Lecture Notes in Computer Science

Lecture Notes in Artificial Intelligence 16090

Founding Editor

Jörg Siekmann

The series Lecture Notes in Artificial Intelligence (LNAI) was established in 1988 as a topical subseries of LNCS devoted to artificial intelligence.

The series publishes state-of-the-art research results at a high level. As with the LNCS mother series, the mission of the series is to serve the international R & D community by providing an invaluable service, mainly focused on the publication of conference and workshop proceedings and postproceedings.

Sašo Džeroski · Jurica Levatić · Gianvito Pio ·
Nikola Simidjievski
Editors

Discovery Science

28th International Conference, DS 2025
Ljubljana, Slovenia, September 23–25, 2025
Proceedings

Springer

Editors
Sašo Džeroski
Jozef Stefan Institute
Ljubljana, Slovenia

Jurica Levatić
Jožef Stefan Institute
Ljubljana, Slovenia

Gianvito Pio
University of Bari
Bari, Italy

Nikola Simidjievski
Institut Polytechnique de Paris
Télécom Paris
Palaiseau, France

ISSN 0302-9743 ISSN 1611-3349 (electronic)
Lecture Notes in Artificial Intelligence
ISBN 978-3-032-05460-9 ISBN 978-3-032-05461-6 (eBook)
https://doi.org/10.1007/978-3-032-05461-6

LNCS Sublibrary: SL7 – Artificial Intelligence

This Springer imprint is published by the registered company Springer Nature Switzerland AG
The registered company address is: Gewerbestrasse 11, 6330 Cham, Switzerland

Preface

The Discovery Science (DS) conference series provides an open forum for intensive discussions and exchange of new ideas among researchers working in the area of Discovery Science. Its focus is on the use of Artificial Intelligence methods in science. Its scope includes the development and analysis of methods for discovering scientific knowledge, coming from machine learning, data mining, and intelligent data analysis, and big data analytics, as well as their application in various domains.

The 28th International Conference on Discovery Science (DS 2025) was held in Ljubljana, Slovenia, from September 23 to 25, 2025. It was organized in close collaboration between the Jožef Stefan Institute and the University of Ljubljana and took place at the Faculty of Computer and Information Science in Ljubljana. This volume contains the papers selected for oral presentation at DS 2025.

The DS conference series started in 1998. For its first 20 editions, the yearly International Conference on Discovery Science (DS) was co-located with the International Conference on Algorithmic Learning Theory (ALT). In 2018, DS was co-located with the 24th International Symposium on Methodogies for Intelligent Systems (ISMIS 2018). Since the 2019 edition, which took place in Split (Croatia), DS has been a stand-alone event. DS 2020 and DS 2021 were online-only events (due to COVID), while DS 2022, 2023 and 2024 were located in Montpellier (France), Porto (Portugal), and Pisa (Italy), respectively.

Discovery Science 2025 was the centerpiece of a larger event, the International Conference on Artificial Intelligence for Science, which covered the whole week of September 22 to 26, 2025 and included a number of additional tracks. The event facilitated the interaction between the Artificial Intelligence community and the communities of scientists from different domains interested in applying AI to their respective sciences. Some of these tracks were supported by the MSCA COFUND Action SMASH and went under the name the 2nd SMASHing conference: AI and Environmental Science, AI and Life Sciences, AI and Physiscs, and AI and Humanities. Additional tracks included AI and Material Science (supported by the DAEMON and EUMINE COST Actions), AI and Space, and the DaFab Summer School. Finally, the AI factories track, devoted to the intersection of AI and High Performance Computing (HPC), was supported by both SLAIF, the Slovenian AI Factory, and SMASH.

DS 2025 had a single research track and received 86 international submissions. Each submission was reviewed by at least two, and in most cases four, Program Committee (PC) members. The reviewing process was single-blind and used the Microsoft CMT system. The Program Chairs decided to accept 38 papers, with an overall acceptance rate of 44%. One accepted paper was co-authored by a program chair, and four others were co-authored by one of the general chairs (who is a co-editor of the proceedings): These papers underwent regular review, being handled and objectively evaluated by other committee members.

Three keynote talks were given during the conference: Iryna Gurevych (Technical University of Darmstadt, Germany) gave a presentation titled "Please meet AI, our dear new colleague. In other words: can scientists and machines truly cooperate?"; Mikel Landajuela (Lawrence Livermore National Laboratory, United States) contributed a talk titled "Deep Symbolic Optimization: Reinforcement Learning for Equation Discovery"; and Joaquin Vanschoren (Eindhoven University of Technology, Netherlands) contributed a talk titled "Auto-Continual Learning". Abstracts of the invited talks are included in the front matter of these proceedings. Besides the presentation of the regular research papers in the main track, the conference offered poster sessions for Late Breaking Papers featuring posters of recent research results related to Discovery Science. Contributions from the main track were also given the opportunity to be presented during poster sessions, in addition to oral presentations.

We sincerely thank Springer for their continued and long-standing support. Springer publishes the conference proceedings as well as a regular special issue of the prestigious Machine Learning journal dedicated to Discovery Science. This special issue provides authors with the opportunity to publish substantially extended and revised versions of their DS conference papers, while also welcoming submissions on Discovery Science topics from the broader research community.

For DS 2025, Springer also supported a Best Student Paper Award. The award went to the paper "Similarity Learning for Spectral Clustering" authored by Vangjush Komini, Nadezhda Koriakina, Debaditya Roy, and Sarunas Girdzijauskas. Among the presented papers, this work was selected by the Program Chairs due to the excellent evaluations received during the review process.

We would next like to thank the organizations and projects that made DS 2025 and AI for Science 2025 possible. On the institution side, these include the Jožef Stefan Institute, University of Ljubljana, University of Bari and Télécom Paris/Institut Polytechnique de Paris. On the project side, we would like to highlight: "AI for Science", funded by ARIS under grant GC-0001; the MSCA COFUND project SMASH (Machine learning for science and humanities); the COST Actions DAEMON and EUMINE; the Future Artificial Intelligence Research (FAIR) PNRR Project; and the SLAIF (Slovenian AI Factory) project, funded by EuroHPC and Slovenia.

Thanks are finally due to all the people that contributed to DS 2025 and AI for Science 2025. We would like to thank all the authors for their submissions, the Program Committee members for their effort and help in reviewing the papers, and the keynote speakers for their contributions to the program. On the organization side, we would like to thank all the members of the Organizing Committee, in particular Vida Groznik and the entire AI for Science team, Sintija Stevanoska, and Antonio Pellicani for pleasant collaboration and organization of all conference-associated activities. We are also grateful to the people behind Microsoft CMT for developing the conference organization system which has proved to be an essential tool in the paper submission and evaluation process.

August 2025

Sašo Džeroski
Jurica Levatić
Gianvito Pio
Nikola Simidjievski

Organization

General Chairs

Sašo Džeroski	Jožef Stefan Institute, Slovenia
Vida Groznik	University of Ljubljana, Slovenia

Program Committee Chairs

Jurica Levatić	Jožef Stefan Institute, Slovenia
Gianvito Pio	University of Bari, Italy
Nikola Simidjievski	Télécom Paris, Institut Polytechnique de Paris, France

Steering Committee Chair

Michelangelo Ceci	University of Bari, Italy

Local Organization Committee

Ana Čufer	Slovenian Artificial Intelligence Society, Slovenia
Mila Marinković	University of Ljubljana, Slovenia
Aleksander Sadikov	University of Ljubljana, Slovenia
Sintija Stevanoska	Jožef Stefan Institute, Slovenia
Maj Zirkelbach	University of Ljubljana, Slovenia
Laura Šušteršič Zorn	University of Nova Gorica, Slovenia
Jure Žabkar	University of Ljubljana, Slovenia

Website Chair

Antonio Pellicani	University of Bari, Italy

Program Committee

Pedro Henriques Abreu	CISUC, Portugal
Reza Akbarinia	Inria, France
Thiago Andrade	INESC TEC - LIAAD, Portugal
Giuseppina Andresini	University of Bari Aldo Moro, Italy
Fabrizio Angiulli	University of Calabria, Italy
Mário Antunes	Universidade de Aveiro, Portugal
Annalisa Appice	University of Bari Aldo Moro, Italy
Martin Atzmueller	Osnabrück University & DFKI, Germany
Isacco Beretta	Università di Pisa, Italy
Robert Bossy	INRA MaIAGE, France
Henrik Boström	KTH Royal Institute of Technology, Sweden
Jure Brence	Jožef Stefan Institute, Slovenia
Francesca Bugiotti	Laboratoire interdisciplinaire des sciences du numérique (LISN), France
Veronica Buttaro	University of Bari Aldo Moro, Italy
Maik Büttner	Otto-von-Guericke University Magdeburg, Germany
Christian Camacho-Villalón	Jožef Stefan Institute, Slovenia
Riccardo Cantini	University of Calabria, Italy
Michelangelo Ceci	University of Bari, Italy
Mattia Cerrato	JGU Mainz, Germany
Ricardo Cerri	Federal University of São Carlos, Brazil
Martina Cinquini	University of Pisa, Italy
Silvia Corbara	Scuola Normale Superiore, Italy
Roberto Corizzo	American University, USA
Paulo Cortez	University of Minho, Portugal
Margarida Costa	Faculdade de Ciências da Universidade do Porto, Portugal
Bruno Cremilleux	Université de Caen Normandie, France
Pedro C. Vieira	Faculty of Sciences of University of Porto, Portugal
Annunziata D'Aversa	University of Bari Aldo Moro, Italy
Le Doan	Teesside University, UK
Katharina Dost	Jožef Stefan Institute, Slovenia
Haimonti Dutta	University at Buffalo, USA
Tome Eftimov	Jožef Stefan Institute, Slovenia
Fabio Fassetti	University of Calabria, Italy
Luca Ferragina	University of Calabria, Italy
Pedro Ferreira	University of Porto, Portugal
Michele Fontana	Università di Pisa, Italy

Johannes Fürnkranz	JKU Linz, Austria
Takayasu Fushimi	Tokyo University of Technology, Japan
Dragan Gamberger	Rudjer Boskovic Institute, Croatia
Boštjan Gec	Jožef Stefan Institute, Slovenia
Gizem Gezici	Scuola Normale Superiore, Italy
Claudio Giovannoni	Università di Pisa, Italy
Hristijan Gjoreski	Ss. Cyril and Methodius University, North Macedonia
Rui Gomes	Universidade Portucalense Infante D. Henrique, Portugal
Rafael Gomes Mantovani	Federal Technology University - Paraná (UTFPR), Brazil
Alexander Gower	Chalmers University of Technology, Sweden
Howard J. Hamilton	University of Regina, Canada
Kohei Hatano	Kyushu University/RIKEN AIP, Japan
Martin Holena	Institute of Computer Science, Czech Republic
Van Anh Huynh-Thu	University of Liege, Belgium
Dino Ienco	INRAE, France
Roberto Interdonato	CIRAD, France
Xiangjian Jiang	University of Cambridge, UK
Ross King	Chalmers University of Technology, Sweden
Ivan Kitanovski	Ss. Cyril and Methodius University, North Macedonia
Dragi Kocev	Jožef Stefan Institute, Slovenia
Boshko Koloski	Jožef Stefan Institute, Slovenia
Ana Kostovska	Jožef Stefan Institute, Slovenia
Vincent Labatut	Université d'Avignon, France
Cristiano Landi	University of Pisa, Italy
Anne Laurent	University of Montpellier, France
Nada Lavrač	Jožef Stefan Institute, Slovenia
Angelica Liguori	University of Calabria, Italy
Donato Malerba	University of Bari Aldo Moro, Italy
Lorenzo Mannocci	UNIPI, Italy
Andrei Margeloiu	University of Cambridge, UK
Fabrizio Marozzo	University of Calabria, Italy
Matej Martinc	Jožef Stefan Institute, Slovenia
Elio Masciari	University of Naples, Italy, Italy
Florent Masseglia	INRIA, France
Sebastian Mežnar	Jožef Stefan Institute, Slovenia
Biljana Mileva Boshkoska	Institute Jožef Stefan, Slovenia
Marta Moreno	ALGORITMI/LASI Center - Universidade do Minho, Portugal

Nontokozo Mpofu	SPU, South Africa
Julio Muñoz-Benítez	INAOE, Mexico
Mirco Nanni	CNR-ISTI Pisa, Italy, Italy
Francesca Naretto	University of Pisa, Italy
Claire Nédellec	INRAE, MaIAGE, France
Masaaki Nishino	NTT Corporation, Japan
Simona Nisticò	University of Calabria, Italy
Ana Rita Nogueira	INESCTEC, Portugal
Andreas Nürnberger	Otto-von-Guericke-Universität Magdeburg, Germany
Nina Omejc	Jožef Stefan Institute, Slovenia
Chetraj Pandey	Georgia State University, USA
Pance Panov	Jožef Stefan Institute, Slovenia
Vincenzo Pasquadibisceglie	University of Bari Aldo Moro, Italy
Kaustubh Patil	Institute of Neuroscience and Medicine, Brain & Behaviour (INM-7), Research Centre Jülich, Germany
Antonio Pellicani	University of Bari Aldo Moro, Italy
Bernhard Pfahringer	University of Waikato, New Zealand
Simone Piaggesi	University of Pisa, Italy
Francesco Pisani	ICAR-CNR, Italy
Vid Podpecan	Jožef Stefan Institute, Slovenia
Pavlin Poličar	University of Ljubljana, Slovenia
Stefano Polimena	University of Bari Aldo Moro, Italy
Luboš Popelínský	Masaryk University, Czech Republic
Aleksandra Rashkovska Koceva	Jožef Stefan Institute, Slovenia
Mathieu Roche	CIRAD, France
Julian Rodemann	LMU Munich, Germany
Joana Santos	FEUP/ISEP, Portugal
José Luis Seixas Jr.	ELTE - Eötvös Loránd University, Hungary
Kaveh Shahhosseini	Teesside University, UK
Larissa Silva	Universidade Federal do Amazonas - ICOMP, Brazil
Nuno Silva	Polytechnic Institute of Porto, Portugal
Paula Silva	INESC TEC - LIAAD, FEUP, Portugal
Esther-Lydia Silva-Ramírez	Universidad de Cádiz, Spain
Nikola Simidjievski	Télécom Paris, Institut Polytechnique de Paris, France
Aristeidis Sionakidis	University of Cambridge, UK
Tomislav Šmuc	Institut Rudjer Bošković, Croatia
Carlos Soares	U. Porto/Fraunhofer AICOS Portugal, Portugal
Domen Šoberl	University of Primorska, Slovenia

Keynote Talks

Please Meet AI, our Dear New Colleague. In other Words: Can Scientists and Machines Truly Cooperate?

Iryna Gurevych

Technical University of Darmstadt, Germany

Automated Machine Learning (AutoML) and Continual Learning (CL) have emerged as pivotal research domains in advancing artificial intelligence. AutoML seeks to automate the design of neural architectures and machine learning pipelines, yet this process is often computationally expensive. Conversely, CL aims to enable models to learn incrementally from new data streams without forgetting what we learned before, yet often relies on fixed predefined architectures. A synergistic integration of AutoML and CL offers a compelling pathway to surmount both limitations, resulting in AI systems capable of efficient lifelong adaptation by dynamically modifying not only their model weights but also their core architecture as needed. Such architectural plasticity is crucial for models to genuinely evolve in response to novel tasks and changing data distributions. This lecture will survey the state-of-the-art in both foundational fields, spanning large-scale and resource-constrained models. Furthermore, we will present recent breakthroughs at their intersection, demonstrating how this synthesis helps us realize more autonomous AI, democratizing its application, and accelerating scientific discovery.

Deep Symbolic Optimization: Reinforcement Learning for Equation Discovery

Mikel Landajuela

Lawrence Livermore National Laboratory, USA

Symbolic regression seeks concise mathematical formulas to describe scientific data, producing models that are transparent and human-interpretable. However, the space of possible symbolic expressions is combinatorially enormous, making search for optimal equations challenging. Deep Symbolic Optimization (DSO) addresses this by framing equation discovery as a sequential decision process guided by reinforcement learning.

In DSO, an autoregressive neural model generates candidate equations token by token, and each formula is scored by its fit to data. A "risk-seeking" policy keeps only the top-scoring expressions to update the model, biasing learning toward promising symbolic forms. Domain priors or constraints can further focus the search on valid or likely solutions. Furthermore, pre-trained models can be used to initialize the search, leveraging existing knowledge to jumpstart the discovery process.

By leveraging deep networks to guide combinatorial search, DSO navigates huge spaces of symbolic formulas, producing compact, interpretable models that capture data patterns. It has achieved state-of-the-art results on symbolic regression benchmarks and even won international equation-discovery competitions. Because DSO works on any discrete-sequence optimization problem, it can be applied beyond equations—for example, to discover symbolic control policies or decision-tree controllers in science and engineering. Overall, DSO shows how reinforcement learning can automate the discovery of human-readable models, aiding scientific discovery with transparent AI.

Deep Symbolic Optimization: Reinforcement Learning for Equation Discovery

Auto-Continual Learning

Joaquin Vanschoren

Eindhoven University of Technology, Netherlands

How can AI and LLMs facilitate the work of scientists in different stages of the research process? Can technology even make scientists obsolete? The role of AI and Large Language Models (LLMs) in science as the target application domain has recently been rapidly growing. This includes assessing the impact of scientific work, facilitating writing and revising manuscripts as well as intelligent support for manuscript quality assessment, peer-review and scientific discussions. The talk will illustrate such methods and models using several tasks from the scientific domain. We argue that while AI and LLMs can effectively support and augment specific steps of the research process, expert-AI collaboration may be a more promising mode for complex research tasks.

Contents

Model Design and Optimization: Symbolic Learning, Decision Trees, and NNs

Unsupervised and Semi-Supervised Learning

Explainable AI for Science

Physics-Informed Methods for Science

Efficient Probabilistic Modeling of Crystallization at Mesoscopic Scale

Pol Timmer(✉), Koen Minartz, and Vlado Menkovski

Eindhoven University of Technology, Eindhoven, The Netherlands
pol.timmer@gmail.com

Abstract. Crystallization processes at the mesoscopic scale, where faceted, dendritic growth, and multigrain formation can be observed, are of particular interest within materials science and metallurgy. These processes are highly nonlinear, stochastic, and sensitive to small perturbations of system parameters and initial conditions. Methods for the simulation of these processes have been developed using discrete numerical models, but these are computationally demanding. This work aims to scale crystal growth simulation with a machine learning emulator. Specifically, autoregressive latent variable models are well-suited for modeling the joint distribution over system parameters and the crystallization trajectories. However, successfully training such models is challenging due to the stochasticity and sensitivity of the system. Consequently, existing approaches fail to produce diverse and faithful crystallization trajectories. This paper introduces the Crystal Growth Neural Emulator (CGNE), a probabilistic model for efficient crystal growth emulation at the mesoscopic scale that overcomes these challenges. We validate CGNE results using the morphological properties of the crystals produced by numerical simulation. CGNE delivers a factor of 11 improvement in inference time and performance gains compared to recent state-of-the-art probabilistic models for dynamical systems.

Keywords: Neural Simulators · Stochastic Systems · Crystallization

1 Introduction

Solidification is the process of a substance transitioning from a liquid or gaseous state to a solid state, which is evident in various systems, including the crystallization of metals and the growth of snow crystals from water vapor, illustrated in Fig. 1. The interplay of highly sensitive nonequilibrium and nonlinear processes as well as stochastic effects results in complex structures emerging during the solidification process. As such, studying these processes not only has significant practical implications in fields like materials science and metallurgy [14], but can also help us to understand the mechanisms behind emergent complexity in general.

In particular, this complexity manifests itself at the mesoscopic scale: while at the atomic scale one might see atoms arrange themselves into a well-organised

S. Džeroski et al. (Eds.): DS 2025, LNAI 16090, pp. 3–16, 2025.
https://doi.org/10.1007/978-3-032-05461-6_1

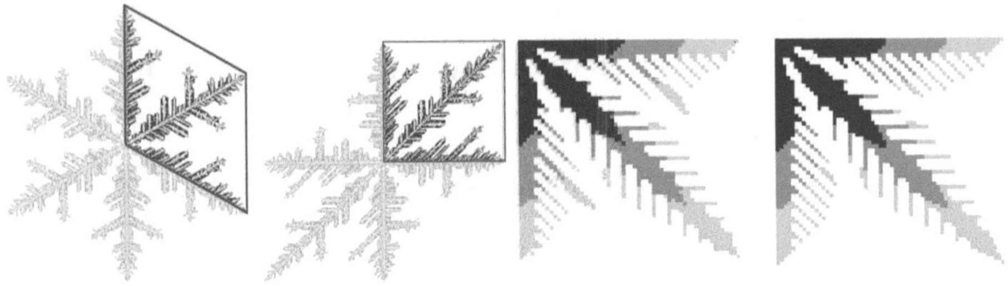

(a) Original crystal representation on a hexagonal grid.

(b) Transformed crystal representation on a square grid.

(c) Crystal growth simulation by numerical LCA model.

(d) Crystal growth simulation by CGNE.

Fig. 1. Illustration of the crystallization simulation process. To account for the crystal's symmetries, we first convert the hexagonal domain on which the snowflake lives to a square grid representation. Then, we simulate a quadrant of the square grid, corresponding to one-third of the hexagonal grid. The crystal growth process produced by CGNE is qualitatively very similar to the one produced by the numerical LCA method.

lattice, and at the macroscopic scale alloys solidify to an ingot, intricate pattern formations caused by the polycrystalline solidification process arise at the mesoscopic scale.

Numerical simulation is a powerful tool to enhance our understanding of these processes, and various methods modeling the stochastic growth of complex solidification processes have been proposed. Computational models based on front-tracking [20,33] and phase-field [3,4,12,17] techniques have provided valuable insights into multigrain and dendritic growth simulations, but occasionally struggle with dynamic and numerical instabilities, especially in cases with both faceted and dendritic growth, such as in snow crystals [1,21]. More recently, local cellular automaton (LCA) based models [10,11,21,27] have advanced the field, capturing previously unseen morphological features for combined faceted and dendritic growth [22]. However, these models demand high computational resources, due to the fine spatiotemporal resolution required to uphold simulation accuracy in the sensitive crystallization process.

Recently, scaling numerical simulations with neural networks has seen a surge in interest from the scientific machine learning community, with successful applications in climate science [5,8,24], nuclear fusion [9,26], and fluid dynamics [13,19]. Similarly, various neural approaches have been proposed in the context of crystal growth modeling. One significant branch of research focuses on Molecular Dynamics (MD) simulations of crystallization [16,34]. While such MD simulations that operate at the microscopic scale have impressively scaled up to 1 million atoms, a simulation at the mesoscopic scale would require many orders of magnitude more atoms to be simulated. Other approaches accelerate phase field simulations directly at the mesoscopic scale [25,28,31]. However, these methods are deterministic and unsuitable for stochastic crystallization processes. More broadly, probabilistic neural simulation methods have been proposed, but these do not straightforwardly apply to the discrete nature of the crystal growth problem [2,7,32], or fail to reliably capture the sensitive and stochastic relationship between environmental parameters and crystal morphology [23].

Although these prior works provide interesting and relevant methods for crystal growth or probabilistic neural simulation in general, no method exists for efficient probabilistic crystal growth simulation at the mesoscopic scale. In this work, we introduce a neural simulator model that greatly accelerates the simulation of mesoscopic-scale crystallization processes. Our main contributions are as follows:

- We propose the Crystal Growth Neural Emulator (CGNE)[1] an autoregressive probabilistic model for simulating crystal growth processes at the mesoscopic scale.
- CGNE is based on PNS [23], a state-of-the-art framework for modeling stochastic dynamical systems. We introduce frozen-state additive sampling, the addition of conditioning on environmental parameters, and temporal subsampling to form CGNE
- We observe that existing training techniques fail to enable the model to reliably capture the joint distribution over environmental parameters and crystal morphologies. We diagnose Latent Variable Neglect (LVN) as the root cause of the problem and furthermore introduce a training technique, called samplewise decoder dropout, that systematically prevents LVN.
- We construct a new snow crystal growth dataset that captures the intricate relationships between environmental parameters and crystal morphology.
- Our evaluations show that CGNE outperforms a state-of-the-art probabilistic neural emulator [23]. Further, we show that decoder dropout significantly enhances simulation diversity and sample quality by preventing LVN, offering broader implications for neural simulation.

2 Approach

2.1 Problem Formulation

At time t, the system state is denoted as (x_t, y), where x_t is a time-varying, binary function on a grid and y is a real-valued vector of static environmental parameters. The system's evolution is represented by the sequence $(x_{0:T}, y)$. We aim to fit a simulation model p to the data by maximizing the log-likelihood over the training set $\{(x_{0:T}, y)_i\}_{i=1}^{N}$:

$$\frac{1}{N}\sum_{i=1}^{N} \log p\left((x_{0:T}, y)_i\right). \tag{1}$$

[1] Source code & dataset: https://github.com/poltimmer/CGNE.

Given the Markovian nature of LCA-based numerical simulators and the known distributions of initial conditions and environmental parameters, this is equivalent to maximizing Eq. 2:

$$\frac{1}{N}\sum_{i=1}^{N}\sum_{t=0}^{T-1}\log p\left((x_{t+1} \mid x_t, y)_i\right). \tag{2}$$

After training on Eq. 2, we can generate new samples by applying $p\left((x_{t+1} \mid x_t, y)_i\right)$ autoregressively.

2.2 Model Design

We aim to develop a model capable of producing realistic simulation trajectories with good sample efficiency. We base our model on PNS [23], a general framework for probabilistic neural simulation of dynamical systems. PNS uses a Conditional Variational Autoencoder (CVAE) structure [30], with a conditional prior $p(z \mid x_t)$ jointly trained with an approximate posterior $q(z \mid x_t, x_{t+1})$ that is informed of the subsequent state of the simulation x_{t+1}. As a result, the conditional prior captures a latent variable z over possible future simulation states, given the current simulation state.

Within the PNS framework, we make CGNE fully convolutional to preserve the data domain's local translation symmetries. This includes a UNet decoder [29] and a spatial latent space, detailed in Fig. 3. The latent variable z passes through an embedding network before concatenation with the input of each residual block in the decoder.

Like the PNS authors, we also incorporate beta annealing [6] and the free bits trick [18] to prevent posterior collapse, a failure mode where the KL term of the VAE loss dominates the reconstruction term, preventing the model encoder from learning useful representations.

To adapt PNS to the problem of mesoscopic scale crystallization, we extend it in a few key areas.

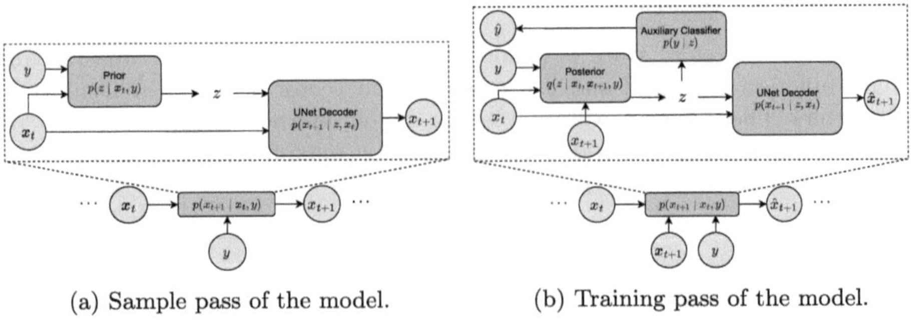

(a) Sample pass of the model. (b) Training pass of the model.

Fig. 2. CGNE model overview.

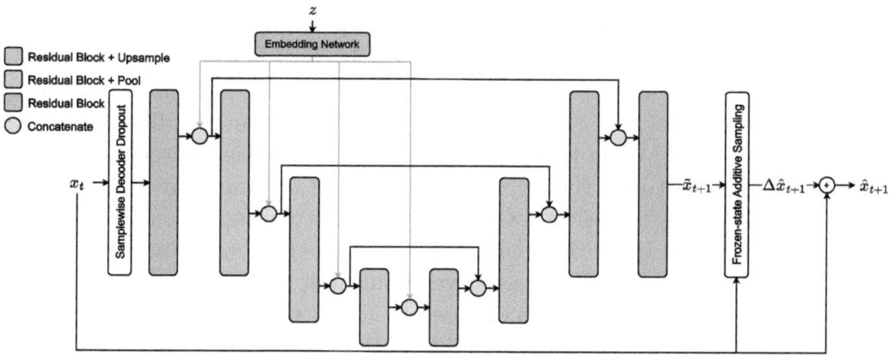

Fig. 3. Overview of CGNE's decoder, including the samplewise decoder dropout and frozen-state additive sampling strategies.

Environmental Parameters. Firstly, since PNS does not support conditioning on environmental parameters, we choose to integrate environmental parameters y into both the prior and posterior of, resulting in a conditional prior $p(z \mid x_t, y)$ and a posterior $q(z \mid x_t, x_{t+1}, y)$, as shown in Fig. 2. An auxiliary classifier $p(y \mid z)$ reconstructs environmental parameters y from the latent variable during training, improving their representation in the latent space [15].

Frozen-state Additive Sampling. Secondly, as we only consider solidification, as opposed to melting and sublimation, we introduce frozen-state additive sampling to guarantee the physical property of monotonic crystal growth. In this approach, the solidified portion from the previous step x_t is preserved and combined with the model's prediction for only the non-solidified region of the prior step $\Delta\hat{x}_{t+1}$. This enforces that only grid cells within the non-solidified region of x_t are eligible for growth. We define this set of grid cells as the "eligible growth area" $\mathcal{E}$. Formally,

$$\Delta\hat{x}_{t+1}^{(i,j)} = \begin{cases} \tilde{x}_{t+1}^{(i,j)} & \text{if } (i,j) \in \mathcal{E} \\ 0 & \text{otherwise,} \end{cases}$$

and

$$\hat{x}_{t+1}^{(i,j)} = \Delta\hat{x}_{t+1}^{(i,j)} + x_t^{(i,j)} = \begin{cases} \tilde{x}_{t+1}^{(i,j)} & \text{if } (i,j) \in \mathcal{E} \\ x_t^{(i,j)} & \text{otherwise.} \end{cases}$$

This strategy ensures a physically accurate monotonic increasing solidification process.

Decoder Dropout. Next, as the decoder is conditioned on the previous simulation state x_t, and differences between consecutive states (x_t and x_{t+1}) are minute, inferring a reasonable estimate for x_{t+1} is possible to a large degree without using z. Consequently, gradients flowing through z, which have a low signal-to-noise ratio, are overshadowed by the gradients flowing through the direct path to x_t. This is amplified by the UNet's skip connections, causing the model to neglect the latent variables. Although the one-step-ahead estimates of such a deterministic model might be reasonable, the lack of accumulation of randomness results in a severe underestimation of sample diversity at the end of the trajectory. While this Latent Variable Neglect is an issue inherent to CVAE-based architectures, the PNS authors did not encounter it in their implementations. LVN only arises in CGNE because of two implementation details, namely the usage of a UNet decoder conditioned on the simulation state x_t, and the small differences between consecutive simulation states.

LVN is different to posterior collapse, as while with posterior collapse the posterior $q(z \mid x_t, x_{t+1}, y)$ collapses to the prior $p(z \mid x_t, y)$, with LVN the environmental parameter y is also ignored, causing the conditional prior to be uninformative.

To remedy this issue, we use a samplewise dropout on the direct connection from x_t to the decoder, similar to word dropout used by Bowman et al. [6]. We schedule the dropout probability linearly from 1 to 0.1 during training only, with this probability being zero during inference. This training strategy reinforces the flow of information through the latent variables, which solidifies the stochastic pathway and consequently significantly improves the model's ability to model realistic and diverse trajectories.

Temporal Subsampling. Another issue arises when modeling a system characterized by gradual differences in subsequent simulation states with a Markov model. When differences between subsequent states are slight, the model can struggle to learn the actual dynamics of the simulation and instead tends to resort to the strong local minimum of an identity function. In this failure mode, model simulations tend to deadlock at some point in the trajectory. Therefore, to improve simulation efficiency and prevent this failure mode, we introduce temporal downsampling of the simulation trajectories, increasing the observed differences between subsequent simulation states.

3 Experiments

We validate our model on snow crystal growth, a highly complex crystallization process stemming from the underlying combination of faceted and dendritic growth. We generate a dataset by implementing Janko Gravner and David Griffeath's stochastic LCA algorithm for snow crystal growth [11]. This algorithm simulates growth from a single seed crystal under seven tuneable environmental parameters that influence the formation and final morphology of the crystal. We fix six parameters and vary ρ, the ambient vapor saturation parameter sampled from a uniform distribution between 0.35 and 0.65. This parameter is most influential on the shape of the snow crystal and therefore leads to a wide range of morphologies when varied.

Table 1. Experiment Configurations.

Method	Temporal Subsampling	Free Bits	Decoder Dropout
PNS	No	No	No
PNS $_{+\mathrm{TS}}$	Yes	No	No
CGNE $_{-\mathrm{DO}}$	Yes	Yes	No
CGNE $_{-\mathrm{FB}}$	Yes	No	Yes
CGNE	Yes	Yes	Yes

Due to the stochastic nature of growth, snow crystals generated from the same parameters can vary drastically at the pixel level. Low ρ values produce thin crystals with few side branches, whereas high values yield thicker crystals with numerous branches; however, the exact branch attachment points can differ even with consistent ρ. This variability makes pixel-wise accuracy an inadequate metric for model validation. Instead, we assess the model based on the high-level morphological properties of the generated structures.

Specifically, we investigate if the joint distribution $p(a, b \mid \rho)$ of the model's generated snowflakes, conditioned on the environmental parameter ρ, matches the joint conditional distribution of the numerical LCA simulator $p_{\mathrm{LCA}}(a, b \mid \rho)$. Here, a is the area of the snowflake and b is the boundary length. We quantify the match of these distributions using the expected value of the Wasserstein distance between the conditional distributions, where the expectation is taken over the environmental parameter ρ:

$$\mathrm{EWD} = \mathbb{E}_{\rho \sim p(\rho)} \left[W_2 \left(p(a, b \mid \rho), \, p_{\mathrm{LCA}}(a, b \mid \rho) \right) \right]. \tag{3}$$

In Eq. 3, W_2 denotes the type-2 Wasserstein distance. EWD values are smaller when model distributions closely match the ground-truth distributions. As a baseline, we take PNS [23], which we configure with the free bits trick disabled, and extended with frozen-state additive sampling, as well as the support for environmental parameters. No other extensions are made to PNS.

We compare CGNE to this baseline, and additionally evaluate three ablations of CGNE to assess impact of temporal subsampling, as well as the independent impact of decoder dropout and the free bits trick, the configuration of which is displayed in Table 1.

3.1 Results

Figure 4 illustrates a set of simulated snowflakes comparing our CGNE method and ablations to the LCA ground truth and simulations using the PNS model. For five values of ρ, we simulate a full snowflake using the five models. We see that PNS fails to produce any meaningful results in its base configuration. Due to lack of temporal subsampling, crystals produced by PNS fail to grow out of

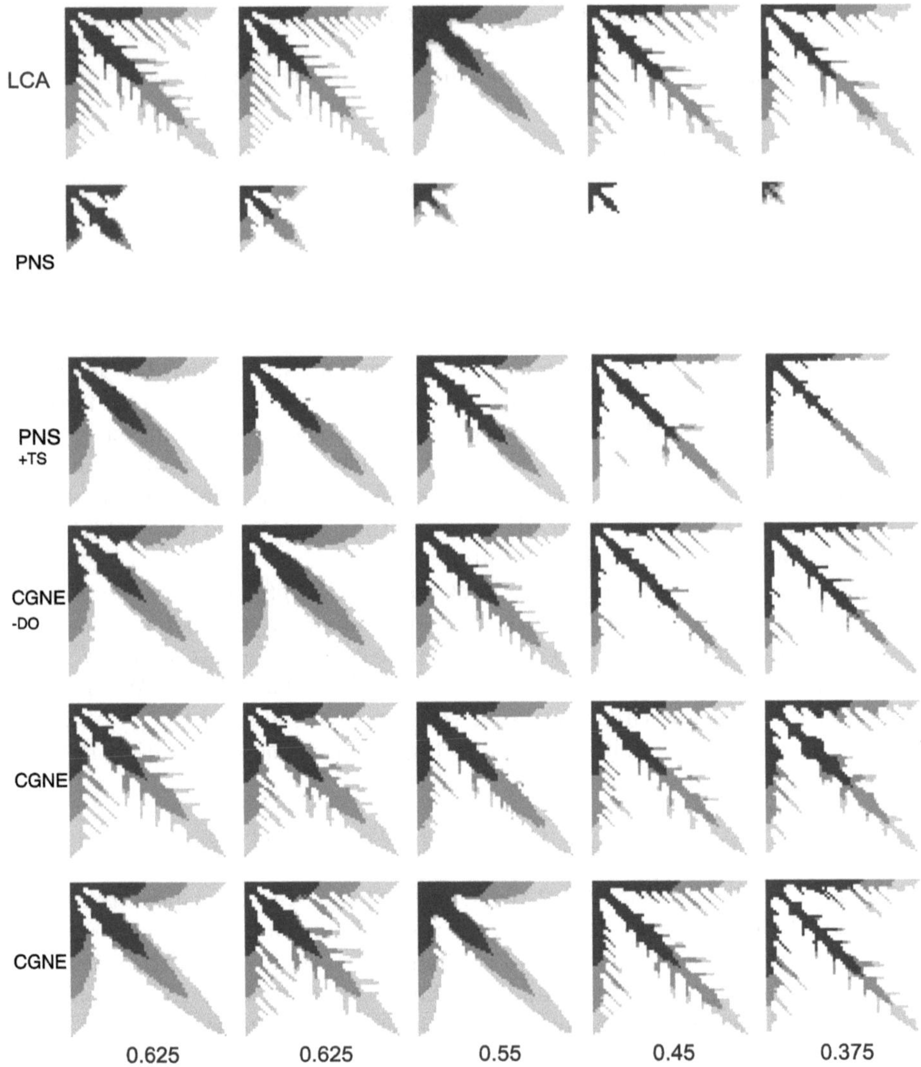

Fig. 4. Qualitative comparison of simulated trajectories for different values of ρ. Three stages of growth are displayed with different shades of blue. (Color figure online)

infancy. The very minute differences in subsequent states cause this Markov process to quickly land in deadlock. We add temporal subsampling to PNS to yield $\text{PNS}_{+\text{TS}}$, allowing better comparison of the effectiveness of other contributions.

In the snowflakes produced by the LCA simulator, we observe a trend of thinner, less voluminous snowflakes as ρ decreases, reflecting lower atmospheric water vapor saturation. Notably, LCA can yield diverse snowflake morphologies for the same ρ, which is an important capability of a neural simulation model for this stochastic ground truth process. While PNS $_{+TS}$ manages to create snowflakes of full size, it struggles to capture this diversity, and generates very similar snowflakes for the same ρ, evident at $\rho = 0.625$. This indicates a tendency towards a deterministic model, caused by the lack of samplewise decoder dropout leading to LVN. Meanwhile, CGNE faithfully reproduces morphological variability, producing distinct snowflakes for the same ρ. Disabling decoder dropout in the CGNE $_{-DO}$ ablation eliminates that variability, confirming the importance of decoder dropout in preventing LVN and preserving the model's ability to model the stochasticity of snow crystal growth.

Not only the diversity, but also sample quality of the simulated snowflakes by PNS $_{+TS}$ is noticeably worse than for CGNE. PNS $_{+TS}$ tends to struggle with the formation of thin branches and finer details. This often results in PNS $_{+TS}$-simulated snowflakes lacking thin side branches, especially for lower values of ρ, and it results in distortions and artifacts in the form of floating crystal pieces, which are physically impossible in the ground truth data-generating process. Conversely, snowflakes generated by CGNE look complete and similar to the ground truth for all values of ρ. The model shows no artifacts and has no issues with thinner branches, even for low values of ρ. The snowflakes visualized in the plot are not identical between the LCA and CGNE models, but this is expected as the models are stochastic. Importantly, the snowflakes generally share a similar style and do not show any distortions. The two ablations of CGNE with free bits and decoder dropout disabled fall qualitatively in between PNS $_{+TS}$ and CGNE, displaying that both decoder dropout and the free bits trick are required to achieve full simulation accuracy.

We further assess how well the snowflake morphology distributions from CGNE, PNS, and their ablations match the LCA model. Figure 5 compares the joint distributions over ρ, crystal area, and boundary length of all five model configurations. CGNE's distributions align closely with the ground truth LCA, capturing all relevant modes as well as overall correlation (Fig. 5e). In contrast, PNS without temporal subsampling has zero overlap between its morphology distributions and the ground truth, due to the deadlocking issue. PNS $_{+TS}$ with temporal subsampling is a drastic improvement, showing some overlap, but its distributions still fail to show the correct overall correlations, and do not capture the modes well. Snow crystals generated by PNS $_{+TS}$ have lower boundary length and lower area comppared to CGNE. This is especially pertinent at lower values of ρ, where fine branch formation occurs.

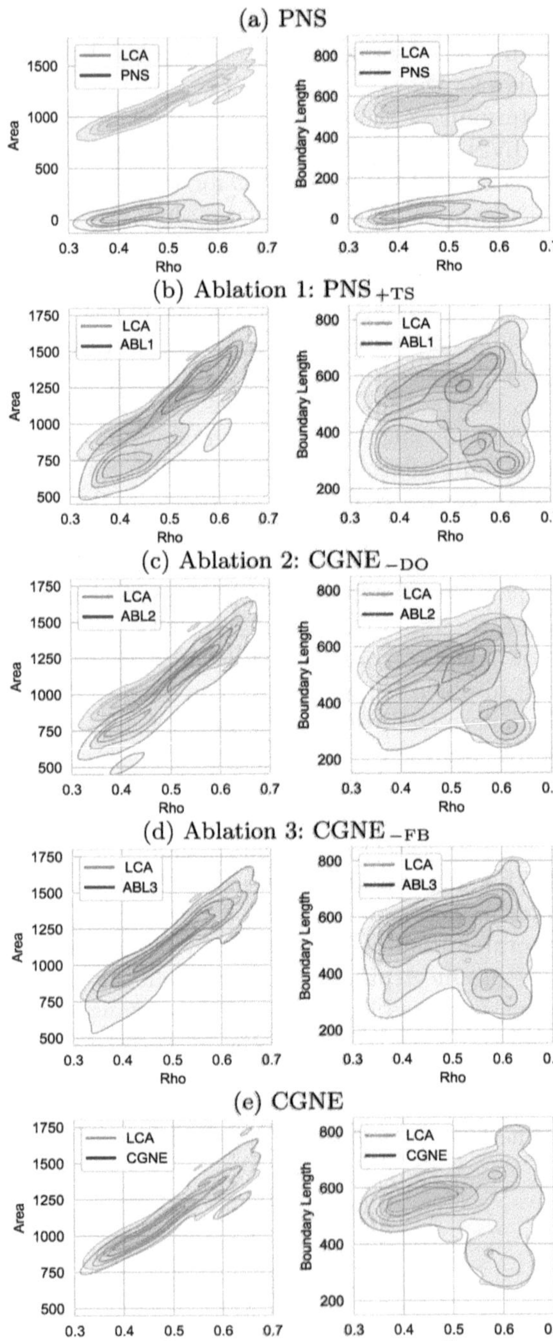

Fig. 5. Joint distributions over crystal morphology properties and environmental parameters compared with ground-truth distribution produced by LCA across five model configurations.

Table 2. Quantitative results. The lower expected Wasserstein distance EWD indicates that simulations generated by CGNE match the ground-truth data-generating process better than PNS.

Method	Temporal Subsampling	Free Bits	Decoder Dropout	EWD ↓
PNS	No	No	No	1087.3
PNS_{+TS}	Yes	No	No	176.1
$CGNE_{-DO}$	Yes	Yes	No	103.8
$CGNE_{-FB}$	Yes	No	Yes	60.4
CGNE	Yes	Yes	Yes	**43.8**

Adding independently the free bits trick or, more substantially, samplewise decoder dropout significantly improves observed distribution alignment. But only in combining the two methods is the full model potential realized, displaying a causal difference between posterior collapse and latent variable neglect.

The quantitative results generally reflect the observations made from the distribution plots (Table 2). The expected Wasserstein distance is significantly better for CGNE compared to PNS and all ablations. This indicates that the distribution over crystal morphologies generated by CGNE overlap better with those generated by LCA. The importance of temporal subsampling to prevent deadlocking is also reflected in the quantitative results table, with PNS_{+TS} yielding a 6x lower EWD score than PNS. Furthermore, we again see that both the free bits trick and decoder dropout independently improve the EWD score, and that both should be used to maximize CGNE's potential.

CGNE significantly enhances inference efficiency, being **11x** faster than our GPU-accelerated LCA model on average. While LCA takes a median of **4.11 s** per simulation trajectory, CGNE completes in just **0.36 s**. This speed advantage is flexible and inversely proportional to temporal resolution, and thus can be even more significant depending on the application.

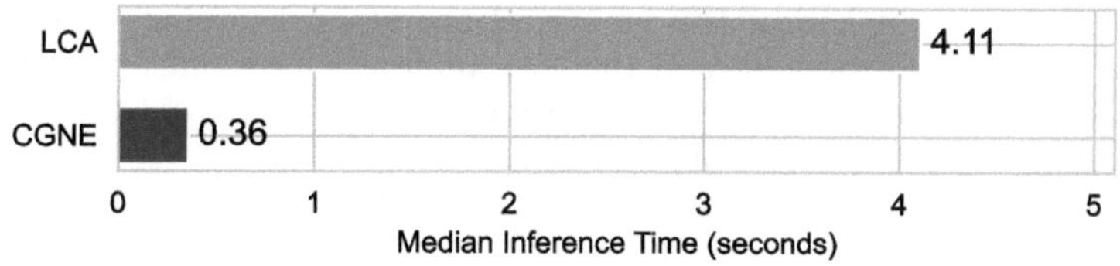

Fig. 6. Median inference time of a full simulation trajectory for LCA and CGNE.

4 Conclusions and Future Work

We propose a model architecture CGNE, that greatly accelerates probabilistic simulations of crystallization processes at the mesoscopic scale while maintaining excellent simulation accuracy. We identify Latent Variable Neglect as a significant challenge for such autoregressive models, which we resolve with sample-wise decoder dropout. As a result, CGNE significantly improves the simulation accuracy and diversity compared to PNS, a recent state-of-the-art model for probabilistic simulation of dynamical systems.

For future research, we consider implementing mechanisms that physically prevent distortions in the form of floating crystal pieces. While these distortions are not an issue in the trained CGNE model, enforcing this property of crystal growth into the model directly could improve training time, or could enable the model to uphold accurate simulations over greater spatial resolutions. Furthermore, we consider investigating how CGNE generalizes to other types of mesoscopic scale crystallization by applying it to datasets of other crystallization processes, such as in metallurgy.

References

1. Barrett, J.W., Garcke, H., Nürnberg, R.: Numerical computations of faceted pattern formation in snow crystal growth. Phys. Rev. E **86**(1) (2012). https://doi.org/10.1103%2Fphysreve.86.011604
2. Bergamin, F., et al.: Guided autoregressive diffusion models with applications to PDE simulation. In: ICLR 2024 Workshop on AI4DifferentialEquations In Science (2024). https://openreview.net/forum?id=1avNKFEIOL
3. Biswas, S., Liu, D., Aagesen, L.K., Jiang, W.: Solidification and grain formation in alloys: a 2d application of the grand-potential-based phase-field approach. Model. Simul. Mater. Sci. Eng. **30**(2) (2022). http://dx.doi.org/10.1088/1361-651X/ac46dc
4. Boettinger, W.J., Warren, J.A., Beckermann, C., Karma, A.: Phase-field simulation of solidification. Ann. Rev. Mater. Res. **32**(1) (2002). https://doi.org/10.1146/annurev.matsci.32.101901.155803
5. Bonev, B., et al.: Spherical Fourier neural operators: Learning stable dynamics on the sphere. In: Proceedings of the 40th International Conference on Machine Learning. Proceedings of Machine Learning Research, vol. 202 (2023). https://proceedings.mlr.press/v202/bonev23a.html
6. Bowman, S.R., Vilnis, L., Vinyals, O., Dai, A., Jozefowicz, R., Bengio, S.: Generating sentences from a continuous space. In: Proceedings of the 20th SIGNLL Conference on Computational Natural Language Learning (2016). https://aclanthology.org/K16-1002
7. Cachay, S.R., Zhao, B., James, H., Yu, R.: DYffusion: a dynamics-informed diffusion model for spatiotemporal forecasting. In: Thirty-seventh Conference on Neural Information Processing Systems (2023). https://openreview.net/forum?id=WRGldGm5Hz
8. Gao, Z., et al.: Prediff: precipitation nowcasting with latent diffusion models. In: Thirty-seventh Conference on Neural Information Processing Systems (2023). https://openreview.net/pdf?id=Gh67ZZ6zkS

9. Gopakumar, V., et al.: Plasma surrogate modelling using fourier neural operators. Nucl. Fusion **64**(5) (2024). https://dx.doi.org/10.1088/1741-4326/ad313a
10. Gravner, J., Griffeath, D.: Modeling snow crystal growth i: rigorous results for packard's digital snowflakes. Exp. Math.**15**(4) (2006). https://doi.org/10.1080/10586458.2006.10128978
11. Gravner, J., Griffeath, D.: Modeling snow crystal growth ii: a mesoscopic lattice map with plausible dynamics. Phys. D Nonlinear Phenom. **237**(3) (2008).https://www.sciencedirect.com/science/article/pii/S0167278907003387
12. Gránásy, L., Pusztai, T., Warren, J.A.: Modelling polycrystalline solidification using phase field theory. J. Phys. Conden. Matter **16**(41) (2004). https://dx.doi.org/10.1088/0953-8984/16/41/R01
13. Gupta, J.K., Brandstetter, J.: Towards multi-spatiotemporal-scale generalized pde modeling. arXiv preprint (2022). https://arxiv.org/abs/2209.15616
14. Idrus-Saidi, S.A., et al.: Liquid metal synthesis solvents for metallic crystals. Science **378**(6624) (2022). https://www.science.org/doi/abs/10.1126/science.abm2731
15. Ilse, M., Tomczak, J.M., Louizos, C., Welling, M.: Diva: domain invariant variational autoencoders. In: Proceedings of the Third Conference on Medical Imaging with Deep Learning. Proceedings of Machine Learning Research, vol. 121 (2020). https://proceedings.mlr.press/v121/ilse20a.html
16. Jarry, P., Jakse, N.: Medium range ordering in liquid al-based alloys: towards a machine learning approach of solidification. In: IOP Conference Series: Materials Science and Engineering **1274**(1) (2023). https://dx.doi.org/10.1088/1757-899X/1274/1/012001
17. Karma, A., Rappel, W.J.: Quantitative phase-field modeling of dendritic growth in two and three dimensions. Phys. Rev. E 57 (1998). https://link.aps.org/doi/10.1103/PhysRevE.57.4323
18. Kingma, D.P., Salimans, T., Jozefowicz, R., Chen, X., Sutskever, I., Welling, M.: Improved variational inference with inverse autoregressive flow. In: Advances in Neural Information Processing Systems. vol. 29 (2016). https://proceedings.neurips.cc/paper_files/paper/2016/file/ddeebdeefdb7e7e7a697e1c3e3d8ef54-Paper.pdf
19. Li, Z., et al.: Geometry-informed neural operator for large-scale 3d PDEs. In: Thirty-seventh Conference on Neural Information Processing Systems (2023). https://openreview.net/forum?id=86dXbqT5Ua
20. Libbrecht, K.G.: Cylindrically symmetric green's function approach for modeling the crystal growth morphology of ice. Phys. Rev. E 60 (1999). https://link.aps.org/doi/10.1103/PhysRevE.60.1967
21. Libbrecht, K.G.: Physically derived rules for simulating faceted crystal growth using cellular automata (2008). https://arxiv.org/abs/0807.2616
22. Libbrecht, K.G.: Snow Crystals: a case study in spontaneous structure formation. Princeton University Press (2022). http://www.jstor.org/stable/j.ctv1qdqztv
23. Minartz, K., Poels, Y., Koop, S.M., Menkovski, V.: Equivariant neural simulators for stochastic spatiotemporal dynamics. In: Thirty-seventh Conference on Neural Information Processing Systems (2023). https://openreview.net/forum?id=CCVsGbhFdj
24. Nguyen, T., Brandstetter, J., Kapoor, A., Gupta, J.K., Grover, A.: Climax: a foundation model for weather and climate. arXiv preprint arXiv:2301.10343 (2023). https://arxiv.org/abs/2301.10343

25. Nomoto, S., Wakameda, H., Segawa, M., Yamanaka, A., Koyama, T.: Solidification analysis by non-equilibrium phase field model using thermodynamics data estimated by machine learning. Model. Simul. Mater. Sci. Eng. **27**(8) (2019). https://dx.doi.org/10.1088/1361-651X/ab3379
26. Poels, Y., Derks, G., Westerhof, E., Minartz, K., Wiesen, S., Menkovski, V.: Fast dynamic 1d simulation of divertor plasmas with neural PDE surrogates. Nuclear Fusion **63**(12) (2023). https://iopscience.iop.org/article/10.1088/1741-4326/acf70d
27. Raghavan, S., Sahay, S.S.: Modeling the grain growth kinetics by cellular automaton. Materials Science and Engineering: A 445–446, 203–209 (2007). https://www.sciencedirect.com/science/article/pii/S0921509306019836
28. Ren, L., Geng, S., Jiang, P., Gao, S., Han, C.: Numerical simulation of dendritic growth during solidification process using multiphase-field model aided with machine learning method. Calphad 78 (2022). https://www.sciencedirect.com/science/article/pii/S0364591622000554
29. Ronneberger, O., Fischer, P., Brox, T.: U-net: convolutional networks for biomedical image segmentation. In: Navab, N., Hornegger, J., Wells, W.M., Frangi, A.F. (eds.) Medical Image Computing and Computer-Assisted Intervention – MICCAI 2015, pp. 234–241. Springer International Publishing, Cham (2015)
30. Sohn, K., Lee, H., Yan, X.: Learning structured output representation using deep conditional generative models. In: Advances in Neural Information Processing Systems (2015). https://proceedings.neurips.cc/paper/2015/file/8d55a249e6baa5c06772297520da2051-Paper.pdf
31. Xue, T., Gan, Z., Liao, S., Cao, J.: Physics-embedded graph network for accelerating phase-field simulation of microstructure evolution in additive manufacturing. npj Comput. Mater. **8**(1) (2022). http://dx.doi.org/10.1038/s41524-022-00890-9
32. Yang, G., Sommer, S.: A denoising diffusion model for fluid field prediction. arXiv preprint (2023). https://arxiv.org/abs/2301.11661
33. Yokoyama, E., Kuroda, T.: Pattern formation in growth of snow crystals occurring in the surface kinetic process and the diffusion process. Phys. Rev. A 41 (1990). https://link.aps.org/doi/10.1103/PhysRevA.41.2038
34. Zhang, L., Yang, M., Zhang, S., Niu, H.: Unveiling the crystallization mechanism of cadmium selenide via molecular dynamics simulation with machine-learning-based deep potential. J. Mater. Sci. Tech. 185, 23–31 (2024). https://www.sciencedirect.com/science/article/pii/S1005030223009623

Score Matching on Large Geometric Graphs for Cosmology Generation

Diana-Alexandra Onuțu[1,2](✉), Yue Zhao[2], Joaquin Vanschoren[1], and Vlado Menkovski[1]

[1] Eindhoven University of Technology, Eindhoven, The Netherlands
{d.onutu,j.vanschoren,v.menkovski}@tue.nl
[2] SURF, Amsterdam, The Netherlands
yue.zhao@surf.nl

Abstract. Generative models are a promising tool to produce cosmological simulations but face significant challenges in scalability, physical consistency, and adherence to domain symmetries, limiting their utility as alternatives to N-body simulations. To address these limitations, we introduce a score-based generative model with an equivariant graph neural network that simulates gravitational clustering of galaxies across cosmologies starting from an informed prior, respects periodic boundaries, and scales to full galaxy counts in simulations. A novel topology-aware noise schedule, crucial for large geometric graphs, is introduced. The proposed equivariant score-based model successfully generates full-scale cosmological point clouds of up to 600,000 halos, respects periodicity and a uniform prior, and outperforms existing diffusion models in capturing clustering statistics while offering significant computational advantages. This work advances cosmology by introducing a generative model designed to closely resemble the underlying gravitational clustering of structure formation, moving closer to physically realistic and efficient simulators for the evolution of large-scale structures in the universe.

Keywords: Galaxy Clustering · Cosmology · Graph Neural Networks · Equivariance · Score Matching · Diffusion Models · Generative modeling

1 Introduction

Influenced by gravity, a nearly uniform distribution of matter evolves into a complex network of clusters, filaments, and voids, known as the large-scale structure of the universe. Analyzing this structure provides valuable information for inferring properties of dark matter and dark energy, advancing our understanding of cosmic structure formation and fundamental physics. Cosmologists study this by comparing observational data collected using telescopes with predictions from theoretical models, such as those produced with N-body simulations. These

Work is partially done during internship in High-Performance ML Team at SURF.

S. Džeroski et al. (Eds.): DS 2025, LNAI 16090, pp. 17–31, 2025.
https://doi.org/10.1007/978-3-032-05461-6_2

comparisons rely on summary statistics that quantify and describe the matter distribution, with the aim of testing and constraining cosmological theories.

One of the leading models for the large-scale structure of the universe is the Lambda Cold Dark Matter (ΛCDM) model. Based on this model, N-body simulations track the dynamics of dark matter particles over cosmic time, each influenced by mutual gravitational interactions and governed by ΛCDM parameters. The simulations result in particle distributions that form the complex large-scale structure characterized by clusters of matter and vast voids. Within these distributions, dark matter particles collapse under gravity to form gravitationally bound structures known as halos, sites for galaxy formation. As such, the spatial distribution of dark matter halos serves as a theoretical counterpart to the observed galaxy distributions. These observations are then compared with galaxy surveys, allowing cosmologists to test the consistency of ΛCDM and extract constraints on model parameters governing the formation of cosmic structures.

Ideally, a large number of high-resolution N-body simulations are required to comprehensively explore the cosmological parameter space, but their high computational cost makes this challenging. For example, Millennium Simulation II [3], which evolved $\sim$10 billion dark matter particles, required approximately one month and 1.4 million CPU hours, while a higher-resolution simulation of $\sim$70 billion particles [17] ran for two months, and the Quijote simulation suite [18], consisting of 44,100 simulations with millions of particles each, required more than 35 million CPU hours.

To overcome the computational challenges of N-body simulations, generative models provide a promising alternative to significantly improve the computational complexity [4,13]. These methods adopt a data-driven approach to approximate the simulation process. Specifically, given a set of large-scale structures, the machine learning-based (ML) models estimate the arrangement of dark matter in space conditioned on cosmological parameters.

A recent approach introduces a diffusion-based generative model to generate samples of large-scale structure based on halo distributions [4]. To mitigate information loss associated with discretized representations such as pixels or voxels, the method uses point clouds limited to the 5,000 most massive halos per cosmology. The model architecture is based on graph neural networks (GNN) or Transformers to account for permutation invariance of point clouds. This approach shows promising results in generating samples that accurately reproduce summary statistics across small to intermediate scales.

However, the previous approach presents several limitations. First, it initializes sampling from a Gaussian prior, which does not reflect the near-uniform matter distribution characteristic of the early universe. This mismatch can make the modeling task more difficult and inefficient. Additionally, the Gaussian prior is defined on an unbounded domain, while cosmological simulations are confined to a bounded, periodic volume. The model must therefore implicitly map unbounded inputs to the bounded simulation space in the absence of an explicit mechanism enforcing spatial boundary constraints. Second, the model does not explicitly encode key domain symmetries: cosmological data are equivariant under the $E(3)$ Euclidean group and exhibit periodic boundary conditions (PBC). Incorporating such symmetries as inductive biases is known to enhance model generalization, as demonstrated by the success of translation-equivariant architectures such as convolutional neural networks. Finally, generated point clouds are limited to only about 1% of the number of halos in simulations, significantly fewer than the number of halos targeted in full simulations. Overall, these limitations hinder the scalability and physical fidelity required for generative models to serve as viable surrogates for computationally expensive N-body simulations.

This work addresses the aforementioned challenges by introducing a score-based generative model [15] to sample large-scale structure conditioned on cosmological parameters. Following [4], we represent the matter distribution as point clouds and construct graphs based on k-nearest neighbors (k-NN). A key difference from diffusion models lies in the noise schedule: instead of interpolating between the data distribution and a Gaussian, random noise is added to the data. This results in the most corrupted state corresponding to a uniform halo distribution, enabling more efficient sampling from an informed prior. PBCs are imposed as in [19] to align the bounded, periodic domain of cosmological simulations. To respect Euclidean symmetries, we employ an $E(3)$ equivariant graph neural network (EGNN) [14]. By incorporating these domain-specific constraints: periodic boundaries, symmetries, and uniform prior, the model's denoising can be interpreted as a simulator of gravitational clustering, providing a physically grounded generative framework for structure formation.

Developing score-based generative models for cosmology involves two main challenges. First, the noise schedule must be carefully designed to map an initially uniform point distribution into the highly clustered halos of large-scale structure. Second, the model must scale to generate point clouds with up to 600,000 halos to capture the full complexity of gravitational clustering. To address the first challenge, we introduce a method for quantifying changes in graph topology across noise levels, providing a principled approach to tuning the noise schedule. To address scalability, we train on smaller sub-volumes sampled from the full simulation box and perform full-resolution generation at inference time.

Prior work [16] shows score-based generative dependence on noise schedule design, with performance degrading when scaling from low- to high-resolution images. In the context of graphs, diffusion and score-based models have been mostly limited to small graphs with 4 to 400 nodes [8,12,19], and generation quality degrades noticeably for larger graphs [8]. For point clouds, existing models typically operate at scales of 1,000 and 2,048 points [9,10]. Within cosmology, where datasets naturally consist of large point clouds, generative approaches have been scaled from 100 points [11] up to 5,000 points [4], highlighting current limitations in generating full-scale point clouds for structure generation.

In this work, we introduce a score-based generative model to sample cosmologies. We summarize our contributions below:

- We introduce a score-based model for cosmologies that generates matter distribution across various ΛCDM configurations, enforcing periodic boundary

conditions, and initializing sampling from a physically motivated uniform prior, thereby aligning the generative process more closely with physical reality compared to Gaussian-initialized diffusion models.
- We incorporate domain symmetries via an equivariant graph neural network, ensuring consistency with cosmic structure and enhancing model generalization and data efficiency.
- We introduce a principled, topology-aware method for designing noise schedules critical for large geometric graphs, addressing a key challenge in scaling score-based models. This is vital because the structure of large graphs is highly sensitive to noise.
- The unique combination of $E(3)$ equivariance, PBCs, physically motivated prior, and a scalable noise schedule tailored for large geometric graphs represents a significant step forward in cosmology simulation, outperforming prior diffusion models. This synergy allows the model to not only generate samples, but do so in a way that reflects the underlying physical processes of structure formation more faithfully than previous approaches.

2 Cosmology Representation

N-body simulations model the gravitational evolution of a large number of dark matter particles from early-universe initial conditions to present epoch (redshift $z = 0$). These simulations run within a periodic box, where particles exiting one side re-enter from the opposite side, mimicking the infinite nature of the universe. Particles interact through gravity, leading to the emergence of non-linear structures from an initially near-uniform density field. The resulting matter distribution reflects the large-scale structure of the universe under a specific ΛCDM model configuration. Thus, each simulation represents a possible instance of the universe consistent with the assumed model parameters, namely, a cosmology.

Table 1. Parameters of the ΛCDM model and corresponding value ranges used in the Latin hypercube sampling of the Quijote simulations [18].

Parameter	Description	Range
Ω_m	Total matter density	$[0.1, 0.5]$
Ω_b	Baryon matter density	$[0.03, 0.07]$
H_0	Hubble constant	$[0.5, 0.9]$
n_s	Scalar spectrum power-law index	$[0.8, 1.2]$
σ_8	Amplitude of density fluctuations at scale 8 h^{-1} Mpc	$[0.6, 1.0]$

Each cosmology is characterized by a distinct set of ΛCDM parameter values, described in Table 1, which influence the formation of large-scale structure. Among these, Ω_m quantifies the total matter content while σ_8 characterizes the clumpiness factor at the 8 h^{-1} megaparsec (Mpc) scale. The two-point correlation function (2PCF), denoted $\xi(r)$, measures clustering strength of cosmologies by quantifying the excess probability, relative to a uniform distribution, of finding pairs of points separated by distance r. Higher 2PCF values indicate stronger clustering; lower values suggest uniformity. Figure 1 shows the resulting large-scale structure and 2PCF for various (Ω_m, σ_8) combinations. Notably, high Ω_m and high σ_8 yield a highly clustered distribution with prominent voids, whereas low Ω_m and low σ_8 produce a more uniform and weakly clustered matter distribution.

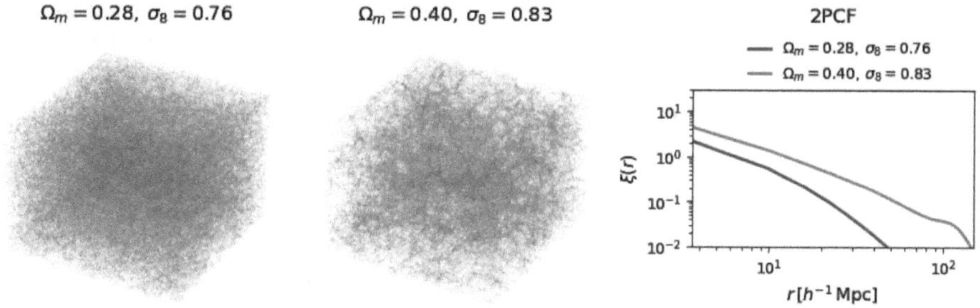

Fig. 1. Large-scale structures and 2PCF of two cosmologies varying (Ω_m, σ_8), drawn from the standard Latin-hypercube sampling at $z = 0$ in the Quijote simulations [18].

3 Methodology

We propose a score-based generative model as an alternative to the diffusion-based approach used in [4] for generating cosmological samples. This formulation offers two key advantages: (i) it enables the explicit enforcement of periodic boundary conditions during both training and inference, and (ii) it adopts a uniform prior, more consistent with the nearly homogeneous matter distribution in the early universe than the commonly used Gaussian prior. Formally, given a set of halos $\mathbf{X}$, representing their spatial positions, the objective is to model the conditional distribution $p(\mathbf{X}|\boldsymbol{\theta})$, where $\boldsymbol{\theta}$ denotes the cosmological parameters.

Score-based [15] and diffusion [7] models follow a similar framework: a forward process $q(\tilde{\mathbf{X}}|\mathbf{X})$ that corrupts data into a noisy version $\tilde{\mathbf{X}}$, and a reverse process where a model $p_\theta(\mathbf{X})$ is trained to iteratively denoise $\tilde{\mathbf{X}}$, generating samples from the data distribution $p(\mathbf{X})$. The forward process follows a noise schedule that progressively perturbs the signal until it becomes pure noise. For matter distribution, this perturbation involves progressively displacing halos from their initial position, leading to a changed large-scale structure. Conversely, the denoising process aims to restore halos to their original position, thereby generating the large-scale structure of the cosmology. To realistically simulate structure formation, the generative model must ensure that halos remain within the simulation bounds, re-entering from the opposite side when displaced beyond a limit.

A subtle difference in the noise schedules makes diffusion models less suitable for periodically bounded data. Figure 2 shows the forward process over T steps for both models, incorporating periodic wrapping at box length S. In diffusion models, the noise schedule gradually transforms the data into a Gaussian, requiring the denoising process to recover the large-scale structure from this prior. Thus, the model does not simulate the clustering of dark matter halos from a physically realistic initial condition. The forward process with wrapping is defined as an interpolation between the data and noise distribution: $\tilde{\mathbf{X}}_t = (\sqrt{\bar{\alpha}_t} \cdot \mathbf{X} + \sqrt{1-\bar{\alpha}_t} \cdot \boldsymbol{\epsilon}) \bmod S$, where the coefficients balance data and noise at timestep t, and $\boldsymbol{\epsilon} \sim \mathcal{N}(0, I)$. The modulo operation keeps halos within the box by wrapping values outside the boundaries back into the domain. As t increases, the signal diminishes, and data approaches a Gaussian centered at 0 when the variance is small. This wrapping effect causes artificial clustering at the boundaries, producing an unrealistic cosmological state.

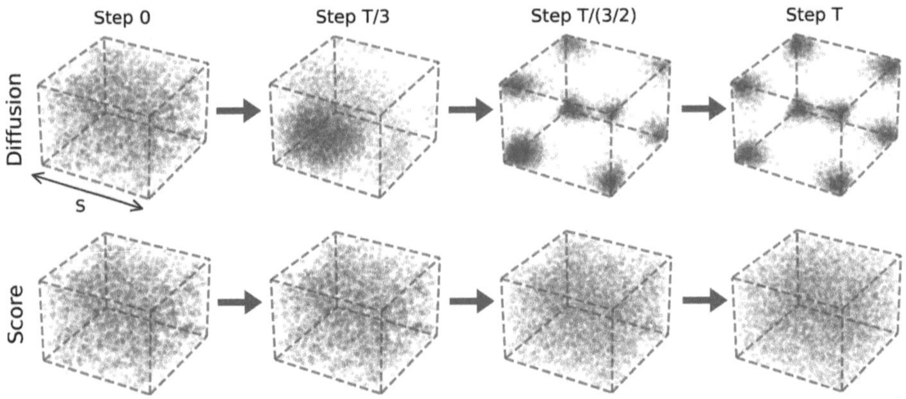

Fig. 2. Forward process visualization of diffusion and score-based models with periodic boundaries. Diffusion models exhibit artificial clustering near boundaries due to wrapping, while score-based models produce a uniform distribution across the domain.

In contrast, score-based models perturb data by displacing halos with some noise. The forward process under periodic boundaries is defined as: $\tilde{\mathbf{X}}_\sigma = (\mathbf{X} + \sigma \cdot \boldsymbol{\epsilon}) \bmod S$, where $\boldsymbol{\epsilon} \sim \mathcal{N}(0, I)$ and σ controls the noise level. As σ increases, halos move further away, and at sufficiently high noise, the perturbed data approximate a uniform distribution due to periodic wrapping, avoiding the boundary clustering seen in diffusion models. Thus, the score-based model must learn an implicit representation of matter clustering to generate large-scale structure.

Next, we formally define the modeling task using score estimation, following the formulation in [15,16], and enforcing periodic boundary conditions as described in [19]. The generative process is defined through the estimation of the *score function*, which is the gradient of the log-probability density of the data corrupted with Gaussian noise. Therefore, the score estimates the direction in which the perturbed data should be moved to recover the data distribution.

Given a cosmology with N halos, each with 3D coordinates, their positions are represented as $\mathbf{X} \in [0, S)^{N\times 3}$, where S is the box length. Let $q_\sigma(\tilde{\mathbf{X}}|\mathbf{X})$ be a noise distribution with periodic wrapping, where $\{\sigma_i\}_{i=1}^L$ is a geometric sequence of L positive noise levels. The objective is to learn the score function $\nabla_{\mathbf{X}} \log q_\sigma(\mathbf{X})$. Since $q_\sigma(\tilde{\mathbf{X}}|\mathbf{X}) = \mathcal{N}(\tilde{\mathbf{X}}|\mathbf{X}, \sigma^2, I)$, we have $\nabla_{\tilde{\mathbf{X}}} \log q_\sigma(\tilde{\mathbf{X}}|\mathbf{X}) = -\boldsymbol{d}_{\min}(\tilde{\mathbf{X}}, \mathbf{X})/\sigma^2$, where $\boldsymbol{d}_{\min} = (\tilde{\mathbf{X}} - \mathbf{X}) - \texttt{round\_to\_nearest\_int}(\tilde{\mathbf{X}} - \mathbf{X})$, following [19].

A conditional score network $\mathbf{s}_{\boldsymbol{\theta}}$ is trained to estimate gradients of the log-density for all noise levels, satisfying: $\mathbf{s}_{\boldsymbol{\theta}}(\tilde{\mathbf{X}}, \sigma) \approx \nabla_{\tilde{\mathbf{X}}} \log q_\sigma(\tilde{\mathbf{X}}|\mathbf{X})$. The training objective is **denoising score matching**, defined for a specific noise level σ as:

$$\ell(\boldsymbol{\theta}; \sigma) \triangleq \frac{1}{2}\mathbb{E}_{p_{\text{data}}(\mathbf{X})}\mathbb{E}_{\tilde{\mathbf{X}}\sim\mathcal{N}(\mathbf{X},\sigma^2 I)}\left[\left\|\frac{\mathbf{s}_{\boldsymbol{\theta}}(\tilde{\mathbf{X}}, \sigma)}{\sigma} + \frac{\boldsymbol{d}_{\min}(\tilde{\mathbf{X}}, \mathbf{X})}{\sigma^2}\right\|_2^2\right], \tag{1}$$

conditioning $\mathbf{s}_{\boldsymbol{\theta}}$ on σ by scaling its prediction, avoiding the need for explicit noise level conditioning parameters as used in diffusion models. The overall loss aggregates the per-noise objectives: $\mathcal{L}\left(\boldsymbol{\theta}; \{\sigma_i\}_{i=1}^L\right) \triangleq \frac{1}{L}\sum_{i=1}^L \sigma_i^2 \ell(\boldsymbol{\theta}; \sigma_i)$ [15]. Furthermore, sampling is performed using annealed Langevin dynamics, as described in [15], with PBCs and a fixed step size $\epsilon > 0$ applied at each update step:

$$\tilde{\mathbf{X}}_t \leftarrow (\tilde{\mathbf{X}}_{t-1} + \frac{\epsilon}{2} s_\theta(\tilde{\mathbf{X}}_{t-1}, \sigma_i) + \sqrt{\alpha_i}\mathbf{z}_t) \bmod S, \qquad \mathbf{z}_t \sim \mathcal{N}(0, I). \tag{2}$$

3.1 Score Matching on Large Geometric Graphs

Noise levels $\{\sigma_i\}_{i=1}^L$ must be chosen such that σ_L is small enough to closely approximate the true data distribution, while σ_1 is large enough to enable score estimation in low-density regions, as discussed in [15]. For image data, these levels are often selected via visual inspection: low noise level preserves most of the image, while high noise produces unrecognizable images.

However, for graph-structured data, such heuristic approaches prove inadequate. As such, our main aim is to directly address the challenge of systematically defining a noise schedule suitable for graphs of arbitrary size. This requires a principled notion of structural change and a metric to quantify the deviation of a perturbed graph from its original form. Motivated by the insight from [16] that noise schedule affects generative performance in high-resolution images, we propose a method to quantify change in graph topology to define noise levels. We define the similarity between a perturbed and original graph in terms of topological structure, measured by the percentage of preserved edges. Specifically, we compute the proportion of shared k-nearest neighbor connections between the original and perturbed graphs across various noise levels. This approach allows for consistent and scalable noise scheduling across diverse graph sizes.

3.2 Equivariant Graph Neural Networks for Cosmology Modeling

The score network employs a GNN to process the perturbed graphs and estimate scores. We adopt the message-passing framework [2,6], leveraging the $E(n)$ equivariant graph neural network [14], with modifications tailored to our task.

Given node features $\mathbf{h}_i \in \mathbb{R}^n$, coordinates $\mathbf{x}_i \in [0, S)^3$, edges $e_{ij} \in \mathcal{E}$, scores $\mathbf{s}_i \in \mathbb{R}^3$ (initialized at zero), and global graph conditioning $\boldsymbol{\theta} \in \mathbb{R}^d$, where n is the feature dimension and d is the number of cosmological parameters, the equivariant graph convolutional layer (EGCL) is defined as: $\mathbf{h}^{l+1}, \mathbf{s}^{l+1} = \text{EGCL}[\mathbf{h}^l, \mathbf{s}^l, \mathbf{x}, \mathcal{E}, \boldsymbol{\theta}]$, where l refers to the messaging passing layer. We retain the original message, aggregation and feature update functions from [14], but change their input: add global conditioning $\boldsymbol{\theta}$ and use relative differences in features and coordinates, as it improves training compared to concatenation or Euclidean distances. Unlike the original model, which predicts coordinates, we predict scores as follows:

$$\mathbf{s}_i^{l+1} = \mathbf{s}_i^l + \frac{1}{|\mathcal{N}_i|} \sum_{j \in \mathcal{N}_i} (\mathbf{x}_i - \mathbf{x}_j) \phi_x(\mathbf{m}_{ij}), \quad \mathcal{N}_i \text{ is neighborhood of node } v_i. \quad (3)$$

The scores are iteratively refined through message-passing, progressively incorporating information from a larger neighborhood. This separation between coordinates and predicted scores ensures that scores remain defined in physical space while preserving all necessary symmetries. To enforce global translation invariance, we adopt the method of [5], subtracting the center of gravity from the predicted scores. This constrains the predictions to a translation-invariant subspace by ensuring a zero mean. As a result, the model's output depends only on the relative arrangement of the points, independent of the absolute position of the system in space. Implementation details are provided in Appendix A.

4 Experiments

We conduct experiments using halo catalogs from the standard regular-resolution Latin Hypercube at redshift $z = 0$, part of the Quijote N-body simulations suite [18]. The dataset includes 2,000 simulated halo catalogs varying configurations of the ΛCDM model, as shown in Table 1. Each simulation is initialized with a unique random seed and runs within a fixed periodic volume $(h^{-1}\text{Gpc})^3$.

For evaluation, we use the 5,000 most massive halos per simulation from [4] to ensure a fair comparison. The dataset includes 1,800 cosmologies for training and 200 for testing. For large graphs training and generation, we use the full halo catalogs from [18]. Each halo catalog represents a cosmology and is treated as a point cloud. The 3D coordinates are normalized to $[0, 1]$ for training. To model the spatial structure, each point cloud is converted into a graph using k-NN with $k = 20$, incorporating PBCs via the Minimum Image Convention. This construction preserves local geometry while respecting the periodicity of the simulation volume.

4.1 Noise Schedule

To design the noise schedule, we quantify topological change by the percentage of preserved edges under perturbation. Specifically, we compute the average edge difference across all cosmologies for varying noise levels σ and the impact of σ on 2PCF of a random cosmology. Results are illustrated in Fig. 3. The common setting, $\sigma_L = 0.01$, causes over 80% edge difference on average and low 2PCF values, indicating that even low noise substantially alters graph topology and clustering strength. This is problematic for score-based models, which require the perturbed graphs at low noise to remain close to the data distribution. The sensitivity arises from graph size: in large graphs, even minor perturbations can significantly disrupt local neighborhoods, as the k-NN structure is highly sensitive to small displacements. In contrast, smaller graphs are more robust to the same level of noise, which may explain why previous work has not required a systematic approach to noise scheduling. Motivated by this analysis, we adopt $\sigma_L = 0.001$, which preserves roughly 30% of the original edges and yields a closer match to the clustering strength of the underlying cosmology. Additionally, we set the maximum noise level to $\sigma_1 = 0.2$, where the graph's topology and clustering strength are degraded, ensuring complete signal removal.

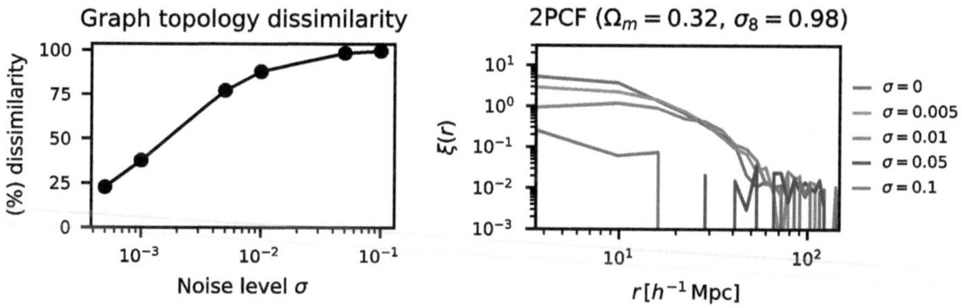

Fig. 3. Effect of noise on graph topology and clustering strength (via 2PCF).

4.2 Quantitative Analysis

We evaluate diffusion and (non)equivariant score-based generative models for cosmology generation. One model is conditioned on all cosmological parameters $\boldsymbol{\theta} = (\Omega_m, \Omega_b, H_0, \sigma_8, n_s)$, the other ones only on (Ω_m, σ_8). We also investigate the impact of a different noise schedule. Performance is evaluated by calculating the mean absolute error (MAE) and mean squared error (MSE) of the 2PCF over 10 generated samples for each of the 200 validation cosmologies. Results are then averaged over all scales and reported in Table 2.

Conditioning on the full parameter set yields the best performance, highlighting the importance of incorporating domain-specific information. Equivariance provides only marginal improvements over the data-augmented non-equivariant

Table 2. Overall model performance comparison. ScoreEGNN (Ω_m, σ_8)** uses a noise schedule commonly found in literature. We retrained the diffusion models from [4].

Model	MAE 2PCF ↓	MSE 2PCF ↓
DiffusionGNN (Ω_m, σ_8)	0.4625	2.7809
DiffusionTransformer (Ω_m, σ_8)	0.2297	0.7147
ScoreGNN (Ω_m, σ_8)	0.2366	0.7092
ScoreEGNN (Ω_m, σ_8)	0.2250	0.6638
ScoreEGNN $\boldsymbol{\theta}$	**0.1764**	**0.3837**
ScoreEGNN (Ω_m, σ_8)**	0.8606	7.2882

ScoreGNN. This can be explained by two factors. First, the 2PCF metric only allows for limited discrimination among cosmology structures, masking some of the advantages of an equivariant model. Second, the ScoreGNN model is able to learn symmetry properties implicitly given sufficient data [14]. Naive noise schedules significantly degrade generative quality, as expected. Score-based (equivariant) GNNs outperform DiffusionGNN, while DiffusionTransformer shows comparable performance, indicating that the Transformer architecture may have sufficient capacity to model the large-scale structure without explicit inductive biases.

4.3 Qualitative Analysis

As shown in Fig. 4, the sampling process begins from a uniform prior and progressively evolves into a clustered large-scale structure, indicating that the model estimates scores that effectively guide halo dynamics. Additionally, Fig. 6 demonstrates that the model captures the dependence on cosmological parameters, reproducing the trend observed in the simulations.

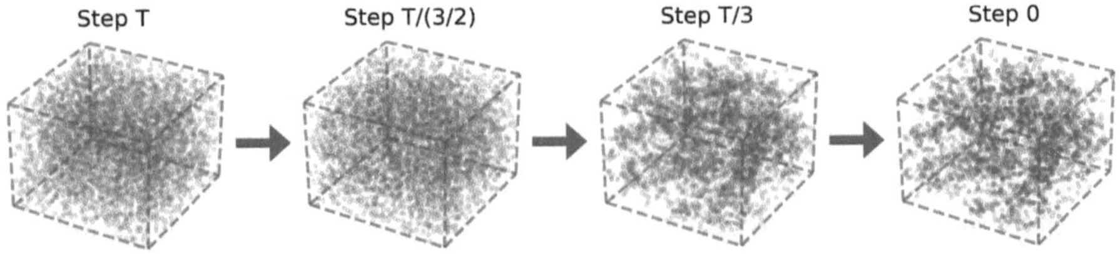

Fig. 4. ScoreEGNN(Ω_m, σ_8) sampling process for ($\Omega_m = 0.10$, $\sigma_8 = 0.92$). Generated sample presents dense regions and voids, as expected for this configuration.

Furthermore, in Fig. 5 we evaluate the clustering strength of generated samples with both ScoreEGNN and DiffusionGNN models, conditioned on (Ω_m, σ_8). Both models generally reproduce the overall trend of the simulated data 2PCF.

However, both models generate samples with high clustering variance at large scales, consistent with the findings of [4], indicating that both generative models struggle to reproduce the clustering patterns on large scales. This reinforces the idea that the problem may be the graph representation and message-passing mechanism, known to struggle with long-range correlations [1].

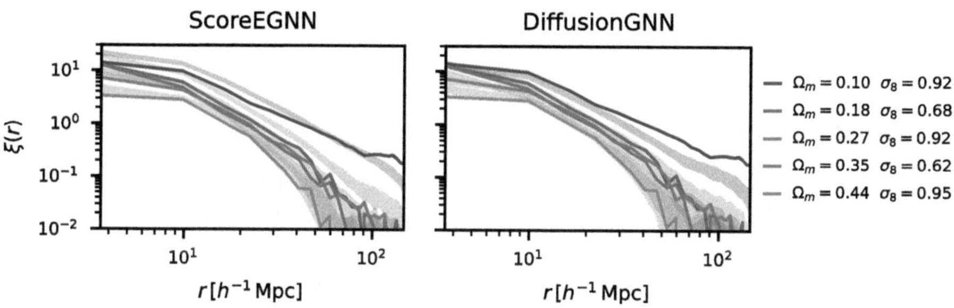

Fig. 5. 2PCF for 5 Ω_m equally spaced cosmologies. Solid line represents the 2PCF of the simulated cosmologies, while the contour is based on the mean and standard deviation of 20 generated samples per configuration.

4.4 Efficiency Analysis

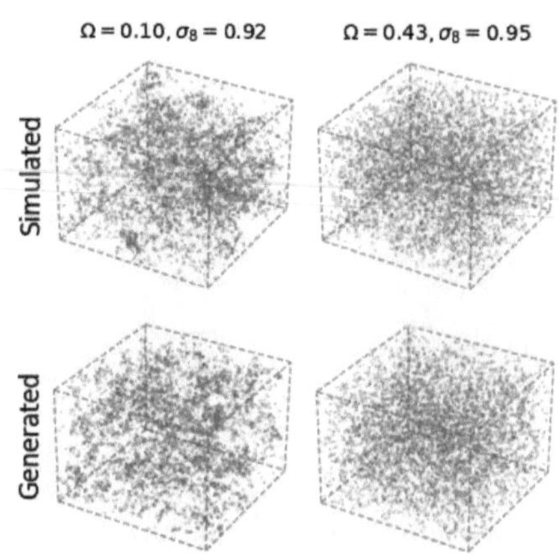

Fig. 6. Simulated and generated cosmologies. Parameter dependence is evident. High-density regions indicate strong clustering; sparse regions correspond to voids.

We also assess computational efficiency. Table 3 shows score-based models outperform diffusion models across all metrics. Score-based models generate the same number of samples roughly twice as fast as diffusion models. Reducing the computational cost of N-body simulations is a major goal in cosmology. Our model generates 2,000 halo catalogs in approximately one hour on 1 H100 GPU, whereas the Quijote simulations required 1.6 million CPU hours. However, a direct comparison is hindered by the limited details of the Quijote setup. For a rough energy use estimate, assuming 30W per CPU core, Quijote consumed 1.6×10^6 core-hours $\times$ 30W $\times$ 3600s/hr $= 1.73 \times 10^{11}$J ($\approx 4.8 \times 10^4$kWh). In comparison, assuming 400W for 1 H100 GPU, our method uses 400W $\times$ 3600s $= 1.44 \times 10^6$J ($\approx$ 0.4kWh). This implies an energy efficiency improvement of approximately five orders of magnitude.

Table 3. Model efficiency comparison. Inference time is time to generate 2,000 samples.

Model	# parameters	# iterations	Training time	Inference time
DiffusionGNN	629K	192K	12h	3h30min
DiffusionTransformer	4.8M	265K	12h	2h30min
ScoreGNN	570K	28K	5h30min	1h5min
ScoreEGNN	417K	28K	5h45min	1h10min

4.5 Large Graphs Generation

Previously, we generated cosmologies containing only the 5,000 most massive halos, omitting the vast majority of halos in a catalog. In this section, we scale the ScoreEGNN (Ω_m, σ_8) model to generate cosmologies with up to 600,000 halos. Training on full graphs of this scale is computationally prohibitive, primarily due to the time and memory cost of our custom PBC-preserving k-NN computations. To mitigate this, we train on smaller sub-volumes obtained via box subsampling. During inference, we operate on the full graph. This means we can only generate one cosmology at a time on 1 GPU due to memory limitations.

Figure 7 shows that generated samples closely resemble the simulated matter distribution and follow the conditional cosmological dependence. The corresponding 2PCF in Fig. 8 confirms that the large-scale structure is well reproduced, though clustering strength at large scales remains underestimated, consistent with observations on smaller graphs in Sect. 4.3.

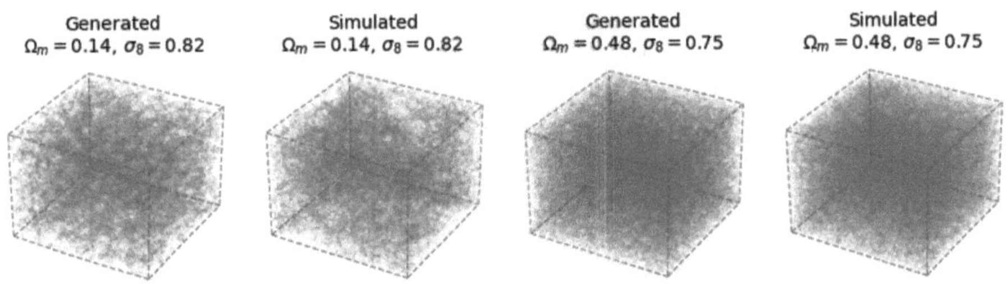

Fig. 7. Comparison of two simulated and two generated cosmologies. The generated samples exhibit clustering patterns similar to those in the simulations.

5 Limitations and Future Work

This work demonstrates the potential of score-based generative models for generating large-scale cosmological structure, but several limitations remain. Score-based models introduce more hyperparameters for the inference process. Developing a theoretical framework for optimizing these settings in graph-based

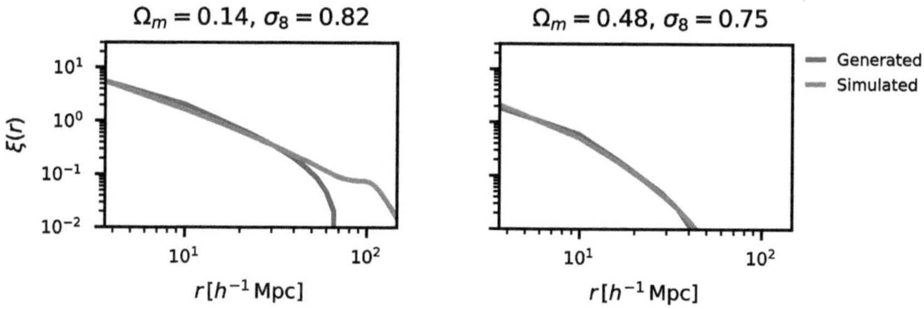

Fig. 8. 2PCF comparison between two simulated and generated cosmologies. Generated data faithfully reproduces 2PCF on smaller scales but deviates on larger scales.

domains could speed up the generation process. Alternatively, flow matching presents a promising direction for future work, simplifying the inference process by avoiding Langevin dynamics.

While our model captures variations in clustering strength across cosmologies, it struggles to reproduce long-range correlations. This difficulty in capturing long-range correlations, evident in the underestimation of the 2PCF at large scales in our generation experiments (Fig. 5 and Fig. 8), may stem from limitations inherent to vanilla graph neural networks, which are known to have difficulty modeling long-range dependence. Future investigations could explore multi-scale or hybrid GNN-Transformer architectures, aiming to leverage the GNN's efficiency for local interactions and the Transformer's capacity for global context, potentially mitigating the observed underestimation of large-scale clustering.

Finally, techniques such as exponential moving average could stabilize training and improve sample quality [8,16], particularly for large graphs.

6 Conclusion

We introduced a score-based generative model for generating large-scale structure conditioned on cosmological parameters. Our approach addresses key limitations of diffusion models by enforcing periodic boundary conditions during both training and inference, initializing sampling from a physically motivated uniform prior, and incorporating Euclidean symmetries. To effectively apply score-based models to large graphs, a topology-aware noise schedule was designed, guided by a quantitative measure of graph perturbation. We demonstrate the scalability of score-based generative modeling to graphs consisting of up to 600,000 nodes, marking the first such application in this context. Empirical results show that model performance is highly sensitive to the noise schedule and that our topology-aware scheduling significantly enhances generative quality. The model outperforms the diffusion-based baselines while offering substantial gains in computational efficiency and improved adherence to physical constraints.

Despite limitations in capturing long-range correlations, our model generates full cosmological samples with promising fidelity. The combination of physically grounded priors, symmetry preservation, and a novel topology-aware noise schedule constitutes a principled advancement in the design of generative models for cosmology, bringing such models closer to capturing the underlying physical processes that govern the evolution of the universe. Most notably, the introduction of the topology-aware noise schedule and the demonstrated scalability to realistic cosmological graph sizes constitute important developments toward machine learning-based simulators of matter clustering. Overall, these contributions move the field closer to developing viable, efficient, data-driven alternatives to computationally expensive N-body simulations.

Acknowledgments. The project has been supported by the Dutch national e-infrastructure with the support of SURF (grant no. EINF-10541). We thank SURF (www.surf.nl) for the support in using the National Supercomputer Snellius.

Disclosure of Interests. The authors have no competing interests to declare that are relevant to the content of this article.

A Implementation Details

All models were trained for 1,000 epochs on 4 Nvidia H100 GPUs on the Dutch national supercomputer Snellius, using data parallelism with micro batch size 4. We used the AdamW optimizer with weight decay 1×10^{-4}, gradient clipping at 0.5, and a learning rate schedule with linear warm-up (epochs 0–100) from 1×10^{-6} to 5×10^{-4}, followed by cosine decay. Most models use a noise schedule with 100 steps, $\sigma_L = 0.001$ and $\sigma_1 = 0.2$; for ScoreEGNN (Ω_m, σ_8)** we used $\sigma_L = 0.01$, $\sigma_1 = 1$. Langevin dynamics used $T = 2$ and $\epsilon = 3 \times 10^{-7}$ across all models.

The equivariant GNN comprises 4 message-passing layers. Node and conditional features are projected to 16-D space and updated via 4-layer MLPs (128 hidden units, SiLU). Node messages are summed, added via residual connection, and normalized. Coordinates are updated via a 2-layer MLP (128 hidden units, SiLU, output 1). Attention scores, computed as in [4], scale messages using a 2-layer MLP (128 hidden units, output 1, SiLU). The non-equivariant GNN omits the coordinate model. Halo coordinates are node features, processed with initial embeddings via a 4-layer MLP (128 hidden units, output 16, SiLU), and final positions are decoded through another 4-layer MLP to 3D space.

References

1. Alon, U., Yahav, E.: On the bottleneck of graph neural networks and its practical implications. arXiv preprint arXiv:2006.05205 (2020)
2. Battaglia, P.W., et al.: Relational inductive biases, deep learning, and graph networks. arXiv preprint arXiv:1806.01261 (2018)

3. Boylan-Kolchin, M., Springel, V., White, S.D., Jenkins, A., Lemson, G.: Resolving cosmic structure formation with the millennium-ii simulation. Mon. Not. R. Astron. Soc. **398**(3), 1150–1164 (2009)
4. Cuesta-Lazaro, C., Mishra-Sharma, S.: Point cloud approach to generative modeling for galaxy surveys at the field level. Phys. Rev. D **109**(12), 123531 (2024)
5. Garcia Satorras, V., Hoogeboom, E., Fuchs, F., Posner, I., Welling, M.: E (n) equivariant normalizing flows. Adv. Neural. Inf. Process. Syst. **34**, 4181–4192 (2021)
6. Gilmer, J., Schoenholz, S.S., Riley, P.F., Vinyals, O., Dahl, G.E.: Neural message passing for quantum chemistry. In: International Conference On Machine Learning, pp. 1263–1272. PMLR (2017)
7. Ho, J., Jain, A., Abbeel, P.: Denoising diffusion probabilistic models. Adv. Neural. Inf. Process. Syst. **33**, 6840–6851 (2020)
8. Jo, J., Lee, S., Hwang, S.J.: Score-based generative modeling of graphs via the system of stochastic differential equations. In: International Conference On Machine Learning, pp. 10362–10383. PMLR (2022)
9. Luo, S., Hu, W.: Diffusion probabilistic models for 3d point cloud generation. In: Proceedings of the IEEE/CVF Conference On Computer Vision And Pattern Recognition, pp. 2837–2845 (2021)
10. Luo, S., Hu, W.: Score-based point cloud denoising. In: Proceedings of the IEEE/CVF International Conference on Computer Vision, pp. 4583–4592 (2021)
11. Makinen, T.L., Charnock, T., Lemos, P., Porqueres, N., Heavens, A.F., Wandelt, B.D.: The cosmic graph: optimal information extraction from large-scale structure using catalogues. Open J. Astrophys. **5**(1), 18 (2022)
12. Niu, C., Song, Y., Song, J., Zhao, S., Grover, A., Ermon, S.: Permutation invariant graph generation via score-based generative modeling. In: International Conference on Artificial Intelligence and Statistics, pp. 4474–4484. PMLR (2020)
13. Riveros, J.K., Saavedra, P., Hortua, H.J., Garcia-Farieta, J.E., Olier, I.: Conditional diffusion-flow models for generating 3d cosmic density fields: applications to f (r) cosmologies. arXiv preprint arXiv:2502.17087 (2025)
14. Satorras, V.G., Hoogeboom, E., Welling, M.: E (n) equivariant graph neural networks. In: International Conference on Machine Learning, pp. 9323–9332. PMLR (2021)
15. Song, Y., Ermon, S.: Generative modeling by estimating gradients of the data distribution. Adv. Neural Inf. Proc. Syst. **32** (2019)
16. Song, Y., Ermon, S.: Improved techniques for training score-based generative models. Adv. Neural. Inf. Process. Syst. **33**, 12438–12448 (2020)
17. Teyssier, R., et al.: Full-sky weak-lensing simulation with 70 billion particles. Astron. Astrophys. **497**(2), 335–341 (2009)
18. Villaescusa-Navarro, F., et al.: The quijote simulations. Astrophys. J. Suppl. Ser. **250**(1), 2 (2020)
19. Xie, T., Fu, X., Ganea, O.E., Barzilay, R., Jaakkola, T.S.: Crystal diffusion variational autoencoder for periodic material generation. In: International Conference on Learning Representations

Maximum Entropy Models for Unimodal Time Series: Case Studies of Universe 25 and St. Matthew Island

Sabin Roman(✉)

Department of Knowledge Technologies, Jožef Stefan Institute, Ljubljana, Slovenia
sabin.roman@ijs.si

Abstract. We present a maximum entropy modeling framework for unimodal time series: signals that begin at a reference level, rise to a single peak, and return. Such patterns are commonly observed in ecological collapse, population dynamics, and resource depletion. Traditional dynamical models are often inapplicable in these settings due to limited or sparse data, frequently consisting of only a single historical trajectory. In addition, standard fitting approaches can introduce structural bias, particularly near the mode, where most interpretive focus lies. Using the maximum entropy principle, we derive a least-biased functional form constrained only by minimal prior knowledge, such as the starting point and estimated end. This leads to analytically tractable and interpretable models.

We apply this method to the collapse of the Universe 25 mouse population and the reindeer crash on St. Matthew Island. These case studies demonstrate the robustness and flexibility of the approach in fitting diverse unimodal time series with minimal assumptions. We also conduct a cross-comparison against established models, including the Richards, Skewnormal, and Generalized Gamma functions. While models typically fit their own generated data best, the maximum entropy models consistently achieve the lowest off-diagonal root-mean-square losses, indicating superior generalization. These results suggest that maximum entropy methods provide a unifying and efficient alternative to mechanistic models when data is limited and generalization is essential.

Keywords: Maximum entropy · Time series · Population collapse

1 Introduction

Unimodal patterns, characterized by a single dominant peak, frequently arise in diverse natural, social, and economic systems, often signaling phases of growth followed by decline or reflecting intrinsic distributional structures. Understanding and modeling these phenomena is essential for predicting critical transitions and characterizing complex systems. While two, three or higher dimensional time series can be effectively modeled through dynamical systems such as differential

S. Džeroski et al. (Eds.): DS 2025, LNAI 16090, pp. 32–44, 2025.
https://doi.org/10.1007/978-3-032-05461-6_3

equations or chaotic maps [27–31], or agent-based models [4], in the case of sparse data for a single variable time series these methods can prove overly complex, under-constrained, and prone to overfitting - making simple, shape-constrained models like the maximum entropy functions presented here a more robust and interpretable alternative.

In this work, we develop and apply maximum entropy models to characterize unimodal data, encompassing both dynamic time series and static distributions. We illustrate the approach through two prominent case studies: the population dynamics of Universe 25 and the reindeer population collapse on St. Matthew Island. Our results demonstrate that maximum entropy formulations effectively capture the essential features of unimodal rise-and-fall behavior as well as stable unimodal distributions, providing a unified statistical framework. This approach highlights the potential of entropy-based models to serve as universal descriptors across ecological, resource, and socio-economic systems.

2 Methods

The intersection of machine learning and time series modeling has witnessed significant advances through the development of structured and multi-target learning approaches. Techniques such as predictive clustering trees (PCTs) and model trees have emerged as powerful tools for capturing complex temporal dynamics, especially in settings involving multivariate or structured outputs [23, 36, 38]. These methods excel in modeling tasks where temporal data is not only univariate but involves interconnected outputs evolving over time - scenarios often encountered in environmental and ecological domains [2, 12, 18].

Multi-target regression has been particularly influential for time series applications that require simultaneous prediction of several dependent variables [15, 25, 37]. In environmental monitoring, for instance, forecasting the simultaneous progression of temperature, humidity, and pollutant concentrations benefits from algorithms capable of exploiting inter-target correlations. Model trees tailored for multi-target prediction allow for accurate and interpretable models, especially when combined with hierarchical or taxonomically structured outputs, as seen in ecological and biological time series [5, 21, 24].

Symbolic regression, particularly via evolutionary approaches, has further extended the ability to discover functional forms underlying temporal trends. These methods are capable of capturing the underlying generative processes of time series data, producing models that are both predictive and interpretable [7, 11, 14, 20, 22, 26, 34, 35, 40, 41, 44]. Such capabilities are valuable in scientific discovery settings where understanding the temporal dynamics is as critical as accurate forecasting.

In real-world applications, hybrid modeling frameworks that integrate empirical data with domain knowledge have shown substantial promise. These approaches are particularly suited for time series exhibiting regular patterns, such as seasonal unimodal curves in phenology, hydrology, and agriculture. The use of hybrid models supports generalization across systems with similar structural characteristics but differing quantitative parameters.

Furthermore, the emphasis on interpretable modeling has driven the adoption of rule-based and tree-based approaches for temporal analysis. These models not only achieve high accuracy but also support model transparency, enabling domain experts to validate and refine predictive insights, which is essential in high-stakes applications like environmental risk assessment or medical prognosis.

While structured and multi-target machine learning approaches have greatly enhanced our ability to model complex temporal systems, they often involve considerable algorithmic complexity and require large volumes of data. Tree-based models and symbolic regression frameworks, for instance, offer flexibility and interpretability, especially in multivariate or relational contexts. However, their application to relatively simple univariate time series—particularly those characterized by a single peak—can introduce unnecessary model complexity and risk overfitting, especially when domain knowledge suggests a well-defined temporal shape.

In contrast, an emerging class of models focuses on fitting parametric unimodal functions - such as the generalized logistic (e.g., Richards) [1,10,39,43,45], skew-normal [3,17,32,33,46] or generalized gamma [6,9,16,19,42] - to describe the evolution of a univariate quantity over time. These functions are inherently designed to capture bell-shaped or skewed temporal profiles with a single peak, making them particularly well-suited for phenomena such as seasonal biological activity, pollutant concentrations, or temperature responses, where a rise-and-fall pattern is expected.

The key advantage of such parametric approaches lies in their simplicity and parsimony. By reducing a time series to a small number of interpretable parameters (e.g., peak time, amplitude, spread, skewness), these models provide compact summaries of temporal behavior. This stands in contrast to decision trees or symbolic models, which may require dozens of nodes or terms to achieve similar predictive performance for simple unimodal dynamics.

Moreover, the novelty of these unimodal function-based models lies in their potential for domain-specific interpretability and model regularization. Parameters can often be directly mapped to physical or biological concepts (e.g., growth rates, onset timing), making them highly suitable for scientific applications where insight into temporal dynamics is as important as prediction. These models also offer robust performance in low-data regimes and naturally handle extrapolation, especially near the tails, where data-driven methods may struggle.

Another benefit is computational efficiency. Parametric curve fitting is typically orders of magnitude faster than training complex machine learning models and can be implemented with standard optimization techniques. This makes unimodal function models ideal for applications requiring real-time analysis, automated monitoring, or integration into embedded systems. While complex machine learning models excel in high-dimensional and relational settings, the use of unimodal parametric functions offers a lightweight, interpretable, and domain-aligned alternative for modeling univariate time series with a single prominent peak.

3 Results and Discussion

Building on the advantages of parametric unimodal functions, an additional compelling approach arises from information theory: using functions derived from the maximum entropy principle as candidate models for unimodal time series. The principle of maximum entropy selects the probability distribution that best represents the current state of knowledge while making the fewest assumptions beyond the given constraints - typically moments like the mean, variance, or known support. When applied to temporal modeling, this leads to least-biased functional forms that encode only what is known (e.g., the existence of a single peak, finite time support, skewness) and nothing more.

We aim to identify candidate functions on a bounded interval that maximize the entropy functional. Without loss of generality we can consider the bounded interval to be $[0, 1]$. Let $p(x)$ be a function that maximizes the entropy functional:

$$H = \int_0^1 p(x) \log p(x) dx \tag{1}$$

under the following constraints:

$$\int_0^1 p(x) = C_1 \tag{2}$$

and

$$p(0) = 0 \text{ and } p(1) = 0 \tag{3}$$

where C_1 is a finite constant (typically $C_1 = 1$ for probability distributions). We can reformulate the constraints in (3) by considering functions $f(x), g(x)$ such that $\lim_{x\to 0} f(x) = \infty$ and $\lim_{x\to 1} g(x) = \infty$ and enforcing the following properties:

$$\lim_{\epsilon\to 0} \int_\epsilon^1 f(x)p(x)dx = C_2 \text{ and } \lim_{\epsilon\to 0} \int_0^{1-\epsilon} g(x)p(x)dx = C_3 \tag{4}$$

where C_2 and C_3 are also finite constants. The conditions in (4) imply that $p(x) = 0$ when $x \to 0$ or 1, which are the desired constraints in (3).

Using Lagrange multipliers, we can maximize the entropy functional H under the constraints imposed in (2) and (4). Different choices of functions $f(x)$ and $g(x)$ lead to different maximum entropy models. If $f(x) = \log x$ and $g(x) = \log(1-x)$ then the solution to the Euler-Lagrange equation derived from (1) with constraints (2) and (4) is (up to a normalization constant):

$$p(x) \sim x^{a-1}(1-x)^{b-1} \tag{5}$$

where $a \geq 1$ and $b \geq 1$ are parameters. Upon normalization this yields the $\beta(a, b)$ distribution, which is a well-known, classic probability distribution. For specific parameter choices (such as $a = 1, b = 1$), the β distribution is known to correspond to a maximum entropy distribution (e.g., the uniform distribution).

However, we have shown that the beta distribution is generally a maximum entropy distribution if the constraints (4) are imposed.

If $f(x) = \frac{1}{x}$ and $g(x) = \frac{1}{1-x}$ then:

$$p(x) \sim \exp\left(-\frac{a}{x} - \frac{b}{1-x}\right) \tag{6}$$

where $a, b > 0$ are parameters. We refer to (6) as the maximum entropy (MaxEnt) function (or distribution). Its mode (peak) is located at $x = \frac{\sqrt{a}}{\sqrt{a}+\sqrt{b}}$. While other choices of maximum entropy models are possible, we focus in this paper on applications of (5) and (6).

Table 1. Root Mean Squared (RMS) loss and standard deviations from cross-comparison of different unimodal models used to fit synthetic time series generated by each method. Diagonal values represent each model fitting its own data, while off-diagonal values highlight cross-model fitting performance.

Methods	Richards	Skewnormal	GenGamma	MaxEnt	Beta
Richards	0.0018(98)	0.04(7)	0.05(8)	0.04(1)	0.08(5)
Skewnormal	0.12(8)	0.07(4)	0.3(1)	0.28(6)	0.47(8)
GenGamma	0.31(9)	0.33(3)	0.4(1)	0.4(1)	0.3(2)
MaxEnt	**0.03(2)**	**0.03(1)**	0.06(9)	0.0002(30)	**0.05(10)**
Beta	**0.03(2)**	0.03(2)	**0.03(2)**	**0.011(5)**	0.001(3)

3.1 Method Comparison

Table 1 presents a cross-comparison of five different methods used to model unimodal time series: Richards, Skewnormal, Generalized Gamma (GenGamma), Maximum Entropy (MaxEnt), and Beta functions. Each row corresponds to the method used for fitting, while each column corresponds to the method that generated the synthetic time series data. The values represent the root mean squared loss in fit accuracy, with standard deviations shown in parentheses. Lower values indicate better performance.

Richards, Skewnormal, and Generalized Gamma functions form the core set of reference models in our comparative analysis because they offer flexible representations of asymmetry - a crucial feature for accurately modeling unimodal time series. Each of these functions can express both left-skewed and right-skewed shapes, allowing them to accommodate a wide range of peak timings and growth-collapse dynamics. The Richards function, as a generalized logistic model, adjusts its curvature through a shape parameter that governs asymmetry. The Skewnormal distribution introduces skew through a shape parameter modifying the Gaussian form, while the Generalized Gamma offers even greater flexibility by encompassing a family of distributions that include both symmetric and highly skewed profiles. These three models serve as the primary benchmarks against which we evaluate the performance of the two-parameter maximum entropy functions.

In contrast, other distributions commonly used in practice - such as the Lognormal and Gompertz - exhibit important limitations. The Lognormal is inherently right-skewed, making it unsuitable for processes with early peaks or left-skewed dynamics. The Gompertz function is in fact a special case of the Richards curve, and as such, it is implicitly included in our comparison through the Richards model itself.

As expected, the diagonal entries - where the model is applied to data generated by the same functional form - in general show the best performance across the board. This is intuitive: each method is best suited to reproducing its own characteristic shape and temporal dynamics. For instance, Richards performs best on Richards-generated data (0.0018), and similarly for the Skewnormal (0.07), MaxEnt (0.0002), and Beta (0.001) models. GenGamma is an exception, showing similarly poor performance when fitting both its own data and data from other models. This suggests that, contrary to its theoretical flexibility, the model may be more rigid in practice, or its parameter estimation may be especially prone to becoming trapped in local minima during optimization. As a result, GenGamma fails to capture its own characteristic dynamics as effectively as the other methods.

What is more revealing are the off-diagonal values, particularly the second-best scores, which highlight the versatility of models in capturing other functional forms. These are emphasized in bold within the table. A key finding from this analysis is the consistently strong cross-fit performance of the MaxEnt model. When applied to time series generated by other methods, MaxEnt achieves the lowest off-diagonal loss in three out of four cases: Richards (0.03), Skewnormal (0.03), and Beta (0.05). This suggests that the MaxEnt-derived models, which are based on the principle of maximizing entropy under shape constraints, provide a highly flexible and unbiased framework for modeling diverse unimodal time series shapes.

As we have shown, the Beta model is also a maximum entropy model and performs competitively in cross-fits, notably yielding the second-best loss on GenGamma-generated data (0.03) and a strong performance on MaxEnt-generated data (0.011). This reflects the Beta function's capacity to adapt to various peak asymmetries and supports its use in bounded time interval modeling.

In contrast, the GenGamma and Skewnormal models show poorer generalization to other types, with higher cross-fit errors - especially when fitting MaxEnt or Richards-generated data. In addition, the Richards, GenGamma and Skewnormal methods enjoy three parameters (or degrees of freedom) and should have a better capacity at fitting diverse patterns. However, the results in Table 1 illustrate that maximum entropy-based models offer the best overall versatility, capturing a wide range of unimodal dynamics with minimal loss. Their data-agnostic nature, guided only by moment constraints, makes them ideal candidates for general-purpose modeling where prior knowledge of the underlying process is limited.

3.2 Universe 25

Universe 25 is a behavioral experiment from the 1970 s by American ethologist John B. Calhoun, designed to study the effects of overpopulation on social behavior in rodents [8]. It involved a population of mice placed in a "utopia"—an enclosed space with unlimited food, water, nesting material, and no predators. The only limiting factor was space. The experiment aimed to simulate conditions of density-dependent stress.

Initially, the population grew rapidly, but over time, social breakdown emerged. Dominant males became aggressive, maternal behavior deteriorated, and many mice withdrew from social interaction entirely. Eventually, the population collapsed entirely, despite the continued presence of resources. This phenomenon, which Calhoun called the "behavioral sink", suggested that social and psychological stress from overcrowding could override even ideal environmental conditions.

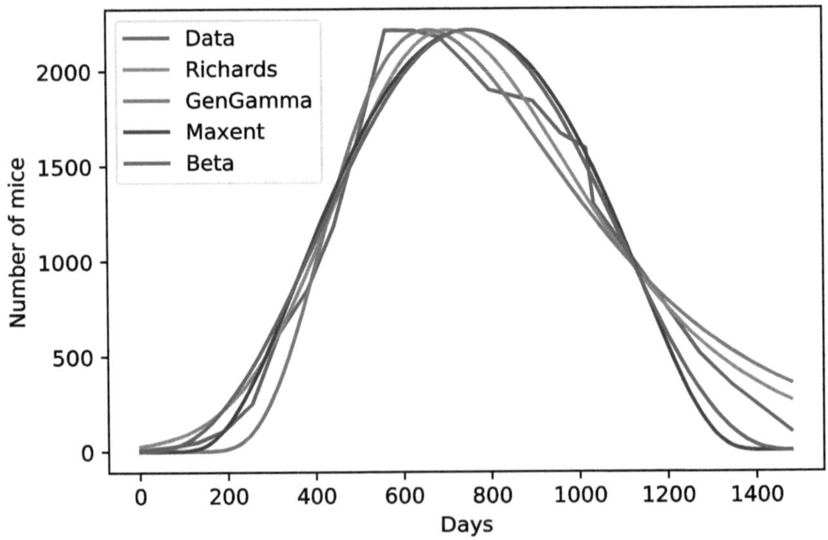

Fig. 1. Mouse population trajectory from the Universe 25 experiment, fitted with four unimodal models: Richards, GenGamma, MaxEnt, and Beta.

Figure 1 shows the mouse population curve over time from the Universe 25 experiment, plotted alongside fitted curves using five different unimodal modeling approaches: Richards, Generalized Gamma (GenGamma), Maximum

Entropy (MaxEnt), and Beta functions. The population data exhibits a characteristic unimodal shape - a rapid growth phase, a population peak, followed by a decline and collapse. The Richards function is a sigmoid-derived model that captures the general shape well, particularly the early growth and peak region. However, it slightly underperforms in capturing the asymmetry of the decline, showing a more gradual descent than observed in the data. The Generalized Gamma distribution is flexible, but in this case, it overestimates the early growth and smooths over some of the sharper transitions in the decline. It offers a reasonable fit but appears to lag in accurately modeling the abrupt changes post-peak.

The MaxEnt-derived curve, based on maximizing entropy under moment constraints, shows a strong fit across the entire curve, especially in the peak and decline regions. It follows the asymmetric descent more closely than most others, making it a strong candidate for capturing the essential features of the population trajectory. The Beta function also fits well, particularly around the growth and peak regions.

The fitted models each provide unique insights into the population dynamics of Universe 25. While all manage to reproduce the general unimodal shape, MaxEnt and Beta functions stand out for their flexibility and fidelity to the real data, particularly in modeling the asymmetric shape of the rise and collapse. The MaxEnt model's performance highlights the power of information-theoretic approaches in modeling complex natural phenomena with minimal assumptions, capturing both symmetry and skewness effectively. Overall, this figure supports the conclusion that modeling with flexible, peak-constrained functions like MaxEnt or Beta distributions can offer more accurate, data-consistent reconstructions of complex behavioral phenomena like those observed in Universe 25.

3.3 Saint Matthew Island

Saint Matthew Island, a remote landmass in the Bering Sea, is home to one of the most dramatic cases of wildlife population collapse ever recorded. In 1944, 29 reindeer were introduced to the island by the U.S. Coast Guard as a food source for a wartime weather station crew [13]. The station was soon abandoned, leaving the reindeer without predators or human interference. With abundant vegetation and no natural predators, the reindeer population grew exponentially. By 1963, their numbers had ballooned to approximately 6,000. However, the island's environment could not sustain such a large population. Overgrazing led to severe habitat degradation, particularly of the slow-growing lichens that formed their primary winter food source. The population crashed catastrophically during the harsh winter of 1963–64. By 1966, only 42 reindeer remained—41 females and a single male—none of which were reproductively viable. Eventually, the population went extinct. Figure 2 illustrates the reindeer population dynamics on Saint Matthew Island from their introduction in the 1940 s until their catastrophic collapse in the mid-1960 s. The empirical data shows a slow initial growth followed by a rapid exponential increase, peaking around 1963, and then a sudden, steep collapse.

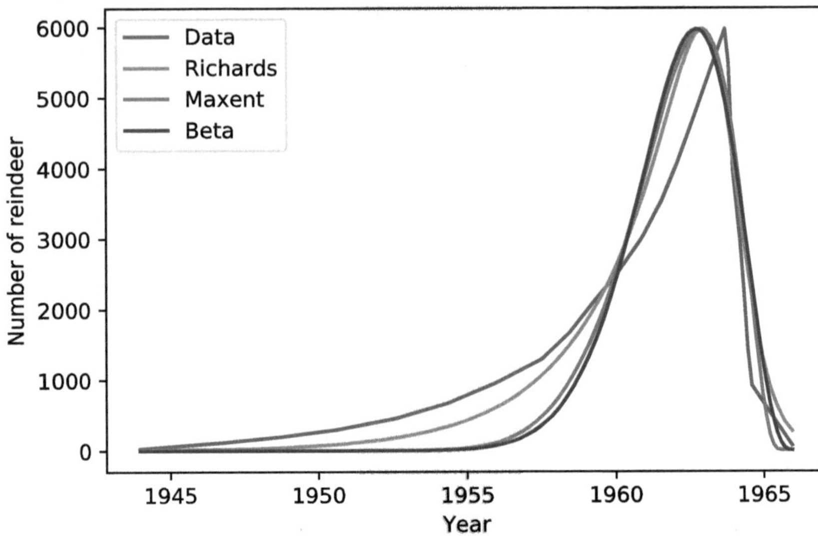

Fig. 2. Reindeer population on Saint Matthew Island (1944–1966) with fitted curves from three unimodal models: Richards, MaxEnt, and Beta. The data show rapid population growth followed by a sharp collapse.

As a generalized logistic function, the Richards curve captures the early and mid-phase population growth fairly well. However, it underestimates the sharpness of the collapse. Its symmetry and gradual decline make it less suited to modeling abrupt ecological breakdowns. The MaxEnt-derived curve shows good adaptability to the data, particularly in reflecting the asymmetrical shape of the population curve—rapid growth followed by an abrupt drop. This reflects the model's strength in capturing real-world ecological constraints without assuming specific underlying mechanisms, making it well-suited for modeling systems with sudden transitions or tipping points. The Beta function also performs well, closely matching the data through the peak and collapse phase. Its bounded nature makes it inherently suitable for modeling finite-time processes, such as a population rising and falling within a constrained time frame.

The extreme dynamics of the Saint Matthew Island reindeer population are a classic example of overshoot and collapse due to resource overexploitation. The ability of MaxEnt and Beta models to closely fit this trajectory underscores their usefulness in ecological modeling, especially when dealing with systems prone to nonlinear behavior and irreversible transitions. These models, grounded in shape and peak constraints, provide a powerful alternative to traditional growth models, offering better sensitivity to abrupt ecological changes such as extinction or collapse events.

4 Conclusion

This study presents a comparative analysis of various unimodal models - Richards, Skewnormal, Generalized Gamma, Beta, and a novel class of Maximum Entropy (MaxEnt) functions - for modeling time series that exhibit a single peak followed by a decline. Through both synthetic data experiments and real-world case studies, including the Universe 25 mouse population collapse and the Saint Matthew Island reindeer die-off, we evaluated each model's ability to fit and generalize complex unimodal dynamics.

A key result from the cross-comparison table of RMS losses is that, while each model performs best when fitting data generated from its own functional form, the maximum entropy models (Beta and MaxEnt) consistently achieve the lowest off-diagonal loss values. This indicates superior flexibility and generalization across different types of unimodal behavior. Importantly, the maximum entropy models accomplish this with only two degrees of freedom, in contrast to the three-parameter structure of the other models evaluated. This makes them not only more parsimonious but also less prone to overfitting, a critical advantage when data are sparse or noisy.

The empirical fits further support this conclusion. In both the Universe 25 and Saint Matthew Island datasets, the maximum entropy models offered closer tracking of asymmetric growth and collapse patterns, outperforming traditional models that tend to impose more symmetric or smoothed shapes. The Beta function is also competitive in accuracy and shares a similar peak-constrained form.

Taken together, these results demonstrate that maximum entropy-based models represent a novel, principled, and efficient framework for modeling unimodal time series. Their ability to provide strong fits using only shape-based constraints - without reliance on parametric assumptions or domain-specific tuning - makes them especially promising for applications in ecology, epidemiology, demography, and beyond. They serve not only as accurate approximators of known dynamics but also as least-biased candidates when functional forms are unknown or poorly understood.

Acknowledgments. This publication is supported by the European Union's Horizon Europe research and innovation programme under the Marie Skłodowska-Curie Postdoctoral Fellowship Programme, SMASH co-funded under the grant agreement No. 101081355. The operation (SMASH project) is co-funded by the Republic of Slovenia and the European Union from the European Regional Development Fund.

Data Availability Statement. The code to reproduce the results is available at: https://doi.org/10.5281/zenodo.16088580 of Interest Statement

Conflict of Interest Statement. The authors declare that the research was conducted in the absence of any commercial or financial relationships that could be construed as a potential conflict of interest.

References

1. Aljarrah, M.A., Famoye, F., Lee, C.: Generalized logistic distribution and its regression model. J. Stat. Distrib. Appl. **7**(1), 1–21 (2020). https://doi.org/10.1186/s40488-020-00107-8
2. Andonovikj, V., Boškoski, P., Džeroski, S., Boshkoska, B.M.: Survival analysis as semi-supervised multi-target regression for time-to-employment prediction using oblique predictive clustering trees. Expert Syst. Appl. **235**, 121246 (2024)
3. Azzalini, A.: The skew-normal distribution and related multivariate families. Scand. J. Stat. **32**(2), 159–188 (2005)
4. Bertolotti, F., Roman, S.: Balancing long-term and short-term strategies in a sustainability game. Iscience **27**(6) (2024)
5. Bogatinovski, J., Todorovski, L., Džeroski, S., Kocev, D.: Comprehensive comparative study of multi-label classification methods. Expert Syst. Appl. **203**, 117215 (2022)
6. Bourguignon, M., Lima, M.d.C.S., Leão, J., Nascimento, A.D., Pinho, L.G.B., Cordeiro, G.M.: A new generalized gamma distribution with applications. Am. J. Math. Manag. Sci. **34**(4), 309–342 (2015)
7. Brence, J., Džeroski, S., Todorovski, L.: Dimensionally-consistent equation discovery through probabilistic attribute grammars. Inf. Sci. **632**, 742–756 (2023)
8. Calhoun, J.B.: Death squared: the explosive growth and demise of a mouse population (1973)
9. Cordeiro, G.M., Ortega, E.M., Silva, G.O.: The exponentiated generalized gamma distribution with application to lifetime data. J. Stat. Comput. Simul. **81**(7), 827–842 (2011)
10. Cortés, I., Reyes, J., Iriarte, Y.A.: A weighted skew-logistic distribution with applications to environmental data. Mathematics **12**(9), 1287 (2024)
11. Gec, B., Omejc, N., Brence, J., Džeroski, S., Todorovski, L.: Discovery of differential equations using probabilistic grammars. In: International Conference on Discovery Science, pp. 22–31. Springer (2022)
12. Ispirova, G., Eftimov, T., Džeroski, S., Seljak, B.K.: Msgen: measuring generalization of nutrient value prediction across different recipe datasets. Expert Syst. Appl. **237**, 121507 (2024)
13. Klein, D.R.: The introduction, increase, and crash of reindeer on st. matthew island. J. Wildlife Manag. pp. 350–367 (1968)
14. Kompare, B., Dzeroski, S.: Two artificial intelligence methods for knowledge synthesis from environmental data, vol. 6. WIT Press (2025)
15. Kostovska, A., Doerr, C., Džeroski, S., Kocev, D., Panov, P., Eftimov, T.: Explainable model-specific algorithm selection for multi-label classification. In: 2022 IEEE Symposium Series on Computational Intelligence (SSCI), pp. 39–46. IEEE (2022)
16. Lahcene, B.: A new extended-gamma family of distributions: properties and applications. J. Appl. Math. Comput. **5**(1) (2021)
17. Lee, S.X., McLachlan, G.J.: On mixtures of skew normal and skew-distributions. Adv. Data Anal. Classif. **7**(3), 241–266 (2013)
18. Levatić, J., Ceci, M., Kocev, D., Džeroski, S.: Semi-supervised predictive clustering trees for (hierarchical) multi-label classification. Int. J. Intell. Syst. **2024**(1), 5610291 (2024)
19. Li, H.C., Hong, W., Wu, Y.R., Fan, P.Z.: On the empirical-statistical modeling of SAR images with generalized gamma distribution. IEEE J. Sel. Top. Signal Proc. **5**(3), 386–397 (2011)

20. Lukšič, Ž, Tanevski, J., Džeroski, S., Todorovski, L.: Meta-model framework for surrogate-based parameter estimation in dynamical systems. IEEE Access **7**, 181829–181841 (2019)
21. Novak Babič, M., Gunde-Cimerman, N., Breskvar, M., Džeroski, S., Brandão, J.: Occurrence, diversity and anti-fungal resistance of fungi in sand of an urban beach in slovenia–environmental monitoring with possible health risk implications. J. Fungi **8**(8), 860 (2022)
22. Omejc, N., Gec, B., Brence, J., Todorovski, L., Džeroski, S.: Probabilistic grammars for modeling dynamical systems from coarse, noisy, and partial data. Mach. Learn. **113**(10), 7689–7721 (2024)
23. Osojnik, A., Panov, P., Džeroski, S.: isoup-symrf: symbolic feature ranking with random forests in online multi-target regression and multi-label classification. Mach. Learn. **114**(2), 34 (2025)
24. Petković, M., Ceci, M., Pio, G., Škrlj, B., Kersting, K., Džeroski, S.: Relational tree ensembles and feature rankings. Knowl.-Based Syst. **251**, 109254 (2022)
25. Petković, M., Levatić, J., Kocev, D., Breskvar, M., Džeroski, S.: Clusplus: a decision tree-based framework for predicting structured outputs. SoftwareX **24**, 101526 (2023)
26. Radinja, M., Škerjanec, M., Šraj, M., Džeroski, S., Todorovski, L., Atanasova, N.: Automated modelling of urban runoff based on domain knowledge and equation discovery. J. Hydrol. **603**, 127077 (2021)
27. Roman, S.: Historical dynamics of the chinese dynasties. Heliyon **7**(6) (2021)
28. Roman, S.: Theories and models: understanding and predicting societal collapse. In: The Era of Global Risk: An Introduction to Existential Risk Studies, pp. 27–54. Open Book Publishers (2023). https://doi.org/10.11647/OBP.0336.02
29. Roman, S., Bertolotti, F.: A master equation for power laws. Royal Soc. open sci. **9**(12), 220531 (2022)
30. Roman, S., Bertolotti, F.: Global history, the emergence of chaos and inducing sustainability in networks of socio-ecological systems. PLoS ONE **18**(11), e0293391 (2023)
31. Roman, S., Palmer, E.: The growth and decline of the western roman empire: quantifying the dynamics of army size, territory, and coinage. Cliodynamics **10**(2) (2019)
32. Said, K.K., Ning, W., Tian, Y.: Likelihood procedure for testing changes in skew normal model with applications to stock returns. Commun. Stat. Simul. Comput. **46**(9), 6790–6802 (2017)
33. Shafiei, S., Doostparast, M., Jamalizadeh, A.: The alpha-beta skew normal distribution: Properties and applications. Statistics **50**(2), 338–349 (2016)
34. Simidjievski, N., Todorovski, L., Džeroski, S.: Modeling dynamic systems with efficient ensembles of process-based models. PLoS ONE **11**(4), e0153507 (2016)
35. Simidjievski, N., Todorovski, L., Kocijan, J., Džeroski, S.: Equation discovery for nonlinear system identification. IEEE Access **8**, 29930–29943 (2020)
36. Stevanoski, B., Kostovska, A., Panov, P., Džeroski, S.: Change detection and adaptation in multi-target regression on data streams. Mach. Learn. **113**(11), 8585–8622 (2024)
37. Stoimchev, M., Kocev, D., Džeroski, S.: Deep network architectures as feature extractors for multi-label classification of remote sensing images. Remote Sensing **15**(2), 538 (2023)
38. Stoimchev, M., Levatić, J., Kocev, D., Džeroski, S.: Semi-supervised multi-label classification of land use/land cover in remote sensing images with predictive clustering trees and ensembles. IEEE Trans. Geosci. Remote Sens. (2024)

39. Stroup, W.W., Ptukhina, M., Garai, J.: Generalized linear mixed models: modern concepts, methods and applications. Chapman and Hall/CRC (2024)
40. Tanevski, J., Simidjievski, N., Todorovski, L., Džeroski, S.: Process-based modeling and design of dynamical systems. In: Machine Learning and Knowledge Discovery in Databases: European Conference, ECML PKDD 2017, Skopje, Macedonia, September 18–22, 2017, Proceedings, Part III 17, pp. 378–382. Springer (2017)
41. Tanevski, J., Todorovski, L., Džeroski, S.: Combinatorial search for selecting the structure of models of dynamical systems with equation discovery. Eng. Appl. Artif. Intell. **89**, 103423 (2020)
42. Thurai, M., Bringi, V.: Application of the generalized gamma model to represent the full rain drop size distribution spectra. J. Appl. Meteorol. Climatol. **57**(5), 1197–1210 (2018)
43. Tjørve, E., Tjørve, K.M.: A unified approach to the richards-model family for use in growth analyses: why we need only two model forms. J. Theor. Biol. **267**(3), 417–425 (2010)
44. Tolovski, I., Kostovska, A., Simidjievski, N., Todorovski, L., Džeroski, S., Panov, P.: Towards reusable process-based models of dynamical systems: a case study in the domain of aquatic ecosystems. In: 2019 42nd International Convention on Information and Communication Technology, Electronics and Microelectronics (MIPRO), pp. 1110–1115. IEEE (2019)
45. Wang, X.S., Wu, J., Yang, Y.: Richards model revisited: validation by and application to infection dynamics. J. Theor. Biol. **313**, 12–19 (2012)
46. Zadkarami, M.R., Rowhani, M.: Application of skew-normal in classification of satellite image. J. Data Sci. **8**(4), 597–606 (2010)

Learning from Complex Data

Who Drives Misinformation? Key Node Detection with Heterogeneous Graph Neural Networks

Liliana Martirano(✉), Francesco Scala, Carmela Comito, and Luigi Pontieri

Institute of High Performance Computing and Networking (ICAR-CNR), Via P. Bucci, 87036 Rende, CS, Italy
{liliana.martirano,francesco.scala,carmela.comito, luigi.pontieri}@icar.cnr.it

Abstract. Misinformation propagation in online networks involves multifaceted interactions between users, contents, and engagement mechanisms (likes, shares, comments). Addressing this issue entails both understanding how information spreads and identifying influential users driving the dissemination process. To tackle these challenges, this paper proposes a framework based on a Graph Attention Network model, applied to a heterogeneous graph representing social interactions and context-aware dynamics. Targeting the binary classification of real vs fake news, it offers insights into both propagation patterns and influential users in the dissemination process. A core contribution is the adoption of two post-hoc mechanisms for uncovering such users: uncertainty-based Active learning-like and GNN-Explainer. A detailed comparative analysis reveals that nodes where the model exhibits the highest confidence often lack rich content information; nevertheless, combining both high-confidence and content-rich nodes grasps complementary aspects and better aligns with influential users in information propagation. The framework is benchmarked against traditional centrality measures, widely used to identify influential users in social networks. A comparative evaluation on two heterogeneous, real-world, social networks confirms that the proposed method both achieves compelling accuracy in finding influential nodes and shows a potential to scale-up to densely-connected graphs on which classic approaches may fail.

Keywords: Graph Neural Networks · GNN-Explainer · Active Learning · Misinformation · Influential nodes

1 Introduction

The proliferation of misinformation in online social networks presents a significant challenge, influencing public opinion, shaping societal discourse, and even affecting democratic processes. With the increasing reliance on digital platforms for information consumption, the rapid and widespread dissemination of misleading content has become a pressing issue. This phenomenon is driven by

S. Džeroski et al. (Eds.): DS 2025, LNAI 16090, pp. 47–62, 2025.
https://doi.org/10.1007/978-3-032-05461-6_4

complex interactions among users, content, and engagement mechanisms such as likes, shares, and comments. Consequently, understanding how misinformation spreads and identifying the key users responsible for its propagation are crucial for developing effective mitigation strategies.

Traditional approaches to influence analysis in social networks often rely on graph-based centrality measures such as degree centrality, closeness centrality, betweenness centrality [16] and PageRank [2], or influence-spreading heuristics related to centrality like VoteRank [17]. While these methods provide valuable insights into network structures, they are not directly applicable to heterogeneous networks, i.e., graphs with multiple node and/or edge types. Moreover, computational constraints make them less suitable for large-scale social media datasets, necessitating the development of more scalable and adaptive solutions. Finally, existing methods often assume uniform importance among neighboring nodes and fail to account for the structural and semantic diversity of relationships in heterogeneous networks.

To address these limitations, we propose a meta-path embedding-based Graph Attention Network (GAT) model [6] to identify influential nodes in heterogeneous networks. The framework leverages GAT and meta-paths applied to heterogeneous social graphs to capture multi-typed dynamics and model misinformation propagation patterns. The proposed model performs binary classification of real and fake news while simultaneously uncovering key users driving information dissemination. A key contribution of this work is the integration of post-hoc techniques to identify influential nodes in the network. Specifically, we explore two complementary approaches: (1) Uncertainty-based Active Learning-like (AL-like) approaches, which prioritize nodes where the model exhibits high confidence, and (2) GNN-Explainer, which identifies critical nodes for the classification task.

To validate our approach, we conduct an extensive comparative analysis against traditional centrality measures, evaluating the efficacy and efficiency of our model in identifying influential users.

Our findings demonstrate that the proposed framework not only achieves competitive accuracy but also demonstrates promising scalability to densely connected graphs, making it well-suited for real-world, large-scale misinformation detection tasks. Experimental validation on two heterogeneous, real-world social graphs exhibiting dissimilar topologies further substantiates the framework's robustness, offering a scalable solution for influence analysis in online networks.

By bridging graph-based deep learning with explainability techniques, our work contributes to a deeper understanding of misinformation propagation, paving the way for more effective countermeasures against the spread of false information in digital environments.

The rest of the paper is structured as follows. Section 2 discusses major related approaches. Section 3 introduces the proposed methodology and describes its base components. Section 4 illustrates the experimental investigation. Finally, Sect. 5 concludes the work and provides pointers for future research.

2 Related Work

Methods for identifying influential nodes fall into two categories: traditional and deep learning-based.

Traditional methods rely on centrality metrics, such as local (e.g., degree), global (e.g., betweenness), and hybrid (e.g., PageRank, VoteRank). Examples include DC [10], WSLC [12], and LGC [11], which use various combinations of edge weighting, subgraph extraction, and path-based analysis. However, they often neglect influence variability, node attributes, and heterogeneity—key limitations in complex or heterogeneous networks.

Deep learning-based approaches better handle high-dimensional data and complex structures but often focus on homogeneous graphs or rely on rigid meta-paths in heterogeneous ones. Models such as those by Yu et al. [15], Kou et al. [5], Zhao et al. [18], Keikha et al. [4], and Ahmad et al. [1] use CNNs, GCNs, or embeddings for influence prediction. MEGA [13], the most relevant to our work, targets academic networks using meta-path-based GAT aggregation. Unlike MEGA, our method dynamically combines active learning and GNN-Explainer to rank nodes based on learned influence patterns, not just predefined structures. It also offers interpretability and is tailored for misinformation detection, broadening its applicability across real-world contexts.

3 Proposed Methodology

The methodology presented in this study is designed to detect key actors, such as influencers, within complex social media misinformation networks. At its core, the approach involves constructing a heterogeneous graph that captures the interactional and structural dynamics of the network, focusing on entities such as news articles, users, tweets, and hashtags, and the relationships among them. This graph serves as the foundation for applying a Graph Attention Network, which leverages attention mechanisms to classify news articles as fake or real while learning from the contextual and social interaction signals embedded in the graph. The final step of our framework involves extracting model-driven insights to identify the most influential (top-k) nodes driving misinformation propagation. In the following, we delve into the details of the proposed methodology (see Fig. 1 for the workflow overview).

3.1 Network Modeling

In the first step we propose two feature-rich network models able to capture the complex semantics underlying the news propagation process within the social media ecosystem. We first construct a heterogeneous attributed graph for the fake news detection task and then derive a homogeneous user-centric meta-path-based weighted graph enabling the conventional application of well-known network analysis techniques while preserving the prominent information.

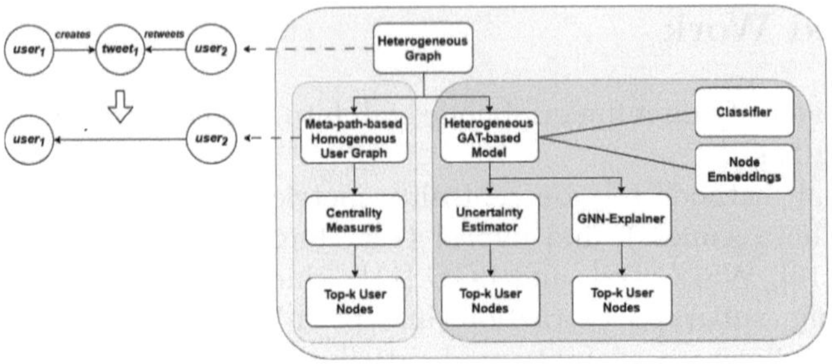

Fig. 1. Logical schemes for detecting influential nodes in a given heterogeneous graph: baseline approach based on computing centrality measures on a homogeneous-node view of the graph (yellow) vs the proposed heterogeneous GAT-based approach (blue). For the yellow block, we extract the (homogeneous) meta-path based graph of user nodes on which we compute the traditional centrality measures to extract top-k nodes. For the blue block, we apply on the original (heterogeneous) graph a GAT-based model, trained at classifying news/claim nodes (e.g., using labels obtained via fact checking), which learns embeddings for all node types; from the trained model we extract top-k user nodes based on uncertainty estimation (AL-like approaches) and explainability techniques.(Color figure online)

The heterogeneous graph captures the intricate relationships among different types of entities involved in the propagation process, like news articles, social media posts, users and hashtags. Each edge type represents a specific type of interaction or association, including retweets, mentions and co-occurrence within posts. The user-centric weighted graph focuses on a single target node type for further analysis using as-is traditional techniques, e.g., to detect the most influential users leading the spread of information.

Formally, we define a heterogeneous attributed network as $G = \langle \mathcal{V}, \mathcal{E}, A, R, \phi, \varphi, \mathcal{X} \rangle$, where $\mathcal{V}$ is the set of nodes, $\mathcal{E} = \bigcup_{r \in R} \mathcal{E}_r \subseteq \mathcal{V} \times \mathcal{V}$ is the set of (directed) edges, A is the set of node types, R is the set of relation types, $\mathcal{E}_r$ is the subset of edges associated with any relation $r \in R$, $\phi : \mathcal{V} \rightarrow A$ is the node-type mapping function, $\varphi : \mathcal{E} \rightarrow R$ is the edge-type mapping function, and $\mathcal{X}$ is a set of matrices $\mathcal{X} = \{\mathbf{X}^{(a)} \mid a \in A\}$ storing node attributes.

More complex interactions in a heterogeneous graphs can be modeled via meta-paths. A *meta-path type* (or simply *meta-path*) in a heterogeneous network is a composite relation modeling high-order proximity induced by a path $a_1 \xrightarrow{r_1} a_2 \xrightarrow{r_2} \ldots \xrightarrow{r_x} a_{x+1}$ between two node types, $a_1 \in A$ and $a_{x+1} \in A$, which are expected to share information – see Fig. 1 for a concrete meta-path example. Let σ_m be a meta-path of type m, and let $\mathcal{M}$ be the set of all the chosen meta-path types in the heterogeneous graph. Notation $\sigma_m(v, v')$ will denote the fact that nodes v and v' (such that v and v' belong to the starting node type and final node type in meta-path σ_m) are connected via σ_m.

A *meta-path instance* is a sequence of connected nodes matching the node and edge types in the meta-path, able to make two distant nodes in the network reachable. A *meta-path-based graph* is a graph comprised of nodes connected via meta-path instances of one or more meta-path types. Given a target node type $\overline{a} \in A$ and a meta-path $\sigma_m \in \mathcal{M}$, with terminal nodes of the same type $\overline{a}$, the resulting meta-path-based graph is a homogeneous weighted graph with one node type $\overline{a}$ and one relation type σ_m, obtained from all meta-path instances of σ_m by removing the intermediate nodes and establishing a direct link between the terminal nodes, weighted on the number of meta-path instances connecting them. Many meta-paths connecting nodes of type $\overline{a}$ can be combined into a single meta-path-based graph.

3.2 Fake News Detection

The second step applies a GAT-based architecture enhanced by meta-paths to the constructed heterogeneous graph to classify news articles as fake or real by relying on both the network topology and the node attributes as in [6]. The GAT operates on the heterogeneous graph at node-level by computing attention scores for each incident edge and meta-path instance[1], dynamically weighing the contributions of neighbors. The attention mechanism allows to prioritize influential relationships, such as a widely retweeted tweet or a frequently mentioned user or frequently co-occurring hashtags.

The model architecture includes an embedding layer to encode the initial features of each node, followed by attention layers to refine the embeddings based on the influence of neighboring nodes. The final output layer classifies nodes by their likelihood of being fake news. The graph is processed in a type-aware manner, with distinct attention mechanisms applied to different edge types (e.g., users retweeting tweets vs. hashtags used in tweets). This ensures that the model captures the semantics of each relationship. Information from neighboring nodes is aggregated based on the computed attention scores, updating the node embeddings to reflect their local context. Please note that, as well as to tell fake and real news apart, the learned embeddings can be used for different downstream tasks, at node, edge and (sub)graph level. By making the GAT-based classifier learn all node embeddings within a shared latent space, we propose to leverage it for identifying influential user nodes.

3.3 Key Node Identification

Identifying key nodes within an online misinformation network is critical for understanding how false information propagates and how it can be mitigated. In the third step, node rankings are determined and key (top-k) nodes are detected based on their roles in either amplifying or suppressing misinformation. Influential users act as amplifiers by driving the spread of misinformation, while suppressors counter false narratives and help curb their dissemination. Understanding these roles is essential for designing effective intervention strategies.

[1] We treat meta-paths as additional relationships between nodes of a given type.

Traditional approaches to identify influential nodes rely on centrality measures such as degree centrality, which highlights well-connected nodes; betweenness centrality, which identifies nodes serving as network bridges; closeness centrality, which measures how efficiently a node can reach others; PageRank, which evaluates node importance based on both the quantity and quality of incoming connections. Additionally, influence-spreading heuristics like VoteRank iteratively select key nodes to maximize diffusion in a network. While these measures offer valuable structural insights, they may fail to account for the contextual aspects of misinformation spread and suffer from high computational costs, hence struggling in dense graphs where high connectivity complicates the distinction of truly influential nodes and becoming impractical for large-scale networks.

To address these limitations, our framework integrates model-driven techniques by leveraging the trained GAT and advanced post-hoc methods. Specifically, two complementary approaches are employed to uncover influential nodes: uncertainty-based AL-like techniques and GNN-Explainer. The AL-like approach prioritizes nodes where the model exhibits high confidence, ensuring that the most informative and structurally significant nodes are analyzed. On the other hand, a GNN-Explainer module provides a score for each instance allowing to identify which nodes and edges contribute the most to the model's predictions. This dual approach ensures that both structural influence and contextual significance are accounted for, approximating the traditional techniques while remaining scalable as the graph grows. Each technique applied on the heterogeneous GAT deal with the multiplicity of types in the graph and results in a ranking of nodes, allowing top-k nodes of a given type to be selected as most relevant.

Uncertainty-Based Active Learning-Like (AL-Like) Approach. Active Learning (AL) techniques aim to reduce labeling costs by iteratively selecting a small number of informative instances from a large unlabeled dataset [8]. In this work, AL-like instance ranking criteria are exploited to identify key user nodes, which deem higher attention in the analysis and management of fake news propagation.

This method prioritizes nodes for further analysis by selecting those where the GAT model exhibits the highest prediction confidence. The rationale is that user nodes associated with highly confident predictions are likely to feature some properties that contribute significantly to (mis-)information propagation. These techniques were selected based on their energy efficiency, as they require only a single inference pass per data sample. The employed methods are simple measures computed directly from the model's output.

The AL-like strategy begins by sampling nodes with high classification certainty. These nodes are examined in greater detail to assess their structural and behavioral significance in the network. Unlike traditional AL involving iterative querying of labels, this AL-like method focuses on identifying nodes that the model deems most influential without requiring additional labeling efforts.

In the current implementation of the approach, we use the following three uncertainty-based ranking criteria, in a opposite way w.r.t. typical AL settings (i.e., the lower the uncertainty, the higher the rank):

- *Least Confidence Sampling* (for short, *AL LCS*): Let p be the probability of the most likely class for a data instance x. Then the least confidence score assigned to x is simply computed as $1 - p$;
- *Margin Sampling* (for short, *AL Margin*): This criterion focuses on the difference between the probability of the most likely class and the second most likely class. If, for a data instance x, p_{top1} and p_{top2} are the probability of the most likely class and of the second most likely class, respectively, then the margin score of x is computed as $p_{top1} - p_{top2}$;
- *Entropy Sampling* (for short, *AL Entropy*): Entropy measures the overall uncertainty across all classes. A high entropy value means the model is unsure about the correct class. For a data instance x, if there are C classes and p_i is the probability of the i-th class, then it is calculated as $-\sum_{i=1}^{C} p_i \log p_i$.

GNN Explainer Approach. GNN-Explainer [14] is a post-hoc interpretability method designed to uncover the most important structural and feature-related factors influencing Graph Neural Network's predictions. It can be extended to heterogeneous graphs with minimal effort. We use a Captum-based explainer which optimizes a mask over the input graph, assigning importance scores to node features and relations based on their impact on the gradient, highlighting key edges and node features that drive the GAT classification outcomes.

To identify highly central nodes responsible for amplifying fake news, we compute the node-level score according to two complementary strategies:

- *feature-based score* (for short, *Expl. Feats.*), computed by summing along the feature dimension, i.e., by aggregating the contributions of all its features;
- *relation-based score* (for short, *Expl. Rel.*), computed by summing along the incident edges, i.e., by aggregating the contributions of all its neighbors.

The disjunction of the two node-level scores enable the understanding of the predominance of one contribution over the other in analyzing the most critical actors in the network. For user nodes, we assign an importance score based on node features —such as engagement metrics, credibility scores, and content characteristics —as well as an importance score based on edges representing interactions between users, allowing for a better understanding of how misinformation spreads and which users play central roles in its propagation.

Unlike the AL-like approach, which focuses on confident nodes, GNN-Explainer provides a broader view by identifying nodes that play a pivotal role in shaping misinformation spread, regardless of their classification confidence.

4 Experimental Evaluation

The goal of this evaluation is to systematically assess the efficacy of the proposed approach by analyzing its ability to identify influential nodes and maximize information diffusion. The objective is to generate a ranking of nodes such that selecting the top-k nodes ensures the highest possible spread of influence on the remaining nodes in few steps.

The evaluation focuses on two performance dimensions: (i) Influence Spread Capacity, i.e., evaluating how many nodes are influenced when selecting top-ranked nodes, and (ii) Computational Efficiency, i.e., comparing the execution time of each method to determine scalability.

4.1 Baselines

To assess the effectiveness of the proposed approach in detecting the most influential nodes, we compared its performance against traditional centrality measures, which evaluate structural significance within the network while offering the advantage of being highly interpretable:

- *Degree Centrality.* Reflects the number of direct connections a user has, capturing their immediate influence;
- *Betweenness Centrality.* Measures how often a user acts as a bridge between other nodes, highlighting their role in spreading information by facilitating communication between otherwise disconnected groups;
- *Closeness Centrality.* Indicates how quickly a user can reach others in the network, capturing their ability to spread information across the network;
- *Pagerank.* Quantifies a user's overall importance within the graph, based on the connectivity with other influential nodes;
- *Voterank.* Identifies influential nodes in a network based on an iterative voting process by selecting influential nodes based on their ability to influence others while avoiding redundancy.

4.2 Performance Metrics

Influence Spread Metrics. To evaluate the influence spread effectiveness, in terms of the extent to which selected influential nodes propagate information, two widely used metrics are considered. Given N the total number of nodes in the network, K the number of selected top-k influential nodes (seed nodes) and R the number of nodes reached (influenced) by the K selected nodes using Breadth-First Search (BFS) algorithm within a fixed small, empirically determined depth, we define our reachability and coverage metrics as follows:

- $Coverage(K) = \frac{R+K}{N}$: measures the proportion of covered nodes, including both seed nodes and influenced nodes, out of the total number of nodes in the graph;
- $Reachability(K) = \frac{R}{N-K}$: evaluates how well the selected seed nodes propagate influence among non-seed nodes, by quantifying the total number of influenced nodes w.r.t the maximum number of nodes that can be influenced.

Table 1. MuMiN statistics, in terms of no. of nodes, edges and meta-paths. In bold the ranked node type and the relations held in the meta-path-based graph.

# Nodes	Claim (C)	2168
	Tweet (T)	4340
	Reply (R)	195459
	User (U)	153168
	Hashtag (H)	28091
	Image (I)	1020
	Article (A)	1453
# Meta-paths	C-T-U-T-C	28867
	C-T-H-T-C	21577
	C-T-R-T-C_r	2859
	C-T-R-T-C_q	3042
	U-T-U	11412
	U-R-U	146056

# Edges	T *discusses* C (C-T)	5081
	R *reply_to* T (T-R_r)	90196
	R *quote_of* T (T-R_q)	101216
	T *has_hashtag* H (T-H)	2289
	T *has_article* A (T-A)	1898
	T *has_image* I (T-I)	1028
	T *mentions* U (U-T_m)	1119
	U *posted* T (U-T_p)	4091
	U *posted* R (U-R)	179247
	U *retweeted* T (U-T_r)	13402
	U *follows* U (U-U_f)	18379
	U *mentions* U (U-U_m)	2797
	U *has_hashtag* H (U-H)	50451

As K increases, i.e., the number of starting influencial nodes grows, Coverage(K) is expected to increase, as it accounts for both the K influential nodes selected at start (based on centrality-like rankings) and those that they can potentially influence; however, this may not necessarily hold for Reachability(K), since a higher K reduces the set of reachable nodes.

Ranking Similarity Metrics. To compare two rankings over the same set, multiple rank correlation coefficients measuring the overall similarity and overlap measures are considered:

- *Spearman's Rank Correlation Coefficient* (ρ): measures how well the rankings match through the correlation between the rank positions of elements in both lists. Given d_i the difference between the ranks of element i in both lists: $\rho = \frac{6\sum d_i^2}{N(N^2-1)}$;
- *Kendall's Tau* (τ): counts the number of concordant (C) and discordant (D) pairs between the two lists: $\tau = \frac{C-D}{N(N-1)/2}$;
- *Jaccard Similarity Index:* measures the percentage of common nodes in the two top-K rankings A_K and B_K: $Jaccard@K = \frac{|A_K \cap B_K|}{|A_K \cup B_K|}$;
- *Precision@K*: calculates how many of the top-K items in one list appear in the other list: $Precision@K = \frac{|A_K \cap B_K|}{K}$;
- *Normalized Discounted Cumulative Gain* (NDCG@K): measures ranking quality by considering the relevance and order of ranked nodes: $NDCG@K = \frac{DCG@K}{IDCG@K}$ where DCG is the Discounted Cumulative Gain and measures the effectiveness of a ranking method by considering both relevance scores and positioning of ranked elements, while $IDCG$ is the ideal discounted cumulative gain, where nodes are perfectly ranked.

4.3 Datasets

The experimental evaluation has been conducted on two real-world datasets, i.e., MuMiN and PolitiFact datasets. Both are modeled as a heterogeneous information network, with multiple node and edge types and external information

Table 2. PolitiFact statistics, in terms of no. of nodes, edges and meta-paths. In bold the ranked node type and the relations held in the meta-path-based graph.

# Nodes	News (N)	696
	Tweet (T)	268306
	User (U)	169106
	Hashtag (H)	18631
# Meta-paths	N-T-H-T-N	44682
	N-T-U-T-N	46025
	N-T-U-U-T-N	20056
	U-T-U	533

# Edges	T *discusses* N (N-T)	276676
	T *has_ hashtag* H (T-H)	59782
	U *posted* T (U-T_p)	285124
	U *retweeted* T (U-T_r)	539
	U *mentions* U (U-U)	84093

associated with nodes available as a set of attributes (for detailed information on node features, please refer to the corresponding publication). The former is multi-topic; a complete description is provided in [7]. The latter is extracted from the FakeNewsNet data repository [9], which fact-checks news pertaining to the US political system.

In the following, we use the terms Claim and News interchangeably, as the MuMiN dataset refers to news items as Claims, while the PolitiFact dataset uses the term News. Classification is performed on nodes of type Claim (C) in the MuMiN dataset and nodes of type News (N) in the PolitiFact dataset.

To enhance classification performance by capturing richer structural and semantic relationships, we handcrafted some *meta-paths* toward the Claim/News node type. Two types are common to both datasets, including connections between pairs of claims (news) discussed in tweets by the same user or associated with the same hashtag. Additionally, dataset-specific meta-paths are defined. For MuMiN, we identify pairs of claims belonging to the same conversation thread through reply or quote relationships, respectively. For PolitiFact, we include pairs of news discussed in tweets posted by users who mention each other.

To construct the meta-path-based user graph we look at all the edges between users and additionaly integrate the U-T-U meta-path for both datasets, which connects users if one has retweeted the other's post, and the U-R-U meta-path for MuMiN, which connects users who have commented on the same tweet.

Tables 1 and 2 show the statistics in terms of number of nodes, number of edges, and number of meta-path instances for each type, for the two datasets.

4.4 Results

To assess the effectiveness of the proposed methodology in identifying influential user nodes, we compare individual and combined rankings versus the baselines using the two influence spread metrics defined in Sect. 4.2. Before delving into this comparative study, let us first show the results of some preliminary analyses.

Table 3. Results of the Fake News Detection task averaged over 5 runs. Execution time is expressed in seconds.

Dataset	F1-micro	F1-macro	Precision T	Recall T	Precision F	Recall F	Time
MuMiN	0.954 ± 0.007	0.788 ± 0.117	0.721±0.139	0.647 ± 0.191	0.967 ± 0.014	0.985 ± 0.009	189 ± 21
Politifact	0.859 ± 0.039	0.845 ± 0.054	0.848 ± 0.066	0.943 ± 0.044	0.909 ± 0.063	0.736 ± 0.153	332 ± 18

Table 4. Agreement between node importance rankings produced by pairs of post-hoc techniques, using multiple metrics.

Dataset	Pair of techniques	Spea. ρ	Kend. τ	Jacc. index		Precision		NDCG	
				@$\frac{N}{2}$	@$\frac{N}{4}$	@$\frac{N}{2}$	@$\frac{N}{4}$	@$\frac{N}{2}$	@$\frac{N}{4}$
Politifact	Expl feats - AL entr	-0.001	-0.001	0.333	0.133	0.500	0.235	0.935	0.868
	Expl feats - AL lcs	0.029	0.020	0.344	0.154	0.512	0.266	0.9400	0.880
	Expl feats - AL marg	-0.001	-0.001	0.333	0.133	0.500	0.235	0.935	0.868
	Expl rel - AL entr	-0.003	-0.002	0.331	0.140	0.497	0.246	0.936	0.870
	Expl rel - AL lcs	0.032	0.021	0.343	0.144	0.511	0.252	0.94	0.873
	Expl rel - AL marg	-0.003	-0.002	0.33	0.140	0.497	0.245	0.936	0.870
	Expl feats - Expl rel	0.638	0.531	0.622	0.502	0.767	0.669	0.978	0.959
	AL entr - AL lcs	0.319	0.218	0.431	0.206	0.602	0.341	0.938	0.884
	AL entr - AL marg	1.000	0.999	1.000	0.999	1.000	0.999	1.000	1.000
	AL lcs - AL marg	0.319	0.218	0.430	0.206	0.602	0.341	0.957	0.894
Mumin	Expl feats - AL entr	-0.031	-0.021	0.313	0.157	0.476	0.272	0.936	0.874
	Expl feats - AL lcs	-0.049	-0.034	0.289	0.143	0.449	0.25	0.934	0.87
	Expl feats - AL marg	-0.031	-0.022	0.312	0.157	0.476	0.271	0.936	0.874
	Expl rel - AL entr	-0.012	-0.008	0.323	0.15	0.489	0.260	0.936	0.873
	Expl rel - AL lcs	-0.039	-0.026	0.32	0.141	0.485	0.247	0.935	0.87
	Expl rel - AL marg	-0.008	-0.005	0.327	0.151	0.493	0.262	0.937	0.873
	Expl feats - Expl rel	0.184	0.201	0.384	0.177	0.555	0.301	0.953	0.889
	AL entr - AL lcs	0.815	0.649	0.719	0.545	0.836	0.706	0.991	0.973
	AL entr - AL marg	1.000	0.988	0.978	0.976	0.989	0.988	1.000	1.000
	AL lcs - AL marg	0.816	0.650	0.721	0.546	0.838	0.706	0.987	0.979

Preliminary tests: GAT accuracy and correlation analysis Table 3 shows a number of accuracy metrics for the fake news classification task. These results confirm that the proposed GAT-based architecture performs well on both datasets ([6]).

Here, we report the main parameters used in the experimentation. As regards the graph neural network model, we employed a l-layer GAT [3] architecture with dropout set to 0.4, hidden channels dimension set to 64 and out channel dimension set to 2 as the number of classes. The number of layers l was set to 2 for the PolitiFact dataset and to 3 for the MuMin dataset, based on the empirical performance and to balance representation capacity with over-smoothing avoidance. We employed a weighted cross entropy loss function in a fully supervised setting, with weights inversely proportional to the class frequency. We trained the model over 200 epochs and used the Adam optimization algorithm. The learning rate was set to 0.005 while the weight decay to 0.001.

We exploit AL-like techniques and GNN-Explainer on the trained model to identify high-confidence and key influential nodes, respectively. These methods

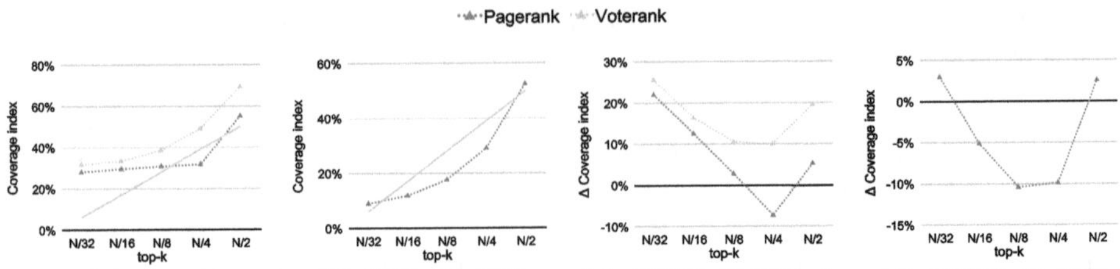

(a) $C(K)$, PolitiFact (b) $C(K)$, MuMiN (c) $\Delta C(K)$, PolitiFact (d) $\Delta C(K)$, MuMiN

Fig. 2. Qualitative analysis of the baseline approach, configured with ranking methods Pagerank and Voterank, when varying the size of the core set K, with dataset and maximum depth $d = 2$ fixed. (a) and (b) show the coverage index $C(K)$, while (c) and (d) show the increment of $C(K)$ given by the influenced nodes and measured as $\Delta C(K) = C(K) - \frac{K}{N} = \frac{R}{N}$.

generate independent node rankings, which are compared in pairs by means of the ranking similarity metrics presented in Sect. 4.2.

Table 4 reveals low correlation across ranking techniques of different post-hoc approaches, indicating that node importance varies across methods. However, the Jaccard Index suggests moderate agreement, especially for top-half-ranked nodes. Notably, consistently high NDCG scores confirm that key influential nodes remain prominent across methods, despite variations in ranking positions.

Among variants of the AL-like approach, rankings align closely, particularly for entropy- and margin-based methods, which show near-identical orderings. Differences across datasets, particularly for the two GNN-Explainer outcomes, reflect the impact of graph sparsity, where high connectivity graphs emphasize structural importance, while more sparse ones rely more on feature-based influence. These findings led us investigate a combined approach in addition to individual rankings. To create the combined ranking, we merge top-K/2 nodes from both AL and explainability-based lists, iteratively combining them to capture complementary aspects of nodes importance.

Coverage/reachability Analysis. The top row in Fig. 3 is meant to help get a general understanding of how the coverage power of baseline methods depend on the number K of core influential nodes retrieved with these methods, on both datasets. For the sake of presentation, let us focus only on methods Pagerank and Voterank –indeed, as shown later on, these two methods were found to excel among classical ranking ones in coverage metrics. Sub-figures (c-d) show the trend of the share $\Delta C(K) = C(K) - \frac{K}{N} = \frac{R}{N}$, of $C(K)$ corresponding to the influenced nodes (i.e. the nodes reached from the initial K ones), i.e., the proportion of nodes that are successfully reached.

Notably, in denser graphs like MuMiN, VoteRank and betweenness centrality – the most competitive baselines in sparser graphs like PolitiFact – fail to compute within a reasonable time, while PageRank shows high coverage capabilities. The Coverage scores align as the core set increases, but the dependency from K

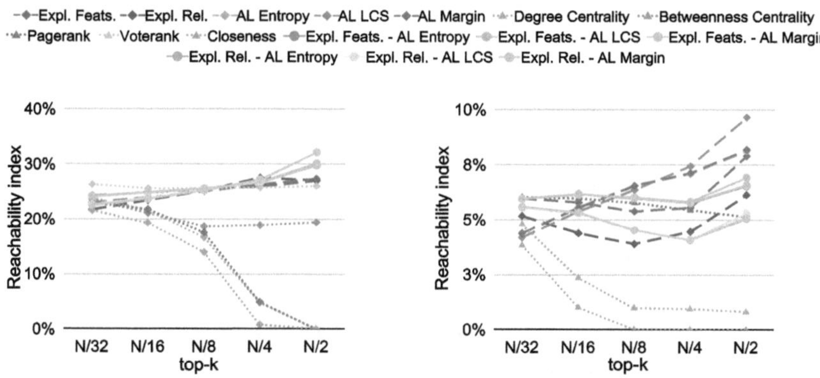

(a) PolitiFact (b) MuMiN

Fig. 3. Reachability index $R(K)$ when varying the number K of nodes in the initial core set, with maximum depth $d = 2$. Different kinds of markers (namely, diamonds, circles and triangles) and styles (namely, dashed, solid and dotted) are used to denote the proposed individual and combined post-hoc techniques and the baselines, respectively. For the sparser dataset (a), few differences are noticed as the initial core set size changes and the best performance is achieved by combining AL margin with an explainability technique, outperformed by voterank only for smaller seed size; for the denser dataset (b), for large initial core set size the individual Active Learning techniques work best, while decreasing the size the combination of AL margin with an explainability technique still achieves the best result.

is non-monotonic. Excluding the core set from the computation (Fig. 2 (c) -(d)) emphasizes the need of a proper selection of K.

Figure 3 shows the Reachability index values of individual, combined and baseline rankings as the size of the core set K varies within a maximum depth $d = 2$. This results clearly reveal that conventional measures of centrality often fail to maximize influence for short distances as the core set increases. On the more sparse graph, VoteRank remains flat while post-hoc techniques slightly increase, with predominance of combinations of explainability outcomes and AL margin. On the higher connectivity graph, no significant changes emerge, with uncertainty-based techniques prevailing.

The effectiveness of our approach is coupled with scalability in denser graphs exhibiting a higher number of connections. Table 5 shows the elapsed time for the computation of node importance scores for individual techniques. Increasing the number of edges slows down the computation and exceeds 40 h for techniques most dependent on the number of edges, such as VoteRank and betweenness centrality. An exception is closeness centrality which benefits from its implementation. Degree centrality and PageRank are efficient but fail in spreading information in few steps in more sparse graphs. Although we do not achieve the best absolute runtimes, our execution times remain reasonable while yielding superior performance results. Our analysis highlights the advantages of combining AL-like and explainability techniques for identifying influential nodes.

Table 5. Computation time of node importance scores, expressed in seconds, including training time.

Technique	politifact	mumin
GNN Explainer	44.81	51.71
AL Entropy	1183.74	3054.21
AL LCS	1352.85	3063.36
AL Margin	1183.74	3093.11
Degree Centrality	0.09	0.33
Closeness Centrality	1542.41	237.86
Betweenness Centrality	11844.87	n/a
Pagerank	1.79	2.45
VoteRank	711.75	n/a

The proposed hybrid approach achieves competitive accuracy while significantly reducing computational overhead, making it an effective solution for large-scale social media networks. Additionally, our influence spread simulations confirm that hybrid rankings outperform traditional centrality measures in identifying nodes that maximize reachability within minimal steps.

5 Conclusion

In this work, we have addressed the critical challenge of misinformation propagation in online social networks by proposing a novel attention-based framework which leverages a trained GAT model to feed advanced post-hoc techniques. Our approach enhances the identification of influential nodes within heterogeneous networks, effectively capturing the faceted nature of information dissemination. By integrating post-hoc techniques, such as uncertainty-based AL-like methods and graph neural network explanations, we provide deeper insights into the mechanisms of misinformation spread while improving model interpretability. Through extensive comparative analysis, our findings demonstrate that the proposed framework outperforms traditional graph-based centrality measures in both accuracy and scalability, . successfully identifying key users responsible for misinformation diffusion even where conventional approaches often struggle or fail. The validation on two heterogeneous, real-world social graphs further highlights the robustness and adaptability of our approach.

Building on these findings, future work will explore the extension of our framework to dynamic social networks, where the continuous evolution of user interactions and misinformation patterns poses additional challenges. Notably, integrating continual learning techniques could enable real-time adaptation to emerging trends, enhancing both the efficiency and responsiveness of misinformation mitigation strategies in large-scale, rapidly changing environments.

Test reproducibility. The code necessary to replicate our experiments is available at: https://github.com/Franco7Scala/ActiveLearningFakeNewsDetection.

Acknowledgments. This work has been partially supported by: (i) project SERICS (PE00000014) under the NRRP MUR program funded by the EU - NGEU; (ii) project MIRFAK (P2022C23K9), ERC field: PE6, funded by the EU - NGEU. (iii) project FAIR (PE00000013), funded by the EU - NGEU.

References

1. Rezaei, A.A., Munoz, J., Jalili, M., Khayyam, H.: A machine learning-based approach for vital node identification in complex networks. Expert Syst. Appl. **214**,119086 (2023)
2. Brin, S., Page, L.: The anatomy of a large-scale hypertextual web search engine. Comput. Netw. ISDN Syst. **30**(1–7), 107–117 (1998)
3. Brody, S., Alon, U., Yahav, E.: How attentive are graph attention networks? *CoRR*, abs/2105.14491 (2021)
4. Keikha, M.M., Rahgozar, M., Asadpour, M., Abdollahi, M.F.: Influence maximization across heterogeneous interconnected networks based on deep learning. Expert Syst. Appl. **140**, 112905 (2020)
5. Kou, J., Jia, P., Liu, J., Dai, J., Luo, H.: Identify influential nodes in social networks with graph multi-head attention regression model. Neurocomputing **530**, 23–36 (2023)
6. Martirano, L., Comito, C., Guarascio, M., Pisani, F.S., Zicari, P.: M3dusa: a modular multi-modal deep fusion architecture for fake news detection on social media. Soc. Netw. Anal. Min. **15**(1), 1–17 (2025)
7. Nielsen, D.S., McConville, R.: MuMiN: a large-scale multilingual multimodal fact-checked misinformation social network dataset. In: Proceedings of the 45th Internationl ACM SIGIR Conference on research and development in Information Retrival, pp. 3141–3153 (2022)
8. Settles, B: Active learning literature survey. Computer Sciences Technical Report 1648, University of Wisconsin–Madison (2009)
9. Shu, K., Sliva, A., Wang, S., Tang, J., Liu, H.: Fake news detection on social media: a data mining perspective. ACM SIGKDD Explor. Newsl **19**(1), 22–36 (2017)
10. Ugurlu, O.: Comparative analysis of centrality measures for identifying critical nodes in complex networks. J. Comput. Sci. **62**, 101738 (2022)
11. Ullah, A.: Identifying vital nodes from local and global perspectives in complex networks. Expert Syst. Appl. **186**, 115778 (2021)
12. Wang, X., Othman, M., Dewi, D.A., Wang, Y.: WSLC: weighted semi-local centrality to identify influential nodes in complex networks. J. King Saud Univ. Comput. Inf. Sci. **36**(1), 101906 (2024)
13. Xie, J., Yu, J., Chen, Z.: Mega: identifying influential nodes in heterogeneous networks based on meta-path and attention. J. Stat. Mech: Theory Exp. **2025**(2), 023401 (2025)
14. Ying, Z., Bourgeois, D., You, J., Zitnik, M., Leskovec, J.: GNNExplainer: generating explanations for graph neural networks. Adv. Neural Inf. Process. Syst. **32** (2019)

15. Yu, E.-Y., Wang, Y.-P., Fu, Y., Chen, D.-B., Xie, M.: Identifying critical nodes in complex networks via graph convolutional networks. Knowl. Based Syst. **198**, 105893 (2020)
16. Zafarani, R., Abbasi, M.A., Liu, H.: Social media mining: an introduction. Cambridge University Press (2014)
17. Zhang, J.-X., Chen, D.-B., Dong, Q., Zhao, Z.-D.: Identifying a set of influential spreaders in complex networks. Sci. Rep. **6**(1), 27823 (2016)
18. Zhao, G., Jia, P., Zhou, A., Zhang, B.: InfGCN: Identifying influential nodes in complex networks with graph convolutional networks. Neurocomputing **414**, 18–26 (2020)

Learning Tree-Structured Distributions

Fadwa Idlahcen(✉)

Faculty of Electrical Engineering, Czech Technical University in Prague, Karlovo namesti 13, Prague 12000, Czech Republic
idlahfad@fel.cvut.cz

Abstract. Tree-structured probabilistic models offer a balance between interpretability and tractability, making them a cornerstone in density estimation and graphical model learning. In this work, we present an empirical evaluation of three recent approaches for learning tree distributions, benchmarking them against the classic Chow-Liu algorithm. Although Chow-Liu is decades old, our results show that it continues to outperform or match the newer methods in most scenarios, including both synthetic and real-world datasets.

A key contribution of this study is the implementation and practical evaluation of two online learning algorithms, which had not previously been tested in practice or compared directly with the other methods we use. We analyze performance across multiple dimensions: realizability of the data distribution, alphabet size, and dataset scale. Our findings reaffirm the competitiveness of the Chow-Liu algorithm while also offering a comparison among the three methods, shedding light on their respective strengths and limitations across different regimes.

Keywords: Tree-structured distribution · Chow-Liu algorithm · Online learning

1 Introduction

Bayesian networks (BNs) are widely used across various domains due to their ability to model causal relationships among random variables—for instance, in medicine [9] and gene regulatory networks [10]. However, learning the structure of BNs from data is an NP-hard problem [6]. Additionally, performing inference with BNs is computationally non-trivial [8].

To address the challenges of learning and inference in BNs, numerous approaches have been proposed. While many of these methods yield strong results, they often involve a trade-off between computational tractability and model expressiveness. Certain BN structures, such as Naive Bayes and tree-structured networks, offer computational efficiency but at the cost of reduced expressiveness. Nonetheless, these simpler distributions are frequently employed in machine learning—for example, in variational inference—due to their practicality. Tree-structured models are particularly common, as they enable efficient probabilistic inference [15]. Furthermore, more expressive models can be

S. Džeroski et al. (Eds.): DS 2025, LNAI 16090, pp. 63–77, 2025.
https://doi.org/10.1007/978-3-032-05461-6_5

constructed by building on tree structures, such as through mixtures-of-trees approaches [14].

In this paper, we compare several methods for learning tree-structured distributions. A foundational approach in this domain is the Chow-Liu algorithm [7], introduced in 1968, which efficiently identifies the tree structure and parameters that maximize the likelihood of the observed data. Our primary objective is to evaluate the efficiency of recent methods relative to the Chow-Liu baseline. These methods are theoretically state-of-the-art, offering theoretical guarantees, and we aim to assess their empirical performance through experimentation:

- Distribution Learning Meets Graph Structure Sampling [2]
- Online Forest Density Estimation (OFDE) [11]
- Mixture of All Trees (MoAT)[1] [16]

The main contribution of this work is a practical comparison of these algorithms[2] We aim to highlight their empirical strengths and limitations, particularly in light of the lack of experimental results in [2] and the unavailability of implementation code for [11].

1.1 Tree-Structured Distributions

In this section, we introduce the notation used throughout the paper and outline the learning framework for tree-structured distributions. A Bayesian network over n discrete variables, each taking values from an alphabet size k, is represented as a directed acyclic graph (DAG) that encodes causal relationships among the variables. These dependencies are quantified through conditional probability distributions for each variable, conditioned on its parent(s) in the network [15].

Tree-structured distributions—referred to here interchangeably as trees or tree distributions—are a subclass of BNs in which the underlying DAG has a maximum in-degree of 1. Alternatively, such distributions can be represented by an undirected graph $G = (V, E)$, also known as a Markov Random Field (MRF), where V and E denote the sets of vertices and edges, respectively. The normalized joint distribution defined over this graph, as described in [14], is given by:

$$Pr_G(\mathbf{x}) = \frac{\prod_{(u,v)\in E} P_{uv}(x_u, x_v)}{\prod_{u\in V} P_u(x_u)^{deg(u)-1}} \tag{1}$$

Here, $\mathbf{x}$ denotes an assignment of values to all variables; $deg(u)$ is the degree of vertex u; $P_{uv}(x_u, x_v)$ is the joint distribution over variables u and v; and $P_u(x_u)$ is the marginal distribution of x_u, obtained from $P_{uv}(x_u, x_v)$ or $P_{vu}(x_v, x_u)$ by marginalizing over x_v for any $v \neq u$.

[1] While this method does not provide theoretical guarantees, we include it due to its strong empirical performance and conceptual similarity to [2].

[2] Our Python implementation is available at https://github.com/Fadwaid2/Tree-distributions..

The task of learning tree-structured distributions involves estimating both the structure and the parameters of the model that maximize the log-likelihood of a given dataset. Specifically, given a set of observations $\{x^{(1)}, x^{(2)}, \ldots, x^{(T)}\}$, where T denotes the number of samples, the goal is to identify the tree model that best fits the data. In this setting, the structure is constrained to be a spanning tree over the n random variables (nodes). A spanning tree is an acyclic, connected subgraph that includes all n nodes and exactly $n-1$ edges. The collection of all possible spanning trees over n nodes forms the spanning tree polytope, a convex hull of incidence vectors of all spanning trees. The number of such trees is exponential in n, and is given by Cayley's formula, which states that there are exactly n^{n-2} spanning trees for a complete graph with n nodes. Learning a tree-structured distribution can also be viewed as minimizing the Kullback–Leibler (KL) divergence between the empirical distribution $\hat{P}$ of the data and a candidate tree-structured distribution Q from the model class. That is, the objective becomes:

$$\min_{Q \in \mathcal{T}} KL(\hat{P} || Q) \tag{2}$$

where $\mathcal{T}$ is the set of all tree-structured distributions. In the realizable case, when the true distribution lies within $\mathcal{T}$, this corresponds to exact recovery of structure; in the agnostic case, the goal is to find the closest tree-structured approximation to a potentially non-tree distribution.

The large combinatorial space of possible spanning trees underscores the computational efficiency of algorithms like Chow-Liu, which cleverly use maximum-weight spanning tree algorithms to optimize over this exponentially large space efficiently, by reducing the KL divergence to a sum over mutual information terms.

1.2 Online Learning

Both [2] and [11] adopt an Online Learning (OL) framework. In this setting, the model learns sequentially from data by sampling from a class of candidate structures [2]. We assume that the observations $\{x^{(1)}, x^{(2)}, \ldots, x^{(T)}\}$ are independent and identically distributed (i.i.d.) samples drawn from an unknown distribution P^*. At each time step t, the learner observes a new data point $x^{(t)}$ and produces an estimate $\hat{\theta}_t$, based on the previous observations $x_{1:(t-1)} := \{x^{(1)}, \ldots, x^{(t-1)}\}$. The performance at time t is measured by a loss function $l(x^{(t)}, \hat{\theta}_t)$, which we denote simply as $l(\hat{\theta}_t)$ for brevity. The overall quality of the sequential estimation is evaluated using regret, which measures the cumulative discrepancy between the learner's predictions and the best possible fixed predictor in hindsight. Formally, the cumulative regret of the sequence $\{\hat{\theta}_t\}_{t=1}^T$ is defined as follows:

$$\sum_{t=1}^{T} l_t(\hat{\theta}_t) - \inf_{\theta \in \Theta} \sum_{t=1}^{T} l_t(\theta) \tag{3}$$

The objective, therefore, is to minimize the cumulative regret, with the loss function defined as the negative log-likelihood. In this context, the minimax regret quantifies the best possible worst-case performance of a forecasting algorithm across all possible classes of N experts and all outcome sequences of length T [4]. Shtarkov [17] introduced a forecasting method based on the Normalized Maximum Likelihood (NML) distribution, which achieves the minimax regret when the time horizon T is known in advance. However, computing the NML distribution is intractable in practice, as it requires summing over an exponential number of terms [4]. As a result, mixture-of-experts approaches are often preferred due to their computational efficiency and the fact that they do not require knowledge of the time horizon T. Moreover, their theoretical regret bounds are comparable to those of the NML forecaster [4], which further motivates their widespread use.

Another widely studied forecasting strategy is Follow-the-Leader (FTL) and its randomized variant, Follow-the-Perturbed-Leader (FPL). These algorithms achieve regret bounds that grow logarithmically with the number of iterations [12].

Outside the online learning framework, the MoAT method provides an exact computation of the normalized likelihood by averaging over all n^{n-2} possible tree structures. This approach enables the use of a full mixture-of-trees model with tractable inference. Further details on each of the selected methods are provided in the following section. One approach to comparing these methods is through their sample complexity—the amount of data required to guarantee a specified level of performance. Below, we summarize the theoretical bounds as reported in the respective works.

For OFDE, the algorithm achieves minimax regret bounds of $O(nk^2 lnT + n^2 lnT\sqrt{T})$ in the horizon-dependent case, and $O(nk^2 lnT + n^2(lnT)^2\sqrt{T})$ in the horizon-independent case [11]. The method in [2] is analyzed under the Probably Approximately Correct (PAC) learning framework. It provides a sample complexity of $\tilde{O}(nk^2\epsilon^{-1})$ (ignoring logarithmic factors), ensuring that the learned mixture-of-trees distribution $\hat{P}$ satisfies:

$$KL(P^*||\hat{P}) \leq 3 \min_{P \in \mathcal{T}}(KL(P^*||P)) + \epsilon \tag{4}$$

with probability at least 2/3, where $\mathcal{T}$ denotes the class of all tree-structured distributions. For the Chow-Liu algorithm, the realizable-case PAC sample complexity has been shown in [3] to be $\tilde{O}(nk^3\epsilon^{-1})$, indicating a factor of k worse dependence compared to the bound in [2]. As far as we are aware, no formal sample complexity guarantees have been established for MoAT. Nonetheless, MoAT remains of practical interest due to its empirical performance.

In this paper, we empirically examine how these theoretical guarantees correspond to observed performance, particularly in data-limited settings.

2 Methods

In this section, we describe the four algorithms employed in this work and provide a summary of their underlying techniques.

2.1 Chow-Liu

The Chow-Liu algorithm [7] is a classical and widely-used method for learning joint probability distributions with tree-structured dependencies. It leverages the key insight that the tree structure maximizing the likelihood corresponds to the maximum-weight spanning tree of the complete graph. In this graph, the weight of each edge represents the mutual information between the two connected random variables.

2.2 Distribution Learning Meets Graph Structure Sampling

In [2], a sampling algorithm embedded within two online learning algorithms, where the set of experts corresponds to a mixture-of-trees, is proposed. The sampling of spanning trees leverages the matrix tree theorem.

Since we employ only one of these OL algorithms in this work—the Randomized Weighted Majority (RWM)—we will refer to the method from [2] simply as RWM hereafter. This approach requires discretization to finitely represent the class of distributions. Although [4] generalizes regret bounds to infinitely large expert classes—making discretization theoretically unnecessary—this step is used in [2] to directly apply regret bounds from that work.

2.3 Online Forest Density Estimation (OFDE)

The OFDE algorithm [11] addresses the problem of learning tree-structured distributions in the OL setting. At each time step, the algorithm maintains and updates a probabilistic estimate over the space of spanning trees, and predicts a tree structure and its associated parameters to model the data.

The structure update step is performed using Follow-the-Perturbed-Leader algorithm. This method perturbs the cumulative loss associated with each potential edge using random noise and then selects a spanning tree that minimizes the perturbed loss. Importantly, this optimization is carried out over the spanning tree polytope using a greedy matroid algorithm—which in the case of trees is equivalent to Kruskal's algorithm.

However, the edge weights in the FPL algorithm are fractional, and generating a valid tree structure requires converting these fractional solutions into integral spanning trees. To accomplish this, OFDE employs swap rounding [5], a randomized rounding procedure that samples a spanning tree from a convex combination while preserving the marginal probabilities of edge inclusion.

Following structure selection, the parameter update step estimates univariate and pairwise marginal distributions using an approach similar to Bayesian mixtures with Dirichlet priors. This leverages the additive nature of the log-likelihood function, allowing the algorithm to incrementally update sufficient statistics and compute a valid likelihood estimate for each tree.

By combining efficient combinatorial optimization with online probabilistic estimation, OFDE provides a scalable approach to learning tree-structured models.

2.4 Mixture of All Trees (MoAT)

The MoAT algorithm [16] introduces a generative model that considers all n^{n-2} possible spanning trees over n variables, operating outside the OL framework. By using a mixture-of-trees representation, the structure is fixed during training, while parameters are learned efficiently.

This tractability is achieved by assigning each tree a weight proportional to its probability within the spanning tree distribution (see Eq. 2 in [16]). Importantly, parameters are shared across all trees in the mixture since the univariate and bivariate probabilities remain consistent for each random variable.

Similar to [2], the matrix tree theorem is employed, but here it is combined with the concept of probability generating polynomials (Eq. 3 in [16]). This yields a closed-form, tractable expression for the likelihood, where the primary computational bottleneck is the calculation of the determinant of the Laplacian matrix—also encountered in the RWM algorithm [2].

MoAT is trained by maximizing the likelihood through stochastic gradient descent, with deterministic parameter initialization. For the latter, mutual information between variables is incorporated following Chow-Liu's method, which improves the model's performance.

3 Results and Discussion

In this study, we aim to rigorously evaluate the performance of recent structure learning methods in comparison to the classical Chow-Liu algorithm. Our goal is to understand how these modern approaches behave under varying conditions, both synthetic and real. To this end, we formulate four key research questions that guide our experimental design and analysis:

- Q1. Comparative Performance: Do the three recent learning methods outperform the classical Chow-Liu algorithm?
- Q2. Robustness Across Data Generating Processes: How does the performance of OL methods differ between the realizable case, where data are generated from a true tree-structured distribution, and the agnostic case, where data arise from more complex non-tree graphs?
- Q3. Scalability with Alphabet Size: Which of the OL methods demonstrates greater robustness and performance as the alphabet size (i.e., the number of discrete variable states) increases?
- Q4. Behavior on Large Datasets: As the number of samples grows, which method scales more effectively in terms of accuracy and computational efficiency?

To address these questions, we design a comprehensive collection of experiments that includes both controlled synthetic datasets—allowing us to systematically vary underlying structures, alphabet sizes, and sample sizes—and real-world datasets, which provide a test of generalization and practical utility.

3.1 Synthetic Datasets

To systematically evaluate the performance of the learning methods, we construct synthetic datasets using discrete random variables across two primary regimes: low-dimensional ($T \gg n$) and high-dimensional. The specific combinations of the number of variables n and the number of samples T used in our experiments are:

- (n,T)=(5,20),(5,50),(10,20),(10,50),(20,50)(20,100),(50,500),(50,1000), (100,500),(100,1000)

We generate datasets under two generative models:

- Tree-structured distributions
- General Markov Random Fields (MRFs) based on the ErdősâĂŞRényi random graph model.

To introduce additional complexity and simulate real-world imperfections, we inject 50% noise by randomly shuffling the observed values of the variables.

We consider discrete alphabets with sizes $k \in \{2, 5, 10\}$. However, since the MoAT algorithm supports only binary variables, it is excluded from experiments involving $k \in \{5, 10\}$. Additionally, to ensure a fair comparison with other methods that recover single tree structures, we adapt the use of MoAT: rather than leveraging the full mixture of trees, we retrieve (using Kruskal's algorithm) and evaluate a single tree structure from the learned mixture.

Each configuration is run 5 times to account for variability in algorithms with stochastic components (namely, OFDE and RWM). We report the mean and standard deviation of the results across these runs.

For the binary case ($k = 2$), we present results for both tree-structured and MRF-generated data using two metrics:

- Structural Hamming Distance (SHD): measures structural accuracy; lower values indicate better alignment with the ground-truth graph.
- Log-likelihood (LL): evaluated on held-out test data sampled from the true generative model.

These results are summarized in Tables 1 and 2, respectively.

For the non-binary cases ($k \in \{5, 10\}$), we report SHD and log-likelihoods in Tables 3 and 4. Across all tables, bolded values indicate the best-performing method for a given metric and setting.

3.2 Real-World Datasets

In addition to synthetic data, we evaluate the performance of the structure learning methods on real-world datasets drawn from the Twenty Datasets benchmark suite for density estimation [18]. These datasets cover a range of domains and data characteristics, providing a robust testbed for empirical validation. A summary of the dataset properties (number of variables and samples) is provided in Table 5.

Table 1. SHD and average LL for $k = 2$ - data from tree.

	Chow-Liu		RWM		OFDE		MoAT	
	SHD	LL	SHD	LL	SHD	LL	SHD	LL
(5,20)	4	-2.76	7.2 ± 1.10	-2.92 ± 0.11	4.8 ± 1.10	-2.85 ± 0.11	**2**	**-2.74**
(5,50)	**2**	-2.86	5.2±3.03	-2.92 ± 0.05	4.8 ± 2.68	-2.92 ± 0.07	**2**	**-2.85**
(10,20)	12	**-6.61**	13.6 ± 2.97	-7.22 ± 0.16	15.6 ± 1.67	-6.88 ± 0.33	**10**	-6.74
(10,50)	14	**-6.24**	13.2 ± 1.79	-6.44 ± 0.19	14.4 ± 0.89	-6.44 ± 0.06	14	-6.35
(20,50)	**32**	**-13.27**	32.8 ± 3.35	-13.78 ± 0.21	35.6 ± 0.89	-13.59 ± 0.17	38	-13.48
(20,100)	36	**-13.16**	**32.8 ± 1.79**	-13.37 ± 0.07	33.2 ± 3.35	-13.41 ± 0.23	36	-13.17
(50,500)	**94**	**-34.02**	93.6 ± 1.67	-34.12 ± 0.05	94.0 ± 2.45	-34.08 ± 0.00	**94**	-34.06
(50,1000)	98	**-34.00**	94.4 ± 2.61	-34.05 ± 0.02	95.2 ± 3.03	-34.02 ± 0.02	**90**	-34.01
(100,500)	194	**-68.68**	194.0 ± 1.41	-68.99 ± 0.07	194.4 ± 2.61	-68.81 ± 0.06	**184**	-68.83
(100,1000)	194	**-68.67**	196.0 ± 2.00	-68.76 ± 0.02	195.2 ± 1.10	-68.72 ± 0.01	**192**	-68.68

Table 2. SHD and average LL for $k = 2$ - data from MRF.

	Chow-Liu		RWM		OFDE		MoAT	
	SHD	LL	SHD	LL	SHD	LL	SHD	LL
(5,20)	5	-3.17	5.0 ± 0.00	-5.25 ± 2.97	5.8 ± 1.10	-7.16 ± 2.33	**3**	**-3.07**
(5,50)	5	-2.90	5.0 ± 0.00	-3.03 ± 0.11	**4.2 ± 1.10**	**-2.86 ± 0.05**	7	-2.97
(10,20)	28	-6.81	26.0 ± 2.00	-6.72 ± 0.19	**25.6 ± 3.58**	-6.77 ± 0.35	26	**-6.59**
(10,50)	29	**-6.56**	24.6 ± 1.67	-6.60 ± 0.04	27.4 ± 2.97	-6.58 ± 0.19	**23**	-6.65
(20,50)	**101**	**-13.33**	101.0 ± 4.00	-13.78 ± 0.12	101.4 ± 3.58	-13.40 ± 0.18	105	-13.35
(20,100)	114	**-13.31**	**108.4 ± 2.97**	-13.50 ± 0.05	110.0 ± 5.83	-13.40 ± 0.08	110	-13.41
(50,500)	755	**-34.00**	751.8 ± 7.16	-34.06 ± 0.02	**745.4 ± 9.74**	-34.02 ± 0.03	749	-34.02
(50,1000)	746	**-33.98**	**727.2 ± 3.90**	-34.04 ± 0.01	734.0 ± 9.38	-34.02 ± 0.03	728	-34.00
(100,500)	2996	**-68.69**	2996.0 ± 8.60	-68.82 ± 0.03	2994.4 ± 11.61	-68.83 ± 0.02	**2994**	-68.71
(100,1000)	**2930**	-68.71	2935.2 ± 9.44	-68.73 ± 0.01	2933.6 ± 5.73	-68.76 ± 0.02	2938	**-68.68**

Table 3. SHD and average LL for alphabet size $k = 5$ and $k = 10$ - data from tree structure.

		Chow-Liu		RWM		OFDE	
		SHD	LL	SHD	LL	SHD	LL
(5,20)	k=5	**4**	**-30.22**	4.4 ± 2.61	-41.94 ± 6.40	6.0 ± 2.00	-34.09 ± 1.89
	k=10	6	**-49.45**	4.4 ± 2.19	-66.25 ± 10.54	**3.6 ± 1.67**	-71.54 ± 4.63
(5,50)	k=5	4	**-11.56**	4.4 ± 1.67	-14.03 ± 1.13	**3.6 ± 1.67**	-13.56 ± 1.38
	k=10	6	**-32.11**	5.6 ± 1.67	-49.18 ± 7.33	**5.2 ± 1.10**	-45.09 ± 4.92
(10,20)	k=5	**12**	**-57.99**	15.2 ± 1.10	-83.80 ± 12.63	14.0 ± 1.41	-83.29 ± 11.95
	k=10	16	**-105.09**	**14.8 ± 2.28**	-130.10 ± 9.83	15.2 ± 1.79	-140.47 ± 6.95
(10,50)	k=5	16	**-22.99**	16.4 ± 1.67	-34.85 ± 5.12	**14.0 ± 2.00**	-32.74 ± 2.34
	k=10	16	**-54.79**	**13.6 ± 2.61**	-99.77 ± 5.83	16.4 ± 2.19	-100.95 ± 12.17
(20,50)	k=5	36	-92.15	33.2 ± 4.15	-93.47 ± 7.04	**33.2 ± 2.68**	**-88.79 ± 2.70**
	k=10	**32**	**-137.09**	36.8 ± 1.79	-238.88 ± 12.75	34.8 ± 2.68	-228.73 ± 9.94
(20,100)	k=5	34	**-32.42**	33.6 ± 1.67	-42.97 ± 3.22	**32.8 ± 3.63**	-43.07 ± 3.10
	k=10	**30**	**-74.47**	35.6 ± 1.67	-156.76 ± 9.21	33.6 ± 2.19	-159.94 ± 4.53
(50,500)	k=5	98	**-78.73**	96.0 ± 2.45	-79.11 ± 0.07	**93.6 ± 3.85**	-79.09 ± 0.16
	k=10	90	**-113.95**	96.4 ± 1.67	-131.30 ± 2.03	**89.6 ± 3.29**	-131.39 ± 1.04
(50,1000)	k=5	**94**	**-78.48**	94.4 ± 1.67	-78.64 ± 0.03	94.4 ± 1.67	-78.74 ± 0.07
	k=10	96	**-109.38**	**94.0 ± 2.83**	-111.25 ± 0.58	94.4 ± 3.85	-111.79 ± 0.35
(100,500)	k=5	194	**-159.33**	**192.0 ± 2.45**	-160.19 ± 0.14	193.6 ± 2.61	-160.38 ± 0.16
	k=10	194	**-227.98**	195.2 ± 1.79	-262.33 ± 4.49	**192.8 ± 3.03**	-263.89 ± 5.14
(100,1000)	k=5	196	**-158.53**	**193.2 ± 1.10**	-158.97 ± 0.06	193.2 ± 2.28	-159.12 ± 0.06
	k=10	**194**	**-221.77**	194.4 ± 3.58	-223.68 ± 0.56	194.0 ± 2.45	-226.48 ± 0.48

Table 4. SHD and average LL for alphabet size $k = 5$ and $k = 10$ - data from MRF.

		Chow-Liu		RWM		OFDE	
		SHD	LL	SHD	LL	SHD	LL
(5,20)	k=5	8	-29.59	5.6 ± 0.89	**-26.28 ± 1.34**	**3.6 ± 0.89**	-27.78 ± 1.50
	k=10	5	**-40.1**	**3.8 ± 1.10**	-46.44 ± 5.17	5.4 ± 1.67	-50.40 ± 4.60
(5,50)	k=5	7	**-12.49**	6.2 ± 1.10	-15.27 ± 1.77	**4.2 ± 1.10**	-14.55 ± 1.29
	k=10	5	**-17.83**	6.6 ± 0.89	-25.40 ± 3.51	**4.6 ± 1.67**	-29.37 ± 4.91
(10,20)	k=5	25	-89.73	26.6 ± 2.19	**-83.16 ± 8.02**	**24.6 ± 2.61**	-85.45 ± 6.35
	k=10	**24**	**-52.45**	26.4 ± 1.67	-77.62 ± 13.83	26.4 ± 1.67	-111.77 ± 8.88
(10,50)	k=5	25	**-21.5**	22.6 ± 0.89	-32.31 ± 2.58	**21.8 ± 1.10**	-36.07 ± 2.27
	k=10	**22**	**-58.82**	23.2 ± 2.28	-69.35 ± 6.69	24.8 ± 1.79	-85.51 ± 6.76
(20,50)	k=5	112	**-68.88**	110.8 ± 2.28	-84.72 ± 4.17	**110.4 ± 3.58**	-93.80 ± 9.16
	k=10	**108**	**-135.44**	109.6 ± 6.23	-210.82 ± 9.53	109.2 ± 5.76	-253.01 ± 11.51
(20,100)	k=5	120	**-33.46**	120.0 ± 3.16	-40.95 ± 2.09	**116.4 ± 7.13**	-38.65 ± 2.57
	k=10	**110**	**-75.38**	111.2 ± 1.79	-120.70 ± 8.03	112.4 ± 3.29	-147.14 ± 12.75
(50,500)	k=5	715	-78.5	699.4 ± 3.29	**-77.29 ± 0.19**	**696.2 ± 7.69**	-78.11 ± 0.29
	k=10	740	**-114.81**	**737.2 ± 6.72**	-124.54 ± 1.13	742.0 ± 7.75	-132.01 ± 1.79
(50,1000)	k=5	723	-77.6	703.4 ± 4.34	**-76.49 ± 0.11**	**699.8 ± 4.82**	-77.42 ± 0.14
	k=10	750	-110.12	**740.4 ± 3.58**	**-109.05 ± 0.40**	742.8 ± 3.63	-111.94 ± 0.39
(100,500)	k=5	2966	-158.83	**2948.4 ± 14.79**	**-157.88 ± 0.11**	2949.2 ± 8.44	-159.20 ± 0.26
	k=10	**2858**	**-230.23**	2858.0 ± 10.10	-250.15 ± 2.05	2862.0 ± 9.70	-258.87 ± 2.65
(100,1000)	k=5	3001	-158.78	2991.4 ± 4.77	**-157.22 ± 0.07**	**2991.0 ± 5.10**	-158.26 ± 0.17
	k=10	2874	-226.09	**2863.2 ± 6.10**	**-224.05 ± 0.53**	2866.0 ± 6.00	-227.63 ± 0.54

We apply all four methods under study to each dataset. To account for variability due to algorithmic randomness (particularly for stochastic methods such as OFDE and RWM), we repeat each experiment five times, and report the average log-likelihood on held-out test data.

The results, presented in Table 6, reflect the quality of the learned models in terms of generalization performance. Higher log-likelihood values indicate better fit to the underlying data distribution.

Table 5. Real-world datasets.

Dataset Name	Train set	Valid set	Test set	# vars	Density
accidents [18]	12758	1700	2551	111	0.291
baudio [13]	15000	2000	3000	100	0.199
bnetflix [13]	15000	2000	3000	100	0.541
dna [18]	1600	400	1186	180	0.253
jester [13]	9000	1000	4116	100	0.608
kdd [13]	180092	19907	34955	65	0.008
nltcs [13]	16181	2157	3236	16	0.332
plants [13]	17412	2321	3482	69	0.180
pumsb star [18]	12262	1635	2452	163	0.270
tretail [18]	22041	2938	4408	135	0.024

Table 6. Average LL of real data. ML: memory limit; TL: time limit.

	Chow-Liu	MOAT	OFDE	RWM
accidents	**-35.72**	-39.43	-38.32 ± 1.53	-44.57 ± 0.74
baudio	**-46.18**	-50.28	-47.38 ± 0.74	-48.46 ± 0.44
bnetflix	-64.13	**-62.88**	-63.08 ± 0.42	-63.56 ± 0.21
dna	-99.22	-95.85	-100.02 ± 0.16	**-88.93 ± 0.53**
jester	-63.48	-64.18	**-61.01 ± 0.34**	-61.96 ± 0.03
kdd	-1.89	**-1.38**	-2.43 ± 0.18	ML
nltcs	-7.63	-7.76	**-7.20 ± 0.25**	-8.01 ± 0.00
plants	-30.85	-32.02	**-28.46 ± 1.11**	-31.38 ± 0.11
pumsb star	**-43.21**	-48.99	-52.16 ± 2.42	TL
tretail	-7.12	**-6.83**	-7.28 ± 0.51	ML

3.3 Hyperparameters

For both synthetic and real-world datasets, we adopt established hyperparameter settings from prior work, if existing, to ensure a fair, reproducible, and unbiased evaluation. No additional tuning was performed on a per-dataset basis to avoid overfitting and to maintain comparability with previously reported results.

- **MoAT**: We follow the recommended settings from [16]. For datasets with fewer than 500 variables, we use a batch size of 1024 and a learning rate of 0.05. For larger datasets ($\geq$ 500 variables), we use a batch size of 64 and a learning rate of 0.01. All models are trained for up to 50 iterations with early stopping enabled. To enable comparison with other single-tree methods, we use Kruskal's algorithm on the final edge weight matrix of the learned mixture to obtain a single tree structure.
- **RWM**: We use the theoretical parameters from [2]. Specifically, we set $\epsilon =$ 0.9, which controls the approximation guarantee, and compute parameters: $\eta = \sqrt{\frac{8 \log N}{T}}$, $\tau = \frac{\epsilon/2}{4k^2}$, and $\gamma = \frac{\epsilon/2}{4}$.
- **OFDE**: We use the default hyperparameters from the horizon-independent setting in [11], with $\alpha_t = \frac{1}{4\sqrt{2t}}$ and $\beta_t = \frac{1}{n}\sqrt{2/t}$, where t is the time step.

3.4 Discussion and Conclusions

To compare the performance of the evaluated methods across synthetic and real-world datasets, we adopt Bayesian signed-rank tests as described in [1]. Unlike traditional null hypothesis testing, Bayesian analysis quantifies the posterior probability that one method outperforms another, providing richer insights into relative algorithmic performance.

We compute the posterior distribution of the mean difference in average log-likelihoods between methods and visualize the results as histograms. Each histogram represents the distribution of sampled differences between two methods, where each bar corresponds to the relative frequency (or probability mass) of a specific range of performance differences. The height of a bar reflects how likely it is that the true difference in performance falls within that interval, based on the posterior distribution. A Region of Practical Equivalence (ROPE), set to 0.5, marks the threshold below which performance differences are considered negligible. Histograms are shown in Figs. 1–5.

Our observations support the following conclusions:

Q1: Do the three methods outperform Chow-Liu, given that they are more recent?

Overall, the Chow-Liu algorithm outperforms the more recent methods on the synthetic datasets. It consistently achieves the highest log-likelihood and the lowest structural error (SHD), especially when the underlying data is generated from tree-structured models (Fig. 1, 3(a,b,c), 4(a,b,c)).

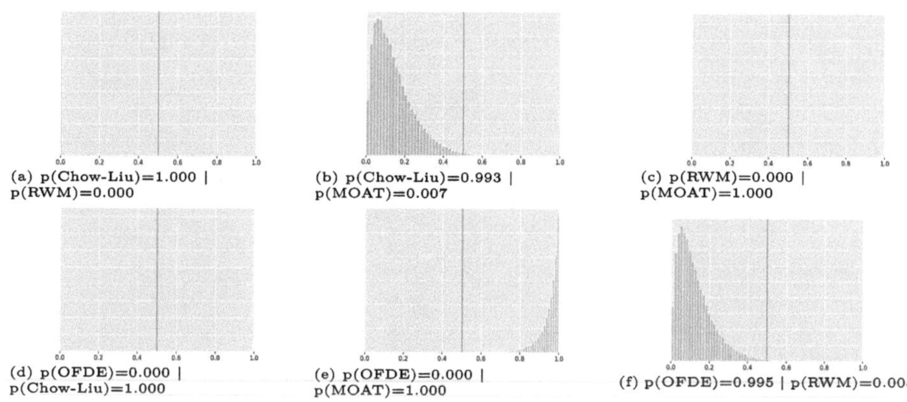

(a) p(Chow-Liu)=1.000 | p(RWM)=0.000
(b) p(Chow-Liu)=0.993 | p(MOAT)=0.007
(c) p(RWM)=0.000 | p(MOAT)=1.000
(d) p(OFDE)=0.000 | p(Chow-Liu)=1.000
(e) p(OFDE)=0.000 | p(MOAT)=1.000
(f) p(OFDE)=0.995 | p(RWM)=0.005

Fig. 1. Histograms for data from tree - k=2

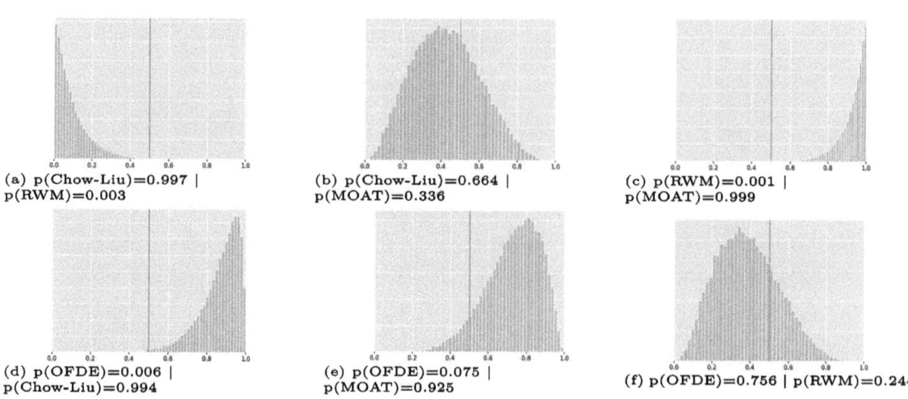

(a) p(Chow-Liu)=0.997 | p(RWM)=0.003
(b) p(Chow-Liu)=0.664 | p(MOAT)=0.336
(c) p(RWM)=0.001 | p(MOAT)=0.999
(d) p(OFDE)=0.006 | p(Chow-Liu)=0.994
(e) p(OFDE)=0.075 | p(MOAT)=0.925
(f) p(OFDE)=0.756 | p(RWM)=0.244

Fig. 2. Histograms for data from MRF - k=2

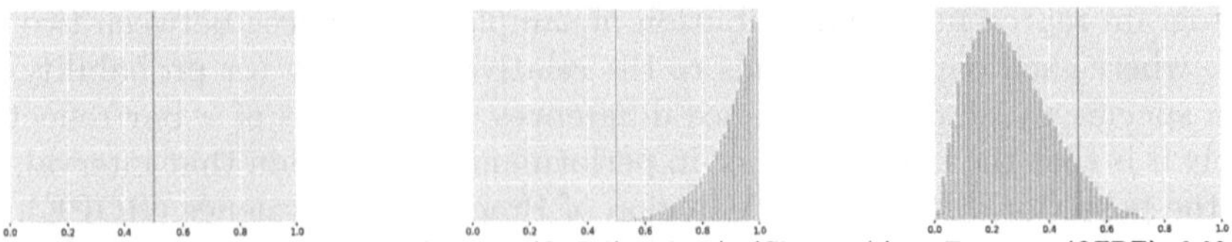

(a) Tree: p(Chow-Liu)=1.000 | p(RWM)=0.000

(b) Tree: p(OFDE)=0.004 | p(Chow-Liu)=0.996

(c) Tree: p(OFDE)=0.924 p(RWM)=0.076

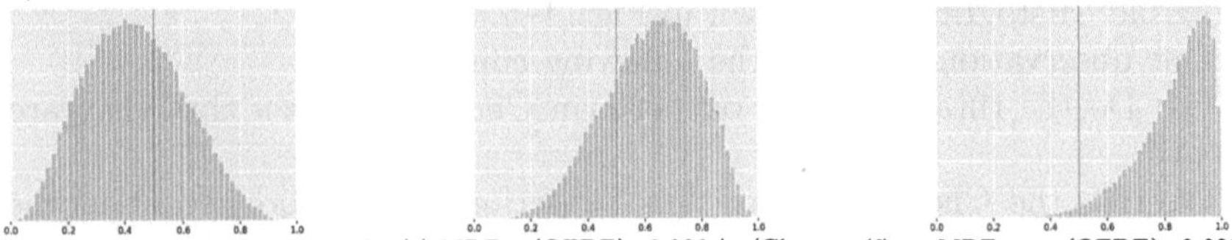

(d) MRF: p(Chow-Liu)=0.648 | p(RWM)=0.352

(e) MRF: p(OFDE)=0.229 | p(Chow-Liu)=0.771

(f) MRF: p(OFDE)=0.021 p(RWM)=0.979

Fig. 3. Histograms for $k = 5$. Top row: data from tree; bottom row: data from MRF.

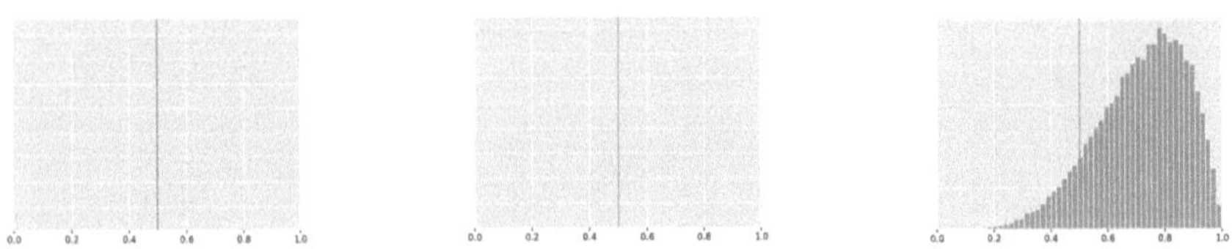

(a) Tree: p(Chow-Liu)=1.000 | p(RWM)=0.000

(b) Tree: p(OFDE)=0.000 | p(Chow-Liu)=1.000

p(RWM)=0.893

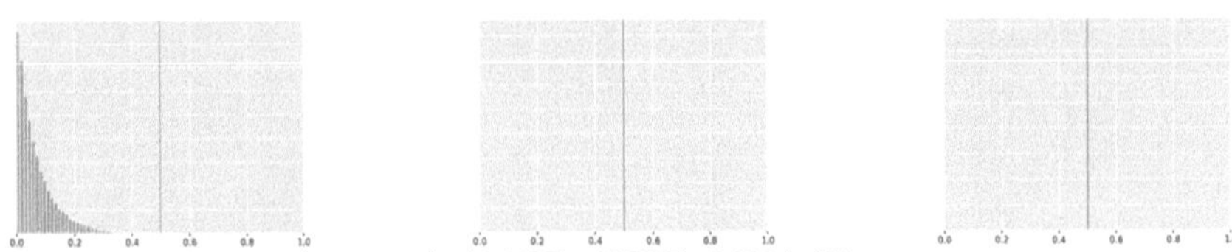

(d) MRF: p(Chow-Liu)=0.999 | p(RWM)=0.001

(e) MRF: p(OFDE)=0.000 | p(Chow-Liu)=1.000

p(RWM)=1.000

Fig. 4. Histograms for $k = 10$. Top row: data from tree; bottom row: data from MRF.

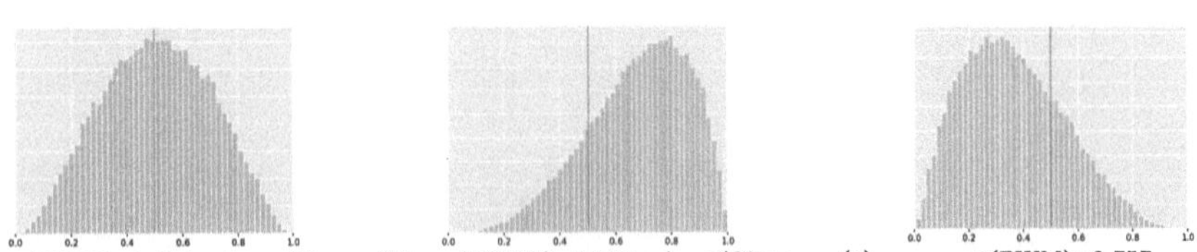

(a) p(RWM)=0.484 | p(Chow-Liu)=0.516

(b) p(MOAT)=0.174 | p(Chow-Liu)=0.826

(c) p(RWM)=0.757 p(MOAT)=0.243

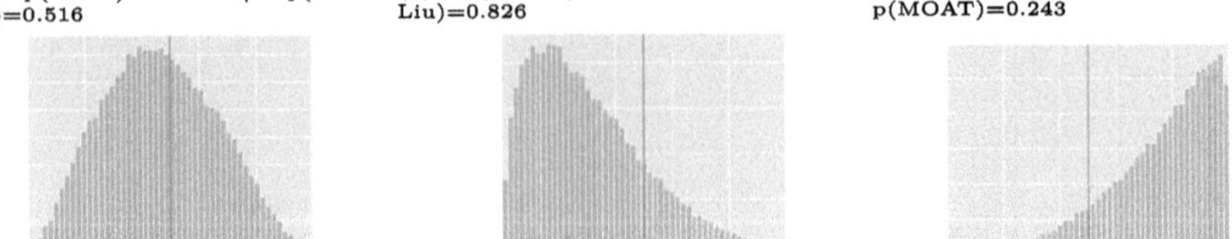

(d) p(OFDE)=0.583 | p(Chow-Liu)=0.417

(e) p(OFDE)=0.854 p(MOAT)=0.146

| (f) p(RWM)=0.108 | p(OFDE)=0.892

Fig. 5. Histograms for real data

While Chow-Liu's performance degrades slightly on real-world datasets (Fig. 5), it still remains highly competitive. Notably, in some real data settings, OFDE slightly outperforms Chow-Liu, indicating strong generalization performance for OFDE in practical scenarios. This result suggests that Chow-Liu remains a robust baseline, but OFDE may be preferable when working with noisy real-world data.

Q2: What are the differences in performance between the realizable and agnostic cases in OL?

In synthetic experiments, we distinguish between the realizable case and the agnostic case. Here, the performance gap between OFDE and RWM becomes more evident.

- In realizable settings, OFDE consistently outperforms RWM (Figs. 1, 3(a,b,c)), except when the alphabet size is larger (Fig. 4(a,b,c)).
- In agnostic settings, where the true distribution may lie outside the hypothesis class, RWM performs better than OFDE (Figs. 3(d,e,f), 4(d,e,f)), except for small alphabet sizes (Fig. 2).

These findings confirm that OFDE is more reliable when model assumptions match the data, while RWM may be better suited for more flexible settings, addressing Q2.

Q3: Which of the OL methods performs better as the alphabet size grows?

Our results indicate that RWM scales better with increasing alphabet size. At $k = 5$ and especially $k = 10$, RWM surpasses OFDE in most settings. This trend is reversed at $k = 2$, where OFDE maintains a consistent advantage.

This suggests that RWM's design, which incorporates exploration and probabilistic updates, adapts more effectively to larger state spaces.

Both Q2 and Q3 justify why OFDE performed better than RWM on real datasets in Table 6, given that the alphabet size is small ($k = 2$) and the structures are more complex—suggested by high densities (Table 5). An interesting observation is that RWM outperformed the other methods on one real-world dataset ("dna"), which stands out as the only high-dimensional dataset in the benchmark, characterized by $n/T \geq 0.1$. A future avenue could be to test on more high dimensional real datasets to check this hypothesis.

Q4: Which method is better when datasets are large?

In very high-dimensional or large-scale settings, the online learning methods (OFDE and RWM) struggle, particularly in terms of scalability and stability. Consequently, we exclude large datasets (with more than 200 variables) from the Twenty Datasets benchmark due to computational costs in memory and time.

In contrast, Chow-Liu and MoAT handle large datasets more effectively, especially in terms of runtime and convergence. Chow-Liu, being a greedy algorithm with a closed-form mutual information computation, scales linearly with the number of variables. MoAT, although slower, remains tractable with appropriate batching.

Therefore, for large-scale applications, Chow-Liu or MoAT are significantly more advantageous, effectively addressing Q4. To pursue the proposed future direction—namely, testing on more high-dimensional real-world datasets—it would be necessary to adapt both OFDE and RWM to better handle such settings.

In conclusion:

- Chow-Liu remains the strongest performer across most settings, especially for tree-structured and moderate-sized data.
- OFDE performs best in realizable and low-alphabet scenarios.
- RWM shines in agnostic settings and scales better with alphabet size.
- Although MoAT is theoretically rich, its practical performance declines when only a single tree is extracted from the mixture. Nonetheless, it remains a strong candidate for scalable applications.

These insights offer practical guidance for selecting learning algorithms based on data characteristics such as distributional assumptions, alphabet size, and dataset scale.

Acknowledgments. This work is supported by the Czech Science Foundation project 23-07299S ("Statistical Relational Learning in Dynamic Domains").

Disclosure of Interests. The author has no competing interests to declare that are relevant to the content of this article.

References

1. Benavoli, A., Corani, G., Demšar, J., Zaffalon, M.: Time for a change: a tutorial for comparing multiple classifiers through Bayesian analysis. J. Mach. Learn. Res. **18**(77), 1–36 (2017). http://jmlr.org/papers/v18/16-305.html
2. Bhattacharyya, A., Gayen, S., John, P.G., Sen, S., Vinodchandran, N.V.: Distribution learning meets graph structure sampling (2024). https://arxiv.org/abs/2405.07914
3. Bhattacharyya, A., Gayen, S., Price, E., Vinodchandran, N.V.: Near-optimal learning of tree-structured distributions by chow-liu (2021). https://arxiv.org/abs/2011.04144
4. Cesa-Bianchi, N., Lugosi, G.: Prediction, Learning, and Games. Cambridge University Press (2006)
5. Chekuri, C., Vondrak, J., Zenklusen, R.: Dependent randomized rounding via exchange properties of combinatorial structures. In: 2010 IEEE 51st Annual Symposium on Foundations of Computer Science, pp. 575–584 (2010). https://doi.org/10.1109/FOCS.2010.60
6. Chickering, D.M., Meek, C., Heckerman, D.: Large-sample learning of Bayesian networks is NP-hard (2012). https://arxiv.org/abs/1212.2468
7. Chow, C., Liu, C.: Approximating discrete probability distributions with dependence trees. IEEE Trans. Inf. Theory **14**(3), 462–467 (1968). https://doi.org/10.1109/TIT.1968.1054142
8. Cooper, G.F.: The computational complexity of probabilistic inference using Bayesian belief networks. Artif. Intell. **42**(2), 393–405 (1990). https://doi.org/10.1016/0004-3702(90)90060-D, https://www.sciencedirect.com/science/article/pii/000437029090060D

9. Heckerman, D., Nathwani, B.: Toward normative expert systems: Part II probability-based representations for efficient knowledge acquisition and inference. Methods Inf. Med. **31**, 106–16 (1992) https://doi.org/10.1055/s-0038-1634868
10. Imoto, S., Miyano, S., Matsuno, H.: Chapter 11 - gene networks: estimation, modeling, and simulation. In: Kriete, A., Eils, R. (eds.) Comput. Syst. Biol., 205–228. Academic Press, Burlington (2006). https://doi.org/10.1016/B978-012088786-6/50030-7, https://www.sciencedirect.com/science/article/pii/B9780120887866500307
11. Koriche, F.: Online forest density estimation. In: UAI (2016)
12. Kotłowski, W., Grünwald, P.: Maximum likelihood vs. sequential normalized maximum likelihood in on-line density estimation. In: Kakade, S.M., von Luxburg, U. (eds.) Proceedings of the 24th Annual Conference on Learning Theory. Proceedings of Machine Learning Research, vol. 19, pp. 457–476. PMLR, Budapest, Hungary (09–11 Jun 2011). https://proceedings.mlr.press/v19/kotlowski11a.html
13. Lowd, D., Davis, J.: Learning markov network structure with decision trees. In: Proceedings of the 2010 IEEE International Conference on Data Mining. p. 334–343. ICDM '10, IEEE Computer Society, USA (2010).https://doi.org/10.1109/ICDM.2010.128, https://doi.org/10.1109/ICDM.2010.128
14. Meila, M., Jordan, M.I.: Learning with mixtures of trees. J. Mach. Learn. Res. **1**, 1–48 (2001). https://doi.org/10.1162/153244301753344605
15. Pearl, J.: Probabilistic Reasoning in Intelligent Systems: Networks of Plausible Inference. Morgan Kaufmann Publishers Inc., San Francisco, CA, USA (1988)
16. Selvam, N.R., Zhang, H., den Broeck, G.V.: Mixtures of all trees (2023). https://arxiv.org/abs/2302.14202
17. Shtar'kov, Y.M.: Universal sequential coding of single messages. Problemy Peredachi Informatsii **23**(3), 3–17 (1987)
18. Van Haaren, J., Davis, J.: Markov network structure learning: a randomized feature generation approach. In: Proceedings of the AAAI Conference on Artificial Intelligence, vol. 2, pp. 1148–1154 (2012). https://doi.org/10.1609/aaai.v26i1.8315

From Reviews to Results: Generative AI for Review-Driven Product and Service Comparisons

Cristian Cosentino[1], Merve Gündüz Cüre[2], Fabrizio Marozzo[1](✉), and Şule Öztürk Birim[2]

[1] University of Calabria, Rende, Italy
{ccosentino,fmarozzo}@dimes.unical.it
[2] Manisa Celal Bayar University, Manisa, Turkey
{merve.gunduz,sule.ozturk}@cbu.edu.tr

Abstract. In today's digital world, user-generated reviews represent invaluable insights reflecting authentic experiences, preferences, and perceptions regarding products and services. Such reviews play a critical role both for consumers seeking informed purchasing decisions and businesses aiming to optimize their offerings and strategies. Recent advancements in machine learning, particularly the emergence of Large Language Models (LLMs) have significantly enhanced the processing and interpretation of this rich, unstructured textual data. Traditional platforms for comparing products and services focus primarily on structured specifications, often neglecting the detailed, experience-based information contained in user reviews. To address this limitation, we propose a novel framework that leverages Generative AI and advanced LLMs to extract and interpret user feedback, enabling more informed, experience-aware comparisons. Our approach involves three main phases: targeted review and metadata collection; topic modeling and sentiment classification using fine-tuned BERT models; and structured comparisons powered by user reviews, featuring attribute-level scores and natural language explanations generated by advanced GenAI tools such as GPT-4. Evaluated on real-world scenarios, including comparisons of similar Amazon products and nearby hotels, our framework outperforms traditional aggregation methods by generating more precise comparative scores and context-aware explanations.

Keywords: Large Language Models · Natural Language Processing · BERT · GPT · ChatGPT · Interpretable Models · Explainability

1 Introduction

In today's digital ecosystem, user-generated reviews have become indispensable, significantly influencing both consumer behavior and business strategy. Platforms such as Amazon, Booking.com, TripAdvisor, and Trustpilot host millions

S. Džeroski et al. (Eds.): DS 2025, LNAI 16090, pp. 78–93, 2025.
https://doi.org/10.1007/978-3-032-05461-6_6

of user opinions that capture real-world experiences, satisfaction levels, and individual preferences. This vast amount of textual feedback has been shown to shape purchasing decisions, build or damage brand trust, and affect market competitiveness. Beyond guiding consumers [5], reviews also provide organizations with actionable insights to improve product design, customer support, and marketing strategies [14]. Furthermore, recent studies suggest that analyzing sentiment and thematic patterns within user reviews can help predict product performance and identify emerging customer concerns before they escalate [15].

Despite their richness, user reviews remain underexploited in most product and service comparison systems. Traditional comparison platforms—such as shopping aggregators, travel booking sites, and consumer review portals—typically prioritize structured attributes like technical specifications, pricing, or expert scores, while ignoring the nuanced content of user-written reviews. When user input is included, it is often reduced to simplified summaries such as star ratings or sentiment scores, which fail to reflect the diversity and specificity of user experiences. Some leading platforms, including Amazon and Tripadvisor, have introduced concise summaries of user feedback for individual products or hotels, but they do not provide mechanisms to directly compare competing options based on this textual evidence.

To address these limitations, we introduce a unified framework that places user feedback at the core of product and service comparisons, enabling more meaningful and transparent evaluations. The process begins by systematically collecting reviews and accompanying structured data—such as technical specifications, product details, and helpfulness votes—for the specific products or services to be compared, sourced from the most relevant online platforms. We then apply BERTopic to extract dominant topics across the combined review corpus, and perform topic-specific sentiment analysis using fine-tuned BERT models to assess whether each theme is discussed positively, negatively, or neutrally for each product [2]. Finally, GPT-4 uses the topic–sentiment signals to generate structured comparison tables along with concise explanatory narratives that reflect both technical information and user opinions.

We conducted extensive experimental evaluations to assess the effectiveness of our review-informed comparison framework. The experiments involved multiple pairs of comparable Amazon products and Tripadvisor hotels, analyzed using two prompting strategies: a baseline that leverages technical information about the products or services, and an advanced method that integrates structured topic–sentiment information extracted from user reviews. Results show that the advanced approach produces more critical and user-aligned evaluations, particularly in identifying product weaknesses and supporting decisions. Evaluations confirm that the explanations generated through our framework are more nuanced and actionable. An ablation study further validates the contribution of each component—topic annotation, sentiment filtering, and structured prompting—by demonstrating consistent performance improvements as each element is introduced.

The remainder of the paper is structured as follows. Section 2 offers a concise review of related works. Section 3 describes the proposed framework. Section 4 presents in-depth comparisons of Amazon products and Tripadvisor hotels. Section 5 analyzes the impact of different prompting strategies on scores and explanations. Finally, Sect. 6 concludes the paper.

2 Related Work

Comparing products based on online reviews is essential for understanding customer preferences and guiding product development. Reviews offer rich, real-world insights into user satisfaction, expectations, and recurring issues. Prior studies have highlighted their value: Mudambi and Schuff [14] emphasized the role of big data in capturing user experience; Yang et al. [21] showed reviews can support satisfaction estimation; and Kessler et al. [11] proposed extracting comparative statements for product ranking.

Advancements in machine learning—particularly deep learning—have enabled automatic extraction of comparative insights. Arora et al. [1] a deep learning approach using LSTMs to extract product comparison information—product names, user opinions, and comparison aspects—from e-commerce reviews, while Sharma et al. [18] presents an RPA system that collects and processes e-commerce data to support personalized product recommendations and Li et al. [12] explored how e-WOM and review helpfulness influence user decisions.

Several frameworks have emerged to support structured product comparison. MatrixMiner [16] extracts feature-value pairs from unstructured text, and SRSS [10] combines topic detection and sentiment analysis to compare opinions. Recent work includes i-SPC [17], a Shopee-integrated tool using Naïve Bayes for multi-criteria evaluation; LDA-based sentiment-topic models for networked comparisons [9]; and web scraping systems for real-time product ranking [7]. Vedula et al.'s ReBARC [19] and HCPC [20] integrate catalog data with reviews to generate explainable, human-centered comparisons.

In addition to academic research, commercial platforms like *Consumer Reports*[1] and *CNET*[2] offer automated or semi-automated product comparisons. While widely used and informative—leveraging expert testing, editorial reviews, and user opinions—these services are typically limited to selected product categories and primarily rely on structured data and editorial insights, often lacking the granularity provided by topic-level sentiment analysis.

Our framework fills this gap by combining advanced topic modeling, sentiment classification, and natural language generation techniques to extract and synthesize insights from user reviews. By integrating with major online platforms and their APIs, it enables the comparison of virtually any product or service available online. The system filters out unreliable or outlier reviews to reduce sentiment distortion and constructs structured overviews that highlight

[1] https://consumerreports.org.
[2] https://cnet.com.

strengths, weaknesses, and possible design rationales—directly grounded in user experiences. Unlike traditional ranking systems, our approach supports in-depth, user-driven comparisons based on authentic feedback rather than curated specifications.

3 Proposed Framework

Our framework transforms technical specifications, consumer reviews, and platform metadata into structured insights that support informed product and service comparisons. It integrates classification and generation techniques within a three-phase pipeline, summarized in Fig. 1.

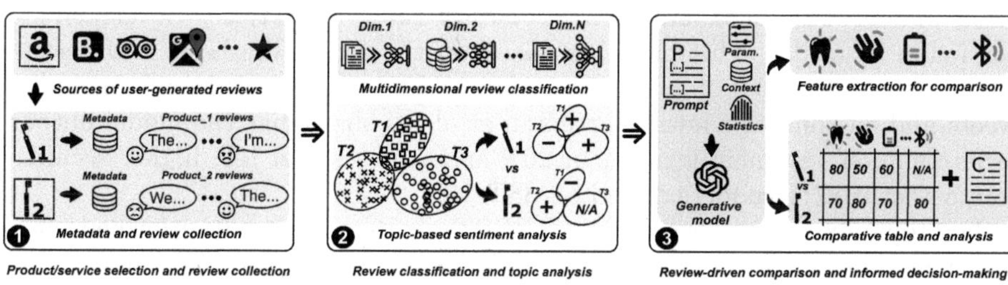

Fig. 1. Execution flow of the proposed framework.

The initial phase, *Product/service selection and review collection*, involves identifying relevant products or services and gathering consumer reviews from prominent platforms offering user-generated content, such as Amazon, TripAdvisor, Booking.com, and Trustpilot. Reviews are collected along with pertinent metadata—including product specification sheets and user feedback indicators (e.g., helpfulness ratings). To complement user opinions, technical specifications are also retrieved either directly from the web or inferred using language models when structured data is unavailable.

The second phase, *Review classification and topic analysis*, employs advanced language models, specifically fine-tuned BERT variants, to classify and analyze collected reviews. Initially, topic modeling techniques (e.g., BERTopic) identify dominant topics across the aggregated reviews. Subsequently, each review undergoes topic-specific sentiment analysis to ascertain whether user opinions related to a particular topic lean positively or negatively. This step includes rigorous outlier detection to ensure reliability; topics lacking sufficient review volume for specific products are excluded from detailed sentiment analysis, preventing unreliable conclusions based on sparse data.

In the final phase, *Review-driven comparison and informed decision-making*, the framework first defines the comparative features based on the most salient topics identified across the review corpus. It then combines technical information and model knowledge with aggregated statistics and individual review annotations—including topic and sentiment labels—to assign scores and generate structured comparison tables. These tables highlight how each product or service is evaluated on key aspects by real users, supporting transparent and evidence-based decision-making. The framework also interprets user criticisms in the context of potential design trade-offs, offering explanations for recurring issues. This approach supports a more grounded understanding of strengths and limitations as perceived by users.

3.1 Implementation Details of Classification Algorithms and Topic Detection

Each review in our pipeline is annotated along two orthogonal dimensions: *sentiment* and *topic*. To compute sentiment scores, we use the pretrained transformer-based model `siebert/sentiment-roberta-large-english`[3], developed by Hartmann et al. [8]. This model is based on the RoBERTa-large architecture [13] and was fine-tuned on a diverse corpus of sentiment-labeled texts, including tweets and customer reviews. Hartmann et al. [8] show that the model outperforms common sentiment lexicons by 20% and achieves over 15% higher accuracy than a DistilBERT model trained on SST-2.

For topic extraction, we adopted *BERTopic* [6], which outperformed alternative methods in both consistency and diversity of topics. The model clusters review embeddings—generated using a pretrained RoBERTa model—into coherent thematic groups. The number of topics is selected by maximizing topic coherence, following the protocol of Chen et al. [3], to balance granularity and generality. Each review is then assigned a dominant topic label, corresponding to the theme with the highest prominence score.

Once topics are identified, we aggregate sentiment within each topic to construct a structured representation of user opinion. Specifically, for each product/service, we compute a topic–sentiment vector containing one entry per shared topic, where each value represents the aggregated sentiment of all reviews associated with the corresponding topic, expressed on a five-level scale (*strongly negative*, *slightly negative*, *neutral*, *slightly positive*, *strongly positive*). This representation enables the comparison of multiple products (two or more) within a unified topic space. By preserving sentiment granularity, the framework captures subtle differences in user perception across competing items. These structured vectors are then used to condition the prompt, guiding the generation of concise and balanced comparative narratives informed by topic-level sentiment signals and representative review excerpts.

3.2 Implementation Details of Prompt Strategies

We evaluate four prompts of increasing complexity, each adding more contextual information to guide the model's judgments. While our implementation relies on OpenAI-4o accessed via the API, the same framework is compatible with other

[3] https://huggingface.co/siebert/sentiment-roberta-large-english.

Table 1. Prompt variants and included input components. Checkmarks indicate which elements are provided at each stage of prompt refinement.

Prompt	Product names	Generate comp. features	Reviews	Topics	Sentiment +outliers	Explanation
`base`	✓	✓				
`intermediate1`	✓	✓	✓			
`intermediate2`	✓	✓	✓	✓		
`advanced`	✓	✓	✓	✓	✓	✓

generative models such as Gemini or Claude, provided they support structured input and instruction-following capabilities. Table 1 summarizes the input components provided with each prompt.

The simplest variant, referred to as the `base` prompt, relies solely on the model's general knowledge and does not incorporate any additional user-provided information. It provides the names of multiple products or services stored in the variable `$products`, and instructs the model to generate a list of dictionaries—each evaluating a specific *product-feature* pair with a *score* in the range `[1, 5]` and a *description*. While the prompt supports comparisons involving any number of products, our experiments focus on scenarios involving two products or services. Although prompts can generate features directly, this often leads to inconsistent or non-overlapping sets across items, hindering comparability. To ensure stable and aligned evaluations, we extract features externally—combining domain knowledge, topic modeling, and product specifications—and supply them as structured input to the prompt.

> `Base prompt`: Your task is to evaluate and compare multiple products listed in {`$products`} using a predefined set of features given in {`$features`}
>
> For each product and each feature, generate an individual evaluation in the form of a dictionary with the following fields: `product` (a short name or identifier of the product), `feature` (the name of the evaluated feature), `score` (a numerical rating in the range `[1, 5]`), and `description` (a concise and objective justification for the assigned score). The response must be a list of such dictionaries, each evaluating a single product-feature pair.
>
> Do not include any introductory or summary text. Do not compare products directly within descriptions. Ensure that each dictionary is self-contained and independently describes the evaluation.

To improve grounding, the next variant—`intermediate1`—supplements the prompt with the full set of user reviews for each product. This allows the model to rely on authentic customer feedback rather than general knowledge alone. Building on this, `intermediate2` introduces the dominant topic labels discovered by BERTopic from the reviews, helping to focus the model's attention on the most salient and recurrent themes.

A further enhancement in the **advanced** prompt is the structured representation of each review as a JSON object, including the review text, its sentiment, and its dominant topic. This enriched format offers a more granular view of product perception across topics. The prompt also incorporates average *topic–sentiment* distributions to guide the model in generating feature-level evaluations. The model considers both the presence and polarity of each topic and receives explicit instructions for handling missing or sparse information.

> Advanced prompt: Your task is to evaluate and compare multiple products listed in {$products} using a predefined set of features in {$features}. For each product, {$<product>_reviews} supplies a list of enriched reviews, each represented as a JSON object with the fields {*text, sentiment, topic*}. Aggregate data are also provided for every <product>, <topic> pair:
>
> – {$<product>_<topic>_sentiment} — overall sentiment label (e.g., high negative, low positive);
> – {$<product>_<topic>_info} — number of reviews in that topic.
>
> For every feature of every product, combine (i) technical knowledge, (ii) the review texts, and (iii) the aggregate statistics. Produce a list of dictionaries; each dictionary contains:
>
> – Product – product name;
> – Feature – feature name;
> – Score – a value in [1,5] or "N/A" if the feature is absent;
> – Description – a concise explanation blending technical and review-based insights.
>
> Do not include introductory or summary text. Do not compare products within descriptions. Ensure each dictionary is self-contained and independently explains its evaluation.

Finally, we introduce an auxiliary prompt that asks the model to explain low-scoring or negatively weighted features by identifying plausible causes—such as design trade-offs, usability flaws, or service gaps—thereby enhancing the interpretability of the generated evaluations.

4 Experimental Results

The experimental evaluation of our framework is based on two curated datasets, publicly available at https://github.com/SCAlabUnical/UserReviewDatasets, designed to support comparative analysis of similar products and services using user-generated reviews. The first dataset contains approximately 10,000 Amazon reviews of functionally equivalent electronic items (e.g., electric toothbrushes, USB cables), with structured metadata such as ratings, titles, verification status, and timestamps. The second dataset includes Tripadvisor reviews of hotels in New York City, with rich metadata such as sub-ratings (e.g., service, value), trip types, and management responses. It comprises 150 hotels and several hundred reviews per hotel. Together, these datasets enable systematic, feature-level comparisons grounded in both user sentiment and thematic relevance.

4.1 Use Case: In-Depth Comparison of Two Electric Toothbrushes via Amazon Reviews

For demonstration purposes, we illustrate our framework using two electric toothbrushes available on Amazon, referred to here as `Model-A` and `Model-B`, as a case study. Although these represent real commercial products, we anonymize their identities to maintain neutrality, avoid brand bias, and focus solely on the methodological aspects of the review-based analysis.

`Model-A` is the newer-generation, professional-grade toothbrush, designed with a more compact and ergonomic form. It features a hard plastic body that gives it a modern, durable look and feel. In addition to improved cleaning performance, it includes a built-in pressure sensor and timer, catering to users who value efficiency and simplicity in oral hygiene. `Model-B`, by contrast, belongs to the previous product generation. It features a more classic design, is slightly bulkier in size, and uses smooth plastic for a softer tactile finish. Its main differentiator is the inclusion of Bluetooth connectivity—specifically in the version of `Model-B` analyzed—which enables integration with a mobile app for brushing feedback and usage tracking.

These two products, which differ in design era, materials, form factor, and functional focus, provide an ideal setting to demonstrate our framework's ability to extract and compare multidimensional insights from Amazon reviews. Specifically, we analyze sentiment and referenced topics to reveal how users perceive and evaluate each product.

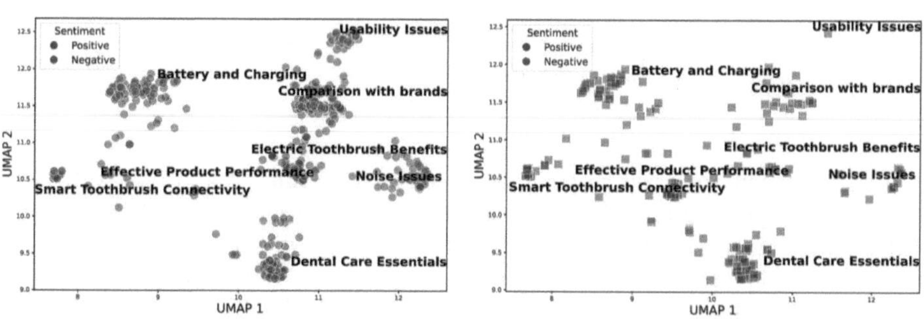

(a) Model-A. (b) Model-B.

Fig. 2. UMAP visualization of BERTopic clusters for `Model-A` and `Model-B` reviews, with sentiment polarity (green = *positive*, red = *negative*) and topic labels generated from representative keywords.

We performed sentiment classification for each review and extracted discussion topics using BERTopic. Figures 2(a) and 2(b) show the topic clusters for `Model-A` and `Model-B`, respectively, visualized in a two-dimensional latent space using UMAP. Each cluster represents a distinct discussion theme, with topic names generated by ChatGPT based on the top representative keywords in each group. Green points denote *positive reviews*, while red points indicate *negative ones*. Finally, we assign each cluster an average sentiment label on a five-level scale (*strongly negative*, *slightly negative*, *neutral*, *slightly positive*, or *strongly positive*) to consistently capture differences in user sentiment across topics.

For `Model-A`, the sentiment distribution across topics is relatively balanced. Clusters such as *Effective Product Performance* and *Dental Care Essentials* exhibit a predominance of positive reviews. In contrast, negative sentiment is more concentrated in topics like *Battery and Charging*, *Usability Issues*, *Comparison with Brands*, and *Noise Issues*. `Model-B`, on the other hand, shows generally less negative sentiment across most clusters. However, negative feedback still appears, particularly within *Usability Issues*, *Comparison with Brands*, and *Noise Issues*, albeit with slightly lower intensity than in `Model-A`.

Table 2. Evaluation scores (1–5 scale) for `Model-A` and `Model-B` using the `base` and `advanced` approaches (1 = very poor, 5 = excellent).

	Base		Advanced	
Feature	Model-A	Model-B	Model-A	Model-B
Battery and Charging	4.0	4.5	3.0	3.5
Comparison with brands	4.0	4.5	3.5	4.5
Dental Care Essentials	4.0	4.5	4.0	4.0
Noise Issues	3.5	4.0	3.5	4.0
Electric Toothbrush Benefits	4.5	4.5	4.5	4.5
Usability Issues	4.0	4.5	2.5	4.0
Effective Product Performance	4.5	4.5	4.5	4.0
Smart Toothbrush Connectivity	1.0	5.0	N/A	3.5

Table 2 shows the average evaluation scores for `Model-A` and `Model-B`, computed using the `base` and `advanced` prompting approaches. Scores are on a 1–5 scale, following a Likert-style evaluation where 1 indicates very poor performance and 5 denotes excellent performance. Under the `base` approach, `Model-B` exceeds `Model-A` on nearly every feature, driven chiefly by its Bluetooth connectivity and higher usability rating. In the `advanced` evaluation—where review evidence tempers the purely technical view—the gap narrows: `Model-B` still leads overall, but several of its ratings drop once user criticisms are considered, while `Model-A` retains a clear edge in *Effective Product Performance*. The `advanced` scores therefore surface trade-offs (e.g., connectivity reliability, grip comfort) that the `base` summary overlooks, giving a more critical, usage-oriented picture of both brushes.

Due to space constraints, we do not include full feature descriptions in the comparison table. Instead, we focus on those aspects where evaluation differences between the two models are most significant. For example, in the case of *Usability Issues*, notable contrasts emerge in both the `base` and `advanced` evaluations:

[Model-A] Usability Issues description (base) = Simple one-button interface is very intuitive but provides no feedback on pressure level or brushing time beyond the built-in timer. [Score = 4.0]
[Model-B] Usability Issues description (base) = Offers LED mode/pressure feedback, but initial Bluetooth setup can be cumbersome and may confuse non-tech-savvy users. [Score = 4.5]

[Model-A] Usability Issues description (advanced) = Intuitive design, but lacks interactive feedback. Frequent user complaints include slippery handle, stiff brush heads, awkward button layout, and unclear pressure signals, making daily use frustrating for many. *These issues likely stem from design trade-offs: the hard-plastic shell may improve waterproofing and hygiene by avoiding textured surfaces or seams, but this reduces grip and tactile comfort.* [Score = 2.5]
[Model-B] Usability Issues description (advanced) = Offers LED guidance but has complex setup. Users do not often mention setup problems, but the Bluetooth connection and settings may still be confusing for people not used to technology. [Score = 4.0]

A second example involves *Smart Toothbrush Connectivity*:

[Model-A] Smart Toothbrush connectivity description (base) = No Bluetooth or app connectivity available, limiting interactive guidance and progress tracking. [Score = 1.0]
[Model-B] Smart Toothbrush connectivity description (base) = Bluetooth connectivity with the official app provides real-time brushing tracking, coaching, and personalized reports. [Score = 5.0]

[Model-A] Smart Toothbrush connectivity description (advanced) = Lacks Bluetooth or connectivity features. Not discussed in reviews due to absence. [Score = N/A]
[Model-B] Smart Toothbrush connectivity description (advanced) = Offers app-based brushing feedback and reports. Reviews praise guidance but mention frequent disconnects and feedback gaps. [Score = 3.5]

The paired excerpts for *Usability Issues* and *Smart Toothbrush Connectivity* show why the review-aware advanced prompt is more diagnostic than the technical-only base version. In the first case, the `base` summaries simply note an intuitive one-button layout for *Model A* and LED feedback with a fussy Bluetooth setup for *Model B*. The `advanced` summaries, however, bring the user voice to the forefront by detailing the slippery handle, stiff brush heads, awkward button placement and unclear pressure signals that frustrate *Model A* owners, while also acknowledging that *Model B*'s guidance works yet can confuse the less tech-savvy. In the second case, the `base` layer records the presence or absence of Bluetooth and assigns extreme scores, but the advanced layer adds context: *Model B*'s app is applauded for coaching yet criticised for frequent disconnects, and *Model A* is marked "N/A" rather than a very low value to signal that the feature is missing, not implemented. By incorporating user experiences and highlighting possible design trade-offs, the `advanced` approach helps contextualize complaints and offers a more detailed and balanced foundation for comparison.

4.2 Use Case: Comparative Analysis of Two Downtown Hotels via TripAdvisor Reviews

To demonstrate the portability of our framework to the hospitality domain, we analyse two large New-York-City properties—hereafter `Hotel-A` (a riverside all-suite brand) and `Hotel-B` (a business-class high-rise belonging to the same global chain). Both are located within Lower Manhattan's financial district and share similar price points, yet they differ in age, room configuration and service philosophy, making them a suitable pair for review-based comparison.

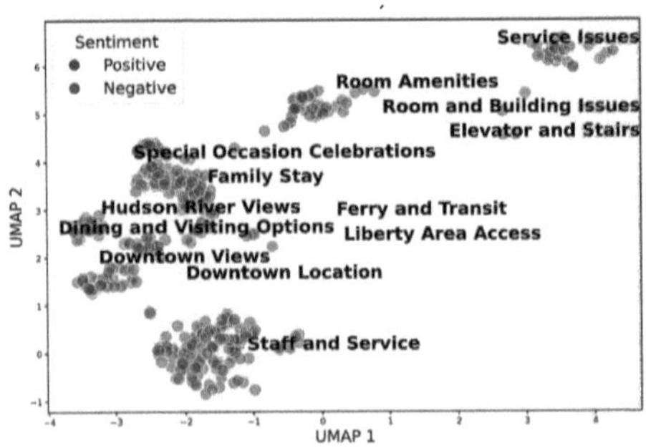

(a) Hotel-A.

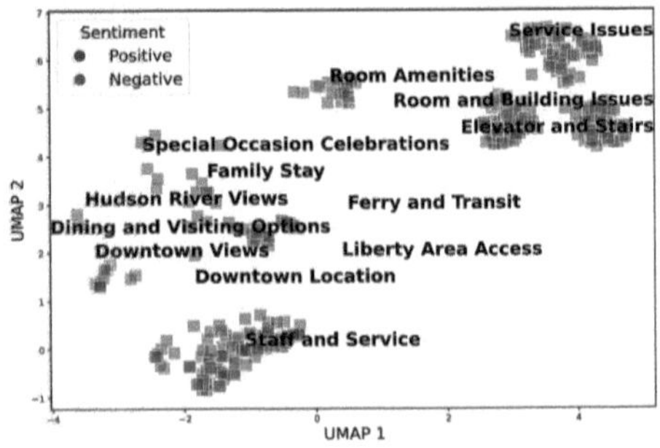

(b) Hotel-B.

Fig. 3. UMAP visualization of BERTopic clusters for `Hotel-A` and `Hotel-B` reviews, with sentiment polarity (green = positive, red = negative) and topic labels generated from representative keywords.

Figures 3(a) and 3(b) show the UMAP projection of BERTopic clusters extracted from the TripAdvisor reviews. For `Hotel-A` (Fig. 3(a), the majority of clusters are dominated by positive sentiment. However, negative feedback is concentrated in a compact *Service Issues* cluster in the upper-right corner. In contrast, `Hotel-B` (Fig. 3(b)) exhibits a broader and more dispersed pattern of negative sentiment. Several clusters—most notably *Elevator and Stairs*, *Service Issues*, *Room Amenities*, and *Room and Building Issues*—show a higher concentration of negative reviews, indicating recurring dissatisfaction across multiple dimensions of the guest experience. This distribution suggests more widespread critical feedback, despite the otherwise similar thematic structure.

Table 3. Evaluation scores (1–5 scale) for `Hotel-A` and `Hotel-B` using the `base` and `advanced` approaches (1 = very poor, 5 = excellent).

	Base		Advanced	
`Feature`	`Hotel-A`	`Hotel-B`	`Hotel-A`	`Hotel-B`
Staff and Service	4.5	4.0	4.5	4.0
Service Issues	5.0	3.0	3.0	2.0
Family Stay	3.5	4.0	4.0	4.0
Elevator and Stairs	4.5	4.0	4.5	2.5
Room Amenities	5.0	4.0	5.0	3.0
Room and Building Issues	4.5	4.0	4.5	3.0
Downtown Location	5.0	5.0	5.0	5.0
Hudson River Views	4.5	4.0	4.5	4.0
Downtown Views	4.5	4.0	4.5	4.0
Special Occasion Celebrations	4.5	4.0	4.5	4.0
Liberty Area Access	5.0	5.0	5.0	5.0
Ferry and Transit	5.0	4.5	5.0	4.5
Dining and Visiting Options	4.5	4.5	4.5	4.5

Table 3 reports 1–5 scores for each feature under two evaluation schemes: a `base` layer that covers tangible facilities and location descriptors, and an `advanced` layer enriched with sentiment-weighted insights from guest narratives.

Three features illustrate how the review layer reshapes the assessment. For *Elevator and Stairs*, both hotels earn strong `base` marks for having modern lifts, yet `Hotel-B`'s `advanced` score falls sharply once guests' complaints about long waits and crowding are considered, whereas `Hotel-A` remains unchanged. In *Service Issues*, operational metrics give `Hotel-A` a perfect score, but the review layer lowers that value and pushes `Hotel-B` even lower after noting delays, limited breakfast hours, and patchy room service. Conversely, the *Family Stay* score for `Hotel-A` rises after positive family experiences in the reviews offset its modest `base` value, while `Hotel-B` remains stable. These shifts show how the descriptive layer grounds the numbers in real usage contexts and reveals friction points that raw specifications alone can overlook.

5 Performance Comparison

To assess explanation quality across domains, we applied both prompting strategies—`base` and `advanced`—to ten Amazon product pairs and ten hotel pairs. For each pair, we generated a feature-comparison table for both items under both strategies. These tables contain attribute-level scores as well as concise descriptions that justify each score, and the evaluation focuses on these explanations rather than on the products themselves. This automated "LLM-as-a-judge" protocol is now common in the literature because it scales well, enforces a consistent rubric, and correlates closely with expert judgment when manual evaluation is impractical [4,22].

Two reasoning-oriented language models, *OpenAI-o3* and *DeepSeek-R1*, served as independent evaluators. Each model rated every explanation on a five-point Likert scale (1 = very poor, 5 = excellent) according to five criteria relevant to product and service comparison:

- *Feature Coverage*: extent to which all salient characteristics are mentioned;
- *Strength Identification*: clarity and completeness in highlighting advantages;
- *Weakness Identification*: accuracy and balance in exposing flaws or trade-offs;
- *Comparative Insight*: effectiveness in contrasting the item with its alternatives;
- *Decision Support*: usefulness of the explanation in guiding a justified choice.

Table 4 confirms the advantage of the review-aware `advanced` prompt. Across both datasets and for both evaluators, the advanced explanations obtain higher scores on the three dimensions most relevant to decision-making—*Weakness Identification*, *Comparative Insight*, and *Decision Support*. This indicates that injecting review evidence makes the text more critical and more useful when weighing alternatives. Results for *Feature Coverage* and *Strength Identification* are mixed. With *OpenAI-o3*, the advanced prompt covers features slightly better, whereas *DeepSeek-R1* shows no clear gain and even a small edge for the base version in a few cases. Strength identification also favours the base summaries in most Amazon pairs and under *DeepSeek-R1* overall. These modest reversals are expected: when the prompt devotes more space to user criticism, it can leave fewer tokens for exhaustive feature lists or positive framing. Even so, the advanced prompt consistently surfaces user-reported issues more clearly

Table 4. Average LLM-as-a-judge scores (1–5) for Amazon product pairs and Tripadvisor hotel pairs, judged by *OpenAI-o3* and *DeepSeek-R1*. Bold marks each evaluator's best metric.

Evaluator	Metric	Amazon dataset		Tripadvisor dataset	
		Base	Adv	Base	Adv
OpenAI-o3	Feature Coverage	4.1	**4.4**	4.1	**4.5**
	Strength Identification	**4.2**	4.1	4.0	**4.1**
	Weakness Identification	2.6	**4.5**	2.5	**4.3**
	Comparative Insight	3.5	**4.1**	3.6	**4.2**
	Decision Support	3.8	**4.2**	3.9	**4.3**
	Mean	3.5	**4.3**	3.6	**4.3**
DeepSeek-R1	Feature Coverage	**4.1**	4.1	**4.2**	4.1
	Strength Identification	**4.2**	4.0	**4.1**	4.0
	Weakness Identification	3.2	**5.0**	3.6	**4.9**
	Comparative Insight	4.0	**4.5**	4.0	**4.4**
	Decision Support	3.9	**5.0**	3.9	**4.8**
	Mean	3.9	**4.5**	4.0	**4.4**

and offers stronger comparative guidance, outweighing the small loss in upbeat content.

The values reported in Table 4 are means computed from ten fully independent runs for each evaluator, with each run executed in isolation to avoid cross-session contamination. The results were stable across repetitions, with low run-to-run variance observed for both base and advanced prompts. Paired-samples t-tests conducted on the per-run scores revealed that advanced prompts consistently outperformed base prompts across most dimensions ($p < 0.05$), confirming that the observed differences were statistically significant across runs.

Table 5 reports composite mean scores for each prompt variant. As additional review-grounded signals are introduced—moving from the `base` prompt, through two intermediate variants, to the fully review-aware `advanced` prompt—the quality of the generated explanations generally rises. *OpenAI-o3* improves monotonically on both datasets, while *DeepSeek-R1* shows a brief dip at `inter1` on the Amazon items before climbing again. The overall pattern is clear: richer inputs that blend topic labels and sentiment statistics help the models produce more informative and actionable comparisons, with the `advanced` prompt delivering the highest scores across evaluators and domains.

Table 5. Composite ablation scores. Values are the mean of the five LLM-as-a-judge criteria for each prompt variant, computed per dataset and evaluator.

Dataset	Evaluator	Base	Inter1	Inter2	Advanced
Amazon	OpenAI-o3	3.5	3.8	3.9	4.3
	DeepSeek-R1	3.9	3.9	4.2	4.5
Tripadvisor	OpenAI-o3	3.6	3.8	4.0	4.3
	DeepSeek-R1	4.0	4.1	4.2	4.4

6 Conclusions

In today's digital landscape, user-generated reviews are a rich source of consumer insight that increasingly shapes purchasing behavior, product design, and business strategy. While modern language models such as BERT and GPT have enabled more powerful tools for extracting information from this unstructured content, effective comparison still requires going beyond raw classification or rating aggregation.

Our framework addresses this need by combining topic-specific sentiment analysis with structured generative summarization to deliver more accurate, balanced, and informative comparisons. By grounding evaluations in both user feedback and product metadata, the system supports decision-making with context-aware insights rather than opaque scores. Results across both Amazon and Tripadvisor domains show that review-aware prompting consistently outperforms traditional specification-based methods. The ablation analysis further highlights the incremental benefits of topic annotation, sentiment filtering, and structured input design.

The architecture is modular and can evolve with new inputs or deployment contexts. Future directions include incorporating multimodal signals (e.g., images or video), adapting to real-time review streams, and supporting interactive or user-personalized comparison tools. Addressing challenges such as opinion bias, review sparsity, and domain transfer will also be essential to ensure the robustness and scalability of the approach.

Acknowledgements and Competing Interests. This work was supported by the research project "INSIDER: INtelligent ServIce Deployment for advanced cloud-Edge integRation" granted by the Italian Ministry of University and Research (MUR) within the PRIN 2022 program and European Union - Next Generation EU (grant n. 2022WWSCRR, CUP H53D23003670006). The authors declare that they have no competing interests.

References

1. Arora, J., et al.: Extracting entities of interest from comparative product reviews. In: CIKM, pp. 1975–1978 (2017)
2. Cantini, R., Cosentino, C., Marozzo, F.: Multi-dimensional classification on social media data for detailed reporting with large language models. In: 20th International Conference on Artificial Intelligence Applications and Innovations, pp. 100–114 (2024)
3. Chen, Y., Peng, Z., Kim, S.H., Choi, C.W.: What we can do and cannot do with topic modeling: a systematic review. Commun. Methods Meas. **17**(2), 111–130 (2023)
4. Chiang, C.H., Lee, H.y.: Can large language models be an alternative to human evaluations? arXiv preprint arXiv:2305.01937 (2023)
5. Cosentino, C., Gündüz-Cüre, M., Marozzo, F., Öztürk-Birim, Ş.: Exploiting large language models for enhanced review classification explanations through interpretable and multidimensional analysis. In: International Conference on Discovery Science, pp. 3–18. Springer (2024)
6. Grootendorst, M.: BERTopic: neural topic modeling with a class-based TF-IDF procedure (2022)
7. Harikirshnan, K., et al.: Intelligent online shopping using ml-based product comparison engine. In: 6th International Conference on Inventive Computation Technologies, ICICT 2023 - Proceedings, pp. 174–179 (2023)
8. Hartmann, J., Heitmann, M., Siebert, C., Schamp, C.: More than a feeling: accuracy and application of sentiment analysis. Int. J. Res. Mark. **40**(1), 75–87 (2023)
9. He, Z., Zheng, L., He, S.: A novel approach for product competitive analysis based on online reviews. Electron. Commer. Res. **23**, 2259–2290 (2023)
10. Jin, J., Ji, P., Yan, S.: Comparison of series products from customer online concerns for competitive intelligence. J. Ambient Intell. Humanized Comput. **10**, 937–952 (2019)
11. Kessler, W., Klinger, R., Kuhn, J.: Towards opinion mining from reviews for the prediction of product rankings. In: WASSA, pp. 51–57 (2015)

12. Li, Y., Zheng, J., Yue, S., Zhi-ping, F.: Capturing and analyzing e-WOM for travel products: a method based on sentiment analysis and stochastic dominance. Kybernetes **51**, 3041–3072 (2021)
13. Liu, Y., et al.: Roberta: A robustly optimized BERT pretraining approach. arXiv:1907.11692 (2019)
14. Mudambi, S.M., Schuff, D.: Research note: What makes a helpful online review? a study of customer reviews on amazon. com. MIS quarterly, pp. 185–200 (2010)
15. Myers, D., et al.: Foundation and large language models: fundamentals, challenges, opportunities, and social impacts. Clust. Comput. **27**(1), 1–26 (2024)
16. Nasr, S.B., et al.: Automated extraction of product comparison matrices from informal product descriptions. J. Syst. Softw. **124**, 82–103 (2017)
17. Samah, K.A.F.A., et al.: Intelligence shopee product comparison (i-SPC) and visualization of product information via naïve bayes adaptation. J. Adv. Res. Appl. Sci. Eng. Technol. **37**, 179–190 (2024)
18. Sharma, D.K.: E-commerce product comparison portal for classification of customer data based on data mining. Materials Today: Proceedings **51**, 166–171 (2022)
19. Vedula, N., Collins, M., Agichtein, E., Rokhlenko, O.: What matters for shoppers: Investigating key attributes for online product comparison. AI and Lecture Notes in Bioinformatics, pp. 231–239 (2022)
20. Vedula, N., Collins, M., Agichtein, E., Rokhlenko, O.: Generating explainable product comparisons for online shopping. WSDM (2023), pp. 949–957 (2023)
21. Yang, C., Wu, L., Kun, T., Yu, C., Zhou, Y., Tao, Y., Song, Y.: Online user review analysis for product evaluation and improvement. J. Theor. Appl. Electron. Commer. Res. **16**, 1598–1611 (2021)
22. Zhang, X., Li, Y., Wang, J., Sun, B., Ma, W., Sun, P., Zhang, M.: Large language models as evaluators for recommendation explanations. In: 18th ACM Conference on Recommender Systems, pp. 33–42 (2024)

Symbolic Regression for Science

Prompting Neural-Guided Equation Discovery Based on Residuals

Jannis Brugger[1,2(✉)], Viktor Pfanschilling[1,5], David Richter[1], Mira Mezini[1,2,3], and Stefan Kramer[4]

[1] Technical University of Darmstadt, Darmstadt, Germany
jannis.brugger@tu-darmstadt.de
[2] Hessian Center for Artificial Intelligence (hessian.AI), Darmstadt, Germany
[3] National Research Center for Applied Cybersecurity ATHENE, Darmstadt, Germany
[4] Johannes Gutenberg-Universität Mainz, Mainz, Germany
[5] German Research Center for Artificial Intelligence, Darmstadt, Germany

Abstract. Neural-guided equation discovery systems use a data set as prompt and predict an equation that describes the data set without extensive search. However, if the equation does not meet the user's expectations, there are few options for getting other equation suggestions without intensive work with the system. To fill this gap, we propose Residuals for Equation Discovery (RED), a post-processing method that improves a given equation in a targeted manner, based on its residuals. By parsing the initial equation to a syntax tree, we can use node-based calculation rules to compute the residual for each subequation of the initial equation. It is then possible to use this residual as new target variable in the original data set and generate a new prompt. If, with the new prompt, the equation discovery system suggests a subequation better than the old subequation on a validation set, we replace the latter by the former.

RED is usable with any equation discovery system, is fast to calculate, and is easy to extend for new mathematical operations. In experiments on 53 equations from the Feynman benchmark, we show that it not only helps to improve all tested neural-guided systems, but also all tested classical genetic programming systems.

Keywords: Symbolic Regression · Disentanglement of Equations

1 Introduction

Equation discovery is the task given a data set $D \in \mathbb{R}^{m \times n}$ to find for m examples a free-form equation $f(\mathbf{x}, \mathbf{c}) = \mathbf{y}$ mapping the independent variables $x_i | 1 \leq i \leq n-1$ and the constants $\mathbf{c}$ to the dependent variable y. To find this equation, pretrained equation discovery systems (EDSs) have become popular in recent years [1,8,9,18,21,23]. These systems use a neural architecture to embed the data set and are trained to predict the equation that generated the data set in a zero-shot way. The method proposed in this paper, *Residuals for equation*

S. Džeroski et al. (Eds.): DS 2025, LNAI 16090, pp. 97–112, 2025.
https://doi.org/10.1007/978-3-032-05461-6_7

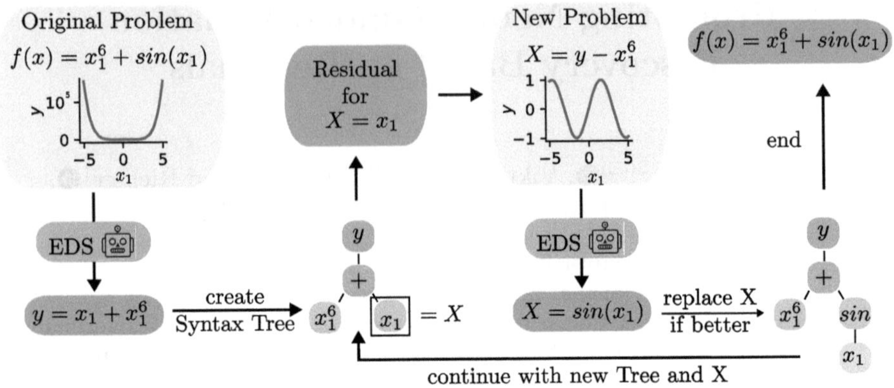

Fig. 1. Overview of how RED helps to disentangle the initial problem $f(x) = x_1^6 + \sin(x_1)$. RED can be combined with every Equation Discovery System (EDS).

discovery (RED), calculates and optimizes the residuals of the data set for a subequation X of the initial equation; that is, we compute what that subequation should have yielded for each data point such that the entire formula predicts the output correctly. These residuals $\mathbf{y}'$ formulate a new problem $f'(\mathbf{x}, \mathbf{c}) = \mathbf{y}'$, and the equation discovery system can predict a solution. If the new solution's error is lower than that of the old solution, the new solution can replace X in the original equation. In Fig. 1, RED is applied to the example equation $f(x) = x_1^6 + sin(x_1)$. RED can help EDSs find a better solution after an initial prediction for two reasons: First, it can find a new representation of the original problem. The chance of finding a representation similar to an example seen during training increases by generating different representations of the original task with the residuals. Second, it can disentangle the original problem and divide it into simpler problems. The initial equation may already contain parts of the solution. By calculating the residual, the correct part of the initial equation is factored out of the original problem, leaving the unsolved subproblem for the EDS to address directly. Based on this, we address the following research questions in this work: How can the residuals be calculated without high computational overhead (**R1**)? Can RED help equation discovery systems to find better equations (**R2**)? How and when should RED be applied to refine an equation (**R3**)? What are the limitations of RED (**R4**)?

2 Method

To explain how RED works, we first introduce the concept of syntax trees to represent an equation f. Subsequently, we use the syntax tree to formulate an equation system with only one operation per equation. Rearranging the equation system towards a specific node X gives us the residual for the node. Next, we answer **R1** and show how to calculate the residual given a syntax tree directly

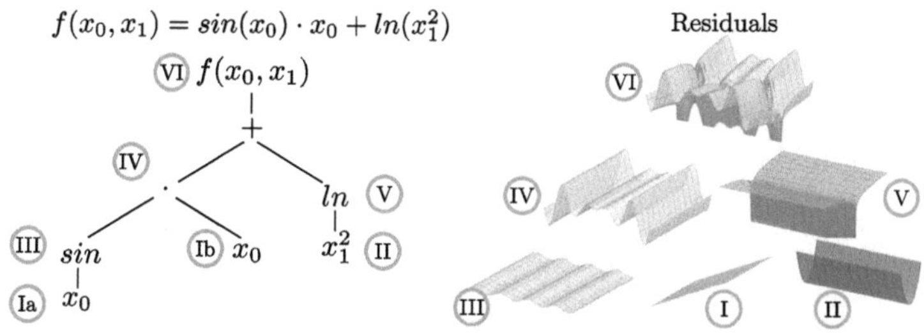

Fig. 2. left Example for a syntax tree. **right** Plots of the subequations composed to $f(x_0, x_1) = sin(x_0) \cdot x_0 + ln(x_1^2)$

by defining the behavior of an operator node based on which adjacent node is calling. Finally, we introduce our method, RED, which uses residuals to generate a new prompt for an EDS and recursively improve an initial equation.

Syntax Tree as Equation System. To represent an equation, we use a syntax tree in which the inner nodes are operators and functions, and the leaf nodes are variables and constants. If a leaf node is called, it returns the corresponding value or the column from the data set. Formally, a syntax tree $\mathcal{S}$ consists of nodes $\mathcal{N}$ and edges $\mathcal{E}$. The edges connect two nodes n_i and n_j where $n_i, n_j \in \mathcal{N} \wedge i \neq j$. Unlike the usual definition of a syntax tree, we add an Y-node as a parent node to what would normally be the root node of the equation. The Y-node returns the y-column of the data set and is used to calculate the residuals. In a syntax tree, an equation is a composition of simple basic components as constants $\mathbf{c}$ and variables $\mathbf{x}$, which are combined into complex expressions by mathematical operations. In Fig. 2, one possible syntax tree for $f(x_0, x_1) = sin(x_0) \cdot x_0 + ln(x_1^2)$ is given. For each node in the syntax tree, the equation equivalent to the corresponding subtree is plotted. Using the example $f(x_0, x_1) = sin(x_0) \cdot x_0 + ln(x_1^2)$, we show how a syntax tree can be translated into an equation system in Eq. (1). We use the Roman numerals from Fig. 2 to represent the result of a subtree.

$$VI = IV + V, \quad IV = III * Ib, \quad Ib = x_0, \quad III = \sin(x_0), \quad V = \ln(x_1^2). \quad (1)$$

The above equation system can be rearranged to any node and defines what the subequation below a node must result in, so that the total equation is equal to y. The result of the calculation we name y_{res}. The only requirement is that all operators on the path between the node for which y_{res} is searched and the root node are invertible.

Calculating the Residual Based on a Syntax Tree. To answer **R1**, we can use the idea of an equation system to calculate the residuals for any node in the syntax tree. For an operator node, the mathematical operation it performs depends on which adjacent node is calling it. An overview of the most essential

Table 1. The mathematical behavior of a node type depends on the caller node. N/A indicates that the inverse is not applicable, while NI indicates that a node is not invertible.

Binary Ops	Caller			Other Nodes	Caller	
	Parent (p) (p)	Child 0 (c_0)	Child 1 (c_1)		Parent (p)	Child 0 (c_0)
Plus	$c_0 + c_1$	$p - c_1$	$p - c_0$	Logarithm	$ln(c_0)$	e^p
Minus	$c_0 - c_1$	$p + c_1$	$c_0 - p$	Sine	$sin(c_0)$	NI
Product	$c_0 * c_1$	p/c_1	p/c_0	Constant	Value	N/A
Division	c_0/c_1	$p * c_1$	c_0/p	Variable	x_i	N/A
Power	$c_1^{c_0}$	$ln(p)/ln(c_1)$	p^{1/c_0}	Y	N/A	y

operator nodes is in Table 1. As the calculation is defined on the node level, new mathematical operations can be added as new node types.

If we want to calculate the residuals for a node it calls its parent node. The parent node calls its other adjacent nodes, and they propagate the call recursively. When the results of the adjacent nodes are returned to the parent node, the parent node performs its defined mathematical operation and returns the result to the node that called it in the first place. Operators that are not bijective (e.g., *sin*) cannot be inverted. Thus, for their child nodes, the residual cannot be computed.

RED. We can use the ability to calculate the residual for each node in a syntax tree to develop RED. The pseudocode of RED is given in algorithm 1. An example for applying RED is shown in Fig. 3.

RED needs an initial equation *tree*, which we want to improve, and a current error $error^{val}$ as a threshold to check if the suggested new subequations improve the initial equation. The idea for RED is to calculate in an ordered manner for each node in the syntax tree whether the initial equation can be improved by using residuals. If a syntax tree is updated, RED is restarted for the new tree until an iteration limit is reached or no further improvement is possible in the tree (line 3).

The Residual List maintains the order of the nodes for which the residual is calculated. As in a breadth-first search, the ordering is from left to right within the nodes of the same depth and then goes to a level deeper. The child node of the Y-node is not added to the list, as using the residual for this node is identical to the original problem. The method *tree*.GETNEXTRESIDUALNODE in line 5 returns the first node from the Residual List. For the residual calculated in line 6, the EDS predicts a new subtree (line 7). This subtree replaces the current node in the original tree and creates a new tree, $tree_{res}$ (line 8). The error of $tree_{res}$ is calculated on the validation data set (line 9). If the error is smaller than the previous best error, we update the best error, and the tree to be optimized is set equal to $tree_{res}$ (line 11 – 12). Finally, *lastUpdatedNode* in line 13 is set to the current *nodeId*. If *tree.GetNextResidualNode* is called again in line 5, the Residual List for the new tree is created. Since we are traversing the tree from

Algorithm 1. RED

Global: Equation Discover Model $\mathcal{M}$, Train Data $\mathcal{D}_{train}$, Validation Data $\mathcal{D}_{val}$
Require: Equation as Syntax Tree *tree*, Residual Max Iteration i_{max}, Error Threshold T, Error of Current Equation $error^{val}$.

function RED($tree$, i_{max}, T, $error^{val}$)
  $i, lastUpdatedNode \leftarrow 0, 0$
  while $error^{val} > T$ and $i < i_{max}$ and LEN($tree.residualList$) > 0 **do**
    $i \leftarrow i + 1$
    $nodeId \leftarrow tree.$GETNEXTRESIDUALNODE($lastUpdatedNode$)
    $res \leftarrow tree.$GETRESIDUAL($nodeId, \mathcal{D}_{train}$)
    $subtree \leftarrow \mathcal{M}.$FIT($\mathcal{D}_{train}$, res)
    $tree_{res} \leftarrow$ UPDATESUBTREE($nodeId$, $tree$, $subtree$)
    $error^{val}_{res} \leftarrow$TESTEQUATION($tree_{res}$, $\mathcal{D}_{val}$)
    if $error^{val}_{res} < error^{val}$ **then**
      $error^{val} \leftarrow error^{val}_{res}$
      $tree \leftarrow tree_{res}$
      $lastUpdatedNode \leftarrow nodeId$
  return *tree*

top to bottom, and only the tree below the *lastUpdatedNode* has changed, we delete the *lastUpdatedNode* and all its direct parent nodes from the Residual List. For the removed nodes, the new subtree does not influence the calculation of their residual, and the problem is equivalent to a previously considered problem.

3 Experiments

We compare[1] RED for five EDSs with six other post-processing methods on the Feynman benchmark [21] as reported in SRBench [4]. For *NeSymReS* [1], we use a pretrained model designed for a maximum of three independent variables. Therefore, we use data sets with three or fewer variables for all models. This results in 53 Feynman data sets. We first present the EDS models and post-processing methods in the following. Subsequently, we answer **R2** by comparing the performance of the post-processing methods. Finally, we examine **R3** and analyze how the number of iterations, noise, and data set size impact RED.

3.1 Models

NeSymReS [1] uses large-scale pretraining on generated data which are encoded with the Set Transformer architecture [13]. A beam-search is used to sample candidates from the decoder. *SymbolicGPT* [23] uses a transformer architecture too, but with a permutation-invariant data set encoder PointNet [17] which interprets the data sets as point clouds. *E2E* [8] is a transformer-based model directly predicting the full mathematical expression, constants included. In a

[1] Code is available: https://github.com/kramerlab/RED_Residuals_For_Equation_Discovery.

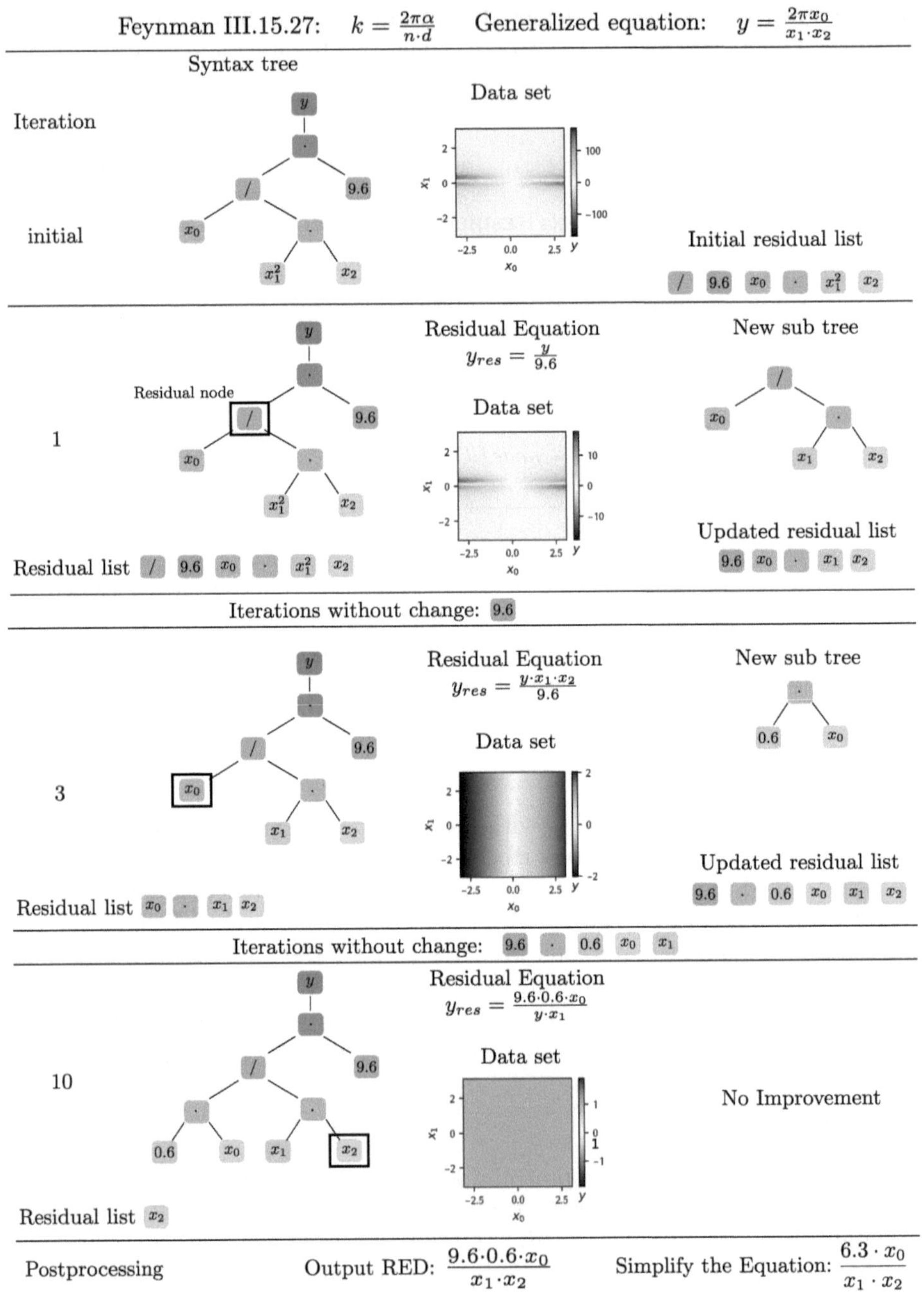

Fig. 3. Example how RED can help to find a better equation using residuals in multiple iterations. For the visualization of the data set, x_2 is set to 1.

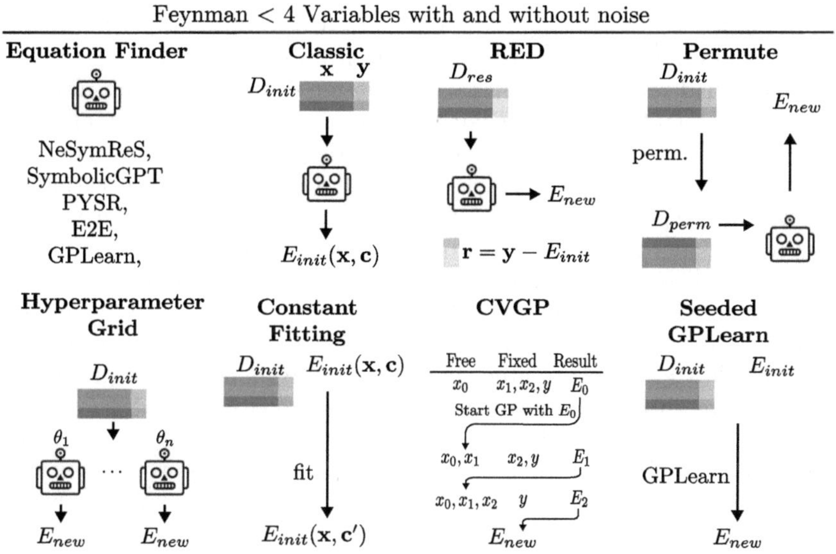

Fig. 4. Overview of the tested post-processing methods.

subsequential step, the constants are refined by feeding them to an optimizer as an informed initialization. The data set is processed as sequential text. *PySR* [5] is a genetic programming approach tying the selection probability of the fittest individual (equation) to the increase in fit quality via simulated annealing. PySR outputs the discovered equations as a Pareto frontier, where for each complexity level the equation with the least error is returned. In our experiment, we use the equation returned with the strategy "best", balancing equation complexity and error. *GPLearn* [20] is an equation discovery library implementing genetic programming and is written in Python.

3.2 Methods for Post-processing Equations

In the following, we describe each post-processing method, and Fig. 4 gives a visual overview of the methods.

Classic: The EDS is applied once with the default parameter on the data set. The result is used as the initial equation for the other approaches. *RED*: Described in detail in Sect. 2. It uses residuals to refine an initial equation. Based on our experiments in Sect. 3.4, we select a maximum of 10 iterations to balance runtime, overfitting, and accuracy. *Permute*: The EDS is applied n-times on the permuted data set. Through the permutation, the EDS gets a new representation of the original data set with each permutation. n is set equal to the performed iterations in RED. *Hyper Parameter Grid*: In addition to the default parameter, two other sets of parameters are defined for each EDS. The sets differ in parameters like the probability of mutations in genetic approaches or the beam size in neural-guided EDSs. *Fitting*: All constants in the initial equation are refitted with the Levenberg-Marquardt algorithm from sklearn [15]. *CVGP*: Control Variable Genetic Programming [6] starts by fitting simple expressions involving a small set of independent variables with all other variables held constant. It then extends expressions learned in previous generations by adding new independent variables, using new control variable experiments in which these variables are allowed to vary. Applying control variable experiments can be interpreted as a post-processing method, which can be combined with every EDS as long the EDS can be seeded with an initial equation. For this reason, we group CVGP as a post-processing method. To do this, we run CVGP on all data sets of the benchmarks and then filter the results according to the data sets in which the current EDS has an error greater than 0.001. CVGP needs the ground truth expression to perform the control variable experiments. *Seeded GPLearn*: We replace 50% of the 1000 candidates in the start population of GPLearn with a subtree of the initial equation. We start a run for each node in the syntax tree. Similar to RED, the subtree is the complete tree without the subtree under the selected node.

We overall employ the following test protocol: In the first phase, we run the Classic approach on each data set in the benchmark. If the Mean Squared Error (MSE) is greater than 0.001, we consider the data set as unsolved and test the post-processing methods. Unless otherwise specified, 300 rows are sampled from each data set. 60% of the data is used as the training set, 20% as the validation set, and 20% as the test set. If a post-processing method does not use a validation data set, the validation data set is added to the training data set. Each equation proposed by a post-processing method is tested on the test data set. We use the MSE as the test metric.

3.3 Performance of RED

We apply each post-processing method as described in Sect. 3.2. All post-processing methods, except Constant Fitting, Classic and CVGP, propose multiple equations. To evaluate RED and answer **R2**, we run two types of experiments. First, we compare all equations the post-processing methods predict for one EDS. Next, we compare the best equations each method predicts for all five EDSs. If a method cannot return an equation, we use one found by Classic.

Win-Loss Statistics. We compare for the EDS NeSymReS for each data set all equations a method predicted vs. the equations of another method. If the equation of the method in the row of Fig. 5 has a lower MSE, we note a win, if equal a draw, and for a higher MSE a loss. In Fig. 5 we present the *win-ratio* = $^{wins}/_{wins+draws+loss}$ for the AI Feynman benchmark. Figure 5 demonstrates that Seeded GPLearn wins more often than it loses against all methods. However, the difference compared to RED is only minimal. In contrast, the equations predicted by RED are of higher quality than the other methods, resulting in a higher win ratio.

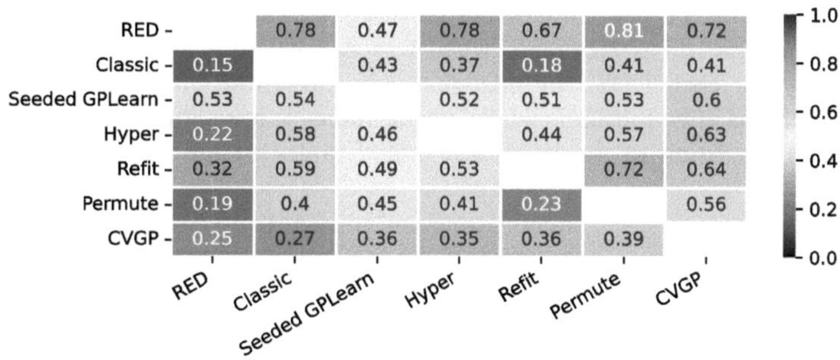

Fig. 5. Win-Ratio for NeSymReS on the AI Feynman Benchmark. We compare two post-processing methods, one vs. one. For each data set we consider all equations both methods suggested and count how often one is better than the other. A ratio of 1 means that all equations of the method in the row are better than the equations of the method in the column

Accuracy for the Best Run. In the following, we analyze the post-processing methods given the best equation they find for a data set in Table 2. To reduce the proposed equations to one, we consider all proposed equations for a method and test them on the test data set. The method is then represented by the equation with the smallest error. First, we summarize the results for each EDS (columns in the table) and afterward we review each post-processing method (rows).

For each EDS, we report the average **Running Time** to run all experiments over 3 seeds. This time is proportional to the time needed to query an EDS once and the length of the proposed equation. The length of the equation has an influence as the number of iterations of Seeded GPLearn and RED depend on it. The metric **# Completed** states for how many of the 53 equations the EDS returns an initial solution.

Only PySR can provide a solution for all and NeSymReS for almost all of these equations. Both EDSs are also the fastest ones. The failures of the other EDSs can have several reasons: The neural-guided approaches are token-based. In the extreme case, these tokens are individual letters. Even if the approaches are trained to predict semantically sound formulas, this is not guaranteed. The EDSs sometimes predict expressions such as $sinI$. Another reason is that searching for a candidate equation results in over- or underflows, divisions by zero, undefined exponentiation in $\mathbb{R}$, or a negative logarithm. If the EDS or post-processing method does not handle the error, we log the error for the current process, and the program continues with the next method or data set. Finally, residuals cannot be applied if the EDS predicts an equation without an operator. The EDS E2E was tested with only one seed and then excluded from further experiments. Although E2E predicts equations with a small MSE, it has to be restarted multiple times, and has the longest suggested equations and running time. We now analyze the performance of the post-processing methods in detail.

The metric **# MSE>0.001** in Classic specifies the number of equations on which we test the post-processing methods. Each post-processing method has the

Table 2. Results of the post-processing methods for multiple EDSs on the Feynman benchmark. We ran all experiments with 3 seeds, except E2E, which is only tested with one seed. For each EDS, the post-processing method with the lowest median is highlighted in **bold**. The methods whose performance is not significant ($p \geq 0.01$) compared to RED according to the Wilcoxon signed-rank test are marked with underline.

	NeSymReS	SymbolicGPT	PySR	GPLearn
Running Time [sec]	5295	7784	2840	12309
Classic				
# Completed	52	33	53	46
# MSE >0.001	22	32	24	32
MSE Q2 , Q3	0.63 , 10.24	1.88 , 28.58	0.08 , 0.35	0.04 , 0.96
# Operators Q2	4	6	1	3
Runtime Q2 [sec]	3.22	1.66	1.93	17.86
Seeded GPLearn				
# Completed	22	28	18	25
MSE Q2 , Q3	<u>**0.02**</u> , 0.39	**0.02** , 0.24	0.01 , 0.09	<u>**0.02**</u> , 0.19
# Operators Q2	4	2	3	3
Runtime Q2 [sec]	140.66	196.25	72.04	109.96
Hyper				
# Completed	22	28	23	30
MSE Q2 , Q3	0.21 , 1.79	1.40 , 18.61	<u>0.01</u> , 0.08	<u>0.03</u> , 0.22
# Operators Q2	5	8	2	3
Runtime Q2 [sec]	8.71	4.23	3.85	35.27
Refit				
# Completed	19	14	22	29
MSE Q2 , Q3	0.48 , 5.38	1.39 , 9.04	0.02 , 0.18	<u>0.03</u> , 0.51
# Operators Q2	4	4	1	2
Runtime Q2 [sec]	0	0	0	0
Permute				
# Completed	22	28	19	24
MSE Q2 , Q3	0.46 , 10.24	<u>0.84</u> , 10.48	<u>0.01</u> , 0.07	0.03 , 0.51
# Operators Q2	4	8	4	4
Runtime Q2 [sec]	29.69	23.06	18.18	113.55
CVGP				
# Completed	16	27	18	25
MSE Q2 , Q3	4.12 , 23.02	<u>0.93</u> , 16.92	0.21 , 8.07	0.04 , 8.86
# Operators Q2	2	2	2	2
Runtime Q2 [sec]	7.78	7.55	7.28	7.52
RED				
# Completed	19	22	19	22
MSE Q2 , Q3	0.06 , 0.57	0.81 , 12.95	**0.00** , 0.05	**0.02** , 0.13
# Operators Q2	12	12	6	8
Runtime Q2 [sec]	28.58	13.71	10.75	123.03

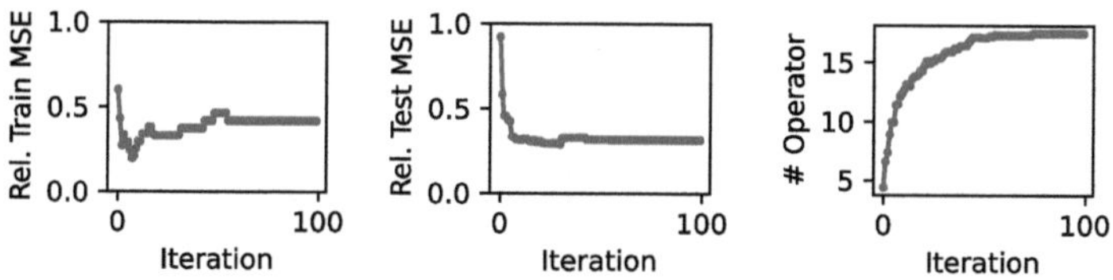

Fig. 6. Influence of the number of iterations in RED on the average relative MSE for train and test set and number of operators in the equations found with NeSymReS on the Feynman benchmark.

metric **# Completed**, which gives the number of data sets for which an equation is returned successfully. **MSE Q2** states the median of the mean squared error across all seeds and data sets. **MSE Q3** gives the third quartile correspondingly. To give an estimate of the complexity of the generated equations, **# Operator Q2** shows the median number of mathematical operators (e.g. $+,-,\cdot, /$, sin, ...) in the proposed equations. Finally, **Runtime Q2 [sec]** indicates the median duration each post-processing method ran per equation.

For Classic, the test error of the neural-guided EDSs are higher, and the equations found are more complex. The post-processing approach Seeded GPLearn can close this gap and improve all examined EDSs. RED also leads to an improvement, while for SymbolicGPT, it is not as drastic as for the other EDSs. One possible explanation is that RED heavily depends on the initial equation, and the initial equations of Symbolic GPT have the highest error. If this initial equation does not help to decompose the original problem, the residual problem can become even more challenging than the original problem. A drawback of RED is that the average length of the equations increases significantly. We will analyze this issue in Sect. 4 in detail. Hyper Parameter Grid, Fitting, and Permute show promising results, especially with the genetic algorithms, but with NeSymReS, they only lead to medium improvements. CVGP can only keep up with the other approaches regarding the median value for GP and is otherwise significantly worse. This is presumably because the control variable approach does not help much in simplifying the equation finding for the benchmark. The extension Racing-CVGP [7] tries multiple permutations of the control variable order, which results in a significant improvement.

In conclusion, and with regard to **R3**, we observe that RED helps improve the results of all the tested EDSs. The combination of neural-guided EDS with genetic EDS, as in Seeded GPLearn, also shows promising results. However, it is less targeted, and requires much more computation time.

3.4 Properties of RED

To answer **R3**, how and when RED should be used, we first study the influence of iterations in RED and then consider the influence of noise and data set length on its performance. Missing equations in RED are not replaced.

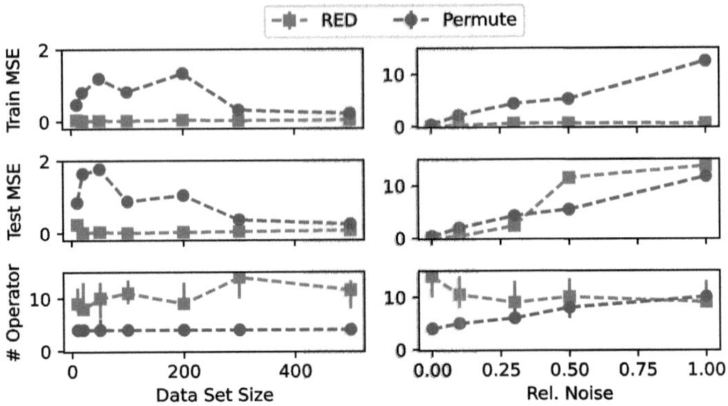

Fig. 7. Influence of **data set size** (left) and **noise** (right) on the median number of operations, train, and test MSE for NeSymReS on the Feynman benchmark.

Effect of Maximum Number of Iteration on RED. To analyze the effects of the number of iterations RED has, we set the maximum number of iterations to 100. We store the train and test error and number of operations in the current equation per iteration for each data set. If we do not have the value for an iteration, we use the value from the last available iteration. We normalize the error for each data set so that the maximum error is 1. Subsequently, the mean value is calculated for all data sets per iteration and plotted in Fig. 6. The figure shows that for NeSymReS (other EDSs behave similarly), the relative test error drops sharply within a few iterations and reaches a minimum. The average number of operators in the equation increases with the number of iterations.

Effect of Noise on RED. To quantify the effect of noisy data on RED, a relative uniform noise (U) is added to each value in the data set D: $D_{noise} = D * (1 + U(-Rel.Noise, +Rel.Noise))$. For D_{noise}, the EDS NeSymReS is used to search for suitable equations with the approaches Permute and RED, and the best equation for each approach is logged for train and test error. For the Feynman benchmark, the median values for train and test MSE error and number of operations in the test equation are given as a function of the relative noise 0, 0.1, 0.3, 0.5 and 1 in Fig. 7. The evaluation shows that the training error for the Permute approach increases linearly with noise. The training error for RED remains almost constant despite the increase in noise. This overfitting on the training data leads to similar values on the test data as with the Permute approach. It is interesting to note that increasing noise leads to a decrease in the median number of operations for RED, whereas an increase can be observed for Permute.

Effect of Data Set Size on RED. The number of data points available for the search can be another source of overfitting. We plot the median values for train, test error, and number of operations in Fig. 7 for the Feynman benchmark.

We run the experiments for data sets of size 10, 20, 50, 100, 200, 300 and 500. For Permute, we see the train and test error decrease with an increasing number of data points. The inability of NeSymRes to fit a few data points is surprising and is probably because the neural model was not trained for such small data sets. However, combined with RED, the MSE remains at a constant low level regardless of the number of samples.

Conclusion: Properties of RED. To summarize our sensitivity analysis, we can state for **R3** that RED can already improve the results with a small number of iterations. In particular, it can strongly improve the results in cases where the EDS has not seen a lot of training data. From a certain level of noise on, RED tends to overfit.

4 Limitations

The conducted experiments show that RED is a promising method to improve the results of any EDS. Nevertheless, to answer **R4**, we want to summarize the challenges here and discuss possible solutions.

First, we address the challenges of RED with neural-guided EDSs and then with genetic EDSs. While RED is independent of the functionality of the pre-trained EDS, it depends on an initial solution, which has to enable the disentanglement. The initial solutions from SymbolicGPT have the highest median error of all EDSs, and accordingly, the results for RED are less substantial than for NeSymReS. Further improvements for neural-guided EDSs can be expected, in which case RED can be even more effective. The second challenge is that RED leads to longer and thus more complex equations. We want to improve the search in future research to tackle this issue. Using not only the validation error but also the length of a new subequation as a criterion if the subequation is accepted will attenuate the tendency of RED towards long equations. A parallel search combined with a dynamic ranking will achieve improvements within the same running time.

In the experiments with genetic EDSs, we show that RED can achieve improvements there. Since genetic approaches perform a random search, the advantage of decomposing the original problem is not so evident here. One possible explanation for the good results of RED with genetic EDSs, is that RED effectively increases the number of search iterations for a data set, as a previously found equation can be further improved. However, an experiment in which we consider the median error of PySR depending on the number of iterations on the Feynman benchmark shows no decrease in error with more iterations. Another explanation is that we restart the genetic algorithm multiple times within RED; by resetting the hyperparameter, we find a solution outside the local maximum, which is not possible by simply extending the runtime of the genetic algorithm. The mutual influence of genetic EDSs and RED and using our method as an operator in genetic algorithms is a promising approach for the future.

5 Related Work

Brute force approaches in equation discovery quickly reach their limits because the search space increases exponentially as the number of variables or operators increases. Since the beginning, equation discovery systems have been using metrics and methods to structure the search space.

The most prominent method is to exploit the reward of a suggested equation as a metric to concentrate the search in promising areas. This procedure can be applied with genetic programming [3,11], reinforcement learning [9,16] or large language models [19].

An orthogonal approach tries to modify the data set that describes the problem and makes it easier to solve. *BACON* [12] adds new variables to the data set based on predefined rules to make the problem solvable by a simple combination of the new set of variables. *CVGP* [6] starts with a few control variables and refines the initial solution by iteratively taking more variables into account. Similar to our work, because the target variable changes, is *additive regression*. While RED makes no assumption about the underlying structure, additive regression [2] uses an additional superposition of continuous single-variable functions. The next term added to the existing one is fitted to minimize the error between the target variable and the equation so far found.

RED tries to decompose the original problem by considering subtrees of the initial solution. *AI Feynman* [21] and *AI Feynman 2.0* [22] follow the same idea by training a neural network on the data set and then using the gradients of this network to decompose the problem into smaller sub-problems whose solutions are combined to a final equation.

Generating different representations of a given problem to get better results from a pretrained neural system is already successfully applied in other domains. In the visual domain, Test-Time Augmentation (TTA) is used to create multiple versions of an image by, e.g., cropping and making decisions based on an ensemble [10]. In prompt engineering for LLMs [14,24], we iteratively refine the prompt for querying the LLM to produce multiple answers.

6 Conclusion

In this research, we proposed RED, a method using residuals to refine an initial equation. Given a syntax tree, we showed that we can calculate the residuals as fast as evaluating the tree using node-based calculation rules (**R1**). In experiments with two benchmarks, we demonstrated that RED improves all equation discovery systems that we tested (**R2**). In a sensitivity analysis, we showed that 10 iterations of RED already led to significantly improved results, and RED helped in cases where the neural EDS has not seen a lot of training data (**R3**). Finally, we gave an outlook on how the increase in equation complexity when using RED can be countered and how the approach can be further improved in the future by an optimized search strategy (**R4**).

Acknowledgments. This research project was partly funded by the Hessian Ministry of Science and the Arts (HMWK) within the projects "The Third Wave of Artificial Intelligence - 3AI" and hessian.AI.

Disclosure of Interests. The authors have no relevant financial or non-financial interests to disclose.

References

1. Biggio, L., et al.: Neural symbolic regression that scales. In: ICML 2021, vol. 139. PMLR (2021)
2. Breiman, L., Friedman, J.H.: Estimating optimal transformations for multiple regression and correlation. J. Am. Stat. Assoc. **80**(391) (1985)
3. Burlacu, B., Kronberger, G., Kommenda, M.: Operon C++: an efficient genetic programming framework for symbolic regression. In: GECCO 2020 Companion. ACM (2020)
4. Cava, W.G.L., et al.: Contemporary symbolic regression methods and their relative performance. In: NeurIPS Datasets and Benchmarks (2021)
5. Cranmer, M.: Interpretable machine learning for science with PySR and SymbolicRegression.jl (2023). https://doi.org/10.48550/ARXIV.2305.01582
6. Jiang, N., Xue, Y.: Symbolic regression via control variable genetic programming. In: ECML PKDD. Springer Nature Switzerland (2023)
7. Jiang, N., Xue, Y.: Racing control variable genetic programming for symbolic regression. AAAI **38**(11) (2024)
8. Kamienny, P., d'Ascoli, S., Lample, G., Charton, F.: End-to-end symbolic regression with transformers. In: NeurIPS (2022)
9. Kamienny, P., Lample, G., Lamprier, S., Virgolin, M.: Deep generative symbolic regression with monte-carlo-tree-search. In: ICML 2023. PMLR, vol. 202 (2023)
10. Kim, I., Kim, Y., Kim, S.: Learning loss for test-time augmentation. In: NeurIPS (2020)
11. Koza, J.R.: Genetic programming as a means for programming computers by natural selection. Statist. Comput. **4** (1994)
12. Langley, P.W., Simon, H.A., Bradshaw, G., Zytkow, J.M.: Scientific Discovery: Computational Explorations of the Creative Process. The MIT Press (1987)
13. Lee, J., Lee, Y., Kim, J., Kosiorek, A., Choi, S., Teh, Y.W.: Set transformer: a framework for attention-based permutation-invariant neural networks. In: ICML 2019. PMLR (2019)
14. Liu, P., et al.: Pre-train, prompt, and predict: a systematic survey of prompting methods in natural language processing. ACM Comput. Surv. **55**(9) (2023)
15. Pedregosa, F., et al.: Scikit-learn: mchine Learning in Python. J. Mach. Learn. Res. **12** (2011)
16. Petersen, B.K., et al.: Deep symbolic regression: recovering mathematical expressions from data via risk-seeking policy gradients. In: ICLR (2021)
17. Qi, C.R., Su, H., Mo, K., Guibas, L.J.: Pointnet: deep learning on point sets for 3d classification and segmentation. In: CVPR 2017. IEEE Computer Society (2017)
18. Shojaee, P., Meidani, K., Farimani, A.B., Reddy, C.K.: Transformer-based planning for symbolic regression. In: NeurIPS (2023)
19. Shojaee, P., et al.: LLM-SR: scientific equation discovery via programming with large language models. In: ICLR (2025)

20. Stephens, T.: Welcome to gplearn's documentation! — gplearn 0.4.2 documentation. https://gplearn.readthedocs.io/en/stable/
21. Udrescu, S.M., Tegmark, M.: AI Feynman: a physics-inspired method for symbolic regression. Sci. Adv. **6**(16) (2020)
22. Udrescu, S., et al.: AI feynman 2.0: pareto-optimal symbolic regression exploiting graph modularity. In: NeurIPS (2020)
23. Valipour, M., You, B., Panju, M., Ghodsi, A.: SymbolicGPT: a generative transformer model for symbolic regression (2021). https://doi.org/10.48550/ARXIV.2106.14131
24. Wang, L., et al.: Plan-and-solve prompting: Improving zero-shot chain-of-thought reasoning by large language models. In: ACL (2023)

Bayesian Grammar Refinement for Efficient Equation Discovery

Jure Brence[1(✉)], Sašo Džeroski[1], and Ljupčo Todorovski[1,2]

[1] Department of Knowledge Technologies, Jožef Stefan Institute, Ljubljana, Slovenia
jure.brence@ijs.si

[2] Department of Mathematics, Faculty for Mathematics and Physics, University of Ljubljana, Ljubljana, Slovenia

Abstract. Equation discovery, or symbolic regression, aims to uncover closed-form mathematical expressions from data, with the goal of facilitating scientific discovery in fields like physics and biology. Probabilistic context-free grammars (PCFGs) provide a structured, interpretable framework for encoding domain knowledge and defining distributions over candidate expressions. While sampling from a fixed PCFGs can provide a simple and effective approach to equation discovery, it struggles to identify complex expressions due to the vastness of the search space. To address this obstacle to successful equation discovery, we consider a novel Bayesian approach that iteratively refines the production rule probabilities within a PCFG, effectively guiding the search toward more promising regions of the space based on observed data. Our computational experiments demonstrate that this approach substantially increases the probabilities assigned to true equations, directly improving both the performance and computational efficiency of the discovery process. However, the lack of context sensitivity in PCFGs limits the effect of tuning production rule probabilities and consequently limits the gains in computational efficiency. This highlights an important challenge for future work in this area. The introduced approach is implemented as part of the Python package ProGED: https://github.com/brencej/ProGED.

Keywords: equation discovery · symbolic regression · probabilistic context-free grammar · PCFG · Bayesian

1 Introduction

Scientific discovery has long been driven by the search for underlying mathematical relationships that capture the structure and dynamics of natural phenomena. From Newton's laws of motion to the Schrödinger equation, many of the most profound scientific breakthroughs have emerged from identifying concise, interpretable equations that accurately describe experimental data. While equations are often derived from theory and verified with experiments, with advances in computational power and data collection capabilities, a data-driven approach is becoming increasingly popular.

S. Džeroski et al. (Eds.): DS 2025, LNAI 16090, pp. 113–127, 2025.
https://doi.org/10.1007/978-3-032-05461-6_8

This approach, known as equation discovery, aims to automate the challenging task of extracting symbolic relationships from data. Equation discovery combines elements of machine learning, symbolic computation, and probabilistic modeling to identify the governing equations of complex systems. However, despite its potential, equation discovery remains a formidable challenge due to the vastness of the search space, the difficulty of incorporating prior knowledge, and the sensitivity of symbolic models to noise and overfitting.

The field of equation discovery has progressed significantly in recent years. Genetic programming, popularized by Eureqa [23], and sparse linear regression methods such as SINDy [3], remain popular approaches. Hybrid methods like AI Feynman [26] incorporate dimensional analysis and symmetry detection to reduce the search space. Deep learning has also become a central theme in recent research. Approaches like Deep Symbolic Optimization (DSO) leverage deep reinforcement learning to directly optimize equation structures, allowing for efficient exploration of the equation space [20]. Transformers [27] have been employed in end-to-end approaches that attempt to directly predict symbolic expressions from data without separate structure and parameter estimation steps [15]. Probabilistic approaches have also gained prominence, with generative methods based on variational autoencoders (VAEs) that map mathematical expressions into structured latent spaces, enabling the use of numerical optimization algorithms for searching the space of expressions. Notable advancements include hierarchical VAEs, which address the challenge of ensuring syntactic correctness in generated expressions [17].

Another class of methods emphasizes the explicit use of background knowledge in equation discovery to facilitate interpretable models that map to concepts in domain theory [8,24]. Context-free grammars have been used as a basis for equation discovery [25] and to incorporate background knowledge in genetic programming [22]. Recent developments in the use of probabilistic context-free grammars (PCFGs) [1,2], build on this tradition, offering more precise and expressive ways to encode background knowledge and define a distribution over the space of candidate mathematical expressions. However, so far, only random sampling from a fixed PCFG has been considered.

To improve the computational efficiency of probabilistic grammar-based equation discovery, we introduce an algorithm that iteratively updates the probabilities of production rules to guide the search towards more promising areas in the space of mathematical expressions. Production probabilities impose soft constraints on the space of expressions and can be used as parameters of the grammar that the algorithm optimizes. Furthermore, since a PCFG defines a probability distribution over the space of mathematical expressions, this procedure has a Bayesian interpretation in that the resulting distribution is an approximation of the posterior distribution.

Bayesian methods, a statistical framework for updating beliefs or probabilities by combining prior knowledge with new evidence, have already been used in equation discovery. A large group of Bayesian methods share similarities with sparse linear regression approaches like SINDy. Bayesian tools are used to estimate parameter uncertainties and to accommodate measurement noise and missing data [18,21], as well as to better control expression complexity through sparsity-inducing prior distributions, such as the spike-and-slab prior [16]. Despite algorithmic innovations such as kernel regression [16] and relevance vector machines [9], these methods are limited to mathematical expressions, linear in parameters. Our grammar-based approach has no such limitation, but uses Bayesian tools only to improve the identification of expression structure, without parameter uncertainties.

On the other hand, the Bayesian machine scientist [13] explores the space of mathematical expressions through a set of operations on symbolic expression trees, similar to mutations in genetic programming. The approach leverages Bayesian statistics with Markov chain Monte-Carlo (MCMC) to directly sample mathematical expressions from the posterior distribution. The posterior probability naturally combines goodness of fit with a prior over expressions, which is estimated from a corpus of mathematical expressions. A very similar approach is used by Bayesian Symbolic Regression [14], which enhances genetic programming with Bayesian MCMC. The advantage of both methods is that they directly sample the true posterior distribution. In contrast, our approach only approximates the posterior by iteratively manipulating the parameters of the distribution. However, the use of PCFGs as priors offers a more powerful and flexible way of expressing background knowledge.

In Sect. 2 we first summarize the approach of PCFG-based equation discovery, then detail the Bayesian refinement algorithm. In Sect. 3 we report on experiments that showcase the method and analyze its performance. We discuss its strengths in shortcomings in Sect. 4 and offer concluding thoughts in Sect. 5.

2 Method

2.1 Probabilistic Grammar-Based Equation Discovery

Probabilistic context-free grammars (PCFGs) provide a structured way to encode domain knowledge for equation discovery, effectively constraining the space of mathematical expressions. A PCFG is defined as a tuple $(\mathcal{N}, \mathcal{T}, \mathcal{R}, \mathcal{S})$, where $\mathcal{N}$ is the set of nonterminal symbols, $\mathcal{T}$ is the set of terminal symbols, $\mathcal{R}$ is the set of production rules, and $\mathcal{S}$ is the start symbol. Production rules have the general form $A \rightarrow \alpha$, where $A \in \mathcal{N}$ and α is a sequence of terminals and nonterminals [19].

In a PCFG, each production rule has an associated probability, and the probabilities of all production rules with the same left-hand nonterminal sum to 1. This defines a probability distribution over all possible derivations of the grammar [6,11]. For example, a simple grammar for generating polynomials can be written as:

$$\begin{aligned} P &\rightarrow P + c \cdot M \; [p_{P1}] \mid c \cdot M \; [p_{P2}] \mid c \; [p_{P3}] \\ M &\rightarrow M \cdot V \; [p_{M1}] \mid V \; [p_{M2}] \\ V &\rightarrow x_1 \; [p_{V1}] \mid x_2 \; [p_{V2}] \mid \; \ldots \; \mid x_m \; [p_{Vm}]. \end{aligned} \tag{1}$$

Here, we used a compact notation that presents all production rules with the same left-hand side in a single line, separated by |. Nonterminal symbols $\mathcal{N} = \{P, M, V\}$ represent different components of a polynomial (polynomial, monomial, variable), and terminal symbols $\mathcal{T} = \{c, +, \cdot, x_1, \ldots, x_m\}$ include numerical parameters and variables. The probability, given in brackets, of a particular derivation or parse tree, is the product of the probabilities of the production rules used. For instance, the expression $c \cdot x_1 \cdot x_2$ might be derived using the rules $P \rightarrow c \cdot M$, $M \rightarrow M \cdot V$, $M \rightarrow V$, $V \rightarrow x_1$, and $V \rightarrow x_2$, yielding a probability $p_{P2} \cdot p_{M1} \cdot p_{M2} \cdot p_{V1} \cdot p_{V2}$.

PCFGs naturally incorporate a form of the parsimony principle: more complex expressions require more production steps, reducing their overall probability due to the multiplicative nature of derivation probabilities. This characteristic makes PCFGs particularly well-suited for equation discovery, where simpler, more interpretable expressions are often preferred.

The simplest way to generate mathematical expressions with a PCFG is a simple Monte-Carlo algorithm that samples the distribution over expressions defined by the PCFG. The algorithm begins with a string, composed of only the start symbol. The algorithm recursively applies production rules to the string until only terminal symbols remain. Whenever more than one choice of production rule is available, the algorithm chooses one randomly according to the probabilities of the production rules [2].

Expressions generated this way contain free parameters, denoted c. We fit their values to the data using numerical optimization. In the case of differential equations, we either compute the derivatives numerically in advance or solve the initial value problem at each step of the optimization [10]. In any case, the error-of-fit, typically mean-squared error, is the primary measure of fitness of generated expressions.

2.2 Bayesian Refinement Algorithm

The Bayesian algorithm, summarized in Fig. 1, uses a probabilistic context-free grammar to discover mathematical equations from data, iteratively refining grammar probabilities based on Bayesian updates.

The initial PCFG is constructed by encoding background knowledge as production rules, defining how complex mathematical expressions are constructed. These define the initial probability distribution over possible mathematical expressions. In each iteration, the algorithm generates a sample of expressions using the current grammar probabilities. The free parameters in these expressions are fitted to the data using numerical optimization, minimizing the root mean square error (RMSE) to assess equation quality. The algorithm then updates the grammar's production rule probabilities based on the best-performing equations. This Bayesian update, implemented using the m-estimate, adjusts the grammar to favor more promising equations over time, effectively approximating a posterior distribution. Over successive iterations, the algorithm refines the grammar, allowing it to discover increasingly accurate and interpretable equations. The final output includes the tested expressions with the

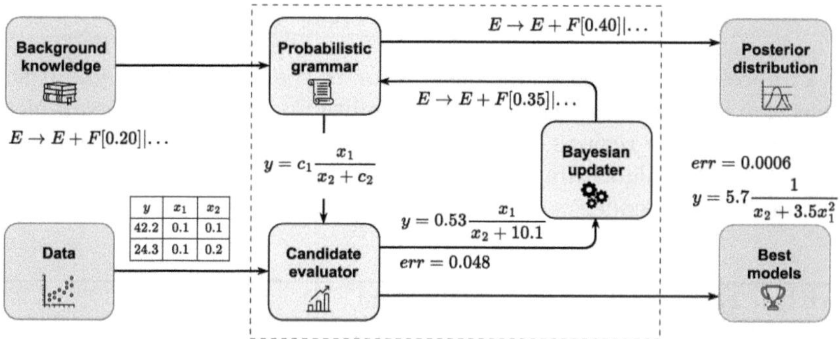

Fig. 1. Overview of the Bayesian PCFG-based equation discovery algorithm, with inputs colored blue, the algorithm components enclosed in a dashed line, and the outputs colored green. Background knowledge is used to construct the initial PCFG. At each iteration, candidate expressions are sampled, evaluated against the data, and used to update the grammar's production rule probabilities via Bayesian refinement.(Color figure online)

lowest error, as well as the PCFG with the final values of its probabilities, which define the posterior distribution.

The **m-estimate** [4] is a flexible Bayesian approach for estimating probabilities, balancing observed evidence with prior knowledge. It generalizes common methods like relative frequency and Laplace's rule of succession, providing a simple way to adjust the influence of prior beliefs.

For a problem with k possible outcomes and N trials, where outcome c occurs n times, the relative frequency estimate is simply $p(c) = \frac{n}{N}$, which relies purely on observed data and ignores prior knowledge. In contrast, Laplace's law of succession adds a uniform prior, resulting in $p(c) = \frac{n+1}{N+k}$, which effectively adds one vitual observation per outcome, smoothing the estimate [12]. The m-estimate further generalizes this by introducing a prior distribution $p_a(c)$ with a tunable parameter m, yielding:

$$p(c) = \frac{n + p_a(c)m}{N + m}, \tag{2}$$

where m controls the relative weight of prior knowledge versus observed data. For $m = 0$, this reduces to the relative frequency estimate, while for large m relative to N, the estimate approaches the prior $p_a(c)$. For integer values of m, the m-estimate generalizes Laplace's rule of succession by effectively adding to the evidence m virtual observations that follow the prior distribution [5,7].

In iterative PCFG-based equation discovery, each production rule used in the derivation of an expression represents a trial and can either be used or not, forming a two-outcome problem ($k = 2$). In each iteration, the algorithm samples a number of expressions and selects a subset of expressions with the lowest error, which are used to update probabilities. Thus, the m-estimate is applied to production rule probabilities, using only the most accurate expressions sampled in each iteration. An expression is included if its error meets the criterion:

$$\text{error} - \text{best\_error} < \epsilon \cdot \text{best\_error}, \tag{3}$$

where ϵ represents tolerance and best_error is the lowest error found so far. The selection criterion conditions the probability updates on good examples and focuses the search on regions of the equation space with high potential, refining the grammar as the algorithm progresses. Algorithm 1 details Bayesian grammar refinement in pseudocode.

Algorithm 1. Bayesian grammar refinement

Require: PCFG $G = (\mathcal{T}, \mathcal{N}, \mathcal{R}, S)$, data D, target v, iterations N_{iter}, samples N_{sample}, m-estimate parameter m, tolerance ϵ
Ensure: Updated PCFG G

$\quad best\_error \leftarrow \infty$
for $i = 1$ to N_{iter} **do**
$\quad es \leftarrow$ sample N_{sample} expressions from G
$\quad$ **for all** $e \in es$ **do**
$\quad\quad$ fit parameters of e do D, compute e.error
$\quad best\_error \leftarrow \min(best\_error \cup \{e.\text{error} : e \in es\})$
$\quad selected \leftarrow \{e \in es : e.\text{error} \leq (1 + \epsilon) \cdot best\_error\}$
$\quad$ **for all** $A \in \mathcal{N}$ **do**
$\quad\quad$ count uses of each rule $r : A \rightarrow \alpha$ in *selected*, store in $r.count$
$\quad\quad total \leftarrow \text{sum}(\{r.\text{count} : r : A \rightarrow \alpha \in \mathcal{R}\})$
$\quad\quad$ **for all** $r : A \rightarrow \alpha \in \mathcal{R}$ **do**
$\quad\quad\quad r.p \leftarrow (r.\text{count} + m \cdot r.p)/(A.\text{count} + m)$ ▷ update prob. of r in G
$\quad\quad\quad$ normalize the values in $\{r.p : r : A \rightarrow \alpha \in \mathcal{R}\}$ so their sum equals 1
return updated grammar G

3 Empirical Evaluation

We study the performance of the proposed Bayesian grammar refinement algorithm with a custom, demonstrative equation discovery task. The purpose of this experiment is to verify whether the algorithm works as intended, improving the performance of PCFG-based equation discovery, refining the grammar according to our expectations and approximating a posterior that maximizes the probability of the correct expression.

3.1 Experimental Setup

The experiment tests the ability of the algorithm to discover the following three equations:

$$y = x_1 - 3x_2 - x_3 - x_5, \tag{4}$$

$$y = x_1^5 x_2^3, \tag{5}$$

$$y = \sin(x_1) + \sin\left(\frac{x_2}{x_1^2}\right). \tag{6}$$

For each of the three equations, we generate 100 data points by uniformly sampling the variables in the $(-10, 10)$ interval. We use a variant of the universal grammar for mathematical expressions with the following initial production rule probabilities:

$$\begin{aligned} E &\rightarrow E + F\ [0.2] \mid E - F\ [0.2] \mid F\ [0.6] \\ F &\rightarrow F * T\ [0.2] \mid F\ /\ T\ [0.2] \mid T\ [0.6] \\ T &\rightarrow (E)\ [0.2] \mid \sin(E)\ [0.2] \mid V\ [0.6] \\ V &\rightarrow x_1\ [1/n_v] \mid x_2\ [1/n_v] \mid \ldots \mid x_{n_v}\ [1/n_v], \end{aligned} \tag{7}$$

where n_v is the number of variables. For simplicity, the grammar does not generate numerical parameters and the three equations do not feature real constants.

Each of the three test equations poses unique challenges for PCFG-based equation discovery. The first equation, despite its simple form, is difficult to generate because it involves five variables but uses only four, requiring the grammar to effectively ignore one variable and rely on repeated addition or subtraction to approximate integer constants. The second equation, while using just two variables, involves high exponents that the grammar must construct through repeated multiplication, further reducing its sampling probability. In contrast, the third equation presents a comprehensive challenge, combining multiple operations, including division and a trigonometric function, without a clear optimization path, providing a more rigorous test of the algorithm's capacity to adapt its production probabilities effectively.

In the experiment, we compare four versions of the Bayesian algorithm based on m-estimate updates (MEU) with varying values of m: 1, 2, 5, 10, and the baseline Monte-Carlo algorithm [2], where we keep production probabilities constant.

We give each tested method a budget of 10^5 expressions to evaluate for each of the three problems. In the case of the Monte-Carlo algorithm, this simply means sampling 10^5 random expressions with the grammar and computing their errors. We run all four versions of MEU in 2000 iterations, sampling 50 random expressions in each iteration. If no expression in an iteration passes the relative threshold selection criterion, we do not update probabilities and move to the next iteration. In the experiment, we use $\epsilon = 0.1$ as the relative threshold. To account for the randomness involved in the algorithms, we run each of the five evaluated algorithms 10 times with different random seeds.

3.2 Results

We summarize the results in Table 1 by considering how often each equation was discovered exactly among 10 runs of each compared method. The first equation proved the most difficult, as it was discovered only in a single run of Bayesian refinement for $m = 1$. The contrast between the Bayesian refinement and the baseline algorithms was most prominent for the second equation – the Bayesian

Table 1. The number of equation discovery successes (exactly recovered equation) among 10 runs with different random seeds for the four variants of Bayesian refinement and the baseline of sampling from a grammar with constant probabilities (constant).

	$m=1$	$m=2$	$m=5$	$m=10$	constant
$y=x_1-3x_2-x_3-x_5$	1	0	0	0	0
$y=x_1^5x_2^3$	10	10	10	9	1
$y=\sin(x_1)+\sin\left(\frac{x_2}{x_1^2}\right)$	0	1	0	3	1

algorithm was able to discover it in all but one run, while Monte-Carlo discovered it only in one out of ten runs. The third equation also proved troublesome for most algorithms.

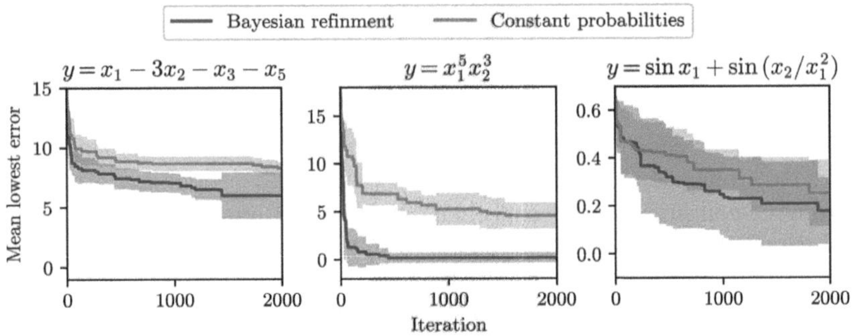

Fig. 2. Optimization curves of Bayesian refinement ($m = 2$) and the baseline of sampling from a grammar with constant probabilities for each of the three equations: a) $y = x_1 - 3x_2 - x_3 - x_5$, b) $y = x_1^5 x_2^3$, c) $y = \sin x_1 + \sin(x_2/x_1^2)$. The horizontal axis depicts the iteration, whereas the vertical axis depicts the lowest error (RMSE) achieved for a given number of expressions. The line depicts the mean and the shaded area the standard deviation across 10 runs with different random seeds.

To further compare the different methods, we calculate and plot optimization curves by averaging the cumulative minimum RMSE across the ten runs at each iteration of the procedure. We compare the optimization curves of Bayesian grammar refinement and the baseline approach of sampling from a grammar with constant probabilities for each of the three equations in Fig. 2. The plots mirror the findings from Table 1, with the mean error of Bayesian approaches dropping to zero quickly for the second equation and remaining relatively high for the first and third equation.

Most importantly, however, the mean error of Bayesian approaches is lower than the mean error of the baseline approach for almost the entire length of the optimization procedure for all three equations. In other words, the Bayesian approach clearly outperformed random sampling in this experiment. The choice of m did not prove to have a notable impact on the performance of the Bayesian refinement algorithm, with the optimization curves for all four options being relatively close. We chose $m = 2$ for visualization as the option with the most consistent results across the three equations in the experiment.

3.3 Production Rule Probabilities

We can further investigate the behavior of the Bayesian grammar refinement algorithm by studying how the probabilities of production rules evolve throughout the optimization procedure.

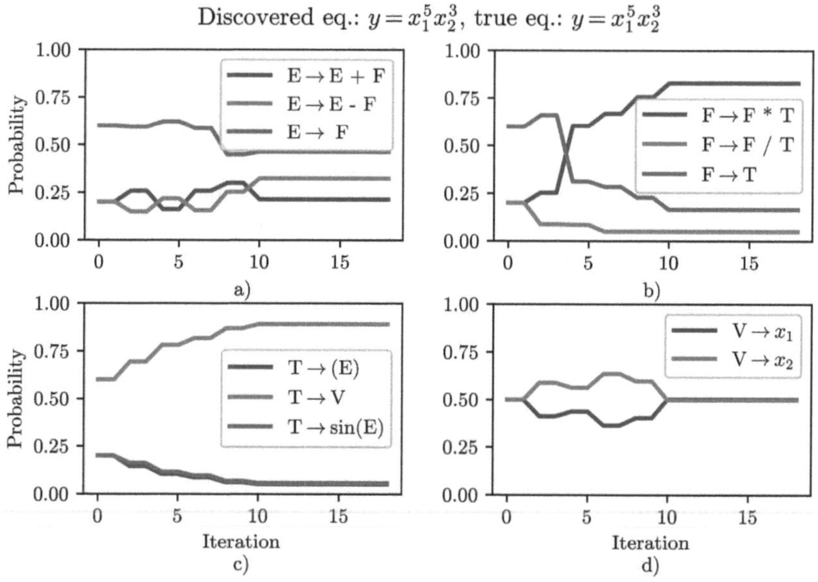

Fig. 3. The probabilities of production rules with the nonterminal a) E, b) F, c) T, d) V on the left-hand side, plotted at each iteration of Bayesian grammar refinement ($m = 2$) for the second equation. In this run, the algorithm was able to exactly recover the target equation with an error of RMSE $= 0$, resulting in early stopping.

Figure 3 depicts the evolution of production rules throughout the refinement procedure in one of the successful runs for second equation, using $m = 2$. This equation features no addition, subtraction, division or special functions, but requires a lot of multiplication. We also expect x_1 to be preferred over x_2. Plot **a** reveals that, contrary to our expectations, the algorithm did not learn to ignore addition and subtraction, but even increased their probability slightly. On the other hand, our expectations are met in plot **b**, which shows the probability of multiplication rising over 0.8 and of division dropping to the minimum.

Similarly, in plot **c**, we see that the probability of (E) and sine dropped to zero, fulfilling expectations. Finally, plot **d** shows a failure of the algorithm to learn any variable preferences. All in all, the behavior of the algorithm when discovering the second equation makes sense, as it successfully learned the crucial importance of multiplication and eliminated excessive recursion.

3.4 Posterior Probabilities

One of the advantages of Bayesian grammar refinement is that it approximates the posterior distribution over the space of mathematical expressions. Once we have completed the optimization of grammar probabilities, we can calculate the posterior probability of individual expressions by parsing them using the PCFG with updated probabilities. Note, however, that because each mathematical expression can be derived by the grammar in a number of different ways, the probability obtained this way is only an approximation of the true probability. On the other hand, this is not an issue when comparing expression probabilities, obtained by parsing the expression in the same way using the same grammar, with different values of production rule probabilities, such as when comparing the prior and posterior expression probabilities.

To further understand the behavior of Bayesian grammar refinement, we compute the probability of the correct expression at various points in the procedure, including the initial and final iteration. To obtain more accurate approximations of the probability, we rewrite the first equation $y = x_1 - 3x_2 - x_3 - x_5$ following the commutativity of summation and generating every possible ordering of the terms:

$$y = x_1 - 3x_2 - x_3 - x_5 = x_1 - x_2 - x_2 - x_2 - x_3 - x_5 = -x_2 + x_1 - x_2 - x_2 - x_3 - x_5 = \dots$$

In total, we generate 120 different expressions, parse each using the initial PCFG and the final PCFG to obtain its prior and posterior probabilities, respectively, and sum the individual probabilities to obtain an approximation of the prior and posterior probabilities of the correct expression. We treat the second equation $y = x_1^5 x_2^3$ in a similar way, generating 56 equivalent representations, whereas the third equation $y = \sin(x_1) + \sin\left(\frac{x_2}{x_1^2}\right)$ allows for only four simple ways of rewriting. We estimate the probability of the target expression by summing over the probabilities obtained by parsing each of its equivalent representations. In Table 2, we summarize the prior and posterior probabilities obtained this way for one of the experiment runs for each of the three equations using production rule probabilities corresponding to the initial and final iterations, respectively.

The posterior probability of the correct expression is higher than the prior probability for all three equations, showing that the Bayesian algorithm indeed leads the distribution towards the correct expression. The improvement is the lowest for the third equation, with the posterior being three times the prior, and the highest for the second equation, improving the probability by 5 orders of magnitude.

Using production rule probabilities from different iterations for parsing, we can plot the evolution of the probability of the correct equation throughout the optimization procedure. We plot the evolution of the minimum, median and

Table 2. Approximated prior and posterior probabilities of the correct expression for each of the three equations from the experiment, obtained by parsing the many equivalent mathematical expressions using the PCFG with the initial and final values of production rule probabilities.

	Equation 1	Equation 2	Equation 3
Prior probab.	$5.3 \cdot 10^{-10}$	$1.7 \cdot 10^{-8}$	$1.6 \cdot 10^{-7}$
Posterior probab.	$1.7 \cdot 10^{-7}$	$1.3 \cdot 10^{-3}$	$4.8 \cdot 10^{-7}$

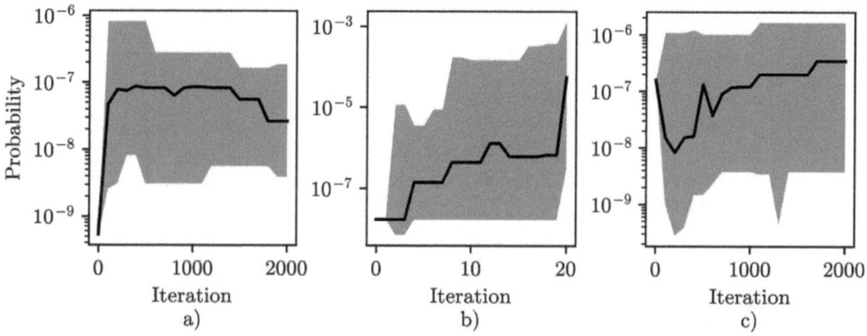

Fig. 4. The approximated probability of the correct expression at each iteration of Bayesian grammar refinement ($m = 2$) for each of the three equations: a) $y = x_1 - 3x_2 - x_3 - x_5$, b) $y = x_1^5 x_2^3$, c) $y = \sin x_1 + \sin(x_2/x_1^2)$. The black line depicts the median probability and the shaded area depicts the region between the minimum and maximum probability among the 10 runs.

maximum probability across the ten runs in Fig. 4. We see that the median probability of the correct expression increases for all three equations, although not monotonically. In the best cases, the probability of the target expressions increases by several orders of magnitude. In the worst cases, the final probability for the first equation is still an order of magnitude higher than the initial. For the second equation, the lowest final probability is the same as the initial probability, and lower than the initial probability for the third equation. We can therefore conclude that in an average run of this experiment, the Bayesian algorithm will result in an increased probability of the correct expression. Runs in which the Bayesian algorithm achieves a counterproductive effect are possible, but unlikely.

4 Discussion

The proposed Bayesian grammar refinement algorithm demonstrates promising results in guiding the search for interpretable mathematical expressions. By iteratively refining production rule probabilities based on the performance of sampled equations, the method effectively approximates a posterior distribution over the space of possible expressions, favoring those that better fit the observed data. Our experiments show that this approach outperforms random sampling from a grammar with constant probabilities, particularly for equations requiring repeated operations (e.g., $y = x_1^5 x_2^3$), where the algorithm successfully amplifies the probability of critical rules like multiplication.

4.1 Limitations

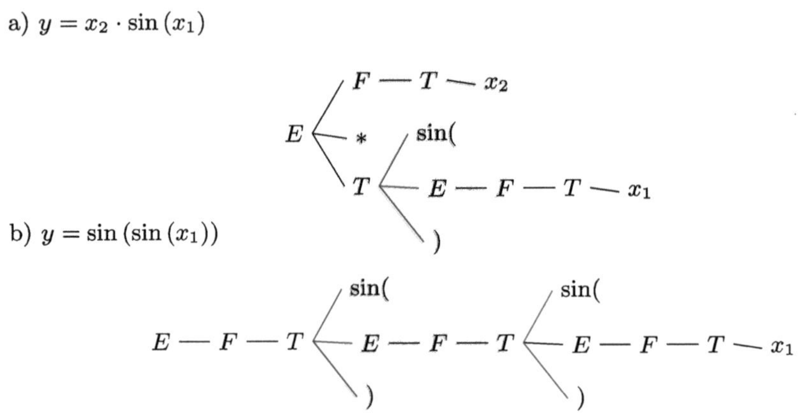

Fig. 5. Parse trees of two expressions, demonstrating the structural limitations of PCFGs. Highlighted in red and blue is the production rule for sine, used in different contexts. In blue, the function has a similar context in both parse trees and makes sense mathematically. In red, the context is different and mathematically implausible, since trigonometric functions typically do not appear within other trigonometric functions. Nevertheless, the probability of the production rule is the same in all three cases. (Color figure online)

Bayesian grammar refinement inherits a fundamental limitation from the underlying PCFG framework: its lack of context sensitivity. The same production rules are used in different contexts, both within the same parse tree and across different parse trees. We demonstrate this limitation in Fig. 5, which depicts the parse trees of the expressions a) $x_2 \cdot \sin(x_1)$ and b) $\sin(\sin(x_1)$. The production rule for sine appears in three different contexts, two of which make sense mathematically, and one that does not. However, the probability of the production rule in a PCFG is the same, regardless of context. There is no way of increasing the probability of a) without also increasing the probability of b), while keeping the structure of the grammar unchanged.

Similarly, using the grammar from our experiment, it is not possible to make x^2 more likely than x. This bias towards simpler expressions is intrinsic in PCFGs due to the multiplicative nature of derivation probabilities—each additional derivation step reduces the overall likelihood of the expression. The consequence of these limitations is notable in Table 2—even though optimized posterior probabilities are substantially higher than the prior probabilities, they are still very small. The expected number of expressions to sample to discover the first equation is $6 \cdot 10^6$ and $2 \cdot 10^6$ for the third equation, which represents a significant computational cost if parameter estimation is required.

The context insensitivity and simplicity bias can be mitigated, though not eliminated, through careful grammar design. For example, introducing dedicated rules for common complex patterns (e.g., $T \to V^n$) can reduce the sampling penalty for exponents, while a separate, more limited version of the nonterminal E can be introduced as an argument for sine. However, this requires domain expertise and risks over-constraining the search space.

A second practical limitation stems from the iterative nature of Bayesian grammar refinement. Unlike Monte-Carlo sampling, which evaluates expressions independently and is trivially parallelizable, our method requires sequential updates to the grammar probabilities after each batch of evaluations. This dependency between iterations hampers computational efficiency, especially when scaling to large datasets or high-dimensional problems, although techniques like asynchronous updates or distributed sampling can mitigate the issue.

4.2 Future Directions

In contrast to purely data-driven methods, the presented method explicitly encodes domain knowledge through the grammar and leverages Bayesian updates to focus the search. While our experiments used a general-purpose grammar to demonstrate the baseline capabilities of the approach, the method's true strength emerges when incorporating domain-specific constraints. In practical applications where background knowledge can be encoded through tailored production rules (e.g., known physical laws or dimensional constraints), the grammar can significantly reduce the search space while maintaining interpretability. Since the Bayesian refinement algorithm demonstrates a clear advantage over sampling from a fixed PCFG, future work should evaluate the method on domain-specific problems where prior knowledge can be rigorously encoded.

On the other hand, to address the discussed limitations, we will consider taking inspiration from the Bayesian Scientist [13] and employing Markov-Chain Monte-Carlo to sample directly from the posterior. If designed correctly, such an approach should avoid the structural and simplicity bias of the current sampling algorithm, while still guaranteeing that generated expressions conform to the grammar.

5 Conclusions

We presented a Bayesian refinement approach for probabilistic context-free grammars, aimed at improving the efficiency and accuracy of grammar-based equation discovery. By iteratively updating production rule probabilities based on observed performance, our method approximates a posterior distribution over mathematical expressions and successfully guides the search toward more promising regions of the expression space. Empirical results show clear advantages over sampling from a constant PCFG, particularly for complex equations requiring repeated operations.

The proposed method inherits a contextual insensitivity intrinsic to PCFGs and faces limitations in scalability due to its sequential nature. Nevertheless, this work represents an important stepping stone in the development of knowledge-based and interpretable equation discovery. Future work will explore more expressive grammar structures and sampling methods that can mitigate these limitations, potentially enabling a broader range of applications across different scientific domains.

Acknowledgements. The authors acknowledge the financial support of the Slovenian Research Agency via the projects GC-0001 and J2-4452. This work was also supported by the EC via the project ELIAS, grant No. 101120237.

Competing Interests. The authors declare that they have no competing interests.

References

1. Brence, J., Džeroski, S., Todorovski, L.: Dimensionally-consistent equation discovery through probabilistic attribute grammars. Inf. Sci. **632**, 742–756 (2023)
2. Brence, J., Todorovski, L., Džeroski, S.: Probabilistic grammars for equation discovery. Knowl.-Based Syst. **224**, 107077 (2021)
3. Brunton, S.L., Proctor, J.L., Kutz, J.N.: Discovering governing equations from data by sparse identification of nonlinear dynamical systems. Proc. Natl. Acad. Sci. **113**(15), 3932–3937 (2016)
4. Cestnik, B.: Estimating probabilities: a crucial task in machine learning. In: Proceedings of the 9th European Conference on Artificial Intelligence, pp. 147–149 (1990)
5. Cestnik, B., Bratko, I.: On estimating probabilities in tree pruning. In: Kodratoff, Y. (ed.) EWSL 1991. LNCS, vol. 482, pp. 138–150. Springer, Heidelberg (1991). https://doi.org/10.1007/BFb0017010
6. Chi, Z.: Statistical properties of probabilistic context-free grammars. Comput. Linguist. **25**(1), 131–160 (1999)
7. Džeroski, S., Cestnik, B., Petrovski, I.: Using the m-estimate in rule induction. J. Comput. Inf. Technol. **1**(1), 37–46 (1993)
8. Džeroski, S., Langley, P., Todorovski, L.: Computational discovery of scientific knowledge. In: Computational Discovery of Scientific Knowledge, pp. 1–14. Springer (2007)
9. Fuentes, R., et al.: Equation discovery for nonlinear dynamical systems: a bayesian viewpoint. Mech. Syst. Signal Process. (2021). https://doi.org/10.1016/j.ymssp.2020.107528
10. Gec, B., Omejc, N., Brence, J., Džeroski, S., Todorovski, L.: Discovery of differential equations using probabilistic grammars. In: Proceedings of the 25th International Conference on Discovery Science, pp. 22–31. Springer (2022)
11. Geman, S., Johnson, M.: Probabilistic grammars and their applications. Int. Encyclopedia Soc. Behav. Sci. **2002**, 12075–12082 (2002)
12. Good, I.J.: Probability and the weighing of evidence. Charles Griffing (1950)

13. Guimerà, R., et al.: A bayesian machine scientist to aid in the solution of challenging scientific problems. Sci. Adv. **6**(5), eaav6971 (2020)
14. Jin, Y., Fu, W., Kang, J., Guo, J., Guo, J.: Bayesian Symbolic Regression (2020). https://doi.org/10.48550/arXiv.1910.08892
15. Kamienny, P.A., d'Ascoli, S., Lample, G., Charton, F.: End-to-end symbolic regression with transformers. Adv. Neural. Inf. Process. Syst. **35**, 10269–10281 (2022)
16. Long, D., Xing, W.W., Krishnapriyan, A.S., Kirby, R.M., Zhe, S., Mahoney, M.W.: Equation discovery with bayesian spike-and-slab priors and efficient kernels (2024). https://doi.org/10.48550/arXiv.2310.05387
17. Mežnar, S., Džeroski, S., Todorovski, L.: Efficient generator of mathematical expressions for symbolic regression. Mach. Learn. **112**, 4563–4596 (2023)
18. North, J.S., North, J.S., Wikle, C.K., Wikle, C.K., Schliep, E.M., Schliep, E.M.: A bayesian approach for data-driven dynamic equation discovery. J. Agric. Biol. Environ. Stat. (2022). https://doi.org/10.1007/s13253-022-00514-1
19. Parkes, A.P.: A Concise Introduction to Languages and Machines. Springer (2008)
20. Petersen, B.K., Larma, M.L., Mundhenk, T.N., Santiago, C.P., Kim, S.K., Kim, J.T.: Deep symbolic regression: Recovering mathematical expressions from data via risk-seeking policy gradients. arXiv:1912.04871 (2019)
21. Ram, P.R.M., Römer, U., Semaan, R.: Bayesian Dynamical System Identification With Unified Sparsity Priors And Model Uncertainty (2021). https://doi.org/10.48550/arXiv.2103.05090
22. Ratle, A., Sebag, M.: Grammar-guided genetic programming and dimensional consistency: Application to non-parametric identification in mechanics. Appl. Soft Comput. **1**(1), 105–118 (2001)
23. Schmidt, M., Lipson, H.: Distilling free-form natural laws from experimental data. Science **324**(5923), 81–85 (2009)
24. Tanevski, J., Todorovski, L., Džeroski, S.: Combinatorial search for selecting the structure of models of dynamical systems with equation discovery. Eng. Appl. Artif. Intell. **89**, 103423 (2020)
25. Todorovski, L., Džeroski, S.: Declarative bias in equation discovery. In: Proceedings of the Fourteenth International Conference on Machine Learning, ICML, pp. 376–384. Morgan Kaufmann (1997)
26. Udrescu, S.M., Tegmark, M.: AI Feynman: a physics-inspired method for symbolic regression. Sci. Adv. **6**(16) (2020)
27. Valipour, M., You, B., Panju, M., Ghodsi, A.: SymbolicGPT: a generative transformer model for symbolic regression. arXiv:2106.14131 (2021)

EDC: Equation Discovery for Classification

Guus Toussaint(✉) and Arno Knobbe

LIACS, Leiden University, Leiden, Netherlands
{g.toussaint,a.j.knobbe}@liacs.leidenuniv.nl

Abstract. Equation Discovery techniques have shown considerable success in regression tasks, where they are used to discover concise and interpretable models (*Symbolic Regression*). In this paper, we propose a new ED-based binary classification framework. Our proposed method EDC finds analytical functions of manageable size that specify the location and shape of the decision boundary. In extensive experiments on artificial and real-life data, we demonstrate how EDC is able to discover both the structure of the target equation as well as the value of its parameters, outperforming the current state-of-the-art ED-based classification methods in binary classification and achieving performance comparable to the state of the art in binary classification. We suggest a grammar of modest complexity that appears to work well on the tested datasets but argue that the exact grammar – and thus the complexity of the models – is configurable, and especially domain-specific expressions can be included in the pattern language, where that is required. The presented grammar consists of a series of summands (additive terms) that include linear, quadratic and exponential terms, as well as products of two features (producing hyperbolic curves ideal for capturing XOR-like dependencies). The experiments demonstrate that this grammar allows fairly flexible decision boundaries while not so rich to cause overfitting.

Keywords: Machine Learning · Binary Classification · Equation Discovery · Explainable Machine Learning · Symbolic Classification

1 Introduction

Equation discovery, the task of discovering analytical functions that model data, is well known for its use in regression settings, where it is then often referred to as Symbolic Regression [2,4]. Many real-world applications require a model to be interpretable, particularly in scenarios where a decision could have substantial consequences. Therefore, there is a movement in machine learning that tries to build models that can be easily interpreted and, where needed, corrected by the practitioners dealing with these models daily. The framework of equation discovery fits this trend perfectly. The result generated does not consist of a large, opaque model that calculates the value for the new data points but rather a simple equation that describes the nature of the data. Furthermore, a domain expert can adjust the pattern language of equations considered by the algorithm,

S. Džeroski et al. (Eds.): DS 2025, LNAI 16090, pp. 128–142, 2025.
https://doi.org/10.1007/978-3-032-05461-6_9

thus allowing the algorithm to adapt to different scenarios where, for example, a non-linear translation of the data makes sense.

As mentioned, the focus of equation discovery (ED) has largely been on regression problems. In this work, we use the ED framework to tackle problems in the classification domain. The central idea is to generate concise and readable equations that define the decision boundary between classes, which domain experts can then evaluate. We show that a classification-oriented ED algorithm may perhaps not beat the state-of-the-art of well-balanced algorithms such as Random Forests or Multi-Layer Perceptrons but be in the same ballpark in terms of classification performance and certainly produce more transparent models than, e.g. Random Forests would. Next to more interpretable models, we also show more expressive power where some standard mathematical functions may capture rather complex class distributions.

When used in a regression setting, the equations considered are candidate functions to model the (numeric) target as a function of several input attributes. These equations typically involve arithmetic operators as well as a set of standard analytical functions (log, exp, sin, ...). Such an approach could then be used to automate scientific discovery (such as Boyle's law, which relates the pressure of a gas in a vessel to its volume and temperature [17]). However, the approach can also be used in more noisy statistical settings, where a regression model captures the influence of certain parameters on the target value. In broad strokes, ED has two separate challenges: first, finding the structure of the equation, and second, optimising the parameters of the candidate equation relative to the available data. Many standard optimisation algorithms are available for the latter, e.g., Gradient Descent [5]. ED then primarily becomes a discrete optimisation problem defined by a search space of allowable equations and a strategy to traverse this search space, either exhaustively or heuristically. Like [26], we define the pattern language, and hence the search space, by means of a configurable grammar that produces a language of equations, typically up to a certain complexity. We provide an example of a grammar that works quite well on datasets involved in our experiments, but we would like to stress that the exact grammar is a parameter of the proposed method. A knowledgeable domain expert may add certain functions to the grammar in order to enrich or reformulate the classification problem in terms of the functions provided.

We need a slightly different setting for ED-based classification since the target values are no longer numeric. Similar to Schwab et al. [23], we model the decision boundary of the classifier in the following basic manner: $T = \textit{true iff } f(x) >= 0$, where $f(x)$ is the equation discovered by ED. In other words, the function f defines a landscape in terms of the available attributes, and whenever the landscape is above 0, the classification is true. It is important to note that this paper focuses on *binary classification*; extensions to multi-class classifications are possible but outside the scope of this paper. For simple formulae, such as linear equations (hyperplanes), this setting is equivalent to existing methods such as logistic regression or SVMs with a linear kernel [10,12]. However, ED comes into its own when richer expressions are involved.

The contributions of our paper can be summarized as follows:

- We propose a method called Equation Discovery for Classification (EDC), that is both interpretable and, with the correct building blocks for the domain at hand, is able to draw complex decision boundaries such that the performance is comparable to the current state-of-the-art in classification.
- With the aim of avoiding redundancy in the equation search space, we suggest a specifically-designed grammar that works in generic cases, and can be refined with application-specific constructs.
- We demonstrate that the algorithm is able to reconstruct a hard-coded decision boundary in artificial data and is able to model challenging XOR-like problems.
- We demonstrate the performance of our model on UCI datasets, where we compare it to state-of-the-art algorithms in terms of area under the curve (AUC), interpretability and model size.

Along with this paper, we publish the code[1] and experimental setup[2] so that the work presented here can be completely reproduced.

2 Related Work

The core idea of equation discovery, as it has been presented, is to find a function that fits the data. To achieve this, a search strategy has to be defined; previous works broadly identify two types of search strategies: exhaustive and heuristic search. Exhaustive search focuses on defining a manageable search space so that all possible equations within that search space can be evaluated. One common approach to exhaustive search for equation discovery is to use a context-free grammar [26]. Implementing a heuristic search algorithm that traverses this search space more efficiently eliminates the need for restricting the initial search space. Genetic programming has been successfully implemented as a heuristic approach in equation discovery [1,2,11]. These come in two flavours. Some try to optimise the structure and the parameters of the equation simultaneously, while others first try to find the structure, and optimise the parameters at a later stage.

Regression is the dominant data mining task in equation discovery (in which case people speak of *symbolic regression*). The aim here is to discover an analytical expression involving various arithmetic operations and closed-form functions (such as *log*, *sin*, etc.) that model a supervised dataset. The target expression typically exhibits a tree-shaped structure (which needs to be discovered) and involves one or more parameters (that need to be 'fit' to the data). For a complete overview of the symbolic regression approach, we refer to the survey paper

[1] The code of the core algorithm can be found in the following repository: https://github.com/GuusToussaint/EDC-core.

[2] The code for reproducing the experiments can be found at the following link https://github.com/GuusToussaint/EDC-experiments datasets can be found at the following link: https://anonymous.4open.science/r/EDC-datasets-B9F7/.

by Makke et al. [19]. Beyond modelling explicit relationships, equation discovery has also been applied to uncovering latent monotonic features in time-series data [27].

In recent years, a handful of efforts have been made to apply the machinery of symbolic regression to the classification task, sometimes called Symbolic Classification (SC). Schwab et al. [23] introduced the topic by describing a simple yet elegant alteration of the classic symbolic regression approach using genetic programming that can solve binary classification problems. The algorithm, AMAXSC, produces a symbolic expression using Genetic Programming, which is then thresholded to produce a discrete decision boundary (much like in our setting). While AMAXSC produces interpretable equations, the experiments show that the method struggles to discover concise and accurate models, often choosing to improve intermediate expressions by adding additional terms, rather than extensively optimizing the model's parameters. This tendency can be mitigated by severely limiting the allowed operators, variables and functions to occur in the expressions.

To improve upon the initial design of AMAXSC and extend it to a multi-label classification setting, Ingalalli et al. [13] introduced the Multi-dimensional Multi-class Genetic Programming algorithm (M_2GP). They argued that a single expression is not informative enough to solve multi-class classification problems. Therefore, they use a multi-expression representation, where each sample is mapped to a custom number of dimensions d, where each dimension is a learned equation. Thus, this approach can be seen as mapping any sample to a latent feature space using d transformations. The final classification is done by calculating the distance between the sample and the centroid of each class in the latent feature space. The initial experiments showed promising results with performance comparable to state-of-the-art classification methods. However, as a result of having an unspecified number of latent features, interpretability is compromised (which is understandable, given the challenges of multi-class classification).

The original method has since been improved by eliminating the need to manually select the number of dimensions d and instead learning it as a parameter within the method. M3GP introduced by Muñoz et al. [21] extends M_2GP by allowing the search process to progressively search for the optimal number of new dimensions that maximize the classification accuracy. In the most recent development, La Cava et al. [16] introduced M4GP, which improves upon M_2GP and M3GP by simplifying the program encoding, using advanced selection methods, and archiving solutions during the run. In our experiments, we demonstrate that M4GP is not able to reach state-of-the-art performance on *binary* classification in terms of model accuracy.

Korns et al. [14] have developed the Multilayer Disciminant Classification (MDC) algorithm, a computationally efficient alternative to the M_2GP. However, the same performance and interpretability drawbacks remain. Our approach aims to improve the interpretation and ease of use in binary classification by having only a single expression, while achieving competitive accuracy, comparable to the state-of-the-art in classification.

3 Background

Assume that we have a dataset X consisting of numeric attributes. A data point $x \in X$ consists of d numeric features such that $x \in \mathcal{R}^d$. When faced with a dataset that contains categorical features, one-hot encoding transforms them into a set of numeric 0,1-features. As this may explode the number of features when dealing with high-cardinality categorical features, we take the pragmatic approach to group one-hot encoded features whose frequency of occurrence is smaller than or equal to 2%.

To evaluate the fitness of an equation, we use the Log Loss [7], also known as the logistic loss or the cross-entropy loss. Since the input for the Log loss function represents probabilities, only values between 0 and 1 are allowed. Therefore, the output of the equation is passed through the logistic function before calculating the loss.

Once an equation is found, the results are ranked based on their probability scores. This ranking is used to calculate the area under the receiver operating characteristic curve (AUC), which determines the final performance of the model. The equation does not assign direct labels to input data but rather a probability score [10]. An appropriate threshold can be chosen based on the costs of mislabeling false cases as true and vice versa. In this work, we will choose a threshold that maximizes the accuracy (misclassification costs are assumed to be equal). This is done by selecting a threshold value that maximizes the true and false positives ratio.

4 Method

The proposed framework for equation discovery in a classification setting consists of two main components. First, candidate equations are constructed during the *search* step. Second, the constants in these candidate equations are optimised during the *optimisation* step.

4.1 Search

We will define the extent of the search space by means of a context-free grammar, further defined below. However, it is safe to say that, in most cases, the defined search space will be too large to examine exhaustively. Even when the search depth is limited, with the growing number of attributes in the dataset, the number of possible equations will explode quickly. For this reason, we will adopt a heuristic search strategy based on *beam search*, which provides an attractive balance between exploration and exploitation.

The notion of using a context-free grammar to define the search space of possible equations in equation discovery was first introduced by Todorovski et al. [26]. The overall search space is limited by declaring a grammar that describes all possible valid equations. However, the grammar can be defined so that all relevant equations remain in the constrained search space. Ideally, a domain expert defines the grammar for each domain the algorithm applies to. A domain expert can give insight into translation combinations of relevant features for the domain.

Our generic, application-independent grammar should include expressive and effective equations that are at the same time interpretable. It includes equations that consist of basic arithmetic operations, including addition, multiplication, and exponentiation. It is designed with an eye on limiting the level of redundancy in the search space such that two equations that are syntactically different but semantically equivalent are only considered once. As a simple example, $x + c \cdot y$ is equivalent to $x - c \cdot y$, since subtraction can be achieved by negating c, so the latter is not part of the grammar. Other opportunities for pruning the search space are explained below the definition of the grammar.

The grammar is defined as follows: $G = (\mathcal{N}, \mathcal{T}, \mathcal{R}, \mathcal{S})$, which contains the following sets of symbols and operations:

- $\mathcal{N}$ contains all non-terminal symbols.
- $\mathcal{T}$ contains all terminal symbols.
- $\mathcal{R}$ contains the rewrite rules in the form $A \rightarrow \alpha$ where $A \in \mathcal{N}$ and $\alpha \in (\mathcal{N} \cup \mathcal{T})^*$.
- $\mathcal{S}$ contains the start symbols.

For our experiments, we have constructed the following grammar G_s:

- $\mathcal{N}_s = \{V, B, X\}$
- $\mathcal{T}_s = \{x_1, x_2, c, +, \cdot, exp\}$
- $\mathcal{R}_s = \{V \rightarrow c \mid V + B,$
  $B \rightarrow c \cdot X \mid c \cdot \exp(c \cdot X) \mid c \cdot X \cdot X,$
  $X \rightarrow x_1 \mid x_2 \}$
- $\mathcal{S}_s = \{V\}$.

Note that this grammar only applies to datasets containing two features, i.e., x_1 and x_2; for our experiments, the non-terminal set X is expanded to fit the number of features available in the dataset.

When dealing with real-world data, the numerical inputs might have an arbitrary scale, which affects the range of the constants to optimize, something we would like to avoid. For this reason, prior to actually starting the EDC algorithm, we normalise the data by means of linear scaling to the interval $[0, 1]$. Our choice of grammar is based on a number of design principles. First off, to keep the search space to a manageable size, we require our equation to be a set of summands: a sequence of building blocks that are connected through addition. As an additional benefit, this keeps derived equations similar to their ancestors: the derived equation just has a summand added and remains unchanged otherwise. Furthermore, we only use addition, not subtraction, between summands because each summand has a constant c that the numeric optimiser can set to a negative number if necessary. As for the summands, we allow a linear term x, multiplication of two features $x \cdot y$ (note that this includes quadratic terms x^2), and exponentiation $\exp(x)$.

For each such term, one or more constants c_i are introduced in specific locations to further parameterize the equation. For the exp function, we add a constant with multiplication 'inside' since that cannot be rewritten into a simpler form. Note that the addition of a constant inside exp is superfluous because that can be rewritten as follows: $\exp(a + x) = \exp(a) \cdot \exp(x) = c \cdot \exp(x)$

The grammar is not set in stone; thus, when deploying the proposed classifier in a domain with different characteristics, it is possible to design a different grammar with the required characteristics.

While the equation language Λ, and thus the search space, is defined by the context-free grammar defined above, we use *beam search* to heuristically traverse a relevant but limited portion of this search space. This search algorithm has successfully been applied in many mining settings, e.g., subgroup discovery [20] or multi-label learning [15]. This approach builds equations within Λ by iteratively adding summands to the equation, starting from a starting constant c_0. For example, the equation $c_0 + c_1 \cdot x_1 + c_2 \cdot \exp(c_3 \cdot x_2)$ is found at search depth $d = 2$ by refining the starting equation c_0 twice, as follows:

$$\begin{aligned}
&c_0 \\
&c_0 + c_1 \cdot x_1 \\
&c_0 + c_1 \cdot x_1 + c_2 \cdot \exp(c_3 \cdot x_2).
\end{aligned}$$

4.2 Optimisation

Each equation encountered in the search space contains constants that need to be optimised for the target at hand. Our choice of optimiser is dependent on the structure of the equation. If the equation is partially differentiable with respect to the constants c_i, we employ stochastic gradient descent (SGD) as our optimisation algorithm. On the other hand, when the equation is not differentiable, which is the case for the exponentiation summand $\exp(c \cdot x)$, we employ a *Hill Climber*-based approach.

The Hill Climber allocates a portion f of its budget to sample random configurations. Subsequently, the top k configurations are chosen as the starting points for the hill climber. The remaining budget of $n \cdot (1 - f)$ evaluation is then evenly distributed among these top k configurations, for $n \cdot (1 - f)/2km$ evaluations each (where m is the number of features). Each evaluation involves taking a step of size α in both directions for each feature, i.e., moving up and down. These potential steps are then evaluated, and the one that results in the most significant decrease in loss is selected as the next step.

5 Experiments

This section describes the experiments conducted to evaluate the EDC algorithm. The experiments are divided into two parts. First, we evaluate the performance of the EDC algorithm under various conditions using a series of artificial datasets. Second, the EDC algorithm is evaluated on a collection of UCI datasets and compared to other common classifiers.

5.1 Artificial

To evaluate the performance, we have developed four sets of artificial data, increasing in degree of difficulty. Each set will evaluate the performance of the EDC algorithm under different conditions. First, we evaluate the algorithm's performance when the target equation is within the search space, and no noise is introduced to the decision boundary. Next, we introduce noise to the decision boundary to assess the resilience of our approach under more realistic conditions. Then, we assess the performance of artificial datasets generated using a more complex grammar, where the decision boundary lies beyond the provided search space for the EDC algorithm. Finally, artificial datasets are generated using random Gaussian clusters. This scenario tests whether adequate classifiers can be induced in the absence of an explicitly defined decision boundary.

Within Search Space. This scenario aims to evaluate the algorithm's performance in the most optimal, structured setting, i.e. the target decision boundary is within the search space, and no noise is added. To achieve this, 100 random equations are sampled from within the search space provided to the algorithm. These equations are then fitted with random constants to create an artificial dataset.

Within Search Space with Noise. In this scenario, equations are sampled from within the algorithm's search space, similar to the first scenario. However, noise is added to the output, making the decision boundary fuzzy. We take the following approach to achieve this 'fuzzy' decision boundary. An equation is sampled from within the search space and fitted with random values for the constants. Then, 2000 data points are drawn uniformly from $[-10, 10] \times [-10, 10]$, and labelled true or false using the sampled equations as the decision boundary (see Figs 1 and 2). After a label is assigned, the points are randomly perturbed using Gaussian noise according to distribution $\mathcal{N}(0, 2^2)$ (a mean of 0 and a standard deviation of 2). The AUC is calculated for both to compare the proposed decision boundary to the target. Note that, due to the noise added to the data points after label assignment, the target decision boundary will not achieve an AUC of 1, which is expected. Since all real-world datasets are affected by noise, it is crucial to assess EDC's ability to identify a suitable decision boundary despite the presence of noise.

Beyond Search Space with Noise. In this scenario, decision boundaries are sampled from a more complex grammar than the one available to EDC. We have extended the original building blocks presented in Sect. 4 with the following:

$$c \cdot X^3, \quad c \cdot X^4$$

Only equations that contain at least one of these building blocks were selected during the random sampling. Since the EDC algorithm is not provided with these extended building blocks, it must approximate the resulting decision boundary with the building blocks present in its search space. As in the previous paragraph, noise is added to the data points to achieve a fuzzy decision boundary. This experiment provides insights into EDC's performance when attempting to approximate equations outside its search space, which is important since, in real-world scenarios, some interactions present in the data might be missing as building blocks in the search space.

Table 1. This table shows the results of the EDC algorithm on our sets of artificial datasets. The results are presented in terms of mean AUC with the standard deviation in parentheses.

Dataset	AUC	
	EDC	Original DB
Within search space	0.999 (±0.00)	1.000 (±0.00)
Within search space with noise	0.951 (±0.03)	0.943 (±0.03)
Beyond search space with noise	0.962 (±0.03)	0.956 (±0.03)

Guassian Clusters. This last scenario creates 100 datasets using a mixture of six randomly generated Gaussian clusters. The labels are assigned by randomly assigning two Gaussian clusters to the *true* class and the remaining four to the *false* class. No explicit target equation exists since the Gaussian intersection determines the dataset's decision boundary. This experiment is designed to evaluate the performance of the EDC algorithm in a setting where the data generated does not come directly from an equation. This setting is the closest to the real world of all the artificial experiments presented in this paper. We compare the results on the 100 generated datasets with the following algorithms. First, we compare against the competing symbolic classification algorithms AMAXSC and M4GP. Second, we compare against other interpretable classification algorithms, namely the decision tree (Tree) and the LDA algorithm. Finally, we also compare against the state-of-the-art in terms of classification performance, namely the multi-layer perception (MLP), the random forest algorithm (RF), and the support vector machine (SVM) with a radial basis function kernel. The results are presented in terms of the mean AUC with the standard deviation on the 100 generated datasets.

Results. Table 1 shows the results of our experiments on the artificial datasets. We note that in the first setting, where the equation is from within the search space, and no noise is added to the decision boundary, the EDC algorithm is able to find approximate equations with an average AUC of 0.999. We can conclude

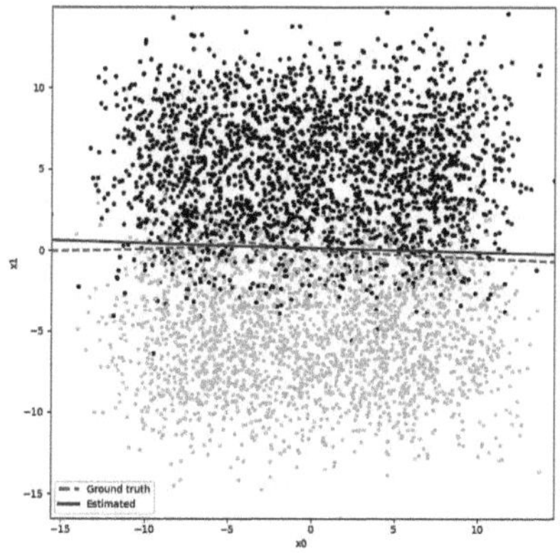

Fig. 1. Target decision boundary (dashed red line), and decision boundary found by EDC (solid blue line). Noise is added to the data, and as a result, the target achieves a lower AUC (0.952) compared to the EDC algorithm (0.978). (Color figure online)

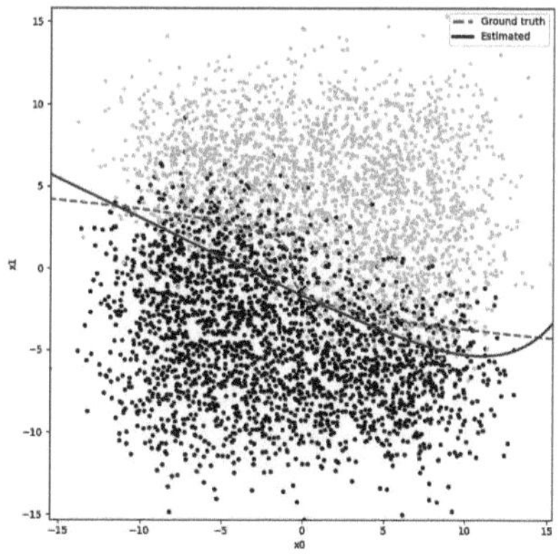

Fig. 2. Target decision boundary (dashed red line), and decision boundary found by EDC (solid blue line). The target equation is sampled from a richer grammar than available to EDC. The target achieves a lower AUC (0.959) compared to EDC (0.967). (Color figure online)

that the EDC algorithm can identify the correct structure of the decision boundary. Any remaining error can likely be attributed to suboptimal optimisation of the constants. However, we note that the obtained errors are very small.

When observing the results of a paired t-test for the second setting, we note a significant difference (in favour of EDC) between the original decision boundary (M = 0.9, SD = 0.03) and EDC (M = 1, SD = 0.03) AUC scores, with a $t(99) = 5.4$ and $p < .001$. Similar behaviour is observed for the third setting, where the target (M = 1, SD = 0.03) and the EDC (M = 1, SD = 0.03) AUC scores also report a significant difference, $t(99) = 5.9$, $p < .001$, for the paired t-test. Interestingly, when adding noise to the decision boundary, the EDC algorithm can find a new decision boundary that fits the data better than the original decision boundary. We observe that for complex decision boundaries, the process of adding noise results in a new decision boundary being formed. This is perhaps somewhat counter-intuitive since one would expect adding Gaussian random noise would not change the patterns hidden in the data. Judging from the results, we note that the EDC algorithm can identify this *new* decision boundary and subsequently achieves a marginal but significantly higher score than the original decision boundary in terms of AUC. From the third setting, we observe similar behaviour, again showcasing that the EDC algorithm can identify a suitable decision boundary that significantly outperforms the original decision boundary in terms of AUC. Figures 1 and 2 show examples of cases where the EDC algorithm outperforms the original decision boundary for the *within search space with noise* and *beyond search space with noise* datasets respectively.

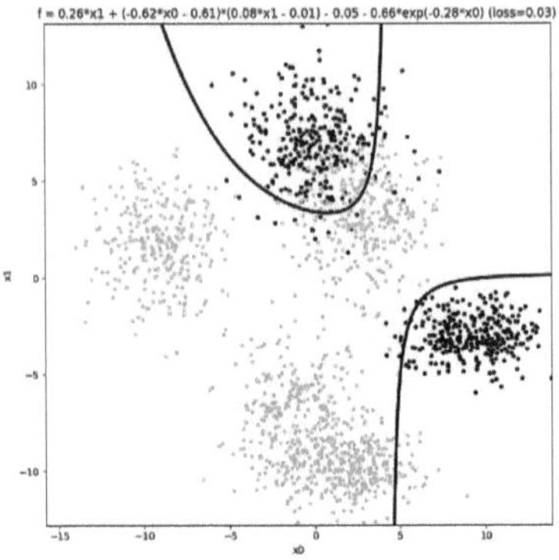

Fig. 3. The proposed decision boundary for the Gaussian clusters artificial dataset. Note that the discovered equation indicated above the figure is produced after translating the equation back to the non-normalised space. This introduces two additional constants for each feature x_i.

Table 2. This table shows the results of the artificial Gaussian clusters experiment. The results are presented in terms of mean AUC with the standard deviation in parentheses. The proposed EDC algorithm performs comparable to the state-of-the-art classification and outperforms all other explainable methods

Algorithm	AUC
MLP	0.972 (±0.034)
SVM	0.970 (±0.038)
RF	0.967 (±0.037)
EDC	0.965 (±0.043)
Tree	0.899 (±0.079)
M4GP	0.894 (±0.082)
LDA	0.811 (±0.139)
AMAXSC	0.802 (±0.181)

Figure 3 shows the proposed decision boundary for a single Gaussian clusters dataset.We observe that the EDC algorithm is capable of defining a decision boundary that fits the data, even when the data-generating process is not guided by a *target* decision boundary. Table 2 shows the results of the comparison between the different classifiers. The state-of-the-art classification algorithms perform the best, with mean AUC values around 0.97. Furthermore, we observe that the explainable methods represent the lower half of the ranking, with a notably high standard deviation across the 100 generated datasets. This lower accuracy is likely caused by the limits on equation complexity introduced to promote explainability. In this respect, EDC appears to strike a good balance in the trade-off between accuracy and explainability. We note that EDC performs on par with the state-of-the-art and outperforms the existing explainable methods.

5.2 UCI

The final set of experiments in this paper involves the evaluation of the EDC algorithm on real-world datasets. A list of UCI datasets that involve binary classification is selected. For each dataset, a 10-fold cross-validation approach is used, and the mean AUC scores (and standard deviations in parentheses) are reported.

For comparison to EDC, we have chosen the same set of algorithms as presented in Sect. 5.1.

Table 3 shows the results of the different classifiers in terms of average AUC on the test set for 10-fold cross-validation. When comparing the results of the

EDC algorithm to the results of the other ED-based classifiers, we observe that for all datasets, the EDC algorithm achieves a higher AUC. This shows that our approach outperforms the current state-of-the-art ED-based approaches on binary classification. Furthermore, as noted in the introduction, the produced model, which is a single equation, is more interpretable than the results obtained from the M4GP algorithm.

To illustrate the interpretability of our approach, one of the equations from the ADULT dataset is shown here:

$$0.75 - 1.27 \cdot \texttt{own-child} \cdot \texttt{education-num} + 3.37 \cdot \texttt{capitalgain} \\ +8.01 \cdot \exp(8.18 \cdot \texttt{married-civ-spouse})$$

Note that `own-child` and `married-civ-spouse` are essentially binary (from the one-hot encoding of nominal attributes). This makes the first summand a conditional negative term depending on `education-num` (a form of if-statement). Similarly, the last summand adds a constant term that depends on the binary value of `married-civ-spouse`: 8.01 if F, and $8.01 \cdot \exp(8.18) = 28,586$ if T.

When comparing the results to the other interpretable classification algorithms, we note that LDA achieves a higher score for 5 of the 9 datasets. This indicates that these datasets contain a linear decision boundary. We note that our EDC algorithm outperforms the simple decision tree algorithm in all datasets. When comparing the results to the state-of-the-art classification methods, we observe that the random forest method outperforms our approach for 6 of the 9 datasets. Furthermore, for the datasets IONOSPHERE and SONAR, the random forest, MLP, and SVM algorithms show substantially higher AUC scores compared to EDC. This would indicate a relationship between the input features and class that is not present in our building blocks.

Figure 4 shows a critical distance plot, as described by Demar [8], of the ranks for the different classifiers evaluated in this work. We identify three overlapping groups of classifiers: the first group consists of the MLP, RF, SVM, LDA and EDC algorithms. The second group contains the LDA, EDC, AMAXSC, and M4GP algorithms. The third and final group contains the AMAXSC, M4GP and decision tree algorithms. This plot shows that the current state-of-the-art classifiers outperform the prior classification algorithms. However, current state-of-the-art classification algorithms do not significantly outperform our proposed EDC algorithms.

Finally, it is important to acknowledge the considerable run time involved with EDC, compared to the other algorithms. The run time of the algorithm depends strongly on the number of examples in a dataset (in a linear fashion), as well as on the number of features. As an indication of the ballpark efficiency of EDC, the run time for the DIABETES dataset (a medium-size dataset) for the EDC algorithm is ~ 167 minutes.

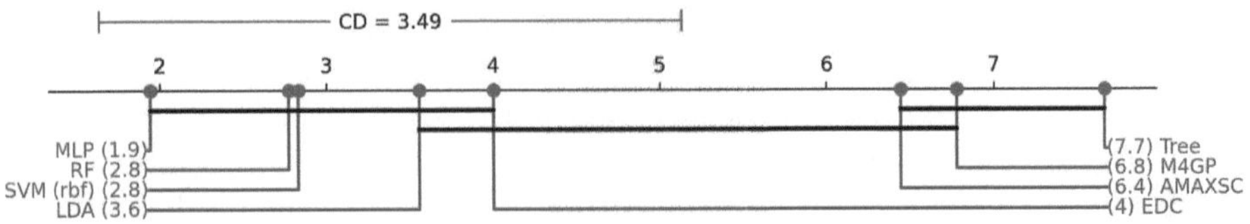

Fig. 4. Critical distance plot of the ranks for the different classifiers for the UCI datasets. The top bar shows the critical distance (CD), which in our setup equals 3.49. EDC outperforms AMAXSC, M4GP, and the decision tree, although not statistically significant. Similarly, MLP, RF, SVM, and LDA perform better but not statistically significantly.

Table 3. Results of the classifiers on the UCI datasets. All scores are the mean Area Under the Receiver Operator Curve (AUC) across 10 folds.

Dataset	EDC	AMAXSC	M4GP	LDA	Tree	MLP	RF	SVM
ADULT [3]	0.889	0.807	0.770	0.902	0.730	0.901	0.880	0.898
BANKNOTE [18]	1.000	0.982	0.999	1.000	0.979	1.000	1.000	1.000
BREAST [28]	0.670	0.617	0.614	0.636	0.590	0.701	0.683	0.709
CREDIT [22]	0.918	0.896	0.869	0.924	0.812	0.910	0.935	0.920
CYLINDER [9]	0.735	0.547	0.703	0.778	0.594	0.844	0.870	0.759
DIABETES	0.830	0.799	0.724	0.829	0.673	0.843	0.826	0.836
IONOSPHERE [25]	0.894	0.888	0.855	0.901	0.892	0.985	0.979	0.979
OCCUPANCY [6]	0.996	0.994	0.990	0.994	0.989	0.997	0.999	0.993
SONAR [24]	0.780	0.762	0.767	0.803	0.731	0.927	0.917	0.916
Average Score	0.857	0.810	0.810	0.863	0.777	0.901	0.899	0.890
Average Rank	4.00	6.44	6.78	3.56	7.67	1.89	2.78	2.78

6 Discussion and Conclusion

We have proposed a new classification algorithm based on equation discovery and thoroughly tested the method on artificial and real-life data (from the UCI Repository).

We evaluated whether the EDC algorithm was able to identify the correct equation on various artificial datasets, where the target equation is known; this was indeed the case. In experiments where the target equation comes from the same grammar as EDC uses, AUC scores on datasets free of noise were in the order of 0.1% from the perfect score. Further experiments with artificial data that *does* include noise showed an interesting phenomenon, where EDC statistically significantly outperformed the score of the target equation. We deliberately avoid the term ground truth, since apparently, as soon as you introduce noise (data noise, not label noise), the implicit decision boundary embedded in the data may shift slightly from the original boundary. For linear decision boundaries, this is very limited and only due to the random sample. However, in curved decision boundaries, it appears that the implicit boundary shifts towards the enclosed region. As a result, our algorithm was almost 1% more accurate than the original equation.

Further experimentation with more challenging target equations outside EDCs grammar (including cubic and quartic terms) demonstrated that EDC is able to approximate the intended decision boundary, again outperforming the original equation in the presence of noise. The results so far confirm that EDCs beam search, which essentially ignores a large part of the discrete search space of equations, is able to focus on those candidate equations that are the most promising, at least on the artificial data. In fact, the current settings of beam width $w = 10$ could perhaps be reduced in favour of shorter run times.

It should be noted that the discovered equations, despite scoring well, may not necessarily take on the same structure as the original. There apparently is some redundancy in our selected pattern language, such that different equations can still produce roughly the same decision boundary. Random effects in the sampling of the dataset or in the Gaussian noise allow some fuzziness in the boundary, which EDC sometimes exploits. For this reason, we have not included a grammatical validation between equations.

The final test concerned experiments on real-world data, for which typically no explicit decision boundary is known. The results indicate that EDC is able to compete with state-of-the-art methods, such as MLPs and RFs, and outperforms all other ED-based classification methods. Given EDC's demonstrated efficacy in a low-dimensional (artificial) dataset and its limited search depth, we hypothesise that EDC occasionally struggles when confronted with datasets that require a combination of many features, more than the maximum of six features our grammar imposes (e.g., $c_0 + c_1 \cdot x_0x_1 + c_2 \cdot x_2x_3 + c_3 \cdot x_4x_5$). Still, the scores in Table 3 demonstrate that this limitation does not frequently occur (most classification tasks apparently can be solved with relatively few features).

Finally, we must address the run time of the EDC algorithm. As stated, this is significantly longer than the algorithms used in our comparison, which have run times of < 1 second. The cause of this is the high number of function evaluations required in the search. Especially the building block $x_i \cdot x_j$ introduces a quadratic number of building blocks. This, combined with the feature expansion from one-hot encoding categorical features, results in long run times for the EDC algorithm.

Acknowledgements. This work was financially supported by research project "*Ship system expanded energy storage devices lifetime via AI-empowered control (SEANERGETIC)*" granted by Dutch research funding agency NWO.

References

1. Antonov, K., Kalkreuth, R., Yang, K., Bäck, T., Stein, N., Kononova, A.: A functional analysis approach to symbolic regression, p. 859–867. GECCO '24, Association for Computing Machinery (2024)

2. Augusto, D.A., Barbosa, H.J.: Symbolic regression via genetic programming. In: Proceedings. Vol. 1. Sixth Brazilian symposium on neural networks, pp. 173–178 (2000)
3. Becker, B., Kohavi, R.: Adult. UCI Machine Learning Repository (1996)
4. Billard, L., Diday, E.: Symbolic regression analysis. In: Classification, clustering, And Data Analysis: Recent Advances And Applications, pp. 281–288 (2002)
5. Boyd, S., Vandenberghe, L.: Convex Optim. Cambridge University Press, Cambridge, England (2004)
6. Candanedo, L.: Occupancy detection. UCI Machine Learning Repository (2016)
7. Cybenko, G., O'Leary, D.P., Rissanen, J.: The Mathematics of Information Coding. Springer, Extraction and Distribution (1999)
8. Demšar, J.: Statistical comparisons of classifiers over multiple data sets. J. Mach. Learn. Res. **7**, 1–30 (2006)
9. Evans, B.: Cylinder bands. UCI Machine Learning Repository (1994)
10. Flach, P.: Machine learning: the art and science of algorithms that make sense of data. Cambridge University Press (2012)
11. Haeri, M.A., Ebadzadeh, M.M., Folino, G.: Statistical genetic programming for symbolic regression. Appl. Soft Comput. **60**, 447–469 (2017)
12. Hosmer, D.W., Lemeshow, S.: Appl. Logistic Regression. John Wiley and Sons, Hobroken, NJ (2000)
13. Ingalalli, V., Silva, S., Castelli, M., Vanneschi, L.: A multi-dimensional genetic programming approach for multi-class classification problems. In: Genetic Programming, pp. 48–60. Springer (2014)
14. Korns, M.F.: An evolutionary algorithm for big data multi-class classification problems, pp. 165–178. Springer, Cham (2018)
15. Kumar, A., Vembu, S., Menon, A.K., Elkan, C.: Beam search algorithms for multilabel learning. Mach. Learn. **92**(1), 65–89 (2013). https://doi.org/10.1007/s10994-013-5371-6
16. Cava, W., Silva, S., Danai, K., Spector, L., Vanneschi, L., Moore, J.H.: Multidimensional genetic programming for multiclass classification. Swarm Evol. Comput. **44**, 260–272 (2019)
17. Levine, I.: Phys. Chem. University of Brooklyn, McGraw-Hill, New York, NY (1978)
18. Lohweg, V.: Banknote authentication. UCI Machine Learning Repository (2013)
19. Makke, N., Chawla, S.: Interpretable scientific discovery with symbolic regression: a review. Artif. Intell. Rev. **57**(1) (2024)
20. Meeng, M., Duivesteijn, W., Knobbe, A.: ROCsearch — an ROC-guided search strategy for subgroup discovery, pp. 704–712. SIAM (2014)
21. Muñoz, L., Silva, S., Trujillo, L.: M3GP – multiclass classification with GP. In: European Conference on Genetic Programming, pp. 78–91. Springer (2015)
22. Quinlan, J.R.: Credit approval. UCI Machine Learning Repository
23. Schwab, I., Link, N.: Reusable knowledge from symbolic regression classification. In: Proceeedings ICGEC, pp. 106–109 (2011)
24. Sejnowski, T., Gorman, R.: Connectionist bench (Sonar, Mines vs. Rocks). UCI Machine Learning Repository (1988)
25. Sigillito, V., Wing, S., Hutton, L., Baker, K.: Ionosphere. UCI Machine Learning Repository (1989)
26. Todorovski, L., Dzeroski, S.: Declarative bias in equation discovery,pp. 376–384. ICML '97, Morgan Kaufmann Publishers Inc. (1997)
27. Toussaint, G., Knobbe, A.: LMFD: Latent Monotonic Feature Discovery, pp. 19–36. ECML-PKDD, Springer (2025)
28. Zwitter, M., Soklic, M.: Breast cancer. UCI Machine Learning Repository (1988)

A Variational Autoencoder for N-Ary Trees

Martin Perčinić[1(✉)], Sebastian Mežnar[1,2], Ljupčo Todorovski[1,3], and Sašo Džeroski[1]

[1] Department of Knowledge Technologies, Jožef Stefan Institute, Jamova cesta 39, 1000 Ljubljana, Slovenia
martin.percinic@ijs.si

[2] International Postgraduate School Jožef Stefan, Jamova cesta 39, 1000 Ljubljana, Slovenia

[3] Faculty of Mathematics and Physics, University of Ljubljana, Jadranska ulica 21, 1000 Ljubljana, Slovenia

Abstract. We introduce nHVAE, a variational autoencoder designed to encode and decode n-ary trees. It extends the hierarchical variational autoencoder, HVAE, initially developed for binary trees. The elementary components of nHVAE are recurrent neural network cells arranged in a hierarchy that follows the structure of the training n-ary trees. The encoder operates bottom-up, and each encoder cell computes the state of the observed tree node from the states of (an arbitrary number of) its children. The decoder operates top-down and left-to-right: each decoder cell computes the new state of the current cell from the states of its parent and its left sibling and decides when to stop the generation of siblings. We empirically evaluate the performance of nHVAE on generating n-ary trees of mathematical expressions. The results show that nHVAE retains the HVAE performance in terms of reconstruction error, as well as its ability to learn from a modest number of training examples and operate in low-dimensional, smooth latent spaces. The ability of nHVAE to generate trees of arbitrary degree enables its application to various downstream tasks on trees, beyond symbolic regression, where HVAE has been used.

Keywords: Variational autoencoders · N-ary trees · Generative models · Mathematical expressions

1 Introduction

Autoencoders have become widely used for unsupervised learning of efficient vector representations (or embeddings) of data in latent spaces. These representations enable addressing various downstream tasks of supervised machine learning from arbitrarily structured data. The versatility of autoencoders has led to the development of quite a few varieties, from sparse autoencoders, which

S. Džeroski et al. (Eds.): DS 2025, LNAI 16090, pp. 143–157, 2025.
https://doi.org/10.1007/978-3-032-05461-6_10

activate only a few neurons at a time in the hidden layer to learn more meaningful features [4], through denoising autoencoders, which learn to remove noise from the input data [17], to contractive autoencoders that are trained to be robust to small changes in the input data, allowing the model to learn more consistent features [12].

Variational autoencoders (VAEs) [8] encode the input data into a probability distribution from which latent space embeddings of data can be sampled and passed to the decoder. This structure enables VAEs not only to embed and reconstruct the input data efficiently, but also to generate new data. Various types of VAEs have been developed, including the conditional variational autoencoder CVAE [14] for image classification and recognition, and the variational autoencoder by Hu et al. [7] for generating text as a sequence of tokens. VAEs for tree-structured, hierarchical data [11] enable addressing downstream tasks that operate on formal structured representations, such as mathematical expressions. An example of such a task is symbolic regression, also known as equation discovery [13,16]. While VAEs for sequences (text) can be, in principle, used to generate sequence-based representations of mathematical expressions, VAEs for hierarchical structures are far more efficient. They can be trained on smaller data sets and accurately embed input data into lower-dimensional latent spaces. The latter is very important, since it can significantly improve the performance on downstream machine learning tasks [11].

Note, however, that the hierarchical variational autoencoder HVAE can only operate on binary trees. While this simplifies the architecture of the HVAE building blocks, it also represents a serious limitation. First, we cannot apply HVAE to relevant and important downstream tasks, such as code generation [10]. Second, even in the case of tree representation of mathematical expressions, we should treat all operators as binary. While binary trees are commonly considered as a natural representation of expressions, the commutative and associative operators, e.g., addition and multiplication, are n-ary. We can always represent such an n-ary operator as a sequence of binary operations (e.g., a sum is a series of binary additions) with algorithms such as the shunting yard algorithm [1], but this comes at the cost of having deeper and more imbalanced expression trees.

To address this limitation of HVAE, we propose a novel variational autoencoder architecture, nHVAE, for encoding and decoding n-ary trees. Following the recursive architecture of the HVAE, each cell of nHVAE recursively computes the corresponding tree node's hidden state from the neighboring nodes' hidden states. The cell is a generalization of the gated recurrent units (GRUs). The encoding proceeds bottom-up, from the leaf nodes to the tree root, where each GRU cell computes the node's hidden state from the states of its children. nHVAE performs decoding top-down, from the root node to the leaves. The GRU cell computes the node's hidden state from the states of its parent and its immediate left sibling (decoded in the previous step). At the same time, a separate network determines that the generated child node is the last, right-most child of its parent.

2 Related Work

Syntactic validity does away with the need to remove syntactically invalid generated expressions, which, in turn, enables faster generation of expressions. Syntactic validity also enhances the model's suitability for the given task. Consequently, variational autoencoder models have been developed with this property in mind. The grammar variational autoencoder GVAE [9] uses context-free grammars to ensure the syntactic validity of the output, representing each example as a sequence of production rules from the grammar used to derive it. Its decoder is used to generate valid sequences of production rules, which are applied to obtain the generated data. Another example is the syntax-directed variational autoencoder SD-VAE [5], which uses attribute grammars to ensure both the syntactic and semantic validity of the generated output.

For hierarchical data, several models have been proposed, most of which extend existing recurrent neural networks (RNNs) to tree-structured topologies. These models are commonly referred to as recursive neural networks. Tai et al. [15] introduce two such encoder models: the Child-Sum Tree-LSTM and the N-ary Tree-LSTM. Both models are able to encode n-ary trees in a bottom-up manner, with the Child-Sum Tree-LSTM adding the children's hidden states into a single state for the LSTM, while the N-ary Tree-LSTM uses separate parameter matrices for each child. The reduced number of trainable parameters makes the Child-Sum Tree-LSTM more suitable for trees with larger branching factor. On the basis of the Child-Sum Tree-LSTM, Zhou et al. [20] propose a Child-Sum Tree-GRU architecture, which extends the GRU cell similarly to how the Child-Sum Tree-LSTM extends the LSTM.

Looking at decoder/generating models, the TreeLSTM proposed by Zhang et al. [19] is able to generate dependency trees, making it well-suited for dependency parsing and text generation. A key challenge of an n-ary tree decoder is decoding a varying number of different child hidden states from a single parent. The doubly-recurrent neural network (DRNN) [2] answers this challenge by utilising two separate RNNs for handling the ancestral and fraternal flow of information, respectively. The DRNN uses the hidden state from both the parent and the left sibling to compute the hidden state of a node. It also provides mechanisms for topological prediction that stop the generation of further siblings and the generation of children, bypassing the need for stopping symbols. The SEQ2TREE decoder [6] on the other hand, implements symbols for stopping the decoding process and for starting a new branch.

3 Methodology

We introduce the nHVAE, a variational autoencoder tailored for efficient encoding and decoding of n-ary trees. Section 3.1 reviews the architecture of the hierarchical variational autoencoder (HVAE). Section 3.2 presents the nHVAE model, delineates its structural deviations from HVAE, and outlines the corresponding training procedure.

3.1 Hierarchical Variational Autoencoder

Figure 1 illustrates the process of encoding and decoding an example expression binary tree with HVAE. The left-hand side of the figure illustrates the encoder, while the right-hand side illustrates the decoder.

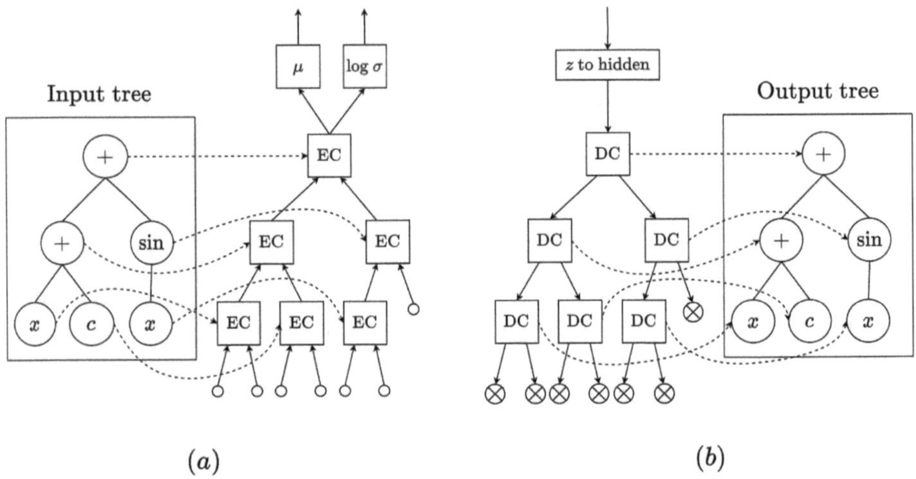

Fig. 1. The encoding (a) and decoding (b) process of HVAE when dealing with the binary expression tree for the expression $x + c + \sin x$.

The encoding proceeds bottom-up through the expression tree, using a single recurrent unit—the encoding cell (EC)—to compute the hidden state at each node. The EC is a modified GRU cell, denoted GRU221, which accepts two hidden state vectors as input rather than one. In the HVAE framework, the EC receives the hidden states of a node's children and the embedding of the node's symbol to compute its hidden state. For leaf nodes or nodes representing unary functions (e.g., sine), zero vectors are supplied as pseudo-children. Upon reaching the root, its hidden state is transformed via two distinct fully connected layers to produce the mean μ and log-variance $\log\sigma$ of a Gaussian distribution. A latent vector is then sampled from this distribution and passed to the decoder.

Decoding proceeds top-down from the root node. The sampled latent vector is first mapped through a fully connected layer to generate the initial hidden state at the root. A decoding cell (DC), implemented as a GRU122 unit, recursively decodes the tree. This cell takes the parent's hidden state as input and outputs a vector twice the hidden size, representing the concatenated hidden states of the left and right child. This output is split and propagated downward to the respective child nodes. Additionally, the DC decodes the symbol for the current node from its hidden state. In the context of expression trees, the decoded symbol determines termination: internal nodes correspond to operators or functions, whereas leaves represent variables or constants, signaling the end of decoding.

3.2 Variational Autoencoder for N-Ary Trees

Expression trees often contain operators with more than two arguments. Most commonly, these are the addition and multiplication operators, which appear in most expressions. Due to their associativity, in the expression tree representation, nodes that represent these operations can have more than two children without causing any ambiguity over which expression the tree represents. For this reason, we are able to easily upgrade HVAE to encode and decode n-ary trees and propose the variational autoencoder for n-ary trees nHVAE as that upgrade.

Encoder The nHVAE encoder, same as the HVAE encoder, performs the encoding from the bottom to the top. Its structure is shown in Fig. 2. The encoder has an encoding cell at each node, capable of taking any number of hidden states as its input. Compared to the HVAE encoder, the encoding cells for the leaves only receive one vector of zeros, and the encoding cells for nodes with unary function symbols do not require any additional states for their inputs.

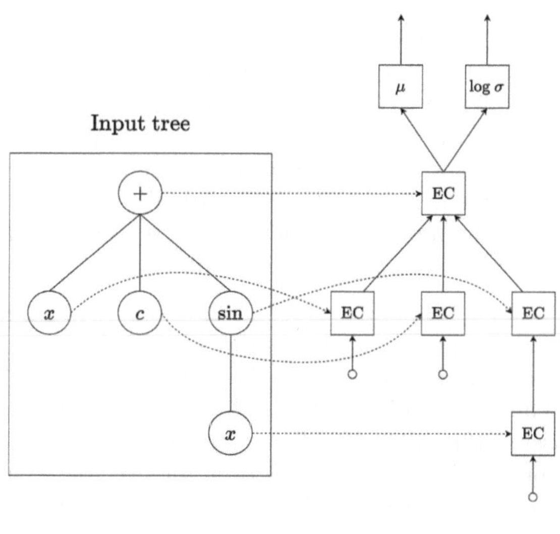

Fig. 2. The nHVAE encoder working on the expression "$x + c + \sin x$".

For the encoding cell, we use a slightly modified version of the Child-Sum Tree-GRU cell, defined with the following equations:

$$\tilde{h} = W_E(\sum_{k=1}^{n} W_k h_k) \tag{1}$$

$$r_k = \sigma(W_{xr}x + W_{hr}h_k), \quad k = 1, \ldots, n \tag{2}$$

$$z = \sigma(W_{xz}x + W_{hz}\tilde{h}) \tag{3}$$

$$n = \tanh(W_{xn}x + W_{hn}(\sum_{k=1}^{n} r_k \odot h_k)) \tag{4}$$

$$h = (\vec{1} - z) \odot n + z \odot \tilde{h}, \tag{5}$$

where $\odot$ denotes element-wise multiplication. The Child-Sum Tree-GRU cell computes the sum of the children's hidden states $h_1, h_2, \ldots, h_n$ and uses this sum and a one-hot encoding x of a node's symbol to compute the output of the update gate (Eq. 3) as well as during the final computation of its hidden state (Eq. 5). The cell also contains separate reset gates for each child, allowing it to selectively store only the hidden states of certain children. Our modification of the Child-Sum Tree-GRU cell adds weight matrices W_E and $W_k, k = 1, 2, \ldots, n$ in Eq. 1, which give different weights to the children. As a result, each child has a different influence on the update gate output. In Eqs. 1- 4, for conciseness, the neural networks' biases are already included by adding another column of parameters to the weight matrices (denoted with W and an index) and adding a row containing 1 to the input and hidden state vectors. In further equations, the biases will also be included in the weight matrix.

Decoder. The decoding process with the nHVAE decoder is performed from the top to the bottom and is shown in Fig. 3. At each node, a decoding cell decodes the node's symbol and the hidden states of the node's children. A node's symbol is decoded by passing the node's hidden state through a fully connected layer and obtaining a probability distribution over all possible symbols. From the distribution, the symbol with the highest probability is chosen as the decoded symbol.

Similarly to the DRNN, the decoding cell receives information from a node's parent and its left sibling, if there is one. The decoding process is therefore not only from the top of the tree to the bottom, but also from left to right. For the decoding cell, we repurpose the GRU221 cell from the HVAE encoder, providing it with both the symbol probability distribution and the hidden state from a node's parent and left sibling. The children's hidden states $h_i^{(d)}, i = 1, \ldots, n$ are computed with the formulas:

$$h_1^{(d)} = W_D(\text{GRU221}(x^{(a)}, \vec{0}, h^{(a)}, \vec{0})) \tag{6}$$

$$h_i^{(d)} = W_D(\text{GRU221}(x^{(a)}, x_{i-1}^{(d)}, h^{(a)}, h_{i-1}^{(d)})), \quad i = 2, \ldots, n. \tag{7}$$

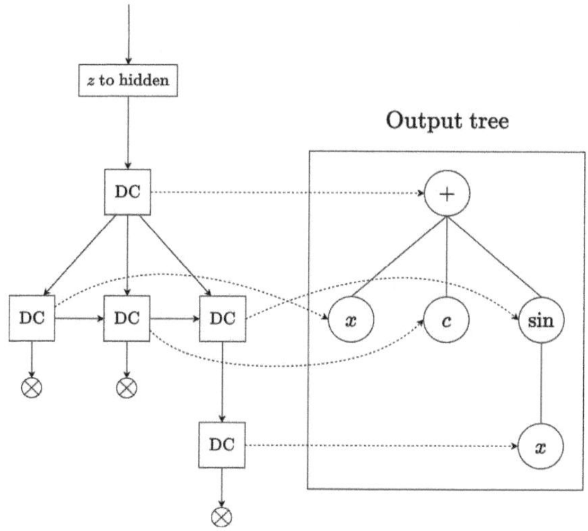

Fig. 3. The nHVAE decoder generating the expression "$x + c + \sin x$".

For each child, the GRU cell receives the symbol distribution $x^{(a)}$ and hidden state $h^{(a)}$ from the parent. For children with a left sibling, the GRU cell also receives the symbol distribution $x_{i-1}^{(d)}$ and hidden state $h_{i-1}^{(d)}$ from the left sibling, whereas for the first child, the GRU cell receives a vector of zeros instead. Our version of the GRU221 cell receives an additional symbol distribution and is defined with the following formulas:

$$r = \sigma(W_{xr}(x^{(a)} \parallel x_{i-1}^{(d)}) + W_{hr}(h^{(a)} \parallel h_{i-1}^{(d)})) \tag{8}$$

$$z = \sigma(W_{xz}(x^{(a)} \parallel x_{i-1}^{(d)}) + W_{hz}(h^{(a)} \parallel h_{i-1}^{(d)})) \tag{9}$$

$$n = \tanh(W_{xn}(x^{(a)} \parallel x_{i-1}^{(d)}) + r \odot W_{hn}(h^{(a)} \parallel h_{i-1}^{(d)})) \tag{10}$$

$$h = (\vec{1} - z) \odot n + \frac{z}{2} \odot h^{(a)} + \frac{z}{2} \odot h_{i-1}^{(d)}, \tag{11}$$

where $\parallel$ is the concatenation operator. For the first child, the symbol distribution and hidden state of the left sibling in the above equations are replaced with vectors of zeros.

We use a separate fully connected layer to halt the generation of siblings. The probability of continuing the generation after i siblings have been generated is computed with the formula:

$$p_{i+1} = \sigma(W_p h_i^{(d)}). \tag{12}$$

If $p_{i+1} \geq 0.5$, another sibling to the right is generated, otherwise the sibling generation process is stopped.

Training. When training a variational autoencoder, we want to maximize the reconstruction accuracy of the decoded tree while keeping the latent space smooth, so that similar trees get encoded to similar points in the latent space and vice versa. Our loss function consists of two terms: cross-entropy for the reconstruction accuracy and the Kullback-Leibler (KL) divergence for regularization.

To compute the reconstruction accuracy, we are interested in two facets of our decoder: correctly decoding the symbol of a node and correctly predicting if a node has a right sibling. For this reason, we also split the cross entropy into two computations. During training, we require the decoded tree to have the same structure as the input tree. This allows us to easily compute the cross-entropy loss for symbol prediction with the formula:

$$\mathrm{CE}_s(p, q) = -\frac{1}{N}\sum_{i=1}^{N}\sum_{s\in S} p_i(s)\log q_i(s), \tag{13}$$

where for each node at the same position in both trees, and for each possible symbol, we calculate the probability of that node having that symbol in the input tree ($p_i(s)$) and in the decoded tree ($q_i(s)$). The sibling prediction accuracy is computed with binary cross-entropy:

$$\mathrm{BCE}(\hat{p}, \hat{q}) = -\frac{1}{K}\sum_{i=1}^{K}\left[\hat{p}_i \log \hat{q}_i + (1-\hat{p}_i)\log(1-\hat{q}_i)\right], \tag{14}$$

where $\hat{p}_i$ is the probability that node i has a right sibling in the input tree and $\hat{q}_i$ is the probability that it has a right sibling in the decoded tree. In this computation, we only look at nodes that have $+$ and $\cdot$ as their parent (since these nodes are the only ones that do not have a fixed number of children), and, out of these, we only look at the nodes that have a left sibling (since the leftmost node always has a right sibling, as addition and multiplication require at least two terms). We compute the reconstruction accuracy by combining both computations:

$$\mathrm{CE}(p, q, \hat{p}, \hat{q}) = \mathrm{CE}_s(p, q) + \alpha \cdot \mathrm{BCE}(\hat{p}, \hat{q}). \tag{15}$$

We selected the value $\alpha = 2$ experimentally. This value leads to a larger focus on sibling prediction, which is necessary for our model, since incorrectly predicting the absence of a sibling generally lowers the reconstruction accuracy more than incorrectly predicting a single symbol.

To compute the Kullback-Leibler divergence, we use the mean and variance parameters from the encoder output with the formula:

$$D_{KL}(\mu, \sigma) = \frac{1}{2}(\sigma + \mu^2 - 1 - \log(\sigma)). \tag{16}$$

The loss function is obtained from the cross-entropy and KL divergence with the formula:

$$RL(p, q, \mu, \sigma) = \mathrm{CE}(p, q) + \beta \cdot D_{KL}(\mu, \sigma). \tag{17}$$

The regularization parameter β is updated at each iteration of the training with the function:

$$\beta = \frac{0.045}{1 + 50 \cdot e^{-10 \cdot \frac{i}{L}}}, \tag{18}$$

where i is the number of completed iterations and L the total number of iterations of the training process. This function is strictly increasing, meaning that at the beginning of the training, more focus is put on maximization of the reconstruction accuracy, and gradually, focus shifts more towards regularization.

4 Evaluation

This section presents the results of our experimental evaluation. Section 4.1 describes the data sets used and the data generation procedure. In Sect. 4.2, we assess the reconstruction error of nHVAE as compared to HVAE, while Sect. 4.3 analyzes the smoothness properties of the learned latent space. The implementation of nHVAE and the data sets used for evaluation can be found at https://github.com/mpercinic/nHVAE.

4.1 Generating Data Sets of Expression Trees

For evaluation, we generated four data sets comprising n-ary expression trees containing 2,000, 15,000, 20,000, and 40,000 samples, respectively, each limited to a maximum of 25 nodes. The trees were constructed by first generating expressions in infix notation using Monte Carlo sampling with a probabilistic context-free grammar, implemented via ProGED [3], and subsequently converting these expressions into tree form. The grammar used for expression generation is defined as follows:

$$\begin{aligned}
E &\rightarrow E + F \;\; [0.2] \mid E - F \;\; [0.2] \mid F \;\; [0.6] \\
F &\rightarrow F \cdot T \;\; [0.2] \mid F/T \;\; [0.2] \mid T \;\; [0.6] \\
T &\rightarrow V \;\; [0.4] \mid C \;\; [0.3] \mid A \;\; [0.3] \\
V &\rightarrow x \;\; [1.0] \\
A &\rightarrow E(P) \;\; [0.1] \mid R(E) \;\; [0.35] \mid (E) \;\; [0.55] \\
P &\rightarrow \hat{}2 \;\; [0.8] \mid \hat{}3 \;\; [0.2] \\
R &\rightarrow \text{sin} \;\; [0.2] \mid \text{cos} \;\; [0.2] \mid e\hat{} \;\; [0.2] \mid \text{sqrt} \;\; [0.2] \mid \text{log} \;\; [0.2].
\end{aligned}$$

The generated expressions are converted into n-ary expression trees using a modified version of the shunting yard algorithm. While the standard algorithm is typically employed to convert infix expressions to postfix notation or, as in HVAE, to construct binary expression trees, our variant generalizes it to produce n-ary trees. In addition to the expression, our algorithm accepts an integer parameter specifying the desired tree arity. For associative operators such as +

and $\cdot$, whose corresponding nodes can admit multiple children, the algorithm greedily forms nodes with as many children as permitted by the specified arity. For instance, the expression $x_1 + x_2 - x_3 + x_4 \cdot x_5$ would be transformed into a tree rooted at + with three children: x_1, $x_2 - x_3$, and $x_4 \cdot x_5$. This heuristic increases the frequency of higher-arity nodes in the data sets.

4.2 Reconstruction Error

To evaluate reconstruction performance, we compute the tree edit distance between the input and decoded expression trees using the algorithm by Zhang and Shasha [18]. We assess nHVAE across all four data sets using five-fold cross-validation, reporting the mean and variance of the reconstruction error. The latent space dimensionality is set to 128, with expression trees of arity 5. Training is conducted over 20 epochs with a batch size of 32. For comparison, HVAE is evaluated under identical conditions. The results are summarized in Table 1 and list data sets by the number of expression trees they contain.

Table 1. The reconstruction error and variance of the nHVAE and HVAE for the four data sets.

Data set	nHVAE	HVAE
2k	5.184 (± 0.203)	4.734 (± 0.186)
15k	3.437 (± 0.090)	2.795 (± 0.081)
20k	3.268 (± 0.127)	2.711 (± 0.061)
40k	2.848 (± 0.083)	2.161 (± 0.080)

Empirically, HVAE achieves slightly lower reconstruction error than nHVAE across all data sets. This performance gap stems from nHVAE's additional objective of predicting sibling structure, whereas HVAE focuses solely on symbol prediction. The sibling prediction task introduces a higher risk of compounding errors—e.g., predicting the absence of a right sibling leads to omission of the entire corresponding subtree—thereby inflating the edit distance. Nevertheless, the observed difference in reconstruction error remains below 1 in all cases, indicating that, on average, nHVAE-generated trees differ from their inputs by fewer than one additional edit operation (insertion, deletion, or substitution) compared to those reconstructed by HVAE.

We further investigate the effect of training set size and latent space dimensionality on reconstruction performance. Specifically, we evaluate the reconstruction error of both HVAE and nHVAE on the 40k data set under varying training set sizes and latent space dimensions. Figure 4 presents the results. Consistent with earlier observations, HVAE marginally outperforms nHVAE across all configurations. Notably, both models exhibit similar trends with respect to data and model size: the reconstruction error difference remains relatively constant as training set size and latent dimensionality vary. The smallest performance gap is observed at lower training set sizes and reduced latent dimensions, suggesting that nHVAE maintains competitive reconstruction capabilities even under data-scarce conditions or when operating with compact latent representations.

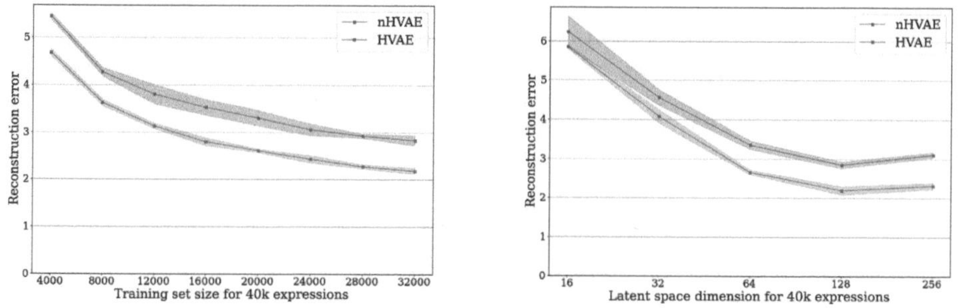

Fig. 4. The reconstruction error obtained with training the autoencoders on the 40k data set for different training set sizes (left-hand side) and different dimensions of the latent space (right-hand side).

4.3 Latent Space Smoothness

Beyond accurately reconstructing the input trees, a key desired property of nHVAE is the smoothness of its latent space, i.e., structurally similar expression trees should be encoded into nearby latent representations. Such continuity facilitates latent space exploration, which is particularly valuable in tasks like symbolic regression, where traversing the latent space can yield semantically meaningful variations of expressions.

To assess this property, we conduct linear interpolation experiments between pairs of expression trees. Given two expressions, we encode their respective trees into latent vectors a and b, and generate intermediate points using the interpolation formula:

$$z_i = (1 - \lambda_i) \cdot a + \lambda_i \cdot b, \qquad \lambda_i = \frac{i-1}{n-1} \tag{19}$$

for $i = 1, 2, \ldots, n$, where n is the number of interpolation steps.

If the latent space exhibits sufficient smoothness, decoding the sequence of interpolated latent vectors—ordered as $a, z_1, z_2, \ldots, z_n, b$—should yield a progression of expression trees that transition gradually from the source expression A (corresponding to a) to the target expression B (corresponding to b). In such a case, intermediate decodings would represent meaningful interpolants between the original expressions.

Table 2 presents representative examples of linear interpolation results for HVAE and nHVAE, illustrating the gradual transformation of expressions across the latent space trajectory. Note that, in Example 2, nHVAE does not correctly decode expression B when $\lambda_5 = 1$, whereas HVAE correctly decodes expressions A and B in all three examples. This result further corroborates the results obtained in our evaluation of the reconstruction error.

Table 2. Examples of linear interpolation between two expressions with nHVAE (on the top) and HVAE (on the bottom). The first and last rows in each half show the source and target expressions, respectively. Each intermediate row corresponds to a decoded expression Z_i obtained from a latent vector $z_i = (1 - \lambda_i) \cdot a + \lambda_i \cdot b$ for the corresponding value of λ_i.

	λ_i	Example 1	Example 2	Example 3
	A	$A:\ x \cdot \frac{x}{c} \cdot \cos(c \cdot x)$	$A:\ \frac{\log x}{x+x}$	$A:\ \log(\frac{c \cdot c}{x}) - x - x^2$
	$\lambda_1 = 0$	$Z_1:\ x \cdot \frac{x}{c} \cdot \cos(c \cdot x)$	$Z_1:\ \frac{\log x}{x+x}$	$Z_1:\ \log(\frac{c \cdot c}{x}) - x - x^2$
	$\lambda_2 = 0.25$	$Z_2:\ \frac{x}{x} \cdot \frac{x}{c} \cdot c$	$Z_2:\ \frac{\log x}{x+x}$	$Z_2:\ c - x - x^2$
nHVAE	$\lambda_3 = 0.5$	$Z_3:\ \frac{x}{c} \cdot \frac{x}{c}$	$Z_3:\ \log(c \cdot c) \cdot x$	$Z_3:\ c - x + x \cdot c - c$
	$\lambda_4 = 0.75$	$Z_4:\ x - \frac{c+x+c}{c} + x$	$Z_4:\ \sqrt{c \cdot c} \cdot x$	$Z_4:\ c + x \cdot c - x \cdot x$
	$\lambda_5 = 1$	$Z_5:\ x - \frac{c+x+c}{c} - x$	$Z_5:\ \sqrt{c \cdot \sin x} \cdot x$	$Z_5:\ c + x \cdot c - x \cdot x$
	B	$B:\ x - \frac{c+x+c}{c} - x$	$B:\ \sqrt{c \cdot \sin x} \cdot x \cdot x$	$B:\ c + x \cdot c - x \cdot x$
	A	$A:\ x \cdot \frac{x}{c} \cdot \cos(c \cdot x)$	$A:\ \frac{\log x}{x+x}$	$A:\ \log(\frac{c \cdot c}{x}) - x - x^2$
	$\lambda_1 = 0$	$Z_1:\ x \cdot \frac{x}{c} \cdot \cos(c \cdot x)$	$Z_1:\ \frac{\log x}{x+x}$	$Z_1:\ \log(\frac{c \cdot c}{x}) - x - x^2$
	$\lambda_2 = 0.25$	$Z_2:\ x \cdot \frac{x}{c} \cdot \cos(c \cdot x)$	$Z_2:\ \frac{\frac{x}{x}}{x+x}$	$Z_2:\ \log(\frac{c}{x}) - x - c$
HVAE	$\lambda_3 = 0.5$	$Z_3:\ \frac{x}{c} - x$	$Z_3:\ \frac{c}{x} \cdot x$	$Z_3:\ \frac{c}{c} - x - x \cdot c$
	$\lambda_4 = 0.75$	$Z_4:\ x - \frac{c+x+c}{c} - x$	$Z_4:\ \sqrt{c} \cdot x \cdot x$	$Z_4:\ c + x \cdot c - x \cdot x$
	$\lambda_5 = 1$	$Z_5:\ x - \frac{c+x+c}{c} - x$	$Z_5:\ \sqrt{c \cdot \sin x} \cdot x \cdot x$	$Z_5:\ c + x \cdot c - x \cdot x$
	B	$B:\ x - \frac{c+x+c}{c} - x$	$B:\ \sqrt{c \cdot \sin x} \cdot x \cdot x$	$B:\ c + x \cdot c - x \cdot x$

Figure 5 shows the binary and n-ary tree representations of the equations in Example 1 for HVAE and nHVAE, respectively. As expected, the n-ary trees are wider than the binary ones and are also not as deep. We see that nHVAE has smooth transitions between the expression trees, with a huge part of the tree at one step being present at the next. In the case of HVAE, while there is no difference between the trees at the first and second, as well as at the penultimate and last steps, there is still a smooth transition across the entire interpolation. We observe that in the case of nHVAE, the interpolation process easily removes a child from an n-ary node in the third step and easily adds an n-ary node with three children to the tree in the fourth step. This behavior confirms that the distance between points in the latent space corresponds to the tree edit distance for n-ary trees in the original space.

From these examples, we observe smooth transitions between successive expressions, indicating that nHVAE learns a continuous and syntactically coherent latent space.

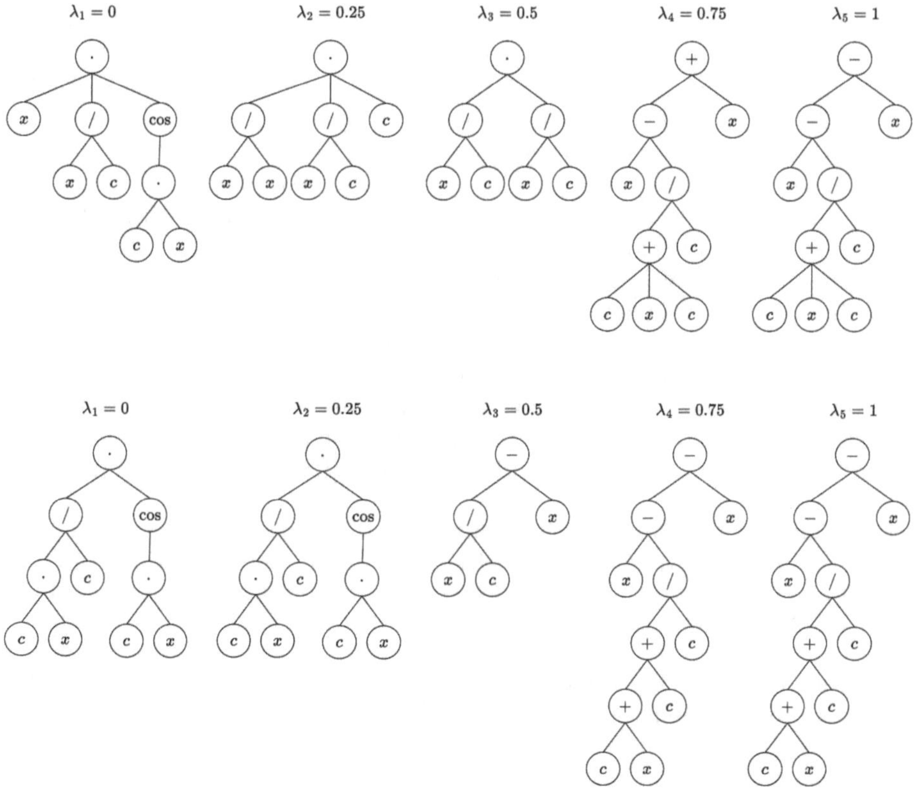

Fig. 5. The first linear interpolation example from Table 2, represented with n-ary expression trees for nHVAE in the top row, and with binary expression trees for HVAE in the bottom row.

5 Conclusion

We introduced nHVAE, a novel variational autoencoder architecture for n-ary trees that extends the hierarchical variational autoencoder (HVAE), previously limited to binary trees. The model incorporates modified gated recurrent units to encode and decode nodes with arbitrary degrees. During decoding, nodes are generated from left to right, conditioned on their left siblings, enabling the model to jointly perform symbol prediction and sibling prediction. For the latter, it is necessary to implement a halting mechanism.

In our experiments, nHVAE demonstrated slightly higher reconstruction error than HVAE, which was an expected outcome due to its additional prediction task. Nevertheless, both models exhibited comparable behavior across various training set sizes and latent space dimensions. Notably, nHVAE maintained robust performance under low-data and low-dimensional conditions. Furthermore, it learns a smooth latent space where similar expression trees map to nearby points, making it amenable to interpolation and latent space search, especially useful in symbolic regression.

The primary advantage of generalizing to n-ary trees lies in the expanded applicability of nHVAE to domains where tree structures of variable arity are prevalent. This includes code generation and completion, where abstract syntax trees (ASTs) frequently involve higher-arity nodes. For example, Liventsev et al. [10] demonstrate a variational autoencoder tailored to code synthesis. Beyond program synthesis, nHVAE may be a foundation for parse tree generation in NLP and representation learning in scientific domains, such as molecular modeling.

As future work, we plan to explore the integration of richer semantic priors into the latent space, refine the halting mechanism for better sibling prediction, and extend the model to accommodate trees with heterogeneous node types and edge annotations. Additionally, evaluating the effectiveness of nHVAE in downstream tasks such as symbolic regression, code completion, and molecular property prediction will further validate its generality and practical impact.

Acknowledgments. This research was funded by the Slovenian Research Agency via the research program "Knowledge Technologies" (P2-0103) and the Gravity project "AI for science" (GC-0001), as well as the European Commission via the project ELIAS (grant No. 101120237).

Disclosure of Interests. The authors have no competing interests to declare that are relevant to the content of this article.

References

1. An algol 60 translator for the x1. Ann. Rev. Autom. Program. **3**, 329–345 (1963). https://doi.org/10.1016/S0066-4138(63)80015-4, https://www.sciencedirect.com/science/article/pii/S0066413863800154, annual Review in Automatic Programming
2. Alvarez-Melis, D., Jaakkola, T.S.: Tree-structured decoding with doubly-recurrent neural networks. In: International Conference on Learning Representations (2017). https://openreview.net/forum?id=HkYhZDqxg
3. Brence, J., Todorovski, L., Džeroski, S.: Probabilistic grammars for equation discovery. Knowl.-Based Syst. **224**, 107077 (2021). https://doi.org/10.1016/j.knosys.2021.107077, https://www.sciencedirect.com/science/article/pii/S0950705121003403
4. Bricken, T., et al.: Towards monosemanticity: decomposing language models with dictionary learning. Transformer Circ. Thread **2** (2023)
5. Dai, H., Tian, Y., Dai, B., Skiena, S., Song, L.: Syntax-directed variational autoencoder for structured data. In: International Conference on Learning Representations (2018). https://openreview.net/forum?id=SyqShMZRb
6. Dong, L., Lapata, M.: Language to logical form with neural attention. In: Erk, K., Smith, N.A. (eds.) Proceedings of the 54th Annual Meeting of the Association for Computational Linguistics (Volume 1: Long Papers), pp. 33–43. Association for Computational Linguistics, Berlin, Germany (2016).https://doi.org/10.18653/v1/P16-1004, https://aclanthology.org/P16-1004/

7. Hu, Z., Yang, Z., Liang, X., Salakhutdinov, R., Xing, E.P.: Toward controlled generation of text. In: Proceedings of the 34th International Conference on Machine Learning - Volume 70, pp. 1587–1596. ICML'17, JMLR.org (2017)
8. Kingma, D.P., Welling, M.: Auto-Encoding Variational Bayes. In: 2nd International Conference on Learning Representations, ICLR 2014, Banff, AB, Canada, April 14-16, 2014, Conference Track Proceedings (2014)
9. Kusner, M.J., Paige, B., Hernández-Lobato, J.M.: Grammar variational autoencoder. In: Proceedings of the 34th International Conference on Machine Learning - Volume 70, pp. 1945–1954. ICML'17, JMLR.org (2017)
10. Tree variational autoencoder for code: Liventsev, V., Bruin, S.d., Härmä, A., Petković, M. IEEE Access **13**, 30262–30273 (2025). https://doi.org/10.1109/ACCESS.2025.3532522
11. Mežnar, S., Džeroski, S., Todorovski, L.: Efficient generator of mathematical expressions for symbolic regression. Mach. Learn. **112**(11), 4563–4596 (2023). https://doi.org/10.1007/s10994-023-06400-2
12. Rifai, S., Vincent, P., Muller, X., Glorot, X., Bengio, Y.: Contractive auto-encoders: explicit invariance during feature extraction. In: Proceedings of the 28th International Conference on International Conference on Machine Learning, pp. 833–840. ICML'11, Omnipress, Madison, WI, USA (2011)
13. Schmidt, M., Lipson, H.: Distilling free-form natural laws from experimental data. Science **324**(5923), 81–85 (2009). https://doi.org/10.1126/science.1165893, https://www.science.org/doi/abs/10.1126/science.1165893
14. Sohn, K., Yan, X., Lee, H.: Learning structured output representation using deep conditional generative models. In: Proceedings of the 29th International Conference on Neural Information Processing Systems - Volume 2, pp. 3483–3491. NIPS'15, MIT Press, Cambridge, MA, USA (2015)
15. Tai, K.S., Socher, R., Manning, C.D.: Improved semantic representations from tree-structured long short-term memory networks. In: Zong, C., Strube, M. (eds.) Proceedings of the 53rd Annual Meeting of the Association for Computational Linguistics and the 7th International Joint Conference on Natural Language Processing (Volume 1: Long Papers), pp. 1556–1566. Association for Computational Linguistics, Beijing, China (2015).https://doi.org/10.3115/v1/P15-1150, https://aclanthology.org/P15-1150/
16. Todorovski, L.: Equation Discovery, pp. 410–414. Springer US, Boston, MA (2017).https://doi.org/10.1007/978-1-4899-7687-1_258
17. Vincent, P., Larochelle, H., Bengio, Y., Manzagol, P.A.: Extracting and composing robust features with denoising autoencoders. In: Proceedings of the 25th International Conference on Machine Learning. p. 1096–1103. ICML '08, Association for Computing Machinery, New York, NY, USA (2008).https://doi.org/10.1145/1390156.1390294, https://doi.org/10.1145/1390156.1390294
18. Zhang, K., Shasha, D.: Simple fast algorithms for the editing distance between trees and related problems. SIAM J. Comput. **18**(6), 1245–1262 (1989).https://doi.org/10.1137/0218082, https://doi.org/10.1137/0218082
19. Zhang, X., Lu, L., Lapata, M.: Top-down tree long short-term memory networks. In: North American Chapter of the Association for Computational Linguistics (2015). https://api.semanticscholar.org/CorpusID:214079
20. Zhou, Y., Liu, C., Pan, Y.: Modelling sentence pairs with tree-structured attentive encoder. In: Matsumoto, Y., Prasad, R. (eds.) Proceedings of COLING 2016, the 26th International Conference on Computational Linguistics: Technical Papers, pp. 2912–2922. The COLING 2016 Organizing Committee, Osaka, Japan (2016). https://aclanthology.org/C16-1274/

Model Design and Optimization: Symbolic Learning, Decision Trees, and NNs

On the Potential of Deep Symbolic Models for Classification Problems

Florian Beck[1](✉), Johannes Fürnkranz[1,2], and Van Quoc Phuong Huynh[2]

[1] LIT Artificial Intelligence Lab, Linz, Austria
fbeck@faw.jku.at
[2] Institute for Application-oriented Knowledge Processing (FAW), Johannes Kepler University Linz, Altenberger Straße 66b/69, Linz4040, Austria
{juffi,vqphuynh}@faw.jku.at

Abstract. With the rise of neural network approaches for machine learning problems, the focus has shifted to learning deep concepts across multiple layers. Although single-hidden-layer networks are sufficient to arbitrarily approximate any function, multilayer networks have proven to be superior in predictive performance for many applications. Unlike for neural networks, in symbolic machine learning approaches like decision tree or rule learning algorithms, the benefits of hidden layers of learned intermediate concepts remain uncertain. In this work, we empirically investigate the potential gains of deep concepts for symbolic approaches from three perspectives. First, we compare the number of possible flat and deep Boolean expressions with limited complexity, underlining the higher expressive power of deep models in such a setting. Second, we use logic minimization algorithms to generate minimal flat and deep Boolean formulas for artificial Boolean classification problems with different numbers of attributes and training examples, showing under which circumstances the use of deep concepts can lead to noticeably less complex models. Third, we compare the predictive performance of flat and deep models with a fixed maximum complexity on these datasets. We interpret these results as evidence that encourages further investigation of algorithms for learning complexity-bounded deep rule sets.

Keywords: deep rule learning · learning in logic · logic minimization

1 Introduction

Deep learning has dramatically improved the state-of-the-art in speech recognition, visual object recognition, object detection and many other domains such as drug discovery and genomics [15]. The key idea is to learn representations of data with multiple levels of abstraction, essentially splitting the task into easier sub-tasks, so that the resulting sub-models can subsequently be combined to the main model.

In contrast to this, symbolic state-of-the-art learners such as rule learners still learn descriptions that directly relate the input features to the target concept, typically as disjunctive normal form (DNF). Such *flat* normal forms can,

S. Džeroski et al. (Eds.): DS 2025, LNAI 16090, pp. 161–175, 2025.
https://doi.org/10.1007/978-3-032-05461-6_11

in principle, represent every Boolean function, in the same way as shallow neural networks with a single hidden layer can represent an arbitrary continuous function by the universal approximation theorem [13]. Nevertheless, deep neural networks are usually easier to train and often yield better performance, presumably because they require exponentially fewer parameters than shallow networks [19]. In analogy, one can argue that even though there is no gain in expressivity when moving from flat to deep rule sets (because every Boolean expression can be converted into an equivalent DNF expression), there may be a practical gain in learning deep rule sets.

In this work, our aim is to analyze this potential gain of deep symbolic models. For this purpose, we briefly present deep rule learning with an illustrative example, highlighting the different sizes of flat and deep rule sets, and cover related work (Sect. 2). The following Sect. 3 compares the size and expressive power of the search spaces for flat and deep Boolean expressions of limited size. We then compare the model shrinkage between flat and deep representations of the same Boolean function in Sect. 4, and the difference in predictive accuracy for flat and deep models of the same size in Sect. 5. Finally, in Sect. 6, we conclude with a summary of the results and a discussion of their impact on future work in deep symbolic learning.

2 Deep Rule Learning

Inductive rule learning has a long history in machine learning, with Michalski's AQ system [20] being one of the earliest learning algorithms, whose legacy can still be found in many state-of-the-art algorithms. With the advent of deep learning, which produces accurate but often inscrutable predictions, interest in the field has been rejuvenated because rules are a natural vehicle for communicating knowledge between man and machine [11].

Modern rule learning algorithms have seen a paradigm shift from a heuristic search for powerful predictive models to an exhaustive search for a model that optimizes a given (regularized) loss function (e.g., [1,29]). As optimality can naturally only be obtained on training data, it remains an open question whether and under what circumstances such algorithms may outperform classical algorithms such as Ripper [8], which make up for the suboptimal search with powerful pruning and out-of-sample optimization techniques, and have a much faster run-time to their credit.[1]

2.1 Flat and Deep Rule Sets

All these algorithms have in common that they are restricted to learning *flat rule models*, i.e., models that directly relate the input features to the target variable.

[1] We note in passing that some popular Python implementations of Ripper do not seem to live up to the full power of the algorithm, and therefore strongly recommend the use of the JRip implementation contained in the Java knowledge discovery workbench Weka.

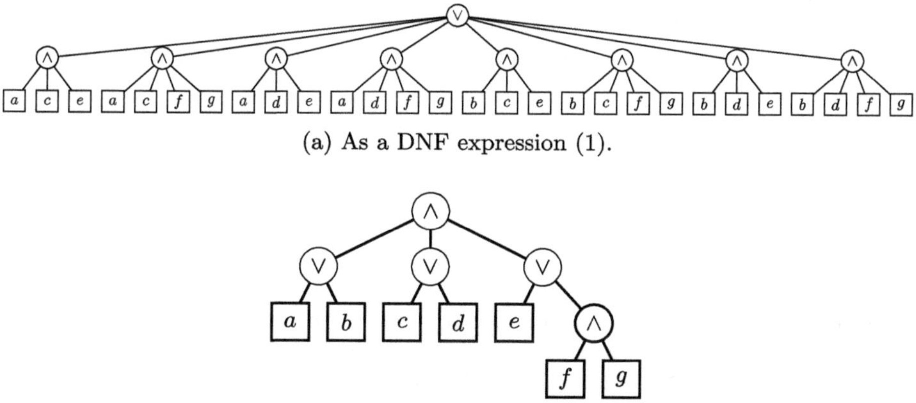

(a) As a DNF expression (1).

(b) As a structured formula (2) .

Fig. 1. Tree representation of two equivalent logical formulas.

While such flat models correspond to logical concepts in disjunctive normal form and therefore allow to represent arbitrary Boolean expressions, deeply structured formulas may be considerably smaller. Consider, for example, the rule set

```
+ :- a, c, e.          + :- b, c, e.
+ :- a, c, f, g.       + :- b, c, f, g.
+ :- a, d, e.          + :- b, d, e.
+ :- a, d, f, g.       + :- b, d, f, g.
```

which defines 8 rules for a positive class over seven input variables a – g. Logically, this corresponds to the following expression in disjunctive normal form (DNF):

$$\begin{aligned}&(a \wedge c \wedge e) \vee (a \wedge c \wedge f \wedge g) \vee (a \wedge d \wedge e) \vee (a \wedge d \wedge f \wedge g) \vee \\ &(b \wedge c \wedge e) \vee (b \wedge c \wedge f \wedge g) \vee (b \wedge d \wedge e) \vee (b \wedge d \wedge f \wedge g)\,.\end{aligned} \tag{1}$$

While this is the smallest possible DNF expression for this Boolean function, a logically equivalent but considerably smaller expression can be found by repeated application of distributivity:

$$(a \vee b) \wedge (c \vee d) \wedge [e \vee (f \wedge g)] \tag{2}$$

Obviously, in this case the structured expression (2) is significantly shorter than its flat DNF counterpart (1).

This can also be seen from the tree representation of both formulas, which is shown in Fig. 1: the DNF formula corresponds to a tree with a single hidden layer, whereas the structured formula needs at least two hidden layers. Note that both representations could be converted into fully connected networks by combining identical nodes (in particular in the input layer), and using binary edge weights, where edges with weight 1 correspond to visible edges, and edges with weight 0 are invisible.

Also note that a structured formula like Fig. 1b can be converted back into a set of rules, where nodes at OR-layers may be interpreted as invented auxiliary concepts:

```
+ :- h21, h22, h23.

h21 :- a.        h22 :- c.        h23 :- e.
h21 :- b.        h22 :- d.        h23 :- f, g.
```

While the benefit of deep rule concepts for the above example is clearly visible, the question arises as to whether this example is only a carefully selected exception or whether similar structured concepts can be found for the majority of datasets. We therefore intend to address the following research question with this work:

RQ : *How likely is it that a deep model is better than the corresponding flat model (and by how much)?*

The qualitative assessment "better" may include multiple aspects such as interpretability, learnability, comprehensibility etc. In this work, we limit ourselves to the quantitative aspect of compactness; we consider smaller models better than larger models. More formally, we define the size l of a rule model as the number of literals used in its representation of a (deep or flat) Boolean expression. Note that many qualitative measures such as interpretability, learnability, and comprehensibility are often associated with compactness as well [7].

2.2 Related Work

In the field of logic circuit design, the observation that deep models are more compact than flat models is not new [10], and theoretical results on logic circuits served as a motivation for deep learning [6]. Finding minimal hierarchical representations of Boolean functions has even been studied since the mid-20th century [2], and has since been extended to machine learning contexts [31].

By drawing inspiration from the analogy to deep neural networks and supported by the theoretical results on logic circuits, several approaches can be taken for learning deep rule models:

The Holistic Approach: A deep rule set is optimized all at once. The structure is built up incrementally, e.g., by adding nodes one-by-one to an initially empty rule network [5].

The Rule Learning Approach: A network structure is built up layer by layer, using a conventional rule learning algorithm that takes the outputs of the previous layers as the inputs of the next layer, with the goal of learning a CNF or DNF description for the next disjunctive or conjunctive layer [4].

The Neural Network Approach: A network structure is fixed and its parameters optimized. This approach is similar in spirit to learning binary [24,26] and ternary neural networks [16,30], sum-product networks [23], or differentiable (propositional) logic [9,22]. In previous work [3], we proposed a layered representation of Boolean networks with alternating conjunctive and disjunctive layers, which seems to be the only approach that aims at learning a deeply structured propositional rule set, albeit only in a greedy fashion.

Traditionally, the problem of learning structured rule sets has been studied in machine learning under various labels, including *constructive induction* [17] and *predicate invention* [27]. However, theses techniques typically aim at an interactive setting where new concept definitions are invented and proposed. For example, in [21] a technique that employs user-provided meta rules for proposing new predicates was introduced, which allows to invent useful predicates from very few training examples. For an excellent recent summary of work in this area we refer to [14]. While most of these systems operate in first-order logic, we aim at inventing auxiliary definitions in propositional logic. Moreover, instead of involving a user to provide templates or assess the outcome, we aim for a purely data-driven process.

3 Search Spaces

As discussed in Sect. 2.1, both flat and deep rule models are capable of representing arbitrary Boolean expressions. Consequently, a perfect representation of the training data (where each example is converted into a single rule) is contained in the hypothesis space of all rule learning algorithms. Obviously, the hypothesis space of a flat rule learner is a subspace of a deep rule learner's hypothesis space, i.e., a deep rule learner can represent more logically equivalent representations of the same Boolean concept—including the DNF representations to which a flat learner is restricted. Without further restriction of the search space, the larger hypothesis space of deep rule learners without a higher expressive power can be considered to be a disadvantage that only adds additional complexity.

However, the size of the search space can be restricted by additional constraints. For example, the size l of the model, i.e., the number of literals in the Boolean expression, can be limited to maintain interpretability of the model. In this case, the search space of a flat rule learner is not only a subspace of a deep rule learner's search space but also has a lower expressive power. We have seen in Sect. 2.1 that structured expressions can be considerably smaller than their logically equivalent counterparts in DNF, so that for a maximum model size l_{max} with $7 \leq l_{max} < 28$ the deep representation (2) can be learned with an adequate learner, whereas the DNF representation (1) would no longer be included in the search space.

In the following, we try to analyze the size and expressive power of the search spaces more systematically by considering the question:

RQ 1: *What is the distribution of flat vs. deep structures for Boolean expressions with l literals?*

To answer RQ 1, we consider a Boolean expression in its tree representation and infer that the tree has l leaf nodes representing a literal, and—by allowing the inner nodes to be of any arity larger than one—up to $l - 1$ inner nodes representing a connector. Without loss of generality, we assume that two adjacent inner nodes do not represent the same connector (otherwise they could be merged into a single inner node of larger arity), i.e. the tree consists of alternating conjunctive and disjunctive layers.

Flat trees therefore consist of either a conjunctive root node connecting to a set of leaf nodes and disjunctive inner nodes (representing a CNF) or vice versa of a disjunctive root node connecting to a set of leaf nodes and conjunctive inner nodes (representing a DNF). We note that by "flipping" the operator of each inner node the same tree "shape" can be used once for representing a CNF and DNF each. The same holds for deep trees with a height $h > 2$, where for any tree shape exactly two ways of alternating conjunctive and disjunctive layers are possible. Therefore, in the following, we do not distinguish between conjunctive and disjunctive inner nodes when determining the number of possible tree shapes.

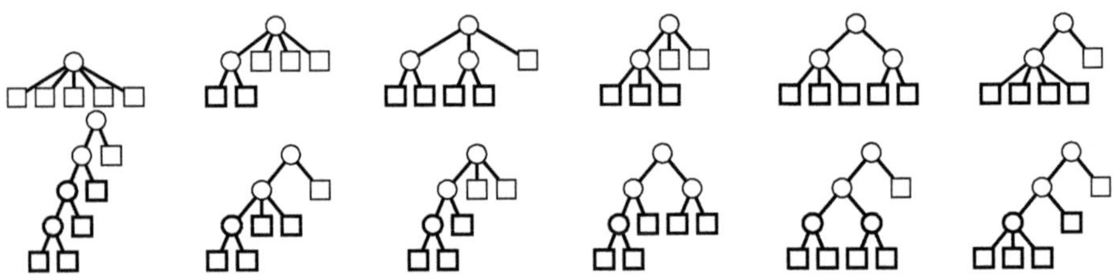

Fig. 2. All 6 flat tree shapes (top row) and all 6 deep tree shapes (bottom row) with exactly $l = 5$ leaves. The leaf nodes are represented as squares and can be connected by 1 to 4 inner nodes, represented as circles.

Finding the number of flat tree shapes is related to the number of partitions.[2] For instance, the integer $l = 5$ has the seven partitions {1, 1, 1, 1, 1}, {2, 1, 1, 1}, {2, 2, 1}, {3, 1, 1}, {3, 2}, {4, 1}, and {5}. Each partition corresponds to a flat tree shape, where each number p_i in the partition denotes a node connected to the root node, being an inner node of arity p_i if $p_i > 1$ or a leaf node if $p_i = 1$, as shown in Fig. 2 (so in case of interpreting the tree as a rule set in DNF, each number corresponds to the length of one rule). The trivial partition consisting of a single number has to be ignored because the arity of the root node violates the precondition of being larger than one. Hence, the number of flat tree shapes with l leaf nodes is defined by the partition numbers with an offset of one.

Similarly, we can describe the shape of deep trees as partitions, whose elements can recursively be split into partitions again. For instance, for $l = 5$ we can find additional deep tree shapes for {{{2, 1}, 1}, 1}, {{2, 1, 1}, 1}, {{2, 1}, 1, 1}, {{2, 1}, 2}, {{2, 2}, 1}, and {{3, 1}, 1}. The number of flat and tree

[2] https://oeis.org/A000041.

Table 1. For each number of leaf nodes ($1 < l < 20$), the number of flat tree shapes (*flat*) and all tree shapes including deep shapes (*total*) is reported. The share of flat tree shapes (*rat.*) among all tree shapes decreases with growing number of leaves l, also if the number of tree shapes is accumulated up to l (*acc.*).

l	flat	total	rat.	acc.	l	flat	total	rat.	acc.
2	1	1	100.00	100.00	**11**	55	7 068	0.78	1.73
3	2	2	100.00	100.00	**12**	76	21 965	0.35	0.80
4	4	5	80.00	87.50	**13**	100	68 954	0.15	0.35
5	6	12	50.00	65.00	**14**	134	218 751	0.06	0.15
6	10	33	30.30	43.40	**15**	175	699 534	0.03	0.07
7	14	90	15.56	25.87	**16**	230	2 253 676	0.01	0.03
8	21	261	8.05	14.36	**17**	296	7 305 788	0.00	0.01
9	29	766	3.79	7.44	**18**	384	23 816 743	0.00	0.00
10	41	2 312	1.77	3.68	**19**	489	78 023 602	0.00	0.00

shapes combined is known as the number of series-reduced planted trees with l leaves,[3] and grows much faster than the partition numbers.

Table 1 summarizes the number of flat and total tree shapes for trees up to 19 leaves. It can be seen that the percentage of flat trees among all tree shapes with l leaf nodes decreases rapidly and is negligible for $l > 10$, where more than 99% of all tree shapes are deep.

The astute reader may have noticed that we only discussed the ratio of flat and deep tree *shapes*, but not of tree *expressions*. The number of expressions also depends on 1) the number of possible features the literals can be picked from, and 2) the number of leaf node permutations resulting in the same expression due to commutativity. While 1) affects both flat and deep trees equally, 2) is arguably considerably higher for flat tree shapes. To illustrate this, consider a flat partition $\{p_1, p_2, \ldots, p_n\}$ with $p_1 + p_2 > 3$ and $n > 2$. By moving a leaf node from p_1 to an additional layer connected to the root node, we obtain the deep partition $\{\{p_1 - 1, p_2, \ldots, p_n\}, 1\}$, which allows $n \cdot \frac{p_1}{p_1 - 1}$ as many different expressions as the flat one. This indicates that the ratio of flat tree shapes over all tree shapes is an upper limit for the ratio of flat tree expressions over all tree expressions.

4 Model Sizes

The previous section showed that the search space of deep rule learners is significantly larger than the search space of flat rule learners and has a higher expressive power when the model size is limited. A consequence of this is that—as every Boolean function can be converted into a flat representation—the same Boolean function will have many equivalent representations. For some concepts,

[3] https://oeis.org/A000669.

the corresponding flat representation is already minimal, so that the extension of the search space to deep concepts is not necessary. To address this issue, we now look at the question from a different perspective:

RQ 2: *For an arbitrary Boolean expression, how much smaller is the smallest equivalent deep expression than the smallest equivalent flat expression?*

The answer to this question might differ depending on how "arbitrary Boolean expressions" are understood. On the one hand, we can generate Boolean expressions by randomly connecting Boolean features with ANDs and ORs. This may also be viewed as "picking" a tree from a deep model's search space. On the other hand, we can randomly select one of the 2^{2^m} possible Boolean functions that can be defined over a fixed number of m Boolean features. Each such function may be viewed as a representative for a class of logically equivalent Boolean expressions which encode this function. We consider these two possible ways in the following subsections.

4.1 Generating Random Boolean Expressions

Given a set of m Boolean features $\mathcal{F} = \{f_1, \ldots, f_m\}$ and a restricted set of operators $\mathcal{O}$, in the simplest case the basic operators $\mathcal{O} = \{\wedge, \vee, \neg\}$, a *Boolean expression* can be generated by repeatedly connecting features with randomly chosen operators. It can be either written as a concatenation of features and operators or represented as a tree. Although there is an infinite number of Boolean expressions, the number of *Boolean functions* is limited by 2^{2^m}.

However, following the above procedure to generate a Boolean expression, not all Boolean functions are obtained equally likely. Gardy [12] introduced natural probability distributions by partitioning the set of Boolean expression trees into 2^{2^m} subsets, each subset gathering all the trees that compute a specific Boolean function. Consequently, the probability of a Boolean function f is proportional to the number of trees that compute it. For example, given only 3 Boolean features $\mathcal{F} = \{a, b, c\}$, generating the tautology *true* (16.5%) is much more likely than generating a (3.14%), $b \wedge c$ (1%) or $(a \wedge b) \vee c$ (0.21%). Overall, "simpler" functions tend to be generated more frequently.

For our experiments on generated expressions, we use various numbers of Boolean features $m \in \{3, 5, 7, 10, 15, 20\}$ and different Boolean expression lengths $l \in \{10, 30, 50, 100, 150, 200\}$. To avoid that large parts of the generated expression evaluate to *true* or *false*, we ignore negations and restrict the set of operators to $\mathcal{O} = \{\wedge, \vee\}$[4]. For each combination of m and l, we construct 1000 expressions by first generating a string of length l from the set of Boolean features, and then randomly combining two adjacent features or subexpressions with $\wedge$ or $\vee$.

For example, for $m = 3$ and $l = 5$, we may generate the following expression:

$$abaac \rightarrow (a \vee b)aac \rightarrow (a \vee b)(a \vee a)c \rightarrow (a \vee b)((a \vee a) \wedge c) \rightarrow ((a \vee b) \wedge ((a \vee a) \wedge c))$$

[4] The omission of negations in models is also common in many rule learners and facilitates the interpretability, especially for nominal attributes.

Table 2. Model size of the minimal DNF and deep representations for various number of Boolean features m and different expression lengths l. The size of the models is highlighted in a color scale from green (small) to red (large).

(a) Flat

l \ m	3	5	7	10	15	20
10	2.35	3.96	5.87	7.78	11.45	13.88
30	2.55	5.28	9.08	15.75	33.51	56.47
50	2.71	5.59	10.05	20.70	52.55	105.10
100	2.80	6.05	11.92	27.28	84.84	219.33
150	2.83	6.48	13.02	33.42	112.08	284.89
200	2.89	6.66	13.58	34.53	130.31	381.14

(b) Deep

l \ m	3	5	7	10	15	20
10	2.25	3.40	4.46	5.42	6.77	7.37
30	2.44	4.30	6.38	9.22	14.39	18.51
50	2.55	4.50	6.95	11.48	20.43	30.07
100	2.65	4.80	7.97	14.40	31.33	54.69
150	2.66	5.12	8.52	16.94	39.62	73.10
200	2.71	5.21	8.84	17.76	45.46	93.53

Subsequently, we use MIS [25], an algorithmic multi-level logic synthesis and minimization program, to convert these random Boolean expressions into minimal DNF and deep representation. Unlike exact minimum logic factoring algorithms (see [28]), MIS' good factor algorithm does not guarantee minimality for the deep representation but provides a better runtime while still providing minimal deep representations for most expressions.

Table 2 shows the model sizes of the minimal DNF and deep representations for different values of m and l. As expected, the minimal deep representation is smaller than the DNF, and both representations grow in size with the number of available features m or the length of the generated expression l. While the difference between the flat and deep model is still small for $m = 3$, it differs considerably for $m = 20$: The DNF is roughly twice as long as the original expression, whereas the deep representation can even halve the original length.

We can also observe in Table 3 that the model size reduction achieved using a deep representation instead of a flat one increases with higher values for m and l. Similarly as in the previous two tables, the inclusion of additional features has a stronger effect than extending the length of the expression. Cases like the one presented in Fig. 1 occur more frequently for $m = 20$ and $l \geq 100$. In this setting, MIS can consistently find deep representations that are 75% smaller than equivalent DNFs.

4.2 Generating Random Boolean Functions

In the following, we ignore the natural probability distributions of Boolean expressions and sample from the space of all Boolean functions with equal probability. One way to achieve this is to consider a truth table consisting of 2^m rows, representing an exhaustive list of (non-contradicting) examples $\{\mathbf{x}_1, \ldots, \mathbf{x}_{2^m}\} = \{0, 1\}^m$. For this, we randomly generate a 2^m-dimensional vector $\mathbf{y} \in \{0, 1\}^{2^m}$ that assigns a Boolean value y_i to each example $\mathbf{x}_i$. Example $\mathbf{x}_i$ is called *positive* if $y_i = 1$, and *negative* if $y_i = 0$. We pass all positive examples as *min terms* to the Quine-McCluskey algorithm [18] implementation in SymPy,[5]

[5] https://www.sympy.org/en/index.html.

Table 3. Reduction of model size (in %) when using a deep instead of a flat representation. The shrinkage is highlighted in a color scale from green (big) to red (small).

l \ m	3	5	7	10	15	20
10	4.01	14.18	23.92	30.40	40.85	46.93
30	4.36	18.59	29.75	41.46	57.06	67.22
50	5.62	19.50	30.88	44.52	61.12	71.39
100	5.53	20.65	33.12	47.19	63.08	75.06
150	5.73	21.02	34.60	49.33	64.65	74.34
200	6.37	21.71	34.87	48.57	65.12	75.46

to obtain a minimal DNF representation of this function. As in the previous section, we then compare its size to the size of the minimal deep representation found by MIS' good factor algorithm.

Figure 3 summarizes the experiments for $m \in \{3, \ldots, 10\}$ Boolean features and 1000 randomly selected Boolean functions.[6] The model size l is plotted in logarithmic scale and grows exponentially for both the smallest flat and deep model. As can be seen in the model shrinkage graph in red, the size reduction of the deep model increases from 8% for $m = 3$ to 49% for $m = 10$. Similarly to the evaluation of generated expressions, a deep representation is more worthwhile the more Boolean features are used in the expression.

We also compare the depths of the deep models for all values $m \in \{3, \ldots, 10\}$. Figure 4 shows, for each value of m, the percentage of the 1000 models having a specific depth (resp. height if interpreted as a tree). Small shares are shown in small blue circles, which become increasingly large and red the larger the share. It can be recognized that the largest share is always for depth $2 \cdot (m-2)$, representing between 48% and 75% of the models. In other words, the addition of a new feature to the Boolean function usually results in an additional conjunctive and an additional disjunctive layer in the deep representation.

So far, we only considered *complete* truth tables, in which a Boolean value y_i is assigned to each example $\mathbf{x}_i$. Thus, all examples are part of the training data. However, the situation in machine learning problems is typically different: The training data covers only a (small) fraction of all possible examples in the feature space, and the model tries to guess the underlying function from such a sample.

To analyze the behavior of the models in this setting, we only label P [N] examples as positive [negative]. For $m = 7$ features, we generate 1000 functions for different combinations of P, N with $P + N < 2^m = 128$. All positive examples are passed as *min terms* and all unlabeled examples as *don't cares* to the Quine-McCluskey algorithm to obtain a minimal DNF, which is then transformed into a deep representation. In addition, we also compute the minimal CNF.

[6] For $m = 3$, this resulted in upsampling, since $2^{2^3} = 256 < 1000$.

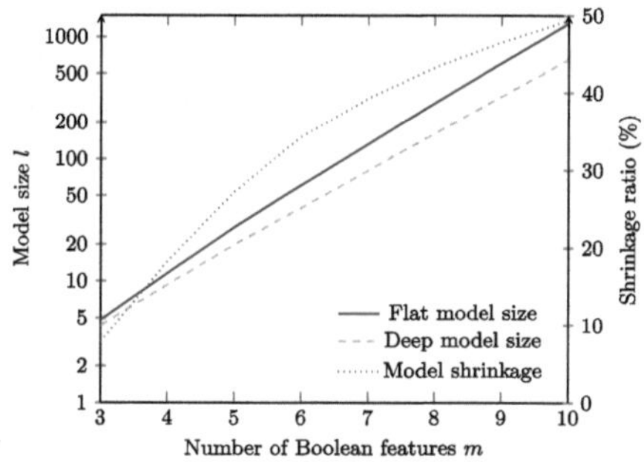

Fig. 3. Average model size l in flat (blue) and deep (green) representation over 1000 Boolean functions with different feature counts m, in logarithmic scale. The percentage of saved literals in the deep model is plotted as model shrinkage (red).

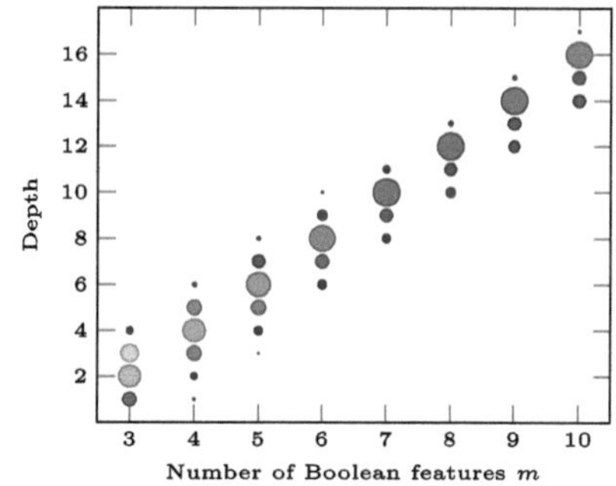

Fig. 4. Depth of the deep model in a scatter plot. For each feature count value m, the share of 1000 Boolean functions with a specific depth is shown as a circle of the corresponding size.

The results are summarized in Table 4. We can see that the more examples are removed from the training data, i.e., the smaller P and N, the smaller the model sizes l_{CNF}, l_{DNF} and l_{deep}. We also observe that for $P > N$ the minimal DNF is more compact than the minimal CNF, and for $N > P$ vice versa. This results in different shrinkage levels if the values for P and N are flipped, since the deep representation was always learned on the minimal DNF. The last row of the table reports the average flat and deep model size in the previous experiment (cf. Fig. 3) where we used exhaustive and roughly balanced data by assigning each example $\mathbf{x}_i$ with 50% chance to the positive or negative class respectively. Overall, the previous reduction of 39% decreases to less than 15% for $P = 10$.

5 Predictive Performance

In the previous section, we analyzed the reduction of the model size when using a deep representation instead of a flat representation of the same Boolean function. In the following, we reverse the approach by comparing flat and deep models of the same size, but representing different Boolean functions, resulting in our last research question:

RQ 3: *If the model size l is limited, how much can a deep model improve the predictive accuracy over a shallow model?*

We approach this question in an online machine learning setting: The training data are passed to the predictor in a fixed sequential order, and each new training example not yet correctly classified by the predictor requires the Boolean function of the model to change accordingly. Thus, the model grows over multiple iterations, allowing one to look for different thresholds l_{max} at the maximal accuracy achieved for models with size $l \leq l_{max}$.

Table 4. Average reduction of model size (in %) when using a deep instead of a flat representation for $m = 7$ Boolean features and different numbers of positive (P) and negative examples (N) to learn the Boolean function. The shrinkage is computed as $1 - l_{deep}/\min(l_{CNF}, l_{DNF})$ and highlighted in a color scale from green (big) to red (small).

P	N	l_{CNF}	l_{DNF}	l_{deep}	**Shr. (%)**
10	10	11.513	11.654	9.927	13.78
10	20	17.217	18.201	14.657	14.87
10	50	27.656	32.411	23.915	13.53
10	100	38.108	48.933	33.361	12.46
20	10	18.243	17.254	14.213	17.62
20	20	28.287	28.155	21.575	23.37
50	10	32.099	27.369	21.565	21.21
50	50	95.363	95.169	61.007	35.90
100	10	48.724	38.032	28.819	24.22
~64	~64		132.726	80.386	39.43

The experiment is carried out on 20 artificial datasets with $m = 10$ Boolean features, which have been previously used in [3], resulting in $2^{10} = 1024$ training examples for each dataset. The training examples are shuffled and then repeatedly removed from the *don't care*-set and added to the *min terms* if positive. In every iteration, it is checked whether the new example contradicts the current model and if so, the minimal DNF is recomputed, and its predictive accuracy on the full dataset is evaluated. Each DNF is also converted into a deep representation, having the same predictive accuracy but potentially a smaller model size l.

The development of the model size l over the percentage of training data seen is shown in Fig. 5. Similarly to previous experiments, the flat model size grows faster than the deep one, so that the shrinkage ratio grows as well. Figure 6 rearranges these data in order to answer RQ 3. In the violet graph, it plots the average accuracy increase of the deep model over the flat model for any maximum model size $1 \leq l_{max} \leq 85$. It rises rapidly and peaks at $l_{max} = 13$ and $32 \leq l_{max} \leq 39$, where the deep model provides an average gain above 4% over the flat model. Furthermore, the orange graph shows that for $15 \leq l_{max} \leq 50$, at least on half of the 20 datasets the best deep model performs better than the best flat one.

For the above experiments, we did an intermediate step by first converting the training data into a minimal DNF, and then refactoring this DNF into a smaller deep representation. Note that because of this, the minimal structured formula found may not be minimal. Consider, e.g., the following two correct

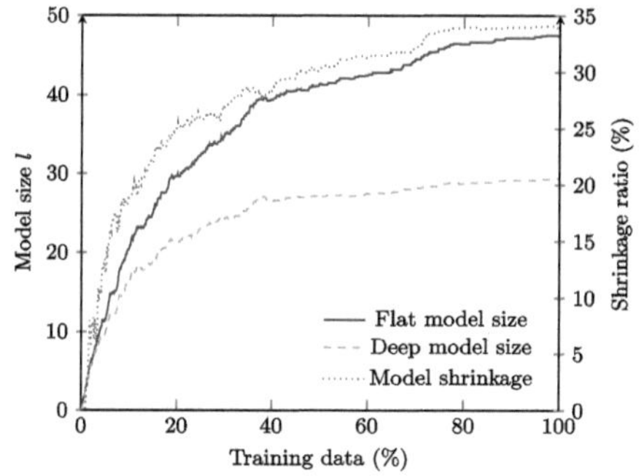

Fig. 5. Average model size l in flat (blue) and deep (green) representation over 1000 Boolean functions with different feature counts m, in logarithmic scale. The percentage of saved literals in the deep model is plotted as model shrinkage (red).

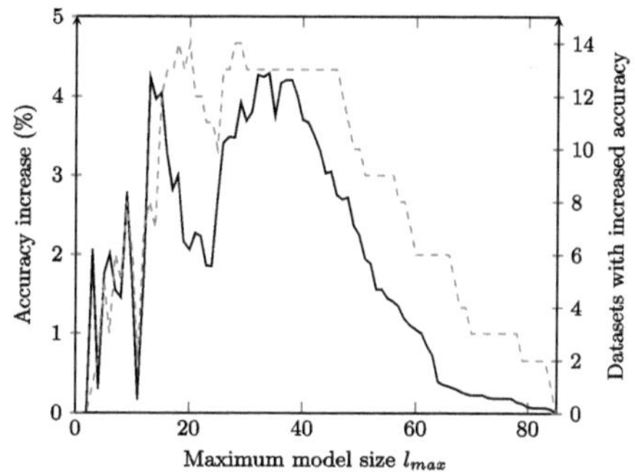

Fig. 6. Average accuracy increase of a deep model over a flat model for a fixed model size over 20 artificial datasets (violet) and the absolute number of datasets with an improved accuracy when using a deep model (orange).

models for given training data:

$$(a \wedge c) \vee (a \wedge d) \vee (b \wedge c) = a \wedge (c \vee d) \vee (b \wedge c) \quad (3)$$

$$(a \wedge c) \vee (a \wedge d) \vee (b \wedge c) \vee (b \wedge d) = (a \vee b) \wedge (c \vee d) \quad (4)$$

The minimal DNF ($l_{DNF} = 6$) is shown in (3) and can be reduced to a deep representation with $l_{deep} = 5$. However, if the deep model was directly learned on the training data, a smaller representation with $l_{deep} = 4$ can be found, as shown in (4).

6 Conclusion

Motivated by the superior performance of deep neural networks over single-hidden-layer networks, we empirically compared flat and deep symbolic models to analyze the potential gains of deep symbolic model from three perspectives. First, we illustrated that the search space of flat models is considerably smaller than the search space of deep models, and less expressive if the size of the model is limited. Second, we calculated the average model shrinkage of saved literals in a deep model compared to the equivalent flat model in different settings, leading to reductions from 4% up to 75%. Third, we showed that the best deep model of limited size outperforms the best flat model of the same size by up to 4% in terms of predictive accuracy.

Note that the presented experiments come with multiple limitations, and the shown advantages of deep symbolic models over flat ones do not necessarily imply the superiority of deep rule learners over flat rule learners. For example, in real-world scenarios, the large search space on deep models demands suitable search strategies, and noisy data requires heuristic adaptations. Nevertheless, in the authors' opinion, the results of this work support the step towards learning deep rule concepts and affirm future work in this direction.

Acknowledgments. We are grateful for the support of the Bilateral AI project and thank the reviewers for their encouraging comments and fruitful pointers to related work.

References

1. Angelino, E., Larus-Stone, N., Alabi, D., Seltzer, M.I., Rudin, C.: Learning certifiably optimal rule lists for categorical data. J. Mach. Learn. Res. **18**, 234:1–234:78 (2017)
2. Ashenhurst, R.L.: The decomposition of switching functions. In: Proceedings of an International Symposium on the Theory of Switching, April 1957, pp. 74–116 (1957)
3. Beck, F., Fürnkranz, J.: An empirical investigation into deep and shallow rule learning. Front. Artifi. Intell. **4**, 145 (2021). https://doi.org/10.3389/frai.2021.689398
4. Beck, F., Fürnkranz, J., Huynh, V.Q.P.: Layerwise learning of mixed conjunctive and disjunctive rule sets. In: Proceedings of the 7th International Joint Conference on Rules and Reasoning (RuleML+RR). Lecture Notes in Computer Science, Springer (2023)
5. Beck, F., Fürnkranz, J., Huynh, V.Q.P.: Learning deep rule concepts as alternating boolean pattern trees. In: Proceedings of the 27th International Conference on Discovery Science (DS). Lecture Notes in Computer Science, Springer (2024)
6. Bengio, Y., et al.: Learning deep architectures for AI. Foundations and trends® in Machine Learning **2**(1), 1–127 (2009)
7. Carvalho, D.V., Pereira, E.M., Cardoso, J.S.: Machine learning interpretability: a survey on methods and metrics. Electronics **8**(8), 832 (2019)
8. Cohen, W.W.: Fast effective rule induction. In: Prieditis, A., Russell, S. (eds.) Proceedings of the 12th International Conference on Machine Learning (ML-95), pp. 115–123. Morgan Kaufmann, Lake Tahoe, CA (1995)
9. Cohen, W.W., Yang, F., Mazaitis, K.: TensorLog: a probabilistic database implemented using deep-learning infrastructure. J. Artifi. Intell. Res. **67**, 285–325 (2020). https://doi.org/10.1613/jair.1.11944
10. Darwiche, A., Marquis, P.: A knowledge compilation map. J. Artifi. Intell. Res. **17**, 229–264 (2002). https://doi.org/10.1613/JAIR.989
11. Fürnkranz, J., Gamberger, D., Lavrač, N.: Foundations of rule learning. Springer-Verlag (2012)
12. Gardy, D.: Random boolean expressions. Discrete Math. Theoretical Comput. Sci. (2005)
13. Hornik, K.: Approximation capabilities of multilayer feedforward networks. Neural Netw. **4**(2), 251–257 (1991)
14. Kramer, S.: A brief history of learning symbolic higher-level representations from data (and a curious look forward). In: Proceedings of the 29th International Joint Conference on Artificial Intelligence (IJCAI), Survey Track, pp. 4868–4876 (2020)
15. LeCun, Y., Bengio, Y., Hinton, G.: Deep learning. Nature **521**(7553), 436–444 (2015)

16. Li, F., Zhang, B., Liu, B.: Ternary weight networks. arxiv **abs/1605.04711** (2016)
17. Matheus, C.J.: A constructive induction framework. In: Proceedings of the 6th International Workshop on Machine Learning, pp. 474–475 (1989)
18. McCluskey, E.J.: Minimization of boolean functions. Bell Syst. Tech. J. **35**(6), 1417–1444 (1956)
19. Mhaskar, H., Liao, Q., Poggio, T.A.: When and why are deep networks better than shallow ones? In: Singh, S.P., Markovitch, S. (eds.) Proceedings of the 31st AAAI Conference on Artificial Intelligence. pp. 2343–2349. AAAI Press, San Francisco, California, USA (2017)
20. Michalski, R.S.: On the quasi-minimal solution of the covering problem. In: Proceedings of the 5th International Symposium on Information Processing (FCIP-69). vol. A3 (Switching Circuits), pp. 125–128. Bled, Yugoslavia (1969)
21. Muggleton, S.H., Lin, D., Tamaddoni-Nezhad, A.: Meta-interpretive learning of higher-order dyadic datalog: predicate invention revisited. Mach. Learn. **100**(1), 49–73 (2015). https://doi.org/10.1007/s10994-014-5471-y
22. Petersen, F., Borgelt, C., Kuehne, H., Deussen, O.: Deep differentiable logic gate networks. In: Koyejo, S., Mohamed, S., Agarwal, A., Belgrave, D., Cho, K., Oh, A. (eds.) Advances in Neural Information Processing Systems 35 (NeurIPS). LA, USA, New Orleans (2022)
23. Poon, H., Domingos, P.M.: Sum-product networks: a new deep architecture. In: Cozman, F.G., Pfeffer, A. (eds.) Proceedings of the 27th Conference on Uncertainty in Artificial Intelligence (UAI), pp. 337–346. AUAI Press, Barcelona, Spain (2011)
24. Qin, H., Gong, R., Liu, X., Bai, X., Song, J., Sebe, N.: Binary neural networks: a survey. Patt. Recogn. **105** (2020)
25. Shapiro, E.Y.: An algorithm that infers theories from facts. In: Proceedings of the 7th International Joint Conference on Artificial Intelligence (IJCAI-81), pp. 446–451 (1981)
26. Simons, T., Lee, D.J.: A review of binarized neural networks. Electronics **8**(661) (2019). https://doi.org/10.3390/electronics8060661
27. Stahl, I.: Predicate invention in inductive logic programming. In: De Raedt, L. (ed.) Advances in Inductive Logic Programming, Frontiers in Artificial Intelligence and Applications, vol. 32, pp. 34–47. IOS Press (1996)
28. Yoshida, H., Ikeda, M., Asada, K.: Exact minimum logic factoring via quantified boolean satisfiability. In: 2006 13th IEEE International Conference on Electronics, Circuits and Systems, pp. 1065–1068. IEEE (2006)
29. Yu, J., Ignatiev, A., Stuckey, P.J., Bodic, P.L.: Learning optimal decision sets and lists with SAT. J. Artifi. Intell. Res. **72**, 1251–1279 (2021). https://doi.org/10.1613/JAIR.1.12719
30. Zhu, C., Han, S., Mao, H., Dally, W.J.: Trained ternary quantization. In: Proceedings of the 5th International Conference on Learning Representations (ICLR). OpenReview.net, Toulon, France (2017)
31. Zupan, B., Bohanec, M., Demšar, J., Bratko, I.: Learning by discovering concept hierarchies. Artif. Intell. **109**(1–2), 211–242 (1999)

Exploring the Design Space of Fair Tree Learning Algorithms

Kiara Stempel(✉), Mattia Cerrato, and Stefan Kramer

Johannes Gutenberg University Mainz, Staudingerweg 9, Mainz 55128, Germany
{stempel,mcerrato}@uni-mainz.de, kramer@informatik.uni-mainz.de

Abstract. Decision trees have been studied extensively in the context of fairness, aiming to maximize prediction performance while ensuring non-discrimination against different groups. Techniques in this space usually focus on imposing constraints at training time, constraining the search space so that solutions which display unacceptable values of relevant metrics are not considered, discarded, or discouraged. If we assume one target variable y and one sensitive attribute s, the design space of tree learning algorithms can be spanned as follows: (i) One can have *one* tree T that is built using an objective function that is a function of y, s, and T. For instance, one can build a tree based on the weighted information gain regarding y (maximizing) and s (minimizing). (ii) The second option is to have *one* tree model T that uses an objective function in y and T and a constraint on s and T. Here, s is no longer part of the objective, but part of a constraint. This can be achieved greedily by aborting a further split in a subspace as soon as the splitting condition that optimizes the objective function in y does not satisfy the constraint on s. The simplest form to examine other possible splits in this subspace is to backtrack in tree construction, once a fairness constraint in terms of s is violated in a subspace of a new split. (iii) The third option is to have *two* trees T_y and T_s, one for y and one for s, such that the tree structure for y and s does not have to be shared. In this way, information regarding y and regarding s can be used independently, without having to constrain the choices in tree construction by the mutual information between the two variables. Quite surprisingly, of the three options, only the first one and the greedy variant of the second have been studied in the literature so far. In this paper, we introduce the above two additional options from that design space and characterize them experimentally on multiple datasets.

Keywords: Fairness · Classification · Decision trees

1 Introduction

As machine learning models increasingly influence critical decision-making processes, addressing fairness has become an essential field of research. Algorithmic fairness involves ensuring that model predictions do not disadvantage individuals based on protected groups such as religion, gender, or race. Among the various algorithms designed or adapted to mitigate discrimination, decision trees

S. Džeroski et al. (Eds.): DS 2025, LNAI 16090, pp. 176–190, 2025.
https://doi.org/10.1007/978-3-032-05461-6_12

belong to the more transparent methods, as they provide a clear and interpretable structure for addressing fairness concerns. Various attempts have been made to incorporate fairness into decision trees, typically focusing on imposing constraints during training, e.g., constraining the search space to discard, discourage, or exclude solutions that lead to unacceptable values of relevant fairness metrics by adjusting the splitting criterion.

To account for both, performance based on the target variable y, and fairness based on the sensitive attribute s, the design space allows for (i) constructing *one* tree T by combining performance and fairness in a single objective function, for instance by weighting the information gain regarding y and s, (ii) constructing *one* tree T by imposing a fairness constraint on the objective function, or (iii) modeling performance and fairness independently in *two* separate trees T_y and T_s. While prior work has mostly focused on the first and, to a limited extent, the second approach, the third design option remains unexplored. However, it allows for an independent use of information regarding y and s, and provides the possibility to examine predictions and decision paths of the two separate trees.

In this paper, we provide a broader perspective on fair tree learning. We experimentally evaluate the above design choices for fair tree learning and identify their characteristics. For this purpose, we introduce a two-tree approach that separates fairness and performance into distinct learning objectives and therefore falls into (iii). Further, we propose a backtracking mechanism for (ii) that actively revises previous splitting decisions when fairness constraints are at risk of being violated. The evaluation is based on the trade-off curves of the tree variants. We obtain a trade-off curve by varying a parameter γ that controls the trade-off between performance and fairness. Several measures derived from the trade-off curve are used to compare the approaches. The area under the trade-off curve (AUTOC), along with the number of unique and local Pareto-optimal points, provides a compact summary of the trade-off curves, while offering different perspectives for evaluation.

2 Related Work

Several methods have been proposed to adapt decision trees to ensure fairness while maintaining predictive performance, typically by modifying the splitting criterion to incorporate fairness considerations: Kamiran et al. [13] introduced the first discrimination-aware decision trees by modifying the splitting criterion to balance information gain on the target and sensitive attribute, combining them via subtraction, addition, or division. The method penalizes splits that reveal too much about sensitive groups and is complemented by post-hoc leaf relabeling. This idea was extended to fair forests by Raff et al. [19], building fair decision trees similar to the ones by Kamiran et al. The difference between the Gini impurity [7] on the class label and the information gain of the protected attribute seeks to balance predictive performance with fairness. Castelnovo et al. [8] modify the splitting criterion to do a constrained search for the best attribute. They maximize the information gain on y while enforcing a threshold for a given fairness criterion. Since most fair tree learning methods optimize for a fixed threshold on the fairness metric, Barata et al. [17] propose a splitting criterion based on AUROC to be independent of the decision threshold.

For online learning, Zhang and Ntoutsi [23] extend Hoeffding trees [10] with a fairness gain, which is defined in terms of the reduction in discrimination caused by a specific split. Zhang and Zhao [24] introduce an adaptive fair information gain to Hoeffding trees. Instead of addition, they multiply the predictive and sensitive gains to account for the possible difference in scales between the gains.

An approach to learning optimal fair decision trees is done by Aghaei et al. [1]. This exact approach unifies the fairness definitions across classification and regression and uses a mixed-integer optimization framework for learning. Their framework incorporates a discrimination regularizer and a corresponding regularization weight that allows tuning the fairness-performance trade-off.

Instead of learning a new, fair model, Zhang et al. suggest repairing an existing unfair tree by flipping or refining specific decision paths in the tree while setting hard constraints such as fairness and semantic difference [22].

Fair classification can also be framed as a multi-objective optimization problem, balancing performance and fairness. Optimization methods fall into exact and meta-heuristic categories, the latter split into single- and multi-objective techniques [20]. Unlike single-objective optimization, multi-objective optimization seeks a set of Pareto-optimal solutions rather than a single optimum [4]. Multi-objective evolutionary algorithms use performance indicators to guide search toward Pareto fronts without computing Pareto dominance relations [20]. Zitzler et al. introduced an indicator-based evolutionary algorithm compatible with various indicators [25]. The hypervolume indicator transforms the problem into maximizing a single metric [3]. Many approaches aim to improve this indicator, to quantify the volume of the region dominated by a set of solutions. [2,5]. La Cava [15] applies the hypervolume to optimize the fairness-performance trade-off via a meta-model that maps protected attributes to sample weights. Freitas and Brookhouse [11] review such multi-objective evolutionary methods in fairness-aware learning.

3 Methods

3.1 Defining the Design Space

In this paper, we address binary classification settings. With n denoting the number of samples and m denoting the number of features, the input variables are represented by matrices $X_{n \times m}$. The corresponding set of class labels is defined as Y, where $y_i \in \{0, 1\}$ for each instance $i = 1, ..., n$. When training a fair decision tree, the objective is to reduce information about the sensitive attribute S, where $s_i \in \{0, 1\}$, to mitigate bias or discrimination based on S. Let $\mathcal{T}$ be the search space of our classification problem, i.e., the set of all potential decision trees that could be constructed based on X, that is

$$\mathcal{T}(X) = \{T \mid T \text{ is a decision tree based on } X\}. \tag{1}$$

Each point in $\mathcal{T}$ corresponds to a unique tree structure. The objective of a standard decision tree algorithm [7] is to find a decision tree $T \in \mathcal{T}$ that maximizes the predictive performance on X. The decision space $\mathcal{D}(T)$ refers to the set of possible outcomes a tree T can produce given any input. That is, $\mathcal{D}(T)$ represents all possible class labels that the fixed tree structure is capable of making. The set of predictions obtained by applying T to input data is denoted by $\hat{Y}$.

Based on this foundation, we can now span the design space of fair tree learning algorithms. When learning a fair tree on the input X, we are given one target variable y and one sensitive attribute s, and the objective is (i) to obtain a good predictive performance while (ii) minimizing the discrimination, or the information about s used for predicting.

One option is to have *one* tree that is built using a single objective function representing both objectives, which in this way is a function of y, s, and T. A typical approach is to adjust the splitting criterion to account for both y and s. The splitting criterion that is optimized during the training process typically corresponds to the information gain G:

$$G_V(A) = H_V(X) - \sum_{a \in \text{Values}(A)} \frac{|X_a|}{|X|} H_V(X_a). \tag{2}$$

It measures the increase in purity regarding a potential splitting attribute A and target attribute V (which can be either y or s), of which we measure the impurity. The impurity of V is given by the entropy $H_V(X)$ of dataset X. Under this design, fairness constraints are integrated into the search space $\mathcal{T}$, since we aim to find the tree model T that optimizes a combined objective function. For accounting for both performance and fairness, one can have two information gains which are linearly combined, where the information gain on y is maximized and the information gain on s is minimized.

The second design option, also defined over $\mathcal{T}$, learns *one* tree using an objective based solely on y and T, with fairness imposed as a constraint on s and T. Instead of weighting fairness in the objective, the trade-off is controlled via thresholds that are part of the constraint. This can be implemented greedily by aborting further splits in a subspace as soon as the split that best optimizes the objective in y violates the constraint on s. A simple extension of this idea is to backtrack during tree construction – i.e., revisiting earlier splitting decisions – when a fairness constraint is violated within a subspace. Beyond the approach of integrating both objectives into a (constrained) single objective function, a third, alternative design involves separating them into two objective functions. This results in learning *two* separate decision trees: one (T_y) optimized for the objective function in y and T, and the other (T_s) for the objective function in s and T. The resulting trees are then combined. In this design, the impact of fairness constraints is effectively shifted from the search space into the decision space $\mathcal{D}(T)$.

Algorithm 1. DTFC

Input: $X \neq \emptyset$, Y, S, γ, max_depth, $min\_samples$, current depth d
Output: A fair decision tree node or **null** if no valid split found
$A \leftarrow$ set of available attributes
$A_{\text{valid}} \leftarrow \emptyset$ ▷ Attributes satisfying the fairness constraint
if A is empty **then**
    return Leaf with majority label in X
end if
for all $A_i \in A$ **do**
    Compute $\text{G}_Y(A_i)$ and $\text{G}_S(A_i)$
    if $\text{G}_S(A_i) \leq \gamma$ **then**
        $A_{\text{valid}} \leftarrow A_{\text{valid}} \cup \{A_i\}$
    end if
end for
Sort A_{valid} by decreasing G_Y
for all $A^* \in A_{\text{valid}}$ **do**
    Create node and candidate node splitting on A^*
    `valid_split` $\leftarrow$ **true**
    for all values a of A^* **do**
        $X_a \leftarrow \{x \in X \mid x[A^*] = a\}$
        if $|X_a| \geq$ `min_samples` and $d < max\_depth$ **then**
            $T_a =$ DTFC(X_a, Y, S, γ, max_depth, $min\_samples$, $d+1$)
            if $T_a =$ **null then**
                `valid_split` $\leftarrow$ **false**
                break
            end if
            Attach T_a to current node under $A^* = a$
        else
            return Leaf with majority label in X
        end if
    end for
    if `valid_split` **then**
        return current node
    end if
end for
return null ▷ No fair and valid split found, i.e., backtrack

3.2 Extending the Constrained Single-Tree Approach with Backtracking

To allow for the examination of more possible splits when having one constrained objective function as design variant (ii) and imposing the fairness constraints in search space $\mathcal{T}$, we introduce DTFC (decision tree with fairness constraint), a variant using a backtracking mechanism that adaptively refines the tree construction process.

The core idea of the backtracking approach is to allow the decision tree algorithm to revisit and actively refine previous splitting decisions whenever the fairness threshold is compromised. In standard fair tree algorithms, splits in tree

T are chosen to balance the objective in y and T with the fairness constraint on s and T. However, this approach may become suboptimal as deeper splits could amplify bias introduced by earlier decisions. DTFC accepts a candidate split only if the information gain on the sensitive attribute s remains below a threshold γ; otherwise, alternative attributes are considered. If no valid split is found, the algorithm backtracks to explore other options at earlier nodes, allowing the algorithm to revisit and correct biased splits. Algorithm 1 provides a formal description of the procedure. The backtracking mechanism of DTFC enables deeper trees by exploring more splitting options than the constrained approach without backtracking. Rather than creating a leaf when fairness thresholds are violated, it searches for alternative splits that may better balance performance and fairness. If no valid configuration is found, DTFC returns an empty tree, ensuring that fairness is not compromised.

However, the backtracking approach is not without limitations. The primary drawback is its computational inefficiency, as the process of revisiting and evaluating alternative splits can introduce a substantial computational overhead compared to conventional tree-growing methods. Although backtracking allows for local adjustments at individual nodes, it does not necessarily optimize the tree globally. Decisions are still made on a greedy node-by-node basis, which may prevent the algorithm from achieving the overall optimal structure for performance and fairness trade-offs.

3.3 Dual-Tree Approach

The dual-tree approach, 2TFT (Dual Trees for Fairness Trade-Offs), separates the training process of fair decision trees into two distinct trees, one optimizing the performance-related objective function, the other optimizing the fairness-related objective function, to shift the trade-off between predictive performance and fairness constraints to $\mathcal{D}(\mathcal{T})$. As T_y, we will denote a tree that is exclusively performance-aware and thus has a fully performance-related objective in y and T_y, whereas T_s refers to a tree that is only fairness-aware and has a fully fairness-related objective in s and T_s. In the training procedure, T_y is trained as a standard decision tree, i.e. the splitting criterion optimizes only the information gain on the target variable, G_Y. In this way, T_y does not address any fairness metrics and is therefore fairness-unaware.

The splitting criterion of T_s is adjusted similarly to the one described in Sect. 2. We want T_s to be fairness-aware, but performance-unaware. Therefore, we do not maximize the information gain on y as we do for T_y. Rather, in each splitting step, we choose the attribute minimizing the information gain G_S we obtain on S, i.e., we maximize $-G_S$. However, also with this tree, we predict with respect to y. The class probability estimates are calculcated as the proportion of samples in a leaf that belong to a certain class. After training, 2TFT combines the class probability estimates of T_y and T_s in $\mathcal{D}(\mathcal{T})$ in a linear combination

$$T(x) = (1 - \gamma) \cdot T_y(x) + \gamma \cdot T_s(x), \tag{3}$$

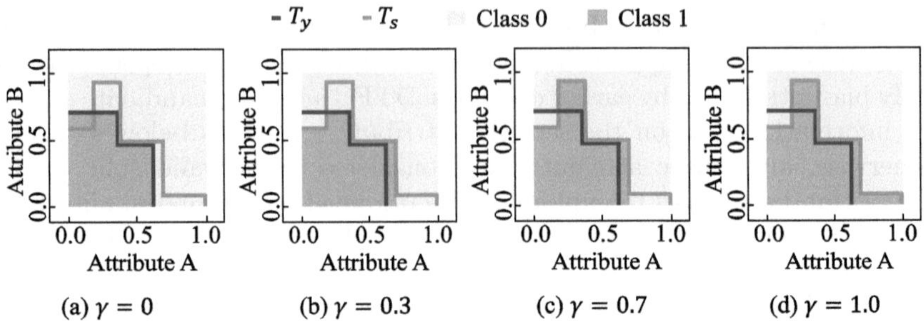

Fig. 1. Model partitions of 2TFT for varying γ in the case of two features. Partitions are colored by predicted class, depending on whether T_y or T_s dominates.

where γ is the parameter that allows for controlling the trade-off. Figure 1 visualizes the combined predictions, resulting in different partitions of $\mathcal{D}(\mathcal{T})$.

In the process of making the tree model fair, two separate trees allow for more flexibility regarding the structure of the decision trees, since their structures can be defined and optimized independently of each other.

3.4 Comparison Methods

The fair tree learning methods implemented for comparison correspond to the following design options: a combined objective optimizing both performance and fairness in one tree for design variant (i) (as introduced by Kamiran et al. [13]), and a constrained objective enforcing threshold on G_S during splitting for variant (ii) (as described by Castelnovo et al. [8]).

The combined-objective approach adjusts the splitting criterion by adding the information gain on Y (G_Y) and the negative gain on S (G_S), weighted by a trade-off parameter γ: $\texttt{max}(1-\gamma) \cdot G_Y - \gamma \cdot G_S$. The constrained approach maximizes G_Y subject to $G_S \leq \gamma$. If no attribute satisfies this constraint, a leaf node is created. By varying γ, we obtain a list of different models representing different possible fairness-performance trade-offs. These trade-offs can subsequently be expressed as sequences of values of performance and fairness measures.

4 Experimental Evaluation of the Trade-Off Curves

The experiments are designed to assess performance–fairness trade-offs across methods spanning the design space of fair tree learning. This section introduces the fairness and performance metrics used to construct trade-off curves, the evaluation criteria for analyzing them, and the experimental setup, including datasets, baselines, and hyperparameters. We also discuss interpretability and the runtime of the methods. Our Python implementation is openly available[1].

[1] https://github.com/kiarastempel/FairTreeExperiments

4.1 Evaluation Measures

We measure the performance of a decision tree T in $\mathcal{D}(\mathcal{T})$ using AUROC. The strength of discrimination can be measured using several different fairness metrics. We focus on the statistical parity difference [21], defined as

$$\text{SPD} = \left| P(\hat{Y} = 1 \mid S = 0) - P(\hat{Y} = 1 \mid S = 1) \right| . \tag{4}$$

The trade-off curve is created with AUROC on the x-axis and $1 - \text{SPD}$ on the y-axis. Based on this, we outline the following measures for analyzing the curve.

AUTOC. Computed as the area under the trade-off curve, AUTOC captures the overall balance between fairness and performance, aiming for points near the upper-right corner of the performance–fairness trade-off space. The curve is represented by a set of points obtained by varying the γ parameter, encoded in two lists: `SPD` (fairness values) and `AUROC` (corresponding performance values). All curves are extended by two points, where we add one point to $(0.5, \max(1 - \texttt{SPD}))$ and another one to $(1, \min(1 - \texttt{SPD}))$, to avoid penalizing steep curves with consistently high AUROC values but limited AUROC range. All points are sorted increasingly by AUROC before approximating the area under the curve using the trapezoidal rule [9]. Consequently, we formally define AUTOC as

$$\begin{aligned} \text{AUTOC} = \frac{1}{2}\Big[& (\texttt{AUROC}_0 - 0.5)\big(\max(1 - \texttt{SPD}) + (1 - \texttt{SPD}_0)\big) \\ & + \sum_{k=0}^{K} (\texttt{AUROC}_{k+1} - \texttt{AUROC}_k)\big((1 - \texttt{SPD}_k) + (1 - \texttt{SPD}_{k+1})\big) \\ & + (1 - \texttt{AUROC}_K)\big((1 - \texttt{SPD}_K) + \min(1 - \texttt{SPD})\big)\Big]. \end{aligned} \tag{5}$$

The measure is motivated by the definition of hypervolumes, as already mentioned in Sect. 2, to evaluate the quality of a set of solutions in multi-objective optimization, particularly in terms of their proximity to the Pareto front, i.e., the optimal trade-offs [3,5]. If we map each solution $T \in \mathcal{D}(\mathcal{T})$ to a point in the objective space $\mathcal{F}$, which represents the set of all possible outcomes of the objective functions evaluated on the models in $\mathcal{T}(X)$, then the hypervolume quantifies the volume of the region dominated by a set of solutions in $\mathcal{F}$.

Number of Points on Pareto Front. The Pareto front represents the set of solutions where no single objective can be improved without worsening the other [14]. In this context, a point is Pareto-optimal within one trade-off curve (we call this locally Pareto-optimal) if no other point in the trade-off curve achieves both higher fairness and better performance. The number of such points reflects the model's ability to generate optimal trade-offs, with higher counts indicating that the model provides users with more flexibility to prioritize fairness or performance as needed. Ideally, all trade-off points are Pareto-optimal, but this may be reduced by redundancy.

Number of Unique Points. This measure counts distinct trade-off combinations, considering identical AUROC and SPD values as duplicates. It reflects how solution diversity evolves under varying fairness constraints, indicating whether changes in the strength of a fairness constraint lead to gradual, smooth performance shifts or sudden, significant impacts.

Variance of Pairwise Distances. The distribution variance evaluates how evenly the points on the trade-off curve are distributed across the fairness-performance space. It is computed as the variance of pairwise Euclidean distances between all points on the curve, with a value of 0 for trade-off curves with a single unique point. Low variance may indicate that the curve is concentrated in specific regions or clusters, while a higher variance suggests a more uniform coverage of the decision space, which makes this measure relevant when assessing the practical usability of the curve and corresponding models.

4.2 Experimental Setup

For each dataset, the class labels Y and sensitive groups S are transformed to being either 0 or 1. Further, all categorical features are one-hot encoded. When training a single tree, for each numerical attribute, 10 equally distributed thresholds are tested for a split. For evaluation, as recommended by Claude and Bengio [16], we perform a 15-times hold-out in the outer loop, splitting the data randomly into $\frac{2}{3}$ of training data and $\frac{1}{3}$ of test data for each hold-out. In each hold-out, the hyperparameters defining the tree size are optimized on the training data using three-times cross-validation in an inner loop. We evaluate all hyperparameter combinations on the three folds by calculating the trade-off curve for all of them. Then, we choose the one where the averaged AUTOC is maximized, and compute all measures as described above. The best found hyperparameters are selected for retraining and evaluating the model on the held-out dataset.

When creating the trade-off curve for a specific hyperparameter combination, the interval of γ is set to $[0, 1]$ for those approaches with weighted combinations or soft constraints (i.e., one tree with combined splitting criterion, and two trees), while $[0, 0.2]$ is used for the approaches using hard fairness-related constraints (i.e., one tree with constrained splitting criterion, the backtracking version, and the relaxed threshold optimizer) to account for the smaller range of the information gain. We test 50 equally distributed values of γ for each run that creates a trade-off curve. To ensure consistency, all models were trained under identical hyperparameter settings. The selected hyperparameters are shown in Table 1. The value of `min_samples` refers to the proportion of instances of the dataset that a leaf node has to obtain at minimum, i.e., if at least one potential child node of the current node gets less than $n\cdot$ `min_samples` training instances, then the current node becomes a leaf. The maximum depth a tree can reach in the training process is set with `max_depth`. Since the backtracking approach is by far the slowest, we set a time limit of 10 h per dataset and evaluate the best hyperparameters out of all combinations that were tested in that time range.

After choosing the 15 best trade-off curves based on AUTOC, we evaluate the found curves subsequently on the other criteria described in Sect. 4.1.

Table 1. Tested hyperparameter combinations for all methods based on trees. For backtracking, we report all hyperparameter combinations that were tested within the time range of 10 h. The value of `min_samples` corresponds to the proportion of instances in the given dataset.

	Dataset	Tested combinations of (`max_depth`, `min_samples`)
General		All combinations with - `max_depth` = {4, 6, 8, 13} - `min_samples` = {0.25, 0.1, 0.01}
DTFC	COMPAS	(4, 0.25), (4, 0.1), (4, 0.01), (6, 0.25), (6, 0.1), (6, 0.01), (8, 0.25), (8, 0.1)
	Adult	(4, 0.25), (4, 0.1), (4, 0.01), (6, 0.25), (6, 0.1)
	Dutch	(4, 0.25), (4, 0.1), (4, 0.01), (6, 0.25), (6, 0.1)

4.3 Datasets

We apply the described methods to the COMPAS dataset, the Adult dataset, and the Dutch Census dataset from 2001 [18].

The COMPAS dataset is widely used to predict criminal recidivism and includes features like prior convictions, age, gender, and race. In this context, race is considered as sensitive attribute S. Another widely studied dataset is the Adult dataset, derived from U.S. Census data, and is commonly used to predict whether an individual earns over \$50K annually. It includes features such as age, education, occupation, race, and gender, the latter typically serving as the sensitive attribute S, highlighting potential gender bias in income prediction. Similarly, the Dutch Census dataset, based on the data from 2001, contains demographic information and is, like Adult, commonly used to explore gender-based discrimination, also in these experiments.

4.4 Baseline

Our baseline is the threshold optimizer, an implementation by fairlearn [6], based on work by Hardt et al. [12]. Unlike tree-based in-processing methods that incorporate fairness constraints during training, this is a post-processing algorithm that adjusts prediction thresholds after training to enforce fairness. It modifies decision thresholds separately for each sensitive group to exactly satisfy a fairness criterion without altering the underlying model. However, in contrast to in-processing methods, it requires sensitive attributes to be available at prediction time. Since this post-adjustment of the threshold is optimal, we expect not to be better than postprocessing. In our implementation, we used a forked version of the repository[2], which includes a relaxed variant of the threshold optimizer, but is open for merge with the original fairlearn library. Here, a tolerance value for the given fairness metric can be specified to create a trade-off curve. As a base model, we pass a decision tree classifier.

4.5 Results and Discussion

In this section, we discuss our experimental results on the datasets described in Sect. 4.3 and the methods introduced in Sect. 3.

Analyzing the trade-off curves of the implemented tree variants provides insights into their ability to balance performance and fairness objectives. The primary focus is to assess how different decision tree models perform under varying constraints, represented by the trade-off parameter γ. This evaluation aims to highlight the strengths and weaknesses of the different approaches and their potential applicability in real-world scenarios.

The trade-off curves illustrated in Fig. 2 and the evaluation criteria listed in Table 2 demonstrate how the performance of each tree variant evolves under different fairness constraints, i.e., increasing γ stepwise in the defined range. The curves are computed by averaging the 15 best curves obtained from the described hold-out procedure. Table 2 gives the measures used in the evaluation process, computed on the optimized curves. For each dataset, we test whether the method with the highest value is significantly better using a pairwise two-sided t-test with $p < 0.05$. This verifies whether the 15 found values of one method differ statistically from those of another. If so, we mark the value with *. For this test, we do not consider the baseline. It serves as a visual comparison, although its AUTOC values might be lower due to the steepness of the curves.

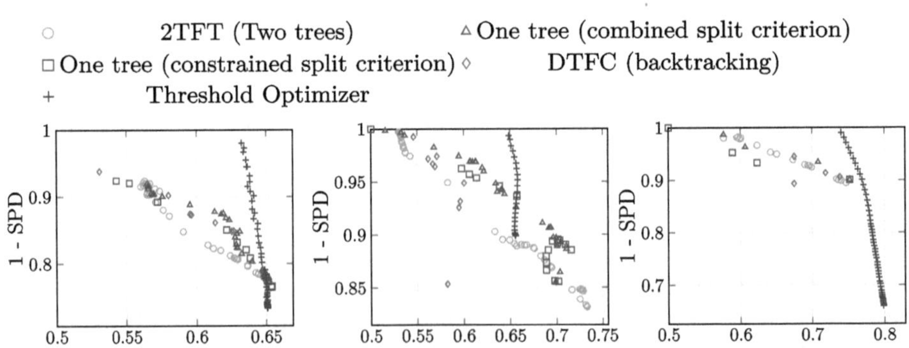

(a) Compas (b) Adult (c) Dutch Census

Fig. 2. We compare the trade-off curves of the described methods on multiple datasets. The 15 best curves were averaged for each dataset to create the shown curves.

---
[2] https://github.com/taharallouche/fairlearn.

Table 2. Evaluation of the trade-off curves for different methods and splitting criteria (SP). For a detailed description of the evaluation metrics we chose, we refer to Sect. 3.1. We print in boldface the best value of each metric for each dataset, where the baseline is not considered. The averaged AUTOC value is marked with * if there was a statistical significance.

Dataset	Method	AUTOC	Number of local Pareto points	Number of unique points	Number of unique Pareto points	Variance of pairwise distances
Adult	2TFT (Two trees)	0.434 ± 0.027	39.3 ± 9.7	**23.7*** ± 8.4	**7.3** ± 8.4	**0.0095*** ± 0.0056
	One tree (combined SP)	**0.451** ± 0.005	25.2 ± 9.4	10.7 ± 4.1	3.3 ± 2.1	0.0041 ± 0.0013
	One tree (constrained SP)	0.448 ± 0.005	29.7 ± 9.5	8.1 ± 1.1	2.1 ± 0.9	0.0021 ± 0.0012
	DTFC (backtracking)	0.220 ± 0.207	**42.9** ± 6.6	3.9 ± 2.7	0.8 ± 0.8	0.0015 ± 0.0016
	Threshold optimizer	0.339 ± 0.205	19.0 ± 18.9	28.5 ± 17.8	22.6 ± 14.6	0.0014 ± 0.0010
COMPAS	2TFT (Two trees)	0.402 ± 0.011	34.5 ± 10.3	**12.5*** ± 4.3	**4.8** ± 3.1	**0.0062*** ± 0.0034
	One tree (combined SP)	0.402 ± 0.009	35.5 ± 11.8	8.0 ± 1.3	3.2 ± 1.2	0.0036 ± 0.0017
	One tree (constrained SP)	0.400 ± 0.008	39.1 ± 17.4	7.9 ± 1.1	2.3 ± 1.4	0.0031 ± 0.0011
	DTFC (backtracking)	**0.430*** ± 0.005	**48.7*** ± 1.1	5.1 ± 0.6	1.1 ± 0.9	0.0009 ± 0.0003
	Threshold optimizer	0.406 ± 0.018	10.5 ± 4.9	36.1 ± 5.6	25.7 ± 7.0	0.0056 ± 0.0014
Dutch	2TFT (Two trees)	0.463 ± 0.004	46.5 ± 4.9	**4.3*** ± 0.9	**0.5** ± 0.7	**0.0053** ± 0.0023
	One tree (combined SP)	**0.464** ± 0.004	49.8 ± 0.4	2.9 ± 0.6	0.2 ± 0.4	0.0025 ± 0.0002
	One tree (constrained SP)	0.459 ± 0.003	**50.0** ± 0.0	3.0 ± 0.0	0.0 ± 0.0	0.0020 ± 0.0007
	DTFC (backtracking)	0.455 ± 0.010	49.5 ± 0.5	2.1 ± 1.0	**0.5** ± 0.5	0.0008 ± 0.0011
	Threshold optimizer	0.419 ± 0.034	30.3 ± 8.1	47.7 ± 4.6	18.3 ± 7.8	0.0090 ± 0.0049

In terms of the AUTOC measure, we notice that the one-tree approach with the combined weighted objective function achieves the best value for two of the datasets (Adult and Dutch), but the difference is not statistically significant. 2TFT, the approach with two objective functions (i.e., with trees T_y and T_s), shows a comparable AUTOC value on Dutch Census, with the combined objective function being only slightly better. The backtracking variation DTFC is significantly better on the COMPAS dataset in terms of AUTOC, even though fewer combinations of hyperparameters were tested. The number of local Pareto points, however, reaches the maximum for Adult and COMPAS when having one constrained objective extended with the backtracking variation. Only for Dutch, the values are rather similar across all methods, with the constrained objective (without backtracking) being slightly better than the others.

Regarding the number of unique points per curve, we obtain substantially different results. 2TFT consistently performs best, making it the most suitable method for producing diverse solutions across the trade-off space and enabling finer granularity in decision-making, as seen in Fig. 2. The combined objective function for one tree can be identified as the second-best. The same pattern holds for distribution variance, indicating that the two-tree method explores the trade-off space more extensively, approaching even the baseline. The number of unique Pareto points—defined as the intersection of Pareto-optimal and unique points—is also highest for the two-tree method, which shares the lead with DTFC

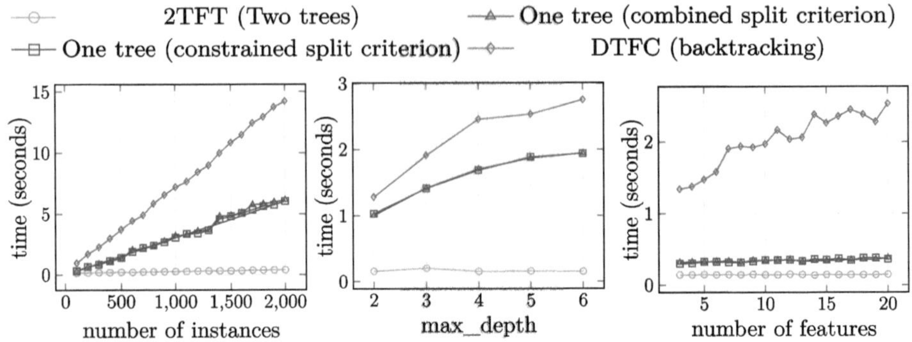

Fig. 3. Runtime analysis on Adult. Each run generates a trade-off curve using 100 instances, depth 3, and 3 features, increasing one of the parameters stepwise.

on the Dutch dataset. However, we note that most of the methods show a very low average number of unique Pareto points.

The results suggest that having one objective may be more suitable when aiming for a smaller set of strong outcomes, although we can identify regions in the trade-off space in Fig. 2, where points of the dual-approach dominate. In contrast, if a broad variety of trade-off solutions is desired, having two objectives, separated into two trees, may be more appropriate. Additionally, in Fig. 2, the backtracking variant appears to fill the gaps in the trade-off curve of the constrained approach without backtracking, even though fewer hyperparameters were tested due to runtime limitations.

The evaluation approach used in this study has certain limitations. One key challenge is the difficulty of reducing an entire performance curve to a single numerical value, as this simplification may not fully capture the complexities present in the trade-off process. Further, the AUTOC metric, while useful for quantitative comparison, may not always align with human judgment in selecting the most suitable trade-off curve.

Figure 3 analyzes running times depending on the number of instances, maximum depth of the tree, and number of features in the dataset. In each run, one trade-off curve is created using 100 instances, a maximum depth of 3, and 3 features, where one of the parameters is increased stepwise. 2TFT runs much faster than one-tree variants, as it builds only two trees reused across the curve. In contrast, one-tree models retrain a tree for each γ value. Among them, the backtracking approach is the slowest due to its extensive search. If performance differences are minimal, the dual-tree variant may be the more practical choice.

From an interpretability perspective, the dual-tree approach offers advantages over the single-tree approach. In the dual-tree method, understanding the trade-off curve requires analyzing only two trees, regardless of the number of trade-off parameter values of γ used to construct the curve. In contrast, the single-tree method produces a separate tree for each γ, complicating interpretation as the trade-off space grows. The approach with two objective functions provides additional transparency by allowing an examination of the predictions separately from the individual trees. This enables a clear understanding of how the model would behave if it was optimized solely for performance or fairness, facilitating a direct comparison of decision paths and decisions between the two objectives. In contrast, the single-tree method lacks this separation, offering only a single set of predictions and decision paths.

5 Conclusion

In this study, we revisited fair tree learning and compared methods for balancing fairness and performance in decision trees across multiple datasets. We examined three different approaches for designing fair tree learning algorithms: (i) one combined objective function in y, s and T, (ii) one objective function in y and T with a constraint in s and T, and (iii) two objective functions, the first being a function only in y and T, and the second a function in s and T, resulting in two separate trees. We introduced a backtracking variant which can be classified into (ii) and an approach for (iii). Methods designed according to (i) or (ii) yielded smaller unique sets of strong outcomes, and the backtracking variant consistently achieved a high number of local Pareto points. The method from design variant (iii) provided broader and more diverse trade-off curves. The two-tree method also offers interpretability advantages, requiring only two trees to understand the full trade-off, and allowing explicit comparison between performance- and fairness-driven decisions.

In future work, it would be intriguing to study, amongst others, the behavior of optimal decision trees in this space and non-linear combinations of models.

Acknowledgments. The research in this paper was supported by the "TOPML: Trading Off Non-Functional Properties of Machine Learning" project funded by Carl Zeiss Foundation, grant number P2021-02-014.

Disclosure of Interests. The authors have no competing interests to declare that are relevant to the content of this article.

References

1. Aghaei, S., Azizi, M.J., Vayanos, P.: Learning optimal and fair decision trees for non-discriminative decision-making. AAAI'19/IAAI'19/EAAI'19 (2019)
2. Auger, A., Bader, J., Brockhoff, D.: Theoretically investigating optimal μ-distributions for the hypervolume indicator: First results for three objectives. In: Parallel Problem Solving from Nature, PPSN XI. vol. 6238, pp. 586–596 (2010)
3. Auger, A., Bader, J., Brockhoff, D., Zitzler, E.: Hypervolume-based multiobjective optimization: theoretical foundations and practical implications. Theor. Comput. Sci. **425**, 75–103 (2012)
4. Bandyopadhyay, S., Saha, S.: Unsupervised Classification: Similarity Measures, Classical and Metaheuristic Approaches, and Applications, chap. Some Single- and Multiobjective Optimization Techniques, pp. 17–58 (2013)
5. Beume, N., Naujoks, B., Emmerich, M.: Sms-emoa: multiobjective selection based on dominated hypervolume. Eur. J. Oper. Res. **181**(3), 1653–1669 (2007)

6. Bird, S., et al.: Fairlearn: a toolkit for assessing and improving fairness in AI. Tech. Rep. MSR-TR-2020-32, Microsoft (2020). https://www.microsoft.com/en-us/research/publication/fairlearn-a-toolkit-for-assessing-and-improving-fairness-in-ai/
7. Breiman, L., Friedman, J.H., Olshen, R.A., Stone, C.J.: Classification and regression trees. Routledge (1984)
8. Castelnovo, A., Cosentini, A., Malandri, L., Mercorio, F., Mezzanzanica, M.: Fftree: a flexible tree to handle multiple fairness criteria. Inform. Process. Manage. **59**(6), 103099 (2022)
9. Dahlquist, G., Åke Björck: Numerical Methods in Scientific Computing, Volume I. Society for Industrial and Applied Mathematics (SIAM), Philadelphia, PA (2008)
10. Domingos, P., Hulten, G.: Mining high-speed data streams. In: Proc. of the Sixth ACM SIGKDD International Conference on Knowledge Discovery and Data Mining, pp. 71–80. KDD '00 (2000)
11. Freitas, A., Brookhouse, J.: Evolutionary algorithms for fair machine learning. In: Handbook of Evolutionary Machine Learning, pp. 507–531 (2024)
12. Hardt, M., Price, E., Srebro, N.: Equality of opportunity in supervised learning. In: Advances in Neural Information Processing Systems. vol. 29 (2016)
13. Kamiran, F., Calders, T., Pechenizkiy, M.: Discrimination aware decision tree learning. In: 2010 IEEE International Conference on Data Mining, pp. 869–874 (2010)
14. Kang, S., Li, K., Wang, R.: A survey on pareto front learning for multi-objective optimization. J. Membrane Comput. (2024). https://doi.org/10.1007/s41965-024-00170-z
15. La Cava, W.G.: Optimizing fairness tradeoffs in machine learning with multiobjective meta-models. In: Proc. of the Genetic and Evolutionary Computation Conference, pp. 511–519 (2023)
16. Nadeau, C., Bengio, Y.: Inference for the generalization error. In: Advances in Neural Information Processing Systems. vol. 12. MIT Press (1999)
17. Pereira Barata, A., Takes, F.W., van den Herik, H.J., Veenman, C.J.: Fair tree classifier using strong demographic parity. Mach. Learn. **113**(5), 3305–3324 (2023)
18. Quy, T.L., Roy, A., Iosifidis, V., Zhang, W., Ntoutsi, E.: A survey on datasets for fairness-aware machine learning. Wiley Interdisciplinary Reviews: Data Mining and Knowledge Discovery **12** (2022)
19. Raff, E., Sylvester, J., Mills, S.: Fair forests: regularized tree induction to minimize model bias. In: Proc. of the 2018 AAAI/ACM Conference on AI, Ethics, and Society, pp. 243–250 (2018)
20. Sharma, S., Kumar, V.: A comprehensive review on multi-objective optimization techniques: past, present and future. Arch. Comput. Methods Eng. **29**, 5605–5633 (2022)
21. Verma, S., Rubin, J.: Fairness definitions explained. In: Proc. of the International Workshop on Software Fairness, pp. 1–7 (2018)
22. Zhang, J., Beschastnikh, I., Mechtaev, S., Roychoudhury, A.: Fair decision making via automated repair of decision trees. In: IEEE/ACM International Workshop on Equitable Data & Technology, pp. 9–16 (2022)
23. Zhang, W., Ntoutsi, E.: Faht: an adaptive fairness-aware decision tree classifier. In: Proc. of the Twenty-Eighth International Joint Conference on Artificial Intelligence, pp. 1480–1486 (2019)
24. Zhang, W., Zhao, L.: Online decision trees with fairness. arXiv preprint https://arxiv.org/abs/2010.08146 (2020)
25. Zitzler, E., Künzli, S.: Indicator-based selection in multiobjective search. In: Parallel Problem Solving from Nature. vol. 3242 (2004)

Neuron-Level Architecture Search for Efficient Model Design

Artem Lunev(✉) and Nikolay Nikitin

ITMO University, Saint-Petersburg, Russia
aelunev@itmo.ru

Abstract. Neural Architecture Search (NAS) methods are used to automate the design of neural network architectures. On many popular datasets, architectures discovered through NAS already outperform manually designed ones. However, most existing approaches rely on predefined sets of common operations, such as fully connected or convolutional layers with a fixed set of their parameters. As a result, these methods are limited in their ability to discover fundamentally new architectures with complex topologies. In this work, we propose an approach that designs architectures at the most fundamental level—by directly manipulating neurons and their connections. In addition, we focus on finding computationally efficient models by eliminating redundant components that may be inherent in manually constructed layers. To achieve this, we introduce an evolutionary algorithm that balances computational cost with architecture performance. Experimental results demonstrate that our approach can discover architectures that match the performance of pruned, manually designed networks while achieving approximately a twofold reduction in FLOPs on image classification datasets. Additionally, the use of specialized tools for sparse model acceleration allows our architectures to achieve up to a 20% reduction in inference time on the CPU compared to manually designed networks.

Keywords: Neural Architecture Search · AutoML · Evolutionary Algorithms · Computational Efficiency

1 Introduction

Modern neural networks achieve state-of-the-art performance across various domains, including image processing, machine translation, speech recognition, and many others. However, designing an efficient neural network architecture remains a challenging task, requiring substantial computational resources and expert knowledge. In recent years, the field of automated neural architecture design, known as Neural Architecture Search (NAS) [1,2], has been actively developing.

NAS methods aim to automate the process of finding optimal architectures, reducing the need for manual tuning. On several benchmark datasets, automatically generated architectures have already surpassed manually designed ones

S. Džeroski et al. (Eds.): DS 2025, LNAI 16090, pp. 191–203, 2025.
https://doi.org/10.1007/978-3-032-05461-6_13

[3,4]. However, automated architectural design has not yet become widespread and the architectures discovered through NAS rarely introduce fundamentally new or innovative design principles. This is primarily because existing NAS approaches rely on predefined sets of commonly used operations from manually designed architectures, such as fully connected layers or convolutional layers with a discrete set of parameters (e.g., number of neurons, kernel size, and stride). Moreover, the ability to explore diverse architectural topologies is essential for capturing the complexity of modern neural network designs [4,5].

Another limitation of current NAS approaches is their task-specific nature. Most methods are designed to optimize architectures for a particular problem, such as segmentation, making it difficult to develop fully automated solutions without human intervention.

In this work, we propose an approach to architecture design that operates at the most fundamental level—manipulating individual neurons and their connections. We hypothesize that this will enable the discovery of architectures that differ significantly from manually crafted ones, featuring more complex topologies and unconventional operations. Furthermore, we expect this method to lead to more computationally efficient models by eliminating redundant components inherent in traditional designs.

To achieve this, we implement a multi-objective evolutionary algorithm that jointly optimizes for model accuracy and computational efficiency. Our method was evaluated on the widely used CIFAR-10 image classification benchmark. The experiments demonstrate that our approach can generate architectures that outperform hand-designed small convolutional networks in terms of computational complexity.

Due to the sparsity of the resulting architectures, their acceleration can be achieved by deploying the models with a specialized CPU runtime, such as DeepSparse [6]. In this setting, our sparse models can outperform traditional convolutional networks in inference speed. These results support the viability of low-level architecture design and highlight its potential advantages in practical deployment scenarios.

Code is available at https://github.com/artemlunev2000/Neuron-Level-NAS.

2 Background

Neural Architecture Search. One of the earliest contributions to the field of Neural Architecture Search is presented in [7]. This seminal work established a foundational framework in which a meta-controller network generates neural network architectures and iteratively refines them using reinforcement learning, typically employing the REINFORCE policy gradient algorithm. The controller constructs architectures by selecting from a predefined set of computational blocks, such as convolutional and fully connected layers with various hyperparameter configurations, and is capable of incorporating skip connections to enhance flexibility. They achieved an error rate of 3.65% on the CIFAR-10 dataset during the architecture search, which was conducted using 800 GPUs.

This idea was further developed in [8], where the authors introduced the concept of a cell-based search space. In this setting, the outer macro-architecture is fixed, composed of repeating normal and reduction cells, and the NAS algorithm only searches for the internal structure of these cells. This significantly reduces the search space, resulting in faster and more efficient discovery of high-performing models.

Subsequent research [3] demonstrated that evolutionary algorithms can discover superior architectures in less time within the same cell-based search space. This is achieved by iteratively applying mutation operations to evolve architectures over generations. One of the key innovations introduced in this work was the use of aging evolution, a strategy where the oldest individuals are removed from the population rather than the weakest, as in classical evolutionary algorithms. This approach prevents premature convergence by encouraging continual exploration of the search space.

A key limitation of these NAS approaches is the computational cost associated with training a large number of candidate architectures to obtain validation accuracy estimates. One-shot NAS methods address this issue by training a single over-parameterized supernet that encompasses all candidate architectures. This supernet is typically optimized once using gradient-based methods. Individual architectures are then evaluated by sampling sub-networks from the trained supernet. Among such approaches, the DARTS family of methods has gained particular prominence [9–11].

Another major challenge in modern NAS research is the design of the search space itself. Formally, the search space is defined as the set of all architectures that can be generated by search algorithm. An effective search space should be expressive enough to contain high-performing architectures, while also being structured in a way that facilitates efficient exploration. Existing types of search spaces include macro search spaces [12,13], chain-structured spaces [14], cell-based architectures [3,8], hierarchical designs [15,16], and grammar-based representations, most notably those based on context-free grammars [4,17]. The latter two are currently receiving increased attention due to their potential to represent highly diverse and unconventional architectures.

Network Pruning. Network pruning is a widely used model compression technique aimed at reducing the size and computational cost of neural networks by removing redundant or less important parameters. This can involve eliminating individual weights, neurons, channels, or even entire layers from a trained model. The primary objectives of pruning are to reduce inference latency, memory footprint, and energy consumption, ideally without significantly degrading model accuracy.

Pruning methods are roughly categorized as either unstructured or structured, depending on the components being removed. In unstructured pruning, individual weights are eliminated based on certain criteria, typically their magnitude. This results in sparse weight matrices with arbitrary patterns of zeros. One of the classical methods in this category is Gradual Magnitude Pruning

(GMP) [18], which iteratively prunes weights with the smallest magnitudes during training or fine-tuning.

In contrast, structured pruning removes entire structural elements of the network, such as layers, channels [19], or attention heads [20]. This results in smaller yet densely connected sub-networks that are better aligned with hardware acceleration libraries and are thus more likely to yield actual speedups during inference.

While unstructured pruning can significantly reduce the number of parameters, it rarely leads to faster execution on GPUs, which are optimized for dense matrix operations and typically ignore sparsity. However, on CPUs, the situation is more favorable. Recent developments have demonstrated that sparsity-aware runtimes, such as DeepSparse [6], can effectively exploit unstructured sparsity by dynamically restructuring the computation graph, leading to measurable speed improvements for pruned models. In some cases, the use of such tools can lead to CPU inference speedups of up to 29x [21].

3 Methods

3.1 Architecture Representation

In this work, directed acyclic graphs (DAGs) are used as the primary representation for neural network architectures. In these graphs, nodes correspond to neurons, while edges represent connections between them. Each connection is associated with a trainable weight. The value of a neuron is computed as the weighted sum of the outputs of all neurons connected to it via incoming edges, similar to fully connected layers. The ReLU activation function is applied to all neurons except for those in the final layer, where the activation function is determined by the task, for example, softmax in the case of classification. Input and output neurons are fixed from the architecture initialization.

When constructing a neural network from a graph, the nodes are first topologically sorted to assign each neuron to a specific network layer. The nodes are arranged such that all edges point from lower-layer neurons to upper-layer neurons. Each layer consists of a one-dimensional set of neurons. Next, weight matrices are constructed. For every pair of layers that are connected, a weight matrix is created with shape (number of neurons in the lower layer × number of neurons in the upper layer). Since the layers are not fully connected, most weight matrices will contain a significant number of zero elements.

A toy example of a graph is shown in Fig. 1. The vertices are grouped into four layers, forming a four-layer architecture. Sparse weight matrices are formed between each pair of layers that are connected: W_{12} (2×3), W_{13} (2×2), W_{23} (3×2), W_{24} (3×2), and W_{34} (2×2).

3.2 Evolutionary Algorithm

The architectural refinement process is driven by an evolutionary search method. During the evolutionary process, candidate architectures are optimized not only

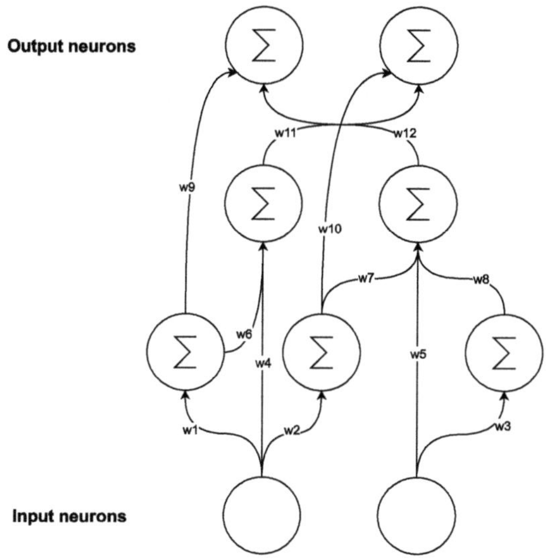

Fig. 1. Architecture representation as a graph. The graph corresponds to a four-layer architecture.

for a primary objective, such as classification accuracy, but also for computational efficiency, measured in FLOPs (floating-point operations). To take several criteria into account, we adopt a modified version of the NSGA-II [22] algorithm for multi-objective optimization. The full procedure is described in Algorithm 1. Our approach incorporates aging evolution [3], meaning that the oldest individuals in the population are removed at each iteration, rather than those with the worst performance.

The initial population size is 60. To ensure diversity in the initial population, individuals are sampled across a range of FLOPs and with varying numbers of layers. For instance, for CIFAR-10 dataset, the range is 3 to 6 layers. In each iteration, 15 parent individuals are selected from the current population via tournament selection with a tournament size of 3. These parents undergo a mutation to generate new children which are added to the population. To maintain a fixed population size of 60, the 15 oldest individuals are then removed. Tournament selection is guided by Pareto dominance: individuals are ranked and then selected based on their non-dominated fronts. This means that individuals in the most preferred first rank are not dominated by any others, the second rank consists of individuals dominated only by the first rank, and so on. In case of a tie, the individual with the largest crowding distance is preferred (i.e., one located in less crowded regions of the objective space).

One of the main challenges in designing mutations is ensuring architectural diversity. If, for example, edges are added uniformly from the list of all possible connections, or if neurons are randomly placed across layers, the resulting architectures tend to be highly similar. They will exhibit approximately the same

Algorithm 1. Multi-Objective Evolutionary Algorithm for Joint Optimization of Accuracy and FLOPs

Require: number of iterations T, number of mutations per iteration M

1: **Initialization:** population P_1, assign rank based on Pareto dominance to all population individuals.
2: **for** $t = 1$ to T **do**
3:     **for** $m = 1$ to M **do**
4:         Select parent from population P_t with tournament selection based on rank and crowding distance.
5:         Mutate parent to generate child, and train resulting model.
6:         Add child to population.
7:     **end for**
8:     Remove M oldest individuals from population.
9:     Assign new Pareto ranks to the population.
10: **end for**
11: **Output:** Architecture from the population with the highest validation accuracy or lowest FLOPs.

number of neurons per layer and similar edge distributions. To address this, two key modifications were introduced: (1) mutations are tied to specific layers, and (2) spatial relationships between neurons are leveraged to form connections, similar to the way convolutions operate. The second aspect is explored in more detail below.

The process of determining the spatial position of a neuron is as follows. The spatial locations of input layer neurons are predefined and correspond to the (x,y) coordinates of the respective pixels. For subsequent layers, the spatial position of a neuron is computed as the average position of the neurons that have outgoing connections to it. This also applies to skip connections. An example of spatial locations, with the first layer positions predefined and the remaining ones computed, is shown in Fig. 2.

These spatial positions are then used to define the probability of an edge forming between neurons. The smaller the Euclidean distance between the spatial positions of two neurons the higher the probability of the emergence of a connection after mutations. This approach results in architectures that resemble convolutional networks, where connections are more likely between spatially close neurons, while still allowing some degree of flexibility. The probability of an edge forming is computed using the following formula:

$$\exp\left(\frac{-\mathrm{dist}}{4 + \mathrm{layer} \cdot 1.5}\right) \tag{1}$$

where dist is the Euclidean distance between the spatial positions of the neurons, and layer indicates the index of the layer that the added edge will lead to. The layer index is included in the formula to progressively weaken the constraint of preferential connectivity between spatially proximate neurons as the network deepens. In the final output layer, where neurons correspond to classification categories, this rule is not applied. In contrast, when removing edges, the oppo-

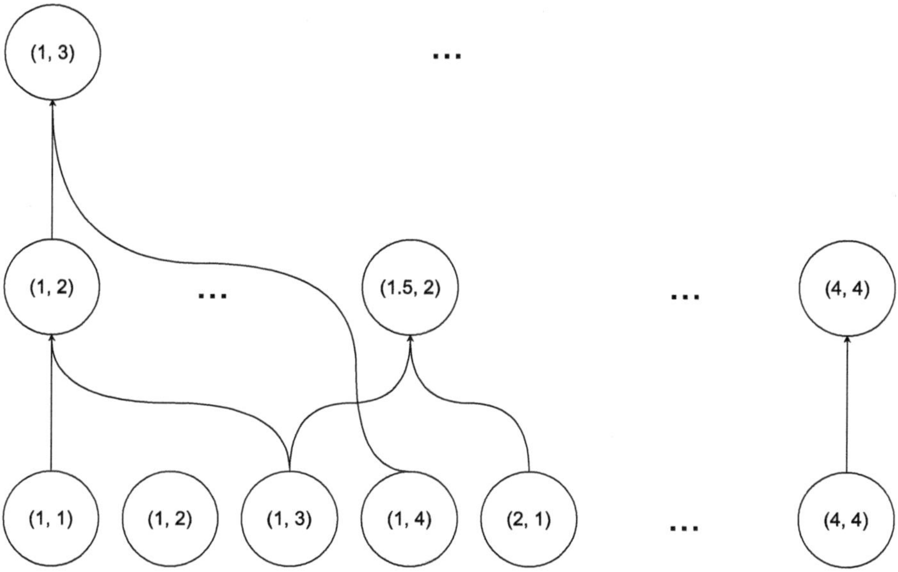

Fig. 2. Example of spatial locations calculation. The first layer spatial locations are predefined. Upper layer nodes calculate spatial locations based on incoming edges.

site principle is applied: the farther apart two neurons are in space, the higher probability of removing their connection. Ultimately, 4 mutation operations were implemented, as described below. All proportions and details of the mutations were selected empirically based on experiments.

1. Adding new nodes to the graph. New nodes are introduced either as a new layer inserted between two randomly selected existing layers or as additions to an existing layer. The number of new nodes varies between 12% and 32% of the current total graph node count. Since nodes cannot exist in isolation, edges are added to connect them to the network. Initially, each new node receives a single random incoming edge from a node in the preceding layer, establishing its initial spatial placement. Additional incoming and outgoing edges are then introduced, connecting the new nodes to adjacent layers in accordance with predefined probability distributions. Furthermore, with a certain probability, skip connections are created between new nodes and non-adjacent layers to enhance network connectivity.
2. Adding new edges to the graph. The number of new edges ranges from 20% to 32% of the current amount of edges. The process involves selecting three pairs of layers to add new edges between them. The probability of selecting a given pair is inversely proportional to the difference in layer indices, meaning that adjacent layers (e.g., layer 1 and layer 2) are more likely to be chosen than more distant pairs (e.g., layer 1 and layer 3). Once three pairs are selected, one is randomly designated as the primary pair, receiving 60% of the new edges, while the other two pairs share the remaining 40%. The primary

pair is the main focus of architectural change, receiving most of the added edges. The other two pairs are auxiliary, ensuring that the mutation affects multiple parts of the architecture. This approach helps speed up the search process by exploring several potential improvements at once. The probability distributions governing edge formation follow the same logic described earlier. Selecting multiple layer pairs rather than a single pair ensures significant modifications to the network, making it possible to receive meaningful structural changes.

3. Removing existing edges. Removes 12% to 22% of existing edges. As in the edge addition process, three layer pairs are selected where edges will be pruned. However, in this case, the probability of selecting a layer pair increases with the distance between layers, promoting the removal of long-range connections.
4. Layer deletion. When deleting the n-th layer, all edges between pairs of layers (n − 1, n) and (n, n+1) are removed, as well as all skipping links involving nodes of the n-th layer. Then edges are added between the pair of layers (n−1, n+1) in an amount equal to 80% of the removed edges between layers (n − 1, n) and (n, n+1). Also, 95% of the edges that corresponded to skip connections are returned. During re-adding these edges, all neurons are assumed to retain their previously assigned spatial positions from before the edges were removed. As a result, this mutation reduces both the number of edges and the number of nodes.

The visualization of all 4 mutations is shown in Fig. 3. Note that in the case of the first and second mutation, initializing the weights of the newly introduced edges with zeros ensures that the neural network continues to represent the same function as it did prior to the mutation. This allows training to continue from the pre-mutation state, leveraging previously learned parameters to accelerate both training convergence and architecture search. While preserving functionality is not feasible for the other mutations, existing edge weights are retained wherever possible, further contributing to improved convergence speed.

4 Experiments

4.1 Experimental Details

Dataset. Experiments were conducted on the CIFAR-10 dataset, a widely used image classification benchmark consisting of 60,000 RGB images of size 32 × 32 pixels, divided into 10 classes such as airplanes, birds, and cats. The standard split includes 50,000 training images and 10,000 test images. For our experiments, we reserved 40,000 images from the original training set for actual training, while the remaining 10,000 images were used as a validation set to monitor the progress of the evolutionary search. No data augmentation was applied during training.

Training Details. All architectures in the initial population were trained for 60 epochs using a fixed learning rate of 1e−3. During mutations, offspring architectures inherited weights from their parents wherever possible. That is, when

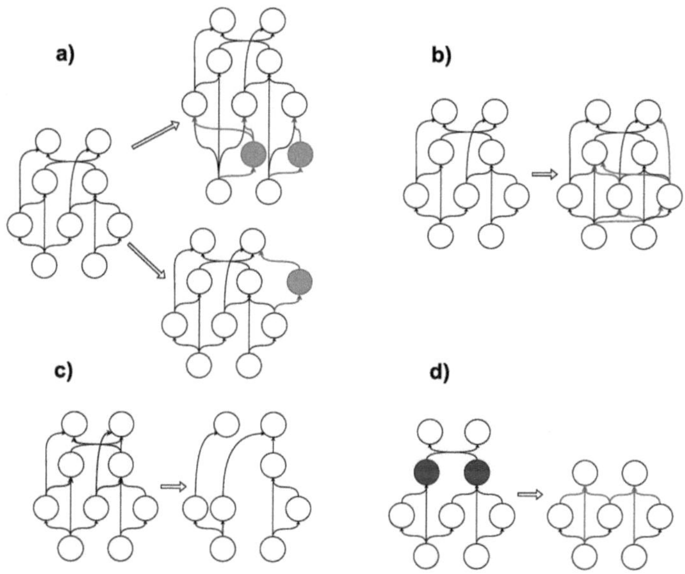

Fig. 3. Mutations visualization. (a) Addition of new nodes and their adjacent edges, two variants: with and without forming a new layer. (b) Addition of edges. (c) Removal of edges. (d) Removal of a layer along with its adjacent edges, followed by insertion of new edges to preserve connectivity.

weights or neurons were added or removed, the existing weights were preserved without reinitialization. For mutations that added new parameters to the architecture, training was performed in two phases. First, only the newly introduced weights were trained for 15 epochs with a learning rate of 5e−4, allowing the architecture to integrate new components while keeping the original structure stable. This was followed by fine-tuning the entire network for 20 additional epochs with a reduced learning rate of 1e−4. For mutations that removed parameters, the entire architecture was trained for 50 epochs, starting from a learning rate of 7e−4 and decaying to 1e−4. In all training procedures, the Adam optimizer was used. The full evolutionary search was performed on a single T4 GPU and completed within approximately two days.

4.2 Evolutionary Search

Figure 4 illustrates the progression of architectures during the evolutionary search process. It compares the first and seventy-fifth generations in terms of accuracy and computational complexity (measured in FLOPs). The red dots represent manually designed architectures, which are relatively simple pruned convolutional networks that were manually tuned to achieve a strong trade-off between accuracy and FLOPs. Pruning was performed progressively during training using the SparseML library [23] with Gradual Magnitude Pruning. The target sparsity levels varied across layers, mostly ranging from 80% to 90%. For example, the manual architecture with the highest accuracy in the figure consists of two convolutional layers followed by a pooling layer and two fully connected layers. By the 75th generation, the evolved architectures (blue dots) outperform the manual ones in both accuracy and computational efficiency.

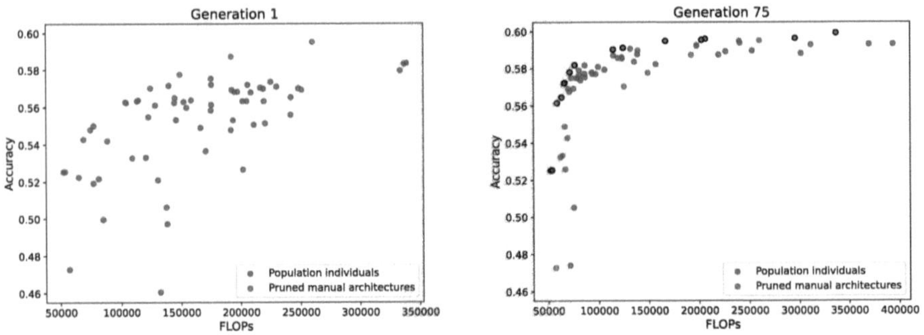

Fig. 4. Comparison of architectures from Generation 1 and Generation 75 with manually designed architectures. Pareto-optimal individuals are highlighted on the right image.

Since FLOPs are not a definitive indicator of actual model speed, we also conducted experiments to explicitly compare CPU inference time between the architectures discovered by our method and manually designed architectures. The main advantage of our models is their sparsity, which is well-leveraged during runtime by inference engines such as DeepSparse [6]. Therefore, we compared the inference time of our models against pruned convolutional models in the DeepSparse runtime. The results are shown in Fig. 5. The inference time was measured using data with a batch size of 128, which was chosen because it provides stable measurements, while smaller batch sizes are more susceptible to measurement noise. One of the best models from the last generation of our search proves to be approximately 20% faster than the manually designed model. However, discovered models do not show better inference performance across all accuracy ranges. A numerical comparison of several models is presented in Table 1.

Additionally, the red dots in Fig. 5 correspond to fully connected variants of the discovered architectures. In other words, if certain weights in these models are zeroed out in a specific way after pruning, the resulting sparse structure would exactly match one of our architectures. However, neither the dense variants nor their pruned versions achieve the same level of accuracy as our discovered models. This suggests that removing the right weights can facilitate the training process, and that pruning does not always find the optimal subset of weights to remove.

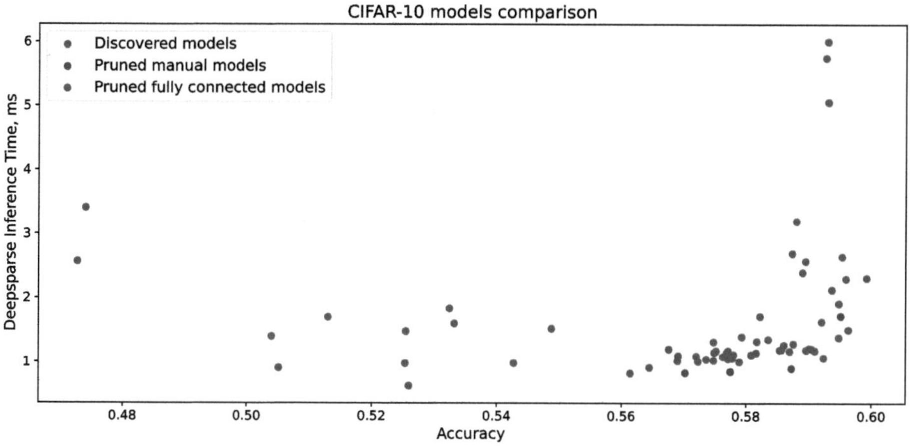

Fig. 5. Comparison of accuracy and inference time of discovered and manual architectures. Inference time is measured on data with a batch size of 128.

Table 1. Comparison of pruned manually designed models and evolutionarily discovered models in terms of Accuracy, FLOPs, and Inference Time.

	Accuracy (%)	FLOPs (K)	Inference Time (ms)
Manual designed models	57.75	149	0.85
	58.72	192	0.9
	59.51	260	1.72
Discovered models	58.17	76	1.32
	59.1	124	1.17
	59.48	166	1.37

5 Conclusions and Future Work

In this work, we introduced a neural architecture search algorithm that explores architectures in the form of graphs, where vertices represent individual neurons and edges correspond to weights. The objective was to investigate whether such a design could discover architectures that require fewer computational resources than manually designed ones. By implementing a multi-objective evolutionary algorithm, we discovered architectures that outperform pruned convolutional networks by a factor of two in terms of FLOPs. To validate the practical potential of our approach, we measured inference time in the DeepSparse runtime and observed a 20% speed-up compared to manually designed models in some scenarios.

The main limitation of the proposed method is its computational cost of the search process, which creates challenges when scaling to larger models. We are currently working on improving the scalability of the algorithm. Another open question is the design of mutation operators that can generate highly diverse architectures capable of performing well across a range of tasks. It is important to find a balance between incorporating prior knowledge (such as the benefit of connecting spatially close neurons), maintaining architectural diversity, and computational cost.

Acknowledgments. This research is financially supported by The Russian Scientific Foundation, Agreement № 25-11-00367.

Competing Interests. The authors declare that they have no competing interests.

References

1. Ren, P., et al.: A comprehensive survey of neural architecture search: challenges and solutions. ACM Comput. Surv. **54**(4), 1–34 (2021)
2. White, C., et al.: Neural architecture search: insights from 1000 papers. arXiv preprint arXiv:2301.08727 (2023)
3. Real, E., Aggarwal, A., Huang, Y., Le, Q.V.: Regularized evolution for image classifier architecture search. In: Proceedings of the AAAI Conference on Artificial Intelligence, vol. 33, no. 1, pp. 4780–4789 (2019)
4. Ericsson, L., et al.: einspace: searching for neural architectures from fundamental operations. In: Advances in Neural Information Processing Systems, vol. 37, pp. 1919–1953 (2024)
5. Cai, H., Yang, J., Zhang, W., Han, S., Yu, Y.: Path-level network transformation for efficient architecture search. In: International Conference on Machine Learning, pp. 678–687 (2018)
6. NeuralMagic: DeepSparse (2021). https://github.com/neuralmagic/deepsparse. Accessed 31 May 2025
7. Zoph, B., Le, Q.V.: Neural architecture search with reinforcement learning. In: International Conference on Learning Representations (2017)
8. Zoph, B., Vasudevan, V., Shlens, J., Le, Q.V.: Learning transferable architectures for scalable image recognition. In: Proceedings of the IEEE Conference on Computer Vision and Pattern Recognition, pp. 8697–8710 (2018)
9. Liu, H., Simonyan, K., Yang, Y.: DARTS: differentiable architecture search. In: International Conference on Learning Representations (2019)
10. Xu, Y., et al.: PC-DARTS: partial channel connections for memory-efficient architecture search. In: International Conference on Learning Representations (2019)
11. Zhang, M., Su, S.W., Pan, S., Chang, X., Abbasnejad, E.M., Haffari, R.: IDARTS: differentiable architecture search with stochastic implicit gradients. In: International Conference on Machine Learning, pp. 12557–12566. PMLR (2021)
12. Kandasamy, K., Neiswanger, W., Schneider, J., Poczos, B., Xing, E.P.: Neural architecture search with Bayesian optimisation and optimal transport. In: Advances in Neural Information Processing Systems, vol. 31 (2018)
13. Dong, X., Liu, L., Musial, K., Gabrys, B.: NATS-Bench: benchmarking NAS algorithms for architecture topology and size. IEEE Trans. Pattern Anal. Mach. Intell. **44**(7), 3634–3646 (2021)
14. Shen, J., Khodak, M., Talwalkar, A.: Efficient architecture search for diverse tasks. In: Advances in Neural Information Processing Systems, vol. 35, pp. 16151–16164 (2022)

15. Christoforidis, A., Kyriakides, G., Margaritis, K.: A novel evolutionary algorithm for hierarchical neural architecture search. arXiv preprint arXiv:2107.08484 (2021)
16. Chen, B., et al.: GLIT: neural architecture search for global and local image transformer. In: Proceedings of the IEEE/CVF International Conference on Computer Vision, pp. 12–21 (2021)
17. Schrodi, S., Stoll, D., Ru, B., Sukthanker, R., Brox, T., Hutter, F.: Construction of hierarchical neural architecture search spaces based on context-free grammars. In: Advances in Neural Information Processing Systems, vol. 36, pp. 23172–23223 (2023)
18. Frankle, J., Carbin, M.: The lottery ticket hypothesis: finding sparse, trainable neural networks. In: International Conference on Learning Representations (2018)
19. He, Y., Zhang, X., Sun, J.: Channel pruning for accelerating very deep neural networks. In: Proceedings of the IEEE International Conference on Computer Vision, pp. 1389–1397 (2017)
20. Michel, P., Levy, O., Neubig, G.: Are sixteen heads really better than one? In: Advances in Neural Information Processing Systems, vol. 32 (2019)
21. Kurtic, E., et al.: The optimal BERT surgeon: scalable and accurate second-order pruning for large language models. In: Proceedings of the 2022 Conference on Empirical Methods in Natural Language Processing, pp. 4163–4181 (2022)
22. Deb, K., Pratap, A., Agarwal, S., Meyarivan, T.A.M.T.: A fast and elitist multiobjective genetic algorithm: NSGA-II. IEEE Trans. Evol. Comput. **6**(2), 182–197 (2002)
23. NeuralMagic: SparseML (2020). https://github.com/neuralmagic/sparseml. Accessed 31 May 2025

Unsupervised and Semi-Supervised Learning

Similarity Learning for Spectral Clustering

Vangjush Komini[1,3](✉), Nadezhda Koriakina[2], Debaditya Roy[3], and Sarunas Girdzijauskas[1]

[1] KTH Royal Institute of Technology, Stockholm, Sweden
{vangjush,sarunasg}@kth.se
[2] Uppsala University, Uppsala, Sweden
nadezhda.koriakina@angstrom.uu.se
[3] Qamcom Research and Technology AB, Stockholm, Sweden
debaditya.roy@qamcom.se

Abstract. Spectral clustering is a widely adopted method capable of identifying complicated cluster boundaries. However, traditional spectral clustering requires the definition of a predefined similarity metric for constructing the Laplacian matrix, a requirement that limits flexibility and adaptability. Instead of predefining this metric upfront as a fixed parametric function, we introduce a novel approach that learns the optimal parameters of a similarity function through parameter optimization. This optimizes a similarity function to assign high similarity values to data pairs with shared discriminative features and low values to those without such features. Previous methods that adapt similarity measures typically treat their parameters as hyperparameters or rely on non-convex optimization strategies. However, these approaches are not well-suited for unsupervised scenarios, as they depend heavily on initial conditions and require labeled data for validation, which is unavailable in such settings. In contrast, our method employs convex optimization to learn the parameters of the similarity metrics directly, rather than treating them as hyperparameters. This enables robust and reliable unsupervised learning, making our approach particularly well-suited for spectral clustering. We validate the effectiveness and adaptability of our method on several benchmark datasets, demonstrating superior performance compared to existing techniques.

Keywords: Spectral Clustering · Similarity Learning

1 Introduction

Spectral clustering (SC) [15] is a very effective method to cluster data with non-convex separation boundaries. In this approach, individual data points can be considered as nodes of a fully connected graph where the edge weights are the similarity values. The underlying assumption for SC is that for every pair of

Code and supplementary appendix are available at https://github.com/Vkomini/SpectraClustering.

S. Džeroski et al. (Eds.): DS 2025, LNAI 16090, pp. 207–221, 2025.
https://doi.org/10.1007/978-3-032-05461-6_14

data in the training set, the similar ones should have a high similarity score and vice versa. Thus, the objective of SC becomes minimizing the graph cut (GC), namely minimizing the sum of edge weights between clusters and maximizing the sum within clusters [15] (see Fig. 1).

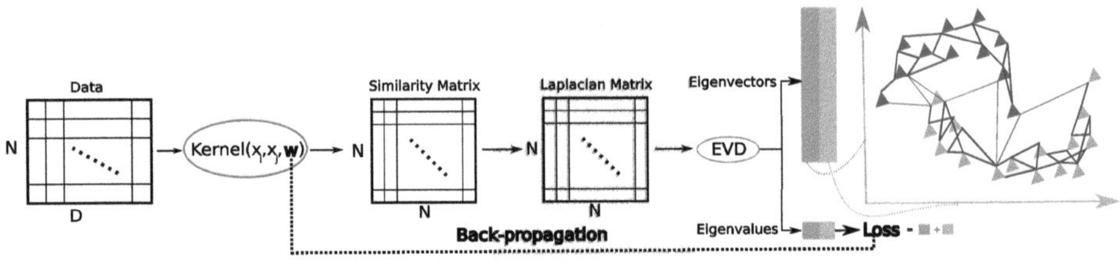

Fig. 1. The proposed kernel function (an alternation of the radial basis function) is optimized using gradient descent to minimize the graph cut (GC).

The performance of SC is primarily attributed to the manually chosen parameters of the *similarity metric*. However, an incorrectly chosen similarity metric can be detrimental to the clustering performance as it might not emphasize enough the similar pairs relative to the dissimilar ones (see Fig. 2c). Hence, instead of hand-picking the parameters of the similarity metric, one can try to learn them instead (see Fig. 1). To do so, we use the GC loss to generate a gradient for the parameter of the kernel function. This gradient is then utilized to optimize the parameter of the kernel function (see Fig. 1). Existing similarity learning methods [2,3,8–14,16,21,25,27–29,31] feature one or more hyperparameters in the objective function. Unlike supervised learning, validating hyperparameter values through an annotated validation set is not possible in an unsupervised learning setup. Given the need for an extensive exploratory and exploitative search in supervised hyperparameter optimization, it is unlikely that optimal parameters will be selected due to the suboptimal range of values being considerably larger than the optimal range in real-world scenarios. If we apply the same principle to an unsupervised setting, randomly selecting hyperparameters would likely result in suboptimal values. This, in turn, would result in incorrect similarity scores affecting the downstream clustering performance [4]. Moreover, existing similarity learning methods employ non-convex optimization, which complicates their application in unsupervised scenarios due to their dependence on the initialization and the challenges associated with validating solutions in the absence of labels. To mitigate these drawbacks, *we propose a novel variation in the radial basis function (RBF) kernel and try to learn its optimal bandwidth through a convex optimization without introducing any hyperparameter in the loss function (see Fig. 1).* In this study, the initial number of clusters (i.e., K) is predetermined; however, the unsupervised estimation of the cluster count is possible by identifying density peaks within the data topology [23]. Building on these insights, the method we introduce offers an unsupervised approach to learning similarities that determine the optimal weights for the edges of a fully connected weighted graph, on which SC is then applied. We assess the clustering performance of our method on various image and text datasets, demonstrating an improvement over alternative approaches. The contributions of this study are outlined as follows:

1. We introduce a novel modification of the RBF kernel that enables the learning of its parameters through convex optimization;
2. We propose a method for updating the RBF bandwidth via gradient descent to minimize GC.

2 Method

(a) (b) (c)

Fig. 2. As shown in Figs. 2a and 2b, each distance is modulated by the exponential kernel, causing larger distances to amplify more strongly. After normalization to a unit vector, smaller distances decrease in magnitude while larger distances increase, as illustrated in Fig. 2c.

The proposed method works directly on the raw data (e.g., in Fig. 2a) and tries to further emphasize any existing separation within the data points (see Fig. 2b). Accordingly, the optimal similarity function would lower the smaller pairwise distances so that all the data points would eventually collapse to a single point (see Fig. 2c). At the same time, the more pronounced distances are amplified (see Fig. 2c). To do so, we choose to exponentiate these pairwise distances in a controlled manner (see Eq. (1) and Fig. 2b), where the smaller the distance between two data points, the higher their similarity and vice versa. As a result, the bigger distances are exponentiated towards higher values much more than the smaller ones (see Fig. 2b). Since the new transformed distances are normalized within a unit vector, increasing the bigger magnitudes would render the smaller ones towards even smaller values and vice versa (see Fig. 2c). The proposed kernel enables a controlled exponential modulation of the distances via its bandwidth σ (see Eq. (1)). To ensure the convexity of the problem, we redefine the distance function as $d(x,y) = u(x,y) + 1$. Given these distances are non-negative (i.e., $u(x,y) \geq 0$) the kernel function in Eq. (1) demonstrates convex behavior over the bandwidth interval $\sigma \in \Sigma = (0, \frac{\sqrt{6}}{3}]$ (see Appendix F). Since SC is mainly influenced by relative distances, uniformly increasing all distances by one unit doesn't change the hierarchy of pairwise distances within a dataset or affect the clustering result. (see Appendix G). Moreover, the proposed modification of the RBF (see Eq. (1)) preserves the intrinsic kernel characteristics of the RBF.

$$K(x,y) = e^{-\frac{d(x,y)}{\sigma^2}} = e^{-\frac{u(x,y)+1}{\sigma^2}} = \sum_{n=0}^{\infty} - \frac{(u(x,y)+1)^n}{(\sigma^2)^n n!}, \text{s.t: } \sigma \text{ is kernel bandwidth.} \tag{1}$$

The superior performance of the proposed kernel is primarily due to its capability to project data into an infinite-dimensional space, as described by the last term in Eq. (1) [15]. By properly tuning its bandwidth σ, one can smoothly truncate the infinite-dimensional expansion (see Eq. (1)) and select the dimensions to be incorporated in the comparison. In other words, the bandwidth σ adjusts the amount of modulation for the similarity metric. Using the proposed kernel as a generic similarity formulation, the task reduces to optimization of σ that yields the most representative distance from a general formulation. To amplify the distance $d(x, y)$ as presented in Eq. (1), the parameter σ must be set to a value less than one (i.e., $\sigma \leq 1$). Note that while $\sigma = 0$ yields minimal GC, this case is not meaningful because all edge values are reduced to zero. As a result, the hierarchical ordering of the edges is lost, and the overall structure of the graph is completely destroyed.

2.1 Learning Operation

In SC, the objective remains identical to other clustering techniques; similar data are projected together while the dissimilar ones are far apart [15]. This objective is equivalent to the GC (see Fig. 1) when data are considered as nodes of the connected graph, while the edges are the pair similarity values. Thus, to construct an objective function that is amenable to the gradient-based method, one has to consider the data projections (i.e., f) and the similarity values (i.e., $A_{i,j} \in R, \forall i,j \in \{1, \cdots, N\}$) simultaneously. Utilizing the general formulation for the GC [15] (see Eq. (2)) as the objective function, it is possible to get an empirical assessment of clustering performance, where L in Eq. (2) is the Laplacian matrix and f are the data embeddings.

$$\text{GC}[A, f] = \sum_{i,j,j<i} A_{i,j}(f_i - f_j)^2 = f^\top L f. \tag{2}$$

The formulation in Eq. (2) reduces the task to learning the similarity matrix (i.e., $A^{[N,N]}$) and the representation of the data (i.e., f). As a result, the computation of GC on the graph reduces to matrix multiplication. Since the utilized similarity is parameterized solely by the bandwidth σ, we optimize the computed similarity matrix by tuning σ (see Eq. (3)).

$$\arg\min_{\sigma,f} \text{GC}[A(\sigma), f(A)] \rightarrow \frac{\partial^2 \text{GC}[A(\sigma), f]}{\partial\sigma\partial f} = 0. \tag{3}$$

Instead of finding both σ and f simultaneously, an alternation between each parameter is adopted (see Algorithm 1). We utilize a Lagrangian multiplier to restrict the domain of the data representation f to a unit magnitude $f^\top f = 1$, which enables the derivation of f and λ as the eigenvectors and eigenvalue of the Laplacian matrix L (see Eq. (4) from [24]). Notice that the Laplacian matrix L is defined as $L = D - A$ where D is the degree matrix. The degree matrix is diagonal, with each diagonal entry (*i.e.*, D_{ii}) representing the sum of the weights of all edges connected to node (i), (*i.e.*, $D_{ii} = \sum_{j=1}^{n} A_{ij}$).

$$\frac{\partial\{\mathrm{GC}[A(\sigma), f] + \lambda(f^\top f - 1)\}}{\partial f} = 0 \rightarrow Lf = \lambda f \tag{4}$$

Since the kernel is symmetric (see Eq. (1)), the resulting Laplacian matrix L is also symmetric upon which eigenvalue decomposition (EVD) produces real and non-negative eigenvalues ($\lambda_i \geq 0, \forall i \in 1, \cdots, N$), and orthogonal eigenvectors ($f^\top f = 1$) [24]. In the case of using the bottom K eigenvector (*i.e.*, $f_{i,\cdots,K}$), the GC reduces to the sum of the corresponding eigenvalues λ as in Eq. (5).

$$\mathrm{GC}[A(\sigma), f_{i,\cdots,K}(A)] = \mathrm{GC}(\sigma)|_{f_{i,\cdots,K}} = \sum_{i=1}^{K} \lambda_i(\sigma) \tag{5}$$

In contrast to Eq. (4), which offers a closed-form solution, finding σ from Eq. (5) requires the use of gradient descent because it involves a nonlinear operation as dictated by the kernel in Eq. (1). Despite the intrinsic non-convex nature of SC [15], we introduce a convex approach for determining the optimal bandwidth σ in the interval $\Sigma = (0, \sqrt{6}/3]$ (see Theorem 2 in Appendix F).

Due to this convexity condition, the parameter σ can be iteratively refined via Newton's method (see Eq. (6)) to achieve a faster than gradient descent convergent progression toward the optimal solution [24].

$$\sigma_i \leftarrow \sigma_i - \frac{\mathrm{grad}_\sigma}{\mathrm{step}_\sigma} \leftarrow \sigma_i - \frac{\sum_{i=0}^{k} f_i^\top \frac{\partial L(\sigma)}{\partial \sigma} f_i}{\sum_{i=0}^{k} f_i^\top \frac{\partial^2 L(\sigma)}{\partial \sigma^2} f_i} \tag{6}$$

The new algorithm adopts a two-stage operation within each iteration, starting computation of the Laplacian matrix and its first and second derivative w.r.t σ, followed by the σ update (see Algorithm 1).

The search for σ remains confined within the convex space in Σ if, and only if, $f_{1,\ldots,K}$ remain restricted as the eigenvectors of L. Moreover, as both σ and the series of eigenvectors $f_{1,\ldots,K}$ are iteratively updated to minimize the GC loss function, the shape of this function gradually evolves towards increasing flatness. Since the proposed kernel provides a convex GC loss $\sigma \in \Sigma$, manually setting σ to a minimal but positive value within Σ would effectively achieve nearly minimal GC values and mitigate the need for the Newton method altogether. However, due to the finite precision, setting σ extremely low could lead to numerical underflow, particularly during the EVD processes, or all similarity weights could be truncated to the smallest positive values. This can compromise the reliability of the eigenvectors in correctly capturing the data clustering structure. For example, forcing the kernel bandwidth σ to zero would provide the minimum possible GC loss equal to zero. However, it would also cause all pairwise node similarities to vanish, leading to a completely disconnected graph and

Algorithm 1. Similarity Learning

procedure LEARNSIMILARITY(Data $\mathbf{X}^{[N,D]}$, Number of clusters K)
  Compute the distance matrix $\Delta^{[N,N]}$.
  Scale and translate the distance matrix: $C^{[N,N]} \leftarrow \frac{\Delta^2}{\max(\Delta^2)} + \mathbf{J}^1$.
  Initialize $\sigma_0 \leftarrow \sqrt{6}/3$.
  for $i = 1$ to n_{iter} **do**
    Compute the affinity matrix: $A \leftarrow e^{-C/\sigma_i^2}$.
    Compute derivatives of A relative to σ_i:

$$\frac{\partial A}{\partial \sigma_i} \leftarrow \frac{2C}{\sigma_i^3} e^{-C/\sigma_i^2} \qquad \frac{\partial^2 A}{\partial \sigma_i^2} \leftarrow \frac{2C}{\sigma_i^6} e^{-C/\sigma_i^2} (2C - 3\sigma_i^2 \mathbf{J}).$$

    Compute the degree and Laplacian matrices and their derivatives:

$$D \leftarrow \text{diag}(\sum A), \qquad L \leftarrow D - A,$$
$$\frac{\partial D}{\partial \sigma_i} \leftarrow \text{diag}(\sum \frac{\partial A}{\partial \sigma_i}), \qquad \frac{\partial L}{\partial \sigma_i} \leftarrow \frac{\partial D}{\partial \sigma_i} - \frac{\partial A}{\partial \sigma_i},$$
$$\frac{\partial^2 D}{\partial \sigma_i^2} \leftarrow \text{diag}(\sum \frac{\partial^2 A}{\partial \sigma_i^2}), \qquad \frac{\partial^2 L}{\partial \sigma_i^2} \leftarrow \frac{\partial^2 D}{\partial \sigma_i^2} - \frac{\partial^2 A}{\partial \sigma_i^2}.$$

    Perform EVD on L and update σ:

$$f_{1,\ldots,K} \leftarrow \text{EVD}(L), \qquad \text{grad}_\sigma \leftarrow \sum_{i=1}^{K} f_i^\top \frac{\partial L}{\partial \sigma} f_i,$$
$$\text{step}_\sigma \leftarrow \sum_{i=1}^{K} f_i^\top \frac{\partial^2 L}{\partial \sigma^2} f_i, \qquad \sigma_i \leftarrow \sigma_i - \frac{\text{grad}_\sigma}{\text{step}_\sigma}.$$

  return $\sigma_i, f_{1,\ldots,K}$

preventing meaningful clustering. Hence, utilizing the Newton update becomes critical because, even with infinite precision and due to the convex nature of the GC loss function and the properties of the Newton update method, achieving a value of σ of zero within a finite number of updates is unattainable (see Appendix H)). In practice, if an underflow occurs as we reduce σ via Newton, the smallest similarity values become truncated to the minimum representable positive values before the bigger ones. Similarly, the GC loss attains the smallest possible values; beyond this point, it ceases to update further because its gradient attains a precisely zero value. However, since the similarity weights remain ordered, the bigger values would not be truncated; thus, the inherent clustering structure remains intact. Hence, this iterative adjustment of σ continues until adjustments to the GC yield negligible changes, indicating that a minimized yet practical representation of the clustering structure within the dataset has been

[1] The matrix J has the same dimensions as C and all of its entries are ones.

attained. If any potential underflow occurs, it would still preserve the inherent cluster in the dataset while preventing further updates on σ.

As the denominator involves the square of σ (see Eq. (1)), both positive and negative values of σ of the same magnitude have the same modulation effect in the proposed kernel. Hence, we consider only positive values for σ, as its negative values do not introduce any additional impact. Furthermore, if the loss at initialization is positive, then the optimal σ is in Σ (see Appendix C). To introduce scale invariance in the distance matrix ($\Delta^{[N,N]}$ in Algorithm 1) and potential numerical underflow, its matrix values are normalized so that its minimum value becomes zero, and its maximum value becomes one. This proposed method works only on the Laplacian matrix setup. Other versions of normalized Laplacian were tried, i.e., symmetric normalized Laplacian $\left(L^{\text{symmetric}}_{\text{normal}} = D^{-\frac{1}{2}}LD^{-1\frac{1}{2}}\right)$, and normalized Laplacian row-wise stochastic version $\left(L^{\text{stochastic}}_{\text{normal}} = D^{-1}L\right)$ but the gradient computation was not updating σ in the anticipated direction.

Complexity Analysis: The Euclidean distance matrix is computed once outside the loop at $O(N^2)$ complexity. Since we are looking solely for the first K eigenvectors, the EVD complexity is $O(K \times N^2)$. Similarly, the complexity of the first and second derivatives is $O(K \times N^2)$ since the output of the computational graph has K eigenvectors. Over nr_{iter} iterations, the total complexity would be $O(nr_{\text{iter}} \times K \times N^2)$.

3 Related Work

The similarity learning domain contains various methodologies established on different fundamental assumptions [21], primarily categorized into empirical heuristics and graph learning approaches.

Empirical Heuristic: The heuristic methods assign a unique σ for each data point by considering local statistics within a selected neighborhood in the raw [27] or embedded space [16]. Yet, scaling this approach is difficult because it requires computing as many σ values as the dataset size, and the choice of neighborhood critically affects performance.

Graph Learning: Rather than relying on heuristics, some studies adopt a data-driven approach to individually learn a scale parameter (i.e., $\{\sigma_k\}_{k=1}^{d}$, where d is the number of dimensions), for each dimension of the dataset [9]. The learning process involves minimizing the cumulative distance between each data point x_i and the weighted mean of its neighbors as in Eq. (7):

$$\min_{\{\sigma_k\}_{k=1}^{d}} \sum_{i=1}^{n} \left\| x_i - \frac{1}{d_j} \sum_{j=1}^{n} E_{ij} W_{ij} x_j \right\|, \text{ s.t: } d_j \rightarrow \text{node degree} \tag{7}$$

In Eq. (7), W_{ij} represents the affinity between data points, defined as[2] $W_{ij} = \exp\left(-\frac{1}{2}\sum_{k=1}^{d} \frac{d(x_{ik}, x_{jk})}{\sigma_k^2}\right)$, and E_{ij} denotes the elements of the connectiv-

[2] $d(x_{ik}, x_{jk})$ is the distance.

ity matrix E (filled with binary values to indicate the presence (i.e., $E_{ij} = 1$) or absence (i.e., $E_{ij} = 0$) of an edge between each data pair), which are determined separately using a heuristic. The values of $\{\sigma_k\}_{k=1}^{d}$ are optimized through a gradient descent procedure, iteratively refining them to improve the representation of the manifold's local geometry. Certain techniques induce locality constraints to evaluate similarity values (i.e., W). This is achieved by minimizing the cumulative difference of individual data relative to neighborhood constraints as in Eq. (8).

$$\min_{W} \sum_{i=1}^{n} \left\| x_i - \sum_{j=1}^{n} E_{ij} W_{ij} x_j \right\|, \text{ s.t: } \sum_{j=1}^{n} E_{ij} W_{ij} x_j = 1 \tag{8}$$

Further enhancements to this method include the imposition of an additional constraint on the similarity values (i.e., $W_{ij} \geq 0$) [25]. To overcome the limitations associated with heuristic-based methods for establishing the connectivity matrix (i.e., E), alternative methods have been developed that optimize the weight matrix W directly, bypassing the need for defining E [29]. The objective (see Eq. (9)) ensures a smooth transition between nearby points, balanced clustering, and uniform edge weight distribution[3].

$$\min_{W \geq 0} \underbrace{\sum_{i,j=1}^{n} \|x_i - x_j\| W_{ij}}_{\text{Smooth transition}} + \underbrace{\gamma_1 \sum_{j=1}^{n} (d_j - 1)^2}_{\text{Balanced clustering}} + \underbrace{\gamma_2 \sum_{j=1}^{n} W_{ij}^2}_{\text{Uniform edge weight distribution}} \tag{9}$$

In comparison to methods that focus on local attributes [29], a global model [2] tries to fit a graph to a vector of data using the following quadratic objective Eq. (10).

$$\min_{W} \sum_{i=1}^{n} \left\| d_j x_i - \sum_{j \neq i}^{n} W_{ij} x_j \right\|^2 \text{ s.t: } d_j \geq 1 \text{ and } \sum_{i} (\max(0, 1 - d_j))^2 \leq \eta n \tag{10}$$

An alternate methodology [3] tries to fit a graph to a specified data set by enforcing sparsity on the learned similarity values via L_1 regularization term (see Eq. (11)).

$$\min_{W} \left(\|X - XW\|_F^2 + \gamma_1 \|W\|_1 \right) \tag{11}$$

In contrast, some researchers have used a low-rank constraint by replacing the sparsity term (i.e., the L_1 norm, $||W||_1$) with a low-rank term (i.e., the nuclear norm, $||W||_*$) [12]. While other approaches have even combined the low-rank

---

[3] γ_1, γ_2, η in Eqs. (9) to (12) are regularizer hyperparameter coefficients.

constraint with sparsity [31], local constraint [14], Markov random walk [13], B-matching [10,11]. Another method, instead of fitting a graph directly on the raw data, tries to alternate between learning a linear transformation matrix (i.e., P in Eq. (12)) of the data such that their pairwise comparison is more robust to noise and learning the similarity values (i.e., W in Eq. (12)) [28] using the locality-preserving principle.

$$\min_{P,W} \sum_{i,j=1}^{n} ||Px_i - Px_j||^2 W_{ij} + \gamma_1 \sum_{i,j=1}^{n} W_{ij} \ln(W_{ij}), \tag{12}$$

$$\text{s.t: } \sum_{i=1}^{n} ||Px_i||^2 = 1, \text{ and } \sum_{j=1}^{n} W_{ij} = 1, W_{ij} \geq 0. \tag{13}$$

Since $||Px_i - Px_j||^2 = ||x_i - x_j||\, P^\top P\, ||x_i - x_j||$ one can consider that this method is unsupervised Mahalanobis distance learning.

4 Experiments

Benchmark Datasets: We ran a series of experiments on benchmark diagnostic datasets [5] (see Fig. 3 and Appendix I) as well as real-world datasets (see Table 1) to showcase the method's capabilities at separating non-convex shapes (see Fig. 3a and Appendices I and J) into linearly separable clusters (see Fig. 3b and Appendices I and J). In all scenarios, the method tries to amplify the big pairwise distances within a confined space while reducing the small pairwise distances. Hence, the method finds a persisting gap within the topology of the dataset and tries to linearize it in the embedding space (see Fig. 3b and Appendix I). The analysis of the optimization trajectory of σ reveals a consistently non-increasing and a non-negative first and second derivative (i.e., $grad_\sigma, step_\sigma$) throughout the trajectory (see Fig. 3d and Appendices I and J). This behavior empirically supports the convex nature of the proposed algorithm. The GC loss of the given Laplacian (i.e., loss $= \sum_{i=1}^{K} \lambda_i$) decreases throughout the training process (see Figs. 3c and 4b and Appendices I and J). Additionally, considering that both the first and second derivatives of the Laplacian matrix (i.e., $\frac{\partial L}{\partial \sigma}$ and $\frac{\partial^2 L}{\partial \sigma^2}$) remain Laplacian matrices (see Lemma 7 in Appendix F)), sharing the same eigenvectors (see Appendix D), their respective GC is defined by the gradient (i.e., $\text{grad}_\sigma = \sum_{i=1}^{K} f_i^\top \frac{\partial L}{\partial \sigma} f_i$) and step update (i.e., $\text{step}_\sigma = \sum_{i=1}^{K} f_i^\top \frac{\partial^2 L}{\partial \sigma^2} f_i$) (see eqs. (14) and (18) in Appendix E). As a result, as σ decreases, the GC decreases for the Laplacian matrix and its first and second derivatives since they all characterize the same graph (see Fig. 3d and Appendices I and J). Eventually, as σ decreases, both the gradient (i.e., $\sum_{i=1}^{K} f_i^\top \frac{\partial L}{\partial \sigma} f_i$) and the step update (i.e., $\sum_{i=1}^{K} f_i^\top \frac{\partial^2 L}{\partial \sigma^2} f_i$) become smaller.

(a)

(b)

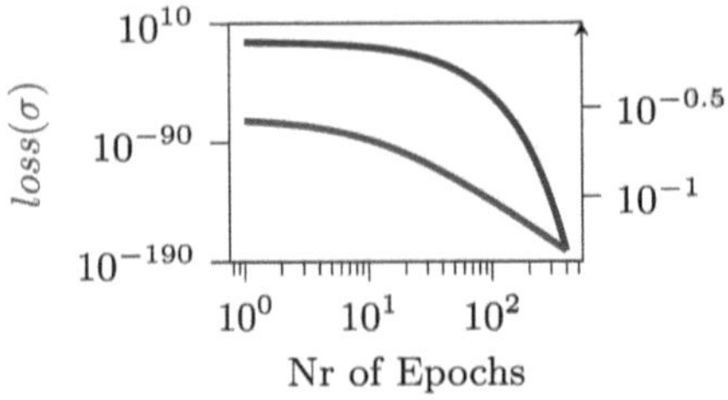

(c)

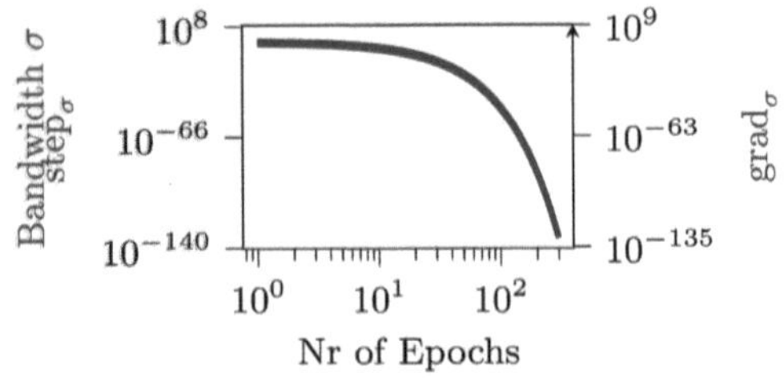

(d)

(e) Epoch 0

(f) Epoch 100

(g) Epoch 200

(h) Epoch 300

Fig. 3. Benchmark (two spirals) dataset in Fig. 3a. Our method learns a linear separation of these two spirals Fig. 3b. In Fig. 3b is the eigen-embedding of the benchmark data. The loss $= \sum_{i=1}^{K} \lambda_i$ along with the bandwidth σ, decreases consistently with the number of iterations (see Fig. 3c). Furthermore, the grad_σ and the step_σ decreases consistently with the number of iterations (see Fig. 3d). A snapshot of the optimized trajectory is in Fig. 3e to 3h.

Image and Text Datasets: Furthermore, this linearization of the separation boundary reduces the complexity of the downstream tasks one can perform upon these embeddings. Hence, to compare the capability of our unsupervised method, we run K-means++ [1,7] on the eigen-embeddings, where the clustering performance would be a proxy evaluator for the learned similarity matrix. We extend to text and image datasets (see Table 1) where the models try to find linear separation between data clusters. In such a real data scenario, the proposed method tries to separate the dataset using persistent low-density (i.e., gap) regions throughout the dataset.

Data Preparation: The image datasets are assumed to be correctly registered beforehand; therefore, no feature selection is needed upfront. To ensure the equivariance of the scale of the features, standardization ($\hat{x}_{\text{stand}} = \frac{x-\hat{\mu}}{\sqrt{\hat{\sigma}^2}}$)

Table 1. Description of the experimented datasets.

Dataset	Size	Dimensions	Clusters
JAFFE	212	177	7
Umnist	575	644	20
YALE	165	105	15
BA	1404	320	36
COIL20	1440	1024	20
ORL	400	1024	36

is performed on non-binary data. Notice that $u(x, y) = \|x - y\|^2$ is used for non-binary data and $u(x, y) = 1 - \cos_{\text{sim}}(x, y)$ for binary data. Once the similarity learning has been completed, we perform K-means++ [1,7] clustering upon the eigen-embeddings. To mitigate the effect of the initialization, K-means++ results are aggregated over 50 different runs.

In Table 2, we compare our approach with two graph-based learning methods: Sparse Subspace Clustering (SSC) [3] and Low-Rank Representation (LRR) [3]. Additionally, we consider six models that integrate graph construction with spectral embedding. These include Clustering with Adapting Neighbors (CAN) [18] and its variations: Projected Clustering with Adapting Neighbors (PCAN) [26] and Self-Weighted Clustering with Adapting Neighbors (SWCAN) [19]. Additionally, we assess several other methods in our evaluation: LAPIN, which optimizes a bipartite graph for subspace clustering [17]; DOGC, which learns the clustering discretization and graph learning simultaneously [6]; JGSED, a recent technique that integrates graph construction, spectral embedding, and cluster discretization [21]; and JSESR, an approach that combines spectral embedding with spectral rotation [20]. Last but not least, self-tuning spectral clustering (SelfT) [27] is also included in the comparison. In contrast to our approach, the effectiveness of other methods depends significantly on selecting appropriate hyperparameters and the type of regularization used. *The aim is to demonstrate the superiority of the proposed method in terms of clustering performance and ease of use.* Recent advancements in deep clustering methods have demonstrated good performance on well-established image datasets. These methods comprise two stages: unsupervised representation learning, followed by the actual clustering step [22,30]. The bulk of the research in this area concentrates on representation learning. This phase involves organizing the data so that similar items are closely embedded, whereas dissimilar items are spaced further apart. This configuration simplifies the task for conventional clustering techniques. However, our proposed method focuses on improving the clustering technique itself.

Evaluation Metrics: Unlike classifiers, where the type of classes and their predictions are kept intact, the prediction cluster indicators are permuted in clustering. Therefore, the clustering evaluation reduces to comparing two different types of sets, i.e., the ground-truth labels set and the cluster indicator. As a result, more than a single metric is required to assess all aspects of the clustering performance; normalized mutual information (NMI) and accuracy (ACC) capture different aspects of the performance. NMI compares two sets using entropy as a criterion. It measures how much entropy is required to describe the second set using the entropy of the first set. Normalization scales the metric from zero to one. In contrast, ACC measures the pairwise exactness of the predicted clusters with the ground-truth clusters. This metric requires a bi-partite set alignment as in an unsupervised learning setup; the predicted cluster index does not correspond to the ground-truth label value at the individual data level. Therefore, the two sets (X and Y of size n) are aligned through the Kuhn-Munkres algorithm (KMA) (*i.e.*, $\hat{Y} = \text{KMA}(Y)$) upfront.

Table 2. ACC and NMI over the different datasets. Comping methods have the average performance while ours is only one estimation.

Data	SelfT	JSESR	LRR	SSC	CAN	PCAN	SWCAN	LAPIN	DOGS	JGSED	Our
ACC											
YALE	47.65	30.18	26.67	24.48	24.24	25.45	26.67	30.45	30.93	32.12	**56.96**
JAFFE	54.28	27.83	29.72	29.72	28.77	28.30	30.19	30.68	31.79	32.08	**86.85**
COIL20	38.17	81.37	73.40	76.18	**83.54**	83.33	79.03	81.96	80.90	82.64	69.79
BA	23.67	46.59	44.59	45.87	42.24	43.59	46.08	46.83	38.53	47.86	**50.21**
UMIST	40.69	61.03	56.17	65.74	64.87	57.74	71.30	71.30	62.17	**71.30**	66.78
ORL	45.78	46.59	44.59	45.87	42.24	43.59	46.08	46.83	38.53	47.86	**67.75**
NMI											
YALE	59.64	33.48	27.53	25.69	20.71	23.54	24.05	32.98	33.42	33.78	**64.01**
JAFFE	69.86	9.97	16.62	14.92	13.33	12.88	15.33	15.91	18.06	17.59	**93.34**
COIL20	63.15	89.25	82.90	90.29	91.09	89.08	89.19	90.03	88.96	**91.17**	85.32
BA	38.96	60.83	57.93	**62.50**	54.24	49.55	57.98	59.04	49.82	60.64	60.85
UMIST	60.93	75.18	70.22	77.68	78.21	70.05	82.95	82.95	78.55	82.95	**85.00**
ORL	72.75	80.37	83.20	79.26	76.45	73.16	76.51	82.74	78.09	81.80	**86.30**

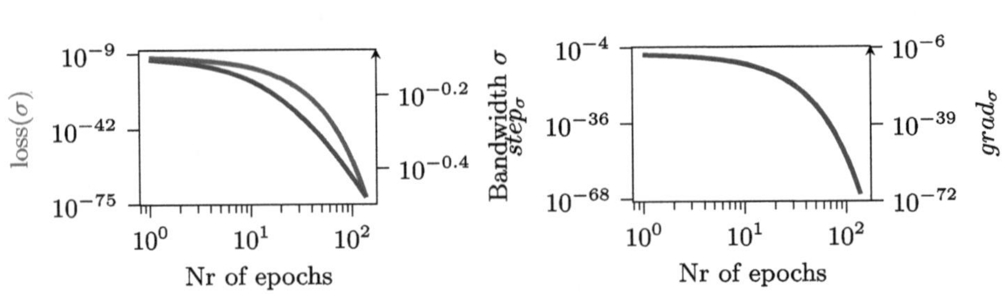

(a) (b)

Fig. 4. The loss $= \sum_{i=1}^{K} \lambda_i$ over the iterations along with the optimization of σ for the JAFFE dataset in Fig. 4a. The $grad_\sigma$ and $step_\sigma$ in Fig. 4b.

Results: The model demonstrates an improvement in the NMI and ACC metrics. This improvement suggests that the entropy of the cluster indices generated by our similarity matrix more closely approximates the actual class indices for the dataset under consideration. Given that NMI provides a broader comparative analysis between true labels and predicted labels, clustering accuracy is computed as a supplementary metric to estimate performance at the individual level. It is important to emphasize that although our method facilitates reproducible similarity learning through its operation within a convex setup and the absence of hyperparameters, competing methods lack these combined attributes. Consequently, the performance of these alternative approaches is heavily dependent upon the initialization of the learning process, which introduces considerable variability and precludes the assurance of even average performance levels. Similarly, optimizing the kernel bandwidth σ follows a similar trajectory as in the benchmark dataset. The method progressively tries to reach an optimal value and decreases the gradient (see Fig. 4b and Appendix K). Likewise, the cumulative magnitude of the first K eigenvalues progresses towards smaller values as in the case of the benchmark data experiments (see Fig. 4a and Appendix K).

5 Conclusion

In this paper, we propose a method for learning the kernel bandwidth by maximizing the linear separation among the K clusters derived from the first K eigenvectors of the Laplacian matrix. We show that arbitrarily choosing the kernel bandwidth σ risks either underemphasizing distances, causing clusters to merge indistinguishably, or overemphasizing distances, isolating individual data points as separate clusters. To address this, we introduce a gradient-based update strategy for σ that explicitly minimizes the GC. Additionally, we propose a modification to the RBF kernel designed to ensure a convex optimization trajectory. The goal of the proposed method is to function without labels in a fully unsupervised manner that takes as input only the number of clusters K. We showcase our method's ability to separate highly non-convex shapes and project the dataset into a linearly separable topology on a series of benchmark datasets. The proposed method is validated on popular datasets and compared with recent similarity learning models. Our proposed method shows an improvement in both normalized mutual information and accuracy. We also highlight a practical limitation of existing approaches: their performance is highly sensitive to the initial settings. *Since these initial settings are not validated in an unsupervised setting, the optimal performance of the competing methods is never guaranteed.* In future work, we plan to explore removing the reliance on a predefined K by integrating automatic cluster number estimation. Leveraging density peak identification within data topology [23] could extend SC from similarity learning to autonomous cluster determination.

References

1. Arthur, D., Vassilvitskii, S.: K-means++: the advantages of careful seeding. In: SODA 2007 (2007)
2. Daitch, S.I., Kelner, J.A., Spielman, D.A.: Fitting a graph to vector data. In: ICML 2009 (2009)
3. Elhamifar, E., Vidal, R.: Sparse subspace clustering: algorithm, theory, and applications. CoRR (2012)
4. Fan, X., Yue, Y., Sarkar, P., Wang, Y.X.R.: On hyperparameter tuning in general clustering problems. In: ICML 2020 (2020)
5. Fränti, P., Sieranoja, S.: K-means properties on six clustering benchmark datasets (2018)
6. Han, Y., Zhu, L., Cheng, Z., Li, J., Liu, X.: Discrete optimal graph clustering. IEEE Trans. Cybern. (2020)
7. Jain, A.K.: Data clustering: 50 years beyond k-means. Pattern Recogn. Lett. (2010), award winning papers from the 19th International Conference on Pattern Recognition (ICPR)
8. Kang, Z., Peng, C., Cheng, Q.: Twin learning for similarity and clustering: a unified kernel approach. CoRR (2017)
9. Karasuyama, M., Mamitsuka, H.: Manifold-based similarity adaptation for label propagation. In: NIPS 2013 (2013)
10. Li, S., Fu, Y.: Low-rank coding with b-matching constraint for semi-supervised classification. In: IJCAI, pp. 1472–1478 (2013)
11. Li, S., Fu, Y.: Learning balanced and unbalanced graphs via low-rank coding. IEEE Trans. Knowl. Data Eng. (2014)
12. Liu, G., Lin, Z., Yan, S., Sun, J., Yu, Y., Ma, Y.: Robust recovery of subspace structures by low-rank representation. CoRR (2010)
13. Liu, R., Lin, Z., Su, Z.: Learning markov random walks for robust subspace clustering and estimation. Neural Netw. **59**, 1–15 (2014)
14. Lu, X., Wang, Y., Yuan, Y.: Graph-regularized low-rank representation for destriping of hyperspectral images. IEEE Trans. Geosci. Remote Sens. **51**(7), 4009–4018 (2013)
15. von Luxburg, U.: A tutorial on spectral clustering (2007)
16. Ng, A., Jordan, M., Weiss, Y.: On spectral clustering: analysis and an algorithm. In: Advances in Neural Information Processing Systems, vol. 14 (2001)
17. Nie, F., Chang, W., Wang, R., Li, X.: Learning an optimal bipartite graph for subspace clustering via constrained laplacian rank. IEEE Trans. Cybern. (2023)
18. Nie, F., Wang, X., Huang, H.: Clustering and projected clustering with adaptive neighbors. In: KDD 2014 (2014)
19. Nie, F., Wu, D., Wang, R., Li, X.: Self-weighted clustering with adaptive neighbors. IEEE Trans. Neural Netw. Learn. Syst. (2020)
20. Pang, Y., Xie, J., Nie, F., Li, X.: Spectral clustering by joint spectral embedding and spectral rotation. IEEE Trans. Cybern. (2020)
21. Qiao, L., Zhang, L., Chen, S., Shen, D.: Data-driven graph construction and graph learning: a review. Neurocomputing (2018)
22. Ren, Y., et al.: Deep clustering: a comprehensive survey (2022)
23. Rodriguez, A., Laio, A.: Clustering by fast search and find of density peaks. Science **344**(6191), 1492–1496 (2014)
24. Strang, G.: Linear Algebra and its Applications. Thomson, Brooks/Cole, Belmont, CA

25. Wang, F., Zhang, C.: Label propagation through linear neighborhoods. IEEE Trans. Knowl. Data Eng. (2008)
26. Yang, X., Li, S., Liang, K., Nie, F., Lin, L.: Structured graph optimization for joint spectral embedding and clustering. Neurocomputing
27. Zelnik-manor, L., Perona, P.: Self-tuning spectral clustering. In: Saul, L., Weiss, Y., Bottou, L. (eds.) NIPS 2004 (2004)
28. Zhang, L., Qiao, L., Chen, S.: Graph-optimized locality preserving projections. Pattern Recogn. (2010)
29. Zhang, Y.M., Huang, K., Hou, X., Liu, C.L.: Learning locality preserving graph from data. IEEE Trans. Cybern. (2014)
30. Zhou, S., et al.: A comprehensive survey on deep clustering: taxonomy, challenges, and future directions. arXiv preprint arXiv:2206.07579 (2022)
31. Zhuang, L., Gao, H., Huang, J., Yu, N.: Semi-supervised classification via low rank graph. In: 2011 Sixth International Conference on Image and Graphics (2011)

Evaluating Clustering Quality in Centroid-Based Clustering Using Counterfactual Distances

Georgios Vardakas(✉), Antonia Karra, Evaggelia Pitoura, and Aristidis Likas

Department of Computer Science and Engineering, University of Ioannina, 45110 Ioannina, Greece
{g.vardakas,a.karra,pitoura}@uoi.gr, arly@cs.uoi.gr

Abstract. Evaluating clustering quality is essential for selecting the optimal number of clusters and for comparing clustering algorithms. Several internal clustering quality indices have been proposed such as the Silhouette score and the Variance Ratio Criterion (VRC) with well-known advantages and limitations. In this work, we present the idea of counterfactuals to quantify cluster separation. Counterfactuals have been recently introduced in the context of clustering to quantify minimum modifications to be applied to a point of one cluster in order to be assigned to another cluster. We exploit counterfactual distances in the context of k-means clustering and define the separation between clusters considering such distances. Then we use the proposed separation measure to define a clustering quality score as the ratio of total separation over total intra-cluster variance of a clustering solution. We evaluated the effectiveness of the proposed score against Silhouette and VRC using various real-world digit datasets.

Keywords: Clustering quality · cluster separation · counterfactual distances · silhouette score

1 Introduction

Clustering is a fundamental task in unsupervised learning, where the goal is to group data points into meaningful clusters based on similarity. Evaluating clustering quality is essential for selecting the optimal number of clusters and for comparing clustering algorithms [5,7]. Among various clustering quality indices, the Silhouette score [9] and the Variance Ratio Criterion (VRC) [1] are the most widely used.

In this work, we focus on the idea of *counterfactuals* to quantify cluster separation. Counterfactuals have been widely adopted to explain classification decisions and have also been proposed recently for clustering problems [8,11]. Roughly speaking, considering a pair of clusters, a counterfactual of a data point in one cluster (source cluster) is its closest point in the other cluster (target cluster). Based on this definition, the counterfactual of point in the source cluster

S. Džeroski et al. (Eds.): DS 2025, LNAI 16090, pp. 222–236, 2025.
https://doi.org/10.1007/978-3-032-05461-6_15

is its closest point that lies in the *boundary* between the source and the target cluster. Thus, counterfactual distances can be used to quantify cluster separation.

This paper introduces a simple and efficient clustering quality criterion based on the ratio of total cluster separation to total intra-cluster variance. The proposed score, called *CFQ*, leverages the *counterfactual-based cluster separation measure* and scales linearly with dataset size, offering a computationally efficient alternative to existing criteria. We have tested and compared our criterion on the problem of selecting the number of clusters in the context of k-means clustering.

The structure of the paper is the following. Section 2 describes related work on clustering quality indices. Section 3 presents counterfactuals in the context of clustering illustrating that they can be used to measure the cluster separation in k-means clustering. Section 4 presents and explains the proposed counterfactual-based measure of cluster separation and defines the CFQ criterion for assessing clustering quality. Section 5 provides details and results of our experimental study with four real digits datasets. Finally, Sect. 6 provides conclusions and future research directions.

2 Internal Quality Indices

Cluster quality indices are used to evaluate the suitability of clustering solutions. A major distinction is between external and internal indices. *External indices* of clustering quality measure the similarity of a clustering solution to a reference solution, usually called ground truth solution. Typical measures of this type are clustering accuracy and normalized mutual information (NMI) [3]. It should be stressed that this similarity comparison involves only the labels of the two compared clustering solutions and does not involve information related to data points or the distance between them.

Nevertheless, clustering is an unsupervised learning problem; therefore, ground truth information is typically not provided to the clustering algorithms. Therefore, the usefulness of external quality indices is rather limited in practice. However, there is an imperative need for assessing the quality of clustering solutions based on the characteristics of the generated clusters in order to be able to select among possible alternative solutions. Such cluster solution evaluation is provided by *internal indices* which take into account only the labels of the clustering solution and information related to the data points that are clustered.

A critical use of internal indices is for the selection of the appropriate number of clusters in cases where this number is not provided by the user. The typical approach to tackle this important problem is to apply a clustering algorithm (e.g. k-means) on the same dataset several times, each time with a different value of number of clusters (k), evaluate the solution produced for each k using an internal quality score and select the solution with highest score.

Internal quality indices usually rely on the combination of two quantities: the first one is *intra-cluster variance* (high compactness) and the second is *separation*. A good clustering solution is typically characterized by high compactness (low intra-cluster variance) and high separation (high inter-cluster variance).

Therefore usually the ratio or the (normalized) difference between inter-cluster and intra-cluster variance is defined as the final quality score. Both compactness and separation can be measured and combined: i) either at point level, where a quality score for each data point is computed and the total score of the solution is the average of individual scores, or ii) at cluster level, where compactness and separation of each cluster are computed and then aggregated for all clusters. Moreover, the computations for compactness and separation could involve either pairwise distances between points or distances between points and cluster centroids or distances between cluster centers. In this work, we propose the use of more informative distance measures between points and clusters that are based on counterfactual distances, i.e. distances from cluster boundaries. In our experimental study, we compared against silhouette and VRC defined as follows.

2.1 Silhouette Coefficient

Assume a dataset X with n points partitioned into k clusters $C = \{C_1, \ldots, C_k\}$. Let $d(x_i, x_j)$ denote the distance between data points x_i and x_j. Silhouette [9] operates at the data point level and computes the distance of a point x to a cluster C_j as the average of distances of x to the points of C_j. A point $x_i \in C_I$ is considered as 'well-clustered' if its distance $a(x_i)$ from its own cluster is small and its minimum distance $b(x_i)$ to any other cluster is large:

$$a(x_i) = \frac{1}{|C_I| - 1} \sum_{x_j \in C_I, i \neq j} d(x_i, x_j) \tag{1}$$

$$b(x_i) = \min_{J \neq I} \frac{1}{|C_J|} \sum_{x_j \in C_J} d(x_i, x_j) \tag{2}$$

where $|C_I|$ and $|C_J|$ are cluster sizes. The silhouette score for x_i is given by:

$$s(x_i) = \frac{b(x_i) - a(x_i)}{\max\{a(x_i), b(x_i)\}}, \quad -1 \leq s(x_i) \leq 1 \tag{3}$$

The overall silhouette score $S(C)$ of clustering solution C is the average over all data points:

$$S(C) = \frac{1}{n} \sum_{i=1}^{n} s(x_i), \quad -1 \leq S(C) \leq 1 \tag{4}$$

Higher $S(C)$ values indicate good cluster assignments, while lower or negative values indicate a bad quality clustering solution.

2.2 Variance Ratio Criterion

The Variance Ratio Criterion [1] (also known as the Calinski-Harabasz Index) evaluates clustering quality by comparing the total between-cluster dispersion (nomimator) to the total within-cluster dispersion (denominator):

$$VRC(C) = \frac{\sum_{j=1}^{k} |C_j| \|\mu_j - \mu\|^2}{\sum_{j=1}^{k} \sum_{x \in C_j} \|x - \mu_j\|^2} \frac{n-k}{k-1} \tag{5}$$

where $|C_k|$ is the size of cluster C_k, μ_k is the center of C_k, and μ *is the overall mean* of the dataset. Higher VRC values indicate better clustering, as they reflect greater between-cluster variability relative to within-cluster dispersion. A key limitation of VRC is that it measures cluster separation (nominator) using the variance of the cluster centers, which is a simplistic approach that may be ineffective for capturing complex cluster structures.

3 Counterfactual Distances

3.1 Counterfactuals for Classification

Counterfactual explanations are local explanations that have been widely used for classification problems. In essence, a counterfactual explanation provides suggestions on how the feature values of an example (called *factual*) should change in order for the modified example (called *counterfactual*) to be classified into a different class [4,12]. More specifically, let f be a classification model and $d(x, y)$ a distance function. Given an example (factual) y, its counterfactual explanation z is the data point closest to y whose outcome $f(z)$ differs from the prediction $f(y)$. More formally:

$$z = \arg\min_{x} d(x, y) \quad \text{s.t.} \quad f(z) \neq f(y). \tag{6}$$

Various search methods have been studied to solve the above problem ranging from gradient-based to genetic algorithms. Those methods are available in popular software libraries for generating counterfactuals in classification tasks supporting various distance metrics (such as L_1 and L_2 norms) and classification models f.

3.2 Counterfactuals for Clustering

We assume a clustering solution $C_1, ..., C_k$ with k clusters on a given dataset X. Given a factual y in cluster C_i and a distance function $d(x, y)$ its counterfactual explanation z is defined as the solution to the following constrained optimization problem [11]:

$$z = \arg\min_{x} d(x, y) \quad \text{s.t.} \quad C_\ell \neq C_i \tag{7}$$

In the above formulation, C_ℓ is the cluster label of z, taking into account the cluster assignment rule (e.g. the distance from cluster centers in the k-means case). Assuming a source and a target cluster, *the counterfactual of a given point (factual) is its closest point lying on the cluster boundary.* Therefore, the counterfactual distance of a point to the target cluster, provides an effective and optimal way to *quantify distance of a point from a cluster.* By summing (or averaging) the counterfactual distances of points, we can compute a measure of separation between two clusters. This idea is exploited in our proposed clustering quality score.

In [11] we have proposed ways for computing counterfactuals for k-means and Gaussian clustering specifically targeting the L_2 distance norm between factual and counterfactual point. The counterfactual computation for k-means clustering is simple and analytical as explained next.

3.3 Counterfactuals for k-Means Clustering

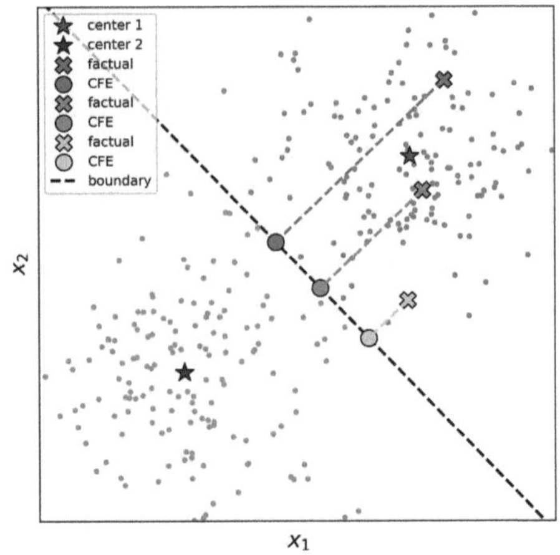

Fig. 1. Counterfactual computation for k-means clustering.

Assuming two clusters C_j and C_l with centers μ_j and μ_l respectively, the cluster boundary set B contains the points z for which:

$$|z - \mu_j|^2 = |z - \mu_l|^2 \tag{8}$$

This equation defines a *hyperplane* that is perpendicular at the middle (point $m_{lj} = (\mu_j + \mu_l)/2$) to the vector $n_{jl} = \mu_l - \mu_j$ connecting the two centers. We refer to the cluster boundary as *separating hyperplane*.

Assuming a point x (factual) of cluster C_j and the L_2 distance measure, its closest counterfactual point $cf_{jl}(x)$ with respect to cluster C_l is computed analytically and corresponds the projection of x to the separating hyperplane:

$$cf_{jl}(x) = x - \frac{(x - m_{lj})^\top n_{lj}}{|n_{lj}|^2} n_{lj} \tag{9}$$

Moreover, the Euclidean distance between x and its counterfactual $cf_{jl}(x)$ is given by the *projection distance*:

$$d_{jl}(x) = \frac{|(x - m_{jl})^\top (\mu_l - \mu_j)|}{||\mu_l - \mu_j||}, \tag{10}$$

Therefore, in the k-means case, the counterfactual is computed *analytically* using the two clusters centers μ_j, μ_l and the factual x. In Fig. 1 we provide illustrative examples of counterfactuals generated considering a synthetic 2-d dataset clustered using k-means. It can be observed, the distances between cluster points and their counterfactuals (which lie on the cluster boundary), can be used to define effective measures of *cluster separation.*

4 Counterfactual-Based Quality Score

Assuming a dataset $X = \{x_1, \ldots, x_n\}$, we propose a new criterion for evaluating the quality of a clustering solution $C = \{C_1, \ldots, C_k\}$ with k clusters:

$$CFQ(C) = \frac{\sum_{j=1}^{k} \text{separation}(j)}{\sum_{j=1}^{k} \text{variance}(j)}, \tag{11}$$

where variance(j) is the total squared distance of points in cluster j to their centroid μ_j (intra-cluster variance), and separation(j) is the minimum distance between cluster j and any other cluster, computed using counterfactual distances as follows.

Given the centroids μ_j and μ_l of clusters C_j and C_l, we can compute the Euclidean counterfactual distance $d_{jl}(x)$ of a point $x \in C_j$ from cluster C_l using Eq. 9. We define the distance $Sep(j,l)$ of cluster C_j to C_l as the sum of squared counterfactual distances of the points of C_j to cluster C_l:

$$Sep(j,l) = \sum_{x_i \in C_j} d_{jl}^2(x_i) \tag{12}$$

Note that $Sep(j,l)$ is not symmetric, i.e. $Sep(j,l) \neq Sep(l,j)$. Then, by transferring the silhouette idea to the cluster level, we define the separation $S(j)$ of cluster C_j as its minimum separation to any other cluster:

$$S(j) = min_{l,l \neq j} Sep(j,l) \tag{13}$$

The total separation $TS(C)$ of the clustering solution is computed as the sum of individual cluster separations:

$$TS(C) = \sum_{j=1}^{k} S(j) \tag{14}$$

It should be noted that the above sum includes one counterfactual distance term for each data point. In order to measure the total variance $TV(C)$ of the clustering solution, we use the typical clustering error minimized by the k-means algorithm:

$$TV(C) = \sum_{i=1}^{n} \sum_{j=1}^{k} I(x_i \in C_j) ||x_i - \mu_j||^2 \tag{15}$$

Note that, as in the case with separation measure, one distance term for each point is also included in the total variance measure. Therefore both TS(C) and TV(C) are of the same scale and we define our clustering quality measure (CFQ) as the ratio of the two terms:

$$CFQ(C) = \frac{TS(C)}{TV(C)} \tag{16}$$

Based on the above definition, higher CFQ values indicate better clustering quality. A notable advantage of CFQ is its *linear computational complexity*: $O(nkd)$ where d is the dimension of the data points. In contrast, the computational complexity of silhouette is $O(n^2)$, thus for large datasets, silhouette is usually approximately computed using randomly selected data subsets.

Figure 2 presents an example where a 2-d synthetic dataset with 5 clusters has been partitioned using k-means with k from 2 to 5. The silhouette, VRC and CFQ scores of each solution are also presented. It can be observed that CFQ selects the correct solution ($k = 4$) evaluating this solution with the highest CFQ score. In contrast, silhouette selects $k = 2$, while VRC selects $k = 5$.

The silhouette score has a tendency to favor simpler partitions with fewer clusters, showing a bias toward *underclustering.* In particular, when well separated groups contain finer internal structures, silhouette may fail to distinguish the subclusters within each group. Instead, it often favors merging them into a single cluster, overlooking meaningful distinctions in locally separated regions.

The proposed counterfactual-based criterion (CFQ) evaluates clustering quality by comparing cluster separability to intra-cluster compactness. While it captures meaningful separations more robustly than traditional metrics, it exhibits a tendency to *overclustering.* In CFQ, the score improves when clusters become more compact (due to splitting) while maintaining sufficient separation. As a result, CFQ tends to favor solutions where large, dense clusters are artificially divided into multiple subclusters, even when such splits do not correspond to meaningful structural differences. This occurs because the decrease of the total variance term is greater than the decrease of the separation term, leading to an inflated CFQ score.

Figure 3 presents another example using a 2-d synthetic dataset with 4 clusters which has been partitioned using k-means with k from 2 to 8. Only the solutions for $k = 2$ and $k = 7$ are presented. CFQ overclusters the dataset selecting the solution with $k = 7$ clusters, while silhouette underclusters the dataset selecting $k = 2$. VRC also selects $k = 7$. Silhouette and CFQ tendencies for underclustering and overclustering, respectively, highlight the need for caution when relying solely on a single internal validation metric to guide clustering decisions.

In what concerns VRC, in practice, it often proves less sensitive to local cluster structures compared to CFQ and Silhouette. Its restrictive reliance on global variance and distances between centers limits its effectiveness when clusters differ significantly in shape, size or density.

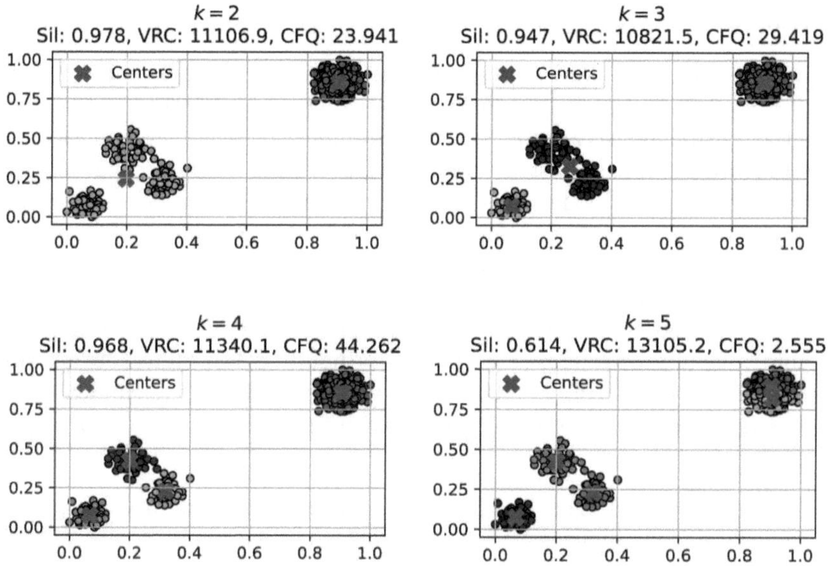

Fig. 2. k-means clustering results for different values of k using a synthetic 2-d dataset. CFQ attains maximum value at the correct solution $k = 4$. Silhouette attains maximum at $k = 2$ (underclustering). VRC attains maximum at $k = 5$ (overclustering).

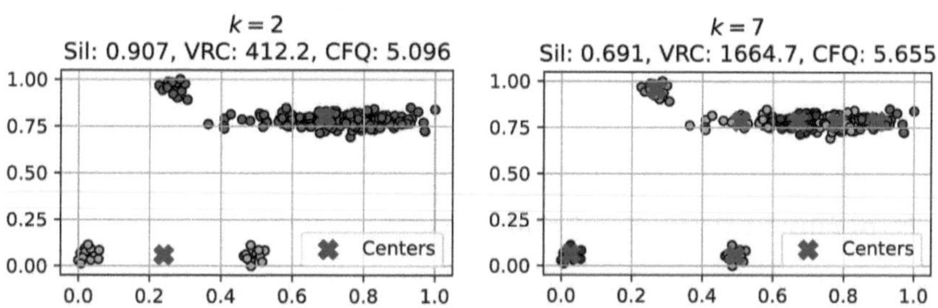

Fig. 3. k-means clustering results for $k = 2$ and $k = 7$ using a synthetic 2-d dataset. Silhouette suggests $k = 2$ (underclustering). CFQ and VRC suggest $k = 7$ (overclustering).

5 Experimental Results

The aim of our empirical study is to comparatively assess the effectiveness of the proposed CFQ criterion in number of clusters estimation: selecting the best among several clustering solutions provided by applying the k-means algorithm on the same dataset for various values of the number of clusters k. We compare CFQ to both silhouette (with squared Euclidean pairwise distances) and VRC.

5.1 Datasets

We considered four well-known 10-class real datasets related to numerical digits (0–9). Those datasets have been selected since they are known to demonstrate substantial clustering structure, i.e. the examples of each digit class tend to form relatively compact clusters that are discernible to some degree. Table 1 summarizes the four digit datasets that we used in our experimental evaluation.

The OptDigits dataset consists of grayscale 8×8 images of handwritten digits. Each 16-dimensional example in the Pendigits dataset contains the (x, y) coordinates of eight pen points on a digital screen. USPS contains 16×16 images of handwritten images. The MNIST embeddings dataset contains 10-dimensional representations of grayscale 64×64 images of handwritten digits, obtained from a trained autoencoder [10]. In all datasets we used min-max normalization as a preprocessing step to map the features of each data point to the $[0, 1]$ interval.

Table 1. The real world digits datasets used in the experiments. n is the number of data instances, d is the dimensionality.

Dataset	n	d	Source
OptDigits	1797	64	[2]
Pendigits	10992	16	[2]
USPS	9298	256	[6]
MNIST (embeddings)	60000	10	[2]

5.2 Experimental Protocol and Results

For each of the four datasets a series of 100 experimental trials we conducted. At each trial, a data subset is first created as follows:

- We randomly select the number of classes k_{actual} as an integer between 2 and 10. Note that this is the ground truth number of clusters of the trial.
- We randomly select k_{actual} different digit classes (from 0–9).
- For each selected digit, we randomly specify the number of examples per class between a minimum value equal to 20 and a maximum value equal to 1000. If the number of available examples of a class is less than 1000, then we set our maximum value equal to this number.
- For each selected class, we randomly select (without replacement) a subset of examples of size equal to the previously specified numbers.

In the above way, a data subset is generated in each trial containing a random number of randomly selected digit classes and with the number of examples per digit varying from possibly very low (near 20) to possibly very high (near 1000). Therefore, some subsets are expected to be highly imbalanced. For each generated subset, the actual number of clusters is known as well as the ground truth labeling of the examples. This information is used to evaluate the clustering solutions selected by the three compared quality scores.

Table 2. NMI statistics over 100 trials for each dataset.

Dataset	Method	Mean	Std
Optdigits	CFQ	0.711	0.118
	SIL	0.725	0.121
	VRC	0.477	0.204
Pendigits	CFQ	0.641	0.135
	SIL	0.629	0.149
	VRC	0.543	0.178
MNIST Embeddings	CFQ	0.712	0.138
	SIL	0.691	0.174
	VRC	0.542	0.194
USPS	CFQ	0.561	0.149
	SIL	0.465	0.207
	VRC	0.394	0.179

After subset generation, the following experiment was conducted at each trial: we applied the k-means algorithm to this subset for values of k from 2 to 15. The solution for each k was evaluated using the three criteria (CFQ, Silhouette, VRC) and for each criterion the solution with highest score was selected as the final solution. The solution selected by each criterion was evaluated i) based on the accuracy of k estimation (absolute difference from actual k) and ii) the NMI [3] of the solution with respect to the ground truth solution.

For each digit dataset, we repeated the above protocol for each of the 100 trials and present two plots. The first plot provides the NMI score for each trial of the k-means solution selected by CFQ (blue line), silhouette (SIL) (orange line) and VRC (green line). The NMI of the k-means solution with ground truth k (k_{actual}) is also presented (black line). The second plot for each dataset presents for each trial the difference ($k^* - k_{actual}$) between the value k^* of the solution selected by each score and the ground truth k_{actual}. The line colors are the same as in the NMI plot. Obviously, the ground truth solution (black line) has a constant difference equal to zero. Therefore solutions above the black horizontal line correspond to overclustering and below it to underclustering. Figure 4 and 5 provide results for Optdigits, Fig. 6 and 7 for Pendigits, Fig. 8 and 9 for MNIST embeddings, while Fig. 10 and 11 provide results for USPS.

Table 2 provides for each dataset the average and standard deviation over the 100 trials of the NMI scores of the k-means solutions selected by each of the three scores. Table 3 presents for each dataset the number of trials for which CFQ solution was found superior, equal or inferior to the silhouette (SIL) or VRC solution in terms of NMI (higher is better) and in terms of estimation of the actual number of clusters $|k^{\star} - k_{actual}|$ (lower is better).

From the results in Table 3 (columns "Tie") it can be observed that in about half of the 400 trials with all datasets, CFQ and SIL provide solutions of equal quality both in terms of NMI and estimation of the true number of clusters. Therefore, they can be considered as equal strength competitors in general. In what concerns VRC, it is clearly inferior to the other scores in all four datasets.

Results for the Optdigits dataset indicate a moderate superiority of SIL compared to CFQ both in terms of NMI and k estimation. while the opposite holds for USPS dataset. In MNIST embeddings CFQ appears to be slightly superior, while in Pendigits SIL is slightly better in terms of k estimation and CFQ in terms of NMI. The inspection of the k estimation plots for all datasets reveals the tendency of CFQ to overestimate the actual number of clusters, while SIL and VRC tend to underestimate this number, with SIL providing estimations closer to ground truth.

Table 3. Number of trials for which CFQ was superior ("CFQ" column), inferior ("Other" column) or equal ("Tie" column) to SIL or VRC in terms of NMI and estimation of true number of clusters k_{actual}.

Dataset	Comparison	NMI			$\|k^{*} - k_{actual}\|$		
		CFQ	Other	Tie	CFQ	Other	Tie
Digits	CFQ vs SIL	20	34	46	9	41	50
	CFQ vs VRC	80	16	4	50	29	21
Pendigits	CFQ vs SIL	26	14	60	13	23	64
	CFQ vs VRC	61	12	27	47	18	35
MNIST	CFQ vs SIL	32	23	45	27	21	52
Embeddings	CFQ vs VRC	76	6	18	63	12	25
USPS	CFQ vs SIL	43	19	38	38	20	42
	CFQ vs VRC	59	5	36	49	9	42

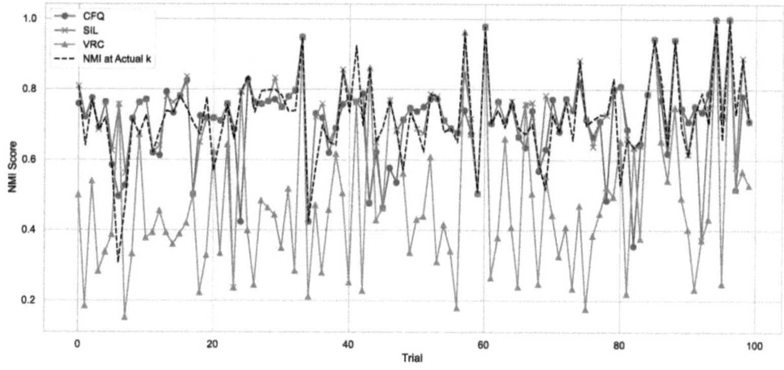

Fig. 4. Optdigits: NMI across trials for different quality scores.

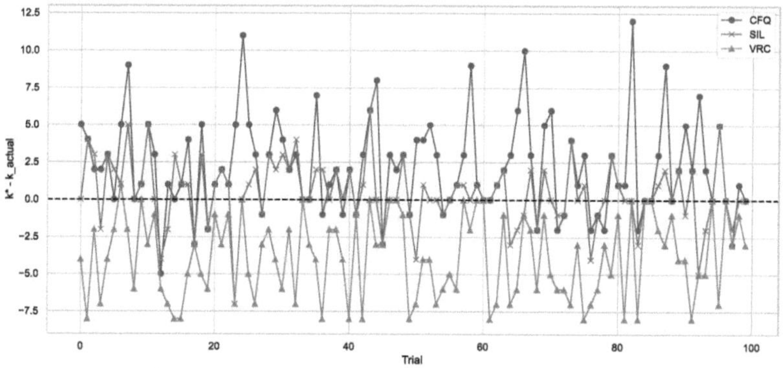

Fig. 5. Optdigits: Error between predicted and true number of clusters ($k^* - k_{actual}$) across trials for different quality scores.

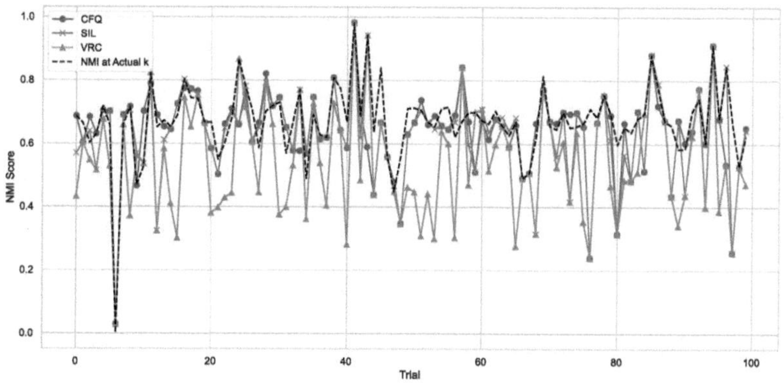

Fig. 6. Pendigits: NMI across trials for different quality scores.

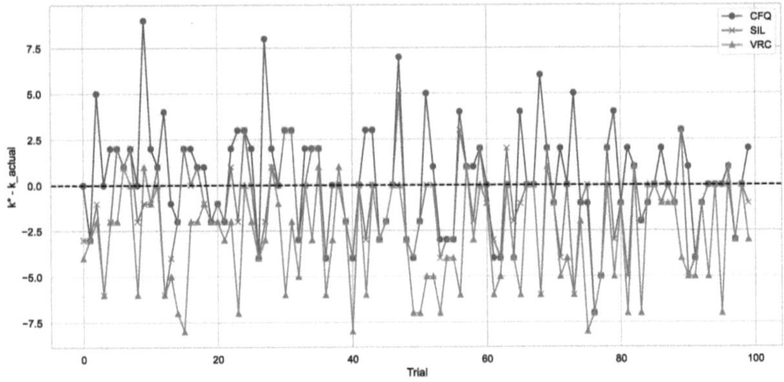

Fig. 7. Pendigits: Error between predicted and true number of clusters ($k^* - k_{actual}$) across trials for different quality scores.

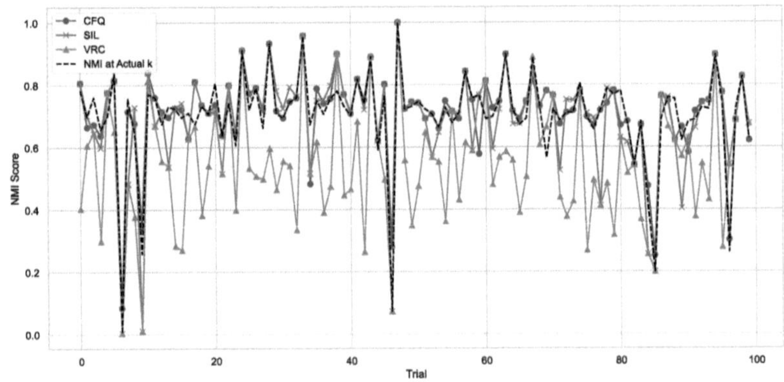

Fig. 8. MNIST (embeddings): NMI across trials for different quality scores.

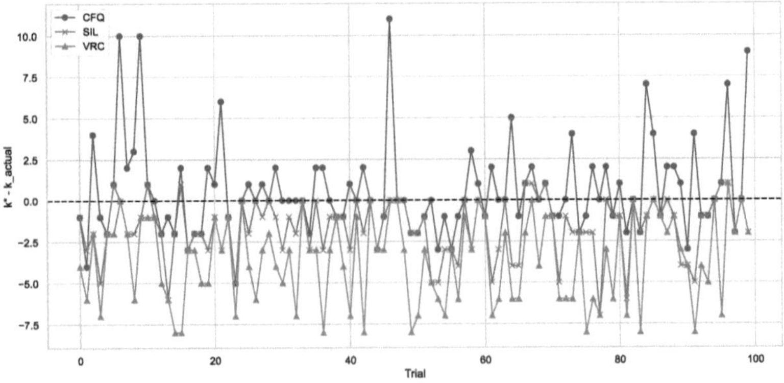

Fig. 9. MNIST (embeddings): Error between predicted and true number of clusters ($k^* - k_{actual}$) across trials for different quality scores.

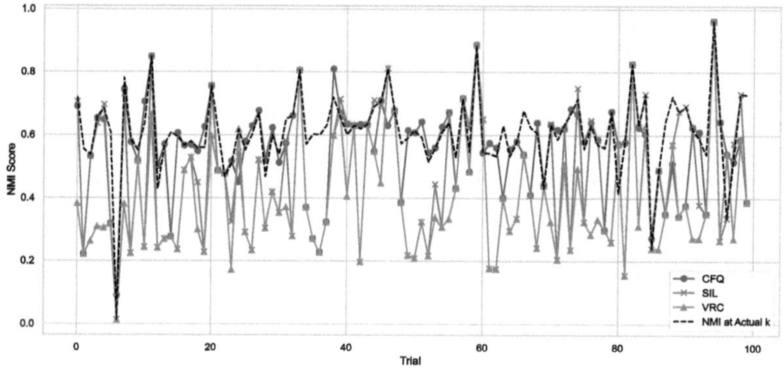

Fig. 10. USPS digits: NMI across trials for different quality scores.

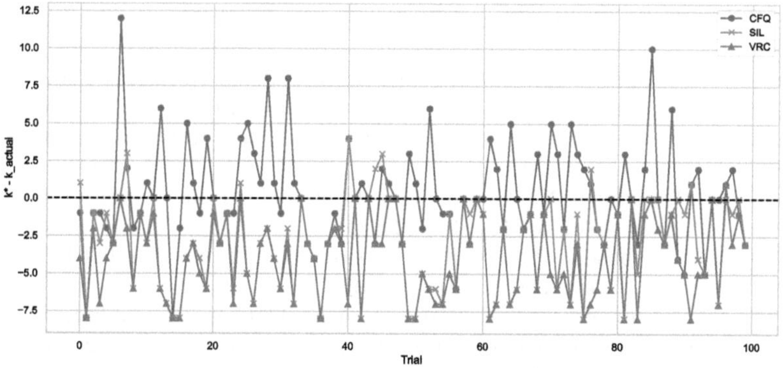

Fig. 11. USPS digits: Error between predicted and true number of clusters ($k^* - k_{actual}$) across trials for different quality scores.

6 Conclusions

We introduced a new approach to quantifying cluster separation by leveraging counterfactual distances. Counterfactuals have been recently used in clustering to represent the minimal changes required for a point from one cluster to be reassigned to another. Building on this idea, we used counterfactual distances within the framework of k-means clustering to define a separation measure between clusters. We also proposed a clustering quality score (CFQ), defined as the ratio of total separation to total intra-cluster variance. We assessed the effectiveness of CFQ in comparison to established metrics such as silhouette and VRC across several digit datasets.

Future work could focus on a more elaborate empirical evaluation of CFQ both in real-world clustering applications (e.g. face clustering, video summarization, text clustering) as well as in assessing its robustness in the presence of noise and outliers. Moreover, we aim to employ counterfactual distances to measure separation in Gaussian clustering. In such a case the cluster boundary is nonlinear and counterfactuals are computed by solving a non-linear equation with a single parameter [11].

Acknowledgment. The research project is implemented in the framework of H.F.R.I. call "Basic research Financing (Horizontal support of all Sciences)" under the National Recovery and Resilience Plan "Greece 2.0" funded by the European Union - NextGenerationEU (H.F.R.I. ProjectNumber: 15940).

Declarations. Conflict of interest The authors declare no conflict of interest.

References

1. Caliński, T., Harabasz, J.: A dendrite method for cluster analysis. Commun. Stat.-Theory Methods **3**(1), 1–27 (1974)
2. Dua, D., Graff, C.: UCI machine learning repository (2017). http://archive.ics.uci.edu/ml
3. Estévez, P.A., Tesmer, M., Perez, C.A., Zurada, J.M.: Normalized mutual information feature selection. IEEE Trans. Neural Netw. **20**(2), 189–201 (2009)
4. Guidotti, R.: Counterfactual explanations and how to find them: literature review and benchmarking. Data Min. Knowl. Disc. **38**(5), 2770–2824 (2024)
5. Halkidi, M., Batistakis, Y., Vazirgiannis, M.: On clustering validation techniques. J. Intell. Inf. Syst. **17**(2–3), 107–145 (2001)
6. Hull, J.J.: A database for handwritten text recognition research. IEEE Trans. Pattern Anal. Mach. Intell. **16**(5), 550–554 (1994)
7. Ikotun, A.M., Habyarimana, F., Ezugwu, A.E.: Cluster validity indices for automatic clustering: a comprehensive review. Heliyon **11**(2), e41953 (2025)
8. Karra, A., Vardakas, G., Pitoura, E., Likas, A.: Generating counterfactual explanations for clustering models based on their equivalence to classification models. In: IFIP International Conference on Artificial Intelligence Applications and Innovations. Springer, Cham (2025)
9. Rousseeuw, P.J.: Silhouettes: a graphical aid to the interpretation and validation of cluster analysis. J. Comput. Appl. Math. **20**, 53–65 (1987)
10. Van Der Maaten, L.: Learning a parametric embedding by preserving local structure. In: Artificial Intelligence and Statistics, pp. 384–391. PMLR (2009)
11. Vardakas, G., Karra, A., Pitoura, E., Likas, A.: Counterfactual explanations for k-means and gaussian clustering. arXiv preprint arXiv:2501.10234 (2025)
12. Verma, S., Boonsanong, V., Hoang, M., Hines, K., Dickerson, J., Shah, C.: Counterfactual explanations and algorithmic recourses for machine learning: a review. ACM Comput. Surv. **56**(12), 1–42 (2024)

Selecting Unlabeled Data for Tabular Self-Supervised Learning

Sintija Stevanoska[1,2(✉)], Katharina Dost[1], Christian L. Camacho Villalón[1], and Sašo Džeroski[1,2]

[1] Jožef Stefan Institute, Ljubljana, Slovenia
{sintija.stevanoska,katharina.dost,christian.camacho.villalon, saso.dzeroski}@ijs.si
[2] Jožef Stefan International Postgraduate School, Ljubljana, Slovenia

Abstract. Self-supervised learning (SSL) offers a promising solution to the problem of label scarcity by leveraging large amounts of unlabeled data to learn transferable representations. However, using extensive unlabeled datasets can introduce substantial computational costs during pretext training and may include noisy or unrepresentative samples that degrade learning. In this work, we explore whether selecting a subset of the unlabeled examples available for the pretext task can reduce computation time while maintaining satisfactory performance of SSL methods for tabular data. In particular, we investigate whether estimates of the uncertainty with which predictions are made for unlabeled data (based on the available labeled data) can inform sampling and lead to superior embeddings. To answer these questions, we carry out large-scale experiments across 28 tabular benchmark datasets with TabNet and SCARF, on varying amounts of labeled data, and with a number of strategies for sampling unlabeled data. Our results show that reducing the pool of unlabeled data yields significant computational gains with only marginal reductions in performance.

Keywords: Self-Supervised Learning · Tabular Data · Unlabeled Data Selection · Active Data Selection

1 Introduction

Current machine learning systems increasingly rely on large volumes of data [6,19,35,41]. While supervised learning has achieved impressive performance across tasks, its reliance on large labeled datasets remains a major bottleneck, as acquiring high-quality labels is often time-consuming, expensive, or infeasible in real-world settings. In contrast, unlabeled data is typically far more abundant and easier to collect. For example, healthcare systems collect abundant patient records, but expert-labeled outcomes are scarce.

S. Stevanoska and K. Dost—These authors contributed equally.

S. Džeroski et al. (Eds.): DS 2025, LNAI 16090, pp. 237–251, 2025.
https://doi.org/10.1007/978-3-032-05461-6_16

Self-supervised learning (SSL) has emerged as a solution by learning useful representations from unlabeled data through pretext tasks. These tasks are learning objectives defined using the structure of the data itself [5,21]. This paradigm has led to significant advances in computer vision (CV) and natural language processing (NLP), where pretrained models such as SimCLR [5], MoCo [14], BYOL [12], BERT [6] and GPT [25] have achieved state-of-the-art performance on downstream tasks.

Compared to images and text, tabular data presents unique challenges for SSL. It lacks the spatial or sequential structure typically utilized in CV or NLP, and often includes heterogeneous features, missing values, and complex interdependencies. Nonetheless, recent work has shown that carefully designed pretext tasks, such as feature masking [1,40], latent space perturbation [22], contrastive learning [2,13], or a combination of these tasks [4,29,33], can improve performance over purely supervised models.

However, a drawback of SSL is its high computational cost. Pretraining often involves processing large volumes of unlabeled data, which can be prohibitive in terms of time and resources [9]. To reduce this burden, several lines of work have examined whether using smaller or more informative subsets of unlabeled data can yield similar performance with reduced computation. For instance, Dhekane et al. [7] demonstrated that for human activity recognition, SSL representations learned from just minutes of sensor data performed comparably to those trained on the full dataset. Similar trends have been observed in vision [5,17], text [15,26] and audio [20,23] domains, where selective strategies like clustering and domain alignment allow models to match full-data performance with a fraction of the unlabeled set.

Building on these findings, a related line of work has explored how confidence or uncertainty estimates can guide data selection for semi-supervised learning. For example, in speech recognition, Drugman et al. [8] showed that selecting unlabeled samples with high model confidence led to substantial gains, while low-confidence examples degraded performance. Notably, simple filtering heuristics performed on par with more complex models. These results raise the question of whether lightweight estimates of model uncertainty can similarly inform the selection of unlabeled data for SSL from tabular data.

In this paper, we investigate whether a small, well-chosen subset of unlabeled data is sufficient for effective pretraining in tabular SSL. Specifically, we explore (i) whether large pools of unlabeled data are necessary to achieve high-quality predictions, (ii) whether reducing the number of unlabeled samples impacts the quality of the learned embeddings, and (iii) whether the time-performance trade-off justifies subsampling the unlabeled space.

To address these questions, we borrow a classic idea from active learning, *query-by-committee (QBC)*, to estimate the uncertainty of any model on unlabeled points [30]. QBC is a lightweight proxy for uncertainty that leverages the disagreement of a "committee" of simple landmark models trained on bootstrap samples of the labeled data. We investigate whether training on subsets with varying levels of disagreement can improve the efficiency of SSL. Unlike in active learning, where uncertainty drives label acquisition, our focus is on data selection for unsupervised pretraining.

We conduct an extensive empirical study across 28 tabular benchmark datasets using TabNet and SCARF as SSL model proxies. Our findings show that even small subsets (100–1000 points) of the unlabeled pool can yield representations and downstream performance comparable to full-data pretraining. However, no single level of committee disagreement consistently outperforms others across datasets. Together, these results suggest that subsampling of unlabeled data offers a practical route to more efficient tabular self-supervised learning.

Our main contributions are the following: (1) We provide the first large-scale evaluation (28 datasets) of unlabeled data subsampling for tabular SSL, including both random and uncertainty-based selection and generalizing beyond specific domains, across multiple labeled and unlabeled data regimes; and (2) We demonstrate that disagreement-based sampling is a simple, model-agnostic strategy that can substantially reduce training costs at similar performance.

The remainder of this paper is organized as follows: Sect. 2 reviews related research. Sections 3 and 4 introduce our experimental methodology and setup, respectively. Section 5 discusses our findings, and Sect. 6 concludes the paper.

2 Related Research

This section reviews prior work in three areas relevant to our study: (1) SSL for tabular data, (2) unlabeled data selection for SSL across domains, and (3) unlabeled data selection in active learning.

Self-supervised Learning for Tabular Data. A recent survey by Wang et al. [36] categorizes SSL methods for non-sequential tabular data into predictive (feature masking/reconstruction), contrastive (view-based similarity), and hybrid approaches. In this work, we evaluate data selection strategies using representatives for both foundational paradigms: (i) TabNet [1] is a predictive transformer-based approach that uses a sequential attention mechanism to select features to use at each decision step. TabNet can be pretrained by masking a subset of the input columns and reconstructing them; (ii) SCARF [2] is a contrastive method that generates positive views by corrupting features with values sampled from the empirical distribution, encouraging invariant representations across views.

Selecting Unlabeled Data for SSL. Recent work across domains suggests full-data pretraining is not always necessary. Studies suggest that selecting informative subsets can improve downstream performance while reducing computational cost. We briefly review approaches across image, text, audio, and tabular domains.

In computer vision, Joshi et al. [16,17] propose augmentation similarity and data distillation strategies to construct compact training subsets. Tripathi et al. [32] dynamically remove less informative samples during training, while Tian et al. [31] cluster data to localize contrastive learning. Other approaches include sampling based on latent diversity [11] or semantic relevance to a target domain

[18]. These methods demonstrate that performance can be preserved or improved even with reduced data volumes.

In language modeling, Hwang et al. [15] and Xie et al. [38] filter by domain alignment or distribution similarity, while Renduchintala et al. [26] and Wang et al. [37] prioritize samples based on training dynamics, such as influence or loss gradients. These approaches yield compact models with strong generalization.

In speech recognition, Li et al. [20] and Park et al. [23] show that focusing SSL training on relevant subsets improves robustness and reduces error rates compared to using the full dataset.

In the tabular domain, research on selective SSL is sparse, with few works specifically addressing unlabeled subset selection. Dhekane et al. [7] find that randomly sampling just 15 min of temporal data is sufficient to match the full-data baseline. Yoon et al. [39] use an ensemble of one-class classifiers trained on disjoint subsets of the unlabeled data to filter outliers from the unlabeled pool, improving downstream anomaly detection. Ghosh et al. [10] propose an SSL framework that focuses training on underrepresented or high-error samples. While their method does not explicitly filter unlabeled data, it prioritizes informative samples during representation learning.

Despite these initial efforts, the literature on unlabeled data selection for tabular SSL remains sparse. Given the larger corpus of work from other domains, this highlights a promising direction for research in making SSL from tabular data both effective and efficient.

Selecting Unlabeled Data for Active Learning. In *active learning*, selecting unlabeled data plays a fundamental role. Here, the goal is to prioritize unlabeled instances that are expected to most improve model performance when labeled. Among the various criteria proposed, *uncertainty-based sampling* remains the most widely used due to its empirical success across domains [24,27]. This strategy selects samples for which the model is maximally uncertain in making a prediction, and thus obtaining the label would be most informative. However, model uncertainty is not always available or reliable, especially for non-probabilistic models. A model-agnostic alternative is query-by-committee (QBC) [28], which measures uncertainty via disagreement among a set of models – each trained on the same labeled data but differing, e.g., by random initialization or bootstrapping. An unlabeled instance is considered uncertain if the committee members predict different outcomes. Unlike single-model confidence scores, disagreement is not tied to a specific model type and can be computed efficiently [30], making it suitable not only for active learning but also for selecting informative unlabeled samples in self-supervised learning.

To our knowledge, no prior work systematically investigates the trade-off of sampling unlabeled data for general-purpose representation learning in tabular data. While strategies such as disagreement-based sampling have shown promise in related contexts (e.g., active learning), they have not been applied to SSL in tabular domains. Our study addresses this gap by providing an extensive empirical comparison of the performance-compute time trade-off when sampling

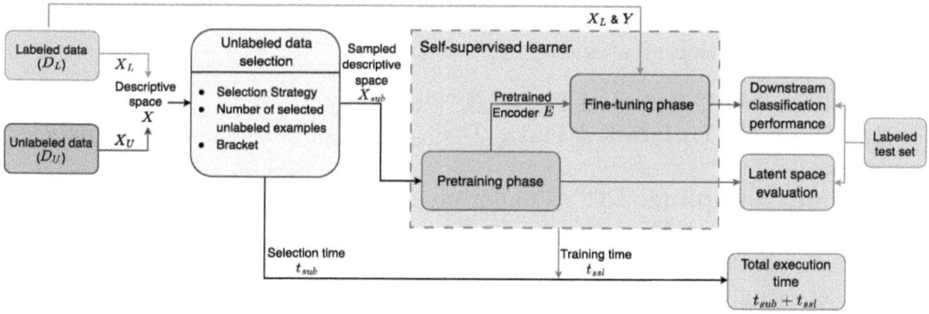

Fig. 1. An overview of the methodology for selecting and using unlabeled tabular data for SSL.

unlabeled data for tabular SSL, using both random and disagreement-based selection across a large and diverse benchmark.

3 Methodology

Our goal is to investigate how the quantity and composition of unlabeled data affect the performance and efficiency of SSL on tabular datasets. Specifically, we examine whether pretraining on a selected subset of unlabeled samples can reduce computational costs while maintaining representation quality and downstream task performance.

Figure 1 summarizes our methodology, which follows a standard SSL pipeline but incorporates two key extensions: (1) a sampling step to subsample the unlabeled pool using different strategies, and (2) a multi-perspective evaluation to assess downstream utility, representation quality, and total execution time.

This section introduces all components in detail.

3.1 Data

Tabular data is organized in a matrix-like structure, consisting of samples $x \in \mathbb{R}^d$, where each row is a data point and each column is a feature (numerical, categorical, or binary). As is common in self-supervised learning, we assume access to a small *labeled dataset* $D_L = \{(x_i, y_i)_{i=1}^{n_L}\}$ consisting of n_L feature vectors x_i with corresponding labels y_i, and a larger *unlabeled dataset* $D_U = \{(x_i)_{i=1}^{n_U}\}$ containing n_U feature vectors, for which no label is known. During pretraining, both feature vectors from D_L to D_U (X_L and X_U respectively) are used to learn the representations, followed by fine-tuning on D_L to solve a downstream task.

3.2 Unlabeled Data Selection

Rather than using the entire unlabeled set (D_U) for pretraining, we first pass D_U through a selection mechanism to construct a smaller, potentially more informative subset $D_U^{\text{sub}} \subset D_U$ of size n_{sub}, yielding a feature vector X_{sub}. We consider the following selection strategies:

- **Baseline: No Unlabeled Data.** D_U^{sub} is empty.
- **Baseline: All Unlabeled Data.** $D_U^{\text{sub}} = D_U$ contains the entire unlabeled dataset.
- **Random Sampling.** D_U^{sub} is uniformly sampled from D_U.
- Committee-Based **Disagreement Sampling.** D_U^{sub} is uniformly sampled from a subset of D_U, whose disagreement scores d fall within a specified bracket (e.g., low, medium, high disagreement). These disagreement scores reflect the uncertainty with which a prediction could be made for this point given the available information (D_L).

To estimate model uncertainty over the unlabeled pool, we implement a committee-based disagreement score borrowed from the active learning strategy *"query-by-committee (QBC)"* [27,30]. QBC offers a model-agnostic and efficient way to estimate uncertainty by measuring disagreement among diverse learners, making it robust to model choice and well-suited for our purpose [27].

We train a committee $\mathcal{C} = \{h_1, \ldots, h_K\}$ of K simple base classifiers (e.g., decision stumps, decision trees, or shallow autoencoders) on bootstrap samples of the labeled dataset D_L. Each classifier is additionally restricted to a random subset of features of size $\sqrt{d}$, where d is the total number of features, to encourage predictive diversity. The sampling of data points and features is in line with that of Random Forests [3].

For each unlabeled instance $x \in D_U$, we collect the set of predicted class labels $\{h_1(x), \ldots, h_K(x)\}$. Let C be the number of unique labels predicted for x, and let p_c denote the fraction of committee members that predicted label c. The disagreement score is computed using the normalized vote entropy:

$$\text{Disagreement}(x) = -\frac{1}{\log_2 C} \sum_{c=1}^{C} p_c \log_2 p_c$$

If all classifiers agree, the entropy is 0. The score increases as predictions are more diverse, reaching 1 when votes are uniformly distributed across C classes.

The disagreement score serves to partition D_U into brackets (e.g., low, medium, high uncertainty) and sample from each bracket to form D_U^{sub}. Particularly, we use percentiles to form those brackets, and bracket $[d_{\text{lower}}, d_{\text{upper}}]$ contains all points x whose disagreement score falls between the d_{lower}'th and the d_{upper}'th percentile of all disagreement scores for D_U.

Selection of D_U^{sub} incurs a time cost t_{sub} that is independent of the choice of bracket as they all contain the same number of points to choose from.

In this study, we opted for 10 Decision Trees and Decision Stumps as the base classifiers in $\mathcal{C}$ due to their computational efficiency. We additionally tested a simple autoencoder in preliminary experiments but omitted it in our benchmarks since its running time outweighed its benefits.

3.3 Self-supervised Training

After selecting D_U^{sub}, we follow a standard two-stage SSL protocol.

In the pretraining phase, an encoder $E : \mathbb{R}^d \to \mathbb{R}^z$ is trained on D_U^{sub} using a fixed pretext task (e.g., TabNet uses masking; Scarf uses contrastive learning). This yields general-purpose representations that capture patterns in the data without labels. The resulting encoder E maps raw inputs X into latent space Z.

In the fine-tuning phase, E is trained end-to-end with a classification head on the labeled dataset D_L, optimizing for predictive performance. This stage evaluates how well the learned representations transfer to the supervised task. The pretraining and finetuning stages incur a combined time cost of t_{ssl}.

3.4 Evaluation

We assess the impact of unlabeled data selection through three complementary evaluation streams, aligned with the research questions introduced earlier. These are visually highlighted in orange in Fig. 1, and provide a comprehensive view of the effects of sampling strategies on SSL performance and efficiency:

i **Downstream Performance.** To evaluate the end-task utility of the learned representations, we fine-tune the encoder on labeled data and report classification performance on a held-out test set using macro-averaged (across different classes) *Area Under the Precision-Recall Curve (AUPRC)*, which accounts for class imbalance. We compare against two baselines: *BaselineNone*, which trains only on labeled data without any SSL pretraining, and *BaselineAll*, which uses the full unlabeled set. These comparisons reveal whether a small, well-chosen unlabeled subset can retain SSL benefits, and how the size of the labeled set, n_L, influences our observations. For each sampling strategy and labeled data setting, we report the best-performing disagreement bracket, averaged across datasets and seeds, to assess the value of disagreement-based over random sampling.

ii **Latent Space Evaluation.** Beyond downstream accuracy, we assess the structure of the learned latent space by training a linear classifier (logistic regression) on top of frozen representations from the pretrained encoder. This linear probing setup is a common SSL diagnostic [5,12,14,21] that evaluates the extent to which semantic class information is linearly encoded in the representation space Z. We again report AUPRC to maintain comparability across evaluation streams. This probe allows us to isolate representation quality from the influence of end-to-end fine-tuning, and serves as a proxy for the usefulness of embeddings in downstream tasks.

iii **Execution Time.** Given our focus on computational efficiency, we track the total wall-clock time $t = t_{\text{sub}} + t_{\text{ssl}}$ required for each configuration, including both the time spent on selecting the unlabeled subset (t_{sub}) and the time spent on SSL pretraining and fine-tuning (t_{ssl}). By jointly analyzing runtime and performance, we assess whether subsampling yields meaningful efficiency gains and whether any added overhead from sampling strategies is justified by improved outcomes.

Together, these three evaluation streams offer a practical assessment of the trade-offs between representation quality, downstream performance, and computational efficiency in selective SSL.

4 Experimental Setup

We evaluate our proposed data selection methodology using a large-scale experimental design across diverse tabular datasets and SSL configurations. This section details the datasets, preprocessing pipeline, model settings, and evaluation protocol.

Datasets. We conduct experiments on 28 tabular classification datasets from the OpenML-CC18 benchmark suite [34], spanning binary and multi-class tasks across domains. To ensure sufficient samples for our experimental setup, we retain only datasets with enough examples to draw at least 500 unlabeled samples from each bracket. Furthermore, CIFAR-10, FashionMNIST, and MNIST are excluded since they are image-based datasets, and our focus is strictly on non-visual tabular data. We split the datasets into training (75%) and testing (25%) sets in a stratified manner, preserving the overall proportions of the class values.

Data Preprocessing. We apply a very minimal data preprocessing pipeline. First, we encode the categorical features and the target, mapping each value within the columns to a unique integer. The numerical features are standardized by subtracting the mean and scaling to unit variance.

Creating Unlabeled and Labeled Datasets. To simulate an SSL scenario, we generate eight labeled subsets per dataset, containing 50, 100, 150, 200, 350, 500, 750, and 1000 labeled examples each. The remaining data points in the training set have their labels removed and are treated as unlabeled examples. The selection of labeled and unlabeled examples is done in a stratified way, ensuring that the class distributions in both the labeled and unlabeled sets match the training data class distribution. We further set 20% of the labeled training dataset aside for validation.

Sampling Unlabeled Data. We include all strategies for sampling unlabeled data mentioned in Sect. 3.2. We draw unlabeled samples of size $n_U^{\text{sub}} \in \{100, 250, 500\}$ for all datasets and additionally sample $n_U^{\text{sub}} \in \{750, 1000\}$ from the sufficiently large datasets. Disagreement sampling uses both decision trees (DT) and stumps (DS) to calculate disagreement scores and then samples from brackets $[0, 40]$, $[20, 60]$, $[40, 80]$, and $[60, 100]$.

Models. We run our experiments on two representative SSL models, TabNet[1] [1] and SCARF[2] [2]. Each model is trained on all combinations of datasets, labeled data amounts, sampling strategies, sample sizes, and disagreement brackets where applicable. We also run the experiments on 10 random seeds, accounting for variation in data splits, model initialization, and sample selection.

[1] https://github.com/dreamquark-ai/tabnet.

[2] https://github.com/clabrugere/pytorch-scarf.

Model hyperparameters are fixed across datasets, as our aim is to evaluate sampling strategies, not to optimize model-specific performance. For SCARF, we extend the original self-supervised-only implementation by adding a downstream classification head, which is a 3-layer MLP, for end-to-end fine-tuning.

Hardware. All experiments are conducted on the Arnes HPC cluster, part of the Slovenian National Supercomputing Network (SLING). Specifically, we use dual-core AMD EPYC Rome 7702P processors running at 2.0 GHz with a 512 KB cache size under Red Hat Enterprise Linux 9.5 (RHEL 9.5). We enforce exclusive CPU usage per run to ensure consistent timing.

Computational Costs. Our benchmark consumed 6,431,116 CPU-seconds and 926 h of wall-clock time across all runs, with average memory usage of 659 MB. These figures reflect the total compute used to evaluate all models, sampling strategies, and dataset splits.

All code, configurations, implementation details, and supplementary results are available in the open-source repository: https://github.com/sintija-s/sudssl/.

5 Results

We evaluate the impact of unlabeled data sampling strategies across the three evaluation streams introduced in Sect. 3.4: (i) downstream classification performance, (ii) representation quality via linear probing, and (iii) execution time.

5.1 Downstream Performance

We start by comparing the downstream classification performance of the three sampling strategies across eight labeled data regimes. Table 1 reports the results as percentages relative to two baselines: using no unlabeled data (BaselineNone) and using the entire unlabeled set (BaselineAll). Values greater than 100% indicate improvements over the respective baselines.

For each sampling strategy and number of labeled examples, we report the performance corresponding to the best-performing configuration, determined by selecting the optimal number of unlabeled samples and, for disagreement-based methods, the best disagreement bracket, averaged over all datasets and random seeds. To simplify the presentation, we report results averaged across the two SSL models (TabNet and SCARF), providing a general view of representation quality irrespective of the underlying architecture.

The results show that sampling just 100–1000 unlabeled examples can retain, or even slightly exceed, the performance of full-data pretraining. In the low-labeled regimes, disagreement-based strategies reach up to 106% of BaselineNone and 101% of BaselineAll, indicating not only improved learning over not using unlabeled examples, but also mild regularization gains over full-data SSL in some cases. As more labeled data becomes available, the relative performance across methods converges to around 98–99% of the full-data baseline.

Table 1. Average downstream performance of sampling methods relative to two baselines: using all unlabeled data (BaselineAll) and using none (BaselineNone). For each method (column Sampling Method) and number of labeled samples (column n_L)), the table reports the best performance as a percentage of each baseline, averaged across datasets. Values above 100% indicate the method outperforms the respective baseline.

n_L	Sampling Method	% of BaselineAll	% of BaselineNone
50	DisagreementSampling(DS)	101	103
	DisagreementSampling(DT)	101	103
	RandomSampling	100	102
100	DisagreementSampling(DS)	100	106
	DisagreementSampling(DT)	101	106
	RandomSampling	99	104
150	DisagreementSampling(DS)	101	106
	DisagreementSampling(DT)	101	105
	RandomSampling	98	102
200	DisagreementSampling(DS)	99	104
	DisagreementSampling(DT)	99	104
	RandomSampling	98	102
350	DisagreementSampling(DS)	99	103
	DisagreementSampling(DT)	99	103
	RandomSampling	98	102
500	DisagreementSampling(DS)	98	102
	DisagreementSampling(DT)	99	102
	RandomSampling	98	101
750	DisagreementSampling(DS)	99	102
	DisagreementSampling(DT)	99	102
	RandomSampling	98	101
1000	DisagreementSampling(DS)	99	101
	DisagreementSampling(DT)	99	101
	RandomSampling	99	101

Random sampling performs very competitively, trailing disagreement-based methods by only 1–3 percentage points in most settings. This suggests that the quantity of unlabeled data may be more critical than the specific sampling heuristic, and that random selection remains a strong baseline in this case.

5.2 Latent Space Probing

Linear probing is a standard technique for evaluating the quality of learned representations by training a simple linear classifier (e.g., logistic regression) on top of the frozen embeddings; high performance indicates that the representations are linearly separable and thus well-structured for downstream tasks. Figure 2 shows the performance of different sampling strategies, measured as a percentage of the performance achieved when using all unlabeled data for pretraining. We report the best-performing (if applicable) disagreement bracket for each strategy and labeled data setting, averaged over both models, TabNet and SCARF.

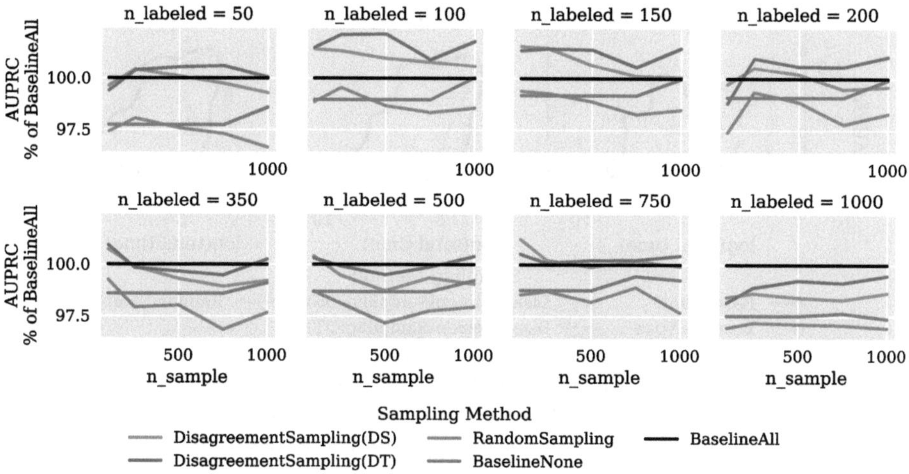

Fig. 2. Linear probing performance (measured as % of BaselineAll) across different amounts of sampled unlabeled data for each labeled data regime (n_L = 50, 100, 150, 200, 350, 500, 750, 1000). For each sampling method, we report the best-performing bracket configuration.

The results show that representations learned from subsampled data are nearly as effective as those learned from the full unlabeled set according to the linear probing evaluation as outlined in Sect. 3.4. For low labeled data regimes (50–200), disagreement-based strategies again reach around 99–102% of the BaselineAll linear probing performance, and they surpass BaselineNone (i.e., pretraining only with the labeled set) in every configuration.

Across all labeled data sizes, we observe that the representation quality plateaus quickly with an increasing number of sampled unlabeled points—often with just 500 examples, suggesting that a large unlabeled pool is not a strict requirement for obtaining useful representations. Additionally, the strong performance of Random Sampling also reinforces that simple sampling strategies can suffice for effective SSL in the tabular domain.

5.3 Execution Time

To understand the trade-off between computational efficiency and model quality, we analyze the relationship between training time and downstream AUPRC for different unlabeled data sampling strategies. Total execution time includes both unlabeled data selection (t_{sub}) and training (t_{ssl}).

Figure 3 shows kernel density contours of AUPRC against the log-transformed runtime for three labeled data regimes: 100, 500, and 1000 examples. Each contour reflects the joint distribution of runtime and performance across all datasets, seeds, and sampling configurations.

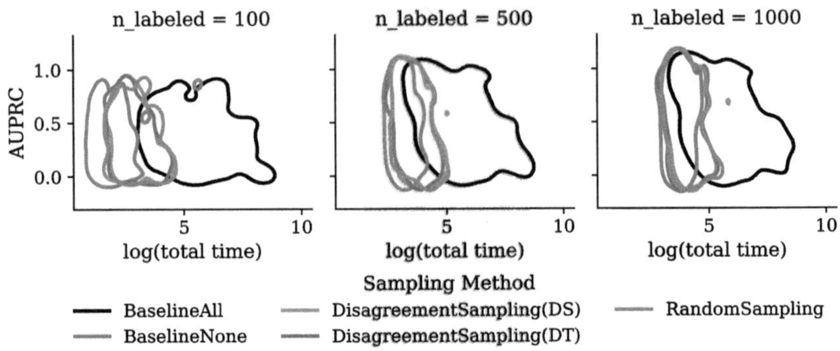

Fig. 3. Joint distribution of downstream AUPRC and total training time (log scale) for different sampling strategies and labeled data sizes. The lines indicate regions where the joint density exceeds 10% of the peak, based on kernel density estimation.

SSL with all unlabeled examples (BaselineAll) is consistently the most expensive strategy, requiring 4–7× the time of sampling-based alternatives, with marginal performance improvement.

Subsampling strategies achieve comparable (or better) AUPRC at a fraction of the cost. This effect is most visible in low labeled data settings, where full-data pretraining is both slow and volatile in performance. On average, the data selection contributes 19/20% of the total time (for DisagreementSampling with decision stumps and trees, respectively), while pretraining and finetuning contribute the 81/80%, highlighting that the overall time savings outweigh the costs of simple selection procedures.

These results strongly support the practical value of unlabeled data reduction: selecting a small but informative subset of unlabeled points enables faster SSL with minimal sacrifice in performance. Moreover, the modest overhead of disagreement-based selection is justified in low-label scenarios, while Random Sampling remains a competitive default.

6 Conclusion

In this work, we investigated whether selectively subsampling unlabeled data for tabular self-supervised learning can reduce computational costs without compromising downstream performance. Through extensive experiments on 28 benchmark datasets using TabNet and SCARF, we evaluated the impact of different sampling strategies across three perspectives: downstream classification performance, representation quality via linear probing, and total execution time.

Our findings show that using only a small subset of unlabeled data (100 to 1000 points) often retains most of the performance benefits of full-data SSL, particularly in low-label regimes. Disagreement-based sampling strategies provide modest gains over random sampling, which proves to be a simple, competitive sampling strategy baseline.

In terms of representation quality, subsampled SSL leads to embeddings that are nearly as linearly separable as those learned from full unlabeled sets, indicating that rich representations can be learned even from partial views of the unlabeled data. Finally, subsampling dramatically reduces training time, achieving up to 7× speedups with minimal losses of performance, which is an important finding for resource-constrained environments.

Overall, these results suggest that careful or even random selection of a small unlabeled subset is a viable alternative to full-data SSL in the tabular domain.

For future work, we plan to explore more data-driven sampling strategies that prioritize informativeness of the unlabeled example set and include a diversity constraint to further reduce its size. Our goal is to surpass current baselines by identifying particularly beneficial unlabeled points for representation learning. We also aim to extend our evaluation to a broader set of SSL methods for tabular data, providing a more complete picture of the sampling-performance trade-off.

We further plan to conduct a meta-analysis linking dataset characteristics to the effectiveness of the sampling strategies. Ultimately, we envision training a meta-learner that can predict the most suitable sampling strategy for a given dataset based on its metafeatures. This will enable automated and adaptive data selection in SSL pipelines.

Acknowledgments. Dost and Camacho are supported by the European Union's Horizon Europe research and innovation programme under the Marie Skłodowska-Curie Postdoctoral Fellowship Programme, SMASH co-funded under the grant agreement No. 101081355. The SMASH project is co-funded by the Republic of Slovenia and the European Union from the European Regional Development Fund. Džeroski and Stevanoska are supported by the Slovenian Research and Innovation Agency under grant P2-0103, with a grant for young researchers provided to Stevanoska and the grant GC-0001. Computations were performed on the ARNES High Performance Computing infrastructure. We gratefully acknowledge the Slovenian National Supercomputing Network for providing computational resources and support.

Disclosure of Interests. The authors have no competing interests to declare that are relevant to the content of this article.

References

1. Arik, S., Pfister, T.: TabNet: attentive interpretable tabular learning. In: AAAI Conference on Artificial Intelligence (AAAI), pp. 6679–6687 (2021)
2. Bahri, D., Jiang, H., Tay, Y., Metzler, D.: SCARF: self-supervised contrastive learning using random feature corruption. In: International Conference on Learning Representations (ICLR) (2022)
3. Breiman, L.: Random forests. Mach. Learn. **45**(1), 5–32 (2001)
4. Chen, S., Wu, J., Hovakimyan, N., Yao, H.: ReConTab: regularized contrastive representation learning for tabular data. arXiv preprint arXiv:2310.18541 (2023)
5. Chen, T., Kornblith, S., Norouzi, M., Hinton, G.: A simple framework for contrastive learning of visual representations. arXiv preprint arXiv:2002.05709 (2020)
6. Devlin, J., Chang, M.W., Lee, K., Toutanova, K.: BERT: pre-training of deep bidirectional transformers for language understanding. In: Conference of the North American Chapter of the Association for Computational Linguistics: Human Language Technologies (NAACL-HLT) (2019)

7. Dhekane, S.G., Haresamudram, H., Thukral, M., Plötz, T.: How much unlabeled data is really needed for effective self-supervised human activity recognition? In: Proceedings of the 2023 ACM International Symposium on Wearable Computers, pp. 66–70. Association for Computing Machinery (2023)
8. Drugman, T., Pylkkonen, J., Kneser, R.: Active and Semi-Supervised Learning in ASR: Benefits on the Acoustic and Language Models (2019)
9. Ericsson, L., Gouk, H., Loy, C.C., Hospedales, T.M.: Self-supervised representation learning: introduction, advances, and challenges. IEEE Signal Process. Mag. **39**(3), 42–62 (2022)
10. Ghosh, S., Xie, J., Kuznetsov, M.: Distributionally robust self-supervised learning for tabular data. In: NeurIPS 2024 Workshop on Table Representation Learning (2024)
11. Ghosh, S., Xu, Y., Zhang, X.: Improving the efficiency of self-supervised adversarial training through latent clustering-based selection. In: ICML 2024 Next Generation of AI Safety Workshop (2024)
12. Grill, J.B., et al.: Bootstrap your own latent a new approach to self-supervised learning. In: International Conference on Neural Information Processing Systems (NeurIPS). Curran Associates Inc., Red Hook, NY, USA (2020)
13. Hajiramezanali, E., Diamant, N.L., Scalia, G., Shen, M.W.: STab: self-supervised learning for tabular data. In: NeurIPS 2022 First Table Representation Workshop (2022)
14. He, K., Fan, H., Wu, Y., Xie, S., Girshick, R.: Momentum contrast for unsupervised visual representation learning. In: Conference on Computer Vision and Pattern Recognition (CVPR), pp. 9726–9735 (2020)
15. Hwang, J., Lee, S., Kim, H., Jeong, Y.S.: Subset selection for domain adaptive pre-training of language model. Sci. Rep. **15**(1), 9539 (2025)
16. Joshi, S., Mirzasoleiman, B.: Data-efficient contrastive self-supervised learning: most beneficial examples for supervised learning contribute the least. In: International Conference on Machine Learning (ICML) (2023)
17. Joshi, S., Ni, J., Mirzasoleiman, B.: Dataset distillation via knowledge distillation: towards efficient self-supervised pre-training of deep networks. In: International Conference on Learning Representations (ICLR) (2025)
18. Kim, S., Bae, S., Yun, S.Y.: Coreset sampling from open-set for fine-grained self-supervised learning. arXiv preprint arXiv:2303.11101 (2023)
19. Krizhevsky, A., Sutskever, I., Hinton, G.E.: ImageNet classification with deep convolutional neural networks. In: Advances in Neural Information Processing Systems (NIPS) (2012)
20. Lu, Z., Wang, Y., Zhang, Y., Han, W., Chen, Z., Haghani, P.: Unsupervised Data Selection via Discrete Speech Representation for ASR (2022)
21. Misra, I., van der Maaten, L.: Self-supervised learning of pretext-invariant representations. In: Conference on Computer Vision and Pattern Recognition (CVPR), pp. 6706–6716 (2020)
22. Nam, J., Tack, J., Lee, K., Lee, H., Shin, J.: STUNT: few-shot tabular learning with self-generated tasks from unlabeled tables. In: International Conference on Learning Representations (ICLR 2023) (2023)
23. Park, C., Ahmad, R., Hain, T.: Unsupervised data selection for speech recognition with contrastive loss ratios. In: IEEE International Conference on Acoustics, Speech and Signal Processing (ICASSP), pp. 8587–8591 (2022)
24. Pullar-Strecker, Z., Dost, K., Frank, E., Wicker, J.: Hitting the target: stopping active learning at the cost-based optimum. Mach. Learn. **113**(4), 1529–1547 (2022)

25. Radford, A., Narasimhan, K., Salimans, T., Sutskever, I.: Improving language understanding by generative pre-training. OpenAI Blog **1**(8), 1–12 (2018)
26. Renduchintala, H.S.V.N.S.K., et al.: INGENIOUS: using informative data subsets for efficient pre-training of language models. In: Findings of the Association for Computational Linguistics: EMNLP 2023, pp. 6690–6705. Association for Computational Linguistics, Singapore (2023)
27. Settles, B.: Active learning literature survey. Technical Report 1648, University of Wisconsin-Madison Department of Computer Sciences (2009)
28. Seung, H.S., Opper, M., Sompolinsky, H.: Query by committee. In: Proceedings of the Fifth Annual Workshop on Computational Learning Theory, pp. 287–294 (1992)
29. Somepalli, G., Goldblum, M., Schwarzschild, A., Bruss, C.B., Goldstein, T.: SAINT: improved neural networks for tabular data via row attention and contrastive pre-training. arXiv preprint arXiv:2106.01342 (2021)
30. Tharwat, A., Schenck, W.: A survey on active learning: state-of-the-art, practical challenges and research directions. Mathematics **11** (2023)
31. Tian, Y., Hénaff, O.J., van den Oord, A.: Divide and contrast: self-supervised learning from uncurated data. In: International Conference on Computer Vision (ICCV 2021), pp. 10043–10054 (2021)
32. Tripathi, A., Shenoy, P., Chakraborty, A.: Dynamic data selection for efficient SSL via coarse-to-fine refinement. In: European Conference on Computer Vision (ECCV 2024), pp. 377–393. Springer, Cham (2025)
33. Uçar, T., Hajiramezanali, E., Edwards, L.: SubTab: subsetting features of tabular data for self-supervised representation learning. In: Proceedings of the 35th International Conference on Neural Information Processing Systems (NeurIPS). Curran Associates Inc., Red Hook, NY, USA (2021)
34. Vanschoren, J., Rijn, J.N., Bischl, B., Torgo, L.: Openml: networked science in machine learning. SIGKDD Explor. **15**(2), 49–60 (2013)
35. Vaswani, A., et al.: Attention is all you need. In: Advances in Neural Information Processing Systems (NIPS) (2017)
36. Wang, W.Y., Du, W.W., Xu, D., Wang, W., Peng, W.C.: A survey on self-supervised learning for non-sequential tabular data. Mach. Learn. **114**(1) (2025)
37. Wang, X., et al.: Farewell to aimless large-scale pretraining: influential subset selection for language model. In: Findings of the Association for Computational Linguistics: ACL 2023, pp. 555–568. Association for Computational Linguistics, Toronto, Canada (2023)
38. Xie, S.M., Santurkar, S., Ma, T., Liang, P.: Data selection for language models via importance resampling. In: Thirty-Seventh Conference on Neural Information Processing Systems (2023)
39. Yoon, J., Sohn, K., Li, C.L., Arik, S.O., Lee, C.Y., Pfister, T.: Self-supervise, refine, repeat: improving unsupervised anomaly detection. Trans. Mach. Learn. Res. (2022)
40. Yoon, J., Zhang, Y., Jordon, J., van der Schaar, M.: VIME: extending the success of self- and semi-supervised learning to tabular domain. In: Advances in Neural Information Processing Systems (NeurIPS) (2020)
41. Yosinski, J., Clune, J., Bengio, Y., Lipson, H.: How transferable are features in deep neural networks? In: Advances in Neural Information Processing Systems (NIPS) (2014)

Explainable AI for Science

Meta Subspace Analysis: Understanding Model (Mis)behavior in the Metafeature Space

Carlos Soares[1,2,3](✉), Paulo J. Azevedo[4], Vitor Cerqueira[1], and Luis Torgo[5]

[1] Laboratory for Artificial Intelligence and Computer Science (LIACC), Faculdade de Engenharia, Universidade do Porto, Porto, Portugal
{vcerqueira,csoares}@fe.up.pt
[2] Fraunhofer Portugal AICOS, Porto, Portugal
[3] BrightFactory, Porto, Portugal
[4] HASLab, INESC TEC, Departamento de Informática, Universidade do Minho, Braga, Portugal
pja@di.uminho.pt
[5] Dalhousie University, Halifax, Canada

Abstract. A subgroup discovery-based method has recently been proposed to understand the behavior of models in the (original) feature space. The subgroups identified represent areas of feature space where the model obtains better or worse predictive performance when compared to the average test performance. For instance, in the marketing domain, the approach extracts subgroups such as: in groups of customers with higher income and who are younger, the random forest achieves higher accuracy than on average. Here, we propose a complementary method, *Meta Subspace Analysis* (MSA), MSA uses metalearning to analyze these subgroups in the metafeature space. We use association rules to relate metafeatures of the feature space represented by the subgroups to the improvement or degradation of the performance of models. For instance, in the same domain, the approach extracts rules such as: when the class entropy decreases and the mutual information increases in the subgroup data, the random forest achieves lower accuracy. While the subgroups in the original feature space are useful for the end user and the data scientist developing the corresponding model, the meta-level rules provide a domain-independent perspective on the behavior of the model that is suitable for the same data scientist but also for ML researchers, to understand the behavior of algorithms. We illustrate the approach with the results of two well-known algorithms, naive Bayes and random forest, on the Adult dataset. The results confirm some expected behavior of algorithms. However, and most interestingly, some unexpected behaviors are also obtained, requiring additional investigation. In general, the empirical study demonstrates the usefulness of the approach to obtain additional knowledge about the behavior of models.

Keywords: Metalearning · Subgroup discovery · Association rules · Model interpretability

S. Džeroski et al. (Eds.): DS 2025, LNAI 16090, pp. 255–269, 2025.
https://doi.org/10.1007/978-3-032-05461-6_17

1 Introduction

Machine Learning (ML) systems are increasingly widespread and applied across varied application domains. Despite important gains in predictive performance, the opacity of state-of-the-art ML approaches remains a bottleneck in the use of these methods, especially in sensitive domains such as healthcare. In this context, understanding the behavior of machine learning methods has become a key goal [8].

As the transparency of models becomes a critical requirement for the responsible use of machine learning methods, several approaches have been developed. For example, recent work aiming to improve model understanding uses subgroup discovery to identify regions of the feature space where performance deviates significantly from the average [21]. This paper as well as other related work focuses on the original feature space, providing insights for end users about specific conditions under which a model works better or worse and understand particular decisions.

However, in the development of models (including the corresponding monitoring and maintenance) as well as in the development of new algorithms, the data scientist/ML researcher could use information about the relation of the data distribution and the performance of the models across different regions of the feature space. The relationship between distributional characteristics and performance variations remains largely unexplored in current interpretability methods.

We bridge this gap by proposing *Meta Subspace Analysis* (MSA), a novel approach that leverages metalearning to analyse the results of predictive models in different subspaces of the input space. By characterizing the data distribution in different regions using subgroup discovery and metalearning, our method provides automated insights into how statistical and information-theoretic properties of the data relate to model performance variations. For example, MSA can identify rules such as: "when the average mutual information between features and target increases and the class entropy decreases in a subgroup, the model achieves higher accuracy compared to its test performance." This type of domain-independent insight helps data scientists/ML researchers understand the statistical properties of data that cause variations in the performance of their models and algorithms.

To validate our proposed method, we conduct experiments using two widely-used classification algorithms: Random Forest (RF) [7] and naive Bayes (NB) (e.g. [9]). We apply these algorithms to the Adult dataset [4], a common benchmark dataset in machine learning for binary classification tasks. Some of the resulting insights from our analysis align well with established knowledge about these algorithms. For instance, we find that the accuracy of Random Forest increases when there is a stronger average relation between quantitative attributes and the target. On the other hand, some rules are harder to explain, which makes them more interesting. For instance, the accuracy of Naive Bayes improves when there is a stronger average relation between **quantitative** attributes and the target, simultaneously with weaker average relation between

qualitative attributes and the target. In the results, we discuss these observations, which clearly show that interesting knowledge can be obtained about models and algorithms with MSA.

In summary, the contributions of this paper are:

- *Meta Subspace Analysis* (MSA), a novel metalearning approach that analyses model performance variations across different subspaces of the feature space by relating them to statistical and information-theoretic properties of the data.
- An experimental evaluation of MSA using RF and NB on the Adult dataset that illustrates how the proposed method can provide insights into model and algorithm behavior.

2 Background

We start by introducing the two tasks that are the basis for the approach, subgroup discovery (Sect. 2.1) and metalearning (Sect. 2.2). Then we describe the previous approach of using subgroup discovery for the interpretability of model errors (Sect. 2.3).

2.1 Subgroup Discovery

Subgroup discovery aims to find subpopulations in datasets that yield values for a predefined property of interest that deviate from the ones observed in a reference population [1,14]. This property of interest (sometimes referred to as the target variable) can be of different sorts, including quantitiative or qualitative. Subgroups that exhibit a notable difference from the population, that is, a significant deviation of the value of the target in the subgroup when compared to the population, are considered interesting. Different proposals make use of variants of different quality measures, and the weighted relative accuracy is the best known. A noticeable extension of this idea is Exceptional Model Mining (EMM) [17], where more than one property of interest can be analyzed in parallel.

2.2 Metalearning

Metalearning is the subfield of machine learning devoted to using historical machine learning experiences to improve the learning process [5].

In many cases, metalearning involves modeling the performance of algorithms or models using metafeatures. Metafeatures are measurable properties of datasets or tasks. These range from basic statistics such as the number of instances or classes to information-theoretic properties such as mutual information or class entropy. The systematic characterization of datasets using metafeatures enables using machine learning to transfer knowledge across different problems efficiently. For a more complete read on metafeatures we refer to the survey by Rivolli et al. [19].

This process is helpful for several tasks, such as algorithm selection [6] or hyperparameter optimization [13], among others [5]. Recent work has used metalearning for understanding model performance, namely to model unusually large errors [15] (cf. Sect. 2.3).

2.3 Subgroup Discovery for Interpretability

A few papers have analyzed model performance using subgroup discovery methods (e.g. [10,11,20,21]). Given that these methods relate performance with the original attributes, while our method uses metafeatures, we describe only the one that we used as part of the proposed method. Here, we also discuss a recent metalearning approach to understand time series forecasting models.

In [21], a specific type of distribution rules is derived from datasets that represent the performance of a classifier. These datasets contain data where the rows are test cases with an additional column to encode the prediction outcome. The rules derived describe situations where the model performance deviates significantly from expectation. The input in this setting is a set of bins representing cells in the confusion matrix of the model. The coding for these cells forms a qualitative distribution. For instance, code "1–2" represents a false negative case in class 1 where class 2 was predicted. Also, code "0" represents a hit. Subgroups with a qualitative distribution that deviates from a reference are considered interesting. The distance between distributions is measured through a χ^2 *goodness of fit* test. An example of such a rule is as follows:

```
Ant sup = 0.01869 pvalue = 0.0010041075378298939000
CM={ 0/0.250,2-1/0.750 } <--
    RI=[1.511 : 1.516] & Na=[13.08 : 13.72] &
    Ca=[7.90 : 8.43]
```

where it shows a Random Forest model performance that highlights a particular bad performance in class 2 of the dataset *Glass* and for the subgroup described in the antecedent. The incidence of this subgroup is 18.69% of the dataset.

Although metalearning has traditionally been used mostly for algorithm selection, recent work has used it for understanding model performance, namely to model unusually large errors in time series forecasting [15]. The approach combines metalearning with data augmentation to generate, characterize and predict when a given model will perform poorly on a given time series.

3 Meta Subspace Analysis

Machine learning models may exhibit varying performance across different regions of the input space, which makes it crucial to understand the conditions under which they perform better/worse [18]. *Meta Subspace Analysis* (MSA) addresses this challenge by relating model performance on different sub-regions of the feature space to metafeatures that characterize the data distribution in those regions.

Figure 1 illustrates the complete workflow of our approach, which consists of three phases: 1) Subgroup discovery; 2) Subgroup Meta-data preparation; and 3) Association rule metalearning, which are detailed next.

An example of a rule that can be obtained is:

```
Sup = 0.15804    Conf = 0.88092
target=pos    <--    PropOutlier=neg
```

This rule means that the accuracy of the corresponding model increases in regions of where the number of outliers reduces (with a support of 15.8% and a confidence of 88%). By analyzing these patterns, we can understand not only where a model performs differently but also what statistical and distributional properties of the data contribute to these performance variations.

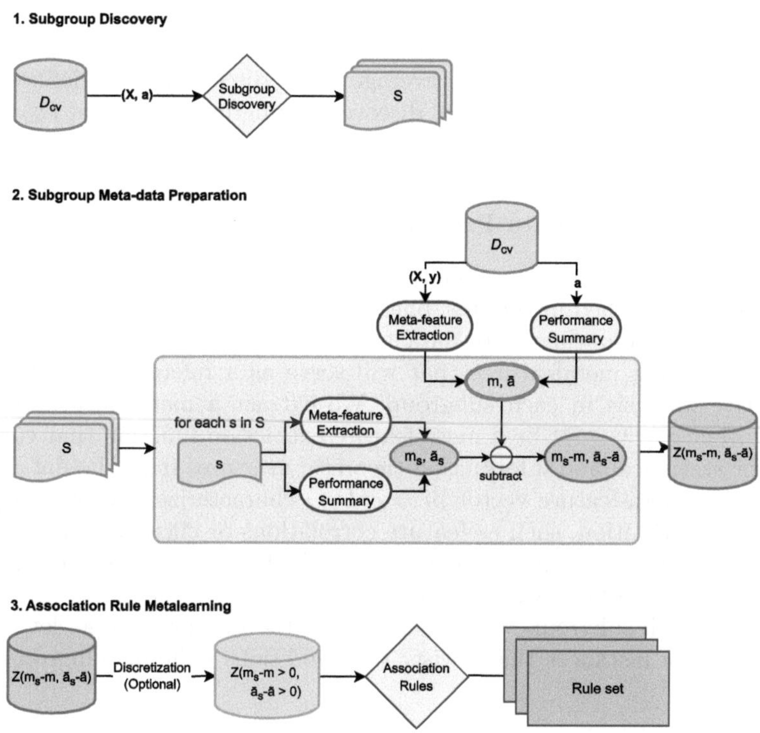

Fig. 1. Workflow behind the proposed method, which is split into three stages: 1) Subgroup discovery to detect subgroups on the original feature space where the performance of the model is significantly better/worse than on average; 2) Subgroup Meta-data preparation, including differencing with respect to the overall metafeatures; 3) Association rule metalearning, by applying an association rule mining algorithm on the meta-dataset.

3.1 Subgroup Discovery

Let D_{train} be a dataset defined as $D_{train} = \{\langle \mathbf{x}_i, y_i \rangle\}_{i=1}^{n}$, where $\mathbf{x}_i$ is a feature vector, y_i is the corresponding target variable, and n is the number of training instances. Without loss of generality, we assume that y is a categorical variable in this work. Given a supervised learning classification model f of the form $y = f(x)$, we can assess its instance-wise performance through cross-validation. Let $D_{cv} = \{\langle \mathbf{x}_i, y_i \rangle\}_{i=1}^{n_{cv}}$ denote the evaluation dataset, where n_{cv} is the number of testing instances. We also denote as $D_{acc} = \{\langle \mathbf{x}_i, a_i \rangle\}_{i=1}^{n_{cv}}$ the performance-annotated evaluation dataset where a_i is a binary variable that denotes the accuracy of the model predictions: $a_i = 1$ if $f(\mathbf{x}_i) = y_i$, or 0 otherwise.

In the first step of our methodology, we use the evaluation data sets obtained from cross-validation, D_{cv} and D_{acc}, to compute base-level statistics and conduct subgroup discovery.

We apply CAREN (Class Association Rule ENgine) [2,3], a subgroup discovery technique, to identify regions in the feature space (X) where the performance of f significantly deviates from the average. Each discovered rule defines a subgroup $s \in S$, where S is the set of all discovered subgroups based on D_{acc}. Rules are evaluated using standard measures such as support and confidence.

3.2 Subgroup Meta-data Preparation

In the second phase of the methodology we create the meta-dataset describing the subgroups identified in the first phase.

It starts by computing meta-data for the test dataset, D_{cv}. This process involves extracting metafeatures that will serve as a reference for subsequent comparative analysis in each subgroup. We extract a meta-feature vector m from D_{cv}. Let $g : D \rightarrow \mathbb{R}^k$ be a meta-feature extraction function that computes k different statistical and information-theoretic measures from the dataset [19]. The resulting meta-feature vector $m = g(D_{cv})$ characterizes various properties of the data distribution, such as feature correlations or class entropy. Then, we compute performance estimates based on a. We denote as $\bar{a}$ the accuracy of f, i.e., the percentage of cases it predicts accurately.

Then, for each subgroup $s \in S$ identified in the previous phase, let $D^s \subset D$ be the subset of instances that satisfy the conditions defining subgroup s. We then perform:

1. Meta-feature extraction: We compute subgroup-specific metafeatures $m_s = g(D^s)$ using the same meta-feature extraction function g defined previously.
2. Performance estimation: We calculate the accuracy $\bar{a}_s$ of model f within the subgroup, computed as the mean of a_j values for all instances in D^s.
3. Difference computation: We compute the differences between subgroup-specific and general measurements to quantify how each subgroup deviates from the reference values. These differences are stored in dataset $Z = \{\langle m_s - m, \bar{a}_s - \bar{a} \rangle\}_{i=1}^{|S|}$. The resulting dataset Z has $|S|$ rows (one per subgroup) and $k + 1$ columns (k meta-feature differences plus one performance

difference). By working with differences rather than absolute values, we can identify how changes in metafeatures from their reference values relate to changes in model performance.

3.3 Association Rule Metalearning

In the final phase, we apply association rule mining to the meta-dataset Z.

This process involves the discretization of the dataset Z. This discretization transforms dataset Z into a binary dataset $Z(m_s - m > 0, \bar{a}_s - \bar{a} > 0)$ containing only binary attributes (including the meta-target) that represent whether each meta-feature and the performance metric are higher or lower in the subgroup compared to the overall dataset. The discretization process aims to create more interpretable rules that are easier to convey to data scientists/ML researchers they are for. This discretization step is optional and the methodology can be applied in either the continuous metafeature differences or to their discretized binary versions, as there are implementations of association rule algorithms that can process quantitative data (e.g., CAREN [3,21]).

In the association rule mining process, each rule has two parts: a set of conditions describing the relationship between metafeatures and their typical values and the resulting impact on model accuracy. If discretization is applied, the conditions and performance changes are binary indicators with the form $(m_s - m) > 0$ and $(\bar{a}_s - \bar{a}) > 0$. Rules can also be extracted from continuous values ($m_s - m$ and $\bar{a}_s - \bar{a}$), which are more expressive and can provide additional information.

The resulting rule set contains patterns such as "when feature correlations increase and class entropy decreases in a subgroup, the model's accuracy improves by X% compared to its overall performance $\bar{a}$", assuming discretization is applied. These rules provide interpretable insights into how changes in data characteristics lead to variations in model performance. In other words, under which conditions the model performs better or worse than its baseline performance on the complete dataset.

The main limitation of the proposed method concerns the dependence in the meta-dataset. As mentioned above, each instance in Z represents meta-feature and performance differences for a given subgroup. Since two subgroups can share data points, their measurements in the meta-level are dependent. This dependence is not considered by the association rule mining algorithm, which works under the assumption that observations are independent.

3.4 Domain-Specific vs Domain-Independent Knowledge

By characterizing algorithm behavior as a function of meta-level features, the language of the knowledge obtained with MSA is independent of the domain. This constrasts with the approaches discussed in the related work, which obtain knowledge in a domain-specific language, because it is uses on base-level features.

However, MSA can generate domain-specific or domain-independent knowledgee, depending on the number of base-level datasets it describes. If the method is applied to subgroups obtained on a single dataset, the knowledge obtained is specific to that problem. On the other hand, if the subgroups describe the performance of the algorithm on multiple problems, than the knowledge obtained in domain-independent.

4 Case Study

We propose an approach that combines subgroup discovery and metalearning to characterize regions of the data space where a given model obtains better or worse predictive performance (namely, accuracy) than the average performance on the test set. Unlike other approaches that characterize those regions in the original feature space, our method uses statistical and other distributional properties of the region (i.e., metafeatures). This provides a domain-independent perspective on the behavior of the algorithms. This perspective can be useful not only for the data scientist addressing a particular problem, but also give insights on the general behavior of algorithms, that is useful for data scientists and machine learning researchers. We note that the method proposed does not replace but rather complements approaches that characterize regions in the base-level feature space.

To illustrate the type of information that the proposed approach can obtain, we apply it to analyze two common algorithms, RF and NB, on a popular benchmark, the Adult dataset [4]. We note that, since MSA is applied to a single dataset, the knowledge obtained is domain-specific, even if represented in a domain-independent language (Sect. 3.4). In the remainder of this section, we describe the setup for the experiments.

4.1 Dataset

The Adult dataset, also known as "Census Income," is a widely used benchmark dataset in machine learning for binary classification tasks. It contains 48,842 instances with 14 features (6 numerical and 8 categorical) extracted from the 1994 Census database. The prediction task is to determine whether the annual income of a person exceeds \$50,000 based on demographic and socioeconomic information such as age, education level, occupation, work hours, and marital status. The dataset has a moderate class imbalance with approximately 24% of instances belonging to the positive class (income $> \$50,000$). The dataset contains some missing values, particularly in the attributes "workclass," "occupation," and "native-country." We handle these with simple methods (mean and mode for quantitative and qualitative attributes, respectively).

4.2 Base-Level Experimental Setup

We reproduced the experimental setup from an earlier publication [21], to ensure that the subgroups analyzed here contain interesting information about the behavior of the selected algorithms on the selected dataset.

Classification Algorithms. Model performance was obtained using the WEKA implementations for RandomForest and NaiveBayes [12]. In both algorithms, standard default parameters were used.

Evaluation Procedure. Performance of RF and NB was measured using accuracy as the evaluation metric, which quantifies the proportion of correctly classified instances. To obtain reliable estimates of model performance, we employed 10-fold cross-validation: the dataset was randomly partitioned into ten equal-sized folds, with each fold used once as the test set while the remaining nine folds served as the training set.

Subgroup Discovery. We apply CAREN (Class Association Rule ENgine) [2,3], a subgroup discovery technique [3,21] using minimal support as 10% and default $\alpha = 0.05$ for significance filtering.

CAREN is a standard implementation of AR learning algorithms and, although the approach is independent of the implementation, we believe that this choice does not affect results significantly.

4.3 Metafeatures

We selected metafeatures based on information theory because it is the basis of the decision tree algorithm used by RF and is related to the conditional probabilities used by NB. The list of metafeatures in presented and briefly described in Table 1.

For more information on these metafeatures, we refer the reader to [19]. We used an implementation in R,[1] which is publicly available, together with the code of this paper.[2]

4.4 Association Rules

CAREN was used to discover relationships between discretized meta-feature differences and performance variations [2,3] in the traditional association rule mining way.

5 Discussion of Selected Rules

Here, we discuss in more detail some of the most interesting rules, starting with the ones obtained for RF and then the ones for NB. We note that the predictive performance of both algorithms is very high (86.5% and 82.8%, respectively), which means that fewer subgroups with below-average performance are expected, when compared to above-average subgroups.

[1] https://github.com/carlospintosoares/mfeR.
[2] https://github.com/carlospintosoares/MLinception.

Table 1. Metafeature used and respective descriptions.

Name	Description
`PropMissing`	Variation in the proportion of missing values; it reflects the ability of the algorithm to deal with them
`ClassEntropy`	Variation in the class distribution, where larger values represent more balanced distributions; it reflects the ability of the algorithm to deal with (un)balanced classes
`AvgMI`	Variation in the amount of discriminative information in the qualitative attributes
`PropOutlier`	Variation in the proportion of outliers; it reflects the ability (or lack of it) of the algorithm to deal with them
`AvgAttrEntropy`	Variation in the average balance in qualitative attribute values; it reflects the ability (or lack of it) of the algorithm to deal with small disjuncts
`NoiseSignalRatio`	Variation in the amount of noise that is not reflected in discriminative power of the qualitative attributes; it reflects the ability of the algorithm to deal with attributes without useful information
`AvgAbsCorrNum`	Variation in the redundancy of the quantitative attributes; it reflects the ability (or lack of it) of the algorithm to deal with them
`AvgMINomPair`	Variation in the redundancy of the qualitative attributes; it reflects the ability (or lack of it) of the algorithm to deal with them
`CanCorrBestLin`	Variation in the canonical correlation of the best linear combination of quantitative attributes; it reflects the ability of the algorithm to extract discriminative linear information from the quantitative data

5.1 Random Forest on the Adult Dataset

The subgroups that were obtained are listed in Table 2. They indicate that RF obtains better performance in subspaces of the Adult dataset when:

RF.A.1 there is more information, either in quantitative or qualitative attributes;
RF.A.2 they are easier, meaning that the attributes have well-behaved distributions (i.e., less outliers in numerical attributes), or are purer (i.e., more skewed class distribution);
RF.A.3 there is more redundancy combined with more information, either in quantitative or qualitative attributes.

The observation **RF.A.1** is based on rules 1 and 2 that indicate that accuracy increases when the amount of information about the label contained on numerical (`CanCorrBestLin`) and nominal (`AvgMI`) attributes, respectively, increases. Rule 3 is a combination of both, but with much less support and without additional confidence. Several other subgroups (11, 12, 14, 15, 16 and 17) are also specializations of these two, without additional value (i.e. small gains in confidence and/or with large reductions in support).

Table 2. Association Rules with Support, Confidence, Class, and Conditions

#	Sup.	Conf.	Effect	Rule
0	0.85	0.85	pos	–
1	0.50	0.87	pos	`CanCorrBestLin` = pos
2	0.51	0.86	pos	`AvgMI` = pos
6	0.16	0.88	pos	`PropOutlier` = neg
7	0.14	0.89	pos	`PropOutlier` = neg & `ClassEntropy` = neg
4	0.64	0.85	pos	`AvgAbsCorrNum` = pos
5	0.63	0.85	pos	`AvgAbsCorrNum` = pos & `AvgMINomPair` = pos
14	0.45	0.87	pos	`AvgMI` = pos & `AvgAbsCorrNum` = pos
15	0.38	0.88	pos	`CanCorrBestLin` = pos & `AvgMI` = pos & `AvgAttrEntropy` = neg
10	0.60	0.86	pos	`AvgAbsCorrNum` = pos & `AvgAttrEntropy` = neg & `AvgMINomPair` = pos
11	0.51	0.86	pos	`AvgMI` = pos & `AvgMINomPair` = pos
12	0.47	0.87	pos	`CanCorrBestLin` = pos & `AvgAttrEntropy` = neg
13	0.45	0.87	pos	`AvgMI` = pos & `AvgAbsCorrNum` = pos & `AvgMINomPair` = pos
3	0.40	0.87	pos	`CanCorrBestLin` = pos & `AvgMI` = pos
16	0.36	0.87	pos	`CanCorrBestLin` = pos & `AvgMI` = pos & `AvgAbsCorrNum` = pos & `AvgMINomPair` = pos
17	0.35	0.87	pos	`CanCorrBestLin` = pos & `AvgMI` = pos & `ClassEntropy` = neg & `AvgMINomPair` = pos
18	0.13	0.89	pos	`PropOutlier` = neg & `ClassEntropy` = neg & `AvgAbsCorrNum` = pos & `AvgMINomPair` = pos
8	0.13	0.89	pos	`PropOutlier` = neg & `ClassEntropy` = neg & `AvgAttrEntropy` = neg & `AvgMINomPair` = pos
9	0.13	0.88	pos	`PropOutlier` = neg & `ClassEntropy` = neg & `AvgAttrEntropy` = neg & `AvgAbsCorrNum` = pos

The following observation (**RF.A.2**) is based on rules 6 and 7, which indicate that accuracy increases when there are fewer outliers in numerical attributes (PropOutlier) and class distribution is more unbalanced (ClassEntropy). Under these conditions, it is easier for RF (and decision trees, in general) to learn purer nodes.

Finally, observation (**RF.A.3**) is more unexpected. Rules 4 and 5 indicate that more redundancy on its own, either in numerical (`AvgAbsCorrNum`) or nominal (`AvgMINomPair`) attributes does not improve the accuracy of RF in the Adult dataset. However, rules 14 and 15 indicate that when combined with more information (`CanCorrBestLin` and `AvgMI`), performance improves. Rule number 10 additionally indicates redundancy combined with lower variance of nominal attributes also leads to improved performance. The justification for this subgroup requires additional analysis.

The remaining rules are all specializations of previous rules, without additional improvement (information).

One interesting observation is that all the rules obtained described conditions where the RF model obtained better performance than the average performance on the test set. This is discussed below.

Table 3. Association Rules with Support, Confidence, Class, and Conditions

#	Sup.	Conf.	Effect	Rule
10	0.66	0.66	pos	
9	0.45	0.67	pos	`AvgAbsCorrNum` = pos
8	0.21	0.69	pos	`CanCorrBestLin` = neg, `AvgAbsCorrNum` = pos
7	0.19	0.70	pos	`AvgMI` = neg, `AvgAbsCorrNum` = pos
6	0.14	0.71	pos	`NoiseSignalRatio` = pos, `AvgAbsCorrNum` = pos
5	0.09	0.71	pos	`CanCorrBestLin` = pos, `AvgMI` = neg
4	0.06	0.71	pos	`NoiseSignalRatio` = pos, `CanCorrBestLin` = pos
1	0.01	0.85	pos	`ClassEntropy` = pos, `AvgMI` = neg, `CanCorrBestLin` = pos
2	0.02	0.77	pos	`AvgAttrEntropy` = pos, `NoiseSignalRatio` = pos
3	0.02	0.77	pos	`AvgAttrEntropy` = pos, `AvgMI` = neg

5.2 Naive Bayes on the Adult Dataset

The subgroups that were obtained are listed in Table 3. They indicate that NB obtains better performance in subspaces of the Adult dataset when there is:

NB.A.1 more information in the quantitative attributes and less information in qualitative attributes;
NB.A.2 more redundancy in quantitative attributes;
NB.A.3 more variability in the qualitative attributes.

Also in the case of the NB model, all the rules obtained described conditions where the performance was better than average. Again, this is discussed below.

The observation **NB.A.1**, based in rules 1, 4 and 5, is very counterintuitive: it states that NB can better capture information in quantitative (`CanCorrBestLin`) than in qualitative attributes (`AvgMI`). Given that the standard NB algorithm has been designed for qualitative attributes, one might expect the opposite to be true: the accuracy of NB only improves when there is more information in the qualitative attributes and is independent of the amount of information in the quantitative ones. However, this result may be explained by the use of Gaussian NB [16], which handles numerical attributes directly (i.e. without the need for discretization). Given that 8 out of the 11 attributes of Adult are numerical, this pattern may simply indicate that NB is able to take advantage of the information in numerical attributes in regions where its usefulness increases.

The observation **NB.A.2** is based on rule 9 (and also rules 7 and 6), which indicates that accuracy increases in subspaces where the correlation between numerical attributes increases (`AvgAbsCorrNum`). This is also very counterintuitive as it seems to contradict one of the basic assumptions of NB: posterior class probabilities can be computed as a product of the conditional probability of the attribute given the class if the attributes are independent. However, it has been shown that [22]:

> In a given dataset, two attributes may depend on each other, but the dependence may distribute evenly in each class. Clearly, in this case, the conditional independence assumption is violated, but naive Bayes is still the optimal classifier. Further, what eventually affects the classification is the combination of dependencies among all attributes. If we just look at two attributes, there may exist strong dependence between them that affects the classification. When the dependencies among all attributes work together, however, they may cancel each other out and no longer affect the classification. Therefore, we argue that it is the distribution of dependencies among all attributes over classes that affects the classification of naive Bayes, not merely the dependencies themselves.

Based on this, our results may indicate that, in subspaces where the correlation between attributes increases, dependencies between attributes are spreading homogeneously over all the classes, thus leading to higher accuracy of NB in that subspace.

The final observation, **NB.A.3**, is based on rules 2 and 3, that indicate that accuracy increases in subspaces where the entropy of attributes (`AvgAttrEntropy`) increases, i.e., when there is more uncertainty about the value of the nominal attributes. The cause for this pattern may be related to observation **NB.A.1**, which indicates that performance decreases with the increase in mutual information between attributes and target. This means that there are subspaces where the nominal attributes are misleading and, thus, with higher uncertainty in their values, there is less risk of overfitting them. The additional conditions, concerning metafeatures `NoiseSignalRatio` (increase) and `AvgMI` (decrease), point in the same direction.

5.3 Subgroups with Below-Average Accuracy

The rules discussed above only cover subgroups with above-average performance. This occurs because the subgroup discovery step returned a much smaller number of negative subgroups (approximately 30%) when compared to the positive ones. Preliminary experiments computing association rules with lower minimum confidence indicate that this hypothesis is correct.

However, this raises an additional question: why does the subgroup discovery step obtain fewer subgroups? We hypothesize that this is due to the high accuracy of both RF and NB models (86.5% and 82,8%, respectively). This means that the number of errors is small, and thus it is difficult to find dense groups with below-average accuracy. Although further work is necessary to investigate this, we believe that the problem could be addressed by using different performance measures in the subgroup discovery step (e.g., the largest difference between the precision in all the test sets and in the subgroup, for all classes).

6 Conclusions and Future Work

In this paper, we presented a novel approach to address model performance using subgroup mining. The derived subgroups represent areas of the feature space where performance deviates from the average. Using metalearning, original derived subgroups are projected into a meta-feature space. In this context, metafeatures describe measurable properties of a given dataset, i.e., information-theoretic properties, basic statistics, etc. Metafeatures iteration is then obtained by association rules generation in a dataset relating metafeatures and performance estimation. Two case studies were used to illustrate the proposed method, using the Adult dataset in NaiveBayes and RandomForest models. These studies enable interesting discussions regarding dataset characteristics and attribute distributions and other properties. Noteworthy the case of numerical attributes in naive bayes and how it contributes to explain counterintuitively the better use of quantitative attributes against qualitative ones. Future work will evolve along in further analysis on different models and datasets.

Acknowledgements. This work was partially funded by the Agenda "Center for Responsible AI", nr. C645008882-00000055, investment project nr. 62, financed by the Recovery and Resilience Plan (PRR) and by European Union - NextGeneration EU. Funded by the European Union NextGenerationEU. Views and opinions expressed are however those of the author(s) only and do not necessarily reflect those of the European Union or the European Commission. Neither the European Union nor the European Commission can be held responsible for them. Also funded by the European Union under the Horizon Europe Framework Programme Grant Agreement No: 101095387. Views and opinions expressed are however those of the author(s) only and do not necessarily reflect those of the European Union or European Commission. Neither the European Union nor the European Health and Digital Executive Agency can be held responsible for them. This work was also funded by: UID/00027 of the LIACC - Artificial Intelligence and Computer Science Laboratory - funded by Fundação para a Ciência e a Tecnologia, I.P./ MCTES through the national funds. This work is also funded by National Funds through the Portuguese funding agency, FCT - Fundação para a Ciência e a Tecnologia, within project LA/P/0063/2020. DOI 10.54499/LA/P/0063/2020—https://doi.org/10.54499/LA/P/0063/2020.

Competing Interests. The authors are unaware of any competing interests.

References

1. Atzmueller, M.: Subgroup discovery. WIREs Data Min. Knowl. Discov. **5**, 35–49 (2015). https://doi.org/10.1002/widm.1144
2. Azevedo, P.J.: Caren-a Java based apriori implementation for classification purposes. Technical report, Université de Mons-Hainaut. Service de Science des Systèmes D'information (2003)
3. Azevedo, P.: Caren—class project association rule engine (2008)
4. Becker, B., Kohavi, R.: Adult. UCI Machine Learning Repository (1996). https://doi.org/10.24432/C5XW20
5. Brazdil, P., Carrier, C.G., Soares, C., Vilalta, R.: Metalearning: Applications to Data Mining. Springer, Cham (2008)
6. Brazdil, P.B., Soares, C., Costa, J.P.: Ranking learning algorithms: using ibl and meta-learning on accuracy and time results. Mach. Learn. **50**, 251–277 (2003)
7. Breiman, L.: Random forests. Mach. Learn. **45**(1), 5–32 (2001)

8. D'Amour, A., et al.: Underspecification presents challenges for credibility in modern machine learning. J. Mach. Learn. Res. **23**(226), 1–61 (2022)
9. Domingos, P., Pazzani, M.: Beyond independence: conditions for the optimality of the simple Bayesian classifier. In: Proceedings of the 13th International Conference on Machine Learning, pp. 105–112 (1996)
10. Duivesteijn, W., Feelders, A.J., Knobbe, A.: Exceptional model mining. Data Min. Knowl. Disc. **30**, 47–98 (2016). https://doi.org/10.1007/s10618-015-0403-4
11. Duivesteijn, W., Thaele, J.: Understanding where your classifier does (not) work - the scape model class for EMM. In: Proceedings - IEEE International Conference on Data Mining, ICDM, vol. 2015-January, pp. 809–814. Institute of Electrical and Electronics Engineers Inc. (2014). https://doi.org/10.1109/ICDM.2014.10
12. Eibe, F., Hall, M.A., Witten, I.H.: The weka workbench. Online appendix for data mining: practical machine learning tools and techniques. In: Morgan Kaufmann. Morgan Kaufmann Publishers San Francisco, California (2016)
13. Feurer, M., Springenberg, J., Hutter, F.: Initializing Bayesian hyperparameter optimization via meta-learning. In: Proceedings of the AAAI Conference on Artificial Intelligence, vol. 29 (2015)
14. Herrera, F., Carmona, C.J., González, P., Jesus, M.J.: An overview on subgroup discovery: foundations and applications. Knowl. Inf. Syst. **29**, 495–525 (2010). https://doi.org/10.1007/s10115-010-0356-2. http://www.springerlink.com/index/10.1007/s10115-010-0356-2
15. Inácio, R., Cerqueira, V., Barandas, M., Soares, C.: Meta-learning and data augmentation for stress testing forecasting models. In: International Symposium on Intelligent Data Analysis, pp. 343–357. Springer, Cham (2025)
16. John, G.H., Langley, P.: Estimating continuous distributions in Bayesian classifiers. arXiv preprint arXiv:1302.4964 (2013)
17. Leman, D., Feelders, A., Knobbe, A.: Exceptional model mining. In: Daelemans, W., Goethals, B., Morik, K. (eds.) Machine Learning and Knowledge Discovery in Databases, pp. 1–16. Springer, Heidelberg (2008)
18. Pimentel, J., Azevedo, P.J., Torgo, L.: Subgroup mining for performance analysis of regression models. Expert. Syst. **40**(1), e13118 (2023)
19. Rivolli, A., Garcia, L.P., Soares, C., Vanschoren, J., de Carvalho, A.C.: Meta-features for meta-learning. Knowl.-Based Syst. **240**, 108101 (2022)
20. Szelazek, M., Hudson, D., Bobek, S., Nalepa, G.J., Atzmueller, M.: Enhanced explanations for knowledge-augmented clustering using subgroup discovery. In: 2023 IEEE 10th International Conference on Data Science and Advanced Analytics (DSAA), pp. 1–11. IEEE (2023). https://doi.org/10.1109/DSAA60987.2023.10302465
21. Torgo, L., Azevedo, P., Areosa, I.: Beyond average performance – exploring regions of deviating performance for black box classification models (2021). https://arxiv.org/abs/2109.08216
22. Zhang, H.: The optimality of Naive Bayes. In: 17th International Florida Artificial Intelligence Research Society Conference, pp. 562–567 (2004). www.aaai.org

tsMIST: Model Sensitivity Analysis with Time Series Morphing

Antónia Brito[1,2(✉)], Moisés Santos[1,2], Duarte Folgado[3], and Carlos Soares[1,2,3]

[1] Faculty of Engineering of the University of Porto, Porto, Portugal
{antoniabrito,mrsantos,csoares}@fe.up.pt
[2] Laboratory for Artificial Intelligence and Computer Science (LIACC), Porto, Portugal
[3] Fraunhofer Portugal AICOS, Porto, Portugal
duarte.folgado@aicos.fraunhofer.pt

Abstract. Ensuring robustness in time series classification remains a critical challenge for safety-sensitive domains like clinical decision systems. While current evaluation practices focus on accuracy measures, they fail to address model stability under semantically meaningful input deformations. We propose tsMIST (*Time Series Model Sensitivity Test*), a novel morphing-based framework that systematically evaluates classifier resilience through controlled interpolation between adversarial class prototypes. By calculating the *switchThreshold* – defined as the minimal morphing distance required to flip predictions – our method reveals critical stability patterns across synthetic benchmarks with tunable class separation and 17 medical time series datasets. Key findings show convolutional architectures (ROCKET) maintain optimal thresholds near 50% morphing (48.2 ± 3.1%), while feature-based models (Catch22) exhibit premature decision flips at 22.7% deformation (±15.4%). In clinical scenarios, tsMIST detected critical ECG misclassifications triggered by ≤12% signal variation – vulnerabilities undetected by accuracy measures. Our results establish that robustness measures must complement accuracy for responsible AI in high-stakes applications. This work advances ML evaluation practices by enabling systematic sensitivity analysis, with implications for model auditing and deployment in safety-critical domains.

Keywords: Model Robustness · Time Series · Classification

1 Introduction

In critical domains such as healthcare, machine learning models must exhibit not only high accuracy but also robustness against subtle input perturbations [2]. Traditional evaluation measures often overlook model sensitivity to semantically meaningful variations, potentially masking vulnerabilities in real-world deployment [23]. This gap is particularly relevant in time series classification tasks, where gradual temporal pattern changes – such as variations in medical signals [16] – may lead to critical diagnostic errors. Recent studies address model

S. Džeroski et al. (Eds.): DS 2025, LNAI 16090, pp. 270–285, 2025.
https://doi.org/10.1007/978-3-032-05461-6_18

robustness through three main perspectives: performance analysis under adversarial scenarios, spatial mapping of model behavior, and resilience to distributional shifts. Work by [9] and [7] emphasizes identifying failure-prone regions, while [23] highlight risks from adversarial attacks. However, few methods explore structured perturbations that preserve temporal semantics, limiting realistic model evaluation.

This paper proposes tsMIST (Time Series Model Sensitivity Test), an innovative sensitivity analysis framework based on time series morphing. Inspired by *tsMorph* [20] originally developed for forecasting, tsMIST adapts the technique for classification by quantifying model resistance to gradual inter-class transformations. Through controlled linear interpolations, the method computes the *switchThreshold* – the minimum morphing required to alter model predictions – providing intuitive robustness measures, $tsMIST_{Avg}$ and $tsMIST_{Std}$.

Experiments encompass synthetic data with sine wave patterns and variable noise, plus 17 real-world medical signal datasets from the Aeon Time Series Classification archive. Three classification algorithms were evaluated: LSTM networks, Random Forest with Catch22 features, and the Rocket model. Results demonstrate that tsMIST identifies vulnerabilities undetected by conventional measures, such as unstable decision boundaries in feature-based models. The code is publicly available at https://github.com/Nia3324/tsMIST for reproducibility and further experimentation.

This work aims to: (1) propose an interpretable robustness measure for temporal classification; (2) validate the effectiveness of the proposal in controlled and real scenarios. This approach reinforces Responsible AI practices by carefully balancing the trade-off between accuracy and stability.

The paper is organized as follows: Sect. 2 reviews related work; Sect. 3 details tsMIST methodology; Sect. 4 describes experimental setup; Sect. 5 presents results; and Sect. 6 discusses implications and future directions.

2 Related Work

Time Series Classification. Time Series Classification (TSCL) methods are often organized according to the type of transformation they apply to the data [1,15], including convolutional approaches [6], deep learning models [10,11], dictionary-based methods [21], distance-based techniques [13], and feature-based classifiers [3,14]. In healthcare, TSCL techniques are used to interpret physiological signals, such as electrocardiograms, and support clinical decision-making [18]. Industrial sectors use TSCL for predictive maintenance by monitoring sensor data to foresee equipment failures, thereby optimizing operational efficiency [19]. The versatility of TSCL across these fields underscores its significance in extracting meaningful insights from temporal data.

Model Robustness. Approaches to model robustness relevant to auditing can be grouped into three main directions: performance-based stress analysis, spatial characterization of model behavior, and resilience to distributional shifts.

First, performance-centric methods aim to identify failure-prone scenarios. Inácio et al. (2025) [9] use meta-learning to estimate the likelihood of large forecasting errors, combining this with targeted data augmentation for stress testing. Similarly, Drapal et al. (2024) [7] define localized performance regions to identify zones where a model behaves reliably or unreliably, offering interpretability aligned with auditing tasks.

Second, spatial mapping of model behavior has been explored through instance space analysis. Muñoz et al. (2018) [17] reduce complex input distributions into two-dimensional spaces that reflect instance difficulty, allowing the visualization of model strengths and weaknesses. While primarily applied to classification, the method generalizes to other domains and supports robustness evaluations by exposing structural failure regions.

Third, robustness under distributional shift has been studied in adversarial and open-world contexts. Sehwag et al. (2019) [23] demonstrate that adversarial OOD inputs can bypass standard detectors, raising concerns about relying solely on non-adaptive robustness tests. Lee et al. (2021) [12] argue that robustness and fairness are aligned in targeting biased or noisy data, and propose unified training schemes to address both. In healthcare, Balendran et al. (2025) [2] categorize robustness into multiple dimensions—input perturbation, domain shift, adversarial resilience—across the ML lifecycle, reinforcing the need for audit methods that cover this range.

These studies collectively inform auditing practices by offering analytical tools, representations, and scenarios to examine model robustness beyond aggregate accuracy.

Dataset Morphing. Dataset morphing has been used to systematically probe model behavior across controlled distributional changes, supporting auditability by exposing gradual failure modes.

One class of methods focuses on real-data interpolation. Correia et al. (2019) [5] transform collaborative filtering datasets to examine how algorithm performance varies with meta-feature transitions. Santos et al. (2023) [20] extend this idea to time series, introducing tsMorph to generate semi-synthetic forecasting datasets along smooth trajectories between real datasets. Both approaches allow the design of structured test scenarios for auditing.

A second group applies morphing in high-dimensional continuous spaces. Schardong et al. (2024) [22] use implicit neural representations to interpolate between face images, creating synthetic variants that challenge facial recognition systems. Golling et al. (2023) [8] use normalizing flows to learn transformations between datasets while preserving likelihood, enabling audits under realistic distributional shifts.

These methods differ in domain and representation, but all support model auditing by generating structured transitions that uncover failure boundaries or distribution sensitivities.

3 Proposed Approach

3.1 Background

A time series represents sequential data points indexed in time order, widely used for temporal pattern analysis. Formally, following [16], we define a time series as $X = \{(x_1, t_1), ..., (x_T, t_T)\}$ where $x_i \in \mathbb{R}^d$ are observations (univariate when $d = 1$, multivariate otherwise). In supervised learning, models learn from labeled datasets $D = \{(X_i, Y_i)\}_{i=1}^N$ to predict target outputs Y_i from input features X_i. The learning objective is to find parameters θ^* that minimize prediction error:

$$\theta^* = \underset{\theta}{\operatorname{argmin}}\, L(Y, f(\theta; X)) \tag{1}$$

where L is a loss function (e.g., cross-entropy) and $f(\theta; X)$ produces predictions.

Time Series Classification (TSC) specializes supervised learning for temporal data by using complete series as inputs. Each instance $X_i = \{(x_{i1}, t_1), ..., (x_{iT}, t_T)\}$ maps to a discrete class label $Y_i \in \{1, ..., C\}$. Unlike forecasting which predicts future values, TSC identifies patterns that discriminate between classes. A classifier $f : \mathbb{R}^{T \times D} \rightarrow \{1, ..., C\}$ must capture temporal dependencies in order to accurately separate classes.

Dataset Morphing for Algorithm Analysis. Dataset morphing systematically probes model behavior through controlled transformations between data distributions. Originally developed for collaborative filtering [5], the method generates intermediate datasets D_j between source $D^{(s)}$ and target $D^{(t)}$ via:

$$\mathcal{D}^{(morph)} : \{D_j \mid D_0 = D^{(s)}, D_n = D^{(t)}, D_j = \tau(D_{j-1})\},\ 1 \leq j < n \tag{2}$$

where τ represents incremental modifications. Santos et al. (2024) [20] adapted this to time series through linear interpolation, creating a method called *tsMorph*:

$$\tau(i) = \alpha_i \cdot \mathbf{X}^{(target)} + (1 - \alpha_i) \cdot \mathbf{X}^{(source)}, \quad \mathbf{X} \in \mathbb{R}^{T \times D} \tag{3}$$

where $\mathbf{X}$ is a matrix of time series with T time steps and D variables and $\tau(i)$ creating smooth transitions between time series pairs (Fig. 1). In their work, this

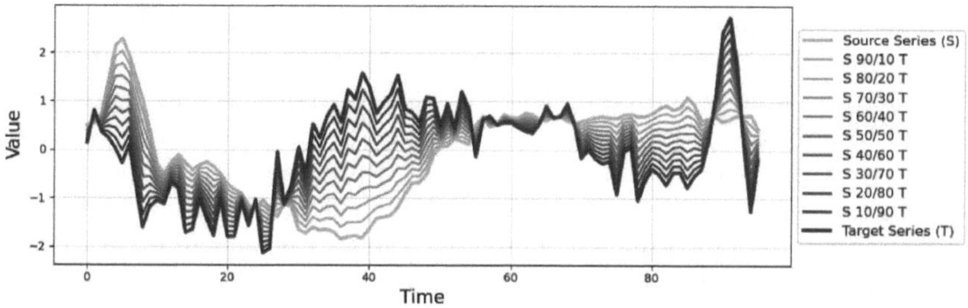

Fig. 1. Gradual morphing of source time series into target.

method was applied for the evaluation of forecasting algorithms. Our tsMIST framework extends the morphing to Time Series Classification (TSC) by analyzing how classifiers respond to these controlled deformations, particularly near decision boundaries.

3.2 Model Sensibility Analysis with Morphing Time Series

In this paper, we present tsMIST, a tool for model sensitivity analysis using morphing. Our approach refines the time series dataset morphing concept, proposed by [20], by shifting the focus to the sample level and adapting it to a TSC paradigm, rather than applying it to entire datasets for Forecasting.

The process begins by identifying source and target samples from distinct classes. Specifically, time series instances that classifiers frequently misclassify due to their high similarity in temporal patterns despite belonging to different classes. To systematically detect these challenging cases, we compute the Dynamic Time Warping (DTW) distance between all cross-class instances and select pairs with the smallest DTW distances as morphing candidates.

Given a labeled time series dataset D, for a given class $c \in \{1, \ldots, C\}$, we define the set of candidate pairs as:

$$P_c = \left\{ \left(X^{(\text{source})}, X^{(\text{target})} \right) \mid X^{(\text{source})} \in D_c,\ X^{(\text{target})} \in \bigcup_{c' \neq c}^{C} D_{c'} \right\} \tag{4}$$

where D_c is the dataset for class c, and $\bigcup_{c' \neq c}^{C} D_{c'}$ is the union of datasets for all classes except c. The similarity between time series pairs is measured using Dynamic Time Warping distance:

$$\text{DTW} : \mathbb{R}^{T \times D} \times \mathbb{R}^{T \times D} \to \mathbb{R}^{+} \tag{5}$$

The final set of morphing candidates M is selected by identifying pairs with minimal DTW distance:

$$M = \underset{\substack{(X^{(\text{source})}, X^{(\text{target})}) \in P_c \\ |M| = \delta}}{\text{argmin}} \left\{ \text{DTW}(X^{(\text{source})}, X^{(\text{target})}) \right\} \tag{6}$$

where P_c is a candidate pair set from Eq. 4, δ is the total fixed number of pairs to select, this value is determined based on the desired number of morphing pairs to be evaluated when computing tsMIST metrics. A low value restricts evaluation to only the most challenging samples, which may underestimate the model's overall robustness. Conversely, a high value may dilute the assessment by including a smaller percentage of ambiguous cases, potentially inflating the perceived robustness performance. Thus, selecting a relevant number of samples is critical: it ensures the model is still tested on hard-to-classify instances while preserving a representative measure of its true robustness. By focusing on these borderline cases – samples from different classes that exhibit high similarity in their temporal patterns – we create a controlled framework to evaluate model robustness.

For each pair of hard-to-distinguish examples, we perform a gradual morphing process between them. Given any classification problem, let $X^{(\text{source})}$ belong to class c (i.e., $Y^{(\text{source})} = c$) and $X^{(\text{target})}$ belong to any other class $c' \neq c$ (i.e., $Y^{(\text{target})} \in \{1, \ldots, C\} \setminus \{c\}$). Using Eq. 3, we generate a sequence of n intermediate time series $X^{(\text{morph})}$, where each subsequent series represents a step in the transformation from $X^{(\text{source})}$ to $X^{(\text{target})}$.

To evaluate model sensitivity, we determine how much morphing is required to change the model's prediction from the source to the target class. Specifically, we calculate:

$$switchThreshold_{P_c} = \frac{i}{n}, \quad 1 \leq i \leq n \tag{7}$$

where i is the index of the first morphed time series in $X^{(morph)}$ that causes the classifier f to change its prediction from the source class, i.e., where $\hat{Y}_i^{(morph)} \neq Y^{(source)}$, and n is the total number of morphed time series. For simplicity, we typically set $n = 100$, so the transition point i can be interpreted as a percentage. However, n can be adjusted to achieve finer or coarser granularity depending on the desired precision of the analysis.

The distribution of *switchThreshold* values across borderline sample pairs provides critical insights into model robustness. Let $tsMIST_{Avg}$ be the average of all *switchThreshold* across all pairs, and $tsMIST_{Std}$, its corresponding standard deviation. Given δ selected morphing candidate pairs:

$$\text{tsMIST}_{Avg} = \frac{1}{\delta} \sum_{k=1}^{\delta} switchThreshold_k$$

$$\text{tsMIST}_{Std} = \sqrt{\frac{1}{\delta} \sum_{k=1}^{\delta} \left(switchThreshold_k - \text{tsMIST}_{Avg}\right)^2}$$

A $tsMIST_{Avg}$ close to 50% indicates that the model maintains stable predictions, suggesting well-defined decision boundaries. In such cases, the model only changes its prediction when the time series has undergone substantial transformation toward the opposing class, reflecting confidence in its decision-making process. Conversely, extreme $tsMIST_{Avg}$ values – either very low or very high – signal potential instability. A low $tsMIST_{Avg}$ implies that minimal changes can trigger a classification switch, indicating fragile decision boundaries. A high $tsMIST_{Avg}$, on the other hand, may suggest that the model is overly rigid towards the source class, failing to adapt even when significant transformations are introduced.

The $tsMIST_{Std}$ further enhances our understanding of model behavior. A low $tsMIST_{Std}$ indicates consistent robustness across the feature space, meaning the model responds uniformly to perturbations regardless of the source-target pair. In contrast, a high $tsMIST_{Std}$ reveals region-dependent stability, where certain areas of the feature space are more vulnerable to perturbations than others. This variability can highlight specific weaknesses in the model's decision surface, guiding targeted improvements.

By aggregating $tsMIST_{Avg}$ and $tsMIST_{Std}$ of *switchThreshold* across all borderline sample pairs, we derive a quantitative measure of model sensitivity to gradual, semantically meaningful transformations. Unlike arbitrary perturbations, the morphing process preserves the temporal structure of the time series while systematically introducing features of the target class. This approach simulates semantically meaningful variations that models might encounter in real-world applications, making the sensitivity analysis both interpretable and practically relevant.

4 Experimental Setup

4.1 Datasets

We conducted experiments to demonstrate how tsMIST can enhance the understanding of classification algorithm robustness. To support the proposed approach, we utilized synthetically generated datasets, which allowed for controlled testing and analysis. Subsequently, we validated our findings using real-life data to ensure the practical applicability and reliability of the results.

Synthetic Datasets. The synthetic time series dataset generation framework is designed to create labeled time-series datasets with distinct patterns for two classes. The process begins by generating time-series data using fundamental waveforms, such as sine, cosine waves. These waveforms serve as the foundation for creating unique patterns for each class.

To simulate real-world variability, random noise is added to the signals. The noise level is adjustable, allowing us to control the degree of variability and study its impact on model performance. The noise was sampled from a Gaussian distribution with zero mean and a user-defined standard deviation.

Additionally, vertical and horizontal shifts are applied to the waveforms to further differentiate the classes. Vertical shifts alter the amplitude of the time series, while horizontal shifts introduce time delays. The difference between classes is evident in Fig. 2.

Once the data for both classes is generated, it is combined into a single dataset. The dataset is then shuffled to ensure randomness and avoid any bias in the order of samples. The synthetically generated datasets used in this study have a shape of (1000, 1, 100), representing 1000 samples, 1 variable, and a time series length of 100. These datasets consist of two balanced classes, each generated from sine curves with varying degrees of shifts.

Real-World Datasets. We evaluated tsMIST on 17 publicly available medical time series datasets from the Aeon archive[1], presented in Table 1. These

[1] https://www.timeseriesclassification.com/dataset.php.

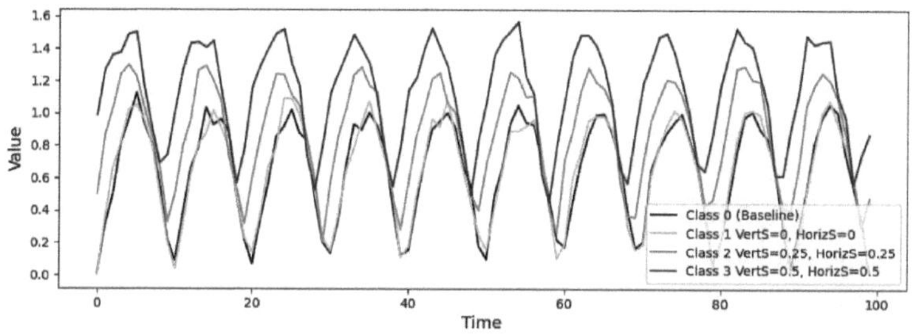

Fig. 2. Classes generated from sine wave with different vertical shifts (vertical and horizontal shifts).

datasets encompass a variety of physiological signals, including electrocardiograms (ECG), electroencephalograms (EEG), and electrooculograms (EOG). The collection includes both univariate and multivariate recordings, with signal lengths ranging from 82 to 2500 time steps, and classification tasks involving between 2 and 12 classes. Additional datasets cover other medical contexts such as Colposcopy, NerveDamage, and MedicalImages. All datasets were processed with an 80/20 train-test split. Full specifications, including sample sizes, dimensions, and class distributions, are available in the code repository.

4.2 Time Series Classification Methods

To evaluate and compare the robustness of different machine learning approaches for time series classification, we selected three very distinct algorithms: (1) a simple recurrent neural network (RNN) with a Long Short-Term Memory (LSTM) module, (2) an ensemble model leveraging extracted features from raw data, and (3) the state-of-the-art Rocket model, which employs randomly generated convolutional kernels.

The neural network classifier was designed with a straightforward architecture. It consists of two LSTM layers with 32 and 16 neurons, respectively, each followed by a dropout layer with a rate of 0.2. These layers are succeeded by a dense layer with 8 neurons and a final output layer with a single neuron using a sigmoid or a softmax activation function, depending on the classification task, binary or multiclass. The model was optimized using the Adam optimizer and trained to minimize binary or categorical cross-entropy loss, ensuring efficient convergence during training.

For the ensemble model, we extracted features using CAnonical Time-series CHaracteristics (Catch22) [14], a compact set of twenty-two statistical time series features designed to capture key temporal patterns. These features were used to train a Random Forest classifier, which combines the predictive power of multiple decision trees to achieve robust performance.

Lastly, we employed the Rocket model [6], a state-of-the-art approach for time series classification. Rocket transforms time series data using 5,000 randomly generated convolutional kernels, which capture diverse temporal patterns

Table 1. Dataset information, including training/testing set shapes and number of classes.

Dataset	Train Shape	Test Shape	# Classes
CinCECGTorso (CET)	(908, 1, 1639)	(228, 1, 1639)	4
Colposcopy (COL)	(128, 1, 180)	(32, 1, 180)	6
ECG200 (ECG)	(128, 1, 96)	(32, 1, 96)	2
ECG5000 (ECG5)	(3200, 1, 140)	(800, 1, 140)	5
ECGFiveDays (ECGFD)	(565, 1, 136)	(142, 1, 136)	2
EMOPain (EMO)	(846, 30, 200)	(212, 30, 200)	3
EOGHorizontalSignal (EOGH)	(463, 1, 1250)	(116, 1, 1250)	12
EOGVerticalSignal (EOGV)	(463, 1, 1250)	(116, 1, 1250)	12
Epilepsy (EPI)	(176, 3, 206)	(44, 3, 206)	4
EyesOpenShut (EOS)	(62, 14, 128)	(16, 14, 128)	2
HandMovementDirection (HMD)	(149, 10, 400)	(38, 10, 400)	4
Heartbeat (HRT)	(261, 61, 405)	(66, 61, 405)	2
MedicalImages (MI)	(729, 1, 99)	(183, 1, 99)	10
NerveDamage (ND)	(130, 1, 1500)	(33, 1, 1500)	3
StandWalkJump (SWJ)	(16, 4, 2500)	(5, 4, 2500)	3
ToeSegmentation1 (TOE)	(171, 1, 277)	(43, 1, 277)	2
TwoLeadECG (TECG)	(743, 1, 82)	(186, 1, 82)	2

without requiring manual feature engineering. The transformed features were then used to train a Ridge classifier, which is a linear model maintaining good performance while being computational efficiency.

Evaluation. In order to calculate $tsMIST_{Avg}$ and $tsMIST_{Std}$ we used $n = 100$ and $\delta = N$, where N is the size of a given dataset. Although the focus of this paper is on assessing robustness of predictions, we also estimated accuracy. To ensure consistent evaluation across all experiments, all models were tested on a 80/20 train-test split in each dataset.

5 Results

5.1 Synthetic Data

Regarding the experiments on synthetic data, we analyzed how different magnitudes of shifts between classes influence the tsMIST values. In Fig. 3, we present the experimental results for all modes, a simple LSTM, a Random Forest trained on extracted Catch22 features, and a Ridge Classifier trained on ROCKET with 5000 random kernels.

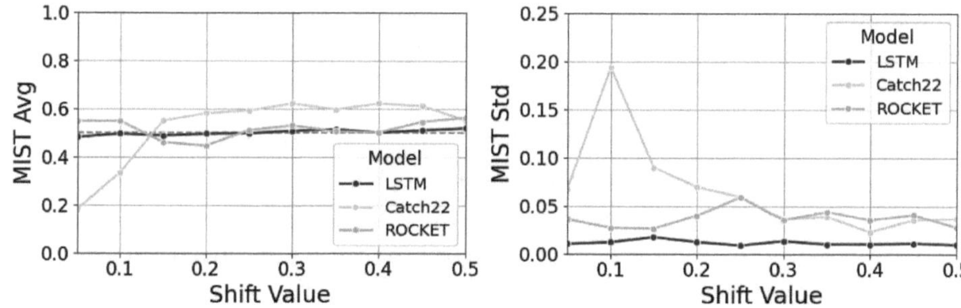

Fig. 3. tsMIST Average and Standard Deviation for different Shifts Variations.

From these results, we observe that the best-performing model is the LSTM, followed by ROCKET, and finally Catch22. It is important to note that the classification task in this example is intentionally straightforward: we systematically generate classes with increasing levels of shift between them, thereby making the classes progressively more distinct.

As shown in Fig. 3, the LSTM consistently achieves $tsMIST_{Avg}$ values near 50% and near-zero $tsMIST_{Std}$, indicating strong robustness even under minimal class separation. ROCKET exhibits similar behavior, albeit with slightly more variability. In contrast, Catch22 shows $tsMIST_{Avg}$ values between 20–30% in low-shift scenarios, implying high sensitivity to small perturbations. This fragility is further evidenced by its erratic $tsMIST_{Std}$, which spikes under low separation—a pattern absent in the more stable architectures. Catch22's statistical descriptors exhibit limited variation in response to the increasing shifts, making it difficult for the model to distinguish between classes that remain substantially overlapping.

5.2 Real World Data

We extended our analysis to real-world data in order to analyze the sensitivity of algorithms to small gradual changes in the input data. In total, this study analyzed seventeen medical-related datasets sourced from the Aeon Time Series Classification Archive.[2]

Table 2 summarizes the average $tsMIST$ scores across all classes for each dataset, along with the corresponding model accuracy. Additional class-specific results are presented in Tables 4 and 5.

In several datasets (e.g., Heartbeat, Colposcopy), Catch22 and LSTM failed to classify any borderline cases, resulting in $tsMIST_{Avg} = 0.0$ or undefined $tsMIST_{Std}$. These anomalies are denoted by dashes (–) in Table 2 and emphasize serious model fragility not captured by accuracy alone.

From Table 2, we observe two distinct scenarios: 1) Cases where all models exhibit similar robustness for a particular dataset, whether good or bad; and 2)

[2] Available at https://www.timeseriesclassification.com/dataset.php?train=&test=&leng=&class=&type==%27ecg%27.

Table 2. Average Score comparison of models across datasets, including tsMIST Average (M_Avg), tsMIST Standard Deviation (M_Std) and model Accuracy (Acc). Bold values represent cases where models had the most inconsistent behavior across metrics.

Dataset	LSTM			Catch22			Rocket		
	M_Avg	M_Std	Acc	M_Avg	M_Std	Acc	M_Avg	M_Std	Acc
CET	0.534	0.09	0.989	0.468	0.157	0.982	0.462	0.068	1.0
COL	0.0	–	0.4	0.21	0.057	0.4	0.213	0.021	0.325
ECG	0.484	0.217	0.9	0.489	0.167	0.8	0.51	0.161	0.9
ECG5	0.418	0.186	0.953	0.29	0.068	0.952	0.488	0.186	0.942
ECGFD	0.51	0.13	1.0	0.48	0.134	0.989	0.51	0.052	1.0
EMO	0.498	0.133	0.823	0.552	0.091	0.917	0.23	0.224	0.785
EOGH	**0.044**	**0.016**	**0.248**	**0.32**	**0.182**	**0.786**	**0.406**	**0.147**	**0.869**
EOGV	**0.0**	–	**0.131**	**0.296**	**0.154**	**0.655**	**0.34**	**0.132**	**0.786**
EOS	0.0	–	0.55	0.35	0.236	0.55	0.51	0.291	0.6
EPI	0.447	0.042	0.745	0.401	0.159	0.982	0.456	0.147	0.982
HMD	0.229	0.121	0.383	0.22	0.018	0.255	0.265	0.173	0.468
HRT	**0.0**	–	**0.622**	**0.0**	–	**0.695**	**0.51**	**0.138**	**0.598**
MI	0.396	0.164	0.716	0.395	0.186	0.769	0.404	0.232	0.825
ND	**0.34**	**0.104**	**0.39**	**0.482**	**0.117**	**1.0**	**0.417**	**0.057**	**1.0**
SWJ	**0.292**	**0.1**	**0.667**	**0.553**	–	**0.5**	**0.0**	–	**0.167**
TECG	0.499	0.169	0.991	0.47	0.195	0.97	0.479	0.134	1.0
TOE	0.51	0.229	0.778	0.453	0.205	0.944	0.51	0.126	0.944

Datasets where different models show clear distinctions in prediction. The latter scenario is more insightful for comparing model robustness, as it provides clear algorithmic behavioral insights.

Table 3. Most changed TSFEL features at the *switchThreshold* for the Heartbeat (HB) dataset tested with ROCKET model.

Rank	1	2	3	4	5
Class Abnormal	39 LPCC 6	33 LPCC 6	23 LPCC 4	23 LPCC 8	36 MFCC 9
Class Normal	39 Mean diff	58 MFCC 5	60 Min	8 Kurtosis	11 Kurtosis

For instance, consider the Heartbeart (HRT) dataset results in Table 2. While all models achieve similar accuracy scores, only ROCKET demonstrates sufficient robustness against adversarial perturbations. This is evidenced by its $tsMIST_{Avg}$ nearing the 50% margin, with a low standard deviation of 0.14%. In contrast, both the LSTM and Catch22 models fail to correctly classify any of

Table 4. MIST Score comparison of models across datasets, including MIST Average (M_Avg), MIST Standard Deviation (M_Std) and model Accuracy (Acc). [1/2]

Dataset	Class	LSTM			Catch22			Rocket		
		M_Avg	M_Std	Acc	M_Avg	M_Std	Acc	M_Avg	M_Std	Acc
CET	1 (24%)	0.701	0.118	0.989	0.629	0.165	0.982	0.549	0.054	1.0
	2 (25%)	0.582	0.085	0.989	0.42	0.16	0.982	0.391	0.056	1.0
	3 (25%)	0.408	0.068	0.989	0.315	0.159	0.982	0.488	0.063	1.0
	4 (24%)	0.447	0.088	0.989	0.51	0.144	0.982	0.422	0.099	1.0
COL	0 (8%)	0.0	–	0.4	0.0	–	0.4	0.0	–	0.325
	1 (11%)	0.0	–	0.4	0.0	–	0.4	0.0	–	0.325
	2 (10%)	0.0	–	0.4	0.308	0.276	0.4	0.0	–	0.325
	3 (25%)	0.0	–	0.4	0.0	–	0.4	0.23	–	0.325
	4 (15%)	0.0	–	0.4	0.83	–	0.4	0.42	–	0.325
	5 (28%)	0.0	–	0.4	0.12	0.067	0.4	0.63	0.128	0.325
ECG	−1 (33%)	0.395	0.206	0.9	0.34	0.144	0.8	0.414	0.161	0.9
	1 (66%)	0.572	0.228	0.9	0.637	0.189	0.8	0.606	0.161	0.9
ECG5	1 (58%)	0.291	0.145	0.953	0.223	0.178	0.952	0.443	0.152	0.942
	2 (35%)	0.272	0.32	0.953	0.15	–	0.952	0.464	0.216	0.942
	3 (1%)	0.779	0.144	0.953	0.616	0.162	0.952	0.599	0.202	0.942
	4 (3%)	0.748	0.32	0.953	0.46	–	0.952	0.453	0.234	0.942
	5 (0%)	0.0	–	0.953	0.0	–	0.952	0.48	0.127	0.942
ECGFD	1 (50%)	0.46	0.13	1.0	0.557	0.157	0.989	0.523	0.052	1.0
	2 (49%)	0.56	0.13	1.0	0.403	0.11	0.989	0.497	0.052	1.0
EMO	0 (77%)	0.231	0.149	0.823	0.14	0.024	0.917	0.4	0.37	0.785
	1 (14%)	0.426	0.113	0.823	0.703	0.174	0.917	0.264	0.286	0.785
	2 (8%)	0.837	0.137	0.823	0.812	0.076	0.917	0.027	0.017	0.785
EOGH	1 (9%)	0.0	–	0.248	0.319	0.281	0.786	0.479	0.201	0.869
	2 (8%)	0.0	–	0.248	0.237	0.049	0.786	0.513	0.126	0.869
	3 (8%)	0.0	–	0.248	0.316	0.237	0.786	0.472	0.118	0.869
	4 (8%)	0.0	–	0.248	0.188	0.145	0.786	0.413	0.2	0.869
	5 (9%)	0.0	–	0.248	0.411	0.171	0.786	0.291	0.116	0.869
	6 (8%)	0.533	0.195	0.248	0.112	0.113	0.786	0.237	0.089	0.869
	7 (7%)	0.0	–	0.248	0.494	0.15	0.786	0.474	0.126	0.869
	8 (6%)	0.0	–	0.248	0.299	0.29	0.786	0.312	0.168	0.869
	9 (8%)	0.0	–	0.248	0.365	0.167	0.786	0.449	0.138	0.869
	10 (7%)	0.0	–	0.248	0.397	0.136	0.786	0.293	0.073	0.869
	11 (8%)	0.0	–	0.248	0.381	0.221	0.786	0.433	0.143	0.869
	12 (7%)	0.0	–	0.248	0.322	0.227	0.786	0.504	0.265	0.869
EOGV	1 (9%)	0.0	–	0.131	0.511	0.148	0.655	0.503	0.221	0.786
	2 (8%)	0.0	–	0.131	0.537	0.253	0.655	0.398	0.102	0.786
	3 (8%)	0.0	–	0.131	0.033	0.012	0.655	0.04	–	0.786
	4 (8%)	0.0	–	0.131	0.353	0.287	0.655	0.603	0.084	0.786
	5 (9%)	0.0	–	0.131	0.333	0.186	0.655	0.267	0.15	0.786
	6 (8%)	0.0	–	0.131	0.306	0.228	0.655	0.504	0.196	0.786
	7 (7%)	0.0	–	0.131	0.0	–	0.655	0.283	0.14	0.786
	8 (6%)	0.0	–	0.131	0.312	0.043	0.655	0.356	0.095	0.786
	9 (8%)	0.0	–	0.131	0.429	0.273	0.655	0.3	0.127	0.786
	10 (7%)	0.0	–	0.131	0.275	0.151	0.655	0.283	0.139	0.786
	11 (8%)	0.0	–	0.131	0.223	0.088	0.655	0.333	0.158	0.786
	12 (7%)	0.0	–	0.131	0.241	0.174	0.655	0.209	0.173	0.786

Table 5. MIST Score comparison of models across datasets, including MIST Average (M_Avg), MIST Standard Deviation (M_Std) and model Accuracy (Acc). [2/2]

Dataset	Class	LSTM			Catch22			Rocket		
		M_Avg	M_Std	Acc	M_Avg	M_Std	Acc	M_Avg	M_Std	Acc
EOS	0 (55%)	0.0	–	0.55	0.222	0.137	0.55	0.463	0.291	0.6
	1 (44%)	0.0	–	0.55	0.478	0.335	0.55	0.557	0.291	0.6
EPI	epilepsy (24%)	0.68	–	0.745	0.524	0.144	0.982	0.458	0.185	0.982
	running (26%)	0.365	0.038	0.745	0.336	0.158	0.982	0.304	0.08	0.982
	sawing (21%)	0.34	–	0.745	0.412	0.197	0.982	0.551	0.189	0.982
	walking (27%)	0.403	0.131	0.745	0.331	0.139	0.982	0.512	0.135	0.982
HMD	backward (22%)	0.02	–	0.383	0.09	–	0.255	0.0	–	0.468
	forward (31%)	0.227	0.2	0.383	0.0	–	0.255	0.235	0.205	0.468
	left (24%)	0.428	0.283	0.383	0.325	0.035	0.255	0.377	0.262	0.468
	right (22%)	0.24	–	0.383	0.465	0.035	0.255	0.448	0.224	0.468
HRT	abnormal (73%)	0.0	–	0.622	0.0	–	0.695	0.6	0.138	0.598
	normal (26%)	0.0	–	0.622	0.0	–	0.695	0.42	0.138	0.598
MI	1 (9%)	0.572	0.292	0.716	0.407	0.227	0.769	0.425	0.252	0.825
	2 (5%)	0.585	0.222	0.716	0.42	0.179	0.769	0.436	0.241	0.825
	3 (6%)	0.486	0.204	0.716	0.515	0.176	0.769	0.485	0.192	0.825
	4 (4%)	0.797	0.147	0.716	0.441	0.292	0.769	0.458	0.255	0.825
	5 (3%)	0.382	0.23	0.716	0.259	0.209	0.769	0.371	0.259	0.825
	6 (2%)	0.0	–	0.716	0.36	0.082	0.769	0.385	0.204	0.825
	7 (4%)	0.722	0.278	0.716	0.426	0.292	0.769	0.35	0.262	0.825
	8 (2%)	0.0	–	0.716	0.259	0.14	0.769	0.234	0.181	0.825
	9 (10%)	0.0	–	0.716	0.389	0.043	0.769	0.47	0.241	0.825
	10 (50%)	0.412	0.271	0.716	0.477	0.221	0.769	0.428	0.233	0.825
ND	0 (16%)	0.574	0.156	0.39	0.764	0.147	1.0	0.58	0.073	1.0
	1 (35%)	0.446	0.156	0.39	0.253	0.145	1.0	0.44	0.073	1.0
	2 (47%)	0.0	–	0.39	0.429	0.061	1.0	0.231	0.025	1.0
SWJ	jumping (28%)	0.375	0.169	0.667	0.34	–	0.5	0.0	–	0.167
	standing (33%)	0.502	0.13	0.667	0.64	–	0.5	0.0	–	0.167
	walking (38%)	0.0	–	0.667	0.68	–	0.5	0.0	–	0.167
TECG	1 (50%)	0.387	0.162	0.991	0.283	0.181	0.97	0.532	0.142	1.0
	2 (49%)	0.61	0.177	0.991	0.657	0.21	0.97	0.426	0.126	1.0
TOE	0 (50%)	0.489	0.229	0.778	0.437	0.224	0.944	0.552	0.126	0.944
	1 (49%)	0.531	0.229	0.778	0.469	0.186	0.944	0.468	0.126	0.944

the selected borderline pairs, resulting in no measurable mean or standard deviation for the morphing. This indicates significantly lower robustness compared to ROCKET.

To provide deeper interpretability for tsMIST results, we leveraged TSFEL (Time Series Feature Extraction Library) [4] to compute the feature-wise differences between the Source series and the *switchThreshold* series. This approach enables interpretability at a meta-feature level, identifying which features most significantly influenced the model's prediction changes. By analyzing these differences, we gain insights into the specific aspects of the time series that the model prioritizes when making predictions.

In Table 3, we present the features exhibiting the largest changes at the *switchThreshold* for the Heartbeat (HB) dataset using the ROCKET model.

For abnormal heartbeats, the predominance of Linear Predictive Cepstral Coefficients (LPPC) features among the top-ranked changes suggests a strong sensitivity to variations in the temporal dynamics of the cardiac signal. LPCCs are typically used to model the resonant structure of signals and may reflect abrupt changes in waveform morphology associated with pathological events such as arrhythmias, ectopic beats, or irregular contraction patterns.

6 Conclusion

This study introduced **tsMIST**, a morphing-based framework for assessing robustness in time series classification models. Our results demonstrate that the proposed measures – $tsMIST_{Avg}$ and $tsMIST_{Std}$ – capture model sensitivity to semantically meaningful input deformations, offering insights not provided by accuracy alone. Experiments on both synthetic and real-world medical datasets revealed that neural models such as LSTM and ROCKET maintain robust decision boundaries, with *tsMISTAvg* values consistently near 50% and low standard deviation. In contrast, feature-based models using Catch22 displayed fragile behavior, with early prediction switches and unstable *tsMISTStd* in scenarios with low class separation. Importantly, tsMIST exposed critical vulnerabilities in clinical datasets that remained undetected by standard evaluation. In datasets such as Heartbeat and Colposcopy, even minor morphing perturbations ($\leq$12%) were sufficient to flip predictions in less robust models – a concerning pattern for safety-critical applications. To enhance interpretability, we applied the TSFEL library to analyze feature-wise changes at the *switchThreshold*. The results highlighted key signal characteristics, such as LPCC, mean difference, and kurtosis, that contribute most to classification decisions. This fusion of robustness quantification and feature-level analysis provides a richer understanding of model behavior. Overall, tsMIST contributes a novel and interpretable approach for auditing time series classifiers. By quantifying prediction instability and identifying failure boundaries, the framework advances responsible AI practices in high-stakes domains. Future work will explore non-linear morphing paths and applications in multivariate and multimodal time series data.

Acknowledgements. This work is a result of Agenda "Center for Responsible AI", nr. C645008882-00000055, investment project nr. 62, financed by the Recovery and Resilience Plan (PRR) and by European Union - NextGeneration EU. Funded by the

European Union – NextGenerationEU. Views and opinions expressed are however those of the author(s) only and do not necessarily reflect those of the European Union or the European Commission. Neither the European Union nor the European Commission can be held responsible for them.

AISym4Med (101095387) supported by Horizon Europe Cluster 1: Health, ConnectedHealth (n.o 46858), supported by Competitiveness and Internationalisation Operational Programme (POCI) and Lisbon Regional Operational Programme (LISBOA 2020), under the PORTUGAL 2020 Partnership Agreement, through the European Regional Development Fund (ERDF).

This work was financially supported by: UID/00027 of the LIACC - Artificial Intelligence and Computer Science Laboratory - funded by Fundação para a Ciência e a Tecnologia, I.P./MCTES through the national funds.

We would like to thank the Aeon Time Series Classification Repository for providing the open-access time series datasets used in this study.

References

1. Bagnall, A., Lines, J., Bostrom, A., Large, J., Keogh, E.: The great time series classification bake off: a review and experimental evaluation of recent algorithmic advances. Data Min. Knowl. Disc. **31**, 606–660 (2017)
2. Balendran, A., et al.: A scoping review of robustness concepts for machine learning in healthcare. NPJ Digit. Med. **8**(1), 38 (2025)
3. Barandas, M., et al.: TSFEL: time series feature extraction library. SoftwareX 11, 100456 (2020). https://doi.org/10.1016/j.softx.2020.100456, https://linkinghub.elsevier.com/retrieve/pii/S2352711020300017
4. Barandas, M., et al.: TSFEL: time series feature extraction library. SoftwareX 11, 100456 (2020). https://doi.org/10.1016/j.softx.2020.100456, https://www.sciencedirect.com/science/article/pii/S2352711020300017
5. Correia, A., Soares, C., Jorge, A.: Dataset morphing to analyze the performance of collaborative filtering. In: Kralj Novak, P., Šmuc, T., Džeroski, S. (eds.) DS 2019. LNCS (LNAI), vol. 11828, pp. 29–39. Springer, Cham (2019). https://doi.org/10.1007/978-3-030-33778-0_3
6. Dempster, A., Petitjean, F., Webb, G.I.: ROCKET: exceptionally fast and accurate time series classification using random convolutional kernels. Data Min. Knowl. Disc. **34**(5), 1454–1495 (2020)
7. Drapal, P., Prudêncio, R.B., Silva Filho, T.M.: Towards explainable evaluation: explaining predicted performance using local performance regions. Appl. Soft Comput. **167**, 112351 (2024)
8. Golling, T., Klein, S., Mastandrea, R., Nachman, B., Raine, J.A.: Morphing one dataset into another with maximum likelihood estimation. Phys. Rev. D **108**(9), 096018 (2023)
9. Inácio, R., Cerqueira, V., Barandas, M., Soares, C.: Meta-learning and data augmentation for stress testing forecasting models. In: Krempl, G., Puolamäki, K., Miliou, I. (eds.) IDA 2025. LNCS, vol. 15669, pp. 343–357. Springer, Cham (2025). https://doi.org/10.1007/978-3-031-91398-3_26
10. Ismail-Fawaz, A., Devanne, M., Weber, J., Forestier, G.: Deep learning for time series classification using new hand-crafted convolution filters. In: 2022 IEEE International Conference on Big Data (Big Data), pp. 972–981. IEEE (2022)

11. Ismail Fawaz, H., Forestier, G., Weber, J., Idoumghar, L., Muller, P.-A.: Deep learning for time series classification: a review. Data Min. Knowl. Disc. **33**(4), 917–963 (2019). https://doi.org/10.1007/s10618-019-00619-1
12. Lee, J.G., Roh, Y., Song, H., Whang, S.E.: Machine learning robustness, fairness, and their convergence. In: Proceedings of the 27th ACM SIGKDD Conference on Knowledge Discovery & Data Mining, pp. 4046–4047 (2021)
13. Lines, J., Bagnall, A.: Time series classification with ensembles of elastic distance measures. Data Min. Knowl. Disc. **29**, 565–592 (2015)
14. Lubba, C.H., Sethi, S.S., Knaute, P., Schultz, S.R., Fulcher, B.D., Jones, N.S.: Catch22: canonical time-series characteristics: Selected through highly comparative time-series analysis. Data Min. Knowl. Disc. **33**(6), 1821–1852 (2019)
15. Middlehurst, M., Schäfer, P., Bagnall, A.: Bake off redux: a review and experimental evaluation of recent time series classification algorithms. Data Min. Knowl. Disc. **38**(4), 1958–2031 (2024)
16. Mohammadi Foumani, N., Miller, L., Tan, C.W., Webb, G.I., Forestier, G., Salehi, M.: Deep learning for time series classification and extrinsic regression: a current survey. ACM Comput. Surv. **56**(9) (2024). https://doi.org/10.1145/3649448
17. Muñoz, M.A., Villanova, L., Baatar, D., Smith-Miles, K.: Instance spaces for machine learning classification. Mach. Learn. **107**, 109–147 (2018)
18. Neves, I., et al.: Interpretable heartbeat classification using local model-agnostic explanations on ECGs. Comput. Biol. Med. **133**, 104393 (2021)
19. Resende, C., et al.: Tip4. 0: industrial internet of things platform for predictive maintenance. Sensors **21**(14), 4676 (2021)
20. Santos, M.R., Carvalho, A.C., Soares, C.: Enhancing algorithm performance understanding through tsMorph: generating semi-synthetic time series for robust forecasting evaluation. In: Proceedings (2024)
21. Schäfer, P.: The boss is concerned with time series classification in the presence of noise. Data Min. Knowl. Disc. **29**, 1505–1530 (2015)
22. Schardong, G., et al.: Neural implicit morphing of face images. In: Proceedings of the IEEE/CVF Conference on Computer Vision and Pattern Recognition, pp. 7321–7330 (2024)
23. Sehwag, V., et al.: Analyzing the robustness of open-world machine learning. In: Proceedings of the 12th ACM Workshop on Artificial Intelligence and Security, pp. 105–116 (2019)

Explaining Deep Neural Networks with Example and Pixel Attribution

Genghua Dong(✉), Henrik Boström, Roman Bresson, and Amr Alkhatib

KTH Royal Institute of Technology, Stockholm, Sweden
{genghua,bostromh,bresson,alkhat}@kth.se

Abstract. Most techniques for explainable machine learning focus on a single modality for the explanations, e.g., using either feature or example attribution. A novel approach, called Hybrid Attribution Network (HAN), is proposed for providing multimodal explanations for image classification. The technique first extracts embeddings from a deep neural network (DNN), which are subsequently used by a KNN classifier to form predictions; example attributions can then be derived from the latter. Based on the example attributions, pixel attributions are further generated to provide complementary feature-level explanations. Results from an empirical investigation show that HAN may provide highly concentrated example attributions, i.e., the predictions can be explained with few training examples, without compromising predictive performance relative to the original deep neural network. Moreover, the pixel attributions are shown to enhance the interpretability of the predictions, by highlighting key pixels in the example attributions. An important finding from the empirical investigation is that the choice of layer to use for the embeddings may have a large impact on both the predictive performance and the generated explanations.

Keywords: Explainable AI · Example attribution · Pixel attribution · Deep Neural Networks

1 Introduction

Deep neural networks (DNNs) are widely applied across various areas, such as autonomous driving, AI-assisted medical diagnosis, and financial fraud detection. However, their 'black box' nature has raised concerns about understanding their internal mechanisms [17]. Explainable artificial intelligence (XAI)[1] focuses on developing models that allow humans to comprehend their predictions. Therefore, such models offer transparency into their insights, aiming to enable users to better trust and manage them [1,17].

Some machine learning models are interpretable by design, leveraging an intelligible architecture, e.g., logistic regression, decision trees and generalized

[1] The term XAI is in this work used as a synonym to explainable machine learning.

S. Džeroski et al. (Eds.): DS 2025, LNAI 16090, pp. 286–300, 2025.
https://doi.org/10.1007/978-3-032-05461-6_19

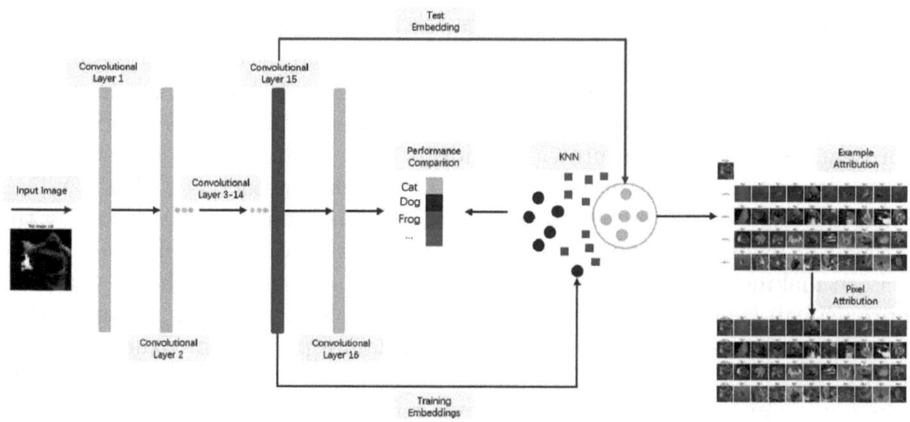

Fig. 1. The structure of HAN.

additive models, while also some neural network architectures have been designed with interpretability in mind [5,20,31]. On the other hand, black-box models, such as most DNNs, are trained without consideration for interpretability, and explained through post-hoc methods, e.g. feature-based, concept-based, rule-based, or example-based explanation methods [31]. Feature-based explanations (FBE) aims at evaluating to what extent each input feature contributes to the model output. In computer vision, this translates to pixel attribution, where the contribution of each pixel to a prediction is computed through methods such as Grad-CAM [41], SmoothGrad [43], Vanilla Gradient [42], or Eigen-Cam [32], allowing the most relevant pixels in a visual explanation to be highlighted. Additionally, FBE also includes the well-known model-agnostic techniques LIME [39] and SHAP [29]. Since input features are not always naturally interpretable, concept-based approaches such as TCAV [23] consider higher-level information, such as a color and an object, to be embedded within the latent space learned by the DNN.

Orthogonal to the previously discussed approaches, example-based explanations (EBE) enhance understanding of model behavior by leveraging specific training examples [31]. Different categorizations of EBE methods have been suggested in the literature; see e.g., [37]. For instance, the prototype and criticisms approach [22] aims to represent the data distribution by selecting the data instances that best represent the majority of data points (prototypes), along with the criticisms that identify instances not well captured by the prototypes. Counterfactual explanations [16] involve applying a minimal alteration to a datapoint so that the prediction is changed. However, it should be noted that these methods do not provide example-level attributions for individual predictions.

In contrast, example attribution techniques allow for identifying and quantifying the influence of training examples on a model's prediction. For example, both exact and approximate example attributions have recently been investigated for random forests [4].

A learning algorithm that by design provides example attributions is the KNN algorithm; a prediction can be exactly explained by the set of data points closest to the input. However, the algorithm is less suited for demanding tasks, such as image classification, unless high-level features have been extracted, e.g., representing edges or textures. To address this, embeddings extracted from DNNs are widely used as inputs to KNN. One such approach is EBE-DNN [11], which directly extracts embeddings from a DNN as input to KNN, allowing the retrieval of example attributions that more accurately and reliably reflect the decision-making process of the DNN. Empirical studies have shown that this approach can generate highly concentrated example attributions, ensuring that limited training examples are sufficient to explain a prediction without sacrificing accuracy. However, EBE-DNN solely depends on example attributions and lacks a mechanism to provide feature-level information, such as highlighting key pixels that influence the model's decisions.

The limitations of relying solely on example attributions emphasize a challenge in XAI; single-method explanations may not always be sufficient for fully understanding model decisions. To address this, multimodal explanation approaches have been explored, which provide a richer framework for interpretability than what is provided by a single method [40], and employ multiple explanation techniques, such as example-based, feature-based, and generated text-based explanations [28,33,35]. For instance, the Pointing and Justification Explanation (PJ-X) model integrates visual and text-based explanations by generating saliency maps that highlight relevant image regions and providing textual justifications for the model's decisions. This combination allows for a deeper understanding of the model's reasoning process [33]. Another example is Disentangled Multimodal Explanations (DIME), which enables an accurate and fine-grained analysis of multimodal models by disentangling local explanations into unimodal contributions and multimodal interactions [30]. Additionally, the Multimodal Text-in-Image Explanation Network (MTXNet) generates multimodal explanations by focusing on the text in the image, producing consistent explanations with human interpretations [38]. However, these multimodal methods face several challenges. For instance, recent studies highlight the disagreement in the output [25] and limited predictability in Visual Question Answering (VQA) explanations [6]. Furthermore, effectively aligning and integrating heterogeneous modalities remains a fundamental challenge, as improper fusion may lead to suboptimal performance [27].

To address the aforementioned limitations, we propose the Hybrid Attribution Network (HAN) method for image classification, which exploits the strengths of both KNN and DNNs to form multimodal explanations. As illustrated in Fig. 1, HAN combines example attributions with pixel attributions to provide a comprehensive explanatory framework. First, embeddings generated from DNN are employed by a KNN classifier to enable the retrieval of training examples that are most similar to the test image in terms of the DNN's learned feature representation. Then, pixel attributions are generated to present the key input pixels that align with the example attribution, which provides complementary insights at the feature level, rather than relying solely on the example-based explanation as in EBE-DNN. Specifically, applying pixel attribution (e.g., Grad-CAM) to retrieved examples enhances interpretability by highlighting consistent visual patterns across similar examples, thus providing an additional interpretive context and highlighting semantically aligned explanations compared to analyzing the test image alone. In contrast to multimodal approaches that require complex data fusion, HAN achieves multimodal-level explanations using only a single data modality, reducing the computational burden and eliminating the need for cross-modal alignment. Our approach aligns with recent research in conceptual interpretation, such as the Cut the Black Box (CB2) method [2], which maps pixel-based representations to a conceptual space for improved model transparency. By integrating both example attribution and pixel attribution, our method aligns most closely to the Unimodal Task, Multimodal Explanation (UM) category [40]. Moreover, HAN bridges the gap between explanation and model performance. As will be observed in the empirical investigation, the accuracy of the underlying DNN may be maintained while providing hybrid explanations.

The remainder of this paper is structured as follows. In the next section, we review related work on the hybrid attribution methods. In Sect. 3, we describe the proposed approach (HAN). In Sect. 4, we present and evaluate HAN on image classification tasks. Finally, in Sect. 5, we summarize the main findings and discuss directions for future work.

2 Related Work

Given the breadth of previous research on explanation methods, we here focus on two main categories that relate closely to the approach proposed in this work, namely similarity-based explanation techniques and prototype-based explanation techniques.

Similarity-based explanation techniques aim to interpret models that rely on the "distance-like" function to quantify the similarity between input instance pairs. Among them, Hamilton et al. [18] unify several explanation techniques under a Shapley-Taylor formalism for visual similarity models and introduce a kernel-based method to extract pairwise correspondences from trained opaque-box models. Salient Attributes for Network Explanation (SANE) [36] addresses the explanation of image similarity models by pairing saliency maps with an attribute that best explains the match, providing more informative insight into similarity models. BiLRP [12] is another explanation method in similarity models which leverages the composition of layer-wise relevance propagation (LRP) explanations [3]. These methods are fundamentally designed to explain similarity models, rather than classification decisions.

Prototype-based explanation techniques embed the learned prototypes into their architectures and use the similarity to these prototypes to guide and interpret predictions. Prototypical part network (ProtoPNet) [7] is a self-explaining classification architecture that integrates a prototype layer in the backbone model, extracting prototypical parts from training images of a given class. During inference, ProtoPNet identifies regions in the input image that closely resemble these learned prototypes, enabling localized visual explanations by highlighting the matching regions that contribute to the final prediction. Prototypical Relevance Propagation (PRP) [14] extends the ProtoPNet architecture by applying LRP, which improves the localization of prototype-based explanations. In particular, both PRP and BiLRP build on the same theoretical foundation as LRP, using model-aware relevance tracking to assign importance scores to input features. K-Means Explainer (KMEx) [13] is an effective method for converting any pretrained model into a prototypical self-explainable model (PSEM) [24] without retraining. By clustering latent representations using k-means, KMEx constructs class-level prototypes that serve as classification references, thereby enabling transparent decision-making in existing networks.

3 Hybrid Attribution Network (HAN)

Building on our previous work [10] for providing example attribution of deep neural networks, we propose a novel hybrid approach, HAN, which provides both example and pixel attribution in a unified, post-hoc setting. In contrast to similarity-based explanation techniques discussed above, HAN focuses on classification models and explains why a test image is assigned to a particular class. In contrast also to the prototype-based methods, which rely on a subset of abstract representations, such as latent patches (ProtoPNet, PRP) or clustered embedding centers (KMEx), to explain predictions, HAN performs post-hoc example-level attribution by retrieving actual training examples based on embedding similarity. When applied to pretrained DNN models with sufficiently discriminative embeddings, HAN can operate as a post-hoc attribution method that explains predictions via nearest-neighbor example retrieval, without requiring any modifications to the backbone model architecture.

The approach is outlined in Algorithm 1. At a given layer of a DNN, HAN leverages the embedding of the training data as the support for a KNN classifier, allowing to make predictions for new examples. In practice, given a set of training examples X with their corresponding labels y, and a deep neural network D with L layers, embeddings are computed for a specified layer $l \leq L$, producing a transformed set of examples X'. This transformation step allows to exploit discriminative properties extracted by the DNN's layers. Similarly, the test example X_{n+1} is converted into its embedding X'_{n+1} using the same transformation function. The k nearest neighbors of the test instance in the transformed space are identified, based on a specified distance metric (δ). From these examples, denoted in the algorithm as $X'(1), \ldots, X'(k)$, where each (i) corresponds to the index of the example ranked i, the resulting example attribution for the test instance is derived. Specifically, the attribution is represented as $N = (X_{(1)}, y_{(1)}), \ldots, (X_{(k)}, y_{(k)})$. Additionally, pixel attributions are derived for the identified examples, providing pixel-level explanation of the prediction. Finally, the predicted label for the test instance is determined by the mode of the labels from the retrieved examples (i.e., the most represented label). The algorithm implicitly assumes equal weighting of the k returned examples, with each weight set to $1/k$, although alternative weighting schemes are possible.

Algorithm 1. HAN

Require:

$X = \{X_1, \ldots, X_n\}$ ▷ objects
$y = \{y_1, \ldots, y_n\}$ ▷ labels
D ▷ deep neural network
$0 < k \leq n$ ▷ no. of examples
$0 < l \leq L$ ▷ layer no.
δ ▷ distance metric (e.g.,cosine)
S ▷ pixel attribution method (e.g., Grad-CAM)
X_{n+1} ▷ test object

$X' \leftarrow \texttt{Embed}(X, D, l)$ ▷ output of the lth layer of D
$X'_{n+1} \leftarrow \texttt{Embed}(X_{n+1}, D, l)$
$\{X'_{(1)}, \ldots, X'_{(k)}\} \leftarrow \texttt{KNN}(X'_{n+1}, X', k, \delta)$
$N \leftarrow \{(X_{(1)}, y_{(1)}), \ldots, (X_{(k)}, y_{(k)})\}$
$\hat{y} \leftarrow \texttt{Mode}(\{y_{(1)}, \ldots, y_{(k)}\})$
`PixelAttribution` $\leftarrow$ `GeneratePixelAttribution`(N, D, S) ▷ Generate pixel attribution for example attributions
return $\hat{y}, N,$ `PixelAttribution`

---

4 Empirical Investigation

In this section, we investigate the hybrid example and pixel attributions generated by HAN and evaluate its predictive performance on two image classification tasks. The main goal is to demonstrate that HAN effectively generates example and pixel attributions for test images, without deteriorating the accuracy of the underlying DNN.

4.1 Experimental Setup

We evaluate HAN using the STL-10 [8] and MNIST [26] datasets. STL-10 is a diverse dataset that consists of 10 classes of 96×96 pixel color images. These classes include animals (e.g., cats, dogs, horses) and vehicles (e.g., cars, trucks, airplanes). The dataset contains 5 000 training images and 8 000 test images, along with 100 000 unlabeled images for unsupervised learning. In our work, we use the labeled images only for the supervised classification tasks.

ResNet18 [19] is chosen as the backbone model for HAN due to its demonstrated effectiveness and efficiency in image processing tasks. ResNet18 consists of 4 blocks, each of which consists of 4 convolutional layers, which are sequentially numbered from shallow to deep layers for reference. In this experiment, a pretrained ResNet18 model from PyTorch [34] is used. The model's weights were pretrained on the ImageNet-1K dataset [9], which has 100 classes and images with a resolution of 224 × 224 pixels. Unlike our previous work, EBE-DNN, we uniformly resize the input images of STL-10 and MNIST to 224×224 to preserve the original ResNet18 architecture as much as possible. For the MNIST fine-tuning task, the first convolutional layer is modified to support single-channel grayscale inputs by changing the number of input channels from 3 to 1, while keeping the kernel size, stride, and padding unchanged. Additionally, the final fully connected layer is modified to match the number of classes of both datasets. For optimization, we use Adam with a learning rate of 1×10^{-4}. A StepLR scheduler with a step size of 10 and a decay factor of 0.1 is applied to adjust the learning rate during fine-tuning, and early stopping is employed to prevent overfitting. After the model is fine-tuned, its weights are frozen to maintain consistency during embedding extraction.

To generate example attributions, embeddings are extracted from each convolutional layer of ResNet18 for both training and test images. These embeddings capture the semantic features learned by ResNet18 and vary across layers due to the network's hierarchical feature abstraction. To address the issue of excessively high-dimensional embeddings flattened from feature maps, particularly in shallow layers, we apply adaptive average pooling with a target output size of 4 × 4 to reduce the spatial dimensions of feature map from each convolutional layer. This pooling operation significantly reduces the embedding length while maintaining a coarse spatial layout in deeper layers. Moreover, it refines the embeddings extracted from shallow layers by suppressing redundant local details, leading to the improved `KNN` accuracy in early layers. We will discuss the accuracy comparison of HAN with and without average pooling in Sect. 4.3.

To investigate the impact of the layer selection, we apply HAN across all layers, i.e., $l \in \{1, \ldots, 16\}$. The algorithm uses the faiss-gpu [21] implementation of `KNN`, with cosine similarity as the distance metric. This metric is chosen as it is well-suited for high-dimensional spaces and is resistant to magnitude scaling caused by operations such as convolution, pooling, and batch normalization.

Once the example attributions are generated, we employ Grad-CAM to form the pixel attribution through the PyTorch library for CAM methods [15]. Specifically, Grad-CAM is applied to test images and the corresponding example attributions, with the target layer being the convolutional layer that generated the embeddings used in the `KNN`. Pixels are then highlighted with an intensity that indicates their contribution to the classification, computed as the gradient of the classification loss w.r.t. each pixel value. In addition, we overlay contour lines directly on the example attributions generated from the last convolutional layers. The contours are derived from the normalized Grad-CAM saliency values. Specifically, we flatten and sort the saliency map in descending order, then compute the cumulative sum until a predefined percentage (e.g. 60%) of the total saliency is reached. Therefore, the derived contours indicate how the model's attention is distributed. Smaller contours highlight focused and salient regions, while larger ones capture areas with less dense information. This method progressively highlights salient regions, offering a clearer visualization of model attention, thereby providing complementary information to the example attributions.

4.2 Illustration of Hybrid Attributions

In this section, we provide illustrations of hybrid attributions, including both example attributions and their pixel attributions, as generated by HAN for MNIST and STL-10 datasets; for the former we consider a particular handwritten digit (*2*) and for the latter an image of a car. The illustrations show how the proposed approach identifies training examples and key pixels that contribute to the black-box model predictions. For each test image, we generate a hybrid attribution leveraging $k = 10$ neighbors obtained from `KNN`. Since ResNet18 naturally consists of 4 blocks and the last layer of each block generates the block's final feature output, we present the hybrid attributions of two test images utilizing embeddings generated from these last layers. Hence, we investigate not only the common properties indicated from hybrid attributions within a single convolutional layer, but also explore how they vary from shallow layers to deep layers of ResNet18.

In Fig. 2, the hybrid attributions of the two selected test images are shown, where the test images are displayed in the first columns, and where the following ten columns display the nearest training images. The first four rows display nearest neighbors based on different embeddings (layer 4, 8, 12 and 16), with pixel attributions (heatmaps) generated by Grad-CAM. The last row displays the nearest neighbors using the last embedding, but with contour plots instead of heatmaps.

In general, Grad-CAM heatmaps exhibit evolving patterns across the convolutional layers, reflecting the hierarchical nature of feature representations in ResNet18. In shallow layers, such as layer 4 and layer 8, the overall gradient magnitudes are relatively small, primarily due to the long backpropagation path from the output logits to the shallow layers, which leads to gradient diminishing. Meanwhile, the sparse regions with salient gradients mostly appear along the object boundaries, reflecting the fact that early convolutional layers primarily capture low-level visual features such as edges and textures. As the depth increases, salient regions increasingly align with structurally meaningful parts of the object. By the final convolutional layer (16), the model relies on high-level semantic features that are strongly associated with the target class, indicating a shift from low-level features to high-level semantic class-specific representations.

As for specific classification task, here represented by the MNIST handwritten digit *2*, we note that due to the simplicity and structural consistency of the handwritten digit dataset, all retrieved examples indicate visual similarity to the test image. Nevertheless, we can still observe a clear progression in feature abstraction across layers. At layer 4 and layer 8, retrieved examples align closely with the test image in terms of global stroke shape and orientation, suggesting that early layers emphasize low-level structural similarity. At intermediate depth (e.g., layer 12), the examples show more variation in local stroke patterns while maintaining class-consistent features, indicating a transition toward localized

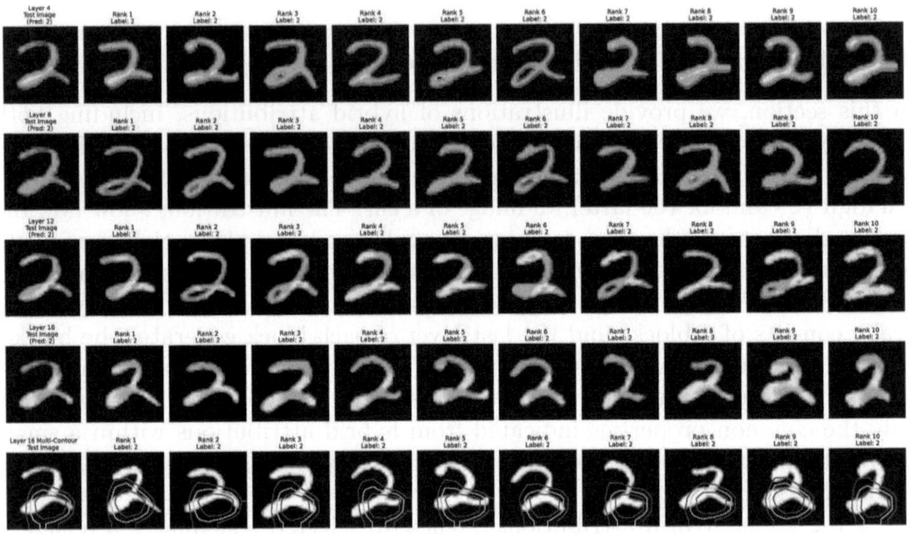

(a) The pixel attribution of a handwritten digit *2* from MNIST dataset

(b) The pixel attribution of a car from STL-10 dataset

Fig. 2. Example attributions are retrieved using KNN based on embeddings from convolutional layers 4, 8, 12, and 16. The final contour map overlays red, green, and blue contours for regions capturing the top 30%, 50%, and 70% of cumulative Grad-CAM saliency, respectively.

semantic representations. At the deepest layer (layer 16), the retrieved images are visually more diverse but consistently preserve key semantic properties, such as the lower curve or closed loop, showing that the model captures class-level properties beyond pixel alignment.

For the STL-10 car classification task, the increased visual complexity of the dataset leads to the inclusion of objects from irrelevant classes (e.g., ships, airplanes) in the example attribution using layer 4. These retrieved images tend to share low-level features with the test image, such as bright or uniform backgrounds. In addition, some of these samples exhibit almost no gradient activation in the Grad-CAM maps, suggesting that the embeddings extracted at layer 4 contribute very little to the final classification. As the network goes deeper, the gradient magnitudes begin to increase. At layer 8, sparse salient regions start to emerge on the main object, indicating that the model begins to focus on the object itself. Some key visual features associated with cars, such as windshields, wheel rims, and tires, become visibly highlighted in the attribution maps. While salient regions in layer 8 begin to focus on the car, especially around structural parts such as wheels and windows, pixel attributions in layer 12 unexpectedly highlight the upper background areas in many images. We suspect that in the training data, cars frequently appear beneath open sky, and this recurring co-occurrence may have led the model at this stage to associate background regions with the car class. Despite this, some saliency remains on semantically meaningful properties, such as the edge of the tire, which still contribute to the model's prediction. In the final layer, the saliency maps show a clear focus on the tires and hubs of the vehicle. Despite the presence of a prominent sky background in the top-ranked example, its contribution to the model's decision appears minimal. The shift of saliency from background in Rank 7 (layer 12) to the tires in Rank 2 (layer 16), despite both images being the same, further illustrates this progression.

Through these observations of hybrid attributions generated by HAN, we arrive at a conclusion consistent with [44]: deeper layers generally produce more discriminative features. Moreover, despite the high diversity of examples retrieved through example attribution, the saliency maps consistently concentrate on the main part of the image, especially in deeper layers. This qualitative consistency indicates that the explanations provided by our HAN framework are reliable and coherent. As a result, this affinity demonstrates that the HAN method through the hybrid attributions offers some insights into ResNet18's classification mechanism and that it provides a more transparent view of the decision-making process, than when relying only on example- or pixel-based explanations.

4.3 Predictive Performance

The accuracy comparison between HAN with and without 4×4 average pooling is presented in Fig. 3. With pooling, the classification accuracy is notably higher in the shallow layers, likely because average pooling reduces local texture fluctuations and aggregates activations into more stable and compact representations,

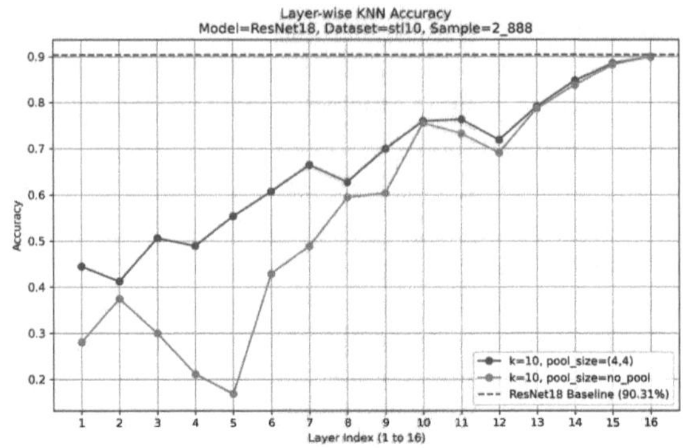

Fig. 3. Comparison of $k = 10$ KNN classification accuracy on STL-10 using convolutional embeddings with and without 4×4 average pooling. Pooling significantly reduces the length of embeddings flattened from feature map while improving the accuracy.

which are suited for distance-based classification. However, in deeper layers, both configurations converge to similar predictive performance, suggesting that high-level semantic features are sufficiently captured regardless of pooling. Additionally, our experiments on both STL-10 and MNIST demonstrate that classification accuracy remains similar across different values of k (i.e., $k = 5$, 10, 15, 20), especially in deeper layers. Therefore, we select $k = 10$ in our work since it provides the best balance between predictive performance and interpretability, where interpretability is measured by the number of training examples included in the explanations.

5 Concluding Remarks

A novel approach to obtaining combined example and pixel attributions from deep neural networks, called HAN, has been introduced. HAN leverages DNNs to extract hierarchical embeddings and employs KNN to retrieve concentrated example attributions. Using these retrieved examples, pixel attributions are further derived to highlight salient regions of them, such as contours with different levels, that contribute to the prediction.

Our investigation demonstrates the progression of properties captured across ResNet18 layers. As shown in Fig. 3, the KNN accuracy per layer increases as the layer becomes deeper, especially when pooling is applied. This indicates that deeper layers produce more discriminative embeddings between classes than shallow layers. A consistent pattern can be found in Fig. 2, especially in the more complex STL-10 dataset illustrated in Fig. 2b. In shallow layers, example attributions sometimes include images from different classes than the test image. As the layer depth increases, the example attributions become more concentrated and exhibits diversity within the correct class. This behavior aligns with the KNN accuracy result presented in Fig. 3. Moreover, in shallow layers, pixel attributions derived from example attributions focus on low-level properties, such as background color and texture, which contribute less to the prediction than the deeper layers. In contrast, pixel attributions from deeper layers highlight the structural and semantic properties of the object. Stronger gradient responses in these regions indicate a higher contribution to the model's decision.

In terms of predictive performance, HAN achieves accuracy comparable to ResNet18 on the STL-10 and MNIST datasets, with the final layer achieving the best results. We compare the model's behavior with and without feature map pooling. Applying pooling leads to more stable performance across layers and improves interpretability, by reducing the spatial noise in the extracted embeddings.

Although HAN is effective, there are still opportunities for further improvement. One area for future research is the quantitative evaluation of HAN, which may involve metrics, such as explanation fidelity, consistency and semantic alignment between the hybrid attribution and the model's prediction. Additionally, future work could explore feature-level explanations based on pixel-to-cosine similarity with example embeddings, offering an alternative to traditional pixel-to-class saliency maps for interpreting instance-level similarity. Other directions concern investigating alternative distance metrics and forming predictions by weighted voting, e.g., by leveraging distances. Finally, further investigation is needed to determine if the HAN framework is equally effective in other domains, such as classification of text or tabular data, thus expanding its applicability across various machine learning tasks.

Acknowledgments. This work was partially supported by the Wallenberg AI, Autonomous Systems and Software Program (WASP) funded by the Knut and Alice Wallenberg Foundation.

Competing Interests. The authors declare that they have no competing interests.

References

1. Adadi, A., Berrada, M.: Peeking inside the black-box: a survey on explainable artificial intelligence (XAI). IEEE Access **6**, 52138–52160 (2018)
2. Atienza, N., et al.: Cutting the black box: Conceptual interpretation of a deep neural net with multi-modal embeddings and multi-criteria decision aid. In: Larson, K. (ed.) Proceedings of the Thirty-Third International Joint Conference on Artificial Intelligence, IJCAI-24, pp. 3669–3678. International Joint Conferences on Artificial Intelligence Organization (2024). https://doi.org/10.24963/ijcai.2024/406
3. Bach, S., Binder, A., Montavon, G., Klauschen, F., Müller, K.R., Samek, W.: On pixel-wise explanations for non-linear classifier decisions by layer-wise relevance propagation. PLoS ONE **10**(7), e0130140 (2015)
4. Boström, H.: Example-based explanations of random forest predictions. In: International Symposium on Intelligent Data Analysis, pp. 185–196. Springer (2024)
5. Bresson, R., Cohen, J., Hüllermeier, E., Labreuche, C., Sebag, M.: Neural representation and learning of hierarchical 2-additive Choquet integrals. In: Bessiere, C. (ed.) Proceedings of the Twenty-Ninth International Joint Conference on Artificial Intelligence, IJCAI-20, pp. 1984–1991. International Joint Conferences on

Artificial Intelligence Organization (2020). https://doi.org/10.24963/ijcai.2020/275, https://doi.org/10.24963/ijcai.2020/275, main track
6. Chandrasekaran, A., Prabhu, V., Yadav, D., Chattopadhyay, P., Parikh, D.: Do explanations make VQA models more predictable to a human? arXiv preprint arXiv:1810.12366 (2018)
7. Chen, C., Li, O., Tao, D., Barnett, A., Rudin, C., Su, J.K.: This looks like that: deep learning for interpretable image recognition. In: Advances in Neural Information Processing Systems, vol. 32 (2019)
8. Coates, A., Ng, A., Lee, H.: An analysis of single-layer networks in unsupervised feature learning. In: Proceedings of the Fourteenth International Conference on Artificial Intelligence and Statistics, pp. 215–223. JMLR Workshop and Conference Proceedings (2011)
9. Deng, J., Dong, W., Socher, R., Li, L.J., Li, K., Fei-Fei, L.: ImageNet: a large-scale hierarchical image database. In: 2009 IEEE Conference on Computer Vision and Pattern Recognition. pp. 248–255. IEEE (2009)
10. Dong, G., Boström, H., Vazirgiannis, M., Bresson, R.: Obtaining example-based explanations from deep neural networks. In: International Symposium on Intelligent Data Analysis. Springer (2025). Accepted, to appear
11. Dong, G., Boström, H., Vazirgiannis, M., Bresson, R.: Obtaining example-based explanations from deep neural networks (2025). https://arxiv.org/abs/2502.19768. Accepted at International Symposium on Intelligent Data Analysis 2025, to appear
12. Eberle, O., Büttner, J., Kräutli, F., Müller, K.R., Valleriani, M., Montavon, G.: Building and interpreting deep similarity models. IEEE Trans. Pattern Anal. Mach. Intell. **44**(3), 1149–1161 (2020)
13. Gautam, S., Boubekki, A., Höhne, M., Kampffmeyer, M.C.: Prototypical self-explainable models without re-training. arXiv preprint arXiv:2312.07822 (2023)
14. Gautam, S., Höhne, M.M.C., Hansen, S., Jenssen, R., Kampffmeyer, M.: This looks more like that: enhancing self-explaining models by prototypical relevance propagation. Pattern Recogn. **136**, 109172 (2023)
15. Gildenblat, J.: Pytorch library for cam methods (2021). https://github.com/jacobgil/pytorch-grad-cam
16. Guidotti, R.: Counterfactual explanations and how to find them: literature review and benchmarking. Data Min. Knowl. Disc. **38**(5), 2770–2824 (2024)
17. Gunning, D., Aha, D.: DARPA's explainable artificial intelligence (XAI) program. AI Mag. **40**(2), 44–58 (2019)
18. Hamilton, M., Lundberg, S., Zhang, L., Fu, S., Freeman, W.T.: Axiomatic explanations for visual search, retrieval, and similarity learning. arXiv preprint arXiv:2103.00370 (2021)
19. He, K., Zhang, X., Ren, S., Sun, J.: Deep residual learning for image recognition. In: Proceedings of the IEEE Conference on Computer Vision and Pattern Recognition, pp. 770–778 (2016)
20. Honda, A., Itabashi, M., James, S.: A neural network based on the inclusionexclusion integral and its application to data analysis. Inf. Sci. **648**, 119549 (2023). https://doi.org/10.1016/j.ins.2023.119549, https://www.sciencedirect.com/science/article/pii/S0020025523011349
21. Johnson, J., Douze, M., Jégou, H.: Billion-scale similarity search with GPUs. IEEE Trans. Big Data **7**(3), 535–547 (2019)
22. Kim, B., Khanna, R., Koyejo, O.O.: Examples are not enough, learn to criticize! Criticism for interpretability. In: Advances in Neural Information Processing Systems, vol. 29 (2016)

23. Kim, B., et al.: Interpretability beyond feature attribution: quantitative testing with concept activation vectors (TCAV). In: International Conference on Machine Learning, pp. 2668–2677. PMLR (2018)
24. Kim, E., Kim, S., Seo, M., Yoon, S.: XProtoNet: diagnosis in chest radiography with global and local explanations. In: Proceedings of the IEEE/CVF Conference on Computer Vision and Pattern Recognition, pp. 15719–15728 (2021)
25. Krishna, S., Han, T., Gu, A., Wu, S., Jabbari, S., Lakkaraju, H.: The disagreement problem in explainable machine learning: a practitioner's perspective. arXiv preprint arXiv:2202.01602 (2022)
26. LeCun, Y.: The MNIST database of handwritten digits (1998). http://yannlecun.com/exdb/mnist/
27. Liang, P.P., et al.: High-modality multimodal transformer: quantifying modality & interaction heterogeneity for high-modality representation learning (2023). https://arxiv.org/abs/2203.01311
28. Liu, Y., Tuytelaars, T.: A deep multi-modal explanation model for zero-shot learning. IEEE Trans. Image Process. **29**, 4788–4803 (2020)
29. Lundberg, S.: A unified approach to interpreting model predictions. arXiv preprint arXiv:1705.07874 (2017)
30. Lyu, Y., Liang, P.P., Deng, Z., Salakhutdinov, R., Morency, L.P.: DIME: fine-grained interpretations of multimodal models via disentangled local explanations (2022). https://arxiv.org/abs/2203.02013
31. Molnar, C.: Interpretable machine learning (2020). Lulu.com
32. Muhammad, M.B., Yeasin, M.: Eigen-CAM: class activation map using principal components. In: 2020 International Joint Conference on Neural Networks (IJCNN), pp. 1–7. IEEE (2020)
33. Park, D.H., et al.: Multimodal explanations: justifying decisions and pointing to the evidence. In: Proceedings of the IEEE Conference on Computer Vision and Pattern Recognition, pp. 8779–8788 (2018)
34. Paszke, A., et al.: PyTorch: an imperative style, high-performance deep learning library. In: Advances in Neural Information Processing Systems, vol. 32 (2019)
35. Patro, B., Namboodiri, V.P.: Differential attention for visual question answering. In: Proceedings of the IEEE Conference on Computer Vision and Pattern Recognition, pp. 7680–7688 (2018)
36. Plummer, B.A., Vasileva, M.I., Petsiuk, V., Saenko, K., Forsyth, D.: Why do these match? Explaining the behavior of image similarity models. In: Vedaldi, A., Bischof, H., Brox, T., Frahm, J.-M. (eds.) ECCV 2020, Part XI. LNCS, vol. 12356, pp. 652–669. Springer, Cham (2020). https://doi.org/10.1007/978-3-030-58621-8_38
37. Poché, A., Hervier, L., Bakkay, M.C.: Natural example-based explainability: a survey. In: Longo, L. (ed.) xAI 2023. CCIS, vol. 1902, pp. 24–47. Springer, Cham (2023). https://doi.org/10.1007/978-3-031-44067-0_2
38. Rao, V.N., Zhen, X., Hovsepian, K., Shen, M.: A first look: towards explainable TextVQA models via visual and textual explanations (2021). https://arxiv.org/abs/2105.02626
39. Ribeiro, M.T., Singh, S., Guestrin, C.: "Why should i trust you?" Explaining the predictions of any classifier. In: Proceedings of the 22nd ACM SIGKDD International Conference on Knowledge Discovery and Data Mining, pp. 1135–1144 (2016)
40. Rodis, N., Sardianos, C., Radoglou-Grammatikis, P., Sarigiannidis, P., Varlamis, I., Papadopoulos, G.T.: Multimodal explainable artificial intelligence: a comprehensive review of methodological advances and future research directions. IEEE Access (2024)

41. Selvaraju, R.R., Cogswell, M., Das, A., Vedantam, R., Parikh, D., Batra, D.: Grad-CAM: visual explanations from deep networks via gradient-based localization. Int. J. Comput. Vision **128**, 336–359 (2020)
42. Simonyan, K., Vedaldi, A., Zisserman, A.: Deep inside convolutional networks: visualising image classification models and saliency maps. arXiv preprint arXiv:1312.6034 (2013)
43. Smilkov, D., Thorat, N., Kim, B., Viégas, F., Wattenberg, M.: SmoothGrad: removing noise by adding noise. arXiv preprint arXiv:1706.03825 (2017)
44. Zeiler, M.: Visualizing and understanding convolutional networks. In: European Conference on Computer Vision/arXiv, vol. 1311 (2014)

Extending Model-Agnostic XAI Methods for Regression Tasks in Spatio-Temporal Domains

Matteo Salis[1,3](✉), Gabriele Sartor[1], Marco Pellegrino[1], Rosa Meo[1], Stefano Ferraris[2], and Abdourrahmane Atto[3]

[1] Computer Science Department, University of Turin, Turin, Italy
{matteo.salis,gabriele.sartor,marco.pellegrino,rosa.meo}@unito.it
[2] Interuniversity Department of Regional and Urban Studies and Planning, University of Turin, Turin, Italy
stefano.ferraris@unito.it
[3] LISTIC Laboratory, Université Savoie Mont Blanc, Annecy, France
abdourrahmane.atto@univ-smb.fr

Abstract. Deep Learning (DL) models have become a game-changer approach for spatio-temporal modeling due to their ability to capture complex relations within data. However, their black-box nature hinders their widespread adoption in critical applications (e.g. environmental monitoring, natural hazard risk assessment, resource management, etc.). Explainable Artificial Intelligence (XAI) techniques have emerged to address this challenge, providing insights into model decisions and the learned influence of input features on predictions. Model-agnostic and perturbation-based approaches, which are particularly valuable due to their flexibility and independence from specific model architectures, are widely used for interpreting vector-based and image-based models. Among these, Local Interpretable Model-agnostic Explanations (LIME), SHapley Additive exPlanations (SHAP), and Randomized Input Sampling for Explanation (RISE) are three prominent methods. In this work, we extend and s RISE with adaptations of LIME, and SHAP for spatio-temporal deep learning models, enabling their application to geospatial image time series producing spatial, temporal, and spatio-temporal explanations. To assess the effectiveness of our approach, we apply the adapted methods to a real-world environmental case study: groundwater level prediction. Our findings demonstrate that RISE-based explanations provide different insights from LIME and SHAP ones, focusing on different aspects of the model behavior. Finally, we discuss the usefulness of producing a spatial, temporal, and spatio-temporal explanation. This comparative analysis highlights the strengths and limitations of each method in explaining deep learning models for dynamic environmental systems.

Keywords: spatio-temporal · XAI · regression task

S. Džeroski et al. (Eds.): DS 2025, LNAI 16090, pp. 301–315, 2025.
https://doi.org/10.1007/978-3-032-05461-6_20

1 Introduction

Over the years, Deep Learning (DL) has proven to be a game-changer across a wide range of applications. Geoscience is one of the most affected fields by this technology, showcasing models with remarkable performance and diverse use cases [4], including semantic segmentation [3,22], object detection [25], extreme weather prediction [16], and more. Although advanced DL models have proven highly effective, they lack interpretable representations to explain their predictions—a key requirement for critical applications. For this reason, the field of the eXplainable Artificial Intelligence (XAI) has recently gained increasing importance [7], for its goal of clarifying the decision-making process and making models compliant with the likely future introduction of safety regulations on AI usage [18].

Among these, model-agnostic XAI methods have been extensively used to analyze any pretrained models as black boxes, study their behavior in making predictions on the dataset, and assess their responses to perturbed instances. These methods (e.g., RISE [15], LIME [17], etc.) have been primarily employed to study the behavior of models that process input vectors and images. Instead, when the input of the model is a video, or in general, a spatio-temporal data structure, these two representations can only be used to synthesize, respectively, the temporal and spatial dimensions as independent aspects. In a spatio-temporal setting, however, a joint examination of both dimensions can provide more accurate explanations of complex behaviors that evolve over space and time. This is particularly relevant in geoscience, which involves numerous applications that rely on image time series [5,13]. Indeed, environmental phenomena occurring at a specific time and location can significantly influence other regions in the future, making it challenging to analyze these interactions when considering spatial and temporal dimensions separately.

For this reason, in this work we study a spatio-temporal extension of existing post-hoc model-agnostic methods (i.e., RISE, LIME, and SHAP) to create interpretable representations giving important intuitions on the causal influences of specific events over time and space jointly. Our focus is directed to the attribution task for regression models dealing with spatio-temporal data structure in the form of videos. Given the importance of creating a representation preserving a physical meaning with a specific unit of measurement [8], misplaced in classification, we concentrate on regression models. The selected methods are based on the perturbation of the model's input, and perturbations can be performed in different ways, producing a different explanation [12]. Defining a perturbation is not a trivial task, especially if data has a physical meaning, as it happens in geoscience. Consequently, to provide a spatio-temporal extension, it is necessary to formalize a process of perturbation and saliency generation, preserving the physical characteristics.

Precisely, the main contributions presented in this work are the following:

1. Formalization and implementation of a spatio-temporal extension for RISE [15] for regression.

2. Analysis of the benefits in terms of explainability of the proposed spatio-temporal extension considering an environmental domain (i.e., the groundwater level prediction).
3. A comparison with other XAI established algorithms, namely LIME [17] and SHAP [10].

In the paper, we first analyze previous works on videos (Sect. 2), introduce the original methods and formalize their extension (Sect. 3), compare spatio-temporal approach with the temporal and spatial one discussing the results (Sect. 4).

2 Related Work

Many XAI methods have been proposed for video-related tasks (video-XAI), in which data, of course, exhibit spatio-temporal relations. For example, algorithms have been proposed to find the most relevant regions to produce a model prediction over the input's temporal and spatial dimensions [9,21]. However, the features of video data are usually RGB channels, which provide the most information when exploited together. Differently, in geoscience, input features could be of very different types and provide distinct information (e.g. weather, satellite image, terrain property, etc.). Thus, depending on the application, salient spatial, temporal, and spatio-temporal regions may be produced for any single feature or for some meaningful group of them. Other inspiring works of video-XAI embodied the optical-flow [14] estimation in their workflow with remarkable results [23] concerning video classification.

However, given the peculiarity of spatio-temporal data, these methods are not easily adaptable. Firstly, because of the questionable usefulness of optical-flow of input features with different physics meanings. Furthermore, adapting a video-XAI algorithm originally proposed for classification to regression tasks is not trivial. The adaptation must be done carefully because the meaning of the produced explanation could change [8]. For example, in the common video classification regime, input and output are usually ratio data with a defined range $\mathbb{R}^3_{[0;255]}$ and $\mathbb{R}_{[0;1]}$ respectively. Instead, for regression, and especially in spatio-temporal domains, features could be either ratio or interval data with different ranges.

Consequently, taking care of the values and the form of the perturbation applied to the input by the XAI algorithms is pivotal. Thus, producing the most useful explanation requires domain knowledge, and it is task-specific [12]. In line with these considerations, recent works proposed and adopted a reformulation of RISE to obtain spatial explanations for their black box models applied to hydrological case studies [2,24]. In these applications, the random binary mask has been replaced by an additive unimodal Gaussian perturbation. The same approach has been adopted by Salis et al. [20], who produced the explanations also along the temporal and spatio-temporal dimensions. However, the original meaning of RISE, which was the *expected output value when the pixel is not perturbed* in [2,20,24] changed in *expected change when the pixel is perturbed*, which is still a proxy of relevance but it is a different definition. Furthermore, the adoption of unimodal Gaussian additive noise makes the masks very different from the original implementation, in which a sparse and smoothed noise is generated from a small binary mask and then upsampled to fit the input extent. This work aims to reformulate RISE while adhering as closely as possible to its original formulation to maintain the same interpretability. Furthermore, we compare it to LIME and SHAP, which can be more easily applied to derive spatial, temporal, and spatio-temporal explanations.

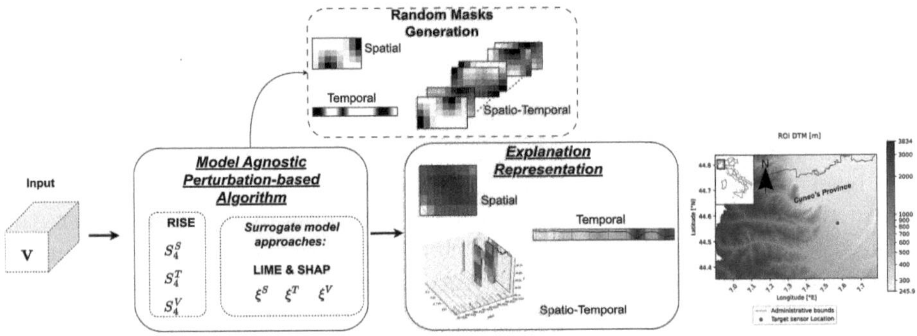

(a) (b)

Fig. 1. (a) Generic pipeline of our approach: given an input video V, random masks are generated based on the desired explanation representation (i.e., 1D for temporal, 2D for spatial, and 3D for spatio-temporal explanations). Depending on the applied method, masks are used to perturb the input of the original model to obtain a saliency representation for RISE, and a surrogate model with its learned parameters for LIME and SHAP. (b) depicts the elevation of the region of interest ROI, with the blue dot indicating the target sensor location.

3 Spatio-Temporal Extension for Model-Agnostic Methods

3.1 RISE

RISE has been proposed by Petsiuk et al. [15] to explain any black-box model f in a classification regime. In particular, RISE ranks the pixels in an input image I[1] to be the most relevant to predict a particular instance (i.e. local explanation). This is achieved firstly by perturbing the input with N random masks, whose generic value is $M_i(\lambda) \in \mathbb{R}_{[0;1]}$ with $i \in [1; N]$, and collecting the corresponding N predictions. Then, a saliency score $S(\lambda)$ for each pixel λ is defined as the expected value of the probability score conditioned by $M(\lambda) = 1$ (Eq. 1).

[1] Formally I is a mapping from the two-dimensional coordinates to the space of colour values (RGB).

$$S_{1(I,f)}(\lambda) = \mathbb{E}_N[f(I \odot M)|M(\lambda) = 1] \tag{1}$$

The underlying assumptions behind the original RISE are: **a)** the truth target class is known and the original output $f(I)$ predicts rightly that class with a high score probability, and **b)** if a pixel is relevant, then $f(I \odot M_i)$ is high even if other pixels have been perturbed, i.e., the model is still able to reproduce the original and truth prediction. In a regression task, the reference point by which evaluating the restoring ability of the model is no longer equal to 1, but it is a real value $y \in \mathbb{R}$. Thus, the original saliency $S_{1(I,f)}(\lambda)$ is no longer positively associated with the relevance of a pixel, but only to the conditional expectation of the output by randomly perturbing some pixels except λ. As a first attempt to recover pixel relevance, we propose to compute the conditional variance given $M(\lambda) = 1$ (Eq. 2). In this way, the saliency is related to the instability of the model by perturbing pixels other than λ, and thus, the lower the conditional variance, the higher the relevance.

$$S_{2(I,f)}(\lambda) = \mathbb{E}_N[(f(I \odot M) - S_{1(I,f)}(\lambda))^2|M(\lambda) = 1] \tag{2}$$

However, S_2 is still not anchored to the original prediction $f(I)$, and thus not related to the restoring ability of the model for a specific pixel. Therefore, we propose to use the conditional bias S_3 (Eq. 3) of the model prediction produced using the perturbed input against the original prediction $f(I)$. Furthermore, to take into account both the bias and the variance, we propose and adopt the conditional root mean squared error S_4 (Eq. 4) as the definitive metric to measure the relevance of pixels for a regression task. For both S_3 and S_4, the lower their values, the higher the relevance.

$$S_{3(I,f)}(\lambda) = \mathbb{E}_N[(f(I \odot M) - f(I))|M(\lambda) = 1] \tag{3}$$

$$S_{4(I,f)}(\lambda) = \sqrt{\mathbb{E}_N[(f(I \odot M) - f(I))^2|M(\lambda) = 1]} \tag{4}$$

Inspired by [20], we extend explanations in the case of a spatio-temporal input video $\mathbf{V} = \{V(h, w, c, \tau) \in \mathbb{R} \mid h \in [1, ..., \mathcal{H}], w \in [1, ..., \mathcal{W}], c \in [1, ..., C], \tau \in [1, ..., \mathcal{T}]\}$, where $\mathcal{H}, \mathcal{W}, \mathcal{C}, \mathcal{T}$ are respectively the height, width, channel and temporal extent, producing spatial, temporal, and spatio-temporal explanations. In the following, we focus on the case of computing explanations for a specific subset $\mathcal{C}$ of the input C features of particular interest for domain experts, adopting the notation $\mathbf{V}_{\mathcal{C}}$.

Spatial Explanation. In this case, the perturbation masks are generated in the 2-dimensional coordinate $(h,\ w)$ space. Then the perturbation is replicated for each frame (or time-step) over the temporal dimension and applied to each channel c of $\mathbf{V}_{\mathcal{C}}$. The resulting saliency map $S^S_{4(\mathbf{V}_{\mathcal{C}},f)}$ is a mapping from the two-dimensional spatial coordinates to $\mathbb{R}$, and for a pixel λ is the root mean squared error by perturbing other pixels for the entire temporal length $\mathcal{T}$.

Temporal Explanation. The same idea of saliency maps could be applied to the temporal dimension. Every random mask $M_{i\in[1;N]}$ is generated in the one-dimensional temporal space and then applied identically to every coordinate point and channel c of $\mathbf{V}_C$. A saliency vector $S^T_{4(\mathbf{V}_C,f)}$ is thus computed as a mapping from the one-dimensional temporal coordinate to $\mathbb{R}$, and for a frame τ is the root mean squared error by perturbing other frames identically over the whole spatial extents.

Spatio-Temporal Explanation. To produce a spatio-temporal explanation, every random mask $M_{i\in[1;N]}$ is generated in the three-dimensional spatio-temporal space and then directly applied to channel c of $\mathbf{V}_C$. In this way, we can compute a silence video $S^V_{4(\mathbf{V}_C,f)}$ as a mapping from the domain of $\mathbf{V}_C$ to $\mathbb{R}$, and for a generic element $V_C(h_\lambda, w_\lambda, c, \tau_\lambda)$, i.e. a pixel λ at frame τ_λ and channel c, it is the root mean squared error by perturbing other $V_C(h \neq h_\lambda, w \neq w_\lambda, c, \tau \neq \tau_\lambda)$ over the whole spatio-temporal extents.

3.2 LIME

LIME [17] takes a different approach. Firstly, given an input instance $x \in \mathbb{R}^d$, a fixed number d' of interpretable representations are extracted from x to represent the input's contiguous and human explicative partition. This could be achieved either by prior knowledge or by unsupervised methods [1]. In general, these representations depend on the specific case study and input data structure (e.g. superpixel for images, supervoxel for video, or more generally partition). Once the interpretable representations have been identified, it is possible to build a presence binary vector of them, named $x' \in \{0,1\}^{d'}$. At this point, N samples are taken from x' by drawing non-zero elements of x' uniformly at random, obtaining perturbed sample $z' \in \{0,1\}^{d'}$, and the corresponding $z \in \mathbb{R}^d$. Given the model f to be explained, a perturbed local dataset $\mathcal{Z}$ is created by considering z' as the input features and $f(z)$ as the target. Finally, an interpretable model g (usually a linear regression with regularization) is trained on the perturbed dataset, and the coefficients ξ of the presence binary vector z' provide the information about input partition relevance. Formally, the model $g(z') : \{0,1\}^{d'} \to \mathbb{R}$ is trained to minimize Eq. 5 which consist in a loss function (Eq. 6) weighted by $\pi_x(z) = e^{\frac{-||x-z||_2^2}{\sigma^2}}$, i.e. a proximity measure between an instance z to x, and a regularization term $\Omega(g)$.

$$\xi(x) = \underset{g\in G}{\arg\min}\ \mathcal{L}(f, g, \pi_x) + \alpha\ \Omega(g), \tag{5}$$

$$\mathcal{L}(f, g, \pi_x) = \sum_{z,z'\in\mathcal{Z}}^{N} \pi_x(z)\ (f(z) - g(z'))^2, \tag{6}$$

Therefore, the relevance of a partition $k \in [1, ..., d']$ is defined as the trained parameter ξ_k of an interpretable model g trained on a neighborhood of x. Thus, ξ_k represents the expected change of the target variable using the partition k

given the other fixed, and the higher the absolute value of ξ_k, the higher the relevance is. One relevant aspect of LIME is defining what using or not using a partition k means (i.e. turning it on or off). For general classification tasks with RGB images, this is often achieved by obscuring superpixels. Differently, in regression tasks, a value has to be set carefully, taking into account the meaning of that value for the feature of interest. Extending LIME to video or spatio-temporal is simpler than RISE. Indeed, it's a matter of defining input feature partitions [6]. Following our proposed approach for RISE, we produce spatial, temporal, and spatio-temporal explanations (ξ^S, ξ^T, ξ^V respectively) by defining spatial, temporal, and spatio-temporal partitions. As in RISE, we focus on a specific subset $\mathcal{C}$ of the C features. To define the spatial superpixels, we adopt SLIC[2] [1] applied to the longitude, latitude, and height of each pixel. Instead, for the temporal dimension, seasons have been used as the separation criterion. Finally, we define spatio-temporal partitions (i.e. supervoxel) as the spatial superpixel extended for each temporal partition.

3.3 Kernel-SHAP

SHAP values [11] is another model surrogate approach and it is based on the game theory grounded Shapley regression values, which assigns an importance value ξ_k to each feature k, of input instance x, that represents the effect on the model prediction of including that feature. It was shown that given a fairly limited set of axioms that an explanation should satisfy, known as Local accuracy ($\sum \xi_k = f(x)$), Missingness, Consistency, Shapley values satisfy all of them [8,11]. Furthermore, it has been demonstrated that using LIME without regularization and with $\pi_x = \frac{(d'-1)}{(d' \; choose \; |z'|)|z'|(d'-|z'|)}$, where $|z'| \in \{0, d'\}$ is the number of non-zero elements in z', leads to an equivalent approximation of the SHAP original definition (Theorem 2 in [11]). Given this connection, we produce spatial, temporal, and spatio-temporal explanations (ξ^S, ξ^T, ξ^V respectively) in the same way as in LIME by defining spatial, temporal, and spatio-temporal input partitions.

Figure 1a sums up the general pipeline to produce explanations using the three proposed approaches (RISE, LIME, and SHAP). We evaluate the methods by using insertion and deletion curves [15] reformulated in terms of mean squared error between the original unperturbed prediction and the one obtained by the perturbed input.

4 Experiments

In this section, we evaluate the spatio-temporal extension in RISE, LIME, and SHAP in the regression task, comparing them with their temporal and spatial versions in the geoscience application of the groundwater level predictions, a

[2] Simple Linear Iterative Clustering (SLIC), which adapts a k-means clustering approach to efficiently generate superpixel and supervoxel.

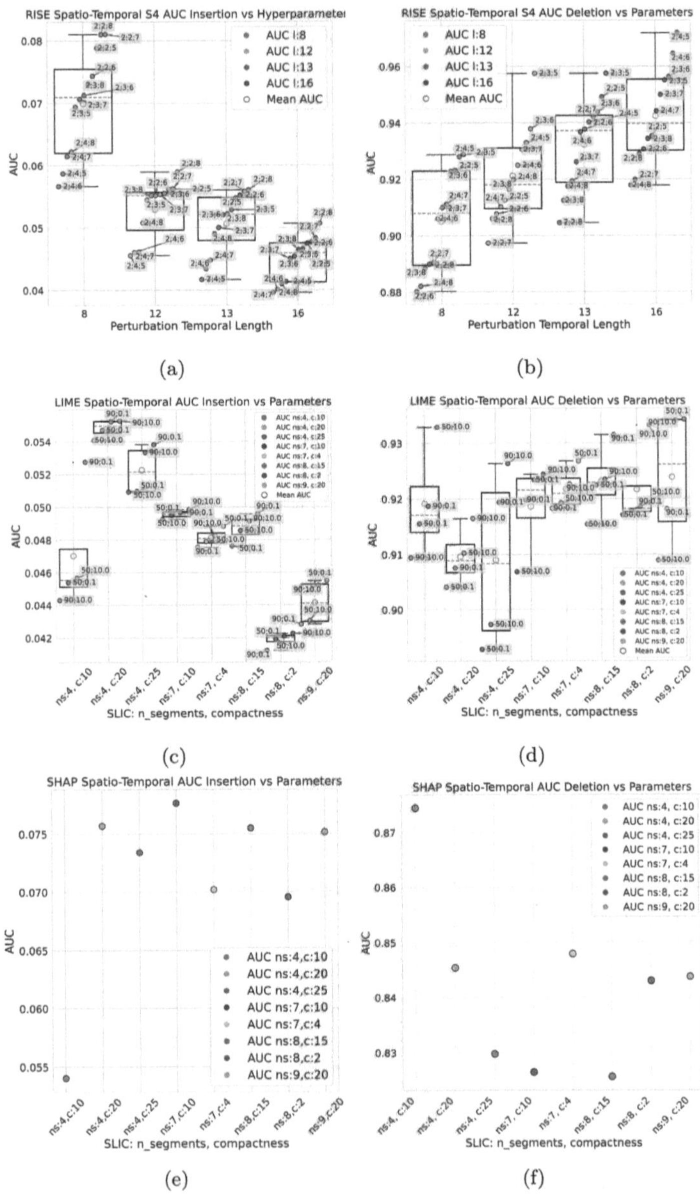

Fig. 2. Insertion and deletion AUC for hyperparameters combinations in the spatio-temporal setting. In (a) and (b), points are labelled by mask extensions and probability of activation (H_M, W_M, p), while the x-axis represents the temporal mask extension L_M. (c) and (d) points report (σ, α) while the x-axis, as in (e) and (f), depicts the number of partitions and compactness used in SLIC.

use case presenting significant spatio-temporal influences. For this purpose, we consider as our black-box the model proposed in [19], a CNN-LSTM taking in input a weekly image time series of weather conditions (i.e., precipitations, min and max temperatures) of two years (weekly 104 time-step) with 5×8 images of 12 km resolution, and predicting the weekly groundwater level in meters for a measurement sensor inside a Region of Interest (ROI) located in the north of Italy (Fig. 1b). We focus on finding explanations for the precipitation channel ($|\mathcal{C}| = 1$). Given that precipitation is a ratio variable, we use 0 (absence of) precipitation as the perturbation value for RISE, LIME, and SHAP. This will highlight differences between current and zero-precipitation scenarios.

4.1 Experimental Setup

The three methods require some hyperparameters. In RISE, the extension of the binary mask has to be chosen: the height H_M and width W_M for the spatial masks, the temporal length L_M for the temporal ones, and all H_M, W_M, and L_M for the spatio-temporal. Furthermore, the probability p of each element of the mask to be 1 or 0 should be tuned. For LIME and SHAP, the hyperparameters consist of the number of partitions to be used and the compactness in SLIC. Moreover, LIME also needs the regularization parameter α in Eq. 5, and the σ in π_x, here set to be a percentile of all the norms $||x - z||_2^2$. To tune these hyperparameters, we performed a grid search selecting the best parameters in terms of AUC of insertion and deletion curves. Figure 2 reports their comparison for the spatio-temporal setting, while the spatial and temporal analyses are omitted due to space constraints and redundancy of results. Since LIME and SHAP require the use of an explainable representation, we decided to vary their SLIC parameters for the spatial dimension, while setting the temporal one to the extent of the seasons to preserve the meaning for an environmental application. Then, the best hyperparameter combinations are adopted to compare RISE, LIME, and SHAP on their insertion and deletion curves in Fig. 3. Finally, Figs. 4, 5, and 6 depict the explainable representations of a dataset's instance for spatial, temporal, and spatio-temporal dimensions, respectively[3].

4.2 Experimental Results and Discussion

Groundwater levels, which measure the elevation of the groundwater body's surface, are highly associated with the elevation of the region under study. That's because water under the terrain flows from the mountain to the valley and then to the plain. From a spatial viewpoint, the explanation in Figs. 4a, 4b, and 4c, fulfils hydrologists' expectations that mountains and valleys are the major recharge locations for groundwater bodies. Looking a the temporal dimension (Figs. 5a, 5b, and 5c), all three methods depict the most relevant time step for the precipitation in predicting the target as the weeks just before the prediction and,

[3] Code and experiments are available at: https://github.com/Matteo-Salis/rise_video.

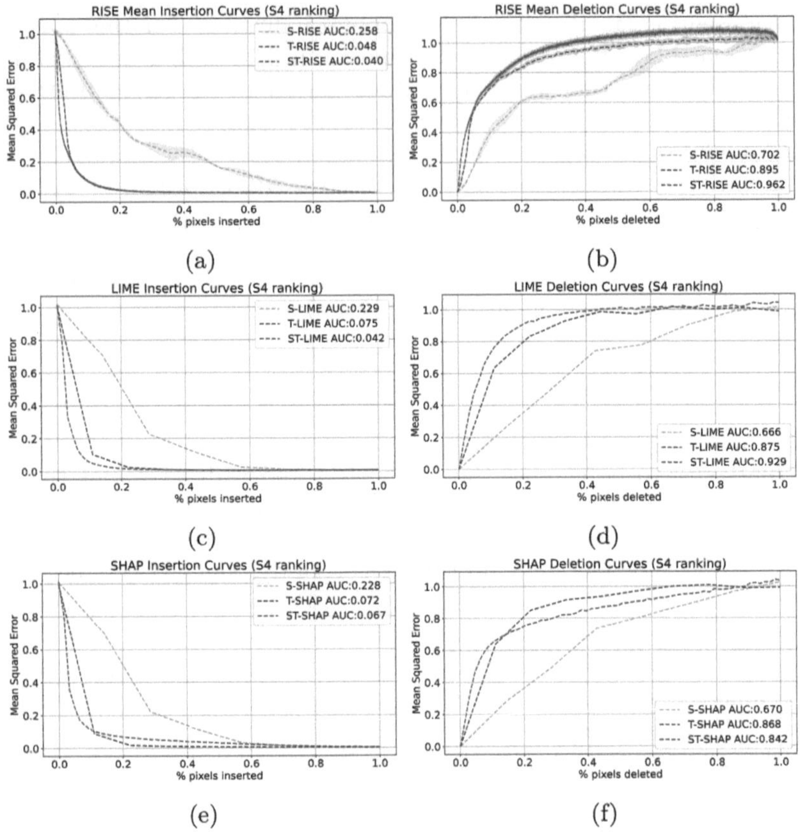

Fig. 3. Means of the insertion and deletion metrics over all the test set for spatio-temporal RISE (a, b), LIME (c, d) and SHAP (e, f) executed with the selected best parameters (10 trials). In the legend, the spatial, temporal, and spatio-temporal settings are identified, respectively, by the prefixes 'S', 'T', and 'ST'.

even more in the case of RISE, the weeks of the previous autumn. This is completely in line with the known long-term memory of the groundwater flow [2,24]. Figures 6a, 6b, and 6c represent the 5th percentile of the most relevant spatio-temporal explanations for the proposed approaches. While SHAP finds the same explanations in the space for the two most relevant time-steps, RISE and LIME detect a spatio-temporal pattern. Indeed, RISE detects the most relevant area in the autumn to be shifted to the north (darker colors) than the area in the spring, just before the prediction. A similar pattern is detected by LIME, in which the vertical supervoxel in autumn disappears in spring time-steps, and then the relevant area is only in the south. Even if the regional groundwater flow understanding involves uncertainties, this spatio-temporal behavior is sensible and supported by the ROI physics knowledge. Indeed, it is possible that the northern region, in which lies the highest mountain in the Cottian Alps (Monviso),

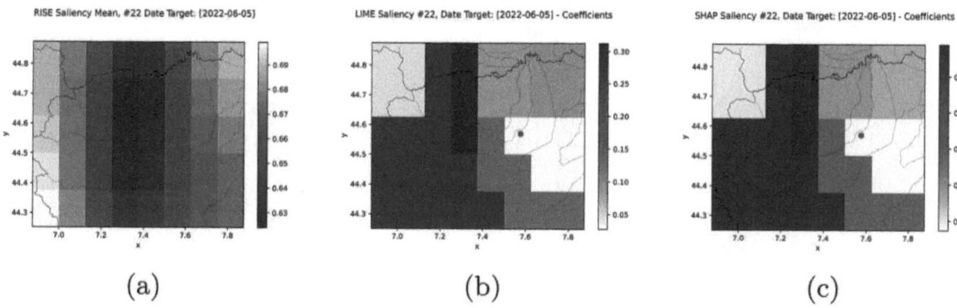

Fig. 4. Spatial explanation for the instance 22 (prediction for 2022-05-05) with RISE (a), LIME (b) and SHAP (c). The saliency is represented by darker red pixels, indicating the importance of each region within the ROI. In RISE, higher importance is given to lower values, while LIME and SHAP highlight higher importance in the regions with higher values (coefficients found by the surrogate model).

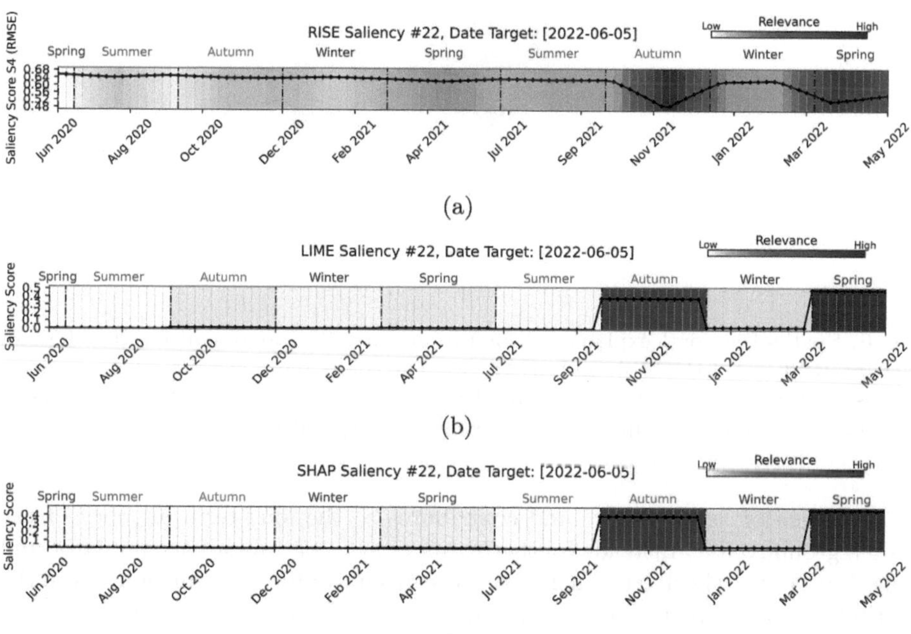

Fig. 5. Temporal explanation for the instance 22 (prediction for 2022-05-05), showing RISE (a), LIME (b), and SHAP (c). The y-axis represents the saliency values, while the x-axis displays the 104 time-steps, with labels indicating the corresponding month, year, and seasonal divisions of the considered temporal sequence. As with the spatial explanation, RISE assigns higher relevance to lower values, whereas LIME and SHAP highlight higher relevance in the regions with higher values.

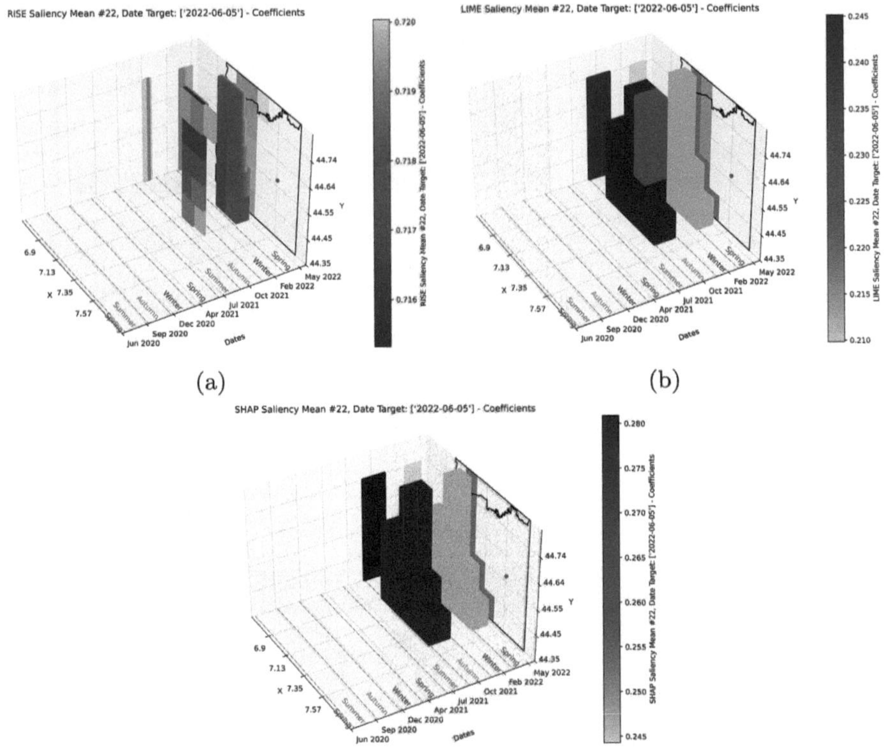

(a) (b)

(c)

Fig. 6. Spatio-Temporal explanation for the instance 22 (prediction for 2022-05-05), showing RISE (a), LIME (b), and SHAP (c). The colors represent the saliency values. Grey-scaled images projected on the front and left are the spatial and temporal mean saliencies respectively, computed on the whole saliency video.

acts as a major snow reserve and contributes to the long-term memory of the plain's groundwater. Instead, southern valleys act as a rainfall accumulator and contribute to the short-term memory of the groundwater phenomenon. From the spatio-temporal explanation, the usefulness of RISE approach emerged. Indeed, without any constraint imposed on the input partition (supervoxel), RISE can provide a more detailed explanation.

Going more into the details of spatio-temporal explanation interpretations, RISE provides, for each element $V_{\mathcal{C}}(h_\lambda, w_\lambda, c, \tau_\lambda)$, the RMSE between the original predictions made with the actual data and the ones obtained by randomly setting precipitations of other elements $V_{\mathcal{C}}(h \neq h_\lambda, w \neq w_\lambda, c, \tau \neq \tau_\lambda)$ to zero. Thus, the RISE's saliency is related to the model's resilience to input perturbations, and restoring ability. While LIME and SHAP for each partition k containing elements $V_{\mathcal{C}}(h, w, c, \tau) \in k$ provide a saliency score that is the expected change in groundwater level by using real precipitation data instead of zero precipitation in $\forall\ V_C \in k$, keeping other elements $V_C \notin k$ fixed. Therefore, their saliency is related to the contribution, in terms of groundwater level, that elements $V_C \in k$ are responsible for when the real data are used instead of zero precipitation. Even if RISE and LIME-SHAP relevance definitions differ, and sound complementary metrics, the actual produced explanations exhibit very similar patterns in space and time. We reserve future investigations on a formal connection between these methods.

Figure 2a depicts a clear positive trend for RISE between the temporal length and the AUC scores. Meaning that a sparser noise helps in capturing the true most relevant input elements. In LIME and SHAP, a slight negative association seems to be present between the number of partitions and the method's performance. This could be because inside a partition, every element is forced to have the same relevance, then with more partitions, a method could achieve a more precise explanation and relevance attribution.

Figure 3 illustrates the advantages of using a spatio-temporal representation in the insertion and deletion metrics. Notably, the AUC values for all methods in the spatio-temporal setting outperform those in other settings. However, it is interesting to observe that the temporal approach remains competitive with the spatio-temporal approach when comparing their AUC values alone. Nevertheless, the spatio-temporal representation provides additional insight by revealing where a significant event occurred in the past, as demonstrated in Figs. 5 and 6. When comparing the spatio-temporal results across different methods, RISE appears to outperform the others, achieving AUC values of 0.04/0.962 for insertion and deletion, respectively, while LIME and SHAP reached 0.042/0.929 and 0.067/0.842, respectively. This superior performance of RISE is likely due to its more precise localization of important regions. In contrast, while superpixels in LIME and SHAP can be beneficial for scaling in high-dimensional spaces, they may not always align well with the physical characteristics of the domain.

5 Conclusion

We presented an extension of RISE to spatio-temporal applications, comparing it with adapted versions of LIME and SHAP algorithms. We analyzed and proved the usefulness of producing spatio-temporal explanations rather than only spatial and temporal ones. Even if RISE, LIME, and SHAP-based explanations differ in the interpretations of scalar values, they found very similar spatial, temporal, and spatio-temporal patterns. The advantage of LIME and SHAP consists in their scaling ability, thanks to the use of input partitions (e.g. superpixels or supervoxels). However, this advantage becomes a disadvantage in the case of missing or incomplete knowledge to leverage for the creation of input partitions. RISE is then able to produce higher-resolution explanations without prior domain information. In general, all three methods provided insights about physics phenomena and thus are potentially helpful not only to improve the trustworthiness of deep learning models but also to discover new spatio-temporal patterns in geospatial applications. In future works, we will analyze in detail the connection between RISE and LIME-SHAP, especially focusing on the impact of different supervoxel generation methods. Furthermore, we will compare these methods on other spatio-temporal use cases.

Acknowledgments. We acknowledge ISCRA for awarding this project access to the LEONARDO supercomputer, owned by the EuroHPC Joint Undertaking, hosted by CINECA (Italy).

Disclosure of Interests. The authors have no competing interests to declare that are relevant to the content of this article.

References

1. Achanta, R., Shaji, A., Smith, K., Lucchi, A., Fua, P., Süsstrunk, S.: SLIC superpixels compared to state-of-the-art superpixel methods. IEEE Trans. Pattern Anal. Mach. Intell. **34**(11), 2274–2282 (2012). https://doi.org/10.1109/TPAMI.2012.120
2. Anderson, S., Radić, V.: Evaluation and interpretation of convolutional long short-term memory networks for regional hydrological modelling. Hydrol. Earth Syst. Sci. **26**(3), 795–825 (2022). https://doi.org/10.5194/hess-26-795-2022
3. Boulch, A., Guerry, J., Saux, B., Audebert, N.: SnapNet: 3D point cloud semantic labeling with 2D deep segmentation networks. Comput. Graph. **71**, 189–198 (2018). https://doi.org/10.1016/j.cag.2017.11.010
4. Camps-Valls, G., Tuia, D., Zhu, X.X., Reichstein, M.: Deep Learning for the Earth Sciences: A Comprehensive Approach to Remote Sensing, Climate Science and Geosciences. Wiley (2021)
5. Castro, R., Souto, Y.M., Ogasawara, E., Porto, F., Bezerra, E.: STConvS2S: spatiotemporal convolutional sequence to sequence network for weather forecasting. Neurocomputing **426**, 285–298 (2021). https://doi.org/10.1016/j.neucom.2020.09.060
6. Gaya-Morey, F.X., Buades-Rubio, J.M., MacKenzie, I.S., Manresa-Yee, C.: REVEX: a unified framework for removal-based explainable artificial intelligence in video (2024). https://doi.org/10.48550/arXiv.2401.11796
7. Höhl, A., et al.: Opening the black-box: a systematic review on explainable AI in remote sensing. IEEE Geosci. Remote Sens. Mag. **12**(4), 261–304 (2024). https://doi.org/10.1109/MGRS.2024.3467001
8. Letzgus, S., Wagner, P., Lederer, J., Samek, W., Müller, K.R., Montavon, G.: Toward explainable artificial intelligence for regression models: a methodological perspective. IEEE Signal Process. Mag. **39**(4), 40–58 (2022)
9. Li, Z., Wang, W., Li, Z., Huang, Y., Sato, Y.: Towards visually explaining video understanding networks with perturbation. In: 2021 IEEE Winter Conference on Applications of Computer Vision (WACV), pp. 1119–1128. IEEE (2021). https://doi.org/10.1109/WACV48630.2021.00116
10. Lundberg, S.M., Lee, S.I.: A unified approach to interpreting model predictions. In: Advances in Neural Information Processing Systems, vol. 30 (2017)
11. Lundberg, S.M., Lee, S.I.: A unified approach to interpreting model predictions. In: Proceedings of the 31st International Conference on Neural Information Processing Systems, NIPS 2017, pp. 4768–4777. Curran Associates Inc., Red Hook (2017)
12. Mamalakis, A., Barnes, E.A., Ebert-Uphoff, I.: Carefully choose the baseline: lessons learned from applying XAI attribution methods for regression tasks in geoscience. Artif. Intell. Earth Syst. **2**(1), e220058 (2023)

13. Miller, L., Pelletier, C., Webb, G.I.: Deep learning for satellite image time-series analysis: a review. IEEE Geosci. Remote Sens. Mag. (2024)
14. Pathak, D., Girshick, R., Dollár, P., Darrell, T., Hariharan, B.: Learning features by watching objects move. In: Computer Vision and Pattern Recognition (CVPR) (2017)
15. Petsiuk, V., Das, A., Saenko, K.: RISE: randomized input sampling for explanation of black-box models. In: Proceedings of the British Machine Vision Conference (BMVC) (2018)
16. Racah, E., Beckham, C., Maharaj, T., Ebrahimi Kahou, S., Prabhat, M., Pal, C.: ExtremeWeather: a large-scale climate dataset for semi-supervised detection, localization, and understanding of extreme weather events. In: Advances in Neural Information Processing Systems, vol. 30 (2017)
17. Ribeiro, M.T., Singh, S., Guestrin, C.: "Why should i trust you?" explaining the predictions of any classifier. In: Proceedings of the 22nd ACM SIGKDD International Conference on Knowledge Discovery and Data Mining, pp. 1135–1144 (2016). https://doi.org/10.1145/2939672.2939778
18. Ruschemeier, H.: Ai as a challenge for legal regulation–the scope of application of the artificial intelligence act proposal. In: Era Forum, vol. 23, pp. 361–376. Springer (2023)
19. Salis, M., Atto, A.M., Ferraris, S., Meo, R.: Time distributed deep learning models for purely exogenous forecasting: application to water table depth predictions using weather image time series. Environ. Model. Softw. (2025). https://doi.org/10.1016/j.envsoft.2025.106568
20. Salis, M., Sartor, G., Pellegrino, M., Ferraris, S., Atto, A.M., Meo, R.: What's behind this water table depth forecasting? RISE application for spatial, temporal, and spatio-temporal explanations. In: Proceedings of AI4IA 2024 - AI4CC: International Workshop on Artificial Intelligence for Climate Change, vol. 3883 (2024)
21. Stergiou, A., Kapidis, G., Kalliatakis, G., Chrysoulas, C., Veltkamp, R., Poppe, R.: Saliency tubes: visual explanations for spatio-temporal convolutions. In: 2019 IEEE International Conference on Image Processing (ICIP), pp. 1830–1834 (2019). https://doi.org/10.1109/ICIP.2019.8803153
22. Tom, M., Aguilar, R., Imhof, P., Leinss, S., Baltsavias, E., Schindler, K.: Lake ice detection from sentinel-1 SAR with deep learning. ISPRS Ann. Photogram. Remote Sens. Spat. Inf. Sci. **V-3-2020**, 409–416 (2020). https://doi.org/10.5194/isprs-annals-V-3-2020-409-2020, https://isprs-annals.copernicus.org/articles/V-3-2020/409/2020/
23. Uchiyama, T., Sogi, N., Niinuma, K., Fukui, K.: Visually explaining 3D-CNN predictions for video classification with an adaptive occlusion sensitivity analysis. In: 2023 IEEE/CVF Winter Conference on Applications of Computer Vision (WACV), pp. 1513–1522 (2023). https://doi.org/10.1109/WACV56688.2023.00156
24. Wunsch, A., et al.: Karst spring discharge modeling based on deep learning using spatially distributed input data. Hydrol. Earth Syst. Sci. **26**(9), 2405–2430 (2022). https://doi.org/10.5194/hess-26-2405-2022
25. Yang, X., et al.: SCRDet: towards more robust detection for small, cluttered and rotated objects. In: Proceedings of the IEEE/CVF International Conference on Computer Vision, pp. 8232–8241 (2019)

Enhancing Local Explanations with GAN-Based Neighborhood Generation

Luca Corbucci(✉), Francesca Naretto, and Anna Monreale

Department of Computer Science, University of Pisa, Pisa, Italy
luca.corbucci@phd.unipi.it

Abstract. The growing deployment of Machine Learning models has increased the demand for interpretability, leading to the development of various explainability methods. Yet, each method has its strengths and limitations, making it challenging to identify a one-size-fits-all solution. This paper introduces FIRE360, a multi-faceted and fast local explanation approach for tabular data. FIRE360 incorporates various explanation types (feature importances, rules, counterfactuals, exemplars, and counter-exemplars), offering a 360-degree model interpretability. FIRE360 also includes a dashboard with indicators for prediction reliability and explanation quality, such as fidelity and robustness. To ensure efficiency, FIRE360 uses GAN-generated synthetic data to approximate the original dataset and selects similar records to the target instance, avoiding costly on-demand neighborhood generation. Multiple local surrogate models are trained to capture different aspects of the data and black-box behavior. FIRE360 achieves higher fidelity than state-of-the-art methods, with improvements up to 95% on multiclass datasets.

Keywords: Explainable AI · Tabular Data · Machine Learning

1 Introduction

In recent years, the rapid advancement and widespread adoption of Machine Learning (ML) techniques across various fields have raised significant ethical concerns. To regulate ML deployment, several legal frameworks [34] have emerged, addressing fairness [22], privacy [15,24], explainability [4,8], and their interplay [6,7,17,23]. The need for explainability has led to the emergence of Explainable AI (XAI), which seeks to enhance the interpretability of black-box (BB) models [4], fostering transparency, trust as well as regulatory compliance.

Several post-hoc explanation methods have been introduced, including Shap [21], Lime [28], and Lore [16]. These differ in mechanisms and outputs: Shap and Lime provide feature importance, while Lore produces rules. Hence, each method offers only one type of explanation at a time.

User studies show no consensus on which explanation type is most interpretable; user preferences often follow patterns depending on expertise and context. For instance, non-expert users tend to favor instance-based explanations,

S. Džeroski et al. (Eds.): DS 2025, LNAI 16090, pp. 316–331, 2025.
https://doi.org/10.1007/978-3-032-05461-6_21

while expert users are more likely to prefer global methods such as feature importance [13]. These preferences can also be influenced by the complexity of the model and the decision-making context [29], leading practitioners to combine multiple explanation methods to meet diverse user needs. However, these methods are computationally expensive, as they require generating synthetic data and training surrogate models for each instance to explain. Others lack reliability in key state-of-the-art metrics, such as fidelity and stability, leading to inconsistent outputs when generating explanations for the same instance multiple times. Providing explanations with such limitations can undermine user trust and raise doubts on the reliability of the system. Furthermore, these issues raise concerns regarding compliance with emerging AI regulations, which emphasize the need for transparent, trustworthy, and stable explanations to ensure fairness, accountability and user protection.

To address these limitations, we introduce `FIRE360` (**F**eature **I**mportance, **R**ules, and **E**xemplars with **360**-degree analysis), a novel XAI method that leverages synthetic data generation to train various white-box surrogate models, enabling the extraction of five explanation types: feature importance, rules, counterfactual rules, exemplars, and counter-exemplars.

Unlike traditional methods, which generate synthetic data for each instance to explain, `FIRE360` constructs a *single, large synthetic dataset* by leveraging the `BB` and its training data. The resulting dataset aims to resemble the original while being denser in feature space. In addition, it enables efficient *similarity search* to retrieve the closest instances and train *local white-box surrogate* models, reducing computational overhead while improving explanation coverage and reliability.

Our main contribution is a new multi-faceted XAI method for tabular data, providing five types of explanations per instance within a unified framework. A key strength of our approach lies in our efficient pipeline design: synthetic data is generated once at the beginning of the process, avoiding the computational overhead of per-instance generation. We validate `FIRE360` on six datasets across four metrics, including `Fidelity`, `Stability`, and `Robustness`, showing improved fidelity over popular explanation techniques, such as `Shap` [21] and `Lime` [28]. We release `FIRE360` source code to ensure reproducibility and enable practitioners to integrate our method into their pipelines[1]. Moreover, we publish a web-based user interface[2], designed to enhance the usability of `FIRE360` by making it easier to interact and visualize the explanations compared to existing solutions.

2 Related Work and Background

We review the relevant literature for our methodology, focusing on XAI techniques, particularly local and post-hoc explanations, in Sect. 2.1. Next, in Sect. 2.2, we discuss synthetic data generation approaches, examining state-of-the-art solutions and their application in `FIRE360`.

[1] Github repository of FIRE360.
[2] Dashboard to visualize the explanations obtained with FIRE360.

2.1 Explainable AI

Nowadays, explainability is one of the most important research directions in AI, as it is crucial to achieve Trustworthy Artificial Intelligence. Bodria et al. [4] provide an overview of existing techniques for ML interpretability, identifying two main types of explanation models: *global* and *local*. Local explainers focus on explaining the results of predictions on single instances [16,21,28], while global ones explain the behavior of the whole ML model [11,12,14]. In addition, explanations can be *post-hoc*, meaning they are generated externally after the ML model has been trained, or *by design*, where both the prediction and the explanation are produced during the model's training process. Lastly, XAI methods can be *agnostic*, if they explain any `BB` model, or *ad-hoc*, specifically designed for a particular ML model. This paper proposes `FIRE360`, a method for generating agnostic, local, and post-hoc explanations. In the following, we present the state-of-the-art methods in this context.

Feature Importance Explanations. These methods produce a feature-importance vector, where each value indicates a feature's contribution to the classification. A well-known example is `Lime` [28], a local, model-agnostic method that explains `BB` model predictions by approximating their behavior near a given instance using randomly perturbed samples.

Given a `BB` and an instance x, `Lime` perturbs x by randomly modifying the features to create synthetic instances, $\{z \in \mathbb{R}^d\}$. These synthetic instances are then passed through `BB` to obtain their predictions, $\theta(z)$. The perturbed dataset is used to fit an interpretable model $E(\cdot)$, which serves as a local approximation of θ. The resulting explanation consists of the learned feature importance values, reflecting how much each feature contributes to the prediction. Several `Lime` variants have been proposed to address its limitations [33,37], but they often increase the computational cost of generating explanations. Therefore, in this paper, we use the classic `Lime` implementation for comparison with `FIRE360`.

In this context, there is also `Shap` [21] (SHapley Additive exPlanations), a local and global, post-hoc, model-agnostic approach that computes feature importance using Shapley values, a concept derived from cooperative game theory. `Shap` provides explanations in the form of additive feature attributions, represented by the equation: $E(z') = \rho_0 + \sum_{i=1}^{M} \rho_i z'_i$, where $z' \approx x$ as a real number, $z' \in [0, 1]$, $\rho_i \in \mathbb{R}$ represents the contribution assigned to each feature, and M is the number of input features. `Shap` satisfies three key properties: (i) local accuracy, ensuring that $E(x)$ matches $\theta(x)$; (ii) missingness, meaning that features with $x_i = 0$ have no impact on the computed `Shap` values; (iii) stability, ensuring that if a model changes so that the marginal contribution of a feature increases, the corresponding `Shap` value also increases. There are various ways to approximate the computation of `Shap` values using different strategies. Among them, we focus on KernelExplainer, the model-agnostic version of `Shap`, which we employ in this work as a competitor. KernelExplainer approximates Shapley values by sampling perturbations around the instance to explain and estimate feature attributions through a weighted linear regression model.

Rule-Based Explanations. This type of explanation, known as a decision or factual rule, follows the form $p \rightarrow y$, where p is a premise consisting of Boolean conditions on feature values and y is the outcome class. The premise p is a conjunction of conditions of the form $x_i \in [v_i^{(l)}, v_i^{(u)}]$, where x_i is a feature, and $v_i^{(l)}$, $v_i^{(u)}$ are its lower and upper bounds, extended with $\pm\infty$. An instance x satisfies a rule r if all conditions in p hold for x. If the instance to explain satisfies p, the rule $p \rightarrow y$ serves as a candidate explanation for the decision $E(x) = y$.

In this setting, we find `Lore` [16] (LOcal Rule-based Explainer) a local, post-hoc, agnostic explainer that provides explanations in the form of rules and counterfactual rules. `Lore` uses a genetic algorithm to generate the neighborhood of the instance to be explained. Given a `BB` and an instance x, where $\theta(x) = y$, `Lore` creates a synthetic set Z of neighbors exploiting a genetic algorithm. It then trains a decision tree classifier ϕ on this set, labelling it with the `BB` predictions $\theta(Z)$. The explanation extracted from ϕ consists of: (i) a decision rule, i.e. the path in the decision tree followed by x to reach the prediction y, and (ii) a set of counterfactual rules, which lead to a different classification than y.

Prototype-Based Explanations. They are representative records that capture the salient characteristics of a group of objects, usually belonging to the same class. They often act as reference points, allowing users to gain insight into the decision-making process of a `BB` by examining instances similar to them. They are usually considered as local explanations, giving information about similarities and differences w.r.t. the record x to explain. A prototype may be (i) a training sample close to the input x, (ii) the centroid of a cluster, or (iii) a synthetically generated record. Different methods adopt varying definitions and criteria for selecting prototypes [20,32]. As an example, Tree Space Prototype [32] is a local post-hoc explainer for tree ensemble models. The goal is to identify class-wise prototypes using a predefined proximity measure between trees.

Explanation Evaluation. In recent years, increasing attention has been given to evaluating the quality of explanations, leading to the introduction of several validation metrics. Among these, `Fidelity`, `Stability`, and `Robustness` are considered the most important for ensuring reliable explanations [4]. `Fidelity` measures how well the explanation function approximates the behavior of the `BB` θ, with its implementation varying depending on the type of explainer being analyzed. `Stability` and `Robustness`, on the other hand, evaluate the consistency of explanations. `Stability` ensures that the same instance receives the same explanation when analyzed multiple times, meaning that if we explain the same record twice, the results should not change. `Robustness` assesses whether similar instances give similar explanations. It is evaluated by analyzing explanations for points in the vicinity of the instance being explained, ensuring that small variations in input do not lead to significantly different explanations.

Overall, `Lime` is computationally efficient, but its reliance on random sampling makes it unstable, leading to different results across runs. `Shap`, on the other hand, is grounded in strong theoretical foundations, yet its approximation process and the choice of the reference dataset can significantly impact both execution time and explanation quality. For `Lore`, the use of a genetic algorithm improves stability at the cost of increased computational time, making it less practical for large-scale applications.

To address this, we propose FIRE360, a method that ensures computational efficiency, stability, and fidelity to the BB's predictions. Unlike existing approaches, FIRE360 offers a comprehensive 360-degree explanation by combining feature importance, rules, counterfactuals, exemplars, and counter-exemplars for a complete and interpretable view of the model's decisions.

2.2 Synthetic Data Generation

In recent years, synthetic data generation has gained increasing attention due to the limited availability of real-world data, as well as ethical considerations, such as fairness and privacy. A widely used techniques employs Gaussian Mixture Models (GMMs) [5], which model multivariate distributions based on the original data. This approach is often combined with copula functions to capture dependencies between variables better. Recently, more effective methods than GMMs, have emerged, particularly Generative Adversarial Networks (GANs) and Variational Autoencoders (VAEs).

GANs [36] are a generative framework based on adversarial training. This approach involves training two models: a *generator* to learn how to approximate the data distribution and a *discriminator*, which evaluates whether a given sample originates from the real training data or is synthetic. When the discriminator can no longer distinguish between real and synthetic data, it indicates that the generator has successfully learned to produce highly realistic synthetic samples. One of the most widely used GAN-based methods for tabular data is CTGAN[3], which employs a conditional generator to preserve the proportions of different attributes and classes present in the real data. This makes it particularly effective for categorical variables, ensuring that synthetic data maintains the original statistical distributions.

TVAE[4] [18], on the other hand, differs from GAN-based approaches in how it models data distributions. Instead of using adversarial training, TVAE (Tabular Variational Autoencoder) encodes input data into a latent space and reconstructs it from latent representations. In practice, a VAE is a generative model that learns a probabilistic latent representation of data. It consists of an encoder that maps input data to a distribution over latent variables and a decoder that reconstructs data from sampled latent representations. In particular, it enables controlled data generation by modeling the latent space as a continuous distribution. TVAE is suited for handling missing values and sparse data but does not include an explicit mechanism for managing categorical variables.

3 FIRE360 Method

FIRE360 is a local, post-hoc and agnostic explanation method for tabular data which provides users with a *multi-faceted view* on the explanation of a model

[3] CTGAN.

[4] TVAE.

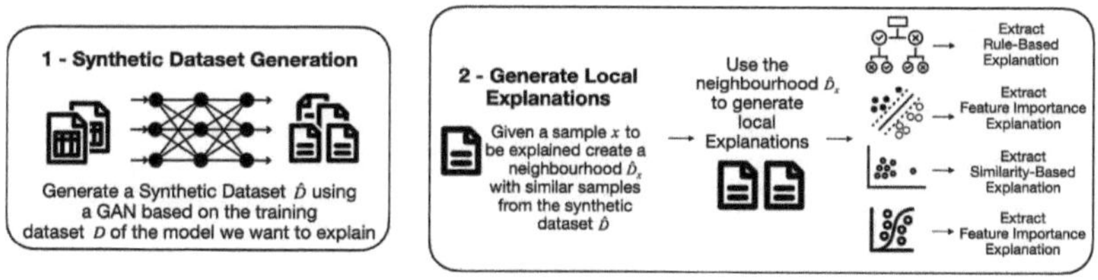

Fig. 1. An overview of our methodology

prediction. It enhances the interpretability by providing explanations in multiple forms: *feature importance*, *decision rules*, *counterfactual rules*, *exemplars* and *counter-exemplars*. By offering diverse perspectives on model predictions, FIRE360 facilitates a comprehensive understanding of the reasoning behind a given decision. As illustrated in Fig. 1, our approach consists of two main phases: (I) the generation of a synthetic dataset and (II) the use of the synthetic dataset to train multiple white-box surrogate models, which are then used to explain the reasoning behind the prediction of each sample made by the BB.

Phase I - Synthetic Data Generation: Given a training dataset D, used to train the BB model θ we aim to explain, FIRE360 generates a synthetic dataset that closely resembles the original data distribution, while ensuring a denser representation of the data space. This allows us to cover a broader range of input variations, enabling the training of white-box models for extracting explanations and analyzing different localities of the feature space more effectively. To achieve our objective, we exploit state-of-the-art synthetic data generation techniques (more information is in Sect. 4). A key aspect of our approach is that the synthetic dataset $\widehat{D}$ serves as the foundation of the explanation pipeline and, differently from the majority of the XAI approaches, is generated only once, ensuring computational efficiency. Once generated, $\widehat{D}$ is labeled using the BB θ, obtaining $\widehat{Y} = \theta(\widehat{D})$. This approach ensures that the synthetic data reflects both the characteristics of the original data and the behavior of the BB.

Phase II - Explanation via White-Box Models: After generating the synthetic dataset $\widehat{D}$ and labelling it using the BB model ($\widehat{Y} = \theta(\widehat{D})$), FIRE360 proceeds with training the white-box surrogate models, from which explanations will later be extracted. Since our method is local, it does not require the entire dataset $(\widehat{D}, \widehat{Y})$. Instead, to better capture the specific patterns in the vicinity of the instance x to be explained, FIRE360 focuses only on the closest points to x. In this way, we ensure that the surrogate models learn the most relevant information for local interpretability. The *neighborhood selection* process extracts a small subset of synthetic samples that are close to x. Mathematically, given an instance x to explain, we select $\widehat{D_x} \subseteq \widehat{D}$, where the number of neighbors $|\widehat{D_x}|$ and the notion of proximity, exploited for the selection of the subset, are user-defined. In particular, for the proximity, depending on the specific dataset, it may be a tailored distance metric (e.g., Euclidean, Cosine, or Jaccard distance).

Fig. 2. A screenshot of the dashboard we developed to allow the practitioners to visualize and explore the explanations generated by FIRE360.

After the extraction of the local dataset $\widehat{D_x}$, FIRE360 trains different white-box models for the explanation extraction: (I) a Decision Tree [19] (DT), (II) a Support Vector Machine [9](SVM), (III) a Logistic Regression [10] (LR) and (IV) a K-nearest neighbors [1](KNN). For obtaining robust and reliable explanations, the surrogate models must achieve high predictive performance, particularly when compared to the BB model. FIRE360 first conducts a grid search to identify the optimal set of hyperparameters for each surrogate model. Further details on the hyperparameter tuning process are provided in Sect. 4. Once all surrogate models are trained, we proceed with extracting explanations. Feature importance explanations are derived from the SVM and LR models, while rule-based explanations, either factual or counterfactual, are obtained from the DT. Finally, similarity-based explanations, in the form of exemplars and counter-exemplars, are extracted using the KNN model. A key advantage of FIRE360 is that it generates the synthetic dataset $\widehat{D}$ once during training, avoiding per-query neighborhood generation and greatly reducing computational overhead. Beyond computational efficiency, FIRE360 also achieves a higher fidelity compared to existing methods, as its surrogate models more accurately replicate the behavior of the BB model, as discussed in the experimental Sect. 2.1. Additionally, it ensures stability, producing consistent explanations across different runs. Finally, by integrating multiple explanation types, FIRE360 provides a comprehensive and multi-faceted analysis of the BB predictions. Our approach simplifies implementation by enabling multiple explanation types through a single, unified method without the need to manage separate techniques.

Visual Dashboard. A key component of FIRE360 is its interactive dashboard, an example of which is shown in Fig. 2. The dashboard displays information about the explained record, including its features and the prediction probability assigned by the BB. Additionally, it presents multiple types of explanations (e.g., feature importance, rules, exemplars, etc.) with their quality scores, such as Fidelity scores. This interface allows end-users to explore the FIRE360 analysis, offering explanations and supplementary information that enhances transparency and fosters trust in the model's decision-making process.

4 Experimental Setting

To validate and evaluate FIRE360, we test it on six diverse tabular datasets[5]: Adult [2], Dutch [35], Covertype [3], House16[6], Letter [30], and Shuttle [31]. These datasets are selected for their distinct characteristics to ensure a comprehensive assessment of our approach. In particular, they include both binary and multi-class classification tasks, numerical and categorical attributes, and vary significantly in both the number of records and of features. This diversity allows us to evaluate the effectiveness and scalability of FIRE360 across different data distributions and learning scenarios. Table 1 summarizes the key characteristics of the datasets used in our experiments and the metrics computed using the BB model on the test dataset. Adult is a binary classification dataset with 48, 842 records and 52 features, including both numerical and categorical attributes. Similarly, Dutch is a binary classification dataset with 60, 420 samples collected in the Netherlands, where the objective is to predict whether an individual's salary exceeds 50, 000$. Covertype is a multiclass dataset with 7 classes and 581, 012 records for classifying forest cover types. House16 contains 22, 784 samples with a binary target. Letter consists of 20, 000 samples and is designed for classifying handwritten English letters into 26 possible categories. Shuttle comprises 43, 500 samples with 7 possible targets. For each dataset, we split it into 80% training D_{train} and 20% test sets D_{test}. The training dataset was further divided into training and validation sets again with an 80-20% split for the hyperparameter tuning done before training the models θ. D_{test} is kept separate from the train dataset throughout the FIRE360 pipeline. In particular, we explain all the samples in D_{test} to evaluate our explanations.

Synthetic Data Generation - Phase 1. To evaluate the effectiveness of the synthetic data generation phase, we test two popular synthesizers: a CT-GAN [36] and a TVAE [18] using the implementation provided by the Synthetic Data Vault library (SDV) [25]. Overall, the results are good and in line with our expectations, confirming the ability of both models to generate high-quality synthetic data. For each synthesiser, we test different configurations, varying the

[5] Details about the pre-processing are reported in the FIRE360 Github repository.
[6] House16 Dataset.

Table 1. A description of the datasets used in the experiments. The number of numerical and categorical characteristics listed in the table is given after pre-processing.

Dataset	Samples	Classes	$\lvert D_{test} \rvert$	Num. Feat.	Cat. Feat.	Accuracy	F1 Score
Dutch	60,420	2	12,084	11	0	0.83 ± 0.01	0.83 ± 0.01
Adult	48,842	2	9,769	9	102	0.85 ± 0.01	0.84 ± 0.01
Covertype	581,012	7	20,000	10	44	0.80 ± 0.01	0.79 ± 0.01
House16	22,784	2	4,557	16	0	0.85 ± 0.01	0.85 ± 0.02
Letter	20,000	26	4,000	16	0	0.82 ± 0.02	0.82 ± 0.02
Shuttle	43,500	7	8,700	9	0	0.98 ± 0.01	0.98 ± 0.01

number of generated samples (from 10,000 to 200,000) and training epochs (1000, 2500, and 5000), following the SDV library's hyperparameter tuning guidelines[7]. After generating the synthetic data $\widehat{D}$, we assess its quality using metrics from the Synthcity library [27], before labelling $\widehat{D}$ with the BB θ. The results highlight that all of the methods have low values of *Common Row Probability* between real and synthetic datasets, meaning that there is a low risk of leaking real records in the synthetic one. In addition, the high performance of *XGB Performance* metric indicates that training XGB on the synthetic dataset yields strong results in classification tasks. Lastly, we also have good results in terms of privacy metrics: none of the methods has less than $k = 23$ for k-anonymity, and the *Probability of re-identification* is always close to zero. Since the results indicate strong overall performance across all models with no significant differences between settings, we opt for using CTGAN for small datasets, e.g., Shuttle, Letter, and House16, and TVAE for Adult, Dutch, and Covertype to leverage its stable training process, especially when handling tabular data with a huge number of records. We train all datasets for 2,500 epochs, as no further improvements were observed beyond that point. The only exception is Covertype, limited to 1,000 epochs due to its large size and computational cost.

Surrogate Model Training - Phase 2. The second phase of our method involves training four surrogate models, with hyperparameter tuning performed using a grid search for each. As discussed in Sect. 3, `FIRE360` is composed of DT, SVM, LR, and KNN. This choice allows us to extract different types of explanations, ensuring a multi-faceted view by capturing various perspectives on the model's decision-making process. Technically, for the implementation of these models, we exploit the *sklearn* library [26]. Specifically, as DT, we use the *DecisionTreeClassifier* performing hyperparameter tuning with a grid search on criterion, max_depth, min_samples_leaf and class_weight. We also use *LogisticRegression* as LR, with liblinear solver, tuning the penalty, C, and class_weight. For SVM, we use a linear kernel and 1000 as the maximum number of iterations, with a grid search to find the best value for class_weight and C hyperparameters. Lastly, for the KNN, we optimize metric, weights, and n_neighbors.

[7] SDV Hyperparameter tuning.

Formally, we denote with ϕ a surrogate model, and with $\phi(\widehat{D}_x, \theta(\widehat{D}_x))$ its training, where $\widehat{D}_x$ is a subset of the synthetic dataset $\widehat{D}$, created at phase 1 and $\widehat{Y}$ the predictions of the BB θ. In the experiments, $\widehat{D}_x$ is created by finding the *top−n* samples from $\widehat{D}$ similar to the sample x we aim to explain. In our experiments, we used cosine similarity as the similarity metric; however, other metrics, such as Euclidean or Jaccard distance, can also be applied in this phase. The value of n is one of the parameters that can determine the quality of the explanations. To evaluate it, we consider three different values: $n = 1000$, $n = 2500$, and $n = 5000$. In addition, since we want $\widehat{D}_x$ to be as representative as possible of the entire input space, we enforce the presence of at least the 30% of samples with a target value different than the target of the sample x. To validate the quality of the surrogate models generated, we compute the `Fidelity` of the different ϕ to assess how well they replicate the behavior of the BB model. `Fidelity` is defined as the proportion of instances to be explained where the model ϕ's prediction matches that of θ. Mathematically, `Fidelity` is defined as: $\texttt{Fidelity} = \frac{1}{N}\sum_{i=1}^{N} \mathbb{I}[\theta(x_i) = \phi(x_i)]$ where $\mathbb{I}$ is an indicator function, ϕ is the surrogate model and N is the total number of test instances. This means that given a sample x, we get 1 if the prediction of the BB is the same prediction obtained with the surrogate, otherwise, we have 0. In addition, we also evaluate $\texttt{Fidelity}_{neigh}$, which is the fidelity of the surrogate model over the neighborhood of each sample x to be explained, that we define as $\texttt{Fidelity}_{neigh} = \frac{1}{N}\sum_{i=1}^{N} \frac{\sum_{j}^{|\widehat{D}_{x_i}|} \mathbb{I}[\theta(x_j)=\phi(x_j)]}{|\widehat{D}_{x_i}|}$. We also evaluate the quality of the explanations provided by computing two metrics: the `Stability` and the `Robustness` of the explanation. Given a sample x and an explainer E, we generate two explanations e and e' using E, and the `Stability` is computed as the differences between the two explanations in terms of the number of shared features. This metric is dependent on the kind of explanations generated. In the case of a rule-based explanation, we measure the distance between e and e', comparing the premises of the rules. For instance, given $e = [age > 60, house\_size > 100]$ and $e' = [age > 30, house\_size > 50]$, the `Stability` is 1 while with $e' = [age > 30, education > 3]$ the `Stability` is 0.5, since they share only 1 feature. Our objective is to determine whether the attributes are shared between the two explanations rather than focusing on the exact threshold, which may fluctuate due to minor variations in the composition of the synthetic datasets and is difficult to track. Formally, we define the `Stability` of the rule-based explanations as: $\texttt{Stability}(e, e') = \frac{|e \cap e'|}{|e \cup e'|}$. For the feature importance-based explanations, instead, we focus on the ranking of the explanations and not on the exact values of the feature importance. In particular, we first sort the features based on the absolute value of the feature importance, and then we measure the `Stability` by comparing the order of the features. Mathematically, we define the stability for the feature importance as: $\texttt{Stability}(e, e') = \frac{\sum_{i=1}^{d} \mathbb{I}(e_i = e_i')}{d}$, where d is the number of features of the dataset. Moreover, we measure the `Robustness` of an explanation. In this case, we evaluate the neighborhood of the record to explain. Hence, given the sample

Table 2. A comparison of FIRE360 (neighbor size $n = 5000$) with Lime, Lore, and Shap across multiple metrics shows that FIRE360 consistently achieves the highest fidelity. The "time" column reports the average explanation time (in seconds) over 100 samples.

Dataset	Method	Time (sec) (↓)	Fidelity (↑)	Fidelity_{neigh} (↑)	Stability (↑)	Robustness k=5 (↑)	Robustness k=10 (↑)	Robustness k=20 (↑)
Adult	FIRE360 (DT)	2.14 ± 0.18	**0.90**	0.94 ± 0.03	**0.92 ± 0.15**	**0.61 ± 0.12**	**0.60 ± 0.11**	**0.59 ± 0.09**
Adult	FIRE360 (SVM)	1.11 ± 0.21	0.54	0.51 ± 0.18	0.73 ± 0.34	0.24 ± 0.10	0.24 ± 0.09	0.24 ± 0.09
Adult	FIRE360 (LR)	2.05 ± 0.63	0.90	0.98 ± 0.01	0.47 ± 0.22	0.25 ± 0.09	0.24 ± 0.08	0.24 ± 0.07
Adult	Lime	**0.08 ± 0.01**	**0.90**	0.73 ± 0.02	0.05 ± 0.02	0.05 ± 0.01	0.05 ± 0.01	0.05 ± 0.01
Adult	Shap	2.44 ± 0.06	-	-	0.41 ± 0.19	0.26 ± 0.11	0.25 ± 0.09	0.25 ± 0.09
Adult	Lore (Random)	13.24 ± 0.56	0.65	**1.00 ± 0.00**	0.34 ± 0.22	0.34 ± 0.14	0.34 ± 0.12	0.34 ± 0.11
Adult	Lore (Genetic)	128.50 ± 2.47	0.64	**1.00 ± 0.00**	0.55 ± 0.17	0.25 ± 0.10	0.25 ± 0.08	0.25 ± 0.07
Covertype	FIRE360 (DT)	2.35 ± 0.58	0.84	0.85 ± 0.03	0.94 ± 0.11	**0.62 ± 0.11**	**0.60 ± 0.10**	0.58 ± 0.09
Covertype	FIRE360 (SVM)	1.19 ± 0.60	0.56	0.52 ± 0.09	0.54 ± 0.38	0.16 ± 0.06	0.16 ± 0.05	0.15 ± 0.05
Covertype	FIRE360 (LR)	17.70 ± 9.38	**0.85**	0.86 ± 0.03	**0.94 ± 0.19**	0.24 ± 0.11	0.23 ± 0.09	0.22 ± 0.09
Covertype	Lime	**0.11 ± 0.11**	0.71	0.26 ± 0.11	0.61 ± 0.05	0.60 ± 0.02	0.60 ± 0.02	**0.60 ± 0.01**
Covertype	Shap	18.63 ± 0.44	-	-	0.18 ± 0.06	0.11 ± 0.03	0.10 ± 0.03	0.09 ± 0.02
Covertype	Lore (Random)	5.07 ± 0.40	0.37	**1.00 ± 0.00**	0.42 ± 0.12	0.41 ± 0.07	0.41 ± 0.06	0.41 ± 0.05
Covertype	Lore (Genetic)	35.11 ± 1.13	0.63	**1.00 ± 0.00**	0.74 ± 0.10	0.44 ± 0.06	0.44 ± 0.06	0.44 ± 0.05
Dutch	FIRE360 (DT)	0.93 ± 0.03	**0.99**	0.99 ± 0.01	0.95 ± 0.11	**0.90 ± 0.14**	**0.87 ± 0.14**	**0.84 ± 0.14**
Dutch	FIRE360 (SVM)	0.35 ± 0.03	0.99	0.98 ± 0.02	0.80 ± 0.26	0.73 ± 0.25	0.66 ± 0.23	0.58 ± 0.20
Dutch	FIRE360 (LR)	6.55 ± 6.83	0.99	0.98 ± 0.01	0.82 ± 0.25	0.75 ± 0.23	0.69 ± 0.22	0.61 ± 0.19
Dutch	Lime	**0.09 ± 0.12**	0.89	0.51 ± 0.19	0.47 ± 0.16	0.46 ± 0.10	0.45 ± 0.09	0.44 ± 0.09
Dutch	Shap	1.72 ± 0.07	-	-	**1.00 ± 0.00**	0.833± 0.20	0.78 ± 0.21	0.72 ± 0.21
Dutch	Lore (Random)	2.13 ± 0.13	0.50	**1.00 ± 0.00**	0.71 ± 0.12	0.68 ± 0.07	0.68 ± 0.06	0.68 ± 0.06
Dutch	Lore (Genetic)	34.37 ± 1.57	0.50	**1.00 ± 0.00**	0.81 ± 0.19	0.59 ± 0.12	0.59 ± 0.11	0.59 ± 0.10
House16	FIRE360 (DT)	2.25 ± 0.08	0.91	0.88 ± 0.03	**0.93 ± 0.11**	**0.68 ± 0.08**	**0.68 ± 0.08**	**0.67 ± 0.07**
House16	FIRE360 (SVM)	0.41 ± 0.05	0.67	0.60 ± 0.13	0.52 ± 0.39	0.22 ± 0.11	0.22 ± 0.10	0.21 ± 0.10
House16	FIRE360 (LR)	1.11 ± 0.33	**0.93**	0.91 ± 0.02	0.75 ± 0.20	0.52 ± 0.13	0.51 ± 0.12	0.49 ± 0.11
House16	Lime	**0.06 ± 0.08**	0.87	0.24 ± 0.14	0.58 ± 0.13	0.47 ± 0.06	0.46 ± 0.04	0.46 ± 0.03
House16	Shap	1.64 ± 0.05	-	-	0.75 ± 0.14	0.19 ± 0.06	0.18 ± 0.05	0.17 ± 0.04
House16	Lore (Random)	2.28 ± 0.10	0.57	**1.00 ± 0.00**	0.51 ± 0.14	0.51 ± 0.08	0.51 ± 0.07	0.51 ± 0.06
House16	Lore (Genetic)	36.76 ± 1.01	0.61	**1.00 ± 0.00**	0.77 ± 0.14	0.55 ± 0.10	0.55 ± 0.08	0.55 ± 0.07
Letter	FIRE360 (DT)	1.21 ± 0.08	0.67	0.61 ± 0.08	**0.97 ± 0.07**	0.66 ± 0.10	0.65 ± 0.09	0.64 ± 0.08
Letter	FIRE360 (SVM)	0.87 ± 0.21	**0.82**	0.77 ± 0.06	0.88 ± 0.27	0.14 ± 0.09	0.13 ± 0.06	0.12 ± 0.05
Letter	FIRE360 (LR)	4.40 ± 1.14	0.77	0.71 ± 0.07	0.92 ± 0.18	0.19 ± 0.11	0.17 ± 0.09	0.15 ± 0.07
Letter	Lime	**0.10 ± 0.12**	0.04	0.05 ± 0.02	0.23 ± 0.11	0.18 ± 0.07	0.17 ± 0.06	0.16 ± 0.05
Letter	Shap	2.43 ± 0.12	-	-	0.52 ± 0.15	0.24 ± 0.10	0.22 ± 0.08	0.19 ± 0.07
Letter	Lore (Random)	2.75 ± 0.07	0.04	**1.00 ± 0.00**	0.61 ± 0.10	0.61 ± 0.05	0.60 ± 0.05	0.60 ± 0.04
Letter	Lore (Genetic)	37.56 ± 0.74	0.04	**1.00 ± 0.00**	0.90 ± 0.06	**0.77 ± 0.04**	**0.77 ± 0.04**	**0.77 ± 0.03**
Shuttle	FIRE360 (DT)	1.18 ± 0.40	0.99	0.98 ± 0.01	0.92 ± 0.15	**0.76 ± 0.15**	0.73 ± 0.13	0.72 ± 0.13
Shuttle	FIRE360 (SVM)	0.47 ± 0.40	**0.99**	0.99 ± 0.05	0.80 ± 0.28	0.43 ± 0.15	0.42 ± 0.14	0.40 ± 0.13
Shuttle	FIRE360 (LR)	1.99 ± 1.83	0.99	0.98 ± 0.01	0.84 ± 0.25	0.50 ± 0.18	0.48 ± 0.16	0.46 ± 0.15
Shuttle	Lime	**0.07 ± 0.09**	0.80	0.07 ± 0.04	0.42 ± 0.19	0.33 ± 0.14	0.32 ± 0.13	0.31 ± 0.12
Shuttle	Shap	1.09 ± 0.05	-	-	**1.00 ± 0.01**	0.69 ± 0.17	0.66 ± 0.15	0.63 ± 0.14
Shuttle	Lore (Random)	1.82 ± 0.18	0.66	**1.00 ± 0.00**	0.71 ± 0.11	0.70 ± 0.07	0.70 ± 0.06	0.70 ± 0.06
Shuttle	Lore (Genetic)	34.70 ± 1.28	0.67	**1.00 ± 0.00**	0.87 ± 0.11	0.75 ± 0.08	**0.75 ± 0.06**	**0.75 ± 0.06**

x and the corresponding explanation e, we find its neighborhood composed of the top-k similar instances from the test dataset. Then, we compare the explanation e with the explanations computed for the samples in the neighborhood, using the same formula of the Stability. We test $k \in [3, 5, 8, 10, 20]$.

5 Experimental Results

To validate the goodness of our results, we compare our solution with three popular explainers: Lime, Shap, and Lore (using both a Random and a Genetic neighborhood generation)[8].

In Table 2, we report the results obtained with FIRE360 and the competitors. For a fair comparison, we use 5,000 samples for training the surrogate models used by FIRE360, Lime, and Lore. For Shap, which has no surrogate model, we apply k-means to obtain the reference points. As suggested by the authors, we use 100 centroids for all datasets, except Letter (200 centroids). In the table, we

[8] Lime source code, Shap source code and Lore source code.

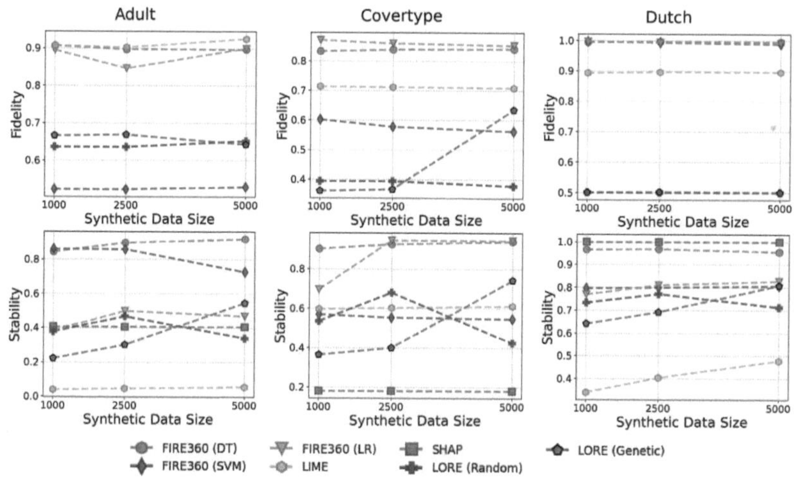

Fig. 3. Ablation study on the number of synthetic records exploited for the explanation generation in the case of Adult, Covertype, and Dutch Datasets, with 2 classes.

report the results on `Fidelity`, $\texttt{Fidelity}_{neigh}$, `Stability` and `Robustness` for `FIRE360` with DT, LR, and SVM, leaving out KNN since, for construction, it does not allow us to measure them. Due to space constraints, we report in the table only the `Robustness` results obtained with $k = 5$, $k = 10$, and $k = 20$; however, we found coherent results with other values of k. From the results reported in Table 2, it is clear that `FIRE360` provides the explanations with the highest `Fidelity`, especially when considering DT and LR models. More specifically, `FIRE360` is stable, particularly when it has to deal with multiclass datasets. Indeed, with Letter `FIRE360` presents a stability that is 95% higher than the competitors, while with Covertype and Shuttle, we observe an improvement of 16% and 19%. Overall, on average, the `Fidelity` obtained with `FIRE360` is 24% higher than the competitors. In addition, even for $\texttt{Fidelity}_{neigh}$, `FIRE360` shows good performance, in line with the ones of `Fidelity`. Regarding the competitors, `Lore` has the lowest fidelity for all the datasets except for $\texttt{Fidelity}_{neigh}$, which is always 1 and hence the highest. `Lime` presents mixed outcomes: it is always better than `Lore`, but it does not reach our `Fidelity` values. High values for `Fidelity` and $\texttt{Fidelity}_{neigh}$ are interesting results since they mean that `FIRE360`'s surrogate models closely mimic the behavior of the `BB`. In terms of `Stability`, i.e., the similarity of two explanations generated for the same record, `FIRE360` usually has the best results, especially using DT. This highlights that our method provides stable explanations and is not affected by randomness. Other competitors, such as `Lore` and `Shap`, present results lower than ours but still acceptable, while `Lime` has the lowest values. The same reasoning applies for `Robustness`, where we evaluate the stability on the k closest records in the test set. The table shows our method yields consistent explanations for sample x as k increases.

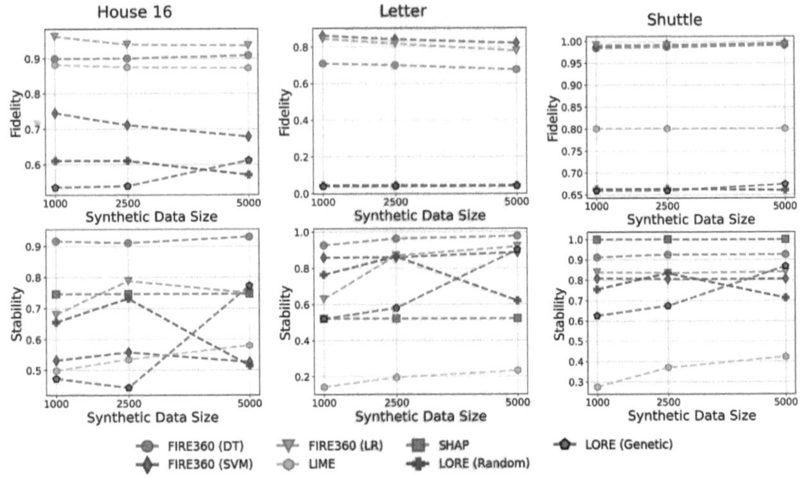

Fig. 4. Ablation study on the number of synthetic records exploited for the explanation generation in the case of House 16, Letter, and Shuttle Datasets, with 2 classes.

We also consider the computational time measured in seconds, reported in Table 2. Depending on the different surrogate models we train, the timing may differ, but overall the results are comparable with those of `Shap`. `Lime`, instead, is always the fastest, even if it is the worst when compared to the other metrics. However, obtaining the same variety of explanations as `FIRE360`, requires running multiple competing methods. Specifically, at least one of `Shap` or `Lime` is needed for the importance of the feature, while `Lore` is necessary to generate rules. This highlights a key advantage of `FIRE360`: it offers multiple explanation types, simplifies implementation, and reduces the need to integrate and manage various separate methods. We also performed an ablation study on the number of synthetic records in $\widehat{D}_x$, evaluating how the quality of the explanations varies as the number of synthetic neighbors for each sample to be explained increases. Results are shown in Figs. 3 and 4. `FIRE360` maintains strong performance regardless of the number of synthetic records. In contrast, other methods perform worse and less consistently, more synthetic data helps some (e.g., `Lore` Genetic) but has little effect on others, especially `Lime`.

6 Conclusion

In this paper, we presented `FIRE360`, a local, post-hoc, and model-agnostic explanation method tailored for tabular data. Its key strength lies in offering diverse explanation forms (*feature importance*, *decision rules*, *counterfactual rules*, *exemplars*, and *counter-exemplars*), to enhance human understanding and streamline explainable AI pipelines without relying on multiple tools.

Experiments show that `FIRE360` outperforms state-of-the-art methods in fidelity, by up to 95% on Letter and 16âĂŞ19% on Covertype and Shuttle.

Beyond fidelity, FIRE360 offers good performance in terms of stability and robustness while maintaining competitive explanation times, thanks to GAN-based neighborhood generation used to learn white-box surrogate models. Future work includes extending FIRE360 to image and time series data and conducting user studies to tailor the dashboard and recommend the most suitable explanations.

Acknowledgements. This research was partially supported by the European Union Horizon 2020 program under grant agreement No. 101120763 (TANGO) and PNRR-"SoBigData.it - Strengthening the Italian RI for Social Mining and Big Data Analytics" - Prot. IR0000013;. Views and opinions expressed are, however, those of the author(s) only and do not necessarily reflect those of the European Union or the European Health and Digital Executive Agency (HaDEA). Neither the European Union nor the granting authority can be held responsible for them.

References

1. Altman, N.S.: An introduction to kernel and nearest-neighbor nonparametric regression. Am. Stat. (1992)
2. Becker, B., et al.: Adult. UCI Machine Learning Repository (1996)
3. Blackard, J.: Covertype. UCI Machine Learning Repository (1998)
4. Bodria, F., et al.: Benchmarking and survey of explanation methods for black box models. Data Min. Knowl. Disc. (2023)
5. Chokwitthaya, C., et al.: Applying the Gaussian mixture model to generate large synthetic data from a small data set. In: Construction Research Congress (2020)
6. Corbucci, L., et al.: Enhancing privacy and utility in federated learning: a hybrid p2p and server-based approach with differential privacy protection. In: SECRYPT (2024)
7. Corbucci, L., et al.: PUFFLE: balancing privacy, utility, and fairness in federated learning. In: 27th ECAI. IOS Press (2024)
8. Corbucci, L., et al.: Semantic enrichment of explanations of AI models for healthcare. In: International Conference on Discovery Science (2023)
9. Cortes, C., et al.: Support-vector networks. Mach. Learn. (1995)
10. Cox, D.R.: The regression analysis of binary sequences. J. Roy. Stat. Soc.: Ser. B (Methodol.) (1958)
11. Craven, M., et al.: Extracting tree-structured representations of trained networks. In: NIPS (1996)
12. Craven, M.W., et al.: Using sampling and queries to extract rules from trained neural networks. In: JMLR. Elsevier (1994)
13. Darias, J.M., et al.: An empirical analysis of user preferences regarding XAI metrics. In: Case-Based Reasoning Research and Development (2024)
14. Deng, H.: Interpreting tree ensembles with inTrees. Int. J. Data Sci. Anal. (2019)
15. Dwork, C., et al.: Fairness through awareness. In: Proceedings of the 3rd Innovations in Theoretical Computer Science Conference (2012)
16. Guidotti, R., et al.: Stable and actionable explanations of black-box models through factual and counterfactual rules. Data Min. Knowl. Disc. (2024)
17. Haffar, R., et al.: GLOR-FLEX: local to global rule-based EXplanations for federated learning. In: 2024 FUZZ-IEEE (2024)

18. Ishfaq, H., et al.: TVAE: triplet-based variational autoencoder using metric learning. arXiv:1802.04403 (2018)
19. Kass, G.V.: An exploratory technique for investigating large quantities of categorical data. J. Roy. Stat. Soc. (1980)
20. Kim, B., et al.: Examples are not enough, learn to criticize! Criticism for interpretability. In: NeurIPS (2016)
21. Lundberg, S.M., et al.: A unified approach to interpreting model predictions. In: NIPS (2017)
22. Mehrabi, N., et al.: A survey on bias and fairness in machine learning. In: ACM Computing Surveys (CSUR) (2021)
23. Naretto, F., et al.: Evaluating the privacy exposure of interpretable global explainers. In: IEEE CogMI (2022). https://doi.org/10.1109/CogMI56440.2022.00012
24. Naretto, F., et al.: Predicting and explaining privacy risk exposure in mobility data. Discov. Sci. (2020)
25. Patki, N., et al.: The Synthetic data vault. In: IEEE International Conference on Data Science and Advanced Analytics (DSAA) (2016)
26. Pedregosa, F., et al.: Scikit-learn: machine Learning in Python. J. Mach. Learn. Res. (2011)
27. Qian, Z., et al.: Synthcity: facilitating innovative use cases of synthetic data in different data modalities (2023). arXiv:2301.07573
28. Ribeiro, M.T., et al.: "Why should i trust you?": explaining the predictions of any classifier. In: ACM SIGKDD (2016)
29. Rong, Y., et al.: Towards human-centered explainable AI: a survey of user studies for model explanations. IEEE Trans. Pattern Anal. Mach. Intell. (2024)
30. Slate, D.: Letter recognition. UCI Machine Learning Repository (1991)
31. Statlog (Shuttle). UCI Machine Learning Repository
32. Tan, S., et al.: Tree space prototypes: another look at making tree ensembles interpretable. In: Foundations of Data Science Conference (2020)
33. Tan, Z., et al.: GLIME: general, stable and local LIME explanation. In: Advances in Neural Information Processing Systems (2023)
34. UE. The Artificial Intelligence Act (2023). https://artificialintelligenceact.eu/the-act/
35. Van der Laan, P.: The 2001 census in the Netherlands: integration of registers and surveys. In: Conference at the Cathie Marsh Centre (2001)
36. Xu, L., et al.: Modeling tabular data using conditional GAN. In: Advances in Neural Information Processing Systems (2019)
37. Zhou, Z., et al.: S-LIME: stabilized-lime for model explanation. In: Proceedings of the 27th ACM SIGKDD (2021)

Few-Shot Learning, Data Streams

Aspect-Based Few-Shot Learning

Tim van Engeland[1], Lu Yin[1,2], and Vlado Menkovski[1(✉)]

[1] Eindhoven University of Technology, Eindhoven, The Netherlands
v.menkovski@tue.nl
[2] University of Surrey, Guildford, UK

Abstract. We generalize the formulation of few-shot learning by introducing the concept of an aspect. In the traditional formulation of few-shot learning, there is an underlying assumption that a single "true" label defines the content of each data point. This label serves as a basis for the comparison between the query object and the objects in the support set. However, when a human expert is asked to execute the same task without a predefined set of labels, they typically consider the rest of the data points in the support set as context. This context specifies the level of abstraction and the aspect from which the comparison can be made. In this work, we introduce a novel architecture and training procedure that develops a context given the query and support set and implements aspect-based few-shot learning that is not limited to a predetermined set of classes. We demonstrate that our method is capable of forming and using an aspect for few-shot learning on the Geometric Shapes and Sprites dataset. The results validate the feasibility of our approach compared to traditional few-shot learning.

Keywords: few-shot learning · aspect-based few-shot learning · representation learning

1 Introduction

Machine learning (ML) methods have been highly successful in image analysis tasks such as image classification [5]. Nevertheless, this formulation poses a significant limitation for the more general goal of recognizing the content of an image or other high-dimensional data. Specifically, to precisely recognize the content of such data, many more labels are typically required. Such labels may not be available during training, may form various hierarchical relationships, the language they use may not be exact, and depending on the context these labels may vary for a given data point. In contrast to this, humans excel at such recognition tasks. We can understand the content of the image data and adjust this understanding to different aspects.

The few-shot learning (FSL) approaches generalize over the classification formulation, as the model is no longer limited to a predefined set of classes, but can generalize from a few training examples [10] to recognize image content beyond what was available during training. The objective of the FSL model is

S. Džeroski et al. (Eds.): DS 2025, LNAI 16090, pp. 335–346, 2025.
https://doi.org/10.1007/978-3-032-05461-6_22

to learn the transferable knowledge from the "base" data (training data) with a small amount of labeled data and to be able to apply it to a "novel" dataset.

There have been multiple developments of the base formulation of FSL [10], however, these formulations are still limited to matching a datapoint to a unique label. We demonstrate the shortcomings of this formulation with the following three examples of four-way-one-shot learning given in Fig. 1. Figure 1a shows a standard example FSL, there is one image of four distinct classes: "horse", "dog", "bunny", and "lion". In this example, we recognize that the query image is in line with the content of the first image and should therefore be labeled as a "horse". However, the second set of images in Fig. 1b, show images of only dogs. Despite this, we can still say that the second image is the most similar to the query image. The label "dog" does not fully describe the content of the image. The first image contains a sitting dog, the second shows a running dog, in the third one the dog lays down, and in the last image, the dog is chewing on a bone. Given this context, we can discern that the physical activity of the dogs is different for each image. So, rather than the presence of the dog being the differentiating factor, we can deduce that the activity of running is a property that both the query and (only) the second image share. The third example in Fig. 1c further demonstrates the point. Each image in the support set displays a running dog. However, each of them runs on a different surface. From left to right, the dogs run in the water, snow, sand and in a field. Therefore, the query image can be matched to (only) the fourth dog as they are both running on grass. In every example, the second image in the support set is the same, as well as the query image, but the matching differs given the rest of the images in the support set. The first support set has images with varying animals. Therefore, the matching was performed on the basis of the species. In the second case, the species was not a factor of variation, but rather the physical activity. Lastly, in the third case, only the background was a factor that allowed us to match a single support set image with the query image. The examples show that the whole support set plays a defining role in how we compare each instance of the support set with the query image.

The current state-of-the-art FSL methods cannot account for those relationships between the instances of the support set because it assumes a single label assignment to each datapoint on which the matching is implemented. As a consequence, only the first case given in Fig. 1b would work with the exact match. For the second case we would need an additional set of labels based on the physical activity. However, adding multiple labels is not efficient and does not address this problem in general.

In this work, we propose an extension of the few-shot learning formulation by incorporating the notion of aspects. Rather than learning to match with a unique label, the model learns the differentiating feature of the support set. Hereby, we will define the *aspect* as the varying feature(s) in the support set and implement the matching based on the values of that feature, which we refer to as properties of the data points. With these definitions in place, we present a framework that

(a) FSL classification based on the predefined label will assign the correct label "horse" to the query image.

(b) Each image in the support set has the label "dog", while the query image is labeled as "horse". Therefore, there is no match for the query image. However, given the support set if we change the aspect from species to physical activity, the query image can be matched based on the label "running".

(c) All the images in the support set are running dogs, but the background of the images is different. If this aspect is considered, a match can be made as only the fourth image shares a common background of a grass field with flowers with the query image.

Fig. 1. Examples of different FSL cases that illustrate how taking different aspects results with different matching between the query and support set data points learns the aspect given by the support set and, as such, identifies the correct assignment between the query and the instances of the support set.

2 Related Work

Few-shot learning (FSL) is a broad research field that can be categorized into two main approaches: non-meta-learning and meta-learning [10]. Non-meta-learning focuses primarily on transfer learning [21], where knowledge from a pre-trained model is transferred to a new task to alleviate the need to train from scratch when data is scarce. On the other hand, meta-learning, often referred to as "learning to learn", has gained significant attention due to its potential to address few-shot learning problems. Meta-learning approaches can be divided into three categories. *Metric-based methods* aim to learn a shared metric space where few-shot predictions can be effectively made by measuring similarities between data points. For example, Euclidean distance-based methods [12,14,15] learn an embedding space and compare query and support examples using Euclidean distance, while hypersphere-based methods [3,8,16] map the embedding space onto a hyperspherical manifold for similarity measurement. Furthermore, psychometric-based metric learning methods take advantage of human perception [17–19] to learn the similarity based on psychometric testing. *Optimization-based methods* focus on adapting gradient-based optimization algorithms to enable rapid learning on new tasks with limited data. Representative works include Model-Agnostic Meta-Learning (MAML) [4], which learns model parameters that are sensitive to rapid adaptation; and Reptile [9], an algorithm that approximates MAML with first-order gradients. In this work, we focus on metric-based approaches for few-shot learning. A closely related study to our research is the Categorical Traversal Module (CTM) [6]. The CTM traverses the entire support set to identify task-relevant features based on both intra-class commonality and inter-class uniqueness. Consider a 2-way 4-shot task with the classes *tiger* and *cheetah*. Each tiger has its recognizable coat of reddish-orange with dark stripes (intra-class commonality) and similarly for cheetahs with their distinct coats. The coats are unique to their species (interclass uniqueness). Both animals belong to the Felidae family (cats) and share characteristics. Therefore, identifying task-relevant features is crucial for correctly distinguishing between a tiger and a cheetah. Nevertheless, as discussed in the introduction, these methods assume a single-class assignment and are unable to deal with the more general setting where class assignment is conditioned on the context given by the support set.

3 Method

3.1 Aspect Definition

In a traditional few-shot classification approach, we are given a query set Q and a support set S of n distinct, previously unseen classes, with k examples each.

The goal of the model is to learn to classify the query instance $x_q \in Q$, into one of the n classes defined by the elements of the support set S. In other words, there exists a data point $x_i \in$ S that matches the query image x_q.

We aim to generalize this class-based assignment by introducing the notion of an aspect. Specifically, rather than assuming that a single class describes the content of a datapoint, we assume that each datapoint is described by a set of properties P_i. In the traditional setting, we can say that a class assignment also has a set of properties P_c, such that if a data point is labeled with the particular class value, then the set of properties P_i is equal to the set of properties of the class P_c. However, in the aspect-based few-shot learning formulation, there is no such assumption, and therefore the properties of the query P_q and the properties of the support set data points P_i may only have a partial overlap.

In aspect-based few-shot learning, we do not assume that each data-point has a single label or class. Instead, we define an aspect as any relevant property (or combination of properties) that both the query and exactly one support instance share, and that the other support instances do not.

For a given element x_m part of the support set S ($x_m \in$ S). The query x_q matches x_m if there exists an aspect A such that A $\subset x_m$ and A $\subset x_q$ and A $\not\subset x_i$ for all i where $i \neq m$.

Therefore, the aspect is not predetermined as the class, but rather becomes specified at query time as a combination of the properties of the query and support set elements.

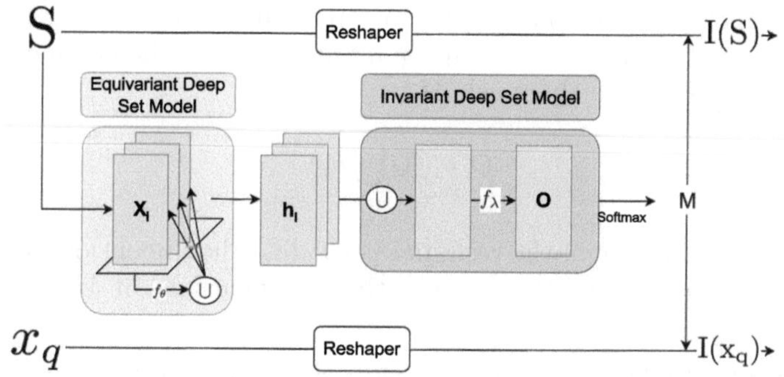

Fig. 2. Detailed depiction of the DSTM components

3.2 Deep Set Traversal Module

The CTM method discussed in the background section aligns with our goal of extracting the aspect from the support set, however, the model does not offer permutation invariance. As the order of the support set does not matter when comparing the support set to the query, the aspects need to be permutationally invariant. To address this, we present a modified architecture named Deep-Set Traversal Module (DSTM).

Permutation-Equivariant Deep Set Model. In Fig. 2, we replace the concentrator with a permutation-equivariant deep set model [20]. The purpose of the model is to enrich the embedding of each element with the information of the other members of the support set:

$$h_i = f_\theta(x_i, \mathrm{P}_{\mathcal{N}(i)}) \tag{1}$$

$$\mathrm{P}_{\mathcal{N}(i)} = \bigcup_{j \in \mathcal{N}(i)} f_\phi(x_j) \tag{2}$$

$\mathcal{N}(i)$ is the neighborhood of x_i, f_ϕ is the neighbor image embedding model.

This model extracts the aspect of each support set image by combining the image representation with the representation of the properties of the neighboring images. The model as specified in Eq. 2 extracts the shared properties in the neighbourhood $\mathrm{P}_{\mathcal{N}(i)}$ by individually processing the representation through embedding function f_ϕ and applying permutation-invariant operation $\bigcup$. The model as specified in Eq. 1 processes the set of properties $\mathrm{P}_{\mathcal{N}(i)}$ together with x_i through the embedding function f_θ to determine the unique properties h_i. This process occurs for each image x_i in the support set S.

Permutation-Invariant Deep Set Model. The second component is the permutation-invariant deep-set model [20]. In Eq. 3 we apply an additional operation $\bigcup$ on aspects h_i to destroy the information regarding the order of the set. The permutation-invariant representation of the set is processed by another embedding function f_λ to have a final output embedding O representing the complete set. As the last step, we apply a Softmax over the channel dimension of the output O to create the mask M.

$$O = f_\lambda(\bigcup h_i) \tag{3}$$

Reshaper. The purpose of the reshaper is to align the dimension of the feature map of the support set and the query with the dimensions of M

Design Choice in DSTM. The last step of the DSTM is to apply M to the support set and query to be able to semantically match. This is done by an element-wise multiplication between the output permutation-variant deep set model M and the reshaper output:

$$\mathrm{I(S)} = M \odot r(\mathrm{S})$$
$$\mathrm{I(x_q)} = M \odot r(\mathrm{x_q})$$

4 Evaluation

To evaluate this approach, our aim is to show that DSTM can deduce the aspect from the support set and use it to match the correct image. Based on our definition of an aspect, the matching is done by the unique discriminating property in the query and the support set. The existing FSL datasets and evaluation do not offer varying aspects as the matching is not done based on different properties, but on the single class assignment assumption. We therefore set out to develop a novel evaluation specific to the aspect-based FSL method.

4.1 Data and Setup

Data. To objectively assess the performance of our approach, we need fine control over the properties of each data point in the dataset such that we can have an unambiguous measure of accuracy. We achieve this by developing a fully synthetic dataset named *geometric shapes*, where 4 factors are being varied to form a set of properties for each data point. We also repeat this evaluation approach on the *sprite* public dataset.

The *geometric shapes* dataset includes two-dimensional geometric shapes with an image size of 112 by 112 pixels. Each image consists of a polygon with a hole in the middle. As such, a polygon can have a number of differentiating properties, namely: color, thickness, and pattern. The *sprites* dataset is derived from the data generated to disentangle sequential data [1,7]. The available sprite sheets [2] offer the possibility of creating custom characters. In this work, we are not interested in the complete sequence, but in individual frames. The selected frames represent different features: physical stance, body type, shirt, pants, and hair color.

Support Sets. Forming the support sets in aspect-based FSL requires certain considerations. Each support set must contain one or more discriminating properties that would allow the query image to be matched with one data point in the support set without ambiguity. Specifically, the query image must match only one data point from the support set with which it shares one or more properties. The values of these properties need to vary in the rest of the data points of the support set. Matching different properties from multiple data points in the support set leads to ambiguity and, while this could be resolved based on different levels or priority or intensity, it is out of scope for this initial version of the method.

Data Split. In FSL the generalization of the model is evaluated on a validation set consisting of unseen classes. As we do not assume a set of pre-existing classes, we cannot employ such a split. In our controlled setting, we can design the dataset so that each image differs in at least one property. Therefore, each image can be classified as a separate class. We refer to this as the *unique data split.* As a consequence, the training data effectively consists of classes that do not appear in the test data ($C_{training} \cap C_{test} = \emptyset$) as there are different images for training compared to testing.

Furthermore, we evaluate the method with a second approach, referred to as the *aspect data split.* In this split, we use all the data points to form the support sets. However, we form unique sets of query images for training and testing.

Therefore, the aspects that are present in training are not present in testing. The goal of this evaluation is not to evaluate the general purpose representation learning properties of the model, but rather to evaluate the capability of the model to correctly extract the aspect, or rather how well it generalizes in doing so.

Loss. The tuplet loss [13] generalizes the triplet loss [11] to explore multiple negative examples simultaneously. Instead of the triplet (x, x^+, x^-), we have a tuplet defined as $(x_a, x^+, x_1^-, .., x_{N-1}^-)$, where N is the size of the support set. The tuplet loss (Eq. 4) brings the anchor embedding to be closer to the positive example, as does the triplet. However, the loss pushes away all negative samples simultaneously instead of only one negative example.

$$L_{tuplet}(x, x^+, x_1^-, .., x_{N-1}^-) = \log\left(1 + \sum_{j=1}^{N-1} e^{\|f(x)-f(x^+)\|_2^2 - \|f(x)-f(x_j^-)\|_2^2}\right) \quad (4)$$

Evaluation Metric. As a metric, we use the Euclidean distance of our aspect-based FSL embedding versus the traditional FSL embedding. Specifically, we use the distance ratio 4.1.

For simplicity, we refer to the distance between the positive example and the query as the positive distance from now on. The same applies to the negative examples as the negative distance. The positive distance within the aspect-based embedding space is expected to be smaller than within a single embedding space due to the focus on the aspect rather than on all properties simultaneously.

We also opt for a relative comparison: the ratio of positive distance to the negative distance. A larger ratio represents a better distinction between the positive and negative example:

$$\text{distance ratio} = \frac{|\text{average positive distance - average negative distance}|}{\text{average positive distance}}$$

Implementation Details. The experiments are implemented in PyTorch (version 2.1.0+cu121). For training, the learning rate is set to 7e-4. We use the ADAM optimizer with L2 regulization set to 1e-2. Each model trains for 50 epochs without any early stopping. The model is saved when the validation loss decreases compared to the best model up to this point. During testing, we generated ten different support sets for each query. Only one support set does not offer enough evidence as the support sets are generated randomly. Therefore, we generate multiple support sets.

4.2 Results

We provide three options for representation learning: a shallow representation model, a small VGG architecture, and a small ResNet architecture. The shallow

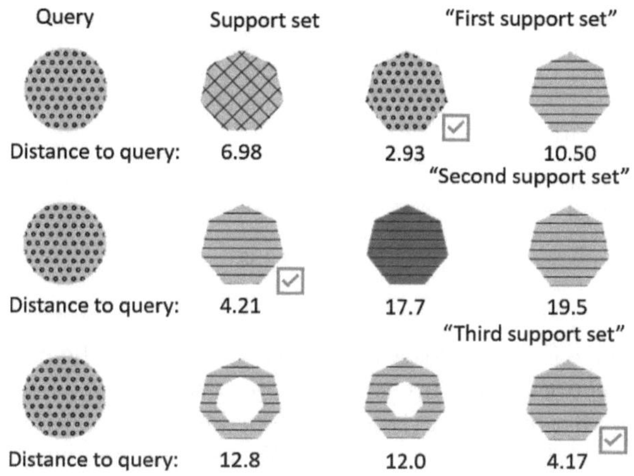

Fig. 3. Example of the distance between the query and support set instances for the geometric shape data. Differentiating aspects leads to different distances between query and the same image.

Table 1. Geometric shapes data set: The average positive and negative distances of the second support set with a 95% confidence interval. The distance ratio interval is calculated based on the 95% confidence interval.

Model	Data Unique Split			Query Split		
	Positive Distance	Negative Distance	Distance Ratio	Positive Distance	Negative Distance	Distance Ratio
Baseline	–	–	–	49.0 ± 0.46	54.9 ± 0.28	0.1–0.14
Shallow + Single Layer DSTM	5.46 ± 0.08	9.72 ± 0.13	0.73–0.84	3.96 ± 0.09	7.57 ± 0.14	0.84–0.99
VGG + Residual Block DSTM	3.23 ± 0.08	6.80 ± 0.12	1.02–1.20	4.03 ± 0.09	8.14 ± 0.15	**0.94–1.10**
ResNet + Residual Block DSTM	3.58 ± 0.08	8.85 ± 0.18	**1.36–1.58**	4.59 ± 0.09	8.06 ± 0.12	0.69–0.81

representation model is a simple architecture of convolutional layer + batch normalization + ReLU followed by a pooling layer. The baseline model would be the shallow representation model without the DSTM. Without the DSTM, we can see the effect of the module on the embedding space. Within the DSTM, we experiment with two types of layers in the permutation equivariant and invariant deep-set models. The first option is to have a single convolutional layer + batch normalization + ReLU for the embedding function $g(;,\mu)$ in the permutation-equivariant model and a single layer for the embedding function $k(;\eta)$ in the permutation-invariant model. The second option is to have a single residual block in those places in the model instead.

Analysis of the Distance Ratio Figure 3 shows the three different support sets for the same query. The upper support set is referred to as the first support set. This support set has two shared features with the query. In this example, the shared features are color and thickness. In the first support set, we see that the second image matches the query on the aspect of the pattern. The third image of the first support set has a distance of 10.5 from the query image. However, we see

Table 2. Geometric shapes data set: The average positive and negative distance of the second support set with a 95% interval. The distance ratio interval is calculated based on the 95% interval.

Model	Data Unique Split			Query Split		
	Positive Distance	Negative Distance	Distance Ratio	Positive Distance	Negative Distance	Distance Ratio
Baseline	–	–	–	40.5 ± 0.56	49.0 ± 0.32	0.19–0.24
Shallow + Single Layer DSTM	4.03 ± 0.07	8.96 ± 0.14	1.15–1.30	2.66 ± 0.06	6.67 ± 0.14	1.40–1.63
VGG + Residual Block DSTM	2.26 ± 0.06	6.19 ± 0.13	1.60–1.85	2.95 ± 0.07	7.84 ± 0.16	**1.54–1.79**
ResNet + Residual Block DSTM	2.54 ± 0.06	8.32 ± 0.19	**2.13–2.42**	3.24 ± 0.07	7.25 ± 0.12	1.15–1.32

that the same image is much closer to the query in the middle or second support set, given the aspect of color. This second support set has only one shared feature in thickness. The distance from the query depends on the aspect of the support set that aligns with the goal of this work. We see the same happening with the third image of the second support set transferred to the lower or third support set. In this support set, we do not have any shared features.

Table 1 shows the average positive distance and negative distance with a 95% interval of the first support set for the best-performing model configuration and the baseline. The boundaries of the distance ratio are calculated on the basis of the interval 95%. We see that the distance ratio of the baseline is small. So, the baseline model cannot differentiate between positive and negative examples well. Each model trained with the DSTM has a better distance ratio. So, the model does a better job of differentiating between positive and negative examples. The VGG and ResNet representation models with the residual block DSTM perform best for the unique split of the data. The VGG representation model with the residual block DSTM also performs well in the case of a query split. However, the shallow representation model with a single-layer DSTM performs better than the ResNet representation model with the residual block DSTM.

Table 3. Sprites data set: The average positive and negative distances of the first support set with a 95% confidence interval. The distance ratio interval is calculated based on the 95% interval.

Model	Data Unique Split			Query Split		
	Positive Distance	Negative Distance	Distance Ratio	Positive Distance	Negative Distance	Distance Ratio
Baseline	–	–	–	14.2 ± 0.05	34.6 ± 0.1	1.42–1.45
Shallow + Single Layer DSTM	1.24 ± 0.01	7.73 ± 0.03	5.16–5.31	1.19 ± 0.01	10.8 ± 0.05	8.01–8.24
VGG + Residual Block DSTM	0.91 ± 0.01	6.67 ± 0.04	**6.20–6.37**	0.76 ± 0.01	10.1 ± 0.04	**12.0–12.46**
ResNet + Residual Block DSTM	1.03 ± 0.01	6.75 ± 0.05	5.39–5.60	1.05 ± 0.01	11.3 ± 0.04	9.58–9.86

Table 2 displays the average positive distance and negative distance with a 95% interval of the second support set for each model configuration and the baseline. The distance ratios are smaller than the first support set in Table 2. There is an increase in the positive distance and the negative distance. It indicates that the DSTM cannot completely filter out the differentiating features between the support set and the query. However, the best model architectures can still differentiate relatively well between negative and positive examples.

Table 3 displays the average positive distance and negative distance with a 95% interval of the first support set for each model configuration and the baseline. The distance ratio of the first support set displays the best results in the case of the query data split. Each DSTM architecture with a query split performs better than every model with a unique data split. However, the unique data split still outperforms the baseline by a large margin. The distance ratios in Table 3 are also an improvement compared to the results of the geometric shapes data set in Table 1.

5 Conclusion

This work addressed the intricate problem of label-based supervision approaches in Few-Shot learning. Traditional Few-Shot learning approaches assume that a single class describes the content of a data point. The class label of the query must match precisely one of the support set class labels. We generalized this class-based assignment by introducing the notion of an aspect.

The experimental results with the DSTM show an improved differentiation between positive and negative examples when we do not consider a single-class assignment to the images. However, this evaluation is still limited to the synthetic cases where we can fully control the properties of the data to form the aspect. Although this is a good initial step in assessing the validity of this approach, to demonstrate the real value of this approach, a user study with natural image data needs to be implemented where the capacity of DSTM to infer the aspect can be contrasted with the innate ability of humans to do the same.

References

1. Liberated pixel cup. http://lpc.opengameart.org/. Accessed 01 Jan 2024
2. Universal LPC spritesheet character generator. https://github.com/sanderfrenken/Universal-LPC-Spritesheet-Character-Generator. Accessed 01 Jan 2024
3. Che, Y., An, Y., Xue, H.: Boosting few-shot open-set recognition with multi-relation margin loss. In: IJCAI, pp. 3505–3513 (2023)
4. Finn, C., Abbeel, P., Levine, S.: Model-agnostic meta-learning for fast adaptation of deep networks. In: International Conference on Machine Learning, pp. 1126–1135. PMLR (2017)
5. He, K., Zhang, X., Ren, S., Sun, J.: Deep residual learning for image recognition. In: Proceedings of the IEEE Conference on Computer Vision and Pattern Recognition, pp. 770–778 (2016)
6. Li, H., Eigen, D., Dodge, S., Zeiler, M., Wang, X.: Finding task-relevant features for few-shot learning by category traversal. In: Proceedings of the IEEE/CVF Conference on Computer Vision and Pattern Recognition, pp. 1–10 (2019)
7. Li, Y., Mandt, S.: Disentangled sequential autoencoder. In: International Conference on Machine Learning (2018)
8. Mettes, P., Pol, E., Snoek, C.: Hyperspherical prototype networks. Adv. Neural. Inf. Process. Syst. **32** (2019)

9. Nichol, A., Schulman, J.: Reptile: a scalable metalearning algorithm. arXiv preprint arXiv:1803.02999, vol. 2, no. 3, p. 4 (2018)
10. Parnami, A., Lee, M.: Learning from few examples: a summary of approaches to few-shot learning. arXiv preprint arXiv:2203.04291 (2022)
11. Schroff, F., Kalenichenko, D., Philbin, J.: FaceNet: a unified embedding for face recognition and clustering. In: Proceedings of the IEEE Conference on Computer Vision and Pattern Recognition, pp. 815–823 (2015)
12. Snell, J., Swersky, K., Zemel, R.: Prototypical networks for few-shot learning. Adv. Neural. Inf. Process. Syst. **30** (2017)
13. Sohn, K.: Improved deep metric learning with multi-class n-pair loss objective. Adv. Neural. Inf. Process. Syst. **29** (2016)
14. Sung, F., Yang, Y., Zhang, L., Xiang, T., Torr, P.H., Hospedales, T.M.: Learning to compare: Relation network for few-shot learning. In: Proceedings of the IEEE Conference on Computer Vision and Pattern Recognition, pp. 1199–1208 (2018)
15. Vinyals, O., Blundell, C., Lillicrap, T., Wierstra, D., et al.: Matching networks for one shot learning. Adv. Neural. Inf. Process. Syst. **29** (2016)
16. Wang, F., Cheng, J., Liu, W., Liu, H.: Additive margin softmax for face verification. IEEE Signal Process. Lett. **25**(7), 926–930 (2018)
17. Yin, L., Menkovski, V., Liu, S., Pechenizkiy, M.: Hierarchical semantic segmentation using psychometric learning. arXiv preprint arXiv:2107.03212 (2021)
18. Yin, L., Menkovski, V., Pechenizkiy, M.: Knowledge elicitation using deep metric learning and psychometric testing. In: Hutter, F., Kersting, K., Lijffijt, J., Valera, I. (eds.) ECML PKDD 2020, Part II. LNCS (LNAI), vol. 12458, pp. 154–169. Springer, Cham (2021). https://doi.org/10.1007/978-3-030-67661-2_10
19. Yin, L., Menkovski, V., Pei, Y., Pechenizkiy, M.: Semantic-based few-shot learning by interactive psychometric testing. arXiv preprint arXiv:2112.09201 (2021)
20. Zaheer, M., Kottur, S., Ravanbakhsh, S., Poczos, B., Salakhutdinov, R.R., Smola, A.J.: Deep sets. Adv. Neural. Inf. Process. Syst. **30** (2017)
21. Zhuang, F., et al.: A comprehensive survey on transfer learning. Proc. IEEE **109**(1), 43–76 (2020)

Learning Fast to Detect Slow: A Few-Shot Neural Approach to Slow DoS Attack Detection

Francesco Scala[1(✉)], Massimo Guarascio[1], Carlo Parrotta[2], and Luigi Pontieri[1]

[1] Institute of High Performance Computing and Networking (ICAR-CNR), Via P. Bucci, 87036 Rende, CS, Italy
{francesco.scala,massimo.guarascio,luigi.pontieri}@icar.cnr.it
[2] Department Computer Engineering, Modeling, Electronics, and Systems Engineering (DIMES), University of Calabria, 87036 Rende, CS, Italy
prrcrl02r14c3521@studenti.unical.it

Abstract. The increasing frequency and sophistication of Denial of Service (DoS) and Distributed Denial of Service (DDoS) attacks, pose significant challenges to modern cybersecurity systems. These threats are further complicated by stealthy variants such as slow DoS attacks, which often evade timely detection. While Deep Learning (DL)-based Intrusion Detection Systems (IDSs) have shown promise in analyzing complex network traffic, their effectiveness is hindered by challenges like limited labeled data, noise, and the presence of Out-of-Distribution (OOD) samples. This paper proposes a hybrid DL-based IDS framework (*ENE4*) that integrates unsupervised and supervised components to improve detection performance under label-scarce conditions. The unsupervised module extracts task-independent features from network traffic, while the supervised one learns task-specific representations. These complementary features are fused to enable robust detection even in few-shot learning settings. Additionally, the model incorporates an adaptation mechanism to leverage knowledge from more frequent and related attack types, enhancing generalization to rare patterns. Experimental results on two standard benchmark datasets demonstrate the effectiveness and robustness of the proposed approach in detecting evasive DoS attacks.

Keywords: Cybersecurity · Deep Learning · Few-Shot Learning · Variational Autoencoder · Ensemble Neural Network

1 Introduction

With attackers increasingly targeting system availability, Denial-of-Service (DoS) and Distributed-Denial-of-Service (DDoS) attacks pose a significant and

S. Džeroski et al. (Eds.): DS 2025, LNAI 16090, pp. 347–362, 2025.
https://doi.org/10.1007/978-3-032-05461-6_23

ongoing cybersecurity challenge. These attacks aim at disrupting service availability by overwhelming systems with traffic or resource requests, often causing severe downtime and financial loss. Highlighting the scale of modern DDoS threats, on July 21st, 2022, the cybersecurity firm Akamai recorded the largest DDoS attack ever observed against a European client, with globally distributed traffic peaking at 853.7 Gbps and 659.6 Mbps.[1]

While traditional DoS and DDoS attacks rely on high volumes of illegitimate traffic, a more covert and harder to detect variant, named *slowDoS* attacks, has emerged [7]. Unlike conventional DoS attacks, slowDoS ones mimic legitimate client behavior by sending low-rate, intentionally delayed requests that exploit server-side connection handling. This subtle approach allows attackers to consume server resources while flying under the radar of standard detection tools. Because the traffic they generate closely resembles that of normal users, slowDoS attacks often evade traditional Intrusion Detection Systems (IDSs), making them particularly challenging to identify and mitigate [4].

In this scenario, Deep Learning (DL)-based approaches represent a highly effective solution [1,2,10,15,18], which does not require labor-intensive feature engineering and can automatically learn hierarchical traffic data representations capturing subtle patterns that may well elude traditional methods [9]. Moreover, DL models can be adapted to process streaming data, enabling real-time analysis and detection. This makes them particularly well-suited for dynamic and evolving threat environments [6], where the ability to learn from new data on the fly is crucial for maintaining robust intrusion detection capabilities.

However, the evasive nature of slowDoS attacks and the scarcity of real-world datasets exacerbate detection efforts. Labeling such data is often impractical, limiting the effectiveness of fully supervised DL methods that rely heavily on large volumes of annotated examples. As a result, DL-based IDSs, while powerful, struggle to maintain accuracy under data-scarce conditions typical of rare or emerging threats like slowDoS attacks.

Contribution. To overcome these limitations, this paper proposes a semi-supervised DL framework based on *ENE4*–an **E**nsemble **N**etwork of feature **E**xtractors for (**4**) few shot learning– that combines the strengths of both supervised and unsupervised learning. Essentially, in this framework, an unsupervised component is used to extract general-purpose features from raw, unlabeled traffic data, while a supervised module learns task-specific patterns from the limited labeled samples available. These representations are then fused to improve the accuracy and robustness of attack detection, even when having very few attack examples using a few-shot learning approach. The framework also includes a knowledge adaptation mechanism that transfers insights from more common attack types to enhance its generalization to stealthier, and less frequent, patterns.

Experimental results on a benchmark dataset demonstrate that the proposed approach achieves improved performance in detecting slow DoS attacks, offering

[1] See https://www.akamai.com/blog/security/largest-european-ddos-attack-ever (accessed on March 30, 2024) for details on this attack.

a promising direction for strengthening intrusion detection in the face of subtle, evasive threats.

Novelty. Our proposed semi-supervised deep learning framework differs from existing approaches tackling the problem to identify in several important ways. Unlike many prior works that focus solely on either supervised or unsupervised methods, our framework combines a variational component, extracting general-purpose features from raw unlabeled traffic data, with a supervised pre-trained classifier, learning task-specific patterns from limited labeled data. This fusion of complementary representations allows the model to achieve improved detection accuracy and robustness, particularly in few-shot learning scenarios.

Moreover, our approach incorporates a knowledge adaptation mechanism that transfers knowledge gained from more common attack types to enhance generalization on stealthier and less frequent slow DoS patterns. While this type of approach is known in the intrusion detection literature, its adoption for slow DoS detection has been scarcely explored.

The classification head of our neural architecture is trained on a joint representation obtained by concatenating two feature sets: *attack-agnostic* features extracted by an unsupervised subnet, and *attack detection-oriented* features extracted by a supervised subnet. This combination enables the model to leverage the strengths of both learning paradigms —unsupervised feature discovery and supervised discrimination— resulting in detection performance that exceeds what either component could achieve independently.

In addition, the proposed framework incorporates a data augmentation module into its backbone, which introduces controlled variability into the training data. This enhances the model's ability to generalize from limited labeled examples—an aspect rarely addressed in existing semi-supervised NIDS solutions.

Finally, it is worth noting that our work explicitly targets the few-shot learning setting for slow DoS detection. Many existing methods require large volumes of labeled data or focus on broad anomaly detection without tackling the challenges posed by limited labeled samples for stealthy slow DoS attacks. By focusing on this realistic and challenging scenario, our framework is particularly well-suited for real-world deployments where labeled attack data are scarce.

Organization. The rest of this paper is structured as follows: After providing an overview of existing approaches in Sect. 2, the proposed method *ENE4* is illustrated in detail, in Sect. 3. The experimental study conducted to evaluate this method, comparatively with previous ones in the field, is then illustrated in Sect. 4. The concluding section finally discusses the main experimental findings and the main contributions of our research work, as well as some limitations of it and future research directions.

2 Related Work

Network Intrusion Detection Systems (NIDSs) play a crucial role in safeguarding network infrastructure from a wide range of attacks, including DoS and, more recently, slow DoS attacks. NIDSs are typically divided into two main categories: *signature-based NIDSs (SNIDSs)* and *anomaly-based NIDSs (ANIDSs).*

Signature-based NIDSs typically operate by comparing observed network traffic against a database of known attack signatures. For example, Snort an open-source Intrusion Prevention System (IPS) tries to detect and prevent malicious network activity using a rule-based engine. These systems are highly effective at detecting well-known attacks for which signatures exist, but they struggle to identify novel or evolving threats, such as new variants of slowDoS attacks, for which no signature is yet available.

Anomaly-based NIDs (ANIDSs) address this challenge by learning a model of normal behavior and flagging significant deviations as potential intrusions. Although they are prone to high false alarm rates (FAR) [3], ANIDSs are commonly employed in intrusion detection scenarios, particularly for their potential to identify previously unseen, or zero-day, attacks.

To address the limitations of traditional approaches, researchers and practitioners have been increasingly resorting to data-driven approaches, leveraging both machine learning and Deep Learning (DL) techniques.

These methods can be applied in both supervised and unsupervised learning paradigms. In addition, some preliminary hybrid approaches have been proposed for the general intrusion detection task without focusing on evasive attacks that are the most challenging to reveal. Here, we provide an overview of the approaches and also discuss recent methods designed for few-shot learning.

Supervised Learning Methods. Supervised approaches require labeled datasets, where each traffic instance is annotated as benign or malicious. In recent years, several supervised ML and DL models have been proposed for slowDoS detection.

In [21], the authors propose the use of ensemble learning methods such as Random Forests (RF) and Gradient Boosting Machines (GBM) to classify slowDoS traffic. Their approach leverages features including connection duration, incomplete HTTP headers, and request rates, demonstrating high accuracy in real-time detection scenarios. The effectiveness of RF and GBM in detecting slowDoS variants like Slowloris is highlighted through extensive evaluation metrics.

An approach based on Convolutional Neural Networks (CNNs) is presented in [19], where the authors show that CNNs excel at extracting spatial features from network traffic data. This enables the model to achieve high precision in distinguishing slowDoS attacks from legitimate traffic flows.

In [17], Multi-Layer Perceptrons (MLPs) are integrated into the ISSDM framework, which is designed to classify traffic in software-defined networks. This method leverages features such as response times and connection patterns, resulting in robust detection rates for slowDoS attacks.

Furthermore, hybrid approaches that combine feature selection techniques with supervised classifiers are explored in [11]. These methods aim to optimize detection performance while reducing computational overhead, thereby enhancing the practicality of slowDoS detection systems.

Despite their effectiveness, supervised methods are limited by the availability and quality of labeled data, which can be costly and time-consuming to obtain, especially for rare or emerging attack types.

Unsupervised Learning Methods. Unsupervised approaches do not require labeled data and are well-suited to detecting novel or previously unseen attacks. In [4], the authors leverage Variational Autoencoders (VAEs) to learn the reconstruction of normal network traffic patterns, flagging deviations as potential anomalies; the probabilistic encoding in the latent space improves generalization and reduces overfitting. The detection of low-rate DoS attacks using clustering algorithms is presented in [28], where the SADBSCAN method is applied to group network traffic based on discrete features such as packet intervals and TCP behavior, while in [27] and [26], the WEDMS and two-step cluster analysis techniques are proposed for similar purposes. In [5], the authors propose the use of sparse U-Net-like autoencoders to extract hierarchical representations of network traffic, supporting the detection of subtle anomalies associated with low-rate DoS attacks. Furthermore, an approach that combines unsupervised autoencoder learning with triplet loss for semi-supervised intrusion detection is presented in [2], showing that limited supervision can enhance detection accuracy in challenging network environments.

A major limitation of unsupervised approaches lies in their tendency to generate a high number of false alarms, which significantly hinders their practical deployment in large-scale networks. While these methods excel at detecting novel or previously unseen attacks without requiring labeled data, the lack of contextual information often causes benign traffic fluctuations to be misclassified as malicious, resulting in excessive false positive rates.

Few-shot Learning Methods. Few-shot learning has recently emerged as a promising direction in network intrusion detection, particularly for scenarios where labeled data are scarce or new attack types appear.

In [13], the authors propose a few-shot intrusion detection method leveraging discriminative representation learning with a supervised autoencoder. Their approach first trains on known samples and then adapts to new attack categories with very few labeled examples; notably, this methodology most closely resembles our own work. The use of prototypical capsule networks combined with attention mechanisms to address data imbalance and enable few-shot learning in network intrusion detection is presented in [25], where strong performance is achieved on imbalanced datasets.

In [12], the authors propose a few-shot network intrusion detection method based on mutual information maximization. Their approach enhances feature representation learning in low-labeled data regimes, enabling effective detection of novel and rare attack types. This method combines unsupervised feature extraction with supervised fine-tuning, improving generalization to unseen intrusions. Finally, in [29], the authors introduce a few-shot learning framework for network intrusion detection based on meta-learning. Their approach employs a balanced resampling method, deep neural networks, convolutional neural networks, and the CenterLoss function to improve performance. Key contributions include the application of FSL to network intrusion detection, the use of balanced resampling, the proposal of network models suitable for sequence data, and the application of CenterLoss.

3 Methodology

In this section, we present *ENE4*, a DL-based solution designed to detect low-rate DoS attacks. The core idea of our approach is to leverage two complementary neural backbones: the first is trained exclusively on legitimate traffic flows to learn *attack-agnostic features*, while the second is pre-trained in a supervised fashion using labeled examples of more common attack types (e.g., standard DoS), yielding *attack-detection-oriented features.*

As depicted in Fig. 1, these two subnetworks are integrated into a single architecture and subsequently fine-tuned using a small number of labeled instances of the specific attack class of interest—namely, slowDoS. This hybrid design enables the model to generalize from both benign and frequent attack behaviors while adapting to the subtle characteristics of slow-rate attacks.

3.1 *ENE4* Overview

Here, we provide an overview of the key components of our neural architecture.

- **Input Layer:** The input layer is the initial component responsible for receiving the input data and forwarding it to the subsequent layers for further processing. In our case, this input consists of a set of statistical features computed

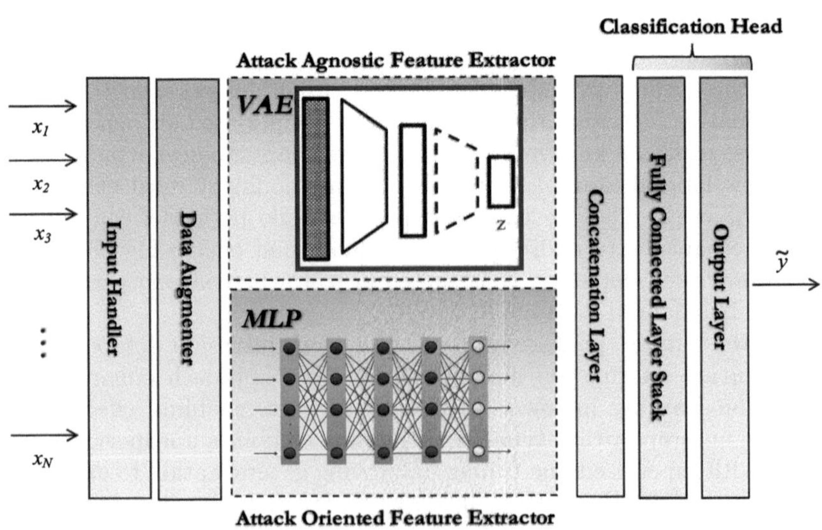

Fig. 1. Overview of the proposed neural architecture *ENE4*.

from network traffic flows, denoted as $\langle x_1, x_2, \ldots, x_N \rangle$ (more details about them can be found in the original papers describing the datasets [22,24]). Each feature corresponds to one node in the input layer, which does not perform any computation but simply passes these values directly to the next layer.

- **Attack Agnostic Feature Extractor:** This is an unsupervised component used to extract task-independent features from raw, unlabeled data. To this purpose, we employed the latent space discovered by a Variational Auto-Encoder (VAE) [14,20]. Unlike techniques like PCA, which yield linear projections, or visualization methods like t-SNE, VAEs offer a more flexible and general method for learning compact, informative latent spaces. Our proposed framework leverages VAEs to extract meaningful representations from the data, allowing us to exploit the natural clustering of similar instances in this latent space.
- **Attack Detection Oriented Feature Extractor.** In our framework it takes the form of a pre-trained Multi-Layer Perceptron (MLP) network that leverages knowledge acquired from a related task, such as detecting DoS attacks. During its initial training phase, the MLP learns to identify patterns and features relevant to this related task. After training, the output layer is removed, and the feature vector produced by the penultimate layer is extracted. These learned features, which capture high-level abstractions of network traffic behavior, are then used as inputs to the classification head of our model. This transfer of learned representations enables the system to benefit from prior knowledge, improving detection accuracy and efficiency when adapting to new or less frequent attack types. Further details about the MLP architecture are provided in the following.
- **Data Augmentation Component:** It is implemented as a Gaussian Noise layer positioned directly after the input layer. This layer adds additive Gaussian noise to the input features during training, which encourages the model to generalize better by simulating natural variability and mitigating overfitting. This process helps the network become robust to small fluctuations and noise inherent in real-world data. The Gaussian Noise layer is active only during training and is disabled during inference to ensure stable and reliable predictions.
- **Classification Head:** This component is trained by combining features extracted from the encoder of the VAE with the features yielded by the pre-trained classifier. Basically, it is composed of a concatenation layer, a stack of m fully connected layers, and an output layer.
- **Class-imbalance aware loss function.** To handle the rare nature of slow-DoS *ENE4* leverages a *Focal loss* [16], defined as follows:

$$FL(p_t) = -w_t(1 - p_t)^{\gamma} \log (p_t)$$

where p_t denotes the predicted probability for the true class, γ is the focusing parameter, and w_t is a class-specific weight (optional).

Training Pipeline. First, we train a Variational Autoencoder (VAE) on a dataset containing a heterogeneous mix of traffic instances, including both benign and malicious traffic, leveraging its ability to capture meaningful representations by organizing similar samples within its latent space.

Simultaneously, we train a supervised model (referred to as the MLP in Fig. 1) on an auxiliary task that is related to—but not the same as—the main detection task. The insights learned during this phase are then transferred to support the final classification task, improving the model's ability to learn effectively from a small amount of labeled data. This training strategy follows a few-shot learning paradigm, enabling the model to adapt to the target task with minimal labeled examples.

Finally, the classification head is trained on the full dataset, which includes a limited number b of labeled instances belonging to the attack class of interest, along with samples of normal traffic.

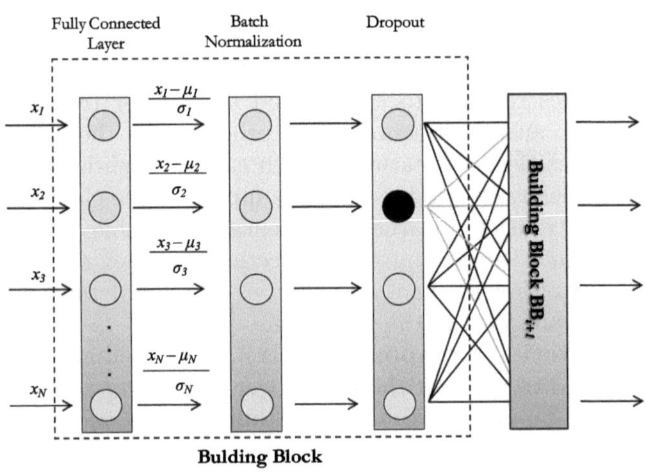

Fig. 2. Building block of the supervised component.

Network Architecture. As presented above, the proposed neural architecture is composed of three interconnected sub-networks. The first sub-network is inspired by the work of [8], where the authors introduce a modular design pattern (see Fig. 2) called a *building block*. Each building block consists of a fully connected (dense) layer followed by a batch normalization layer and a dropout layer. This pattern can be stacked multiple times to increase the depth of the network. The two key hyperparameters controlling this sub-network are the number of repetitions k of the building block and the number of neurons p in each fully connected layer. For all hidden layers, the ReLU activation function is applied to introduce non-linearity, while a softmax is applied to the output layer.

The second sub-network is a VAE with an encoder-decoder structure and a latent code of dimension d; only the encoder is used during inference. Specifically, the encoder maps the input into a d-dimensional latent representation through two linear layers with ReLU activations and batch normalization, followed by separate linear layers that produce the mean and log-variance of the latent distribution.

Finally, the third sub-network, referred to as the *classification head*, combines the latent features from the VAE encoder and the feature extracted by the pre-trained classifier by concatenating them into a joint representation. These combined representations are passed through a three-layer fully connected network with ReLU activations to generate the final prediction.

4 Experimental Evaluation

In this section, we present the experimental investigation conducted to evaluate the performance of *ENE4*. We begin by describing the datasets and parameter settings used in our experiments. Then, we analyze the results obtained, with a focus on answering the following two research questions:

- **RQ1:** Does our proposed solution outperform standard baseline algorithms from the literature in scenarios where only a limited number of labeled samples are available?
- **RQ2:** How does the detection performance of the model vary with the number of labeled samples used during training?

By addressing these questions, we aim to assess both the effectiveness and the robustness of *ENE4* in few-shot learning scenarios, particularly in detecting evasive attacks such as slowDoS under realistic data constraints.

4.1 Evaluation Setting

Datasets and Competitors. In our experimentation, we considered two well-known benchmarks in literature: *CIC-IDS2017* and *5G-NIDD*. The datasets include different types of attacks, but we focused only on the ones belonging to the DoS family. In more detail:

- *CIC-IDS2017* [24]: This dataset contains benign traffic and up-to-date attacks captured over five days, with labeled network flows. It includes diverse attacks and protocols, network configurations, and metadata, addressing shortcomings of previous datasets;
- *5G-NIDD* [22]: This dataset was built on a 5G test network; it contains labeled data from DoS attacks and port scans, extracted from base stations and attacker nodes.

Detailed information about the class distribution of both datasets can be found in their original publications.

We benchmarked *ENE4* against several widely used classical machine learning (ML) techniques, including K-Nearest Neighbors (KNN), XGBoost, Random Forest, and Decision Tree based on the implementation of scikit-learn library[2]. These methods were selected due to their broad adoption and proven effectiveness in the domain of network traffic analysis and intrusion detection. Unlike our proposed framework, which supports a staged and fine-tuned training process, these classical models do not allow for transfer learning or fine-tuning. Therefore, they were trained directly on the datasets containing both DoS and Slow DoS attacks. Each method was evaluated using precision (Prec), recall (Rec), and F1-score as performance metrics.

For training, we used all benign records and all available DoS attack samples from the CIC-IDS2017 dataset. For the 5G-NIDD dataset, we included all instances of HTTP-Flood attacks, as these classes are widely available and easier to retrieve. In contrast, Slow-DoS attacks are significantly more challenging to collect and annotate. To simulate a realistic and difficult few-shot learning scenario, we considered a challenging scenario where the budget b of labelled attack examples is 5. Given the challenging scenario, characterized by a severely limited amount of available data, we deliberately chose not to use a validation set during training.

Furthermore, we trained the model exclusively on the CIC-IDS2017 dataset to evaluate the learning capabilities of our proposed framework under varying quantities of available training instances from the target class.

Settings and Parameters. The network was trained using the Adam optimizer with a learning rate of 0.0001. A dropout rate of 0.05 was applied throughout the fully connected layers to prevent overfitting. The latent space of the variational autoencoder was set to dimension $d = 8$, providing a compact yet expressive representation. The supervised component has been instantiated with $k = 5$ building blocks and $p = 32$ neurons per layer.

Training was conducted in multiple stages: both the predictive model and the VAE were initially trained for 150 epochs using only normal traffic and DoS attack samples. This was followed by a first phase of training the ensemble head for a maximum of 50 epochs, with early stopping applied to prevent overfitting. A subsequent adaptation phase was then carried out for an additional 150 epochs. Due to the extremely limited availability of labeled data, model selection was based on performance on the training set. Specifically, the model achieving the highest Area Under the ROC Curve (AUC) on the training data was selected as the final model, as no separate validation set could be constructed in this few-shot learning scenario.

The loss function used during the early training phases was the Cross Entropy, which was later replaced with Focal Loss [16] to better address class imbalance. To further improve performance under low-data conditions, the gamma parameter was increased as the number of training samples decreased.

[2] https://scikit-learn.org.

During training, Gaussian noise with mean 0.0 and standard deviation 0.05 was injected into the head's input space to promote robustness and improve generalization.

To ensure a fair and reproducible comparison, all baseline machine learning models were configured using standard hyperparameter settings commonly adopted in the literature, with minor adjustments based on preliminary experiments. Decision Tree Classifier was constrained to a maximum depth of 3 to prevent overfitting and maintain model interpretability. The KNN algorithm was configured with 3 neighbors, balancing sensitivity and computational cost. The Random Forest Classifier employed 80 estimators to provide a sufficiently robust ensemble. Similarly, the XGBoost Classifier was set with a base score of 0.5 and 80 estimators, following common practices for binary classification while maintaining efficiency. These configurations were kept consistent across all experiments to ensure that performance differences reflect algorithmic capabilities rather than hyperparameter optimization.

All the experiments were run on an Intel Xeon CPU E5-2698 v4 @ 2.20 GHz, 250 GB RAM, with Tesla V100-DGXS-32GB GPU.

4.2 Test Results

The focus of this analysis is to evaluate each model's effectiveness in identifying rare and underrepresented Slow DoS attacks, particularly under highly imbalanced data conditions. This is a critical scenario in real-world intrusion detection, where new or stealthy attack types may be poorly represented in labeled data.

The results reported in Table 1 demonstrate the effectiveness of the proposed model *ENE4* compared to classical machine learning baselines on both the CIC-IDS2017 and 5G-NIDD datasets. On CIC-IDS2017, while *ENE4* achieves lower precision (0.60) than all baseline methods, it significantly outperforms them in recall (0.39) and consequently in F1 score (0.47), indicating a better balance between detecting true positives and limiting false negatives. The baseline models, particularly KNN, XGBoost, Random Forest, and Decision Tree, show perfect or near-perfect precision but suffer from very low recall, suggesting that they are overly conservative and miss many positive instances. In contrast, on the 5G-NIDD dataset, *ENE4* attains both very high precision (0.95) and recall (0.99), resulting in an F1 score of 0.97, substantially outperforming all baselines. Notably, several baselines such as KNN fail to detect any positive instances on 5G-NIDD, and even the best-performing baseline (Decision Tree) trails behind *ENE4* in overall balance.

Table 2 further illustrates how *ENE4*'s performance scales with the number of training samples from the target class. Even with very limited data (as few as 5 samples), the model achieves an F1 score of 0.47, which gradually improves as more samples become available. Higher performance gains are achieved when the sample size exceeds 600, with the F1 score reaching 0.87 at 9,000 samples. This trend highlights the model's ability to learn effectively in low-data regimes and its continued improvement with increasing data availability (Fig. 3).

Table 1. Comparison of *ENE4* and baseline machine learning techniques on the CIC-IDS2017 and 5G-NIDD datasets. Precision, Recall, and F1 scores are reported for each method to evaluate detection performance across both datasets.

Dataset	Technique	Prec	Rec	F1-Score
CIC-IDS2017	*ENE4*	0.60	**0.39**	**0.47**
	KNN	**1.00**	0.16	0.28
	XGBoost	**1.00**	0.22	0.36
	Random Forest	**1.00**	0.18	0.30
	Decision Tree	**1.00**	0.02	0.04
5G-NIDD	*ENE4*	0.95	**0.99**	**0.97**
	KNN	0.00	0.00	0.00
	XGBoost	**1.00**	0.16	0.28
	Random Forest	**1.00**	0.02	0.04
	Decision Tree	**1.00**	0.62	0.77

Table 2. Performance metrics of the proposed model *ENE4* across different training sample sizes on the CIC-IDS2017 dataset. The table reports Precision, Recall, and F1 scores corresponding to the varying numbers of training instances used, illustrating how model performance scales with data availability.

Budget	Prec	Rec	F1-Score
5	0.60	0.39	0.47
20	0.55	0.41	0.47
125	0.55	0.41	0.47
625	0.94	0.39	0.55
1250	0.99	0.66	0.79
3000	0.98	0.73	0.84
9000	0.88	0.87	0.87

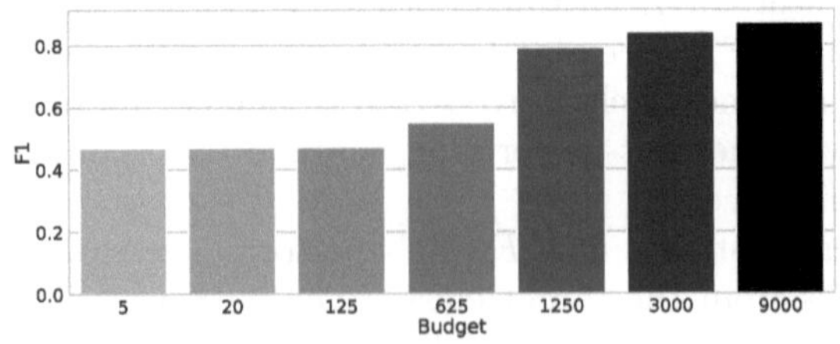

Fig. 3. Variation of the F1 score for the proposed model *ENE4* as a function of the number of training samples. The bar chart displays the F1 score (y-axis) achieved by *ENE4* when trained with varying amounts of data, specifically with 5, 10, 20, 40, 125, 625, 1250, 3000, and 9000 samples (x-axis).

5 Discussion and Conclusion

In this work, we defined a deep learning-based framework for detecting low-rate, slow DoS attacks. The approach leverages a hybrid architecture composed of supervised and unsupervised subnets and employs a pre-training strategy to accelerate the neural model's learning. Based on our findings, *ENE4* demonstrates strong and consistent performance in detecting underrepresented slow DoS attacks within highly imbalanced network traffic datasets. It achieves a more balanced trade-off between precision and recall compared to classical machine learning techniques, significantly outperforming them on both the CIC-IDS2017 and 5G-NIDD datasets. While traditional baselines often exhibit high precision at the cost of poor recall, failing to detect many positive instances, *ENE4* proves effective even when trained with a minimal number of labeled examples, underscoring its robustness in few-shot learning settings.

In addition, the model's performance scales with data availability, yet remains reliable in low-data regimes, reinforcing its practicality for real-world deployment where labeled data is often limited.

Furthermore, by adopting a training paradigm grounded in knowledge transfer, where a pre-trained model serves as background knowledge for subsequent fine-tuning, *ENE4* significantly reduces the computational overhead typically required to train deep models from scratch. This approach not only accelerates the training process but also aligns with the principles of Green AI [23], which advocate for energy-efficient and resource-conscious machine learning. By reusing previously acquired knowledge, *ENE4* achieves robust performance while minimizing energy consumption and environmental impact.

Future Work. Building on the promising results of *ENE4* in detecting underrepresented Slow DoS attacks, several avenues for future research emerge. First, we aim to extend the evaluation of the proposed framework to a broader range of cyber threats, including other forms of low-rate or stealthy attacks, such as data exfiltration, botnet command-and-control traffic, or advanced persistent threats (APTs) we also plan to conduct an ablation study to analyze how individual components—namely the VAE, MLP, and classifier—contribute to the overall performance of *ENE4*. In particular, we aim to quantify the impact of the VAE and transfer learning on the proposed approach through systematic ablation. Additionally, we plan to investigate the usage of *ENE4* in a human-in-the-loop scenario in order to leverage the domain expert to label the most informative unlabeled instance and boost the model performance.

Finally, we plan to explore dynamic fine-tuning strategies and continual learning techniques to better handle evolving threats and data drift over time.

Test Reproducibility. The code necessary to replicate our experiments is available at: https://github.com/Franco7Scala/SlowDOSEarlyDetection.

Acknowledgment. This work was partially supported by research project FAIR (PE00000013), funded by the EU under the program NextGeneration EU, and by project SERICS (PE00000014) under the NRRP MUR program funded by the EU - NGEU.

References

1. Ahmad, Z., et al.: Network intrusion detection system: a systematic study of machine learning and deep learning approaches. Trans. Emerg. Telecommun. Technol. **32**(1), e4150 (2021)
2. Andresini, G., Appice, A., Malerba, D.: Autoencoder-based deep metric learning for network intrusion detection. Inf. Sci. **569**, 706–727 (2021)
3. Axelsson, S.: The base-rate fallacy and the difficulty of intrusion detection. ACM Trans. Inf. Syst. Secur. **3**(3), 186–205 (2000)
4. Bårli, E.M., Yazidi, A., Viedma, E.H., Haugerud, H.: Dos and DDoS mitigation using variational autoencoders. Comput. Netw. **199**, 108399 (2021)
5. Cassavia, N., Folino, F., Guarascio, M.: Detecting dos and DDoS attacks through sparse U-net-like autoencoders. In: Reformat, M.Z., Zhang, D., Bourbakis, N.G. (eds.) 34th IEEE International Conference on Tools with Artificial Intelligence, ICTAI 2022, Macao, China, 31 October–2 November 2022, pp. 1342–1346. IEEE (2022). https://doi.org/10.1109/ICTAI56018.2022.00203
6. Coppolillo, E., et al.: Generative methods for out-of-distribution prediction and applications for threat detection and analysis: a short review. In: Digital Sovereignty in Cyber Security: New Challenges in Future Vision (CyberSec4Europe 2022), vol. 1807, pp. 65–79 (2022)
7. ENISA: ENISA threat landscape 2020 - list of top 15 threats (2020). https://www.enisa.europa.eu/publications/enisa-threat-landscape-2020-list-of-top-15-threats
8. Folino, F., Folino, G., Guarascio, M., Pisani, F.S., Pontieri, L.: On learning effective ensembles of deep neural networks for intrusion detection. Inf. Fusion **72**, 48–69 (2021). https://doi.org/10.1016/J.INFFUS.2021.02.007
9. Guarascio, M., Manco, G., Ritacco, E.: Deep learning. Encycl. Bioinform. Comput. Biol.: ABC Bioinform. **1–3**, 634–647 (2018)
10. Guarascio, M., Cassavia, N., Pisani, F.S., Manco, G.: Boosting cyber-threat intelligence via collaborative intrusion detection. Future Gener. Comput. Syst. **135**, 30–43 (2022)
11. Hosseini, S., Azizi, M.: The hybrid technique for DDoS detection with supervised learning algorithms. Comput. Netw. **158**, 35–45 (2019). https://doi.org/10.1016/j.comnet.2019.04.027, https://www.sciencedirect.com/science/article/pii/S1389128618306881
12. Iliyasu, A.A., et al.: A few-shot network intrusion detection method based on mutual information maximization. Sci. Rep. **15**, 93185 (2025). https://doi.org/10.1038/s41598-025-93185-0, https://www.nature.com/articles/s41598-025-93185-0
13. Iliyasu, A.S., Abdurrahman, U.A., Zheng, L.: Few-shot network intrusion detection using discriminative representation learning with supervised autoencoder. Appl. Sci. **12**(5) (2022). https://doi.org/10.3390/app12052351, https://www.mdpi.com/2076-3417/12/5/2351
14. Kingma, D.P., Welling, M.: Auto-encoding variational bayes. arXiv preprint arXiv:1312.6114 (2013)
15. Kocher, G., Kumar, G.: Machine learning and deep learning methods for intrusion detection systems: recent developments and challenges. Soft. Comput. **25**(15), 9731–9763 (2021)

16. Lin, T.Y., Goyal, P., Girshick, R., He, K., Dollár, P.: Focal loss for dense object detection. In: 2017 IEEE International Conference on Computer Vision (ICCV), pp. 2999–3007 (2017). https://doi.org/10.1109/ICCV.2017.324
17. Mabel, P., Nagappasetty, R.: An intelligent system to detect slow denial of service attacks in software-defined networks. Int. J. Electr. Comput. Eng. (IJECE) **13**, 3099 (2023). https://doi.org/10.11591/ijece.v13i3.pp3099-3110
18. Mishra, P., Varadharajan, V., Tupakula, U., Pilli, E.S.: A detailed investigation and analysis of using machine learning techniques for intrusion detection. IEEE Commun. Surv. Tutor. **21**(1), 686–728 (2019)
19. Nugraha, B., Murthy, R.N.: Deep learning-based slow DDoS attack detection in SDN-based networks. In: 2020 IEEE Conference on Network Function Virtualization and Software Defined Networks (NFV-SDN), pp. 51–56 (2020). https://doi.org/10.1109/NFV-SDN50289.2020.9289894
20. Pinheiro Cinelli, L., Araújo Marins, M., Barros da Silva, E.A., Lima Netto, S.: Variational autoencoder. In: Variational Methods for Machine Learning with Applications to Deep Networks, pp. 111–149 (2021)
21. Rios, V., Inacio, P., Magoni, D., Freire, M.: Detection of slowloris attacks using machine learning algorithms, SAC 2024, pp. 1321–1330. Association for Computing Machinery, New York (2024). https://doi.org/10.1145/3605098.3635919
22. Samarakoon, S., et al.: 5G-NIDD: a comprehensive network intrusion detection dataset generated over 5g wireless network (2022). https://doi.org/10.21227/xtep-hv36
23. Schwartz, R., Dodge, J., Smith, N.A., Etzioni, O.: Green AI. Commun. ACM **63**(12), 54–63 (2020). https://doi.org/10.1145/3381831
24. Sharafaldin., I., Habibi Lashkari., A., Ghorbani., A.A.: Toward generating a new intrusion detection dataset and intrusion traffic characterization. In: Proceedings of the 4th International Conference on Information Systems Security and Privacy - ICISSP, pp. 108–116. INSTICC, SciTePress (2018). https://doi.org/10.5220/0006639801080116
25. Sun, H., Wan, L., Liu, M., Wang, B.: Few-shot network intrusion detection based on prototypical capsule network with attention mechanism. PLOS ONE **18**(4), 1–18 (2023). https://doi.org/10.1371/journal.pone.0284632
26. Tang, D., Dai, R., Tang, L., Li, X.: Low rate dos attack detection based on two-step cluster analysis and UTR analysis. Hum.-Cent. Comput. Inf. Sci. **10**, 1–20 (2020). https://hcis-journal.springeropen.com/articles/10.1186/s13673-020-00235-2
27. Tang, D., Man, J., Tang, L., Feng, Y., Yang, Q.: WEDMS: an advanced mean shift clustering algorithm for LDoS attacks detection. Ad Hoc Netw. **102**, 102145 (2020). https://doi.org/10.1016/j.adhoc.2020.102145, https://www.sciencedirect.com/science/article/pii/S1570870519302653
28. Tang, D., Zhang, S., Chen, J., Wang, X.: The detection of low-rate DoS attacks using the SADBSCAN algorithm. Inf. Sci. **565**, 229–247 (2021). https://doi.org/10.1016/j.ins.2021.02.038, https://www.sciencedirect.com/science/article/pii/S0020025521001808
29. Yu, Y., Bian, N.: An intrusion detection method using few-shot learning. IEEE Access **8**, 49730–49740 (2020). https://doi.org/10.1109/ACCESS.2020.2980136

Adaptive Isolation Forest

Jia Justin Liu[1(✉)], Guilherme Weigert Cassales[1], Fei Tony Liu[2,3], Bernhard Pfahringer[1], and Albert Bifet[1,4]

[1] AI Institute, University of Waikato, Hamilton, New Zealand
{justin.l,guilherme.cassales,bernhard,abifet}@waikato.ac.nz
[2] Artificial General Intelligence Pty Ltd., Victoria, Australia
[3] University of New South Wales, Kensington, Australia
fei.t.liu@unsw.edu.au
[4] LTCI, Télécom Paris, IP Paris, Paris, France

Abstract. Anomaly detection in real-world data streams often struggles with concept drift, where evolving data distributions challenge algorithms to maintain a balance between accuracy, speed, stability, and plasticity. We present Adaptive Isolation Forest (AIF), a novel anomaly detection algorithm designed to effectively adapt to such changes in a resource-efficient and balanced manner. AIF's novelty can be found in the combination of a smart model update mechanism together with a newly developed MinTreeMaxMass (MTMM) criterion, which scores individual trees for replacement. Extensive evaluations on various benchmark datasets demonstrate that AIF significantly outperforms existing state-of-the-art streaming anomaly detection algorithms in terms of detection accuracy. Moreover, AIF achieves linear time and space complexities, providing a robust solution that maintains high accuracy and efficiency, balancing stability and plasticity in dynamic data streams.

Keywords: Anomaly Detection · Streaming Models · Ensembles

1 Introduction

Anomaly detection is one of the key problems in data mining and machine learning, focussing on recognizing abnormal data patterns that significantly deviate from the norm [3]. Practical applications include detecting financial fraud, network intrusions, equipment malfunctions, and disease outbreaks [16].

In many real-world applications, information is generated as continuous data streams, introducing unique complexities that require efficient algorithms capable of managing large volumes of data with limited computational resources [1]. Moreover, real-world data streams often undergo distribution changes over time, violating the IID assumption that most machine learning algorithms rely on [9]. Unlike traditional offline settings, streaming anomaly detection must balance stability and plasticity to maintain high performance over time [6]. Stability ensures that previously learnt normal patterns remain intact, preventing unnecessary false alarms due to minor fluctuations. Plasticity enables the model to

S. Džeroski et al. (Eds.): DS 2025, LNAI 16090, pp. 363–378, 2025.
https://doi.org/10.1007/978-3-032-05461-6_24

adapt to new and evolving data distributions, addressing concept drift [24]. The capacity to achieve this balance is termed adaptability.

We present a novel algorithm for detecting anomalies in evolving data streams using a smart-update mechanism with a novel criterion to manage partial model updates, maintaining good adaptability. This approach overcomes the limitations of existing methods that are either static or computationally expensive.

The main contributions of this work are summarised as follows:

- We introduce Adaptive Isolation Forest (AIF), a novel anomaly detection algorithm that adapts dynamically to changing data distributions. Its adaptivity is achieved by first assessing the ensemble's performance and then strategically replacing less effective trees with newly constructed trees qualified by a novel criterion - MinTreeMaxMass (MTMM).
- We propose a smart update mechanism that updates only part of the model instead of reconstructing the entire ensemble. By adopting this partial update mechanism using trees qualified by the MTMM criterion, AIF balances the resources required for stability, plasticity, accuracy, and detection speed, making it well-suited for online, real-time applications.
- We perform a comprehensive empirical assessment on various benchmark datasets comparing AIF with current state-of-the-art open source anomaly detection algorithms for data streams, demonstrating that AIF exceeds competitors in terms of AUC-ROC score.

The remainder of this paper is structured as follows. We start with the problem formulation in Sect. 2. Then, we give an overview of the recent developments in the literature, with a particular focus on anomaly detection algorithms for data streams. Next, we explain the AIF algorithm and its components. After that, we present the results of the AIF experiment with nine baselines. Finally, we conclude the article and suggest potential future research.

2 Problem Formulation

We consider the task of anomaly detection in data streams, which are assumed to be a continuous flow of d-dimensional examples $\mathbf{x}_i$. Specifically, $S = \{\mathbf{x}_1, \ldots, \mathbf{x}_i\}$, $\mathbf{x}_i \in \mathbb{R}^d$, which contains normal examples $\mathbf{x}_n \in S_n$ and anomalies $\mathbf{x}_a \in S_a$, resulting in the data set S, where $S = S_n \cup S_a$. Furthermore, the flow can be divided into a sequence of equally sized windows $S = \{\mathbf{w}_1, \cdots, \mathbf{w}_n\}$. We also assume that the data stream must be continuously and progressively monitored to assess the abnormality of $\mathbf{x}_i$ without analysing previous observations $\{\mathbf{x}_1, \cdots \mathbf{x}_{i-1}\}$.

We address the issue of identifying all anomalies $S_a \subset S$. Our goal is to create an anomaly score $\alpha : S \rightarrow \mathbb{R}$ that ranks the anomalies higher, meaning $\alpha(\mathbf{x}_a) \gg \alpha(\mathbf{x}_n)$ for all $\mathbf{x}_a \in S_a$ and $\mathbf{x}_n \in S_n$, so that a suitable threshold can separate them.

Algorithm 1. iTree(X, e, l)

Input: X - input data, e - current tree height, l - height limit
Output: an iTree

if $e \geq l$ or $|X| \leq 1$ **then**
  return leafNode($Size \leftarrow |X|$)
else
  let Q be a list of attributes in X
  randomly select an attribute $q \in Q$
  randomly select a split point $p \in [min_q, max_q]$
  $X_l \leftarrow filter(X, q < p)$
  $X_r \leftarrow filter(X, q \geq p)$
  return branchNode($Left \leftarrow iTree(X_l, e+1, l)$,
    $Right \leftarrow iTree(X_r, e+1, l)$,
    $SplitAtt \leftarrow q$,
    $SplitValue \leftarrow p$)
end if

3 Related Work

Anomaly detection algorithms have been extensively researched in the literature [3,23]. This section reviews the most relevant anomaly detection algorithms found in the literature and provides our analysis and insights.

Isolation Forest [12] is an algorithm for anomaly detection, known for its scalability and efficiency. It isolates anomalies through random splitting without assumptions about the data. The algorithm, as discussed in Algorithm 1, has a linear time complexity and uses Isolation Trees, where an anomaly score is based on the path length in trees. However, it cannot adapt to changes in the data streams over time.

To handle streaming data, iForestASD (iFASD) [4] modifies batch learning algorithms using a sliding window approach. The algorithm exemplifies this by updating its model only when concept drift is detected, based on changes in anomaly rates. Although efficient for fixed anomaly rates, real-world applications are challenging due to unpredictable rate changes.

LODA [18] uses sparse random linear projectors in an ensemble to project data onto a one-dimensional space, with histograms used for outputs. The mean log-likelihoods serve as anomaly scores, illustrating that weak detectors can perform well when used together.

RRCF [8] is inspired by the iForest algorithm and is adapted for stream learning. It dynamically maintains tree structures by updating nodes and using the removal impact scoring function. However, it is resource intensive because of frequent tree modifications.

The iMondrian Forest (iMF) [13] algorithm uses random binary trees to isolate data points. In data streams, it checks if a point is within each node's bounding box. An insertion begins at that node if a data point is outside the bounding box. Similarly to RRCF, it has high computational complexity.

Half-Space Trees (HST) [22] is an anomaly detection algorithm that uses randomness and ensemble techniques. It evaluates examples based on the mass of the nodes in the path from root to leaf, not the path length. However, determining feature ranges is challenging for open-ended streams.

xStream [14] is a density-based ensemble anomaly detector for streaming data with evolving examples and feature space. It provides constant time and space processing, suitable for noisy, high-dimensional streaming data.

RS-Hash [20] uses randomized hashing to create an ensemble with hashed data and weak detectors. It has linear computational complexity relative to the dataset size and fixed space requirements.

SDOstream (SDOS) [10] improves SDO for streaming data by incorporating a hyperparameter to manage model ageing and memory, allowing efficient anomaly detection.

Online Isolation Forest (oIFOR) [11] uses dynamic multiresolution histograms for adaptive data distribution, processing data incrementally with a sliding window to integrate new data and discard old data.

In general, these algorithms tend to rapidly discard previous models when exposed to recent examples, frequently disregarding the essential adaptability necessary to maintain a balance between stability and plasticity.

4 Adaptive Isolation Forest

We propose the Adaptive Isolation Forest (AIF) to detect anomalies in streaming data. Specifically, we present a novel model update mechanism that maintains a good adapbility between stability and plasticity to achieve a high level of performance when learning from new data. This mechanism is based on the MinTreeMaxMass (MTMM) method, which combines the weighted normalised tree size and the maximum mass across the ensemble of trees and newly constructed trees with recent examples.

4.1 Streaming Methodology

AIF maintains a set of trees constructed by Algorithm 1. Each tree profiles a set of examples to isolate them. The tree is built randomly to eliminate bias, which requires an ensemble to improve the performance of a single tree. AIF builds trees on a sliding window without subsampling. When a new window of examples is collected, AIF performs a model update mechanism based on MinTreeMaxMass (Algorithm 3) to replace one of the trees in the ensemble.

The AIF stream processing is presented in Algorithm 2. With a given stream S, window size w, ensemble size k, weight β, and a number of new trees m to consider, we initialise the ensemble $\mathbb{F}$ with the first window of w examples (lines 1 to 6). Note that AIF does not use subsampling. Subsampling would need to collect a larger set of examples at the start of a stream, thus delaying initialisation. As the trees select attributes and split points in a completely random fashion, they are still very diverse. Furthermore, for any subsequent window, only

Algorithm 2. AIF(S, w, k, β, m)

Input: S - stream, w - windows size, k - ensemble size, β - weight, m - number of new trees
Output: anomaly scores

```
 1: F ← φ
 2: l ← ceiling(log2|X|)
 3: collect the first window of w examples X from S
 4: for i = 1 to k do
 5:     F ← F ∪ iTree(X, 0, l)
 6: end for
 7: X ← φ
 8: while data stream continues do
 9:     get new data point x from S
10:     s ← Score(x) {output s as anomaly score for x}
11:     output s as anomaly score for x
12:     X ← X ∪ x
13:     if |X| = w then
14:         F ← UpdateForest(X, F, β, m)
15:         X ← φ
16:     end if
17: end while
```

one tree can enter the ensemble, thus achieving a balance between stability and plasticity. After model initialisation, we continue to test the arriving examples while collecting them in a window. When a new example arrives, it is assigned an anomaly score and placed in the window (lines 10–12). If the window is full, the Algorithm 3 is called (lines 13 to 16) to update the model. The anomaly score function is defined as

$$Score(\mathbf{x}) = 2^{-\frac{\Sigma_{T \in \mathbb{F}} h_T(\mathbf{x})}{kc(w)}}, \tag{1}$$

where $h_T(\mathbf{x})$ denotes the path length that x travels within a tree $T \in \mathbb{F}$, and $c(w)$ indicates the average path length for an incomplete search in a binary tree.

4.2 Model Update Mechanism

The proposed mechanism uses MinTreeMaxMass (MTMM) to evaluate the capacity of individual trees to distinguish anomalies from normal examples. After each tree is evaluated, low-scoring trees are replaced with trees built from recent examples. In this section, we introduce the definition of MTMM and the model update mechanism based on it.

To understand the mechanism of the proposed MTMM, we first examine the anomaly discrimination power of trees by measuring the difference between the average path length of normal examples from S_n and that of anomalies from S_a in a dataset S, where $S = S_n \cup S_a$.

$$\mathrm{D}_T(S) = \frac{1}{|S_n|} \sum_{\mathbf{x}_n \in S_n} h_T(\mathbf{x}_n) - \frac{1}{|S_a|} \sum_{\mathbf{x}_a \in S_a} h_T(\mathbf{x}_a), \tag{2}$$

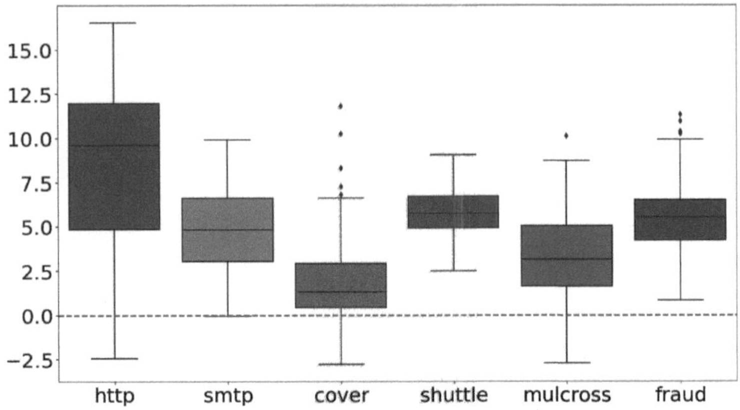

Fig. 1. Distribution of the discrimination power of iTree D(S) between the average path length of normal examples and anomalies.

where T stands for a tree, $h_T(\mathbf{x})$ is the path length of $\mathbf{x}$ traversing the tree T. Figure 1 shows the distribution of the discrimination power for 100 trees in a sample of the chosen benchmark datasets. The low values of D_T suggest that certain trees are ineffective in differentiating anomalies from normal examples. Some trees even mistakenly identify normal examples as anomalies in three of the six datasets, of which the discrimination power of a tree is negative. Thus, our goal is to improve the discrimination power of the forest by optimising the power of individual trees T in the ensemble:

$$\text{Maximise}[\mathrm{D}(\mathbb{F}) = \sum_{T \in \mathbb{F}} \mathrm{D}_T],$$

where $\mathbb{F}$ represents the forest and $\mathrm{D}(\mathbb{F})$ refers to the discrimination power of the ensemble. Since evaluating $\mathrm{D}_T(S)$ requires true labels of $\mathbf{x} \in S$, we cannot directly assess it. In unsupervised learning, we do not know whether $\mathbf{x}$ is a normal point $\mathbf{x} \in S_n$ or an anomaly $\mathbf{x} \in S_a$. To approximate $\mathrm{D}_T(S)$, we formulate two proxy metrics that are found to correlate with D_T. First, the size of a tree $Z_T = |T|$ is defined as the total count of nodes, containing both branch and leaf nodes within the tree. Second, the highest mass among leaves is denoted as $M_T = \max(m_1, m_2, \ldots, m_L)$, with m_i representing the count of data points in the i-th leaf node, and L indicating the number of leaf nodes within the tree.

To verify the MTMM, we first build a hundred trees on a dataset, each containing a sub-sample of 256 examples. Second, we document the length of the travel path for all the examples in the dataset. Using the ground truth label, we can compute the discrimination power $\mathrm{D}_T(S)$ for each of the 100 trees using Eq. 2. Third, we characterise isolation trees with respect to tree size Z_T and maximum mass M_T. Ultimately, we are able to graph the discrimination power (on the x-axis) of each tree in conjunction with the tree size and the maximum

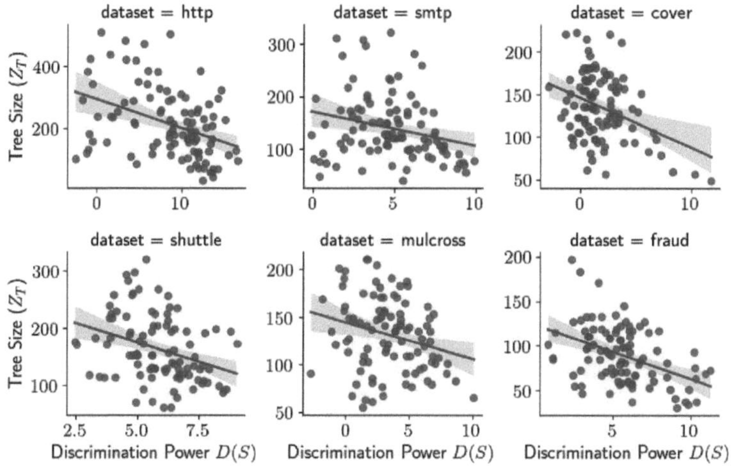

Fig. 2. The negative correlation between the size of a tree(Z_T), and the discrimination power($D(S)$) for 100 trees.

mass (on the y-axis). This entire procedure is carried out on the six chosen datasets in Fig. 1, as shown in Fig. 2 and Fig. 3.

We can confirm that the two proxy metrics correlate well with discrimination power. Generally, the tree with the highest discrimination power is usually smaller than the tree with the lowest, as seen in Fig. 2. Furthermore, the tree with the highest discrimination power generally has a higher maximum mass than the tree with the lowest power, shown in Fig. 3. These two observations demonstrate that the two metrics are related to the distribution of the discrimination power mentioned above.

With these measures as proxies for the discrimination power of individual trees, it is possible to construct a mechanism to replace trees that are ineffective in distinguishing anomalies from normal examples. This mechanism is based on a weighted sum of the normalised tree size and maximum mass over an ensemble of newly constructed trees on a window:

$$I_\beta(T) = \beta(1 - \overline{Z}_T) + (1 - \beta)\overline{M}_T, \quad (3)$$

where β is a weight value, $\overline{Z}_T$ is the normalised tree size of T(i.e., Z_T), $\overline{M}_T$ is the normalised maximum mass of the tree T(i.e., M_T), and I is the MTMM based metric to indicate the ability of a tree to distinguish the two classes of examples. Higher values of I indicate a better discrimination power.

The Algorithm 3 presents details about the update mechanism of the AIF. When it is called, we first initialise an empty set $\mathbb{T}$ for new trees (line 1) and calculate the height limit l (line 2). Second, we build m trees on the set of examples and put them in the set $\mathbb{T}$ (lines 3 to 5). Third, we find the highest scoring tree T_h in the newly built tree set $\mathbb{T}$ and the lowest scoring tree T_l in the ensemble $\mathbb{F}$ (line 6), based on Eq. 3. Finally, we replace the tree T_l in the

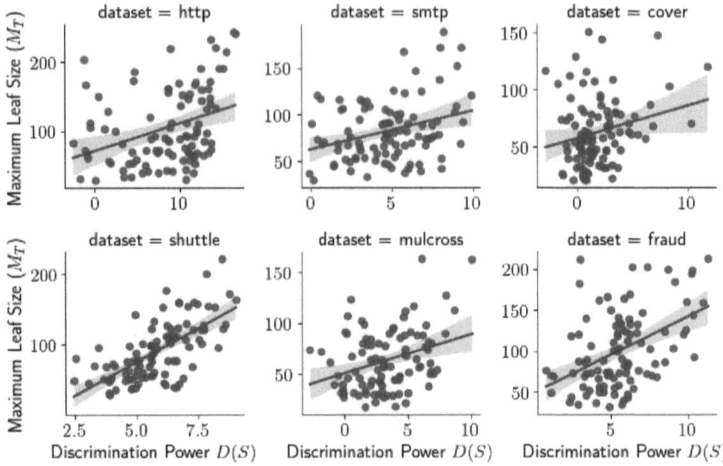

Fig. 3. There is a positive correlation between the maximum mass M_T and the anomaly discrimination power D(S).

ensemble with T_h if it is superior in I_T. Otherwise, we replace the oldest tree in the ensemble with T_h, ensuring that the ensemble is up-to-date with the current distribution of the data (lines 7–11).

It is crucial to emphasise that alternative batch learning algorithms for anomaly detection could be modified similarly to AIF by sequentially structuring base learners. However, the primary technique, MTMM, is specifically designed to be employed with AIF to maintain strong discrimination power under the concept of isolation.

4.3 Complexity Analysis

The primary process within the main loop of Algorithm 2 includes two key operations: scoring the instance (line 10) and updating the model (line 14). In the case of scoring, each instance traverses from the root of the tree to a terminal node (that is, a leaf), which amounts to $O(\log_2(w))$ complexity with the height limit $\log_2(w)$, where w represents the size of the window. The time complexity within the forest is $O(k \log_2(w))$ for scoring, where k refers to the size of the ensemble. Regarding the update of the model, since the construction of a tree has a worst-case time complexity of $O(w \log_2(w))$, the corresponding worst-case time complexity to update the model at each time step (that is, for each window) is $O(mw \log_2(w))$, where m refers to the number of new trees to construct for each update. As a result, the amortised time complexity for each example within the window averages to $O(m \log_2(w))$ to update the model.

Regarding AIF's space complexity, it is easy to see that it is constant: AIF only maintains two different data structures, a sliding window of size w and an ensemble of k trees of size w each. Additionally, AIF also generates m trees of size

Algorithm 3. UpdateForest($X, \mathbb{F}, \beta, m$)

Input: X - examples, $\mathbb{F}$ - The forest, β - weight, m - number of new trees to build on X
Output: The forest $\mathbb{F}$

$\mathbb{T} \leftarrow \phi$
$l \leftarrow ceiling(log_2|X|)$
for $i = 1$ to m **do**
  $\mathbb{T} \leftarrow \mathbb{T} \cup iTree(X, 0, l)$
end for
Normalize tree size and max leaf across all the trees in $\mathbb{F} \cup \mathbb{T}$
$T_h \leftarrow \text{argmax}_{T \in \mathbb{T}} I_\beta(T), T_l \leftarrow \text{argmin}_{T \in \mathbb{F}} I_\beta(T)$ {Eq. 3}
if $I_\beta(T_h) \geq I_\beta(T_l)$ **then**
  Replace T_l in $\mathbb{F}$ with T_h
else
  Replace the oldest tree in $\mathbb{F}$ with T_h
end if
Return $\mathbb{F}$

w every time the sliding window is full, i.e. after every w examples. Therefore, the total space complexity is $O((k+m)w)$, where k, m, and w are constants that define the size of the ensemble, the number of new trees to consider per window and the window size, respectively.

Since baselines are implemented in different languages and use different frameworks, it is unfair to compare runtime and memory consumption. Instead, we compare them on the basis of their time and space complexity. The time complexity is amortised per updating example, making it easier to compare them directly. It is crucial to emphasise that our focus is on comparing the time complexity of model updating, as this process typically requires the most time. Table 1 presents the complexities of all the baseline algorithms and AIF. In terms of time complexity, while most algorithms are influenced by the ensemble size k, AIF is determined by the number of new trees per window m. This allows a balance between speed and accuracy by choosing an appropriate m.

5 Experiments

We evaluated the adaptive isolation forest (AIF) in 23 anomaly detection benchmark datasets and compared with nine baselines. We used area under the receiver operating characteristic curve (AUC-ROC) as the evaluation metric to compare our proposed algorithm with the baselines mentioned in the related work section. Furthermore, we present critical difference diagrams that reflect the ranking of algorithms in the benchmark datasets. In addition, we evaluate the influence of the parameters of the proposed algorithm and offer insights into the limit limits of the AIF.

Table 1. The worst-case time and space complexity for training a model. Parameters include w: window size, k: ensemble size, d: dimensionality, h: tree height, ϕ: subsample size, η: maximum number of examples per leaf, b: number of histogram bins, s: number of random projections, m: cache size, and n: the number of examples have seen.

Algorithm	Time Complexity	Space Complexity
AIF	$m \log_2(w)$	$(k+m)w$
RRCF	kdw	kw
HST	$k(h+w)$	$k2^h$
iFASD	$k\phi \log_2(\phi)$	w
LODA	$kd^{\frac{1}{2}}$	$k(d^{\frac{1}{2}}+b)$
xStream	$ksdh$	$ksbh+md$
RS-Hash	ks	ksb
SDOS	$\log(m)$	m
oIFOR	$k \log_2(\frac{w}{\eta})$	$k\frac{w}{\eta}+w$
iMF	$kd(n+w)log_2(n+w)$	kn

5.1 Datasets

We conducted our experiments on a variety of real-world datasets. Furthermore, we also consider a synthetic dataset from OpenML, Mulcross, which uses a multivariate normal distribution with a selectable number of anomaly groups. The selected datasets have at least 10,000 examples, as our algorithm is designed to handle large amounts of data in the real world. Table 2 presents information on each dataset, including its name with reference, number of features, number of examples, number of anomalies, and outlier percentage.

5.2 Setup

All experiments were performed on an Intel® Core™ i5-10500 CPU @ 3.10GHz with 32 GB RAM under Ubuntu 22.04.4 LTS. The proposed Adaptive Isolation Forest (AIF) anomaly detection algorithm is implemented in Python with CapyMOA [7]. AIF is compared with nine other streaming anomaly detection algorithms: RRCF; HST, iForestASD, and LODA, all obtained from PySAD; xStream, RS-Hash, and SDOstream, by [23]; oIFOR; and iMF. To maintain fairness, the models were configured with similar parameters: RRCF with a tree size of 256 and 100 trees; HST with a window size of 256, eight height levels, and 100 trees; iForestASD with a tree size of 256, a 2048 sliding window size, and 100 trees; LODA with 100 projectors and 100 bins; xStream with 100 projections and Half-Space Chains, a depth of eight, and a window size of 256; RS-Hash with 100 detectors, each containing 4 hash functions, with a hash range of 10000; SDOstream with a window size of 256, 100 elements, and 6 nearest observers; oIFOR with 100 trees, a maximum example size of 32, and a sliding window size of 2048; iMF with 100 trees, a minibatch size of 100. For AIF, we used a window

Table 2. Overview of the 23 benchmark datasets.

Dataset	#Examples	#Features	#Anomalies	%Anomalies
ALOI [5]	49,534	27	1,508	3.04
backdoor [15]	95,329	196	2,329	2.44
campaign [17]	41,188	62	4,640	11.27
celeba [17]	202,599	39	4,547	2.24
census [17]	299,285	500	18,568	6.20
cover [2]	286,048	10	2,747	0.96
donors [17]	619,326	10	36,710	5.93
fraud [17]	284,807	29	492	0.17
http [19]	567,498	3	2,211	0.39
insects [11]	212,514	33	11,701	5.51
Friday [21]	191,033	78	1,966	1.03
magic [5]	19,020	10	6,688	35.16
mammo [25]	11,183	6	260	2.32
mulcross [19]	262,144	4	26,114	10.00
NYC [8]	10,320	1	536	5.19
NYC-S [8]	10,273	48	536	5.22
shuttle [19]	49,097	9	3,511	7.15
skin [5]	245,057	3	50,859	20.75
smtp [19]	95,156	3	30	0.03
Thu-Inf [21]	288,602	78	36	0.01
Thu-WA [21]	170,366	78	2,180	1.28
Tue [21]	445,909	78	13,835	3.10
Wed [21]	692,703	78	252,672	36.48

size of 256, an ensemble size of 100, a model update weight of 0.5, and m set to 10. Each experiment was carried out on ten separate runs, resulting in a total of 230 executions (23 datasets $\times$ 10 runs) for each algorithm. The performance of all the listed algorithms was assessed using the prequential method with the benchmark datasets.

5.3 Results

In this part, we present our findings on keeping dataset examples in their initial order per run. Maintaining the original sequence typically uncovers two main aspects of the stream: data distribution can change over time (non-IID), and the anomaly density may vary greatly between subsequences.

We first report the performance results as the average AUC-ROC score with the standard deviation. Moreover, we rank stream learning algorithms based on their AUC-ROC scores from the 230 executions and illustrate the results in

Table 3. AUC-ROC scores with standard deviations, in percentages (0–100), for the original example sequence. The scores are rounded to one decimal place; the highest scores are in bold. iMF often failed to complete within the allocated time budget.

dataset	AIF	HST	RRCF	iFASD	LODA	xStream	RS-Hash	SDOS	oIFOR	iMF
ALOI	36.5(0.5)	78.7(1.4)	57.9(2.3)	**99.4(0.0)**	46.3(0.3)	33.2(0.7)	69.3(0.7)	70.4(0.1)	63.5(4.8)	71.5(0.7)
backdoor	61.1(2.7)	56.7(7.0)	73.6(1.8)	61.4(1.2)	20.4(6.2)	**76.9(3.4)**	61.1(5.6)	73.1(0.3)	65.7(8.1)	-
campaign	71.1(0.4)	69.3(1.5)	60.2(2.6)	69.3(0.9)	62.3(2.3)	64.0(1.5)	**71.8(1.3)**	59.2(0.1)	66.7(3.9)	69.0(0.4)
celeba	**78.6(0.2)**	74.1(2.5)	61.6(5.9)	69.1(0.9)	57.8(14.8)	67.3(3.7)	72.7(2.1)	57.8(0.1)	74.5(7.2)	-
census	**66.6(0.6)**	60.1(2.9)	54.7(1.7)	59.3(0.9)	51.2(14.0)	59.2(1.2)	60.0(3.7)	49.0(0.1)	60.9(6.4)	-
cover	**88.1(0.6)**	46.3(2.0)	57.3(2.6)	66.5(1.3)	51.7(1.1)	58.7(3.3)	41.8(6.6)	51.2(0.2)	86.1(3.9)	-
donors	73.7(1.0)	73.8(1.9)	60.6(7.2)	77.2(1.0)	33.4(16.6)	76.4(4.1)	74.5(1.9)	79.8(0.1)	**80.4(5.2)**	-
fraud	94.5(0.2)	91.3(0.6)	86.1(2.5)	94.2(0.3)	56.2(7.0)	94.9(0.4)	**95.1(0.5)**	94.7(0.1)	94.9(0.3)	-
Friday	51.2(2.8)	57.4(2.6)	58.9(3.6)	59.8(1.5)	52.8(7.5)	31.2(4.3)	41.3(2.9)	**59.8(0.4)**	46.0(9.4)	-
http	94.4(0.0)	31.3(0.6)	24.1(1.0)	**95.0(0.7)**	29.0(18.9)	32.0(1.4)	7.5(0.3)	8.0(0.3)	23.5(4.0)	-
insects	75.4(0.9)	**76.5(1.0)**	60.4(2.3)	72.2(1.5)	44.8(4.2)	75.2(1.3)	73.6(3.0)	70.5(0.2)	71.7(2.5)	-
magic	60.7(0.9)	**80.6(1.3)**	48.8(6.2)	41.2(1.9)	38.1(0.7)	72.0(0.8)	74.6(0.7)	77.6(0.1)	74.2(2.0)	63.2(0.6)
mammo	86.8(0.9)	80.5(1.0)	74.4(3.1)	72.1(0.5)	26.3(8.7)	**87.7(0.7)**	84.0(0.6)	84.5(0.2)	84.4(1.9)	83.7(0.6)
mulcross	97.9(0.2)	32.4(1.9)	72.5(11.9)	95.2(0.8)	2.5(0.9)	94.9(1.0)	17.9(2.6)	0.9(0.0)	**98.8(0.7)**	-
NYC	60.7(1.2)	61.7(0.0)	48.9(2.7)	**63.4(0.5)**	52.0(1.6)	40.8(0.7)	49.5(1.0)	56.5(0.7)	51.3(1.3)	57.0(1.2)
NYC-S	65.4(2.1)	42.4(0.8)	55.5(3.5)	70.8(0.9)	50.0(0.1)	52.5(3.1)	50.0(0.0)	**79.0(1.0)**	47.2(1.6)	77.6(0.8)
shuttle	**99.7(0.1)**	97.8(0.1)	77.5(9.8)	95.6(0.0)	47.0(5.5)	98.9(0.2)	87.4(3.5)	99.6(0.0)	97.5(1.9)	92.5(0.7)
skin	66.2(0.8)	10.4(0.1)	36.5(1.1)	**69.9(0.5)**	54.3(4.0)	9.3(1.1)	18.3(0.6)	34.7(0.1)	30.6(2.4)	-
smtp	**89.2(1.4)**	83.7(0.5)	80.4(4.2)	88.9(0.5)	65.9(4.9)	81.6(0.4)	83.5(0.2)	85.1(0.5)	87.8(1.9)	86.3(0.8)
Thu-Inf	91.8(0.6)	90.7(0.7)	84.3(2.4)	91.3(0.3)	30.7(3.6)	94.1(0.2)	93.1(1.8)	94.1(0.0)	**94.8(0.6)**	-
Thu-WA	61.6(2.3)	54.2(5.2)	60.6(6.0)	61.8(2.0)	**84.3(5.4)**	72.7(3.4)	19.4(1.6)	73.3(0.2)	45.1(16.0)	-
Tue	63.3(0.8)	64.4(2.2)	45.3(2.7)	39.2(0.9)	53.5(3.8)	**66.6(0.9)**	33.2(1.4)	54.6(0.1)	62.2(1.3)	-
Wed	**87.1(0.7)**	79.4(2.4)	43.1(1.4)	57.2(1.3)	49.5(3.1)	76.9(0.8)	36.1(3.6)	58.8(0.0)	55.5(4.0)	-
Avg AUC(↑)	**74.9**	64.9	60.1	72.6	46.1	66.0	57.2	64.0	68.0	-
Avg Rank(↓)	**3.24**	4.44	6.23	4.14	6.90	4.59	5.72	4.48	4.21	-

a critical difference diagram. Additionally, we explore the influence of the AIF parameters. It is important to note that the iMF is omitted from the critical difference diagram because its experiments were not completed within a reasonable budget time.

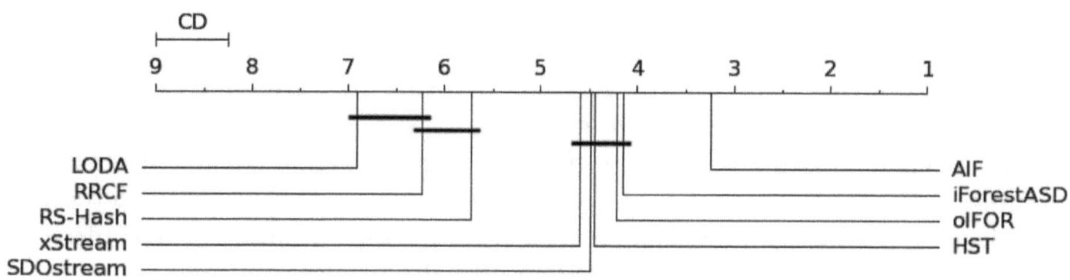

Fig. 4. AIF holds the top position in the CD diagram rankings ($p = 0.05$).

The performance results as shown in Table 3, indicate that our AIF algorithm outperforms the other algorithms in terms of both the average AUC-ROC score and the rank. In addition, we evaluated the algorithms by ranking them for each run and illustrating the results with a critical difference diagram (Fig. 4). The findings indicate that AIF is significantly better than the baseline models.

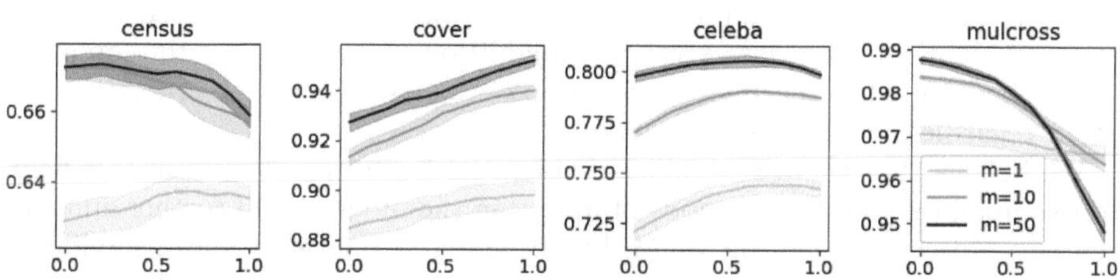

Fig. 5. AUC-ROC with variance(y-axis) of different Datasets for different β(x-axis) and m(legend).

5.4 Parameter Influence

To assess the impact of parameters on our proposed algorithm, we measured the detection performance of AIF on four datasets using the AUC-ROC. The results of the experiments presented in Fig. 5 show that increasing the number of trees built (that is, the parameter m) per window generally improves performance. However, the improvement diminishes with an increasing number of additional trees constructed per window. Thus, the parameter $m = 10$ represents a balance between achieving higher accuracy and maintaining computational efficiency. On the other hand, the parameter β influences the performance differently depending on the dataset. For instance, the cover dataset favours smaller tree sizes, whereas the mulcross dataset prefers larger maximum mass. Consequently, the weight value β should be optimised for each dataset.

5.5 Limitation

Our study revealed that consecutive anomalies at the start of data streams present a significant issue for the AIF, as it forms the ensemble based on the opening data segment (window size of 256 in this study). It then replaces only one tree at each subsequent window. Initially, the ensemble consisted mainly of anomalies rather than normal data points, hindering effective anomaly detection. It required numerous windows before the ensemble appropriately represented the normal data points. By then, the AIF had already misclassified many anomalies as normal points, leading to subpar performance. We ascribe the AIF's poor performance on the ALOI and skin datasets to this underlying issue.

6 Conclusion and Future Work

We presented a novel algorithm, called the Adaptive Isolation Forest (AIF), which is designed to identify anomalies in data streams. In addition, we have developed a criterion-based approach called MinTreeMaxMass (MTMM) to update the model. Our experiments show that AIF provides higher predictive accuracy than eight contemporary (excluding iMF) and widely used anomaly detection algorithms for data streams. In addition, it exhibits greater computational efficiency when learning new examples, as its performance remains consistent regardless of the ensemble size.

We believe that there is significant potential for future studies in this field. For example, refining the methods of updating the model could further enhance accuracy. Moreover, adapting MTMM to other algorithms is a topic of significant interest for further advancement. Furthermore, the specified special cases provide guidance on addressing sequential anomalies within data streams. Another possible direction will be an in-depth analysis of the interaction of stability and plasticity in the context of dynamically changing data streams.

Acknowledgments. TAIAO programme CONT-64517-SSIFDS-UOW funded by MBIE.

References

1. Bifet, A., Gavaldà, R., Holmes, G., Pfahringer, B.: Machine Learning for Data Streams with Practical Examples in MOA. MIT Press (2018)
2. Blackard, J.A., Dean, D.J.: Comparative accuracies of artificial neural networks and discriminant analysis in predicting forest cover types from cartographic variables. Comput. Electron. Agric. **24**(3), 131–151 (1999)

3. Chandola, V., Banerjee, A., Kumar, V.: Anomaly detection: a survey. ACM Comput. Surv. (CSUR) **41**(3), 1–58 (2009)
4. Ding, Z., Fei, M.: An anomaly detection approach based on isolation forest algorithm for streaming data using sliding window. IFAC Proc. Vol. **46**(20), 12–17 (2013)
5. Emmott, A., Das, S., Dietterich, T., Fern, A., Wong, W.K.: A meta-analysis of the anomaly detection problem. arXiv preprint arXiv:1503.01158 (2015)
6. Gama, J., Žliobaitė, I., Bifet, A., Pechenizkiy, M., Bouchachia, A.: A survey on concept drift adaptation. ACM Comput. Surv. (CSUR) **46**(4), 1–37 (2014)
7. Gomes, H.M., et al.: CapyMOA: efficient machine learning for data streams in python (2025)
8. Guha, S., Mishra, N., Roy, G., Schrijvers, O.: Robust random cut forest based anomaly detection on streams. In: International Conference on Machine Learning, pp. 2712–2721. PMLR (2016)
9. Gunasekara, N., Pfahringer, B., Gomes, H.M., Bifet, A.: Survey on online streaming continual learning. In: IJCAI, pp. 6628–6637 (2023)
10. Hartl, A., Iglesias, F., Zseby, T.: Sdostream: low-density models for streaming outlier detection. In: ESANN, pp. 661–666 (2020)
11. Leveni, F., Weigert Cassales, G., Pfahringer, B., Bifet, A., Boracchi, G.: Online isolation forest. In: International Conference on Machine Learning (ICML). Proceedings of Machine Learning Research (PMLR) (2024)
12. Liu, F.T., Ting, K.M., Zhou, Z.H.: Isolation forest. In: 2008 Eighth IEEE International Conference on Data Mining, pp. 413–422. IEEE (2008)
13. Ma, H., Ghojogh, B., Samad, M.N., Zheng, D., Crowley, M.: Isolation mondrian forest for batch and online anomaly detection. In: 2020 IEEE International Conference on Systems, Man, and Cybernetics (SMC), pp. 3051–3058. IEEE (2020)
14. Manzoor, E., Lamba, H., Akoglu, L.: xstream: outlier detection in feature-evolving data streams. In: Proceedings of the 24th ACM SIGKDD International Conference on Knowledge Discovery & Data Mining, pp. 1963–1972 (2018)
15. Moustafa, N., Slay, J.: UNSW-NB15: a comprehensive data set for network intrusion detection systems (UNSW-NB15 network data set). In: 2015 Military Communications and Information Systems Conference (MilCIS), pp. 1–6. IEEE (2015)
16. Pang, G., Shen, C., Cao, L., Hengel, A.V.D.: Deep learning for anomaly detection: a review. ACM Comput. Surv. (CSUR) **54**(2), 1–38 (2021)
17. Pang, G., Shen, C., Van Den Hengel, A.: Deep anomaly detection with deviation networks. In: Proceedings of the 25th ACM SIGKDD International Conference on Knowledge Discovery & Data Mining, pp. 353–362 (2019)
18. Pevný, T.: Loda: lightweight on-line detector of anomalies. Mach. Learn. **102**, 275–304 (2016)
19. Rayana, S.: Odds library (2016). https://odds.cs.stonybrook.edu
20. Sathe, S., Aggarwal, C.C.: Subspace outlier detection in linear time with randomized hashing. In: 2016 IEEE 16th International Conference on Data Mining (ICDM), pp. 459–468. IEEE (2016)
21. Sharafaldin, I., Lashkari, A.H., Ghorbani, A.A., et al.: Toward generating a new intrusion detection dataset and intrusion traffic characterization. ICISSp **1**, 108–116 (2018)
22. Tan, S.C., Ting, K.M., Liu, T.F.: Fast anomaly detection for streaming data. In: Twenty-Second International Joint Conference on Artificial Intelligence. Citeseer (2011)
23. Vázquez, F.I., Hartl, A., Zseby, T., Zimek, A.: Anomaly detection in streaming data: a comparison and evaluation study. Expert Syst. Appl. **233**, 120994 (2023)

24. Wang, L., Zhang, X., Su, H., Zhu, J.: A comprehensive survey of continual learning: theory, method and application. IEEE Trans. Pattern Anal. Mach. Intell. (2024)
25. Woods, K.S., Doss, C.C., Bowyer, K.W., Solka, J.L., Priebe, C.E., Kegelmeyer, W.P.: Comparative evaluation of pattern recognition techniques for detection of microcalcifications in mammography. Int. J. Pattern Recognit Artif Intell. **7**(06), 1417–1436 (1993)

Investigating Delays of Combined Feature and Label Acquisitions on Data Streams

Maik Büttner(✉) and Myra Spiliopoulou

Otto-von-Guericke-University Magdeburg, Magdeburg, Germany
maik.buettner@ovgu.de

Abstract. Missing data is a common occurrence in data streams, where the feature values and label information are not assumed to be available. While acquiring feature values or labels may be possible, doing so is often associated with monetary and time costs. For instance, acquiring labels is time consuming and may require specialized experts or testing equipment to be available. Similarly, acquiring feature values may require access to restricted and or limited sources with limited availability. If one would be able to maximize performance and only acquire necessary information, budget could be saved. In this work we present a combined approach to acquiring feature values and labels under latency in a stream setting. Our method is tested on four data sets under various latencies and missingness conditions to evaluate its usefulness. We also present an ablation study on the impact of individual factors and their impact on the task performance.

Keywords: active feature acquisition · active learning · verification latency · stream mining

1 Introduction

Missing data is to be expected in real-world scenarios and often pose a challenge when evaluating incoming data. While such data may be missing, a variety of domains allow for the acquisition of said missing data when capacity and means are available. For example, quality assurance measures can be taken at different levels of detail with the option to investigate the quality of products using simple automated processes to manual inspections performed by qualified workers. The latter of these processes are presumed more expensive, more limited in availability and often more time-consuming which may be modeled by different types of costs and delays.

It is commonplace to presume that such evaluations mostly concern the target of an instance be it a label or a continuous value. Such cases fall into the domain of active learning [5]. However, it is also possible for feature values to be unknown and to be acquired with similar processes. In such cases we still reamin in the domain of active learning, specifically active feature-value acquisition [10].

S. Džeroski et al. (Eds.): DS 2025, LNAI 16090, pp. 379–391, 2025.
https://doi.org/10.1007/978-3-032-05461-6_25

2 Related Works

The topics of stream learning, active learning, active feature acquisition and verification latency are rarely evaluated together. However, there exist individual works in these fields that investigate aspects of them.

For example, the authors of [9] delve into the topic of active learning with verification delay. They introduce a wrapper for general active learning strategies to project the time points at which such strategies would benefit from the acquisition of labels in order to counteract the delays of verification latency.

Similarly, the work of [6] explores the impact of verification latency and various budgets on active learning in data streams. To do so, they introduce a dynamic budgeting mechanism that distributes more or less budget based on the current stream needs and a propagation mechanism that compensates the oversampling of regions while waiting for labels to arrive and thus ensure a more well-rounded exploration of the feature-space.

Related to the topic of verification latency is the use of semi-supervision to handle the lack of labels when the latency is extensive. The work of [13] is dedicated to compare semi-supervised methods for the use in extreme verification latency scenarios that have to make do without any additional label information. They find that their compared methods perform better or worse depending on the context and requirements of the methods.

More recent variants of semi-supervised extreme verification latency models include the work by [8] in which the authors introduce a Bayesian method to deal with verification latency and choose appropriate models to generate pseudo-labels to bridge the gap until true labels arrive.

The authors of [11] investigated the impact of verification latency on stream classifiers and drift detectors and introduced a new kappa statistic that takes the latency into account. They found that the no change classifier performed best in the null latency scenario.

Data acquisition on partially observed data has been investigated by [1]. The authors perform a pool-based active learning and feature value acquisition strategy based on uncertainty metrics and imputation strategies while considering costs. The work shows impressive results concerning the selection and acquisition of relevant features and labels that enables and closely matches the performance achieved vs. a classifier trained on the fully observed data set.

3 Methodology

In this section we introduce an intelligent method to evaluate and acquire potential instances and their feature values. The overall approach is shown in Algorithm 1.

3.1 Definitions

We assume a stream $\mathcal{S}$ consisting of batches $X_t \in \mathcal{S}$ of feature vectors $x_{t,i} \in F$ and associated labels $y_{t,i}$. F describes the feature space. The batches are separated by indices t which might stand in for fixed time intervals or a fixed number of instances until packaged and presented. Our goal is to train a classifier C who learns a function $f(x_{t,i}) \rightarrow y_{t,i}$ to predict labels. A vector mask $m_{t,i} \in [0,1]^{1+|F|}$ describes if the values of $y_{t,i}$ and $x_{t,i}$ are known (1) or unknown (0).

We further assume a set of experts $|O| = 1 + |F|$ that can be independently tasked to query the true information of a feature or label in exchange for a cost c_k and a response time d_k. We assume the costs c_k for a feature value or label to be consistent based on the type (i.e. feature 2 will always cost the same c_2) while the delay $d_k \in [0, \infty)$ follows an unknown function $delay_k(x_{t,i,k}) = d_k$.

On request, an expert may perform the evaluation of a particular value in parallel, i.e. requests are presumed to be independent of each other. The cost c_k is applied regardless if the request is ever fulfilled. Within this framework, a budget B is tracked and expressed as the ratio B_{used} of spent B_{spent} and given B_{given} budget ($B_{used} = \frac{B_{spent}}{B_{given}}$) and to be used for acquisition requests.

3.2 Budgeting

Budgeting is integral to the selection and acquisition of features and labels, hence the manager determining these acquisition decisions is as well. We follow the budgeting as described in [4] employing the Incremental Percentile Filter that bases its decision making on the most recent goodness values seen. Requests are tracked in two separate windows which store the arrival time of the request and its associated goodness value respectively. Based on whether the rank of a goodness value's placement in the value-ordered window falls below a threshold T, a positive or negative decision is made. The limited capacity of the sliding window ensures the relevance and actuality of the incoming decisions.

The threshold T is determined by the current circumstances in which decisions are to be made with the current budget ratio and potential to overextend the budget having a direct effect on its value. More specifically following [4]:

$$T = \begin{cases} \frac{B_{gain}}{\widehat{A_C} \cdot B_{used}} & \text{if } B_{used} \leq 1 \\ \frac{B_{gain}}{(\widehat{A_C} \cdot B_{used}) \cdot (\lfloor p_c \cdot (B_{used} - 1) \rfloor + 2)} & \text{else} \end{cases} \tag{1}$$

where $p_c = 32$ is a user-specified penalty parameter and $\widehat{A_C}$ is an estimated average cost of queries.

The goodness values the BM sees are the expected improvements in the quality of the task performance were it acquired, i.e. the quality gain from before and after acquisition of the feature values and labels. Tying the quality of this decision necessitates that the quality function considers the delays and costs involved in its current instance and label.

To approach these criteria, we use a combined effort of feature and label importance measures. We quantify the feature importance by a simple feature to label function that only considers the direct relation: For this, we employ the average euclidean distance (AED) [14]. To take a label importance measure, the label is treated as a feature with perfect predictability in order to align the

Algorithm 1. Pseudo-code of the proposed approach

Require: Initial data X_0, the data stream $\mathcal{S}$, a window of batches W, a classifier C, a budget manager BM, an imputation model I, a budget threshold T, a queue of oracle requests Q, A set of oracles O

add X_0 to W
train initial model C_0 on W
for $X_t \in \mathcal{S}$, $t \geq 1$ **do**
  create test data set X'_t by imputing missing values in X_t using I
  test classifier C_{t-1} on X'_t
  for $a_{b,j,k} \in \{a \in Q | a \text{ is available}\}$ **do**
    set value of $m_{b,j,k} = 1$
  end for
  update metrics relevant to acquisition using W // specifically, update AED
  for $x_{t,i} \in X_t$, $i \geq 1$ **do**
    get acquisition set $A_{t,i}$ of $x_{t,i}$ // specifically, all missing information of $x_{t,i}$
    Calculate gain of $A_{t,i}$ // see equation 7
    if BM wants to acquire A **then**
      // see subsection 3.2
      for $a_{t,i,k} \in A_{t,i}$ **do**
        query associated O_k for a_k and add request to queue Q
        if $d_k(a_{t,i,k}) = 0$ **then**
          set value of $m_{t,i,k} = 1$
        end if
        add cost c_k to B_{spent}
      end for
    end if
  end for
  add X_t to W
  update I on W
  create data set D'_t using by imputing missing values in W using I
  train the next classifier C_t on D'_t
  update budget threshold T // see equation 1
end for

importance scales across attribute types. I.e. the label reaches the maximum importance value of AED, that is number of labels - 1. It is important to note that the feature importance is global for all features, independent of the true value.

$$imp(F_k) = AED \tag{2}$$

The cost factor is handled by the known costs for a query and further enhanced by taking the delays into account. The delay is modeled as the observed mean of the previous acquisitions and further described by the variance therein. The greater the mean, the more quality if the value of the feature is known.

$$cost(F_k) = c_k \cdot recency\_bias(\overline{d_k}) \tag{3}$$

To calculate the mean delay of a query, we need to consider how to model the infinite delay case. We assume the extreme verification latency to take on the number equivalent to the size of the sliding window $|W|$ to signify the last chance the acquisition of this value could ever yield a benefit ($d_k \geq |W| \rightarrow d_k = |W|$). Since never getting any information about a query is an important edge, we further add a penalty term that can be used to penalize the chance of no acquisition further.

In order to prioritize recent information, one might alter the linear description of the delays and use a *recency_bias* function instead, that attributes more and more costs the longer the wait for an expected value where r is a user-defined recency bias parameter with $r \geq 1$.

$$recency\_bias(\overline{d_k}) = \begin{cases} r^{\lceil \overline{d_k} \rceil} & \text{exp.} \\ r(1 + \lceil \overline{d_k} \rceil) & \text{lin.} \end{cases} \tag{4}$$

Combined this allows for the calculation of quality q per value and a calculation of the instance quality q. The difference between the quality after an acquisition set $A_{t,i}$ is added to $x_{t,i}$ and before is called quality gain Δq:

$$q(F_k, x_{t,i,k}) = \frac{imp(F_k)}{cost(F_k)} \tag{5}$$

$$q(x_{t,i}) = \frac{\sum_{k=0}^{|F|} q(F_k, x_{t,i,k})}{\sum_{k=0}^{|F|} m_{t,i,k}} \tag{6}$$

$$\Delta q(x_{t,i}, A_{t,i}) = q(x_{t,i} \cup A_{t,i}) - q(x_{t,i}) \tag{7}$$

3.3 Imputation Methods

To keep the methods focused on the delay and selection processes, the imputation performed for feature values takes the mean and mode of the respective feature depending on the feature type. Labels are required for the feature importance measure, hence the classifier itself provides pseudo-labels for instances which are left without a label. The classifier itself is only trained on true label-instance pairs. If there is no true label-instance pair for a particular label available then the last known true label-instance pair is used as a single substitute. For example, in the no acquisition case after the initial batch has been phased out by the sliding window the last instance-label pair for each label from the initial batch is remembered and trained on as the sole member of this particular class.

4 Evaluation

In this section we introduce the experiments, the data sets and evaluate our results.

4.1 Data Sets

We test our method on the four data sets seen in Table 1. The data set *sea* [12] and *cfpdss* [4] are of a synthetic nature. The former aims to learn whether the sum of two randomly chosen features exceeds an unknown threshold. The features and thresholds in question change after 15,000 instances in the form of a sudden drift. The latter data set applies various drift types and feature correlations which change every thousand instances. The so changing concepts test the response of the method to these different changes. Data sets *electricity* [7] and *airlines* [2] are taken from real world observations. The former aims to predict the Australian stock prices for electricity. The latter aims to predict whether a flight will be delayed or not.

Table 1. Data sets used in our experiments

Data set	Type	Instances	Labels	Cat./num. features	Classifier
airlines	real	539383 (10787)	2	4/3	SVC
sea	synthetic	60000 (1199)	2	0/3	SVC
electricity	real	45312 (906)	2	1/7	GNB
cfpdss	synthetic	13000 (259)	2	5/5	SVC

4.2 Test Parameters

In this subsection we briefly describe method parameters that apply based on needs and test scenarios. Each data set had $m\%$ of all feature values uniformly random removed. The removal process was performed for each separate test run so that each test run ran with a different set of removed feature values. We prepared all data sets by splitting them into 50 instance strong batches with the first batch serving as initial data on which to learn the first models. The sliding window was set to 10 batches and an initial budget threshold of 0.25 was set. For the *smraed* method we mandated it to acquire everything of an instance or nothing. We use the kappa plus statistic to evaluate the task performance of the tested methods due to its harsher baseline than regular Cohen's kappa for data streams [3]. Each of the following results presented are the averaged results over the entirety of the data sets and five iterations.

4.3 Method Performance

In this subsection we briefly discuss the performance of our compared methods. To this end Fig. 1 shows the kappa plus statistics of the four compared methods along an increasing number of missing feature values. The *nafa_cat* method shows the absolute worst performance expected on each data set when no acquisitions are performed. In contrast method *all_cat* shows the best case scenario in which all feature values and labels are acquired and immediately made available. Method *label_ cat* highlights a method that only acquires labels for its acquisition. And finally method *smraed_ cat* is the described method that intelligently chooses its instances for acquisition.

Figure 1 further highlights that data set electricity is not well served by the classifiers chosen with a kappa statistic far below the no change classifier. Data sets *airlines* and *cfpdss* show close competitiveness between the introduced method and the acquire all labels method. Since the method is only permitted to acquire all of an instance or nothing its ability to compensate for the low number of true labels is noteworthy. In the very simplistic problem case of the *sea* data set the approach does outperform the all label acquisition strategy. Of further note are the different levels of task performance decline with an increase in missing feature values. All but the *airlines* data set are strongly impacted by the loss of true information, whereas it only shows a modest decline in performance. We suspect that the large number of feature values its "AirportFrom" and "AirportTo" features can take are critical to this difference in behavior.

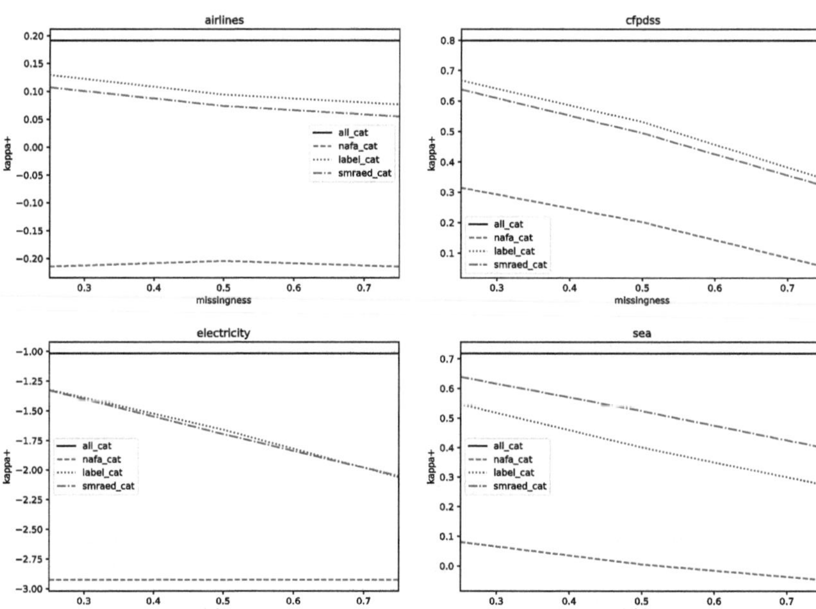

Fig. 1. Method task performances compared

4.4 Impact of Label Availability

In this subsection we take a brief look at the impact a steady provision of labels has on the task performance. These results are shown in Fig. 2. A value of labels

0.2 stands in for a label being made available every fifth instance. Within this figure we can see that even a small amount of 10% to 20% of labels are sufficient to reach close to maximum task performance that can be achieved if all labels were known. Data set *airlines* remains the only data set that is the least impacted by the number of missing feature values while the remaining data sets have clear performance differences.

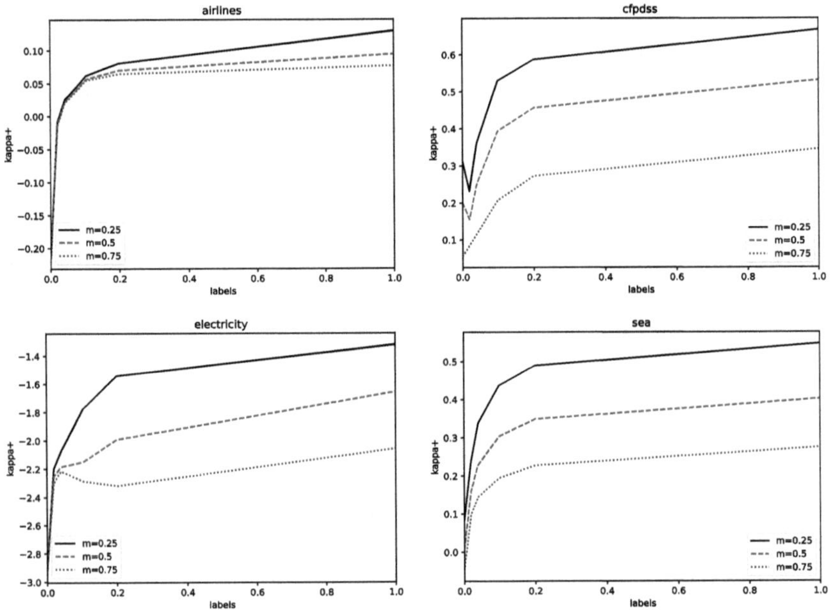

Fig. 2. Task performance compared over different label availability

4.5 Impact of Increasing Acquisition Delays

In this subsection we take a look at the impact consistent acquisition delays of features and labels have on the task performance. The results to these comparisons are seen in Fig. 3 and Fig. 4.

We can see that the biggest drop in performance on the all information acquired method in Fig. 3 occurs from the no delay to minimal delay scenario. Given the nature of streams and the lack of an active drift detection method the delay prolongs the time in which old concepts are held, trained on and evaluated with when new information should already phase out the older information. Surprisingly, performance stays consistent after some delay is introduced hinting that new concepts once learned are not changing quick enough for even more delays to be as impactful. Given that the *cfpdss* data set only switches concepts every thousand instances, the equivalent of a delay of 20, and the *sea* data set switching concepts only every 15,000 instances, the equivalent of a delay of 300, their stability is less surprising. Most affected are the *electricity* and *airlines* data sets in which an increased delay can even lead to a temporary improvement in task performance, especially if the delay inadvertently benefits time lagged effects within the data sets.

Method *smraed* shows a more consistent decline in performance with increased delay as seen in Fig. 4. While *sea*'s simple nature shows similar tendencies to the all acquisitions case, the rest of the data sets show a mostly linear decline in performance with increased delays. We therefore conclude that the inconsistent availability of information already part of the *smraed* method to be the deciding factor in the steepness at which performance declines.

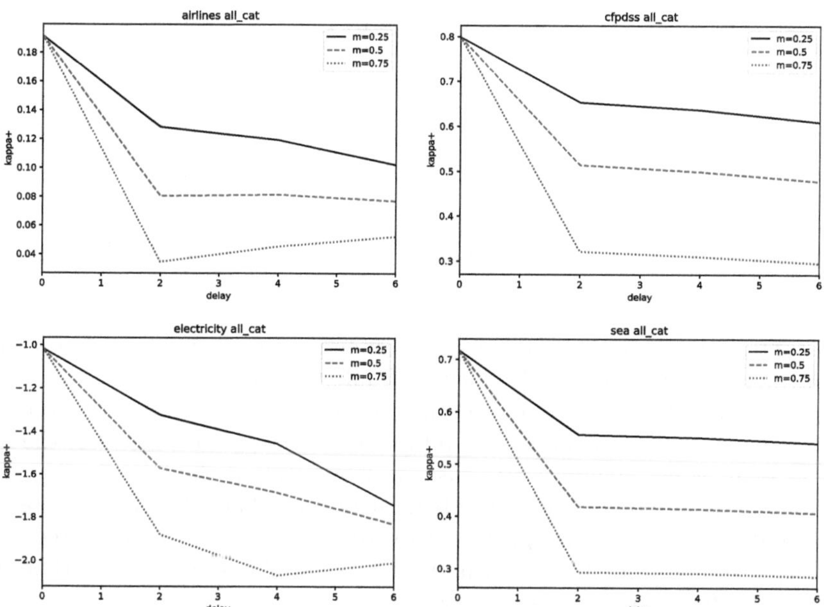

Fig. 3. Task performance of acquiring all information along fixed acquisition delays

4.6 Impact of Delay Variance

In this section we perform experiments about the change in task performance when the delay of acquisitions get more varied and inconsistent. To this end, a mean delay of $\overline{d_k} = 4$ for acquisitions was set and the range in which acquisitions' delays may fall was increased from a $d_k = 4$ to a $d_k = [2, 6]$ (marked as variance four) and $d_k = [0, 8]$ (marked as variance eight). The results are presented in Fig. 5 and Fig. 6.

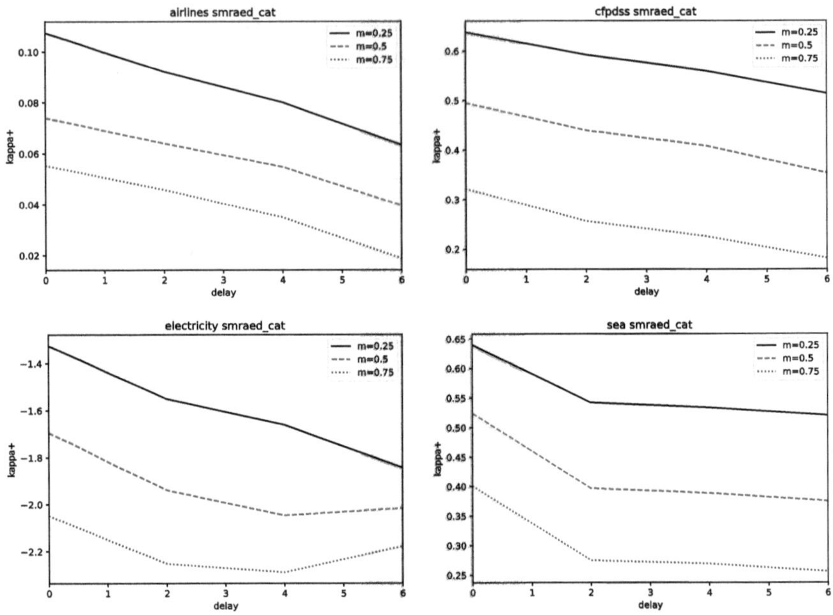

Fig. 4. Task performance of acquiring limited information along fixed acquisition delays

In contrast to the increasing delay, an extended range of delays shows a benefit on the task performance. Acquiring *all* information sees improvements with increasing ranges hinting that the recency of some information at the expense of the delay of other to be a good trade-off for the classifier. In particular, the delay range of eight which includes rare cases in which acquisitions become available immediately after request shows the biggest benefit. The effect can be observed across all data sets and missingness of features. Opposite to the increased delays however, it is now the stable synthetic data sets of *cfpdss* and *sea* that see a lesser impact from the increase in range.

The intelligent acquisition method *smraed* seems mostly unaffected by the change in possible delays. Since it lacks any awareness of the true acquisition delay before making a request it cannot exploit particularly informative acquisitions that would be available immediately. Instead, the mean empirical delay is the same across all use cases which makes beneficial or detrimental effects mostly random.

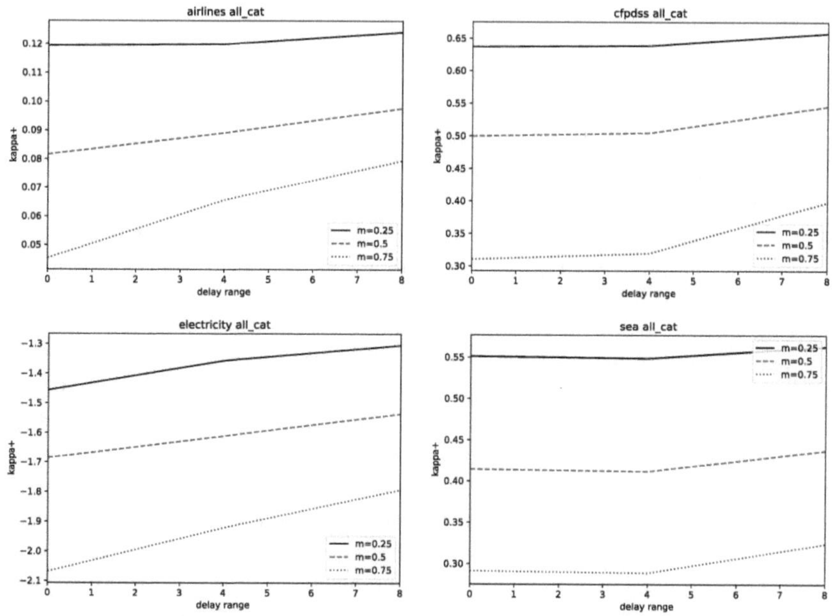

Fig. 5. Task performance of acquiring all information along an increasingly broader range of acquisition delays.

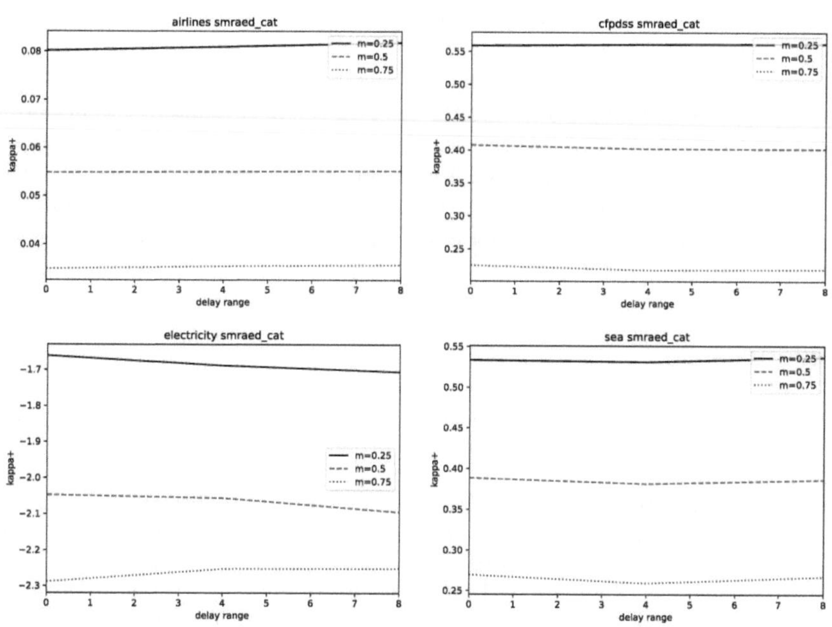

Fig. 6. Task performance of acquiring limited information along an increasingly broader range of acquisition delays.

5 Conclusion

We introduced a batch-based sliding window method for the combined evaluation of acquisitions of incomplete instance-label pairs on a data stream and evaluated them under verification latency and against simple baselines. To do this we compared the impact of acquisition delays on these methods with increasing delays, increasing delay ranges, and increasing provision of labels on two synthetic and two real data sets. We surmised that our method performs similarly to a only-label acquisition method despite its all or nothing acquisition policy. Furthermore, we showed that an increase in acquisition delays to have a particular strong impact from no to some delay across all data sets. We also showed that a wider range of acquisition delays at the same expected delay can lead to an increase in performance if all acquisitions are performed due to a beneficial trade-off of some immediate access to information vs. even more delayed acquisitions while our introduced method was largely unaffected by the greater range of delays.

In future studies active learning methods combined with the same imputation strategies could compete against our introduced method along with a investigation of the impact costs may have on the selection and performance of these methods. We also suggest an investigation into the extreme latency cases and the impact of wasting budget on never returned information has on the task performance. A more thorough study on an exemplary data set or domain could further delve into the aspects of delay and cost modeling with different delays and costs that follow real-world parameters.

Statements and Declarations. The authors have no relevant financial or non-financial interests to disclose. The authors have no conflicts of interest to declare that are relevant to the content of this article. All authors certify that they have no affiliations with or involvement in any organization or entity with any financial interest or non-financial interest in the subject matter or materials discussed in this manuscript. The authors have no financial or proprietary interests in any material discussed in this article.

References

1. Astorga, N., Liu, T., Seedat, N., Schaar, M.: Active learning with llms for partially observed and cost-aware scenarios. Adv. Neural. Inf. Process. Syst. **37**, 20819–20857 (2024)
2. Bifet, A., et al.: MOA: massive online analysis, a framework for stream classification and clustering. In: Proceedings of the First Workshop on Applications of Pattern Analysis, pp. 44–50. PMLR (2010)
3. Bifet, A., Read, J., Žliobaitė, I., Pfahringer, B., Holmes, G.: Pitfalls in benchmarking data stream classification and how to avoid them. In: Blockeel, H., Kersting, K., Nijssen, S., Železný, F. (eds.) ECML PKDD 2013. LNCS (LNAI), vol. 8188, pp. 465–479. Springer, Heidelberg (2013). https://doi.org/10.1007/978-3-642-40988-2_30

4. Büttner, M., Beyer, C., Spiliopoulou, M.: Reducing missingness in a stream through cost-aware active feature acquisition. In: 2022 IEEE 9th International Conference on Data Science and Advanced Analytics (DSAA), pp. 1–10 (2022)
5. Cacciarelli, D., Kulahci, M.: Active learning for data streams: a survey. Mach. Learn. **113**(1), 185–239 (2024). https://doi.org/10.1007/s10994-023-06454-2
6. Castellani, A., Schmitt, S., Hammer, B.: Stream-based active learning with verification latency in non-stationary environments. In: 31st International Conference on Artificial Neural Networks and Machine Learning - ICANN 2022. Lecture Notes in Computer Science, vol. 13532, pp. 260–272. Springer (2022)
7. Harries, M., Wales, N.S.: Splice-2 comparative evaluation: electricity pricing. Technical report (1999)
8. Hofer, V., Krempl, G., Lang, D.: Anticipative bayesian classification for data streams with verification latency. J. Appl. Stat. **51**(14), 2812–2831 (2024)
9. Pham, T., Kottke, D., Krempl, G., Sick, B.: Stream-based active learning for sliding windows under the influence of verification latency. Mach. Learn. **111**(6), 2011–2036 (2022). https://doi.org/10.1007/s10994-021-06099-z
10. Saar-Tsechansky, M., Melville, P., Provost, F.J.: Active feature-value acquisition. Manag. Sci. **55**(4), 664–684 (2009). https://doi.org/10.1287/mnsc.1080.0952
11. de Souza, V.M.A., Pinho, T., Batista, G.: Evaluating stream classifiers with delayed labels information. In: 2018 7th Brazilian Conference on Intelligent Systems (BRACIS), pp. 408–413 (2018)
12. Street, W.N., Kim, Y.: A streaming ensemble algorithm (sea) for large-scale classification. In: Proceedings of the Seventh ACM SIGKDD International Conference on Knowledge Discovery and Data Mining, KDD 2001, pp. 377–382. Association for Computing Machinery, New York (2001). https://doi.org/10.1145/502512.502568
13. Umer, M., Polikar, R.: Comparative analysis of extreme verification latency learning algorithms. CoRR abs/2011.14917 (2020). https://arxiv.org/abs/2011.14917
14. Yuan, L., Pfahringer, B., Barddal, J.P.: Iterative subset selection for feature drifting data streams. In: Proceedings of the 33rd Annual ACM Symposium on Applied Computing, pp. 510–517 (2018)

Foundation Models for Science, LLMs

FoodSEM: Large Language Model Specialized in Food Named-Entity Linking

Ana Gjorgjevikj[1], Matej Martinc[2], Gjorgjina Cenikj[1], Sašo Džeroski[2], Barbara Koroušić Seljak[1], and Tome Eftimov[1](✉)

[1] Computer Systems Department, Jožef Stefan Institute, Ljubljana, Slovenia
{ana.gjorgjevikj,gjorgjina.cenikj,barbara.korousic, tome.eftimov}@ijs.si

[2] Department of Knowledge Technologies, Jožef Stefan Institute, Ljubljana, Slovenia
{matej.martinc,saso.dzeroski}@ijs.si

Abstract. This paper introduces FoodSEM, a state-of-the-art fine-tuned open-source large language model (LLM) for named-entity linking (NEL) to food-related ontologies. To the best of our knowledge, food NEL is a task that cannot be accurately solved by state-of-the-art general-purpose (large) language models or custom domain-specific models/systems. Through an instruction-response (IR) scenario, FoodSEM links food-related entities mentioned in a text to several ontologies, including FoodOn, SNOMED-CT, and the Hansard taxonomy. The FoodSEM model achieves state-of-the-art performance compared to related models/systems, with F1 scores even reaching 98% on some ontologies and datasets. The presented comparative analyses against zero-shot, one-shot, and few-shot LLM prompting baselines further highlight FoodSEM's superior performance over its non-fine-tuned version. By making FoodSEM and its related resources publicly available, the main contributions of this article include (1) publishing a food-annotated corpora into IR format suitable for LLM fine-tuning/evaluation, (2) publishing a robust model to advance the semantic understanding of text in the food domain, and (3) providing a strong baseline on food NEL for future benchmarking.

Keywords: Named-entity linking · Large language model · Food ontology

1 Introduction

Linking food data to ontology entities is critical in aligning it with Findable, Accessible, Interoperable, and Reusable data (FAIR) principles and supporting the European Union (EU) goals for data-driven innovation and sustainability. Ontologies provide standardized terms/structures that make food data discoverable and efficiently accessed from diverse platforms [18]. Ontologies break down the silos between agricultural, health, and nutrition data, facilitating seamless integration in line with initiatives like the European Open Science Cloud

S. Džeroski et al. (Eds.): DS 2025, LNAI 16090, pp. 395–410, 2025.
https://doi.org/10.1007/978-3-032-05461-6_26

(EOSC) [6]. Only in 2025, 91 new projects under Horizon Europe Cluster 6 addressed food systems and sustainability, all requiring FAIR food data management. Many researchers struggle with data interoperability due to the time-consuming linking to ontologies and not having a technical background, highlighting the need to automate this process. Annotating food data with ontology entities ensures consistency and reusability, empowering development of interpretable and reproducible artificial intelligence (AI) applications. It aligns with EU strategies like the Green Deal and Farm to Fork Strategy, promoting sustainability, traceability, and food safety. It underpins the creation of a shared agriculture and food "data space", enabling innovation which adheres to ethical/privacy standards.

Recently, significant progress has been made in predictive healthcare and environmental modeling, enabled by the availability of biomedical semantic vocabularies/standards, which have supported shared tasks in natural language processing (NLP) to produce annotated corpora and advance research [2,27,34,35]. The BioNLP Shared Tasks (2013, 2016) [27] focused on biomedical topics, e.g., gene event extraction, bacteria biotopes, and plant seed regulation. BioCreative challenges [2,34] advanced tasks like gene mention recognition, protein-protein interactions, and chemical/drug entity recognition with machine learning (ML) methods (e.g., Conditional Random Fields, Support Vector Machines, rule-based systems). BioCreative IV [24] focused on chemical named-entity recognition (NER), while BioCreative V focused on disease-related entity relations. Later editions explored crowdsourcing, interoperability, and application in clinics and semantic text similarity. Those resources are vital to interpret health information, along with the abundant biomedical and health-related data.

In contrast, the food and nutrition domain lacks equivalent semantic resources, crucial in the development of domain-specific AI models. Before introducing the FoodBase [32] annotated corpora, food information extraction (IE) relied on rule-based NER methods, such as drNER [17] for dietary recommendations and FoodIE [30] for recipe data, focusing on distinguishing between food vs. non-food entities. With FoodBase, ML approaches emerged, including BuTTER [8], the first bidirectional LSTM for food NER, and FoodNER [36], a fine-tuned BERT model for NER and simultaneous named-entity linking (NEL) tasks (linking recognized mentions of food entities in text to their corresponding ontology entities). FoodOntoMap [31] is a resource designed to normalize and link food concepts to various food ontologies, enhancing data integration. Initially developed with mappings to four ontologies (Hansard Corpus [1], FoodOn [13], OntoFood, and SNOMED Clinical Terms (SNOMED CT) [12]), it was expanded to FoodOntoMapV2 [33] by adding RCD, Medical Subject Headings (MeSH) [11], Systematized Nomenclature of Medicine (SNMI), and National Drug Data File (NDDF) for broader interoperability. The public dataset includes unique food concepts and mappings in CSV and Excel, supporting applications that connect food, health, and environmental research. FoodViz [37] is a web-based tool enabling experts to identify food entity mentions in text and link them to entities from food ontologies, using FoodOntoMap data. Exact matches are linked directly, while ambiguous cases may be linked to multiple entities due to food ontology coverage gaps. FoodKG [19] is a semantic-driven knowledge graph for food recommendation, unifying domain-specific ontologies to support healthy eating. It integrates food and recipe data while preserving provenance, using FoodOn for entity linking.

Recently, the so-called foundation models [4] have revolutionized NLP. The development of large language models (LLMs) has enabled task-agnostic architectures to solve downstream NLP tasks in a zero-shot fashion without the need for labeled data [5]. Researchers are exploring LLMs for biomedical applications, focusing on improving clinical NER through prompt engineering and fine-tuning for biomedical NLP tasks, including NER and NEL [3,40]. LLMs have also been applied to food NER [29] with promising results, but a recent evaluation of GPT models for NEL revealed that while they can identify food entity mentions, they struggle linking them to specific food ontologies [28].

This paper presents FoodSEM, a fine-tuned open-source LLM for food NEL. Through an instruction-response (IR) scenario, FoodSEM is able to link food-related entity mentions (single or multi-word sequences) in a text given in its instruction to entities (represented through their URIs) from three food ontologies (FoodOn, SNOMED-CT, and the Hansard taxonomy). By making all resources publicly available, this paper's contributions are (1) publishing food-annotated corpora in IR format suitable for LLM fine-tuning, (2) publishing a robust model to advance the semantic understanding of text from food domain, and (3) providing a strong baseline on food NEL to serve as a benchmark dataset. FoodSEM achieves high F1 scores, up to 98% for some ontologies. FoodSEM's comparison with 0-shot, 1-shot, and 5-shot prompting baselines demonstrates its superior performance over its non-fine-tuned version. The code and datasets are available on GitHub[1]. FoodSEM is available in Hugging Face's model repository[2].

The paper is organized as follows. Section 2 reviews related work on food NEL and Sect. 3 the used ontologies and corpora. Section 4 details the fine-tuning process. Section 5 presents the results, and Sect. 6 the conclusion.

2 Related Work

NEL [38] is critical in integrating data across diverse domains. By associating entity mentions in text with entities' unique identifiers from ontologies, NEL provides a consistent way to reference those entities, allowing information aggregation for the same entities from multiple sources, regardless of different naming conventions or synonyms. In this direction, StandFood [16] is a semi-automatic system that links food concepts to the FoodEx2 [15] taxonomy, considering lexical, but not semantic similarity. The development of food-annotated corpora like FoodBase [32], CafeteriaFCD [21], and CafeteriaSA [9] enabled the emergence of corpus-based NEL methods. FoodNER [36] is a corpus-based framework

[1] https://github.com/matejMartinc/FoodSEM.

[2] https://huggingface.co/Matej/FoodSEM-LLM.

for food NER and NEL, featuring 15 models, fine-tuned from three pre-trained BERT models on five semantic resources: food vs. nonfood entities, two subsets of Hansard entities, FoodOn, and SNOMED-CT entities. Leveraging BERT's capabilities in IE, FoodNER achieves macro F1 of 93.30%–94.31% for food vs. nonfood classification and 73.39%–78.96% for semantic entity prediction. It supports five tasks, including distinction of entities at the level of food groups using entities from Hansard, FoodOn, and SNOMED-CT, establishing a new benchmark in food NER and NEL. The FoodNER models work well on recipe instruction data, but fail to generalize on scientific articles. SciFoodNER [7] is a scientific food NER and NEL model fine-tuned on scientific abstracts (SAs) annotated with food entities. The model identifies food mentions in text and links them to Hansard, FoodOn, and SNOMED-CT. BioBERT [22] achieved the best performance, with a median macro F1 of 66% for NEL to Hansard, 43% to FoodOn, and 58% to SNOMED-CT. Recent 0-shot evaluations of GPT-3.5 and GPT-4 models have shown low accuracy in food NEL [28], too low for practical use.

3 Food Semantic Resources

This section first describes the used food ontologies, and then the text corpora annotated with ontology entities, used to create the LLM fine-tuning dataset.

FoodOn [13] is an open-source ontology designed to comprehensively describe entities with a "food role", spanning the entire food system from agricultural production to consumption. It offers a structured vocabulary covering food sources, products, and processes, with detailed description of food-related entities. As part of the Open Biological and Biomedical Ontology (OBO) Foundry, FoodOn integrates seamlessly with other life science ontologies, promoting interoperability. It is organized into facets, addressing subdomains like organism parts, food processing methods, and quality attributes. **SNOMED-CT** [12] is a standardized medical ontology that provides a comprehensive vocabulary of clinical concepts (diseases, treatments, anatomy, and procedures). It contains the concept of Food that can be used for food concept normalization. Widely adopted worldwide, it facilitates integration between healthcare systems through its hierarchical structure and interoperability. Maintained by SNOMED International, it supports advanced data analysis and multilingual use, making it essential for effective health information management. The **Hansard taxonomy** [1] is a structured collection of concepts developed in the SAMUELS project (2014–2016) to enable semantically-driven searches within its dataset. This taxonomy organizes concepts into 37 overarching semantic groups, facilitating intuitive information retrieval. Among them, the "Food and Drink" (denoted AG) categorizes culinary/beverage terms and concepts. By structuring data semantically, the taxonomy enhances accessibility/usability of complex textual corpora for researchers/developers, supporting advanced search features and more nuanced analyses.

The **CafeteriaFCD** corpus [21] is a curated text dataset of food consumption data (recipes) annotated with food entities from Hansard, FoodOn, and

SNOMED-CT. Built on the FoodBase corpus (containing 1,000 gold-standard annotated recipes), it has been created with the NCBO Annotator for FoodOn and SNOMED-CT entities, later manually corrected by domain experts. It contains 7,429 annotations. The recipe text varies in length (58–2024 characters, 511.8 on average). Each recipe includes at least one annotation. In relation to Hansard, the recipes contain up to 37 annotations (10.7 on average). From FoodOn, they contain up to 22 annotations (7.6 on average), and from SNOMED-CT up to 17 (5.5 on average). **CafeteriaSA** [9] is a pioneering annotated text corpus of 500 SAs, annotated with food entities from multiple ontologies. It includes 6,407 annotations from Hansard, 4,299 from FoodOn, and 3,623 from SNOMED-CT. The average abstract length is 1,467.6 characters (ranging 236–2,510 characters). From Hansard, there are on average 8.4 annotations per SA (ranging 1–27), from FoodOn 4.8 annotations (1–18), and from SNOMED-CT 3.8 annotations (1–15).

4 FoodSEM

4.1 Fine-Tuning Dataset Creation

```
<document>
␣␣␣␣<id>0recipe1006</id>
␣␣␣␣<infon key="full_text">␣Mix␣the␣cream␣cheese,␣beef,␣olives,␣onion,␣and␣Worcestershire␣sauce␣together␣in␣a␣bowl␣until␣evenly␣
      ↪ blended.␣Keeping␣the␣mixture␣in␣the␣bowl,␣scrape␣it␣into␣a␣semi-ball␣shape.␣Cover,␣and␣refrigerate␣until␣firm,␣at␣
      ↪ least␣2␣hours.␣Place␣a␣large␣sheet␣of␣waxed␣paper␣on␣a␣flat␣surface.␣Sprinkle␣with␣walnuts.␣Roll␣the␣cheese␣ball␣in␣
      ↪ the␣walnuts␣until␣completely␣covered.␣Transfer␣the␣cheese␣ball␣to␣a␣serving␣plate,␣or␣rewrap␣with␣waxed␣paper␣and␣
      ↪ refrigerate␣until␣needed.␣</infon>
␣␣␣␣<annotation id="1">␣<infon
      ↪ key="semantic_tags">http://purl.obolibrary.org/obo/FOODON_03301889;http://purl.obolibrary.org/obo/FOODON_00001013</infon>
␣␣␣␣␣␣␣␣<location offset="3" length="12" />␣<text>CREAM␣CHEESE</text>␣</annotation>
␣␣␣␣<annotation id="2">␣<infon
      ↪ key="semantic_tags">http://purl.obolibrary.org/obo/FOODON_03301704;http://purl.obolibrary.org/obo/NCBITaxon_4679</infon>
␣␣␣␣␣␣␣␣<location offset="10" length="5" />␣<text>ONION</text>␣</annotation>
␣␣␣␣<annotation id="3">␣<infon
      ↪ key="semantic_tags">http://purl.obolibrary.org/obo/FOODON_03305003;http://purl.obolibrary.org/obo/FOODON_03311146</infon>
␣␣␣␣␣␣␣␣<location offset="13" length="20" />␣<text>WORCESTERSHIRE␣SAUCE</text>␣</annotation>
␣␣␣␣<annotation id="4">␣<infon key="semantic_tags">http://purl.obolibrary.org/obo/NCBITaxon_16718</infon>
␣␣␣␣␣␣␣␣<location offset="63" length="7" />␣<text>WALNUTS</text>␣</annotation>
␣␣␣␣<annotation id="5">␣<infon key="semantic_tags">http://purl.obolibrary.org/obo/FOODON_00001013</infon>
␣␣␣␣␣␣␣␣<location offset="67" length="11" />␣<text>CHEESE</text>␣</annotation>
␣␣␣␣<annotation id="6">␣<infon key="semantic_tags">http://purl.obolibrary.org/obo/NCBITaxon_16718</infon>
␣␣␣␣␣␣␣␣<location offset="71" length="7" />␣<text>WALNUTS</text>␣</annotation>
␣␣␣␣<annotation id="7">␣<infon key="semantic_tags">http://purl.obolibrary.org/obo/FOODON_00001013</infon>
␣␣␣␣␣␣␣␣<location offset="78" length="11" />␣<text>CHEESE</text>␣</annotation>
␣␣␣␣<infon key="category">Appetizers␣and␣snacks</infon>
</document>
```

Listing 1.1. BioC-formatted recipe from CafeteriaFCD mapped to FoodOn ontology.

The fine-tuning dataset is created by processing the two Cafeteria corpora in BioC XML format [10], enabling interoperability of biomedical data and annotations. The process starts by converting BioC-formatted data into a format suitable for LLM fine-tuning (converting each BioC-formatted instance into an IR pair). Listing 1.1 shows a BioC-formatted instance from CafeteriaFCD, giving the recipe text and food annotations. The annotations consist of mentions of food entities in the recipe, their location in the text, the actual mention text, and the Uniform Resource Identifiers (URIs) of the linked entities from FoodOn. The mention location in the text is described through its start offset and the number of characters. One recipe has two additional BioC-formatted instances, with all food annotations linked to Hansard and SNOMED-CT entities, respectively. Each BioC instance is converted into an IR pair sequence, including two types of instruction, asking the LLM to either: (1) identify all mentions of food entities in the provided text (which can either be a recipe or a SA), pertaining to the NER task, or (2) link all identified food entity mentions to entities (represented through URIs) from the Hansard taxonomy, FoodOn, or SNOMED-CT, pertaining to the NEL task. The first instruction is concatenated with the actual recipe/SA text in which the entity mentions should be identified, followed by the expected LLM response. The second instruction is repeated three times, each time asking the LLM to link the previously identified entity mentions with entities from one of the three ontologies. Each instruction is followed by the expected LLM response, which contains a generic starting response phrase, followed by a list of entity mentions with their linked ontology entities (represented with URIs). The described transformation of the previously shown BioC-formatted recipe is given in the box below. A sequence of four IR pairs is given, with the instruction given in tags [INST] ... [/INST]. The first pair pertains to the NER task, while the remaining three pertain to the NEL task. There are no duplicate recipe and SA instances in the fine-tuning dataset to prevent data leakage.

[INST] Find all food entities present in the text: Mix the cream cheese, beef, olives, onion, and Worcestershire sauce together in a bowl until evenly blended. Keeping the mixture in the bowl, scrape it into a semi-ball shape. Cover, and refrigerate until firm, at least 2 hours. Place a large sheet of waxed paper on a flat surface. Sprinkle with walnuts. Roll the cheese ball in the walnuts until completely covered. Transfer the cheese ball to a serving plate, or rewrap with waxed paper and refrigerate until needed. **[/INST]** Certainly, the entities connected with food are outlined as follows: cream cheese, beef, olives, onion, Worcestershire sauce, walnuts, cheese ball.
[INST] Is it within your capabilities to link the extracted food entities to a Hansard taxonomy? **[/INST]** Certainly, the entities are associated properly: cream cheese - AG.01.e [Dairy produce]; AG.01.e.02 [Cheese]; AG.01.n [Dishes and prepared food]; AG.01.n.18 [Preserve], beef - AG.01.d.03 [Beef], olives - AG.01.h.01.e [Fruit containing stone], onion - AG.01.h.02.e [Onion/leek/garlic], Worcestershire sauce - AG.01.h [Fruit and vegetables]; AG.01.l.04 [Sauce/dressing], walnuts - AG.01.h.01.f [Nut], cheese ball - AG.01.e.02 [Cheese]; AG.01.n.18 [Preserve].
[INST] Might I trouble you to connect the extracted food entities to a FoodOn ontology, if possible? **[/INST]** Definitely, the entities are linked suitably: cream cheese - FOODON-03301889; FOODON-00001013, onion - FOODON-03301704; NCBITaxon-4679, worcestershire sauce - FOODON-03305003; FOODON-03311146, walnuts - NCBITaxon-16718, cheese - FOODON-00001013.
[INST] Might it be possible for you to link them to a SNOMEDCT ontology, if needed? **[/INST]** Absolutely, the entities are related as such: cream cheese - SNOMEDCT-226849005; SNOMEDCT-255621006; SNOMEDCT-102264005, beef - SNOMEDCT-226916002, olives - SNOMEDCT-227436000, onion - SNOMEDCT-735047000, worcestershire sauce - SNOMEDCT-443701000124100; SNOMEDCT-227519005, cheese - SNOMEDCT-102264005.

To create diverse instruction phrases that precede the recipe or SA text, a pool of 1,000 instruction phrase variations was generated with GPT-3.5. An instruction phrase was randomly selected for each BioC-formatted instance from the corpora and concatenated with the recipe/SA text. Each response that follows the instruction also begins with a generic phrase, followed by a list of food entity mentions and their linked entity URIs from the specified ontology. The beginning response phrase was also randomly selected from a separate pool of 1,000 GPT-3.5-generated phrase variations and concatenated with the linked entity URIs. After converting the BioC-formatted instances into IR pair sequences, we analyzed the entity coverage across the ontologies in the fine-tuning datasets. This analysis was performed only on CafeteriaFCD, as it contains recipe data with a greater diversity of food entities compared to the SAs from CafeteriaSA. The analysis showed underrepresented entities, for which artificially generated IR instances were added to the fine-tuning dataset for a balanced distribution. Figure 1 illustrates the FoodON entity distribution

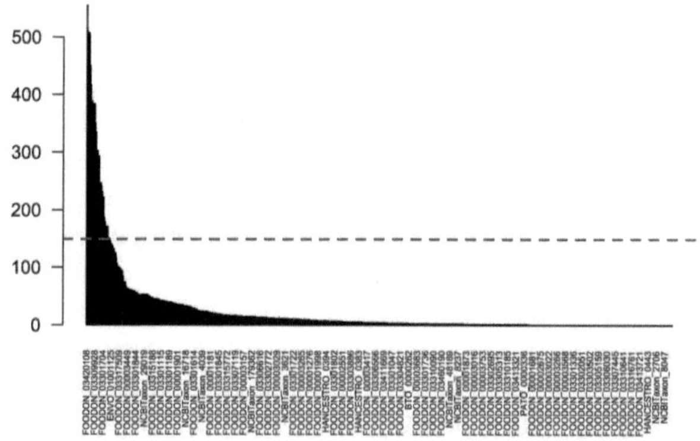

Fig. 1. Distribution of FoodON entities in CafeteriaFCD. Only a subset of entity names is displayed on the x-axis. The red line represents the threshold of 150. (Color figure online)

in CafeteriaFCD. Similar analysis was done for the other two ontologies, but is not shown due to space limitations. Based on the distributions, we set a desired threshold of 150 mentions by entity in the fine-tuning dataset (to be refined in the future through sensitivity analysis). If an entity has fewer than 150 mentions ($k < 150$), we include it in the generation of artificial IR pairs until it reaches the threshold. Unlike the IR pair sequences generated from CafeteriaFCD/SA, which involve one NER instruction and three NEL instructions, the artificially generated IR pairs do not include a NER instruction, only one NEL instruction. To achieve this, we generated a sample of food entity labels for each entity within each ontology, with a size of $150 - k$. Next, we combined all labels across entities within each ontology, shuffled them, and created sets with 7, 9, and 12 labels chosen randomly. Each set was used to generate three IR pairs, each instructing the LLM to link the labels to one of the three ontologies. We generate 13,492 artificial IR pairs for NEL to FoodOn, 1,611 to Hansard, and 4,445 to SNOMED-CT, making sure that no duplicate instances are generated, to avoid data leakage during LLM cross-validation (CV) (see Sect. 5.1). We note again that CafeteriaFCD/SA IR sequences include NER and NEL IR pairs, while the artificially generated sequences include only NEL IR pairs. The final fine-tuning dataset consists of 1,479 Cafeteria IR sequences, 13,492 artificial FoodOn NEL IR pairs, 1,611 artificial Hansard NEL IR pairs, 4,445 artificial SNOMED-CT NEL IR pairs, or a total of 21,027 fine-tuning instances. FoodSEM is expected to correctly do NEL to three ontologies, so it is crucial that it sees all entity URIs from those ontologies during fine-tuning. In such a way, it learns to link single/multi-word mentions to correct entity URIs without hallucinations. The model is not designed to find links to entities absent from the three ontologies, as NEL is an information retrieval task.

4.2 Fine-Tuning Process

We use the instruction-tuned 8B parameter version of the Meta Llama 3 family of LLMs [14]. Llama 3 is an auto-regressive language model using an optimized transformer architecture [39]. The model was pre-trained on a dataset with over 15 trillion tokens, sourced from publicly available content and curated using advanced filtering methods like heuristic filters and semantic deduplication. The 8B model has a knowledge cutoff of March 2023. The instruction-tuned version we use[3] is optimized for dialogue use cases by applying an alignment process including supervised fine-tuning, rejection sampling, proximal policy optimization, and direct preference optimization. It was instruction-tuned on over 10 million manually labeled data points, curated to improve reasoning, code generation, and similar. We selected an open-source LLM since proprietary LLMs (e.g., GPT-4, Gemini, or Claude) cannot be fine-tuned for specific tasks/domains. With our limited GPU resources, it was infeasible to opt for a bigger model (e.g., 70B or 400B parameter model), as it would significantly increase training time/cost. We chose the Llama 3 8B model due to its competitive performance on open-source LLM benchmarks[4] when conducting our experiments. To reduce memory usage and speed up the fine-tuning process, we load the model in 4-bit precision and fine-tune it using Low-Rank Adaptation (LoRA) [20]. We freeze all layers in the backbone model and train only low-rank perturbations to query and value weight matrices in the model. We train the model using a supervised fine-tuning (SFT) trainer, using the next word prediction objective. The fine-tuning hyperparameters are: learning rate of 2e-4, LoRA intrinsic rank of 16, LoRA alpha of 16, LoRA dropout of 0.05, and sequence length of 1024. We fine-tuned the model for one epoch using batch size of 10, and 10 warm-up steps. The input sequences were padded to the longest sequence in the batch and truncated if longer than 1024 tokens. By default, the Llama tokenizer does not have a padding token, therefore we added it by using one of the designated extra token spots, reserved for additional special tokens (we use the "<|reserved_special_token_250|>" spot[5]).

5 Evaluation

5.1 Experimental Design

We evaluate FoodSEM's performance in a CV setting. As per Sect. 4, the fine-tuning uses four IR pair instance datasets: (1) NEL to Hansard taxonomy (consisting of an instruction to link food entity mentions to entities from Hansard

[3] We use the model available on HuggingFace: https://huggingface.co/meta-llama/Meta-Llama-3-8B-Instruct.

[4] https://huggingface.co/spaces/open-llm-leaderboard/open_llm_leaderboard.

[5] Initial experiments showed that a popular practice of setting the "end of sequence" token as a padding token resulted in model generating repetitions and having problems with finishing the generated sequence.

taxonomy and expected response), (2) NEL to FoodON, (3) NEL to SNOMED-CT, and (4) NER of mentions of food entities in recipes/SAs. Each of the datasets is divided in five random chunks with the Scikit-learn library. For each CV fold, we construct a dataset-specific training set by concatenating 4 training chunks and using the remaining fifth for testing. The final training set per CV fold is constructed by concatenating (and shuffling) 4 datasets-specific training sets. The test set for each fold is constructed by concatenating all datasets-specific test folds. Therefore, the training set in each fold consists of 80% of the data and the test from the remaining 20%. We do not use CV folds for hyperparameter search or early stopping. To avoid catastrophic forgetting [23] and prevent overfitting, we add in each CV fold's training set IR instances from the Dolly-HHRLHF subset [25] of the open source MosaicML Instruct-v3 dataset [26]. The Dolly-HHRLHF dataset is a general instruction-tuning dataset generated by Databricks[6] employees, covering different tasks from domains, such as brainstorming, open/closed question-answering, and summarization. Specifically, into the training set of each CV fold, we randomly shuffle in 34,229 IR instances from Dolly-HHRLHF, filtered according to length (all instances shorter or equal to the max sequence length of 1024 tokens). Using the five training sets, we obtain five fine-tuned models, tested on the appropriate test set. During testing, the model is given an instruction part of the IR instance as input, and the generated response is compared to the expected one. Note that for both Cafeteria corpora, the NEL task follows a NER task, so the model is, in the first step, instructed to extract food entity mentions. In the second step, the extracted mentions are fed to the model as input, with an instruction to link them to an ontology (Hansard, FoodON, or SNOMED-CT). Fine-tuning one model instance took ~48 h on an NVIDIA A100 80GB GPU. The entire fine-tuning of five models took about 240 h of GPU time.

The fine-tuned models were compared to the non-fine-tuned on each of the five test folds. Scenarios involving 0-shot, 1-shot, and 5-shot prompting of the non-fine-tuned model were used as baselines. In each, the model was prompted with an instruction part of the test IR instance concatenated with none, one, or five example IR instances sampled randomly from the matching training set, respectively. The IR instance(s) were selected with regard to Hansard, FoodOn, SNOMED-CT, or food NER, depending on the given input test set instruction. An example of a 1-shot instruction prompt is shown in the box below.

[INST] The following are examples of questions (with answers) about nutrition. **Question**: I kindly ask for links to the FoodOn ontology for the mentioned entities: cheddar cheese, cookie dough, flax? **Answer**: Absolutely, the entities are related properly: cheddar cheese - http://purl.obolibrary.org/obo/FOODON_03302458; http://purl.obolibrary.org/obo/FOODON_00001013, cookie dough - http://purl.obolibrary.org/obo/FOODON_03310689; http://purl.obolibrary.org/obo/FOODON_00002466; http://purl.obolibrary.org/obo/FOODON_03301585; http://purl.obolibrary.org/obo/FOODON_03311552, flax - http://purl.obolibrary.org/obo/NCBITaxon_4006.
Respond to the following question in the same manner as seen in the examples above. **Question**: Could you provide links to the FoodOn ontology for these entities: green tea, oil, salmon, flax, lettuce, eggnog, spring, flaxseed meal, maraschino cherry? **Answer**: **[/INST]**

[6] https://www.databricks.com.

Custom LLM response parsers were developed to extract predictions (entity URIs) even when responses do not follow the formatting the LLM was fine-tuned to output. This also applies to the baseline 0-shot, 1-shot, and 5-shot responses of the non-fine-tuned LLM. Only entity mentions for which a URI is given in the true response are taken in consideration during the evaluation. Incorrect entity URI linking, non-meaningful responses (e.g., empty ones, hallucinations that contain no entity URIs for the mentions in the instruction) were counted as errors.

5.2 Results and Discussion

The LLM performance was evaluated with three metrics: macro-weighted NEL precision, macro-weighted NEL recall, and macro-weighted NEL F1. In NEL, precision measures the proportion of correctly linked entities out of all entities linked by the LLM, i.e., its ability to avoid false positives. Recall evaluates the proportion of correctly linked entities out of all ground truth entities, i.e., the ability to link all entities mentioned in the text. The F1 combines precision and recall in a single metric, offering a holistic measure of the LLM's linking accuracy. Although we tried to balance entity distribution across each ontology through artificial IR instances, the dataset is imbalanced overall, as some entities appear more than 150 times in the Cafeteria corpora and were not removed from the dataset. To alleviate this issue, we calculate the macro-weighted average of the metrics, derived by taking the weighted mean of precision, recall, or F1 for each entity, factoring in the actual frequency of each entity in the test dataset.

Table 1 gives FoodSEM's performance on artificially generated, CafeteriaFCD, and CafeteriaSA IR instances. As the artificial IR instances concern the NEL task only, and for both Cafeteria corpora, the NEL task is preceded by an NER task, any entity mentions not recognized during the NER task are excluded from the NEL process, which then affects the NEL performance. For each test dataset (learning task + ontology), results are presented based on five-fold CV. On the **artificially generated** IR instances, FoodSEM has consistently high and robust performance across all metrics. On the artificial instances, on FoodOn the average macro-weighted precision is 0.953(±0.002), recall 0.936(±0.002), and F1 0.942(±0.002). For SNOMED-CT, they are 0.981(±0.003), 0.969(±0.003), and 0.974(±0.002), and for Hansard taxonomy 0.948(±0.004), 0.929(±0.011), and 0.937(±0.007), accordingly. On **CafeteriaFCD**, FoodSEM results for FoodOn are 0.923(±0.004), 0.770(±0.018), and 0.827(±0.012), for SNOMED-CT 0.956 (±0.010), 0.804(±0.024), and 0.863(±0.019), and for Hansard taxonomy, 0.915 (±0.013), 0.791(±0.010), and F1 0.843(±0.010), accordingly. On **CafeteriaSA**, FoodSEM results for FoodOn are 0.804(±0.029), 0.580(±0.035), and 0.654(±0.029), for SNOMED-CT 0.849(±0.019), 0.640(±0.025), and 0.713(±0.022), for Hansard 0.872(±0.016), 0.653(±0.035), and 0.730(±0.026), accordingly. From the results for CafeteriaFCD and CafeteriaSA, it can be concluded that the precision for each ontology is consistently high and robust, although it slightly decreases on CafeteriaSA.

This suggests that most of the entities linked by FoodSEM are relevant. However, the recall is somewhat lower, indicating that not all entity mentions in the SA are successfully linked. This is likely due to errors in the NER task, which are propagated to the NEL task, i.e., if an entity mention is not extracted in the NER task, the NEL task cannot link it, affecting NEL F1.

Table 1. FoodSEM results on artificial, CafeteriaFCD, and CafeteriaSA IR instances.

Artificial NEL Samples					CafeteriaFCD				CafeteriaSA			
Test Fold	Test Set	Macro Weighted Precision	Macro Weighted Recall	Macro Weighted F1	Test Set	Macro Weighted Precision	Macro Weighted Recall	Macro Weighted F1	Test Set	Macro Weighted Precision	Macro Weighted Recall	Macro Weighted F1
1	NEL (FoodOn)	0.954	0.934	0.941	NER+NEL (FoodOn)	0.923	0.769	0.823	NER+NEL (FoodOn)	0.787	0.523	0.609
2		0.953	0.933	0.940		0.928	0.746	0.811		0.825	0.577	0.657
3		0.951	0.937	0.942		0.927	0.759	0.824		0.812	0.608	0.675
4		0.955	0.938	0.944		0.917	0.786	0.836		0.762	0.584	0.645
5		0.952	0.937	0.943		0.921	0.790	0.840		0.833	0.608	0.683
Avg		0.953(0.002)	0.936(0.002)	0.942(0.002)		0.923(0.004)	0.770(0.018)	0.827(0.012)		0.804(0.029)	0.580(0.035)	0.654(0.029)
1	NEL (Snomed)	0.979	0.966	0.972	NER+NEL (Snomed)	0.969	0.826	0.882	NER+NEL (Snomed)	0.827	0.612	0.685
2		0.978	0.971	0.974		0.950	0.782	0.847		0.870	0.660	0.735
3		0.981	0.967	0.973		0.944	0.774	0.840		0.857	0.670	0.736
4		0.984	0.970	0.976		0.955	0.818	0.872		0.831	0.636	0.703
5		0.983	0.972	0.977		0.960	0.821	0.876		0.858	0.621	0.707
Avg		0.981(0.003)	0.969(0.003)	0.974(0.002)		0.956(0.010)	0.804(0.024)	0.863(0.019)		0.849(0.019)	0.640(0.025)	0.713(0.022)
1	NEL (Hansard)	0.950	0.935	0.941	NER+NEL (Hansard)	0.922	0.802	0.854	NER+NEL (Hansard)	0.851	0.594	0.685
2		0.945	0.909	0.925		0.911	0.777	0.834		0.865	0.665	0.731
3		0.944	0.934	0.938		0.929	0.800	0.854		0.878	0.688	0.754
4		0.954	0.935	0.942		0.917	0.786	0.842		0.895	0.654	0.743
5		0.947	0.933	0.938		0.895	0.788	0.833		0.871	0.665	0.736
Avg		0.948(0.004)	0.929(0.011)	0.937(0.007)		0.915(0.013)	0.791(0.010)	0.843(0.010)		0.872(0.016)	0.653(0.035)	0.73(0.026)

Table 2. Non-fine-tuned LLM results in 5-shot prompting baseline scenario.

Artificial NEL Samples					CafeteriaFCD				CafeteriaSA			
Test Fold	Test Set	Macro Weighted Precision	Macro Weighted Recall	Macro Weighted F1	Test Set	Macro Weighted Precision	Macro Weighted Recall	Macro Weighted F1	Test Set	Macro Weighted Precision	Macro Weighted Recall	Macro Weighted F1
1	NEL(FoodOn)	0.368	0.055	0.093	NER+NEL (FoodOn)	0.515	0.249	0.320	NER+NEL (FoodOn)	0.356	0.113	0.160
2		0.367	0.064	0.106		0.504	0.248	0.318		0.382	0.113	0.165
3		0.365	0.061	0.103		0.541	0.246	0.322		0.326	0.106	0.149
4		0.352	0.056	0.095		0.527	0.255	0.326		0.351	0.139	0.191
5		0.358	0.057	0.096		0.540	0.247	0.324		0.319	0.101	0.142
Avg		0.362(0.007)	0.059(0.004)	0.099(0.006)		0.525(0.016)	0.249(0.004)	0.322(0.003)		0.347(0.025)	0.114(0.015)	0.161(0.019)
1	NEL (Snomed)	0.421	0.127	0.189	NER+NEL (Snomed)	0.510	0.242	0.314	NER+NEL (Snomed)	0.442	0.194	0.256
2		0.432	0.134	0.196		0.490	0.249	0.317		0.411	0.164	0.224
3		0.407	0.126	0.185		0.486	0.231	0.301		0.388	0.179	0.235
4		0.400	0.122	0.181		0.461	0.236	0.299		0.331	0.144	0.187
5		0.418	0.127	0.187		0.473	0.219	0.291		0.386	0.153	0.208
Avg		0.416(0.012)	0.127(0.004)	0.188(0.006)		0.484(0.018)	0.235(0.011)	0.304(0.011)		0.392(0.041)	0.167(0.020)	0.222(0.026)
1	NEL (Hansard)	0.396	0.069	0.114	NER+NEL (Hansard)	0.530	0.377	0.425	NER+NEL (Hansard)	0.377	0.232	0.275
2		0.383	0.072	0.116		0.491	0.354	0.395		0.482	0.264	0.314
3		0.421	0.074	0.123		0.512	0.357	0.406		0.487	0.265	0.331
4		0.417	0.074	0.121		0.545	0.387	0.434		0.533	0.291	0.344
5		0.405	0.067	0.111		0.534	0.372	0.427		0.492	0.293	0.341
Avg		0.404(0.016)	0.071(0.003)	0.117(0.005)		0.522(0.021)	0.369(0.014)	0.417(0.016)		0.474(0.058)	0.269(0.025)	0.321(0.028)

Table 3. Non-fine-tuned LLM performance in 1-shot prompting baseline scenario.

Artificial NEL Samples					CafeteriaFCD				CafeteriaSA			
Test Fold	Test Set	Macro Weighted Precision	Macro Weighted Recall	Macro Weighted F1	Test Set	Macro Weighted Precision	Macro Weighted Recall	Macro Weighted F1	Test Set	Macro Weighted Precision	Macro Weighted Recall	Macro Weighted F1
1	NEL (FoodOn)	0.195	0.034	0.055	NER+NEL (FoodOn)	0.347	0.067	0.107	NER+NEL (FoodOn)	0.236	0.039	0.062
2		0.200	0.036	0.059		0.335	0.080	0.124		0.181	0.021	0.034
3		0.202	0.032	0.054		0.344	0.083	0.130		0.18	0.028	0.046
4		0.208	0.036	0.059		0.293	0.070	0.109		0.171	0.022	0.038
5		0.189	0.035	0.057		0.326	0.061	0.096		0.190	0.029	0.047
Avg		0.199(0.007)	0.035(0.002)	0.057(0.002)		0.329(0.022)	0.072(0.009)	0.113(0.014)		0.192(0.026)	0.028(0.007)	0.045(0.011)
1	NEL (Snomed)	0.183	0.045	0.069	NER+NEL (Snomed)	0.277	0.061	0.097	NER+NEL (Snomed)	0.229	0.035	0.055
2		0.208	0.048	0.075		0.254	0.057	0.091		0.177	0.026	0.044
3		0.220	0.055	0.084		0.279	0.059	0.093		0.210	0.036	0.061
4		0.216	0.056	0.085		0.286	0.064	0.101		0.211	0.031	0.051
5		0.217	0.057	0.086		0.286	0.062	0.098		0.235	0.043	0.070
Avg		0.209(0.015)	0.052(0.005)	0.080(0.007)		0.276(0.013)	0.061(0.003)	0.096(0.004)		0.212(0.023)	0.034(0.006)	0.056(0.010)
1	NEL (Hansard)	0.175	0.114	0.130	NER+NEL (Hansard)	0.304	0.131	0.177	NER+NEL (Hansard)	0.206	0.076	0.105
2		0.141	0.102	0.112		0.311	0.128	0.173		0.282	0.071	0.104
3		0.163	0.100	0.116		0.338	0.161	0.210		0.211	0.068	0.098
4		0.147	0.104	0.116		0.313	0.158	0.199		0.357	0.094	0.138
5		0.169	0.116	0.131		0.304	0.130	0.175		0.289	0.077	0.111
Avg		0.159(0.014)	0.107(0.007)	0.121(0.009)		0.314(0.014)	0.142(0.016)	0.187(0.017)		0.269(0.063)	0.077(0.010)	0.111(0.016)

We compared FoodSEM with the non-fine-tuned model using three baseline evaluation scenarios, the same metrics, and the same folds. Results for the 5-shot prompting baseline scenario are presented in Table 2, while for the 1-shot prompting scenario in Table 3. The results from the 0-shot prompting were 0.0 across all metrics/ontologies, so they are not given separately. The tables highlight FoodSEM's superior performance over its non-fine-tuned version. For the 5-shot prompting baseline on the NEL task, the macro-weighted F1 scores are 0.099(±0.006) for FoodOn, 0.188(±0.006) for SNOMED-CT, and 0.117(±0.005) for Hansard. In comparison, the 1-shot baseline yields macro-weighted F1 of 0.057(±0.002) for FoodOn, 0.080(±0.007) for SNOMED-CT, and 0.121(±0.009) for Hansard. For CafeteriaFCD, the 5-shot baseline gives a macro-weighted F1 of 0.322(±0.003) for FoodOn, 0.304(±0.011) for SNOMED-CT, and 0.417(±0.016) for Hansard. For CafeteriaSA, the performance of F1 is 0.161(±0.019) for FoodOn, 0.222(±0.026) for SNOMED-CT, and 0.321(±0.028) for Hansard. In comparison, the 1-shot baseline achieves a macro-weighted F1 of 0.045(±0.011) for FoodOn, 0.056(±0.010) for SNOMED-CT, and 0.111(±0.016) for Hansard on CafeteriaSA. On CafeteriaFCD, the F1 is 0.113(±0.014) for FoodOn, 0.096(±0.004) for SNOMED-CT, and 0.187(±0.017) for Hansard. Overall, the 5-shot prompting baseline achieves a max F1 around 0.40, while the 1-shot baseline scores significantly lower. Figure 2 presents a heatmap of the mean macro-weighted F1, showing FoodSEM's superior results.

The baseline results suggest that the multi-stage reasoning (NER followed by NEL), gives the LLM additional context and leads to better results, a hypothesis to be tested in future work. The results also show that linking to Hansard tends to be easier for the non-fine-tuned model than for the other ontologies. The reason may be mostly syntactic, since linking to FoodOn and SNOMED-CT requires generating long URIs, which do not resemble standard natural language on which LLMs are pre-trained. This claim is also supported by observed examples of deterioration of the syntactic capabilities of the baseline LLM when prompts containing several long URIs (in the randomly sampled 1-shot or 5-shot IR instances) are given to the LLM.

Here we also reference NEL results achieved on CafeteriaFCD with BERT and BioBERT [36]. The macro-weighted F1 for the combined NER+NEL tasks are 0.781 for FoodOn (achieved with BERT), 0.789 for Hansard (with BioBERT-large), and 0.761 for SNOMED-CT (with BERT). The macro-weighted F1 of BioBERT on CafeteriaSA are 0.430 for FoodOn, 0.660 for Hansard, and 0.580 for SNOMED-CT [7]. Although the folds used in those studies differ from ours, the results indicate the expected performance. These methods have been introduced by the authors of this paper in their previous attempt of addressing NEL. The current study overcomes the challenges faced in our earlier attempts to address the NEL task.

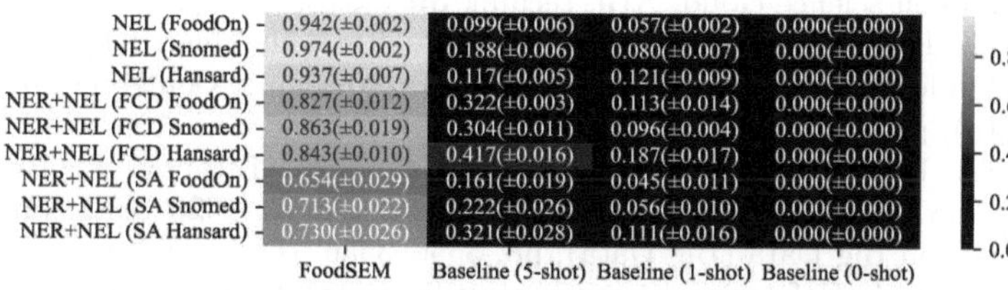

Fig. 2. Heatmap of macro-weighted F1. The benchmarking scenarios are presented in rows, while FoodSEM and baseline models (0-shot, 1-shot, and 5-shot) are shown in columns.

6 Conclusion

We present FoodSEM, a fine-tuned open-source LLM for food named-entity linking (NEL) that achieves state-of-the-art accuracy by linking text entities to multiple food ontologies. Supported by a dedicated instruction-response dataset, FoodSEM outperforms 0-, 1-, and 5-shot baselines, reaching F1 scores up to 98%. Future work will extend its scope with scientific datasets and explore retrieval-augmented generation (RAG) to identify unseen entities in evolving ontologies.

Acknowledgments. The authors acknowledge the support of the Slovenian Research Agency through program grants No. P2-0098 and No. P2-0103, project grant No. GC-0001, as well as the Horizon Europe EU research and innovation framework programme FishEuTrust (101060712), LLMs4EU (101198470), ERA Chair AutoLearn-SI (101187010), and MSCA-PF AutoLLMSelect (101211695).

References

1. Abercrombie, G., Batista-Navarro, R.T.: A sentiment-labelled corpus of hansard parliamentary debate speeches. In: International Language Resource and Evaluation Conference 2018: ParlaCLARIN Workshop, pp. 43–47. Clarin (2018)
2. Arighi, C.N., et al.: Biocreative iii interactive task: an overview. BMC Bioinform. **12**, 1–21 (2011)
3. Biana, J., Zhai, W., Huang, X., Zheng, J., Zhu, S.: Vaner: leveraging large language model for versatile and adaptive biomedical named entity recognition. arXiv preprint arXiv:2404.17835 (2024)
4. Bommasani, R., et al.: On the opportunities and risks of foundation models. arXiv preprint arXiv:2108.07258 (2021)
5. Brown, T., et al.: Language models are few-shot learners. Adv. Neural. Inf. Process. Syst. **33**, 1877–1901 (2020)
6. Budroni, P., Claude-Burgelman, J., Schouppe, M.: Architectures of knowledge: the european open science cloud. ABI Technik **39**(2), 130–141 (2019)
7. Cenikj, G., Petelin, G., Seljak, B.K., Eftimov, T.: Scifoodner: food named entity recognition for scientific text. In: 2022 IEEE International Conference on Big Data (Big Data), pp. 4065–4073. IEEE (2022)
8. Cenikj, G., Popovski, G., Stojanov, R., Seljak, B.K., Eftimov, T.: Butter: bidirectional LSTM for food named-entity recognition. In: 2020 IEEE International Conference on Big Data (Big Data), pp. 3550–3556. IEEE (2020)
9. Cenikj, G., et al.: Cafeteriasa corpus: scientific abstracts annotated across different food semantic resources. Database **2022**, baac107 (2022)
10. Comeau, D.C., et al.: Bioc: a minimalist approach to interoperability for biomedical text processing. Database **2013** (2013)
11. Díaz-Galiano, M.C., García-Cumbreras, M.Á., Martín-Valdivia, M.T., Montejo-Ráez, A., Ureña-López, L.A.: Integrating mesh ontology to improve medical information retrieval. In: Peters, C., et al. (eds.) CLEF 2007. LNCS, vol. 5152, pp. 601–606. Springer, Heidelberg (2008). https://doi.org/10.1007/978-3-540-85760-0_76
12. Donnelly, K., et al.: SNOMED-CT: the advanced terminology and coding system for ehealth. Stud. Health Technol. Inform. **121**, 279 (2006)
13. Dooley, D.M., et al.: Foodon: a harmonized food ontology to increase global food traceability, quality control and data integration. NPJ Sci. Food **2**(1), 23 (2018)
14. Dubey, A., et al.: The llama 3 herd of models. arXiv preprint arXiv:2407.21783 (2024)
15. EFSA: The food classification and description system foodex 2 (revision 2). Technical report, Wiley Online Library (2015)
16. Eftimov, T., Korošec, P., Koroušić Seljak, B.: Standfood: standardization of foods using a semi-automatic system for classifying and describing foods according to foodex2. Nutrients **9**(6), 542 (2017)
17. Eftimov, T., Koroušić Seljak, B., Korošec, P.: A rule-based named-entity recognition method for knowledge extraction of evidence-based dietary recommendations. PLoS ONE **12**(6), e0179488 (2017)
18. Guarino, N., Oberle, D., Staab, S.: What is an ontology? Handbook on ontologies, pp. 1–17 (2009)
19. Haussmann, S., et al.: Foodkg: a semantics-driven knowledge graph for food recommendation. In: Ghidini, C., et al. (eds.) ISWC 2019. LNCS, vol. 11779, pp. 146–162. Springer, Cham (2019). https://doi.org/10.1007/978-3-030-30796-7_10

20. Hu, E.J., et al.: Lora: low-rank adaptation of large language models. arXiv preprint arXiv:2106.09685 (2021)
21. Ispirova, G., et al.: Cafeteriafcd corpus: food consumption data annotated with regard to different food semantic resources. Foods **11**(17), 2684 (2022)
22. Lee, J., et al.: Biobert: a pre-trained biomedical language representation model for biomedical text mining. Bioinformatics **36**(4), 1234–1240 (2020)
23. Luo, Y., Yang, Z., Meng, F., Li, Y., Zhou, J., Zhang, Y.: An empirical study of catastrophic forgetting in large language models during continual fine-tuning. arXiv preprint arXiv:2308.08747 (2023)
24. Mao, Y., et al.: Overview of the gene ontology task at biocreative IV. Database **2014** (2014)
25. MosaicML: Dolly-hhrlhf dataset (2023). https://huggingface.co/datasets/mosaicml/dolly_hhrlhf
26. MosaicML: Mosaicml instruct-v3 dataset (2023). https://huggingface.co/datasets/mosaicml/instruct-v3
27. Nédellec, C., et al.: Overview of bionlp shared task 2013. In: Proceedings of the BioNLP Shared Task 2013 Workshop, pp. 1–7 (2013)
28. Ogrinc, M., Koroušić Seljak, B., Eftimov, T.: Zero-shot evaluation of chatgpt for food named-entity recognition and linking. Front. Nutr. **11**, 1429259 (2024)
29. Pitsilou, V., Papadakis, G., Skoutas, D.: Using LLMs to extract food entities from cooking recipes. In: 2024 IEEE 40th International Conference on Data Engineering Workshops (ICDEW), pp. 21–28. IEEE (2024)
30. Popovski, G., Kochev, S., Korousic-Seljak, B., Eftimov, T.: Foodie: a rule-based named-entity recognition method for food information extraction. ICPRAM **12**, 915 (2019)
31. Popovski, G., Korousic-Seljak, B., Eftimov, T.: Foodontomap: linking food concepts across different food ontologies. In: KEOD, pp. 195–202 (2019)
32. Popovski, G., Seljak, B.K., Eftimov, T.: Foodbase corpus: a new resource of annotated food entities. Database **2019**, baz121 (2019)
33. Popovski, G., Seljak, B.K., Eftimov, T.: Foodontomapv2: food concepts normalization across food ontologies. In: Fred, A., Salgado, A., Aveiro, D., Dietz, J., Bernardino, J., Filipe, J. (eds.) IC3K 2019. CCIS, vol. 1297, pp. 413–426. Springer, Cham (2020). https://doi.org/10.1007/978-3-030-66196-0_19
34. Rastegar-Mojarad, M., et al.: Biocreative/OHNLP challenge 2018. In: Proceedings of the 2018 ACM International Conference on Bioinformatics, Computational Biology, and Health Informatics, p. 575 (2018)
35. Stenetorp, P., Topić, G., Pyysalo, S., Ohta, T., Kim, J.D., Tsujii, J.: BioNLP shared task 2011: supporting resources. In: Proceedings of BioNLP Shared Task 2011 Workshop, pp. 112–120 (2011)
36. Stojanov, R., Popovski, G., Cenikj, G., Koroušić Seljak, B., Eftimov, T.: A fine-tuned bidirectional encoder representations from transformers model for food named-entity recognition: algorithm development and validation. J. Med. Internet Res. **23**(8), e28229 (2021)
37. Stojanov, R., Popovski, G., Jofce, N., Trajanov, D., Seljak, B.K., Eftimov, T.: Foodviz: visualization of food entities linked across different standards. In: Nicosia, G., et al. (eds.) LOD 2020. LNCS, vol. 12566, pp. 28–38. Springer, Cham (2020). https://doi.org/10.1007/978-3-030-64580-9_4

38. Tedeschi, S., Conia, S., Cecconi, F., Navigli, R.: Named entity recognition for entity linking: what works and what's next. In: Findings of the Association for Computational Linguistics: EMNLP 2021, pp. 2584–2596 (2021)
39. Vaswani, A., et al.: Attention is all you need. In: Advances in Neural Information Processing Systems, vol. 30 (2017)
40. Zhu, Z., et al.: Comparative analysis of large language models in chinese medical named entity recognition. Bioengineering **11**(10), 982 (2024)

Multi-domain Validation of LLM-Based Simulators via Interpretable and Latent Representations

Luca Coda-Giorgio, Giacomo Fidone, and Laura Pollacci(✉)

Department of Computer Science, University of Pisa, Pisa, Italy
l.codagiorgio@studenti.unipi.it, giacomo.fidone@di.unipi.it,
laura.pollacci@unipi.it

Abstract. The emergence of simulators powered by Large Language Models (LLMs) has enabled the realistic modeling of complex social phenomena while significantly reducing the costs and challenges of real-world data collection. Despite their promise, assessing the reliability of these simulators remains an open challenge. Existing validation methods often focus on isolated domains and operate at fixed levels of granularity, limiting their generalizability. In this work, we introduce SIMVALE (SIMulator VAlidation with Latent Embeddings), a generalizable, multi-domain framework for quantitatively assessing LLM-based social simulators. SIMVALE leverages both interpretable features and latent representations to yield global and local assessments about the fidelity of a simulator to real-world dynamics, or about the effects of controlled interventions. We demonstrate the effectiveness of SIMVALE through a case study evaluating a simulator's ability to reproduce Online Social Network behavioral patterns and capture the impact of moderation interventions.

Keywords: Simulator Evaluation · Latent Representation · Text-based Simulation

1 Introduction

Large Language Models (LLMs) have significantly advanced Natural Language Processing and Artificial Intelligence by enabling sophisticated tasks such as text summarization, translation, and generation through their capacity to produce contextually coherent and nuanced text. Their recent integration into simulation frameworks has enhanced autonomous agents' capabilities, facilitating realistic modeling of human behavior and social dynamics in controlled environments. LLM-driven social simulators support the investigation of emergent phenomena, hypothesis testing and intervention assessment, while mitigating the challenges and costs associated with real-world data collection.

However, ensuring the fidelity of LLM-based social simulators is critical to prevent biased or misleading observations. This necessitates rigorous validation frameworks that quantitatively evaluate the extent to which simulated outputs

S. Džeroski et al. (Eds.): DS 2025, LNAI 16090, pp. 411–426, 2025.
https://doi.org/10.1007/978-3-032-05461-6_27

replicate the target behavioral patterns observed in real-world scenarios. Existing validation methodologies have notable limitations. Subjective human evaluations, while informative, lack scalability and reproducibility. Many approaches target narrow behavioral traits—like personality—instead of assessing the realism of agent behavior across multiple interacting domains. In addition, high contextual specificity often hampers generalizability.

Starting from these considerations, we propose SIMVALE (SIMulator VAlidation with Latent Embeddings)[1], a quantitative, multi-domain validation framework tailored to LLM-based social simulators. SIMVALE enables validation through a multi-domain, data-driven methodology. The framework processes real and simulated corpora through two complementary pipelines: one extracts transparent, interpretable features related to different domains (e.g., sentiment, readability, personality), while the other derives contextualized representations by fine-tuning BERT models on task-specific traits. A macro-level (global) validation is performed by computing distributional divergences across transparent features, while a micro-level (local) validation is performed by computing the same divergences on latent profiles uncovered by clustering BERT representations. Validation can target realism (real vs. simulated), alteration effects (pre- vs. post-intervention), or both. This approach ensures semantic coherence, scalability and generalizability across different simulation scenarios. We test SIMVALE on COSMOS (COunterfactual Simulations of MOderation Strategies) [5], an Online Social Network (OSN) simulator where the behavior of OSN agents is observed under both unmoderated and moderated conditions. COSMOS is indeed a suitable choice for demonstrating SIMVALE's ability to validate both the simulator's fidelity in reproducing observed OSN behaviors and the effectiveness of intervention mechanisms, both at a global level and across latent toxicity profiles.

2 Literature Review

Simulators enable a controlled environment to investigate complex systems, enabling analysis of emergent behavior and the impact of interventions. Agent-Based-Models (ABMs) simulate autonomous entities' interactions to capture system dynamics [7,13]. Advances in LLMs have produced new ABM simulators with more realistic agents, improving language, memory, and reasoning capabilities [7,8,15,16]. A core challenge is validating these simulators to ensure realistic behavior, which is often overlooked by existing applications. Although no standard framework exists, the literature suggests multi-faceted evaluations, especially at the agent level [3,7,10,16]. Validation splits into subjective (human judgments, annotations, Turing tests) [14,17] and objective (quantitative metrics like behavior similarity, coherence) [2], with hybrid approaches preferred. Evaluations may be agent-oriented—focusing on believability, consistency, controllability, and diversity—or task-oriented—measuring task-specific performance [3]. Both views complement each other. Micro-level evaluation focuses on individual texts or their aggregation (e.g., by agent), while macro-level evaluation examines

[1] The code is publicly available at https://github.com/LauraPollacci/SIMVALE.

emergent properties of the system as a whole [7]. For example, [8] analyze content perplexity (micro) and emotional diffusion (macro). Further, [10] presents a rigorous multi-layer validation of personality in LLM agents, comparing simulated Big Five profiles with human data via human/LLM ratings and LIWC-22 features, showing strong alignment.

Unlike methods that rely on subjective human judgments [14, 17], SIMVALE offers a quantitative, scalable, and automated alternative that mitigates evaluator bias, improves reproducibility, and supports large-scale simulation analysis. Moreover, unlike task-specific objective metrics [2], our framework performs multi-domain validation across linguistic style, sentiment, and psychological traits, using feature-based divergences that are independent of any task. Compared to [6, 10], focused on personality and limited domains, we assess multiple behavioral dimensions. At the macro level, similar to [8], we analyze aggregate trends but via a feature-based approach beyond temporal dynamics. At the micro level, instead of evaluating agents individually, we cluster outputs into latent semantic profiles. This enables targeted and interpretable validation while simplifying analysis. A key contribution is enabling comparative evaluation across simulator settings, assessing impacts of different policies or interventions globally and within content profiles. In sum, SIMVALE offers an objective, agent- and task-oriented, multi-level framework, providing a scalable, general, and multi-domain tool to rigorously assess behavioral realism in LLM-based social simulations, advancing their reliability and design.

3 Methodology

SIMVALE (SIMulator VAlidation with Latent Embeddings) is a general, multi-domain framework for validating LLM-based social simulators. An overview of SIMVALE is described in Algorithm 1. SIMVALE relies on three input sources: 1) *real data*, used as the gold standard; 2) *original simulated data*, the baseline; and 3) *altered simulated data*, used to assess intervention effects (line 1). The real and original simulated data are employed to assess *realism*, i.e., how well the simulator reproduces real-world patterns; while the original simulated data and the altered simulated data are employed to assess *alteration*, i.e., how simulated outputs change under controlled interventions. For enabling comparisons at both global and local levels, SIMVALE implements two extraction paths, targeted at: (i) *transparent features*, capturing interpretable indicators, and (ii) *fine-tuned embeddings*, offering task-specific latent representations.

The transparent feature pipeline defines *domains*—sets of features capturing related aspects of linguistic behavior—and extracts corresponding indicators after preprocessing (e.g., emoji handling, lemmatization) (lines 4–5). Domains include: *a)* **Task-agnostic domains**: linguistic style (e.g., word counts, lexical diversity), readability (e.g., Kincaid, SMOG), and sentiment (e.g., Plutchik, Valence-Arousal-Dominance (VAD), emoji sentiment) (line 4); *b)* **Task-specific domains**: domains considered more relevant for the simulator's objectives, e.g.

Algorithm 1: SIMVALE

Input: D_r, D_s, D_a ; // real, original simulated, and altered simulated datasets
Output: Validation reports and ranked divergences
1 **foreach** $D \in \{D_r, D_s, D_a\}$ **do**
2 $D \leftarrow$ Preprocess(D) ; // clean text
3 **foreach** $D \in \{D_r, D_s, D_a\}$ **do**
4 $F_a \leftarrow$ ExtractAgnostic(D) ; // style, readability, sentiment
5 $F_s \leftarrow$ ExtractSpecific(D) ; // e.g., toxicity, personality
6 $F_D \leftarrow$ Concat(F_a, F_s) ; // complete transparent feature set
7 $E_D \leftarrow$ Embed(D, θ_{BERT}) ; // fine-tuned BERT embeddings
8 $E_D \leftarrow$ Standardize(E_D) ; // normalized embeddings
9 $E_D^+ \leftarrow$ Concat(E) ; // latent representations
10 $C_r \leftarrow$ Cluster($E_{D_r}^+$) ; // cluster real data
11 $C_s \leftarrow$ Cluster($E_{D_s}^+$) ; // cluster simulated data
12 $A \leftarrow$ Align(C_r, C_s) ; // map clusters between D_r and D_s
13 **foreach** *dataset pair* $(D_A, D_B) \in \{(D_r, D_s), (D_s, D_a)\}$ **do**
14 $F_A \leftarrow F_{D_A}$, $F_B \leftarrow F_{D_B}$; // feature matrices
15 **foreach** *domain* M **do**
16 **foreach** *feature* $f \in M$ **do**
17 $p \leftarrow$ Test($F_A[f], F_B[f]$) ; // KS/MWU/Chi-square test
18 **if** $p < \alpha$ **then**
19 $d \leftarrow$ Divergence($F_A[f], F_B[f]$) ; // mean, std, S, etc.
20 LogGlobal(f, M, d) ; // store global divergence
21 RankGlobal(M) ; // rank globally divergent features
22 **foreach** *cluster* $c \in A$ **do**
23 $F_A^c \leftarrow$ Subset(F_A, c) ; // samples in cluster c from D_A
24 $F_B^c \leftarrow$ Subset(F_B, c) ; // aligned samples from D_B
25 **foreach** *domain* M **do**
26 **foreach** *feature* $f \in M$ **do**
27 $p \leftarrow$ Test($F_A^c[f], F_B^c[f]$) ; // local statistical test
28 **if** $p < \alpha$ **then**
29 $d \leftarrow$ Divergence($F_A^c[f], F_B^c[f]$) ; // local difference
30 LogLocal(f, c, M, d) ; // store local divergence
31 RankLocal(M, c) ; // rank features per cluster
32 **return** *Global + local divergence logs and rankings*

personality or toxicity (line 5). Each text instance is thus represented as a vector of normalized features (line 6), enabling multi-domain comparisons across datasets (lines 13–31).

The embedding pipeline builds rich representations for clustering. A subset of task-specific features is used to fine-tune BERT models on real data. For each text, we extract embeddings from the final hidden layer of the BERT models, standardized and concatenated to obtain a single latent representation (lines 7–9). This captures task-specific linguistic information, enabling a deeper profiling of behavioral patterns. Latent content profiles are identified via clustering on the resulting latent representations (lines 10–11). The optimal number of clusters k is selected on real data and reused for simulated datasets to allow alignment. To match clusters across real and simulated data, a one-to-one mapping is computed by comparing the task-specific features of cluster centroids (line 12). A proximity matrix (e.g., cosine or Euclidean distance) is built, and the Hungarian algorithm—also known as the Linear Sum Assignment algorithm [12]—is applied to determine optimal alignment. Each matched pair is labeled with an interpretable name for downstream analysis.

Validation compares real, original, and altered simulated data at global and local levels (lines 13–31). Global validation compares features considering full datasets, while local validation compares features considering a specific latent profile. For each feature (lines 15–16), we assess divergence between distributions (line 17), favoring non-parametric tests (e.g., Kolmogorov–Smirnov (KS), Mann-Whitney, Chi-square) to handle possible non-normality. Significant divergences are then quantified (line 18–19). For continuous features, the pipeline compares means, variances, and percentiles to monitor shifts in different aspects of the distributions; for ordinal features, it evaluates entropy and modal shifts. For each measure m, divergences are computed as: $m_{real} - m_{original}$ (realism) and $m_{original} - m_{altered}$ (alteration).

Features are ranked by divergence magnitude within each domain to highlight key differences (line 21). The same process is applied locally to latent profiles (lines 22–31), enabling finer-grained evaluations of simulator fidelity or alteration effects. We apply SIMVALE to COSMOS, a simulator of online toxic behavior and moderation based on personality traits.

4 Case Study

In this section, we demonstrate the use of the SIMVALE framework to validate COSMOS, an LLM-powered OSN simulator designed to model the emergence of toxic behavior and its interaction with moderation interventions. Toxic behavior is induced via a role-prompting strategy grounded on users' Big Five (OCEAN) personality traits [1] to enhance psychological believability. The case study evaluates COSMOS's realism and alteration effects at both global and local levels.

4.1 Datasets

As *real data* we employ PANDORA [9], which consists of a *User Dataset* of Reddit users annotated with demographics and personality traits, each associated with a set of submissions from a *Comment Dataset.* The *User Dataset* contains 10,295 users described by 38 features. After removing users without OCEAN scores and excluding 28 irrelevant or incomplete features, 1,362 users remain, each with 8 key features: `author`, `gender`, `age` and the five OCEAN traits. Missing `gender` values were imputed using the `is_female_proba` feature with thresholds 0.7 and 0.3. The *Comment Dataset* includes 16,638,070 entries with 16 features. After deduplication and pruning, 4 features remain: `author`, `body`, `date`, and `time_of_day`. Merging user and comment data by `author` yields a *Merged Dataset* of 2,710,671 instances with 11 features, containing both OCEAN scores and cleaned text. Given the size of the *Merged Dataset* compared to the simulated data, we take a sample of 50,000 comments, which will serve as a proxy for the real data in all subsequent analyses. *Simulated data* was generated via COSMOS, configured with 30 agents associated to various demographic and psychological traits. Each agent produced 150 comments under two regimes—unmoderated and moderated—resulting in about 3000 texts per dataset. Moderation applies four different *ex ante* strategies: *One-Size-Fits-All* (OSFA), where the moderator influences agent behavior with a default message; and *Personalized Moderation Interventions* (PMIs), where the moderator influences agent behavior through personalized messages adopting either *Neutral*, *Empathizing* or *Prescriptive* tones. All comments were cleaned by employing regex to remove HTML tags, URLs, mentions, hashtags and expanding contractions with Contractions[2].

4.2 Extraction Pipelines

The task-agnostic features capture general linguistic, emotional, and readability aspects of comments. These include—among others—punctuation counts, uppercase words, emojis, and sentiment-labeled emoticons tagged using resources such as the Emoji Sentiment [11], Emoji and emoticon datasets[3]. Emotional tone is assessed using NRCLex[4] and the NRC VAD Lexicon[5], which provide valence, arousal, and dominance scores. Further, subjectivity scores are computed via TextBlob[6], while readability metrics and counts of difficult words are obtained through Textstat[7]. After preprocessing, NLTK[8] is used to extract POS tags and compute normalized textual indicators like Type-Token Ratio (TTR) and difficult word proportion. Regarding task-specific features, these include personality scores based on the OCEAN model (*openness*, *conscientiousness*, *extraversion*, *agreeableness*, *neuroticism*), which are already available in both real and simulated data; and toxicity-related features computed from raw text using Detoxify[9]. In turn, these latter refer to six subtypes (toxicity traits), i.e., *toxicity*, *severe toxicity*, *obscene*, *threat*, *insult*, and *identity attack*.

To complement feature-based extraction with a deeper semantic view, we employed BERT models fine-tuned on task-specific toxicity traits, training a model for each subtype and using the real dataset as a proxy for true online toxicity patterns. Each model was fine-tuned as a multi-class single-label classifier on toxicity scores in $[0, 1]$ discretized into three ordinal bins—i.e., *low* $[0, 0.33)$, *medium* $[0.33, 0.66)$, and *high* $[0.66, 1]$—allowing the model to capture toxicity intensity. To mitigate class imbalance, random undersampling constrained the majority-to-minority class ratio to a maximum of 4, with this threshold optimized through grid search within the range $[2, 5]$ based on validation performance. Shuffled datasets were split with stratified sampling into fine-tuning (90%) and test (10%) sets, with the fine-tuning set further split (90%/10%)

[2] https://pypi.org/project/contractions/.
[3] Respectively, https://pypi.org/project/emoji/, https://c.r74n.com/faces.
[4] https://pypi.org/project/NRCLex/.
[5] https://saifmohammad.com/WebPages/nrc-vad.html.
[6] https://textblob.readthedocs.io/en/dev/.
[7] https://pypi.org/project/textstat/.
[8] https://www.nltk.org/.
[9] https://pypi.org/project/detoxify/.

into training and validation subsets. Models were implemented with PyTorch[10] and Transformers[11], using the `bert-base-uncased` architecture. The classification head was appended using `AutoModelForSequenceClassification`, with the embedding layer and the first eight transformer layers frozen to limit overfitting. Training employed early stopping with a patience of 2 epochs based on validation Macro-F1 score. At each epoch, average loss and Macro-F1 were computed on both training and validation sets. Hyperparameters—including learning rate, dropout rate, and the number of frozen layers—were optimized via randomized search[12]. The fine-tuned models, i.e., `BERT_obscene`, `BERT_threat`, `BERT_insult`, and `BERT_identity_attack`, produce embeddings encoding toxicity latent spaces that allow behavioral profiling and semantic content analysis.[13]

For each comment, embeddings were extracted from the last hidden layer's `[CLS]` token of each fine-tuned model, yielding four 768-dimensional vectors per comment. These vectors were standardized and concatenated into a single 3,072-dimensional toxicity embedding. Principal Component Analysis (PCA) was applied using Sklearn[14] to reduce dimensionality while preserving 90% of variance. PCA was computed on the embeddings of the real dataset and then used to project both real and unmoderated datasets into a shared 357-dimensional latent space for comparative analysis.

4.3 Toxic Profiles

SIMVALE allows to cluster toxicity-based embeddings, identify latent behavioral profiles and align them across both moderated and unmoderated datasets. These profiles capture underlying patterns of harmful content and enable fine-grained evaluation within COSMOS. Thus, as a first step, we applied clustering to the toxicity embeddings to uncover coherent groups of toxic comments. These embeddings encode both general linguistic patterns from BERT pre-training and toxicity-specific traits learned during fine-tuning. We employed K-Means clustering (via Scikit-learn) due to its scalability and robustness on large, high-dimensional datasets. To determine the optimal number of clusters k, we applied the elbow method to the real dataset—considered a proxy for ground truth—by analyzing the Sum of Squared Errors (SSE) for $k \in [1, 20]$. The elbow point, corresponding to the optimal k, was automatically identified using KneeLocator[15], resulting in $k = 6$. K-Means clustering was applied separately to latent representations derived from the real and unmoderated datasets, resulting in two distinct sets of six clusters. Each comment was assigned a label ranging from 0 to 5, corresponding to its membership in a cluster characterized by similar toxicity

---

[10] https://pytorch.org/.

[11] http://pypi.org/project/transformers/.

[12] Learning rates: 1–5 $\times 10^{-5}$; dropout: 0.1–0.5 (step 0.1); frozen layers: 0, 5–9.

[13] Optimal hyperparameters and Macro-F1s—`BERT_obscene`: 3e-5, 0.3, 8, 0.789; `BERT_threat`: 2e-5, 0.3, 8, 0.758; `BERT_insult`: 4e-5, 0.3, 8, 0.901; `BERT_identity_attack`: 3e-5, 0.3, 8, 0.820.

[14] https://scikit-learn.org/stable/.

[15] https://pypi.org/project/kneed/.

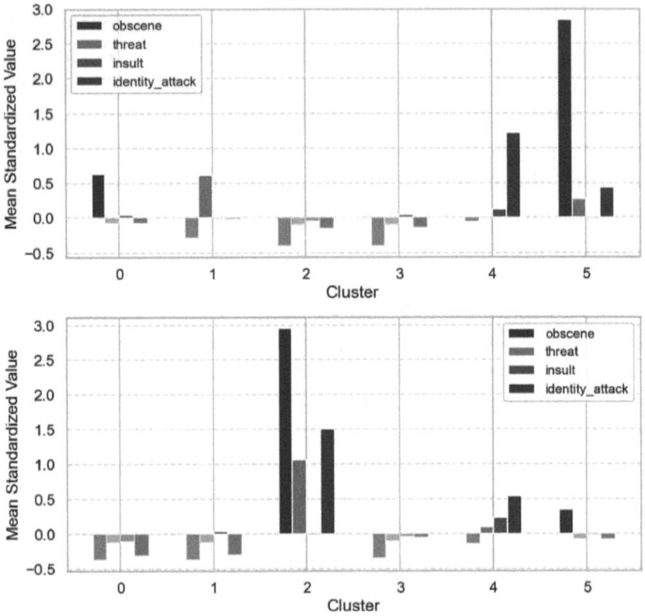

Fig. 1. Standardized toxic centroids by cluster for the real dataset (top) and the unmoderated simulated dataset (bottom).

patterns. To align clusters from the real (Fig. 1, top) and unmoderated (Fig. 1, bottom) datasets for local validation, we compared their toxic centroids. Toxic traits were first standardized to emphasize inter-cluster differences. We then computed pairwise cosine similarities between toxic traits of cluster centroids to build a similarity matrix. Finally, we applied the Hungarian algorithm (via SciPy[16]) to identify the optimal one-to-one mapping that maximizes total similarity. Cluster 0 (real) and cluster 5 (unmoderated) showed high similarity (0.98) and were labeled *Vulgar* due to prevalent obscenity. Cluster 2 (real) matched cluster 0 (unmoderated) with similarity 0.93, both labeled *Healthy* for low toxicity. Cluster 3 (real) and cluster 1 (unmoderated) aligned at 0.94, labeled *Offensive* for high insult scores. Clusters 4 (real) and 4 (unmoderated) scored 0.92, labeled *Discriminatory* for frequent insults and identity attacks, while Clusters 5 (real) and 2 (unmoderated) also scored 0.92 and were labeled *Hostile* for high obscenity, threat, and identity attacks. Cluster 1 (real) and cluster 3 (unmoderated) had low similarity (0.14), indicating no valid match; these were labeled independently as *Aggressive* (real) and *Healthy* (unmoderated). Finally, each comment in both datasets was annotated with its toxic profile for a finer-grained validation of COSMOS. All moderated variants of unmoderated comments inherit their original profile labels to ensure consistent evaluation of moderation effects relative to the original toxic intent.

[16] https://scipy.org/.

Table 1. Global realism: metric differences (simulated – real) for the four most (*), i.e., ($p < 0.01$), significantly different continuous features per domain, plus all personality features. Negative values indicate increases, positive values decreases in the simulated data compared to real data.

Domain	Feature	rel_mean	rel_std	rel_p75	rel_p90
Toxicity	identity_attack	-3.83^*	−1.38	−6.50	−11.43
	obscene	0.20^*	0.12	0.11	0.38
	threat	0.12^*	0.26	−0.76	−0.71
Sentiment	joy	-0.58^*	0.27	−0.30	0.10
	sent_balance	-0.49^*	0.43	0.00	0.27
	sadness	-0.42^*	0.38	−0.43	0.18
	trust	-0.40^*	0.43	−0.125	0.04
Linguistic style	num_words_unique	-0.86^*	0.50	−0.74	−0.12
	num_words	-0.70^*	0.63	−0.87	−0.04
	upper_ratio	0.61^*	0.09	0.00	1.00
	words_sent_ratio	-0.06^*	0.57	0.03	0.22
Readability	smog	-2.38^*	0.28	−0.62	−0.31
	coleman_liau	-0.60^*	0.06	−0.29	−0.14
	ari	-0.37^*	0.00	−0.17	0.02
	flesch_kincaid	-0.34^*	0.42	−0.20	−0.05

		rel_entropy	dominant (real → unmoderated)
Personality	openness	−0.05	very high → high
	conscientiousness	0.20	very low → very low
	extraversion	0.08	very low → very low
	agreeableness	0.10	very low → very low
	neuroticism	0.15	very low → very high

4.4 Validation

The goal of the final phase of our case study is to validate COSMOS by assessing its ability to replicate realistic OSN behavior and reduce harmful content through moderation. The validation is performed by comparing features previously extracted, encompassing three task-agnostic domains (linguistic style, readability, sentiment) and two task-specific ones (personality, toxicity). Comparisons are made between dataset pairs—real vs. unmoderated and unmoderated vs. moderated versions—both globally and locally on subsets defined by latent toxic profiles, i.e., *Healthy*, *Vulgar*, *Offensive*, *Discriminatory*, *Hostile*, and *Aggressive*. We employed the two-sample KS test (via SciPy), with features showing $p < 0.01$ considered significantly different, and features showing $p > 0.01$ considered significantly similar. For significantly different features, we

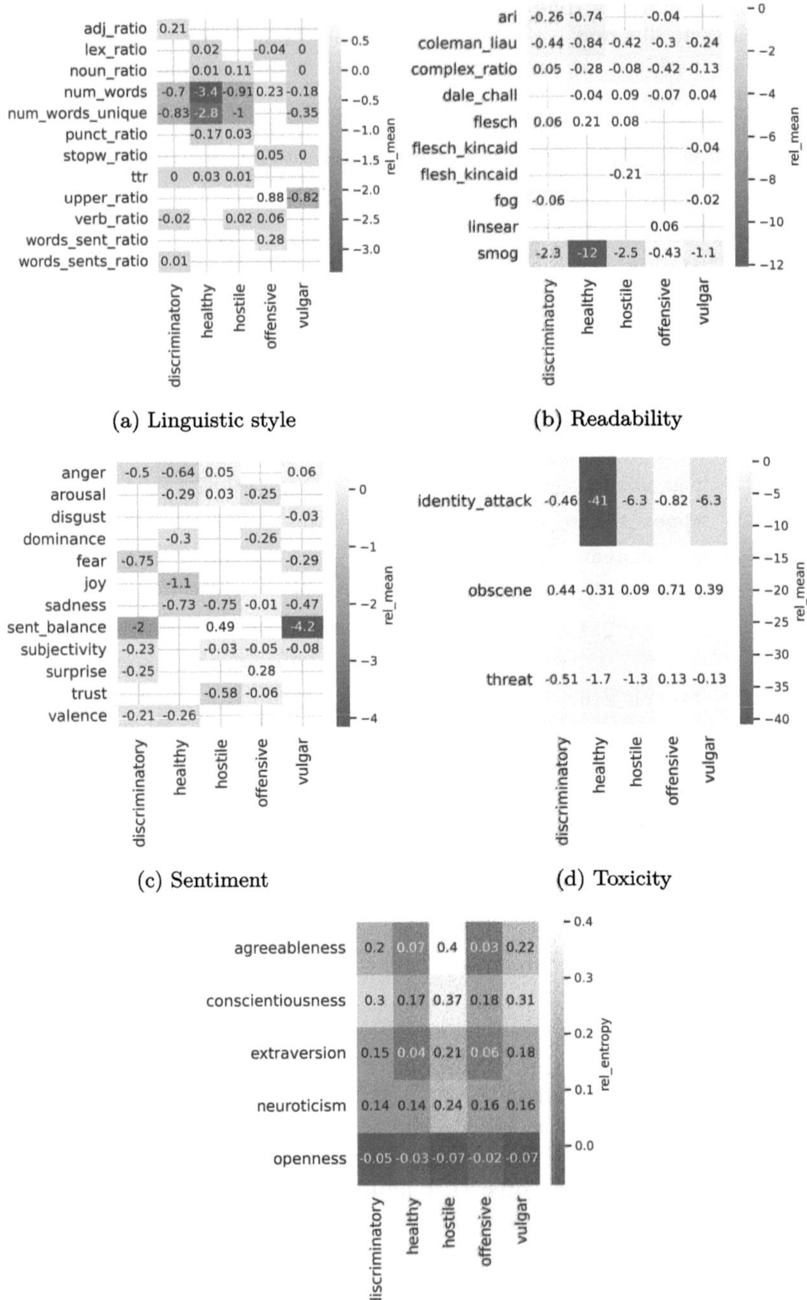

(a) Linguistic style

(b) Readability

(c) Sentiment

(d) Toxicity

(e) Personality

Fig. 2. Local realism: relative mean difference (for continuous features) and entropy difference (for personality). Among continuous features, we select the three most and least significant features per domain.

computed metrics that quantify divergence (e.g., relative changes in mean, standard deviation, and percentiles) for realism and alteration.

Realism. We present first local and global divergences between real and unmoderated data to assess realism in COSMOS.

Global-Level Realism Assessment. Table 1 presents the main differences between real and simulated data, only considering the four most divergent features for each domain based on the relative mean difference. COSMOS reproduces overall behavioral trends but shows systematic deviations. In toxicity, `insult` is well replicated ($p > 0.01$) but `identity_attack` is markedly overrepresented in simulated data, with a substantial relative mean decrease, indicating a bias toward identity-based aggression. Other toxic features such as `obscene` and `threat` show small relative mean increases. For sentiment features, `anticipation` and `sentiment-emoji balance` show no significant differences, while `joy`, `sentiment balance`, `sadness`, and `trust` show negative relative mean difference, suggesting less nuanced emotional expression in simulations. For linguistic style, the `emoji_ratio` is well replicated. However, divergent features indicate longer, more lexically diverse texts with more unique words and fewer uppercase letters, resulting in exaggerated verbosity uncommon in real user language. Readability shows no statistically significant differences; but, metrics suggest increased structural complexity in the simulated data, with relative mean decreases across several indices. Personality also shows no significant differences; most features are realistic, with matching values across the board. Only `openness` and `neuroticism` values show a shift from *very high* to *high* and from *very low* to *very high*, respectively, indicating an elevated emotional reactivity in simulated content. These results suggest that while COSMOS approximates behavioral patterns with reasonable fidelity, it tends to amplify toxic and affective signals and under-represents the natural variability observed in real data.

Local-Level Realism Assessment. To capture latent behavioral patterns across toxic profiles, we analyzed domains individually (Fig. 2), comparing simulated behaviors to real-world distributions. In linguistic style (Fig. 2(a)), *Hostile* and *Vulgar* profiles show increased structural and lexical complexity, marked by greater verbosity and emphatic formatting when compared to real distributions. The *Discriminatory* profile adopts a more repetitive and nominal structure, indicating limited syntactic variability. *Offensive* and *Healthy* remain closer to baseline linguistic characteristics. For readability (Fig. 2(b)), all toxic profiles show reduced values compared to the baseline. This effect is strongest in *Hostile* and *Discriminatory*, indicating denser syntax and more complex phrasing. Conversely, *Healthy* and *Vulgar* content maintain moderate readability deviations. Sentiment (Fig. 2(c)) shows consistent reductions in affective balance across toxic profiles. *Hostile*, *Offensive* and *Discriminatory* display stronger negative affect, particularly in fear, anger, and sadness. The *Vulgar* profile shows an unexpectedly positive sentiment balance, which contrasts with its toxic attitude. Regarding toxicity (Fig. 2(d)), *Hostile* and *Vulgar* profiles demonstrate higher levels of

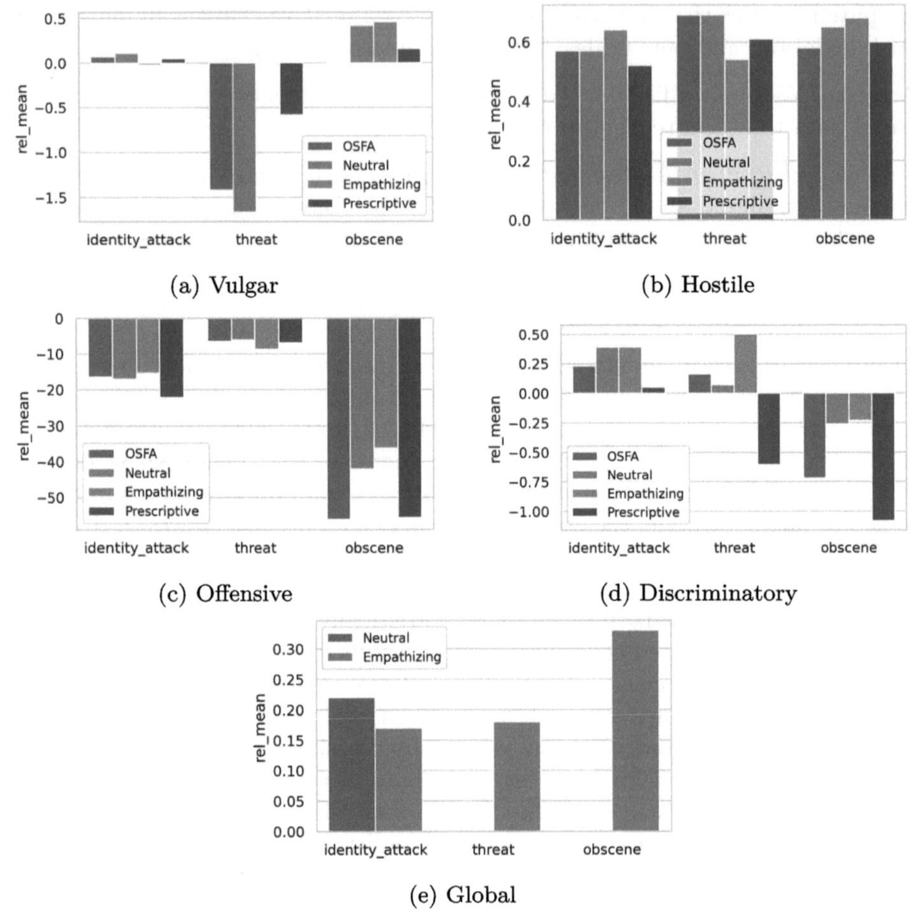

Fig. 3. Moderation effects (relative mean difference) on toxicity at the global level (e) and for each latent profile (a, b, c, d).

identity attack and threat. *Discriminatory* is characterized by higher threat and lower obscene content. Notably, the *Healthy* profile retains some toxic traits, with significant increases in thread and identity attack. Regarding Personality (Fig. 2(e)), *Hostile* and *Discriminatory* profiles display the highest relative entropy across most traits, particularly in conscientiousness, agreeableness, and neuroticism. *Vulgar* and *Discriminatory* show moderate increases in entropy for agreeableness and conscientiousness. *Healthy* maintains low entropy across all traits, indicating minimal deviation from baseline distributions. *Offensive* shows a general alignment with baseline values with a slight decrease in the entropy of conscientiousness and neuroticism.

Moderation Effects. The second validation goal is to assess the effectiveness of moderation strategies in reducing harmful content in COSMOS simulations.

Given their relevance to moderation effects, we focus on analyzing toxicity-related divergences. We compare the original simulated data against each altered simulated set—OSFA, Neutral, Empathizing and Prescriptive—to identify the most effective strategy and determine the best fit for each latent toxic profile.

Global-Level Moderation Effects Assessment. Figure 3(e) compares the global impact of Neutral and Empathizing strategies across all toxic traits. OSFA and Prescriptive are not included as they do not produce statistically significant divergences for any toxic feature, indicating that both approaches are globally ineffective. Similarly, severe toxicity and insult are missing due to the inability of any strategy to yield significant variations. Both Neutral and Empathizing show positive moderation effects, with Empathizing consistently yielding greater reductions in identity attack, threat, and obscene content. In contrast, Neutral results in more limited improvements, particularly for threat and identity attack, and shows no reduction in obscene content. These findings indicate that Empathizing generalizes better across behavior types and substantially reduces global toxicity, especially in high-severity contexts.

Local-Level Moderation Effects Assessment. Figure 3 (a, b, c, d) shows the effectiveness of moderation strategies across latent toxic profiles. The *Healthy* and *Aggressive* profiles are excluded due to the lack of moderated or simulated samples, respectively. For the *Vulgar* profile (Fig. 3(a)), Neutral, Empathizing, and Prescriptive strategies reduce obscene content, with Empathizing having the strongest effect. OSFA and Prescriptive, however, increase threat-related content, suggesting backfiring. In the *Hostile* profile (Fig. 3(b)), all strategies produce clear reductions in toxicity. Neutral and OSFA result in the largest decrease in threat-related content, while Empathizing yields the most substantial reduction in both identity attack and obscene categories. The *Offensive* profile (Fig. 3(c)) shows consistently negative outcomes: all strategies increase toxicity, especially identity attack and threat, indicating a systematic failure to moderate this behavior. For the *Discriminatory* profile (Fig. 3(d)), Empathizing is the most effective strategy, resulting in considerable reductions in identity attack and threat, despite a slight increase in obscene content. Prescriptive leads to a worsening of both threat and obscene metrics, while OSFA and Neutral offer only minor improvements in identity attack while increasing obscene content. These findings underscore the necessity of long-recognized profile-specific, personalized moderation approaches [4], as the efficacy of each strategy is highly dependent on the underlying behavioral patterns.

5 Discussion and Conclusions

SIMVALE introduces a generalizable, multi-domain framework for the validation of LLM-based social simulators, combining the analysis of interpretable features and the clustering of latent representations to assess realism and intervention effects at both global and local levels. Applied to COSMOS, an OSN simulator, SIMVALE allows to unveil a partial fidelity in reproducing real behavioral patterns.

Personality traits and linguistic style align closely with empirical data, whereas toxicity and sentiment dimensions show amplification, particularly in the identity attack dimension. Moreover, COSMOS completely fails to reproduce the Aggressive toxic profile. Moderation analysis through SIMVALE identifies Empathizing as the most effective strategy, especially in reducing vulgar and discriminatory content. Neutral moderation reduces identity-based hate, while OSFA and Prescriptive strategies demonstrate limited effects. The Offensive profile remains mostly unchanged, highlighting a key limitation of current interventions.

SIMVALE has some limitations. It relies on real-world data as a proxy for simulated patterns, which may be scarce or outdated in specialized or fast-changing contexts, affecting validation reliability. The framework also relies on external feature extraction models, whose generalization capabilities may vary and introduce potential biases. Also, current evaluation is based on isolated texts and does not fully capture the emergent properties arising from agents' interactions. Finally, in this work we have tested SIMVALE on a single example—COSMOS—due to the scarcity of LLM-based simulators supporting both realism and alteration assessments. Nonetheless, the tool is inherently designed to enable comparative evaluations, and its effectiveness will be further demonstrated through future extensions as more simulators become available. Despite these limitations, SIMVALE helps identify potential biases in LLM-based simulators by quantifying distributional shifts between real and simulated content across multiple domains. Moreover, SIMVALE is naturally well-suited for offering comprehensive validations of LLM-based simulators operating in a wide range of fields, including not only OSNs, but also education, political debate, and mental health, among others. Moreover, SIMVALE can be easily applied to new interpretable domains, provided their relevance to the validation objectives and the availability of reliable extraction pipelines. Future work should focus on expanding the pool of target simulators, and on domains and behavioral dimensions for providing an even more comprehensive understanding of simulation performance. These developments will further contribute to improving the quality of LLM-based simulators, particularly in high-stakes scenarios demanding high behavioral accuracy.

Acknowledgments. This work was partially supported by the NextGenerationEU – National Recovery and Resilience Plan (Piano Nazionale di Ripresa e Resilienza, PNRR) – Project: "SoBigData.it - Strengthening the Italian RI for Social Mining and Big Data Analytics" – Prot. IR0000013 – Avviso n. 3264 del 28/12/2021; by the PRIN 2022 framework project PIANO under CUP B53D23013290006, by the Italian Project Fondo Italiano per la Scienza FIS00001966 MIMOSA.

Ethical Statement. The application of SIMVALE to COSMOS employs user data from Reddit (PANDORA) anonymized in compliance with EU Regulation 2016/679 (GDPR), ensuring untraceability to real individuals.

Disclosure of Interests. The authors have no competing interests to declare that are relevant to the content of this article.

References

1. Ackerman, C.: Big five personality traits: the ocean model explained. Positive Psychology.com (2017)
2. Aher, G.V., Arriaga, R.I., Kalai, A.T.: Using large language models to simulate multiple humans and replicate human subject studies. In: International Conference on Machine Learning, pp. 337–371. PMLR (2023)
3. Chen, C., Yao, B., Ye, Y., Wang, D., Li, T.J.J.: Evaluating the LLM agents for simulating humanoid behavior. In: Proceedings of the 1st Workshop on Human-Centered Evaluation and Auditing of Language Models (CHI Workshop HEAL) (2024)
4. Cresci, S., Trujillo, A., Fagni, T.: Personalized interventions for online moderation. In: HT, pp. 248–251. ACM (2022)
5. Fidone, G.: Generative simulations of online social networks for evaluating moderation strategies (2024). https://etd.adm.unipi.it/t/etd-10132024-035632/
6. Frisch, I., Giulianelli, M.: LLM agents in interaction: measuring personality consistency and linguistic alignment in interacting populations of large language models. arXiv preprint arXiv:2402.02896 (2024)
7. Gao, C., et al.: Large language models empowered agent-based modeling and simulation: a survey and perspectives. Humanit. Soc. Sci. Commun. **11**(1), 1–24 (2024)
8. Gao, C., et al.: S3: social-network simulation system with large language model-empowered agents. arXiv preprint arXiv:2307.14984 (2023)
9. Gjurković, M., Karan, V.M., Vukojević, I., Bošnjak, M., Snajder, J.: PANDORA talks: personality and demographics on Reddit. In: Ku, L.W., Li, C.T. (eds.) Proceedings of the Ninth International Workshop on Natural Language Processing for Social Media, pp. 138–152. Association for Computational Linguistics, Online (2021). https://doi.org/10.18653/v1/2021.socialnlp-1.12. https://aclanthology.org/2021.socialnlp-1.12/
10. Jiang, H., Zhang, X., Cao, X., Breazeal, C., Roy, D., Kabbara, J.: Personallm: investigating the ability of large language models to express personality traits. arXiv preprint arXiv:2305.02547 (2023)
11. Kralj Novak, P., Smailović, J., Sluban, B., Mozetič, I.: Emoji sentiment ranking 1.0 (2015). http://hdl.handle.net/11356/1048
12. Kuhn, H.W.: The hungarian method for the assignment problem. Naval Res. Logist. Q. **2**(1–2), 83–97 (1955)
13. Macal, C.M., North, M.J.: Agent-based modeling and simulation: ABMS examples. In: 2008 Winter Simulation Conference, pp. 101–112 (2008). https://doi.org/10.1109/WSC.2008.4736060
14. Park, J.S., O'Brien, J., Cai, C.J., Morris, M.R., Liang, P., Bernstein, M.S.: Generative agents: interactive simulacra of human behavior. In: Proceedings of the 36th Annual ACM Symposium on User Interface Software and Technology, pp. 1–22 (2023)
15. Park, J.S., Popowski, L., Cai, C., Morris, M.R., Liang, P., Bernstein, M.S.: Social simulacra: creating populated prototypes for social computing systems. In: Proceedings of the 35th Annual ACM Symposium on User Interface Software and Technology. UIST 2022. Association for Computing Machinery, New York (2022). https://doi.org/10.1145/3526113.3545616

16. Wang, L., et al.: A survey on large language model based autonomous agents. Front. Comp. Sci. **18**(6), 186345 (2024)
17. Ziems, C., Held, W., Shaikh, O., Chen, J., Zhang, Z., Yang, D.: Can large language models transform computational social science? Comput. Linguist. **50**(1), 237–291 (2024)

Fine-Tuning Transformer-Based LLMs in Hierarchical Text Classification

Joana Santos[1,2(✉)], Nuno Silva[2,4], Carlos Ferreira[2,4], and João Gama[3,4]

[1] Faculty of Engineering, University of Porto, Porto, Portugal
[2] School of Engineering, Polytechnic of Porto, Porto, Portugal
{ffs,nps,cgf}@isep.ipp.pt
[3] Faculty of Economics, University of Porto, Porto, Portugal
jgama@fep.up.pt
[4] INESC TEC, Porto, Portugal

Abstract. Hierarchical document classification is essential for structuring large-scale textual corpora in domains such as digital libraries and academic repositories. While recent advances in large language models (LLMs) have opened new possibilities for text classification, their applicability to hierarchical settings under real-world constraints remains underexplored. This study investigates both generative and discriminative transformer-based models, evaluating their effectiveness across multiple inference strategies: zero-shot baseline, local fine-tuning, and a global approach using category-specific models. Experiments on two real-world hierarchical datasets provide a comprehensive comparison of classification accuracy, F1-macro scores, and inference times. The results highlight that, although generative LLMs can deliver competitive (yet variable) performance at higher levels of the hierarchy, their high inference costs hinder their use in time-sensitive applications. In contrast, fine-tuned discriminative models—particularly BERT-based architectures—consistently offer a more favorable trade-off between performance and efficiency.

Keywords: supervised online learning · hierarchical text classification · Large Language Models · generative vs. discriminative

1 Introduction

Hierarchical document classification is essential for organizing and navigating large-scale textual data across diverse domains such as digital libraries, news aggregation, and academic search engines. As the volume of unstructured text grows rapidly, scalable and efficient classification systems become increasingly critical, especially in scenarios demanding timely responses, including online content moderation, recommendation systems, and real-time analytics.

S. Džeroski et al. (Eds.): DS 2025, LNAI 16090, pp. 427–442, 2025.
https://doi.org/10.1007/978-3-032-05461-6_28

Although large language models (LLMs) have shown impressive capabilities across a wide range of natural language processing tasks, their substantial computational demands limit their applicability in latency-sensitive classification scenarios [18].

In this paper, we compare generative and discriminative large language models (LLMs) in the context of hierarchical classification for low-latency scenarios. Our initial experiments with generative models such as LLaMA [1] and Mistral [14] showed competitive performance on classification tasks. However, their inference times without fine-tuning were prohibitively long, and their accuracy varied significantly across datasets, impacting deeper classification at deeper levels of the hierarchy. This variability, combined with high latency, makes them impractical for real-time or high-throughput applications where both responsiveness and consistent sub category performance are critical.

Consequently, we shift our focus to transformer-based discriminative models, including BERT [9], DistilBERT [23], RoBERTa [20], and DeBERTa [12]. Using a set of controlled experiments on hierarchically structured datasets, we evaluate various training and inference strategies, examining the trade-offs between classification accuracy, macro F1-score, and execution time. Our results show that, when fine-tuned, these models provide a compelling balance between performance and computational efficiency, making them well-suited for real-world deployments requiring both reliability and low latency.

The main contributions of this work are as follows:

- a comprehensive empirical comparison between generative and discriminative language models, highlighting the trade-offs between classification performance and computational cost;
- an evaluation of multiple modeling strategies—including local fine-tuning and global category-specific models—across two hierarchically structured datasets;
- an analysis of the operational feasibility of LLMs in resource-constrained, low-latency environments.

To the best of our knowledge, this study is novel because it revisits the inference-time performance of generative and discriminative large language models (LLMs) for hierarchical text classification (HTC) under the lens of deployment efficiency rather than hardware limitations. While modern hardware has made large-scale model training and inference more feasible, real-world applications, such as real-time systems, edge computing, or high-throughput pipelines, still demand latency-aware solutions. Prior work has primarily emphasized architectural innovations or classification performance, such as prompt tuning for discriminative models [26] and generation-based approaches for HTC [13], or focused on modeling choices and error analysis in generative setups [27], leaving the inference-time trade-offs between generative and discriminative approaches underexplored. Our study bridges this gap by quantifying these trade-offs and further enhancing discriminative models through fine-tuning for HTC, offering a nuanced view of both predictive effectiveness and deployment viability.

The paper is organized as follows: Sect. 2 reviews related work; Sect. 3 details the methodology and setup; Sect. 4 presents and discusses results; Sect. 5 concludes and suggests future work.

All the code is available at https://tinyurl.com/yrhx249a and https://tinyurl.com/rce7vfax.

2 Related Work

This section reviews key approaches relevant to the study, including traditional classifiers, LLMs, discriminative transformers, and hierarchical classification methods.

Traditional text classification approaches, such as K-Nearest Neighbors (KNN), Naïve Bayes, Support Vector Machines (SVM), and decision trees, have been widely used in news classification [6–8,15]. While these methods have proven effective [8], they come with several limitations [6,7]. Many require manual feature extraction, which can be labor-intensive and domain-dependent. Additionally, they often struggle with generalization and contextual understanding, limiting their ability to adapt to diverse and evolving news content [7,15].

Moreover, when applied to large datasets, these models can lead to significant computational costs and increased demand for labeled data, making them less scalable in real-world scenarios [16]. Given these challenges, there is a growing need for more advanced models that offer greater generalization capabilities and can dynamically adapt to the context of the documents being classified.

2.1 Large Language Models

Large language models (LLMs) are trained on vast collections of documents, enabling them to process and generate text even in nuanced contexts. These models primarily rely on statistical pattern recognition—matching new inputs with previously learned patterns—rather than true semantic understanding [2]. This often limits their ability to integrate structured or domain-specific knowledge and leads to issues such as hallucinations and logical inconsistencies [19].

LLMs can also be distinguished by their architectural design, which shapes how they process inputs and produce outputs. The transformer architecture [29], enables models to capture long-range dependencies via self-attention mechanisms, as a foundation for both generative and discriminative variants of LLMs:

- Discriminative LLMs – These models, such as BERT [9], RoBERTa [20], and DeBERTa [12], are fine-tuned to map input texts directly to a fixed set of output labels. They are typically used in supervised classification tasks and are highly efficient during inference. Their architecture is well suited for structured tasks like hierarchical text classification (HTC), where intermediate classification steps can be explicitly modeled [5,31];
- Generative LLMs – Models like LLaMA [28] and Mistral [14] are designed to generate free-form text and are repurposed for classification using prompting

techniques. In zero- and few-shot scenarios [24,30], these models are conditioned to generate label names as outputs. While they offer great flexibility and adaptability [3,10], especially for high-level categories, their significantly longer inference times and higher computational cost often render them impractical for latency-sensitive applications [4].

The distinction between generative and discriminative models has long been studied in the context of text classification. Yogatama et al. [32] conducted an early comparative study using recurrent neural networks, showing that generative models tend to perform better in low-resource settings due to faster convergence, while discriminative models typically achieve higher accuracy given sufficient data. More recently, Zheng et al. [33] revisited the theoretical foundations of this distinction, extending classic results to the multiclass setting and demonstrating that generative classifiers can achieve comparable performance with significantly fewer training samples. In the transformer era, Kostina et al. [17] provided a comprehensive empirical review of large language models for text classification, evaluating both discriminative (e.g., BERT-based) and generative (e.g., GPT-based) architectures. Their results confirm that generative LLMs offer strong performance, particularly in low-data or few-shot scenarios, but often at the cost of significantly increased inference time compared to their discriminative counterparts.

2.2 Hierarchical Text Classification

HTC addresses the challenge of assigning documents to categories organized in a taxonomy, often structured as a tree or directed acyclic graph. This form of classification is critical in applications such as news aggregation, digital libraries, medical coding, and scientific literature organization, where labels are inherently structured [25]. HTC introduces complexities beyond flat classification, including data imbalance across levels, the need for consistency across the hierarchy, and increased error propagation from higher to lower levels. The emergence of deep learning and contextual embeddings has significantly improved HTC [5]. Particularly, transformer-based models have enabled end-to-end HTC with better generalization, minimal feature engineering, and improved performance in multi-level prediction settings [5].

Two main strategies are commonly used in HTC [21]:

- Global approaches train a single classifier to predict the entire label space or at least one level of the hierarchy in a unified model. These methods are simpler to manage but may underperform due to class imbalance and lack of specialization;
- Local approaches use a set of models, each focused on a specific node or branch of the hierarchy (e.g., per category or level). These are often more accurate at deeper levels but can be more error-prone if top-level predictions are incorrect.

3 Methodology and Experimental Setup

This section presents the methodology and experimental setup used to evaluate hierarchical document classification. It describes the datasets, and models employed in the study.

3.1 Experimental Design

To assess the operational feasibility of LLMs in resource-constrained, low-latency environments, we designed three experimental setups that reflect different hierarchical classification strategies and their associated computational trade-offs:

1. Generative vs. Discriminative LLMs (Preliminary evaluation): As an initial step in our experimental design, we conducted a zero-shot classification experiment (named *baseline*) comparing generative language models (e.g., LLaMA [1], Mistral [14]) with discriminative transformer-based models (e.g., BERT, RoBERTa). Although the generative models exhibited competitive accuracy at higher levels of the hierarchy, their inference times were significantly longer and their accuracy was highly variable across datasets. These exploratory findings framed the design of our remaining experiments, leading us to focus subsequent evaluations exclusively on discriminative models, which offer more consistent performance and lower latency in online classification settings;
2. Global Top-Level (named *global top*) and Global Sub category (named *global sub*) Classification (Independent Models): In this setup, a single global model is trained to classify top-level categories, and another global model is trained to classify subclasses—independently of each other. This approach assumes no dependency between the two levels, enabling parallel or sequential predictions. While simple to implement, it does not leverage the hierarchical structure and may result in inconsistent sub category predictions relative to the top-level class;
3. Global Top-Level Classification + Local Sub category Models (Conditional Invocation): Here, a global model classifies top-level categories, and then a separate local model, specific to the predicted top class, is used to classify its corresponding subclasses. We explore three variants of this dependency:
   (a) Unconditional Invocation: The subclassification (named *global top+local sub*) model corresponding to the predicted top-level category (*global top*) is always invoked, regardless of classification confidence. E.g., if the top-level category is A, then the local sub category model trained on A1, A2, ..., An is used;
   (b) Confidence-Gated Invocation (10% Threshold): The subclassification step is only performed if the top-level model's confidence in its prediction exceeds 10% (named *global top+local sub 10%*). This introduces a minimal confidence threshold to avoid unnecessary or misleading subclassification when the top-level prediction is uncertain;
   (c) Confidence-Gated Invocation (20% Threshold): Similar to the previous variant, but with a stricter threshold of 20% (named *global top+local sub*

20%). This further reduces the number of subclassification invocations, prioritizing precision and resource savings over coverage.

These experiments assess accuracy-latency trade-offs in hierarchical classification, focusing on how inter-level dependencies and confidence thresholds affect reliability and efficiency in real-time settings.

3.2 Datasets

Two text datasets were created specifically for hierarchical classification, both sourced from their respective APIs.

The Guardian [11] is a British newspaper that publishes daily news. It provides an API allowing access to content in multiple formats, including text, images, and video. For this study, the focus was placed exclusively on news articles in text format. The dataset is available here: https://doi.org/10.5281/zenodo.15552206.

OpenAlex [22] is a free and open catalog of scientific research documents. It offers access to a wide range of scientific publications, as well as detailed information about authors, venues, and research concepts. The dataset is available here: https://doi.org/10.5281/zenodo.15552278.

By collecting data from these two APIs, we constructed the datasets with the characteristics summarized in Table 1.

Table 1. Datasets

	The Guardian	Openalex
Period (date)	2000 - 09/04/2025	2024
Number of categories	37	26
Number of subcategories	325	252
Total documents	1 343 964	2 009 945
Average title size (characters)	46	106
Average abstract size (characters)	131	1411
Class distribution	imbalance	imbalance

Each dataset was then split into 80% for training, 10% for validation, and 10% for testing.

3.3 Models

Two distinct types of models were employed in the system: (i) models based on the BERT architecture, optimized specifically for classification tasks, and (ii) generative models primarily designed for text generation, but adapted here for use as classifiers. The BERT-based models (encoder-only architectures) vary in several aspects, including model size and internal mechanisms (Table 2).

In addition to the BERT-based models, two types of generative models were utilized: LLaMA-3 and Mistral. These models differ in architectural design, which affects aspects such as the number of layers activated during inference. Both models were used in their instruction-tuned configurations, optimized to follow human prompts effectively. Table 3 summarizes their key characteristics.

Table 2. Classifiers models

	BERT-base	DistilBERT	RoBERTa-base	DeBERTa-base
Type	Encoder-only	Encoder-only	Encoder-only	Encoder-only
Parameters	~110 million	~67 million	~125 million	~140 million
Inference Speed	baseline	faster than BERT	slower than BERT	similar to RoBERTa
Hugging Face Reference	google-bert/bert-base-uncased	distilbert/distilbert-base-uncased	FacebookAI/roberta-base	microsoft/deberta-base

Table 3. Generative models

	Llama-3 3B [28]	Mistral 7B [14]
Type	dense-only	dense and Mixture of Experts (MoE)
Parameters	~3.21 billion	~7.25 billion
Hugging Face Reference	meta-llama/Llama-3.2-3B-Instruct	mistralai/Mistral-7B-Instruct-v0.3

3.4 Evaluation Metrics

To evaluate the performance of the models in the hierarchical classification task, we employed a combination of standard and complementary metrics. Accuracy was used as a primary metric to assess the overall correctness of the classification results. However, given the imbalanced nature of the datasets, we also included the F1-macro score, which provides a more balanced evaluation by taking into account both precision and recall across all classes, regardless of their frequency.

In addition to classification performance, we measured training time and inference time to assess the computational efficiency of each model. These metrics are important for understanding the practical feasibility of deploying the models, especially in real-world applications where both accuracy and latency matter.

4 Results and Discussion

The following sections present and analyze the results from three key stages: comparing generative and discriminative LLMs, training and validation, and detailed testing on both top-level and sub-level classes.

4.1 Generative Vs. Discriminative LLMs

In the preliminary experiments, the goal was to assess the models zero-shot classification performance upon the provided datasets. Two separate evaluations were conducted: one focused on classification into top-level categories (*global top*), and the other on subcategory classification (*global sub*). Table 4 summarizes the results of these experiments. To facilitate comparison, the highest accuracy value (indicating the best performance) is marked in bold, while the lowest accuracy value is marked in bold-italic. For execution time, where shorter times are preferable, the lowest execution time is highlighted in bold, and the highest (longest) execution time is marked in bold-italic.

Table 4. Generative vs. Discriminative LLMs results.

Model	Type	The Guardian			OpenAlex		
		Acc (%)	F1-M (%)	Time	Acc (%)	F1-M (%)	Time
BERT-base	global top	1.63	0.11	0:01:39	5.74	0.52	0:02:08
	global sub	0.24	0.01	0:01:40	0.44	0.01	0:02:07
DistilBERT	global top	1.29	0.14	**0:01:05**	4.04	0.54	**0:01:28**
	global sub	0.42	0.01	**0:01:05**	0.47	0.01	0:01:33
RoBERTa-base	global top	1.67	0.09	0:02:01	6.19	0.45	0:02:29
	global sub	0.22	0.00	0:02:01	0.43	0.00	0:02:32
DeBERTa-base	global top	1.67	0.09	0:03:18	6.19	0.45	0:03:29
	global sub	0.22	0.00	0:03:18	0.43	0.00	0:03:30
LLaMA-3 3B	global top	***1.01***	2.01	1:44:09	**8.19**	9.44	2:16:50
	global sub	0.00	0.00	***31:50:48***			
Mistral 7B	global top	**7.11**	10.76	4:13:44	***0.44***	0.44	***6:34:03***
	global sub						

The BERT-family models—BERT-base, DistilBERT, RoBERTa-base, and DeBERTa-base—consistently show low latency (ranging from 1–3 min per dataset) and modest classification performance. Among them, DistilBERT shows the best execution efficiency, achieving the lowest inference times across both datasets while maintaining similar accuracy and F1-macro (F1-M) scores to its larger counterparts.

Notably, all BERT-based models demonstrate substantial performance drops on the subcategory level, with accuracy below 1% and near-zero F1-M scores. This suggests that while discriminative models can generalize moderately at the top level, they struggle to disambiguate fine-grained categories when trained globally without additional hierarchical signals or context specialization.

Generative models like LLaMA-3 3B and Mistral 7B present a clear trade-off: they can deliver strong classification performance at the top level of certain datasets, but with significantly higher inference times.

For instance, LLaMA-3 3B achieves the highest F1-M (9.44%) and accuracy (8.19%) on the top level of OpenAlex—substantially outperforming all BERT-based models—yet requires over 2 h to complete inference. However, their performance varies considerably across datasets, indicating limited robustness and a strong sensitivity to domain-specific data distributions—concerns that are reinforced by LLaMA-3 3B's results in the global sub experiment, where it achieved 0.00% in both accuracy and F1-M. This raises concerns about their generalizability and reliability in diverse, real-world scenarios. Its subcategory classification on The Guardian is especially problematic, requiring over 31 h, with 0% accuracy. This computational cost led to early termination of further global sub-level experiments on OpenAlex.

Mistral 7B also shows strong top-level F1-M (10.76%) on The Guardian, indicating that generative models can better handle broad topic categorization. However, its performance degrades sharply on OpenAlex, with just 0.44% accuracy, while consuming over 6 h for inference. Like LLaMA-3, Mistral's sub-level experiments were not executed due to impractical execution time.

While LLMs demonstrate strong raw performance, their computational demands often outweigh their practical benefits. In contrast, discriminative models such as BERT and DistilBERT—despite lower accuracy and macro F1 scores—offer clear advantages in time efficiency and scalability. These insights suggest that further fine-tuning of BERT-based models is a promising direction for improving hierarchical classification performance in latency-sensitive systems, while the prohibitive inference time of generative models makes them unsuitable for such applications.

Based on these findings, the remainder of this study focuses exclusively on discriminative models, specifically BERT-based architectures.

4.2 Training and Validation

During the training phase, classifier models were trained separately for each dataset using three different approaches: (i) *global top*, (ii) *global sub* and (iii) *local sub*.

Table 5 presents the results obtained on a subset of the validation datasets for each dataset and training approach. Key performance metrics—including accuracy (Acc), macro-averaged F1 score (F1-M), and execution time (Time)—are reported to provide a comprehensive comparison of each model's effectiveness and efficiency. These validation results were used to guide the selection of the final training configurations adopted for subsequent testing, ensuring a balance between performance and computational cost.

Models were trained using a learning rate of 2e-5, weight decay of 0.01, and a batch size of 128 on both training and evaluation, optimized to fully utilize an NVIDIA RTX 4090 GPU. Mixed-precision training (fp16) and an 8-bit optimizer (adamw_bnb_8bit) were employed to accelerate training and reduce memory consumption. The training ran for two epochs with a 5% warmup ratio, and validation accuracy was used to select the best model. This setup provided an effective trade-off between training speed, stability, and classification performance.

Across both datasets, DistilBERT consistently achieved the highest accuracy and F1-macro scores, particularly in the global top-level classification setting, with accuracies of 63.83% (The Guardian) and 55.57% (OpenAlex). This indicates that DistilBERT's architecture offers a good balance of training efficiency and classification performance for top-level categories.

Table 5. Validation results

Model	Type	The Guardian			OpenAlex		
		Acc (%)	F1-M (%)	Time	Acc (%)	F1-M (%)	Time
BERT-base	global top	29.74	9.04	0:56:06	41.62	8.89	1:25:43
	global sub	8.65	0.82	0:56:16	11.10	0.61	1:24:04
	local sub	35.50	5.75	1:01:44	31.94	7.00	1:43:28
DistilBERT	global top	63.83	47.18	0:59:33	55.57	31.44	1:29:09
	global sub	32.39	12.39	0:59:30	22.66	6.34	1:28:52
	local sub	48.42	13.66	1:02:02	42.84	15.40	1:35:30
RoBERTa-base	global top	50.37	19.60	1:57:09	48.79	15.38	2:54:37
	global sub	12.71	0.92	1:57:21	14.30	1.09	2:54:54
	local sub	37.63	6.34	2:02:15	33.89	7.11	3:04:56
DeBERTa-base	global top	11.21	0.54	0:47:34	23.09	1.44	0:45:20
	global sub	3.47	0.02	0:45:30	3.48	0.03	0:43:00
	local sub	34.33	4.89	0:45:45	29.41	5.77	0:41:49

As expected, models trained for *global top* classification outperform those trained for *global sub* classification in both accuracy and F1-macro, reflecting the greater difficulty of correctly classifying finer-grained subcategories.

The *local sub* classification, generally achieves better performance on subcategory classification than the *global sub* classification. This suggests that decomposing the subcategory classification task into multiple category-specific models helps improve discrimination among subcategories within each top-level category.

Different models and configurations result in significantly different training times. RoBERTa-base requires the longest training time, often almost double that of other models, due to its larger size and computational demands. DistilBERT and DeBERTa-base are notably faster, making them preferable in scenarios with limited computational resources.

Training models for *global sub* + *local sub* generally take longer than training for *global top* classification, which is consistent with the increased complexity and the number of models involved in the local approach.

4.3 Test - Top-Level Classes

The top-level classification experiment assesses the performance of various models and strategies, benchmarked against the results obtained using zero-shot classification (Table 6).

The zero-shot results indicate very low accuracy and F1-macro scores across all models and datasets, ranging from approximately 1% to 6% accuracy and near-zero F1-macro values. This confirms our preliminary observations.

In contrast, the *global top* classification shows a significative improvement in all performance metrics. Among the models, DistilBERT clearly stands out, achieving the highest accuracy and F1-macro scores: 63.87% accuracy and 46.99% F1-macro on The Guardian, and 55.61% accuracy and 31.51% F1-macro on OpenAlex. This highlights DistilBERT's strength in efficiently capturing top-level category distinctions in both datasets.

Table 6. Top-level categories classification results

Model	Experiment	The Guardian			OpenAlex		
		Acc (%)	F1-M (%)	Time	Acc (%)	F1-M (%)	Time
BERT-base	zero-shot	1.63%	0.11%	**0:01:39**	5.74%	0.52%	**0:02:08**
	global top	**29.49%**	**8.97%**	0:01:55	**41.52%**	8.84%	0:02:37
DistilBERT	zero-shot	1.29%	0.14%	**0:01:05**	4.04%	0.54%	**0:01:28**
	global top	**63.87%**	**46.99%**	0:01:16	**55.61%**	**31.51%**	0:01:43
RoBERTa-base	zero-shot	1.67%	0.09%	**0:02:01**	6.19%	0.45%	**0:02:29**
	global top	**50.40%**	**19.64%**	0:02:22	**48.70%**	**15.40%**	0:03:05
DeBERTa-base	zero-shot	1.67%	0.09%	**0:03:18**	6.19%	0.45%	**0:03:29**
	global top	**11.21%**	**0.54%**	0:03:41	**23.09%**	**1.44%**	0:04:23

Other models also improve substantially: BERT-base and RoBERTa-base both achieve moderate accuracy gains (around 29–50% accuracy), while DeBERTa-base shows the smallest improvement but still outperforms its zero-shot baseline.

Execution times reflect only the inference duration on the test set. Zero-shot inference times are naturally the shortest, because models are used without additional overhead. *global top* models show a slight increase in inference time, typically within 10–20 seconds more than baseline. For example, DistilBERT's inference time rises from about 1:05 (baseline) to 1:16 (global) on The Guardian. This moderate increase is expected due to the use of trained weights and possibly more complex processing during inference but remains efficient for practical applications.

Based on the analysis, the following conclusions can be drawn:

- *global top* fine-tuning significantly improves classification accuracy and F1-macro compared to zero-shot, confirming the importance of domain/task-specific training;

- DistilBERT delivers the best balance of accuracy, F1-macro, and efficient inference time, making it the strongest candidate for top-level classification in this study;
- Dataset characteristics affect classification outcomes, underscoring the need for dataset-specific strategies.

4.4 Test - *global Sub*

The *global sub* classification experiment assesses the performance of various models and methodologies in distinguishing finer-grained classes, against the results obtained using zero-shot classification (Table 7).

Table 7. Sub classes results

Model	Experiment	The Guardian			OpenAlex		
		Acc (%)	F1-M (%)	Time	Acc (%)	F1-M (%)	Time
BERT-base	zero-shot	0.24%	0.01%	**0:01:40**	0.44%	0.01%	**0:02:07**
	global sub	**8.65%**	**0.83%**	0:01:55	**11.02%**	0.61%	0:02:33
	global top+local	7.04%	0.55%	0:16:17	9.30%	**0.64%**	0:23:52
	global top+local sub 10%	6.60%	0.56%	0:16:33	9.30%	**0.64%**	0:22:40
	global top+local sub 20%	2.12%	0.31%	0:09:18	8.16%	0.57%	0:16:52
DistilBERT	zero-shot	0.42%	0.01%	**0:01:05**	0.47%	0.01%	**0:01:33**
	global sub	**32.09%**	**12.29%**	0:01:17	**22.62%**	**6.29%**	0:01:46
	global top+local	29.80%	9.30%	0:12:52	21.27%	5.76%	0:14:42
	global top+local sub 10%	29.79%	9.27%	0:11:16	21.27%	5.76%	0:13:52
	global top+local sub 20%	29.40%	9.29%	0:10:23	21.21%	5.75%	0:12:52
RoBERTa-base	zero-shot	0.22%	0.00%	**0:02:01**	0.43%	0.00%	**0:02:32**
	global sub	12.70%	0.92%	0:02:23	**14.27%**	1.10%	0:03:03
	global top+local	**15.41%**	**1.57%**	0:21:15	13.38%	**1.21%**	0:26:14
	global top+local sub 10%	15.20%	**1.57%**	0:18:26	13.38%	1.20%	0:25:12
	global top+local sub 20%	12.00%	1.42%	0:16:38	12.93%	1.19%	0:22:00
DeBERTa-base	zero-shot	0.22%	0.00%	**0:03:18**	0.43%	0.00%	**0:03:30**
	global sub	**3.60%**	**0.02%**	0:03:40	**3.46%**	**0.03%**	0:04:27
	global top+local sub	1.85%	0.01%	*1:55:08*	*2.57%*	0.02%	*0:52:32*
	global top+local sub 10%	1.85%	0.01%	2:11:43	2.57%	0.02%	0:49:04
	global top+local sub 20%	0.00%	0.00%	1:23:49	2.57%	0.02%	0:47:50

Across all models and datasets, zero-shot classification yields extremely low accuracy and F1-macro scores—often nearing zero—reinforcing our preliminary observations. As expected, inference times are the shortest, since the models operate without additional fine-tuning or processing overhead.

Instead, *global sub* significantly improves classification metrics:

- Accuracy - for instance, DistilBERT achieves the highest accuracy among the models (32.09% on The Guardian and 22.62% on OpenAlex), showing that fine-tuning greatly enhances the model's ability to classify sub-categories;
- F1-macro - similarly, DistilBERT leads with an F1-macro of 12.29% and 6.29%, which, despite being relatively low, represents a substantial gain compared to the baseline;
- Inference Time - test times increase slightly compared to the zero-shot inference (e.g., from 1:05 to 1:17 for DistilBERT on The Guardian), but remain practical and efficient for inference tasks.

On the other hand, *global top + local sub* yields mixed results:

- in some cases, such as RoBERTa on The Guardian, it achieves the highest accuracy (15.41%) and F1-macro (1.57%), slightly outperforming local models in F1-macro;
- the execution time for inference in the global setting increases significantly, e.g., RoBERTa's inference time rises to over 21 min on The Guardian and 26 min on OpenAlex. This likely reflects the overhead of managing multiple models and predictions per category;
- for DeBERTa, global inference time becomes prohibitively high (over 1 h 55 min on The Guardian), raising concerns about scalability in real-world scenarios.

Probability-based approaches were developed to explore potential advantages in hierarchical classification. They achieved accuracy and F1 scores comparable to the *global top + local sub* strategy, but often incurred higher inference times—highlighting the trade-off between predictive performance and runtime efficiency.

Based on this analysis, the following conclusions can be drawn:

- fine-tuning (local or global) drastically improves sub category classification performance over baseline;
- DistilBERT offers the best balance of accuracy, F1-macro, and inference time;
- dataset characteristics influence performance.

5 Conclusion

This work evaluated the performance of both generative and discriminative transformer-based models for hierarchical text classification, focusing on top-level and sub-level category prediction across two datasets: The Guardian and OpenAlex.

Generative models demonstrated strong accuracy in zero-shot top-level classification, showing promise in scenarios with limited training data. However, their substantial computational cost—particularly at inference—limited their practical applicability, leading us to exclude them from fine-tuning experiments. In contrast, discriminative models such as BERT and DistilBERT performed poorly in baseline settings but significantly improved with local or global fine-tuning. DistilBERT, in particular, achieved the best balance between classification performance and inference efficiency. When accounting for both training and inference times, fine-tuned discriminative models outperformed generative ones in both accuracy and runtime, making them better suited for real-world hierarchical classification pipelines.

Looking ahead, future work will explore integrating ontologies into the classification pipeline to better model and exploit hierarchical relationships among categories. Using structured knowledge representations can help enforce semantic consistency across levels, improve generalization, and support more explainable models. Another direction involves leveraging generative models not for direct classification, but as components in active learning frameworks—helping to identify informative or ambiguous samples, simulate annotations, or guide human-in-the-loop labeling efforts. This approach may reduce annotation costs and increase model adaptability in dynamic environments. Combining these directions with continual learning strategies could enable scalable, self-improving classification systems that evolve alongside the data.

Acknowledgments. This work is funded by national funds through FCT - Fundação para a Ciência e a Tecnologia, I.P., under the support UID/50014/2023 (https://doi.org/10.54499/UID/50014/2023).

References

1. AI@Meta: Llama 3 model card (2024). https://github.com/meta-llama/llama3/blob/main/MODEL_CARD.md
2. Bender, E.M., Koller, A.: Climbing towards NLU: on meaning, form, and understanding in the age of data. In: Jurafsky, D., Chai, J., Schluter, N., Tetreault, J. (eds.) Proceedings of the 58th Annual Meeting of the Association for Computational Linguistics, pp. 5185–5198. Association for Computational Linguistics, Online (2020). https://doi.org/10.18653/v1/2020.acl-main.463
3. Brown, T., et al.: Language models are few-shot learners. Adv. Neural. Inf. Process. Syst. **33**, 1877–1901 (2020)
4. Bucher, M.J.J., Martini, M.: Fine-tuned 'small' LLMs (still) significantly outperform zero-shot generative AI models in text classification. arXiv preprint arXiv:2406.08660 (2024)
5. Chalkidis, I., Fergadiotis, M., Malakasiotis, P., Androutsopoulos, I.: Large-scale multi-label text classification on EU legislation. arXiv preprint arXiv:1906.02192 (2019)
6. Chy, A.N., Seddiqui, M.H., Das, S.: Bangla news classification using Naive Bayes classifier. In: 16th International Conference Computer and Information Technology, pp. 366–371 (2014). https://doi.org/10.1109/ICCITechn.2014.6997369
7. Daud, S., Ullah, M., Rehman, A., Saba, T., Damaševičius, R., Sattar, A.: Topic classification of online news articles using optimized machine learning models. Computers **12**(1) (2023). https://doi.org/10.3390/computers12010016
8. Deb, N., Jha, V., Panjiyar, A.K., Gupta, R.K.: A comparative analysis of news categorization using machine learning approaches. Int. J. Sci. Technol. Res. **9**(1), 2469–2472 (2020)

9. Devlin, J., Chang, M., Lee, K., Toutanova, K.: BERT: pre-training of deep bidirectional transformers for language understanding. CoRR abs/1810.04805 (2018)
10. Fields, J., Chovanec, K., Madiraju, P.: A survey of text classification with transformers: how wide? how large? how long? how accurate? how expensive? how safe? IEEE Access **12**, 6518–6531 (2024)
11. Guardian, T.: Latest news, sport and opinion from the guardian. https://www.theguardian.com/europe
12. He, P., Liu, X., Gao, J., Chen, W.: Deberta: decoding-enhanced bert with disentangled attention. In: International Conference on Learning Representations (2021)
13. Jain, V., et al.: Higen: hierarchy-aware sequence generation for hierarchical text classification. In: Proceedings of the Conference. Association for Computational Linguistics. Meeting, vol. 2024, p. 1354 (2024)
14. Jiang, A.Q., et al.: Mistral 7b (2023)
15. Kareem, I., Awan, S.M.: Pakistani media fake news classification using machine learning classifiers (2019). https://doi.org/10.1109/ICIC48496.2019.8966734
16. Katari, R., Myneni, M.B.: A survey on news classification techniques. In: 2020 International Conference on Computer Science, Engineering and Applications (ICCSEA), pp. 1–5 (2020). https://doi.org/10.1109/ICCSEA49143.2020.9132866
17. Kostina, A., Dikaiakos, M.D., Stefanidis, D., Pallis, G.: Large language models for text classification: case study and comprehensive review (2025)
18. Lee, Y., et al.: Characterizing and efficiently accelerating multimodal generation model inference. arXiv preprint arXiv:2410.00215 (2024)
19. Liu, Y., et al.: Summary of chatgpt-related research and perspective towards the future of large language models. Meta-Radiol. **1**(2), 100017 (2023)
20. Liu, Y., et al.: Roberta: a robustly optimized BERT pretraining approach. CoRR abs/1907.11692 (2019)
21. Plaud, R., Labeau, M., Saillenfest, A., Bonald, T.: Revisiting hierarchical text classification: inference and metrics. arXiv preprint arXiv:2410.01305 (2024)
22. Priem, J., Piwowar, H., Orr, R.: Openalex: a fully-open index of scholarly works, authors, venues, institutions, and concepts (2022)
23. Sanh, V., Debut, L., Chaumond, J., Wolf, T.: Distilbert, a distilled version of bert: smaller, faster, cheaper and lighter. arXiv abs/1910.01108 (2019)
24. Snell, J., Swersky, K., Zemel, R.: Prototypical networks for few-shot learning, vol. 2017-December, pp. 4078–4088 (2017)
25. Sun, A., Lim, E.P.: Hierarchical text classification and evaluation. In: Proceedings 2001 IEEE International Conference on Data Mining, pp. 521–528. IEEE (2001)
26. du Toit, J., Dunaiski, M.: Prompt tuning discriminative language models for hierarchical text classification. Nat. Lang. Process. 1–18 (2024)
27. Torba, F., Gravier, C., Laclau, C., Kammoun, A., Subercaze, J.: A study on hierarchical text classification as a seq2seq task. In: European Conference on Information Retrieval, pp. 287–296. Springer (2024)
28. Touvron, H., et al.: Llama: open and efficient foundation language models. arXiv preprint arXiv:2302.13971 (2023)
29. Vaswani, A., et al.: Attention is all you need. In: Advances in Neural Information Processing Systems, vol. 30 (2017)
30. Veeranna, S.P., Nam, J., Mencía, E.L., Fürnkranz, J.: Using semantic similarity for multi-label zero-shot classification of text documents, pp. 423–428 (2016)
31. Wilkho, R.S., Chang, S., Gharaibeh, N.G.: FF-BERT: a BERT-based ensemble for automated classification of web-based text on flash flood events. Adv. Eng. Inform. **59**, 102293 (2024)

32. Yogatama, D., Dyer, C., Ling, W., Blunsom, P.: Generative and discriminative text classification with recurrent neural networks (2017)
33. Zheng, C., Wu, G., Bao, F., Cao, Y., Li, C., Zhu, J.: Revisiting discriminative vs. generative classifiers: theory and implications. In: Krause, A., Brunskill, E., Cho, K., Engelhardt, B., Sabato, S., Scarlett, J. (eds.) Proceedings of the 40th International Conference on Machine Learning. Proceedings of Machine Learning Research, vol. 202, pp. 42420–42477. PMLR (2023)

Model Uncertainty, Trustworthy and Human-Centered Explainable AI

Beyond the Single-Best Model: Rashomon Partial Dependence Profile for Trustworthy Explanations in AutoML

Mustafa Cavus[1(✉)], Jan N. van Rijn[2], and Przemysław Biecek[3,4]

[1] Department of Statistics, Eskisehir Technical University, Eskişehir, Turkey
mustafacavus@eskisehir.edu.tr
[2] Leiden Institute of Advanced Computer Science, Leiden University, Leiden, The Netherlands
[3] Faculty of Mathematics and Information Science, Warsaw University of Technology, Warsaw, Poland
[4] Informatics and Mechanics, Faculty of Mathematics, University of Warsaw, Warsaw, Poland

Abstract. Automated machine learning systems efficiently streamline model selection but often focus on a single best-performing model, overlooking explanation uncertainty—an essential concern in human-centered explainable AI. To address this, we propose a novel framework that incorporates model multiplicity into explanation generation by aggregating partial dependence profiles (PDP) from a set of near-optimal models, known as the Rashomon set. The resulting Rashomon PDP captures interpretive variability and highlights areas of disagreement, providing users with a richer, uncertainty-aware view of feature effects. To evaluate its usefulness, we introduce two quantitative metrics, the *coverage rate* and the *mean width of confidence intervals*, to evaluate the consistency between the standard PDP and the proposed Rashomon PDP. Experiments on 35 regression datasets from the OpenML-CTR23 benchmark suit show that in most of the cases, the Rashomon PDP covers less than 70% of the best model's PDP, underscoring the limitations of single-model explanations. Our findings suggest that Rashomon PDP improves the reliability and trustworthiness of model interpretations by adding additional information that would otherwise be neglected. This is particularly useful in high-stakes domains where transparency and confidence are critical.

Keywords: Human-centered XAI · Rashomon effect · Partial dependence profile · AutoML

1 Introduction

As artificial intelligence (AI) systems become increasingly embedded in high-stakes domains such as healthcare, finance, and legal applications, the need

S. Džeroski et al. (Eds.): DS 2025, LNAI 16090, pp. 445–459, 2025.
https://doi.org/10.1007/978-3-032-05461-6_29

for explanations that are not only technically accurate but also meaningful to human users has become paramount. Human-centered explainable AI addresses this need by focusing on creating explanations that align with users' values, expectations, and decision-making contexts [1,2]. Unlike traditional explainable AI (XAI) approaches that emphasize algorithmic transparency, human-centered XAI seeks to make AI systems more socially intelligible and trustworthy. However, the diversity in user backgrounds and needs present a fundamental challenge: what constitutes a good explanation varies widely among stakeholders [3]. This necessitates interpretability approaches that go beyond performance metrics to also consider how explanations affect user confidence, understanding, and trust [4]. In this context, automated machine learning (AutoML) systems—while streamlining model selection and optimization—tend to prioritize predictive performance over interpretability, often resulting in a single best-performing model. Even for notable systems that ensemble various well-performing models (such as Auto-sklearn [5] or AutoGluon [6]) it remains non-trivial to utilise these additional models into robust explainations. This can obscure uncertainty and limit the explanatory richness required by human-centered XAI, particularly in sensitive, decision-critical domains.

A persistent obstacle in achieving these goals is explanation disagreement, where different explanation methods—or even different models—offer conflicting explanations of the same prediction [7,8]. Such inconsistencies can undermine user trust and contribute to epistemic uncertainty [9], particularly in complex models where interpretability is already limited [10]. For example, when two explanation methods attribute a model's cancer diagnosis to entirely different features—such as tumor size versus cell irregularity—it becomes unclear which rationale to trust, amplifying epistemic uncertainty. While disagreement is often viewed as a problem, recent studies [11–14] suggest that it can also signal uncertainty and promote more deliberative, cautious decision-making. Thus, rather than eliminating disagreement, it may benefit from frameworks that expose and manage it in ways that support human assessment.

A foundational cause of explanation disagreement lies in the *Rashomon effect*—the existence of many equally accurate models that offer diverging interpretations of the same data [15]. This effect reveals the multiplicity inherent in machine learning (ML): different models may perform similarly well in terms of test set performance according to a given performance measure, yet vary in structure, fairness, and interpretability [16]. This raises ethical and practical concerns about relying on a single best model, as the chosen explanation could have significant consequences depending on which model is selected [17,18]. Moreover, decision stages such as data pre-processing in the modeling process can produce a Rashomon effect [19,20]. Therefore, embracing model multiplicity through the Rashomon lens can help expose this uncertainty and provide users with a richer understanding of the model's behavior and limitations [21,22].

In this paper, we propose a novel framework that integrates the Rashomon effect into model interpretation by leveraging partial dependence profiles (PDP), which is a model-agnostic XAI tool that describes the effect of a feature on the predicted outcome [23]. Rather than relying solely on the top-performing model, our approach considers a set of models with near-optimal performance, known as the Rashomon set. By aggregating the PDPs of these models into what we term the Rashomon PDP, we offer a more comprehensive and uncertainty-aware explanation. This perspective not only highlights where model consensus exists but also where interpretive ambiguity emerges, ultimately supporting more informed and cautious decision-making. Experimental results conducted on 35 real-world datasets using the H2O [24] AutoML show that when the Rashomon ratio (a measure of how many models perform nearly as well as the best one) is high, the coverage of the best model's PDP by the Rashomon PDP tends to be low. In such cases, our proposed framework proves particularly effective, as it better captures the diversity of plausible explanations. The main contributions of this paper are as follows:

1. We propose the Rashomon PDP, a novel aggregation method that reflects model multiplicity and explanation uncertainty.
2. We introduce a metric to quantify the coverage of best-model explanations by the Rashomon PDP, providing a diagnostic for interpretive robustness.
3. Through extensive empirical evaluation on 35 datasets, we demonstrate that our framework reveals hidden interpretive variability, particularly in high Rashomon ratio settings.

The paper is structured as follows. Section 2 details preliminaries and defines the Rashomon PDP. Section 3 covers the experimental setup, datasets, and metrics. Results are presented in Sect. 4. Section 5 provides highlighted examples that demonstrate how the framework performs in both high-agreement and high-divergence settings. Finally, Sect. 6 discusses conclusions, limitations, and future work.

2 Methods

Let $\mathcal{D} = \{(\mathbf{x}_i, y_i)\}_{i=1}^{n}$ be a dataset where each input $\mathbf{x}_i \in \mathbb{R}^p$ represents a p-dimensional feature vector and $y_i \in \mathbb{R}$ is a continuous response variable. We denote the j-th feature by x_{ij} for the i-th instance, and refer to the corresponding feature dimension across the dataset as X_j. The primary goal is to train a diverse ensemble of regression models and analyze their behavior using partial dependence profiles, focusing on models within a Rashomon set.

2.1 Model Training and Rashomon Set Formation

A collection of regression models $\mathcal{M} = \{M_1, M_2, \ldots, M_K\}$ is generated using an AutoML framework. Each model M_k is trained on the same training data and evaluated on a separate test set. The predictive performance of each model is assessed using an appropriate regression metric $\phi(M_k)$, such as mean squared error or mean absolute error. The best-performing model is identified as:

$$M^* = \arg\min_{M_k \in \mathcal{M}} \phi(M_k). \tag{1}$$

To investigate model multiplicity, we define the *Rashomon set* $\mathcal{R}_\varepsilon \subseteq \mathcal{M}$ as the set of models whose performance is within a predefined tolerance $\varepsilon > 0$ of the best model:

$$\mathcal{R}_\varepsilon = \{M_k \in \mathcal{M} \mid \phi(M_k) \leq \phi(M^*) * (1+\varepsilon)\}. \tag{2}$$

For example, reasonable values for ε could be 0.05 or 0.1, depending on the exact problem domain. This set captures models that perform similarly well, allowing us to study functional variability among near-optimal models. Two common metrics used to characterize this set are *Rashomon set size* and *Rashomon ratio* [26].

Rashomon Set Size. refers to the number of models identified within the Rashomon set, i.e., $\mathrm{RSS}_\varepsilon = |\mathcal{R}_\varepsilon|$ quantifies how many models achieve performance comparable to the best-performing model under the tolerance ε.

Rashomon Ratio. is the ratio of the Rashomon set size relative to the total set size of trained models $\mathrm{RR}_\varepsilon = |\mathcal{R}_\varepsilon|/|\mathcal{M}|$. This ratio provides a normalized measure of model multiplicity, indicating the fraction of similarly performing models among all candidates. A higher Rashomon ratio suggests greater diversity of well-performing models, which can be further amplified in the presence of data noise [27], as noisy conditions often allow a wider range of models to achieve similar performance.

2.2 Partial Dependence Profile

For a selected input feature X_j, the *partial dependence profile* (PDP) [23] under a model $M_k \in \mathcal{R}$ quantifies the marginal effect of X_j on the model's output. Formally, the PDP is a function defined over the domain of X_j, denoted $\mathcal{X}_j$, and describes how the model's prediction changes as X_j varies while averaging out the effects of all other features.

The estimated partial dependence function for feature X_j at value $x \in \mathcal{X}_j$ is given by:

$$\hat{f}_j^{(k)}(x) = \frac{1}{n}\sum_{i=1}^{n} \hat{f}^{(k)}(x, \mathbf{x}_{i,-j}), \tag{3}$$

where $\hat{f}^{(k)}$ is the prediction function of model M_k, and $\mathbf{x}_{i,-j}$ denotes the i-th input vector with its j-th feature replaced by the fixed value x.

2.3 Rashomon Partial Dependence Profile

The *Rashomon partial dependence profile* aggregates PDPs across all models in the Rashomon set by averaging them, producing a unified profile that reflects the central trend among near-optimal models:

$$\bar{f}_j(x) = \frac{1}{|\mathcal{R}_\varepsilon|} \sum_{M_k \in \mathcal{R}_\varepsilon} \hat{f}_j^{(k)}(x). \tag{4}$$

To quantify uncertainty arising from model variability, a nonparametric bootstrap is applied over the Rashomon set. Specifically, B bootstrap samples $\mathcal{R}_\varepsilon^{(1)}$, $\mathcal{R}_\varepsilon^{(2)}, \ldots, \mathcal{R}_\varepsilon^{(B)}$ are drawn with replacement, and for each replicate:

$$\bar{f}_j^{(b)}(x) = \frac{1}{|\mathcal{R}_\varepsilon^{(b)}|} \sum_{M_k \in \mathcal{R}_\varepsilon^{(b)}} \hat{f}_j^{(k)}(x). \tag{5}$$

While the Rashomon PDP itself aggregates over multiple near-optimal models, the bootstrap further quantifies the variability of this aggregate due to the finite sample of models in the Rashomon set. This allows us to construct confidence intervals that reflect the stability of the aggregated explanation across different plausible subsets of models.

A pointwise $100(1-\alpha)\%$ confidence interval for $\bar{f}_j(x)$ is then computed using the percentile method:

$$\mathrm{CI}_j(x) = \left[Q_{\alpha/2}\left(\{\bar{f}_j^{(b)}(x)\}_{b=1}^{B}\right),\ Q_{1-\alpha/2}\left(\{\bar{f}_j^{(b)}(x)\}_{b=1}^{B}\right)\right], \tag{6}$$

where $Q_p(\{\bar{f}j^{(b)}(x)\}_{b=1}^{B})$ denotes the empirical p-th quantile of the bootstrap distribution of $\bar{f}_j^{(b)}(x)$. To summarize, the computation of the Rashomon partial dependence profile along with the associated uncertainty quantification via bootstrap is outlined step-by-step in Algorithm 1.

3 Experimental Setup

This section presents a systematic evaluation of our framework through a series of experiments on diverse regression tasks. We aim to investigate how model multiplicity, as captured by the Rashomon set, influences feature effect estimates and how uncertainty can be quantified through partial dependence analysis.

3.1 Datasets

To evaluate the practical utility of our proposed framework, we conduct experiments using the OpenML Curated Tabular Regression benchmark suite 2023 (OpenML-CTR23) [28]. This benchmark includes 35 tabular regression datasets carefully selected to represent a wide range of domains, dataset sizes, and feature

Algorithm 1: Computation of Rashomon partial dependence profile

Input: Dataset $\mathcal{D} = \{(\mathbf{x}_i, y_i)\}_{i=1}^n$, model set $\mathcal{M} = \{M_1, \ldots, M_K\}$, performance metric ϕ, tolerance $\varepsilon > 0$, feature index j, bootstrap iterations B, confidence level α.

Output: Rashomon Partial Dependence Profile $\bar{f}_j(x)$ and confidence intervals $\text{CI}_j(x)$.

Compute performance $\phi(M_k)$ for each model M_k on test data;
Select best model $M^* \leftarrow \arg\min_{M_k \in \mathcal{M}} \phi(M_k)$;
Form Rashomon set $\mathcal{R}_\varepsilon \leftarrow \{M_k \in \mathcal{M} \mid \phi(M_k) \leq \phi(M^*) * (1 + \varepsilon)\}$;
Define grid $\{x_1, \ldots, x_m\}$ for feature X_j;
foreach $M_k \in \mathcal{R}_\varepsilon$ **do**
  foreach $x \in \{x_1, \ldots, x_m\}$ **do**
    Compute $\hat{f}_j^{(k)}(x) = \frac{1}{n} \sum_{i=1}^n \hat{f}^{(k)}(x, \mathbf{x}_{i,-j})$;
foreach $x \in \{x_1, \ldots, x_m\}$ **do**
  Compute $\bar{f}_j(x) = \frac{1}{|\mathcal{R}_\varepsilon|} \sum_{M_k \in \mathcal{R}_\varepsilon} \hat{f}_j^{(k)}(x)$;
for $b = 1$ **to** B **do**
  Draw bootstrap sample $\mathcal{R}_\varepsilon^{(b)}$ with replacement from $\mathcal{R}_\varepsilon$;
  foreach $x \in \{x_1, \ldots, x_m\}$ **do**
    Compute $\bar{f}_j^{(b)}(x) = \frac{1}{|\mathcal{R}_\varepsilon^{(b)}|} \sum_{M_k \in \mathcal{R}_\varepsilon^{(b)}} \hat{f}_j^{(k)}(x)$;
foreach $x \in \{x_1, \ldots, x_m\}$ **do**
  Compute confidence interval
  $\text{CI}_j(x) = \left[Q_{\alpha/2}(\{\bar{f}_j^{(b)}(x)\}_{b=1}^B),\ Q_{1-\alpha/2}(\{\bar{f}_j^{(b)}(x)\}_{b=1}^B)\right]$;
return $\bar{f}_j(x)$, $\text{CI}_j(x)$

complexities. Building on the design principles of the OpenML-CC18 classification benchmark, the CTR23 suite is specifically tailored to support systematic evaluation in regression contexts. Its diversity allows for robust investigation of model performance and interpretability across varied real-world scenarios.

3.2 Setup

We perform all experiments using the H2O AutoML framework [24], an automated machine learning tool that supports a wide range of algorithms, including gradient boosting machines, random forests, generalized linear models, and stacked ensembles. The Rashomon set R_ε is created using models whose performance is close to that of the best model produced by H2O AutoML, with a tolerance of $\varepsilon = 0.05$—a threshold commonly adopted in prior studies [19,25]. For every dataset in the benchmark suite, we use a consistent configuration: the number of models is limited to `max_models = 20` and the total training time is capped at `max_runtime_secs = 360`. This setup ensures fair comparison across datasets while maintaining computational efficiency.

3.3 Metrics

To quantify the variability and reliability of feature effect estimates derived from the Rashomon set, we employ two evaluation metrics. The *mean width of confidence intervals* captures the overall uncertainty across models, while the *coverage rate* assesses how well the Rashomon profile aligns with the partial dependence of the best-performing model.

Mean Width of Confidence Intervals. To summarize the overall uncertainty in the Rashomon partial dependence profile for a feature X_j, we define the *mean width of confidence intervals* (MWCI). Given a grid of input values $\{x_1, x_2, \ldots, x_{n_x}\}$ for X_j, the MWCI is calculated as:

$$\mathrm{MWCI}_j = \frac{1}{n_x} \sum_{\ell=1}^{n_x} \left[Q_{1-\alpha/2} \left(\{ \bar{f}_j^{(b)}(x_\ell) \} \right) - Q_{\alpha/2} \left(\{ \bar{f}_j^{(b)}(x_\ell) \} \right) \right]. \tag{7}$$

Coverage Rate. To assess how well the Rashomon partial dependence profile encompasses the predictions of the best-performing model, we define the *coverage rate* (CR). Let $f_j(x)$ denote the partial dependence profile of the best model M^*. Then the coverage rate is defined as:

$$\mathrm{CR}_j = \frac{1}{n_x} \sum_{\ell=1}^{n_x} \mathbb{I}\left[f_j(x_\ell) \in \mathrm{CI}_j(x_\ell) \right] \tag{8}$$

where $\mathbb{I}[\cdot]$ is the indicator function, which is 1 if the value $f_j(x_\ell)$ lies within the Rashomon-based confidence interval $\mathrm{CI}_j(x_\ell)$, and 0 otherwise. This provides a quantitative measure of how consistently the Rashomon profile captures the behavior of the best model.

4 Results

Table 1 presents key metrics to evaluate model performance and uncertainty across different datasets. Best-performing model performance (BMP) reflects the root mean squared error (RMSE) of the top model on each dataset. Model set size (MSS) denotes the total number of models trained per dataset, as this might vary under the fixed time-limit that we set. Rashomon set size ($\mathrm{RSS}_{0.05}$) indicates how many models achieve performance close to the best-performing model while the tolerance parameter $\varepsilon = 0.05$, thus forming the Rashomon set. The Rashomon ratio ($\mathrm{RR}_{0.05}$), defined as the proportion of the Rashomon set relative to the total model set, captures the extent of model multiplicity. A higher Rashomon ratio indicates that a larger subset of models perform similarly well, implying greater model diversity within the Rashomon set, which is important for assessing explanation diversity.

The results highlight considerable variation in model multiplicity and uncertainty. For example, datasets such as `cars`, `grid_stability`, and four more

Table 1. Set sizes and uncertainty metrics for each dataset. **BMP**: Best-performing model performance in terms of RMSE, **MSS**: Model set size, $\mathbf{RSS}_{0.05}$: Rashomon set size, $\mathbf{RR}_{0.05}$: Rashomon ratio, **MWCI**: mean width of confidence intervals, **CR**: coverage rate.

Dataset	BMP	MSS	$RSS_{0.05}$	$RR_{0.05}$	MWCI	CR
abalone	2.1467	19	13	0.6842	0.5646	0.4214
airfoil_self_noise	1.5680	19	1	-	-	-
auction_verification	371.0463	19	1	-	-	-
brazilian_houses	307.0835	16	1	-	-	-
california_housing	45532.4256	13	6	0.4615	8782.0985	0.2687
cars	2056.7710	22	2	0.0909	538.4280	1
concrete_compressive_strength	4.1028	22	5	0.2272	0.6134	0.6937
cps88wages	443.8287	14	13	0.9285	38.5566	0.6410
cpu_activity	2.1419	8	4	0.5	2.0780	0.5976
diamonds	528.6411	16	7	0.4375	350.7176	0.3416
energy_efficiency	0.7068	22	4	0.1818	0.6040	0.84
fifa	8659.1076	16	9	0.5625	1093.6380	0.7092
forest_fires	112.5122	22	21	0.9545	5.9025	0.1436
fps_benchmark	0.8370	15	2	0.1333	0.2242	0.9626
geographical_origin_of_music	3.8452	22	3	0.1363	0.3885	0.9782
grid_stability	0.0081	19	2	0.1052	0.0003	1
health_insurance	14.6829	19	17	0.8947	1.5426	0.7837
kin8nm	0.1085	18	2	0.1111	0.0119	1
kings_county	110139.3275	16	5	0.3125	77076.2159	0.4087
miami_housing	86377.6786	18	8	0.4444	10383.3733	0.4695
Moneyball	22.6084	22	3	0.1363	6.9054	0.9812
naval_propulsion_plant	0.0006	17	1	-	-	-
physiochemical_protein	3.5789	15	6	0.4	1.2140	0.2888
pumadyn32nh	0.0215	15	8	0.5333	0.0004	0.7437
QSAR_fish_toxicity	0.8382	22	12	0.5454	0.1391	0.5444
red_wine	0.6035	22	9	0.4090	0.0401	0.7136
sarcos	2.0706	6	4	0.6667	0.8788	0.4259
socmob	11.5282	22	1	-	-	-
solar_flare	0.7593	22	9	0.4090	0.0487	0.3750
space_ga	0.1094	19	7	0.3684	0.0440	0.75
student_performance_por	0.9307	22	2	0.0909	0.0312	1
superconductivity	9.2580	6	5	0.8333	0.6905	0.4865
video_transcoding	1.1825	3	2	0.6667	0.4857	1
wave_energy	17589.2242	2	1	-	-	-
white_wine	0.6060	19	6	0.3157	0.0506	0.4454

datasets have small Rashomon sets, indicating that only a few models match the top model's performance, resulting in lower model diversity. Conversely, datasets with higher Rashomon ratios, such as `cps88wages` and `health_insurance`, exhibit many models with similar performance, reflecting greater diversity among well-performing models. Additionally, MWCI and CR reveal higher uncertainty in datasets such as `energy_efficiency` and `fifa`, suggesting more variability in model predictions and wider confidence intervals. These insights emphasize the significance of considering model multiplicity and uncertainty when interpreting model behavior within the Rashomon set.

Figure 1 illustrates the relationship between the Rashomon ratio and coverage rate, along with Spearman's rank correlation significance test [29,30]. We observe a moderate negative correlation, indicating that the coverage rate tends to decrease as the Rashomon ratio increases. This correlation is statistically significant with $\rho = -0.53$, 95% confidence interval $[-0.75, -0.19]$, and $p = 0.003$. The observed relation between Rashomon ratio and coverage rate aligns with the intuition that as the Rashomon set grows larger, there exists a greater diversity of near-optimal models providing different explanations. This model variability naturally leads to increased epistemic uncertainty, which reduces the reliability of explanations. In other words, when many equally good but diverse models fit the data, pinpointing a consistent explanation becomes more challenging, thereby lowering the coverage rate.

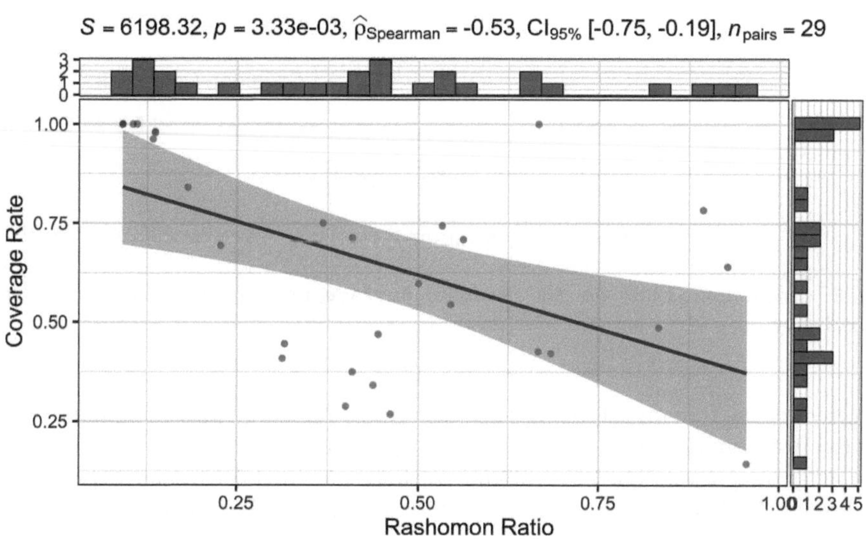

Fig. 1. Visual representation of the results across datasets. Each scatter represents a dataset and reveals the Rashomon ratio and coverage rate of that dataset. Most importantly, we have calculated the Spearman's rank correlation between the Rashomon ratio and coverage rate across datasets.

5 Highlighted Examples

To illustrate the usability of our framework, we highlight two examples. These examples were selected to illustrate the range of behaviors observed in different Rashomon sets. They highlight how our framework adapts to both high-agreement and high-divergence scenarios, revealing when explanation consensus is trustworthy and when caution is warranted.

5.1 Example 1: Surface Area on Energy Efficiency Dataset

An example Rashomon PDP on the variable `surface_area` of the dataset `energy_efficiency` [31] is presented in Fig. 2.

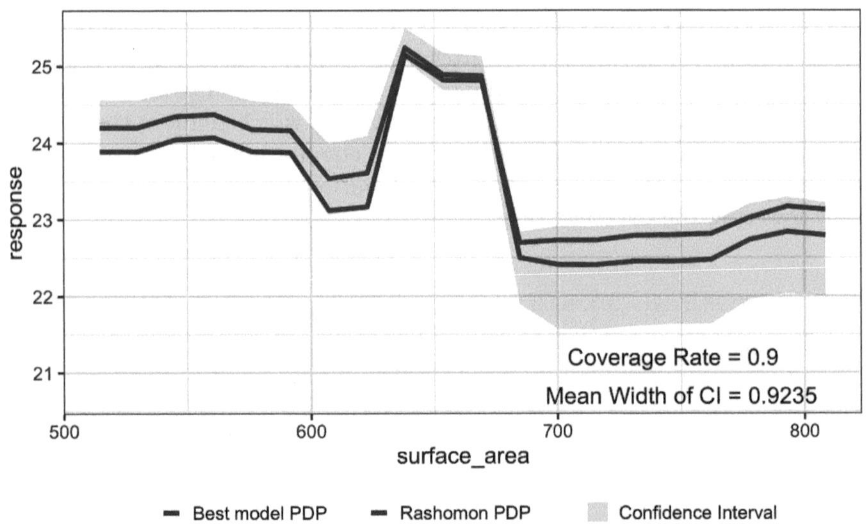

Fig. 2. Rashomon PDP on the variable `surface_area` of the dataset `energy_efficiency`.

Predictions remain relatively high and stable for lower `surface_area` values, but exhibit a sharp increase around 650, followed by a steep decline and stabilization at a lower level. The close alignment between the best model PDP and the Rashomon PDP across most of the `surface_area` range indicates that the top-performing model behaves similarly to the broader set of well-performing models. Minor deviations, particularly around 600, suggest localized disagreement without substantial divergence. The confidence interval illustrates the variability within the Rashomon set and remains relatively narrow, with a mean width of 0.9235, indicating a high level of model agreement. It slightly widens after the sharp drop near 680, reflecting increased uncertainty in that region. A coverage rate of 0.9 implies that 90% of the best model PDP lies within the confidence intervals, further supporting the strong agreement between the best model and the Rashomon PDP. The low mean width of the confidence interval demonstrates that models in the Rashomon set vary little, consistently capturing the relationship between `surface_area` and the response variable.

5.2 Example 2: Temp on Forest Fires Dataset

Another Rashomon PDP example is given on the variable `temp` of the dataset `forest_fires` [32] given in Fig. 3.

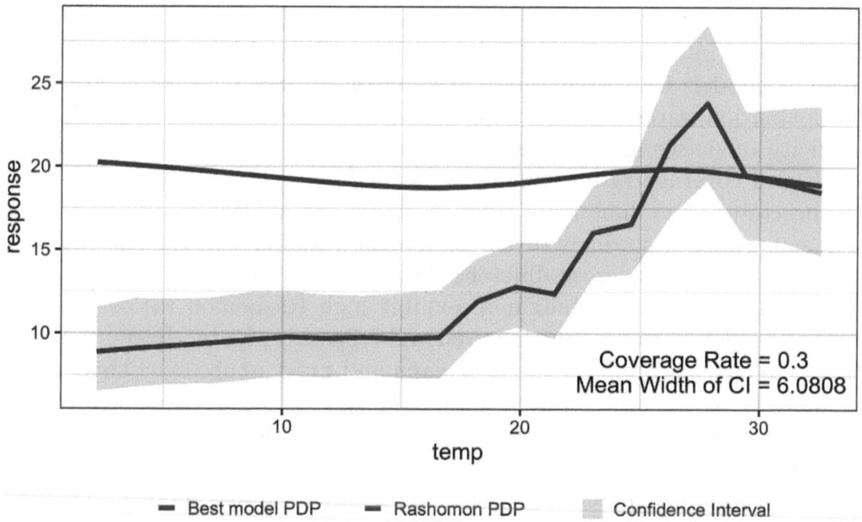

Fig. 3. Rashomon PDP on the variable `temp` of the dataset `forest_fires`.

Predictions for `temp` start low and steadily increase until around 25, followed by greater variability. The best model PDP shows a smooth, slightly decreasing trend, whereas the Rashomon PDP exhibits a pronounced upward trend with higher variance, especially beyond `temp` values of 20. This divergence indicates that the best model behaves quite differently from many other well-performing models in the Rashomon set, particularly in higher temperature ranges.

The confidence interval around the Rashomon PDP is relatively wide, with a mean width of 6.0808, reflecting a high level of disagreement among models regarding how `temp` affects the response. The coverage rate of 0.3, meaning only 30% of the best model PDP lies within the Rashomon PDP's confidence intervals, further underscores this divergence. This suggests that the best model may be overconfident or misaligned with the broader ensemble in this feature dimension. Such a low coverage rate combined with a high interval width signals significant model uncertainty or instability in the learned relationship between `temperature` and the response, possibly due to limited or noisy data in this region or inherent complexity in the variable effect.

Together, these examples demonstrate the practical value of the Rashomon PDP: in cases of strong model agreement, it confirms the reliability of explanations, while in cases of divergence, it surfaces critical uncertainty that would be hidden by relying solely on the best-performing model.

6 Conclusion

This paper introduces the Rashomon PDP, a novel approach to explainability in AutoML systems by leveraging the diversity of near-optimal models. Our findings demonstrate that aggregating partial dependence profiles from a Rashomon set offers a reliable understanding of feature effects than relying solely on the best-performing model. Experimental results across diverse regression tasks reveal that explanation variability is both dataset-dependent and influenced by the multiplicity of performant models. Metrics such as mean width of confidence intervals and coverage rate highlight the added value of our method in quantifying uncertainty and enhancing trust in model explanations.

The experimental findings reveal significant variability in the size and diversity of Rashomon sets across different datasets. For instance, datasets such as `cps88wages` and `health_insurance` exhibit high Rashomon ratios, indicating that many models perform similarly well and reflecting greater diversity in model behavior. In contrast, datasets such as `cars` and `grid_stability` have minimal Rashomon sets, suggesting limited model multiplicity and a lower possibility for diverse explanations. Moreover, the examples further illustrate these dynamics: in the `energy_efficiency` dataset, the Rashomon PDP shows strong alignment with the best model, indicating consistent behavior across models, whereas in the `forest_fires` dataset, a clear divergence is observed, highlighting disagreement among well-performing models. These results underscore the importance of jointly considering model multiplicity and uncertainty when interpreting model behavior and drawing robust conclusions.

Despite these promising results, the study has also several limitations. First, our analysis is restricted to the regression task. While the partial dependece profiles are primarily defined for regression tasks, it would be extremely useful to also gain similar explanations for other task types, such as classification or clustering. Second, the Rashomon set is formed using a fixed tolerance parameter ε, whose optimal value may vary across datasets and use cases. It would be interesting to also explore varying values for epsilon. Third, the approach currently focuses on tree-based model ensembles within the H2O AutoML, limiting generalizability to other model families or tools. Fourth, the runtime of the AutoML was limited to 360 seconds per dataset, which may constrain the exploration of the model space and affect the Rashomon set size. Finally, in terms of explainability, the study exclusively employs PDP, potentially overlooking insights that could be gained from alternative XAI methods.

Future work can extend research towards the Rashomon PDP in several directions. Incorporating theoretical background into the Rashomon framework could yield deeper insights into feature contributions. Additionally, adaptive or data-driven strategies for Rashomon set construction may improve robustness of explanations across domains. For example, one could investigate the effect of learning curves [33], and measure whether additional observations sampled in regions of disagreement can improve the coverage rate. The approach can also be expanded to include other XAI techniques—such as SHAP, ICE, or LIME—to assess whether the Rashomon perspective reveals similar interpretability patterns across different explanation modalities. Finally, user studies are needed to evaluate the impact of Rashomon PDP on human trust, comprehension, and decision-making in real-world applications.

In summary, this work advocates for uncertainty-aware explanations in AutoML, emphasizing that a single-model perspective may be insufficient for trustworthy AI. Rashomon PDP fosters more reliable explanations by accounting for model multiplicity and providing interpretable confidence intervals.

Acknowledgments. This study is funded by the Polish National Science Centre under the SONATA BIS grant 2019/34/E/ST6/00052.

References

1. Ehsan, U., Riedl, M.O.: Human-centered explainable ai: towards a reflective sociotechnical approach. In: Stephanidis, C., Kurosu, M., Degen, H., Reinerman-Jones, L. (eds.) HCII 2020. LNCS, vol. 12424, pp. 449–466. Springer, Cham (2020). https://doi.org/10.1007/978-3-030-60117-1_33
2. Maity, S., Deroy, A.: Human-Centric eXplainable AI in Education. arXiv preprint arXiv:2410.19822 (2024)
3. Ehsan, U., Liao, Q.V., Muller, M., Riedl, M.O., Weisz, J.D.: Expanding explainability: towards social transparency in AI systems. In: Proceedings of the 2021 CHI Conference on Human Factors in Computing Systems, pp. 1–19 (2021)
4. Suffian, M., Stepin, I., Alonso-Moral, J.M., Bogliolo, A.: Investigating human-centered perspectives in explainable artificial intelligence. In: CEUR Workshop Proceedings, vol. 3518, pp. 47–66 (2023)
5. Feurer, M., Eggensperger, K., Falkner, S., Lindauer, M., Hutter, F.: Auto-Sklearn 2.0: hands-free AutoML via Meta-Learning. J. Mach. Learn. Res. **23**, 261:1-261:61 (2022)
6. Erickson, N., et al.: AutoGluon-Tabular: Robust and Accurate AutoML for Structured Data. arXiv preprint arXiv:2003.06505 (2020)
7. Roy, S., Laberge, G., Roy, B., Khomh, F., Nikanjam, A., Mondal, S.: Why don't XAI techniques agree? Characterizing the disagreements between post-hoc explanations of defect predictions. In: 2022 IEEE International Conference on Software Maintenance and Evolution, pp. 444–448. IEEE (2022)
8. Barr, B., Fatsi, N., Hancox-Li, L., Richter, P., Proano, D., Mok, C.: The disagreement problem in faithfulness metrics. arXiv preprint arXiv:2311.07763 (2023)
9. Löfström, H., Löfström, T., Szabadvary, J. H.: Ensured: Explanations for Decreasing the Epistemic Uncertainty in Predictions. arXiv preprint arXiv:2410.05479 (2024)

10. Krishna, S., Han, T., Gu, A., Wu, S., Jabbari, S., Lakkaraju, H.: The disagreement problem in explainable machine learning: a practitioner's perspective. arXiv preprint arXiv:2202.01602 (2022)
11. Gomes, I., et al.: Finding patterns in ambiguity: interpretable stress testing in the decision boundary. In: IEEE/CVF Conference on Computer Vision and Pattern Recognition Workshops (CVPRW) (2024)
12. Reingold, O., Shen, J. H., Talati, A.: Dissenting explanations: leveraging disagreement to reduce model overreliance. In: Proceedings of the AAAI Conference on Artificial Intelligence, vol. 38, no. 19, pp. 21537–21544 (2024)
13. Schwarzschild, A., Cembalest, M., Rao, K., Hines, K., Dickerson, J.: Reckoning with the disagreement problem: explanation consensus as a training objective. In: Proceedings of the 2023 AAAI/ACM Conference on AI, Ethics, and Society, pp. 662–678 (2023)
14. Vascotto, I., Rodriguez, A., Bonaita, A., Bortolussi, L.: When Can You Trust Your Explanations? A Robustness Analysis on Feature Importances. arXiv preprint arXiv:2406.14349 (2024)
15. Breiman, L.: Statistical modeling: the two cultures. Stat. Sci. **16**(3), 199–231 (2001)
16. Rudin, C., Chen, C., Chen, Z., Huang, H., Semenova, L., Zhong, C.: Interpretable machine learning: fundamental principles and 10 grand challenges. Stat. Surv. **16**, 1–85 (2022)
17. Watson-Daniels, J., Parkes, D. C., Ustun, B.: Predictive multiplicity in probabilistic classification. In: Proceedings of the AAAI Conference on Artificial Intelligence, vol. 37, no. 9, pp. 10306–10314 (2023)
18. Watson-Daniels, J., Calmon, F. D. P., D'Amour, A., Long, C., Parkes, D. C., Ustun, B.: Predictive churn with the set of good models. arXiv preprint arXiv:2402.07745 (2024)
19. Cavus, M., Biecek, P.: An experimental study on the Rashomon effect of balancing methods in imbalanced classification. arXiv preprint arXiv:2405.01557 (2024)
20. Cavus, M., Biecek, P.: Investigating the impact of balancing, filtering, and complexity on predictive multiplicity: a data-centric perspective. Inf. Fusion 103243 (2025)
21. Biecek, P., Samek, W.: Position: explain to question not to justify. In: Proceedings of the 41st International Conference on Machine Learning, pp. 3996–4006 (2024)
22. Ganesh, P., Daldaban, I.I., Cofone, I., Farnadi, G.: The cost of arbitrariness for individuals: examining the legal and technical challenges of model multiplicity. arXiv preprint arXiv:2407.13070 (2024)
23. Friedman, J.H.: Greedy function approximation: a gradient boosting machine. Ann. Stat. **29**(5), 1189–1232 (2001)
24. LeDell, E., Poirier, S.: H2O AutoML: scalable automatic machine learning. In: Proceedings of AutoML Workshop at ICML, p. 24 (2020)
25. Müller, S., Toborek, V., Beckh, K., Jakobs, M., Bauckhage, C., Welke, P.: An empirical evaluation of the Rashomon effect in explainable machine learning. In: Joint European Conference on Machine Learning and Knowledge Discovery in Databases, pp. 462–478. Springer, Cham (2023)
26. Semenova, L., Rudin, C., Parr, R.: On the existence of simpler machine learning models. In: Proceedings of the 2022 ACM Conference on Fairness, Accountability, and Transparency, pp. 1827–1858 (2022)
27. Semenova, L., Chen, H., Parr, R., Rudin, C.: A path to simpler models starts with noise. Adv. Neural. Inf. Process. Syst. **36**, 3362–3401 (2023)
28. Fischer, S.F., Feurer, M., Bischl, B.: OpenML-CTR23 – a curated tabular regression benchmarking suite. In: AutoML Conference 2023 (Workshop) (2023)

29. Patil, I.: Visualizations with statistical details: the ggstatsplot approach. J. Open Source Softw. **6**(61), 3167 (2021)
30. Spearman, C.: The proof and measurement of the association between two things. Am. J. Psychol. **100**(3/4), 441–471 (1987)
31. Tsanas, A., Xifara, A.: Accurate quantitative estimation of energy performance of residential buildings using statistical machine learning tools. Energy Build. **49**, 560–567 (2012)
32. Cortez, P., Morais, A.D.J.R.: A data mining approach to predict forest fires using meteorological data. In: Proceedings of 13th Portuguese Conference on Artificial Intelligence (2007)
33. Mohr, F., van Rijn, J.N.: Learning curves for decision making in supervised machine learning: a survey. Mach. Learn. **113**(11), 8371–8425 (2024)

CACTUS: A Context-Aware Framework for Counterfactual Explanations Across Diverse Prediction Domains

Diego García Pérez[1(✉)], Zhendong Wang[2], and José M. Enguita González[1]

[1] University of Oviedo, Gijón, Spain
{garciaperdiego,jmenguita}@uniovi.es
[2] KTH Royal Institute of Technology, Stockholm, Sweden
zhwang@kth.se

Abstract. Counterfactual explanations have been proposed to provide actionable insights for competitive black-box classifiers by suggesting modifications to alter undesirable prediction outcomes into desired ones. Despite the rapid development of counterfactual methods that provide insightful "what-if" explanations in various applications, most existing approaches fail to generate counterfactuals considering whether the contextual integrity of the original instance –such as demographic consistency or user-specific constraints– is explicitly controlled. This article introduces Context-Aware Counterfactual Explanations (CACTUS), a novel framework for generating feasible counterfactuals that can either preserve or change the user-defined contextual features of the original instance. We illustrate CACTUS by conducting the counterfactual search in a context-conditional latent space, using a composite β-VAE model that can learn enhanced disentanglement of context-related factors. Our experimental evaluation indicates that the context-preserving CACTUS outperforms state-of-the-art methods in maintaining contextual consistency, while remaining competitive in compactness, proximity, and validity metrics. Moreover, our qualitative analysis highlights that CACTUS generates plausible counterfactual explanations that can explicitly preserve or modify user-defined contexts in real-world scenarios.

Keywords: Counterfactual Explanations · Context-aware Systems · Deep Learning · Model Interpretability

1 Introduction

Deep learning models have demonstrated their competitive performance across various prediction domains, including tasks like image and time series classification [4,7]. However, these models are often considered "black boxes", which compromises model interpretability and makes it difficult to understand model predictions. To mitigate this issue, previous work in eXplainable AI (XAI) has been

S. Džeroski et al. (Eds.): DS 2025, LNAI 16090, pp. 460–475, 2025.
https://doi.org/10.1007/978-3-032-05461-6_30

introduced, including feature importance methods like SHAP and saliency maps, as well as counterfactual explanations [13,17]. The development of counterfactual explanations has gained significant attention because they align with the "right to explanation" regulations, such as the General Data Protection Regulation (GDPR) [22]. Counterfactual explanations offer post-hoc support in explaining already deployed black-box models without revealing their internal structures, while providing actionable "what-if" insights into the model's reasoning process [5,21]. For example, in a real-world loan application, when the applicant receives a negative prediction (i.e., loan rejection) from the bank, the bank can suggest counterfactual explanations to improve the applicant's profile, such as paying off the student debt. By applying these changes, the applicant can potentially achieve a desired positive outcome and obtain the loan approval.

Recent advancements in counterfactual explanation methods have introduced specific desired properties to guide the generation of counterfactual instances, known as "counterfactual desiderata" [21]. By incorporating *actionability* and *causality* as desiderata, the generated counterfactuals can prevent immutable context changes (e.g., requesting the applicant to become younger) and adhere to the applicant's preferences, which can be achieved by preserving the causal relations learned jointly with the counterfactual explainer [15] or by separating mutable and immutable features through structural interventions [10]. However, these approaches are associated with two challenges: first, training a causal structural model is expensive and often requires additional domain knowledge; second, causal models require tabular data structures, which may not be suitable for all classification scenarios. One alternative to mitigate these challenges is to introduce *context-aware* adaptation, integrating context information (e.g., user-related aspects or image properties) through different modalities to enhance machine learning tasks for interpreting the scenario accurately [8,16]. For example, the word "doctor" can have different meanings in various contexts, such as hospitals and academic institutions. Similarly, in the loan application scenario, explicitly incorporating the applicant's user-related contexts (e.g., age group and socioeconomic situations) into the counterfactual generation processes that can align with the applicant's preference is particularly important. To the best of our knowledge, there has been no prior work on integrating context-awareness into counterfactual generation across diverse prediction domains.

Furthermore, previous works investigated the generation of *plausible* counterfactuals that adhere to the underlying data manifold, including nearest-neighbor estimators [9,12], density estimation measures [2,11,18], and class prototypes [19]. By utilizing plausibility, the counterfactual search guarantees that generated counterfactual instances are in high-density regions of the training data manifold. However, plausibility focuses solely on optimizing the underlying distribution, hence lacking control of explicit context information from user-defined inputs, e.g., the high-density distribution may not necessarily share the same context as the original. Related to plausibility, earlier studies explored leveraging deep generative models to guide the counterfactual search through learned latent representations, including auto-encoders (AEs) and variational auto-encoders

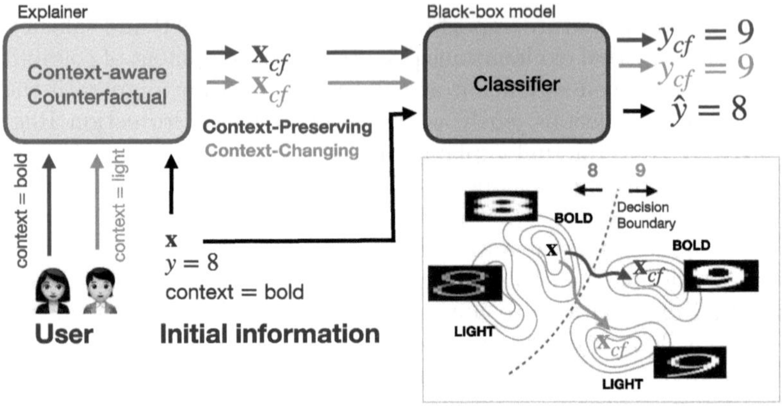

Fig. 1. An illustration of two context-aware scenarios in the CACTUS framework for image counterfactual generation (changing from class "8" to class "9"). CACTUS offers flexibility to generate counterfactuals with either context-preserving (*blue-colored* arrows) or context-changing (*orange-colored*), leading to the counterfactual target class while accounting for user-defined context inputs (e.g., image properties as the "bold" or "light" context).

(VAEs) [18,19,24]. These approaches effectively utilize latent representations to generate counterfactuals following high-density regions across diverse prediction tasks, such as time series and image classification. Nevertheless, similar to plausibility, following the latent data distribution may not explicitly guarantee control of the user-defined contexts information through counterfactual generation [5]. This ongoing challenge underpins the limitations of the current state of counterfactual generation, and it is desirable to have flexibility in either *preserving* or *changing* user-defined context inputs (e.g., following the applicant's preferences) while maintaining coherence at the context level [21].

In this paper, we propose a counterfactual framework for Context-aware Counterfactual Explanations (CACTUS), which provides explicit control over user-defined context information to generate more feasible counterfactuals through latent representations. CACTUS introduces two context-aware scenarios that can be tailored by user preferences: (1) context-preserving and (2) context-changing. To highlight the contributions of CACTUS, we illustrate the framework using an image counterfactual example in Fig. 1. In both scenarios, the CACTUS framework employs a learned context-conditioned latent space to change the input class from "8" to class "9", where the latent representations are organized according to user-defined contexts (i.e., image properties as "bold" and "light" fonts). In the context-preserving scenario, the counterfactual search remains within regions of the latent space that share the same context as the original, i.e., the bold font. This ensures that the generated counterfactuals retain the specific context and avoid unnecessary changes to unrelated features, thereby aligning with user preferences. Alternatively, the context-changing scenario allows the user to explicitly modify their contextual conditions (i.e., the light font), enabling the search to explore latent regions different from the original context. By allowing the explicit control of contextual inputs, this approach can generate feasible counterfactuals that are desirable based on user-specific preferences, hence mitigating the issue of data drift and progression, as well as the expense of extracting a causal model for actionability [21]. In this paper, we quantitatively compare context-preserving scenario using a context-conditional β-VAE model and state-of-the-art counterfactual methods in binary tabular and image classification domains [6,14]. To the best of our knowledge, no state-of-art method is capable of generating context-changing counterfactuals, which makes their evaluation challenging. Nevertheless, our qualitative analysis demonstrates the flexibility of CACTUS in producing counterfactual examples across both context-aware scenarios. Finally, an ablation study reveals a trade-off between validity and context actionability in both scenarios.

Our Contributions. We summarize our contributions as follows:

1. We propose two alternative scenarios in CACTUS for context-aware counterfactual explanations: (1) context-preserving to maintain user-defined contexts within a specific manifold constrained by the latent space; and (2) context-changing, where it is possible to guide the original instance toward a target context and alter its contextual features.
2. We illustrate the CACTUS framework using a context-conditional β-VAE model, where a context learning subtask is conducted to disentangle the context from the modification of feature inputs.
3. We demonstrate the effectiveness of CACTUS using tabular and image binary classification tasks, comparing it to state-of-the-art methods using quantitative metrics and conducting qualitative analysis in credit prediction and typography font scenarios.

2 Related Work

Context-aware adaptation is crucial in machine learning, especially when context information can accurately interpret application scenarios [8]. It integrates context information (e.g., user-related aspects) through various modalities to, for instance, enhance recommendation systems [16] or select relevant features for classification tasks [23]. Recently, researchers proposed context-aware adaptation in post-hoc XAI methods. For instance, context-aware learning can capture feature dependencies for local interpretable model-agnostic explanations (LIME) [1]; context-aware SHAP can complement as a feature selection approach, thereby enhancing model fairness and robustness [20]. To our knowledge, there has not been an adaptation of context-awareness to counterfactual explanations to provide more feasible counterfactuals following the user's preferences.

Counterfactual *actionability* has been investigated to preserve context information, such as age and sex groups, while modifying only mutable features. Mahajan et al. [15] proposed to utilize structural causal models (SCMs) with a causal proximity regularizer, which preserves causal constraints and demonstrates the feasibility of the generated counterfactuals. Karimi et al. [10] highlighted that interventions using causal reasoning can be incorporated to enhance the actionability and minimize the action cost in algorithmic recourse. Nevertheless, incorporating context-awareness as a counterfactual desideratum remains a gap, which can complement actionability and explicitly control context features during counterfactual generation. Another group of counterfactual methods explored *plausibility* to ensure that counterfactuals closely resemble the ground-truth data manifold. Laugel et al. [12] introduced plausibility by employing Local Outlier Factor (LOF) as a distance measure in a distribution-aware approach; the DACE method incorporated LOF through mixed-integer linear optimization, both ensuring counterfactuals aligned with the target class distribution [9]. Recent research has explored latent representations with counterfactual plausibility in various prediction domains. Pawelczyk et al. [18] adapted plausibility and proposed connectedness in the C-CHVAE model to assess the proportion of counterfactuals using correctly classified instances for tabular credit prediction, employing a VAE model and a Gaussian mixture prior to prevent immutable feature changes. The ProtoCF method defined a reconstruction error term from a learned AE and class prototypes to prevent out-of-distribution counterfactual generation, guiding the search for more plausible counterfactuals for image and tabular classification [19]. The NICE model incorporated an AE with nearest-neighbors in optimization to generate plausible counterfactual instances, demonstrating a trade-off between plausibility and proximity [2]. LatentCF++ proposed instantiations using convolutional neural networks (CNN) and long short-term memory (LSTM) AEs to compare the performance trade-offs for time series classification [24]. Both ProtoCF and LatentCF++ serve as suitable baselines compared to CACTUS in our empirical evaluation.

3 Problem Formulation

Let $\mathbf{x} \in \mathbb{R}^n$ denote an input instance characterized by n features, and let $y \in \{0, 1\}$ be the ground-truth label, where $y = 1$ indicates that $\mathbf{x}$ belongs to the positive class, and $y = 0$ to the negative class. We define $f : \mathbb{R}^n \to [0, 1]$ as the trained binary classifier that maps the input $\mathbf{x}$ to the class prediction probability $\hat{y}$, such that $\hat{y} = f(\mathbf{x})$. Each input $\mathbf{x}$ is associated with a multi-label context vector $\mathbf{c} = [c_1, \ldots, c_d]$, where d denotes the total number of contextual conditions. Each component $c_i \in \{0, 1\}$ indicates whether the input instance belongs to the i-th context condition.[1] The context label c_i is predefined using user-interpretable features (e.g., age group or number of dependents), and are independent of the main classification task (i.e., c_i are excluded from $\mathbf{x}$).

[1] An input instance can be assigned to one or more contexts, e.g., a loan applicant $\mathbf{x}$ might simultaneously belong to the contexts `above 50 years` ($c_1 = 1$) and `more than 2 dependents` ($c_2 = 1$).

In this paper, we formulate the problem of generating context-aware counterfactuals that incorporate user-defined contextual information. This objective is formally defined by two problem statements.

Problem 1: Context-preserving Counterfactual Explanations. Given $f(\cdot)$, an input instance $\mathbf{x}$, and its context label $\mathbf{c}$ such that $\hat{y} = f(\mathbf{x})$, we define a context-preserving counterfactual example $\mathbf{x}_{cf}$ as an instance that leads to a different prediction (*validity*) and remains as close as possible to the original instance (*proximity*), while preserving the original context:

$$y_{cf} = f(\mathbf{x}_{cf}) \text{ s. t. } y_{cf} = \hat{y}_{cf} \neq \hat{y}, \ dist(\mathbf{x}, \mathbf{x}_{cf}) \text{ is minimized, and } \mathbf{c}_{cf} = \mathbf{c},$$

where y_{cf} denotes the targeted class by the user, $\hat{y}_{cf}$ is the predicted label of the counterfactual instance $\mathbf{x}_{cf}$, $dist(\cdot, \cdot)$ defines the distance between the original sample and the counterfactual, and $\mathbf{c}_{cf}$ is the context annotation associated with $\mathbf{x}_{cf}$.

Problem 2: Context-changing Counterfactual Explanations. We define the problem for a context-changing counterfactual instance $\mathbf{x}_{cf}$ as one that satisfies:

$$y_{cf} = f(\mathbf{x}_{cf}) \text{ s. t. } y_{cf} = \hat{y}_{cf} \neq \hat{y}, \ dist(\mathbf{x}, \mathbf{x}_{cf}) \text{ is minimized, and } \mathbf{c}_{cf} \neq \mathbf{c}.$$

This formulation follows a similar notation as in Problem 1, except that the context is expected to change, i.e., $\mathbf{c}_{cf} \neq \mathbf{c}$.

These two formulations allow users to explicitly specify whether the original context should be preserved or changed in the counterfactual explanation. We refer to this desired property as *context validity*.

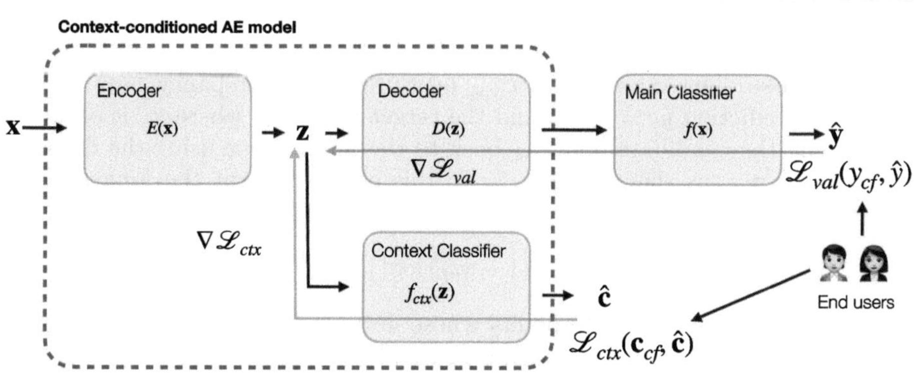

Fig. 2. Overview of the CACTUS framework: a context-conditional AE-based model in which the latent representation $\mathbf{z}$ is guided by a context classification subtask to integrate user-defined context preferences into gradient-based counterfactual generation in the latent space.

4 Context-Aware Counterfactual Explanation Framework

We propose *Context-Aware Counterfactual Explanations* (CACTUS), a framework for generating context-aware counterfactual explanations in binary classification problems. The framework's overview is presented in Fig. 2. In this section, we illustrate the framework in detail, outline its main components, and demonstrate how it can be instantiated with different deep learning models.

4.1 Counterfactual Search in Latent Space

CACTUS ensures plausible counterfactuals and avoid unrealistic outputs by generating them within a latent space learned by an AE-based model. As shown in Fig. 2, the input instance $\mathbf{x}$ is mapped into the latent space by a pre-trained encoder $E(\cdot)$, such that $\mathbf{z} = E(\mathbf{x})$, where $\mathbf{z} \in \mathbb{R}^m$ denotes the representation of $\mathbf{x}$ in a compact m-dimensional latent space, i.e., $m \ll n$. Ideally, $\mathbf{z}$ captures the most relevant and underlying information of the input, forming a manifold representation of the input space. Hence, performing the counterfactual search in this manifold leads to more realistic modifications to the input. Accordingly, we propose an instantiation of CACTUS using a β-VAE model [3] as the AE-based architecture, due to its enhanced disentanglement capabilities, which improve sample generation. Nevertheless, the CACTUS framework is agnostic to the specific latent space model and could be instantiated with other alternative AE variants.

The counterfactual search starts from an initial latent point $\mathbf{z}_0 = E(\mathbf{x}_0)$, and iteratively explores candidate solutions $\mathbf{z}_k$ at the k-th iteration that move toward satisfying the desired counterfactual properties. We propose a multi-objective loss function that evaluates how well a counterfactual candidate in the latent space satisfies the validity, context validity, and proximity requirements, as defined in Sect. 3.

Validity Loss. The validity loss $\mathcal{L}_{val}$ measures the discrepancy between the classifier's prediction $\hat{y}_k = f(\mathbf{x}_k)$ and the target label y_{cf}, where $\mathbf{x}_k$ is obtained by mapping the candidate point $\mathbf{z}_k$ back to the input space using the decoder, i.e., $\mathbf{x}_k = D(\mathbf{z}_k)$, as shown in Fig. 2. To this end, we adopt the *binary cross-entropy* loss function:

$$\mathcal{L}_{val}(\mathbf{z}_k, y_{cf}) = -y_{cf}\log(\hat{y}_k) - (1 - y_{cf})\log(1 - \hat{y}_k), \quad \hat{y}_k = f\left(D(\mathbf{z}_k)\right). \tag{1}$$

This term penalizes candidate points whose decoded instances do not lead the classifier to predict the desired counterfactual label.

Distance Loss. We incorporate a *distance loss* term that penalizes extensive modifications and enforces proximity. Accordingly, we propose using the Euclidean distance between the original input $\mathbf{x}$ and the counterfactual $\mathbf{x}_{cf}$ as a proximity constraint:

$$\mathcal{L}_{dist}(\mathbf{x}_0, \mathbf{z}_k) = \|\mathbf{x}_0 - \mathbf{x}_k\|_2, \quad \mathbf{x}_k = D(\mathbf{z}_k). \tag{2}$$

Context Loss. To enforce context validity, the proposed CACTUS generates counterfactuals within a latent space explicitly structured by contextual information. To this end, we propose a modification to the standard AE architecture by adding an auxiliary classification head that estimates the context annotation $\mathbf{c}$ of the input $\mathbf{x}$ from its latent representation $\mathbf{z}$:

$$\hat{\mathbf{c}} = f_{ctx}(\mathbf{z}), \tag{3}$$

where the classifier $f_{ctx}(\cdot)$ estimates the context label using a shallow neural network composed of fully-connected layers. The overall architecture, including the auxiliary context classifier, is illustrated in Fig. 2. This composite architecture is jointly trained prior to counterfactual generation, guiding the latent space to encode contextual information in its structure.

The estimated context vector $\hat{\mathbf{c}}_k$ at the k-th iteration of the search can be compared to the target context $\mathbf{c}_{cf}$ to evaluate whether it satisfies the user-specified context. As a result, we introduce a context loss function $\mathcal{L}_{ctx}$ based on the binary cross-entropy between $\hat{\mathbf{c}}_k$ and $\mathbf{c}_{cf}$:

$$\mathcal{L}_{ctx}(\mathbf{c}_{cf}, \mathbf{z}_k) = -\mathbf{c}_{cf} \log(\hat{\mathbf{c}}_k) - (1 - \mathbf{c}_{cf}) \log(1 - \hat{\mathbf{c}}_k). \tag{4}$$

This loss guides the optimization toward counterfactuals that satisfy the context validity requirement in both the context-preserving ($\mathbf{c}_{cf} = \mathbf{c}_0$) and context-changing ($\mathbf{c}_{cf} \neq \mathbf{c}_0$) scenarios.

Finally, the validity, distance and context losses are combined into a multi-objective loss function:

$$\mathcal{L}(\mathbf{x}_0, \mathbf{z}_k, y_{cf}, \mathbf{c}_{cf}) = \alpha \mathcal{L}_{val} + (1 - \alpha)\mathcal{L}_{ctx} + \gamma \mathcal{L}_{dist}, \tag{5}$$

where the user-defined hyperparameters α and γ provide the flexibility to control the relative importance of the validity and context terms, and the distance term, respectively. We conduct an ablation study to illustrate the hyperparameter effects of CACTUS in our experimental evaluation in Sect. 5.2.

4.2 Gradient-Based Search

The search in the latent space is conducted by minimizing the loss term in Eq. 5:

$$\mathbf{z}_{cf} = \arg\min_{\mathbf{z}} \mathcal{L}(\mathbf{x}_0, \mathbf{z}_k, y_{cf}, \mathbf{c}_{cf}), \quad \mathbf{x}_{cf} = D(\mathbf{z}_{cf}). \tag{6}$$

We present the pseudo-code of the minimization problem in Algorithm 1, where the initial point $\mathbf{z}_0$ is iteratively updated according to the *gradient descent* rule:

$$\mathbf{z}_k = \mathbf{z}_{k-1} - \eta \frac{\nabla \mathcal{L}(\mathbf{x}_0, \mathbf{z}_k, y_{cf}, \mathbf{c}_{cf})}{\max\left(|\nabla \mathcal{L}(\mathbf{x}_0, \mathbf{z}_k, y_{cf}, \mathbf{c}_{cf})|\right)}, \tag{7}$$

where η is the step size. To ensure stable updates during optimization, the gradient direction is normalized at each step so that updates follow the direction of descent while maintaining a fixed maximum step size, independent of the gradient's magnitude. The update rule in Eq. 7 is executed for a total of `max_epochs` iterations. At each step, the candidate $\mathbf{x}_k$, obtained from $\mathbf{z}_k$ via the decoder, is stored along with its corresponding distance and context score if it improves upon the previous best in terms of prediction and context validity.

Algorithm 1. CACTUS counterfactual generation algorithm

Input: $\mathbf{x}$, y_{cf}, $\mathbf{c}_{cf}$, α, γ, η,max_epochs,$E(\cdot)$,$D(\cdot)$,$f_{ctx}(\cdot)$,$f(\cdot)$
Output: $\mathbf{x}_{cf}$,$\mathbf{z}_{cf}$,$\mathbf{c}_{cf}$,best_dist,best_ctx

$\mathbf{x}_0 \leftarrow \mathbf{x}$
$\mathbf{z}_0 \leftarrow E(\mathbf{x}_0)$
best_dist $\leftarrow 10e10$
best_ctx $\leftarrow 10e10$
prob_error $\leftarrow 1$
for epoch $= 0$ to max_epochs **do**
  $\mathbf{z}_k \leftarrow \mathbf{z}_{k-1} - \eta \cdot \frac{\nabla \mathcal{L}(\mathbf{x}_0, \mathbf{z}_k, y_{cf}, \mathbf{c}_{cf})}{\max(|\nabla \mathcal{L}(\mathbf{x}_0, \mathbf{z}_k, y_{cf}, \mathbf{c}_{cf})|)}$
  $\mathbf{x}_k \leftarrow D(\mathbf{z}_k)$
  $\hat{y}_k \leftarrow f(\mathbf{x}_k)$
  $\hat{\mathbf{c}}_k \leftarrow f_{ctx}(\mathbf{z}_k)$
  if $|y_{cf} - \hat{y}_k| \leq$ prob_error **and** $\mathcal{L}_{ctx}(\mathbf{c}_{cf}, \hat{\mathbf{c}}_k) \leq$ best_context **then**
    $\mathbf{z}_{cf} \leftarrow \mathbf{z}_k$
    $\mathbf{x}_{cf} \leftarrow \mathbf{x}_k$
    best_ctx $\leftarrow \mathcal{L}_{ctx}(\mathbf{c}_{cf}, \hat{\mathbf{c}}_k)$
    best_dist $\leftarrow \mathcal{L}_{dist}(\mathbf{x}_0, \mathbf{x}_k)$
    prob_error $\leftarrow |y_{cf} - \hat{y}_k|$
  end if
end for

5 Empirical Evaluation

5.1 Experimental Setup

Datasets. The proposed CACTUS instantiation is evaluated using TMNIST [14] for image classification and *GiveMeCredit* [6] for tabular data. TMNIST is a publicly available version of the MNIST dataset, in which each image is represented as a vector of pixel intensities. Alongside the digit label, the dataset provides font-style annotations, from which we derive two binary context variables c_i: `bold/light` ($0 =$ bold, $1 =$ light) and `regular/italic` ($0 =$ regular, $1 =$ italic). We design our study for a binary classification setting to distinguish between the digits "8" (negative class) and "9" (positive class). The GiveMeCredit dataset is publicly available and used for credit risk prediction, where each instance corresponds to an individual described by financial and demographic attributes. In addition to the binary label indicating a serious financial situation, we derive two binary context variables c_i from demographic features: `age` ($0 =$ under 50, $1 =$ 50 and above) and `NumberOfDependents` ($0 =$ no dependents, $1 =$ one or more).

Baseline Models. CACTUS is compared against several counterfactual generation methods using latent space representations, including *LatentCF++* [24] and *ProtoCF* [19]. For ProtoCF, we compare with two variants: ProtoCF-C and ProtoCF-D, where ProtoCF-C incorporates the reconstruction error of the AE, whereas ProtoCF-D does not.

Hyperparameters and Model Settings. The classifier and β-VAE models of the proposed framework are trained prior to counterfactual generation. For the TMNIST dataset, we train a CNN classifier consisting of two convolutional blocks, each comprising three consecutive convolutional layers followed by a max-pooling layer, and a two-layer fully connected network that outputs the predicted positive class probability. For the β-VAE used in TMNIST, we define an encoder and a decoder, both composed of a single convolutional block, that map input samples into a 16-dimensional latent space and reconstruct them, respectively. The context output is estimated from the latent representation by a two-layer fully-connected subnetwork. Similarly, for the GiveMeCredit dataset, we train a DNN classifier composed of four fully-connected layers and an output layer that estimates the positive class probability. Regarding the β-VAE for GiveMeCredit, we use an encoder consisting of two fully connected layers, which maps input instances into a 4-dimensional latent space. The decoder reconstructs the input from the latent representation using two fully-connected layers. Finally, the context classifier estimates the context vector using a single fully connected layer.

All experiments presented in Sects. 5.2 and 5.3, except for the ablation study, were performed with: $\alpha = 0.7$, $\gamma = 0.1$, and `max_epochs` = 300. ProtoCF and LatentCF++ models were evaluated using the default settings specified in their respective original papers [19,24].

Evaluation Metrics. The counterfactual examples generated by CACTUS and the baseline models are quantitatively evaluated using the following standard metrics in the literature: 1) *Validity* [15,23], defined as the F1 score between the predicted and target labels; 2) *Compactness* [24], similar to Sparsity, measured as the ratio of features that differ between the counterfactual and the original instance, relative to the total number of input features; 3) *Proximity* [12,15,18], computed as the normalized Euclidean distance[2] between the counterfactual and the original input; and 4) $Context\text{-}LOF_k$[3], similar to LOF [9], computed with $k = 10$ neighbors within the target context, which quantifies how well the counterfactual belongs to the distribution of the desired context. Note that categorical attributes in tabular data are not explicitly handled in the computation of proximity and compactness metrics.

5.2 Quantitative Results

Table 1 reports the quantitative results comparing CACTUS with the baseline models in counterfactual generation, based on five different sets of 100 samples

[2] Euclidean distance between $\mathbf{x}_0$ and $\mathbf{x}_{cf}$, divided by the number of features n.
[3] The LOF is computed using the 10 nearest neighbors in the target context $\mathbf{c}_{cf}$.

Table 1. Quantitative comparison of CACTUS (context-preserving) and baseline models using Validity, Context-LOF$_{10}$, Compactness, and Proximity metrics on the GiveMeCredit and TMNIST datasets. Reported values correspond to the mean and standard deviation over five repeated tests with 100 samples each. For each dataset, method, and metric, the best and second-best values are highlighted in bold and underlined, respectively.

		Validity (↑)	Con.-LOF (↓)	Compact. (↑)	Proximi. (↓)
Credit	CACTUS	<u>0.77 ± 0.03</u>	**1.64 ± 0.07**	**0.08 ± 0.02**	<u>0.20 ± 0.01</u>
	LatentCF++	**1.00 ± 0.00**	<u>2.02 ± 0.10</u>	<u>0.04 ± 0.00</u>	**0.17 ± 0.01**
	ProtoCF-C	**1.00 ± 0.00**	3.91 ± 0.20	0.03 ± 0.00	0.42 ± 0.02
	ProtoCF-D	**1.00 ± 0.00**	2.90 ± 0.42	0.03 ± 0.01	0.33 ± 0.02
TMNIST	CACTUS	<u>0.98 ± 0.01</u>	**1.51 ± 0.03**	0.64 ± 0.00	**0.51 ± 0.01**
	LatentCF++[a]	0.16 ± 0.03	1.35 ± 0.02	0.57 ± 0.01	0.18 ± 0.01
	ProtoCF-D	**1.00 ± 0.00**	2.27 ± 0.04	<u>0.71 ± 0.00</u>	0.57 ± 0.01
	ProtoCF-C	**1.00 ± 0.00**	1.70 ± 0.04	**0.78 ± 0.01**	0.65 ± 0.05

[a] Due to a low validity score, the LatextCF++ results on the TMNIST dataset were excluded from the analysis.

each, created through random splits of the test data to minimize the impact of sample-specific noise. All counterfactuals are generated under the context-preserving setting (i.e., $\mathbf{c}_{cf} = \mathbf{c}_0$). The table summarizes the results across all trials, reporting the mean and standard deviation for each evaluation metric.

We observed that CACTUS consistently achieved the second-best validity score across both datasets, closely following the ProtoCF-C and ProtoCF-D models. This result suggests that adding context constraints makes the generation of valid counterfactuals more challenging, potentially highlighting a trade-off between prediction and context validity, which was further investigated in the ablation study.

Besides, CACTUS outperformed all other models in terms of Context-LOF$_{10}$, highlighting its ability to better preserve the original context. According to compactness, CACTUS obtained the highest score over ProtoCF and LatentCF++ on GiveMeCredit, and remained competitive on the TMNIST dataset. Moreover, CACTUS achieved consistently competitive proximity scores (second best in GiveMeGredit and best in TMNIST (see footnote a in Table 1)), demonstrating that the generated counterfactuals remained close distances to the original instances.

Second, we performed an ablation study to examine the effects of the hyperparameter α in the counterfactual search defined in Eq. 5 to explore the interactions between context and validity loss, as shown in Fig. 3. We measured the Validity and Context-LOF$_{10}$ in both context-preserving and context-changing scenarios when α was gradually increased from 0 to 1 with a step of 0.25. In Fig. 3, we observed a consistent upward trend in both Validity and Context-LOF$_{10}$ as α increases, across all datasets and both context-aware scenarios. This emphasizes

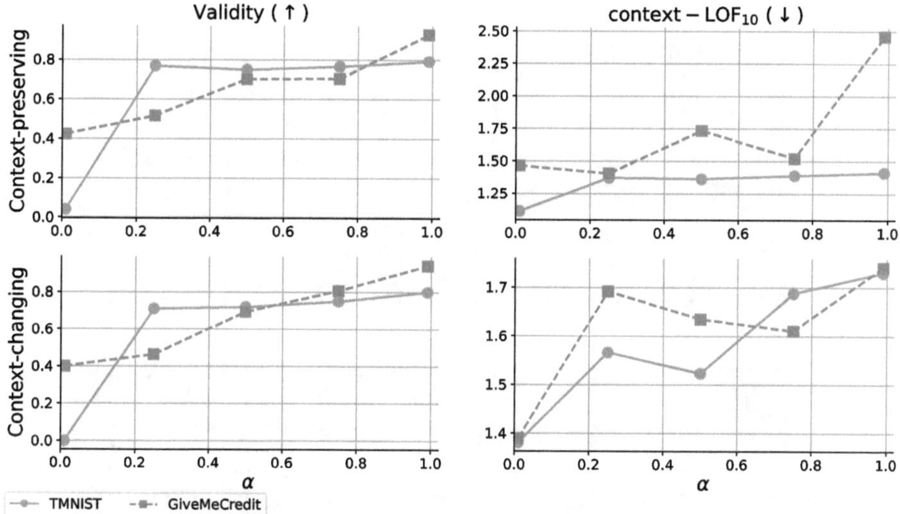

Fig. 3. Effect of the parameter α on CACTUS validity and Context-LOF$_{10}$ across TMNIST and GiveMeCredit datasets under context-preserving and context-changing conditions.

the trade-off between prediction and context validity, with $\alpha \approx 0.7$ representing a balanced choice that achieved strong performance in both evaluation metrics and context-aware scenarios.

5.3 Qualitative Analysis

We present several counterfactual examples to qualitatively examine CACTUS in both context-preserving and context-changing settings. Figure 4 (a) shows the counterfactuals generated by CACTUS and the baseline models for two TMNIST instances with $\mathbf{c}_0$ = `Bold` and $\mathbf{c}_0$ = `Light`, respectively. CACTUS generated more plausible and realistic digits while effectively preserving the initial context, whereas the ProtoCF models exhibited spike noise, and the outputs of LatentCF++ generally showed reduced validity. Figure 4 (b) presents a comparison between context-changing and context-preserving counterfactuals for two representative TMNIST instances, demonstrating that CACTUS could effectively control contextual features –thickness of the digits in this case– while maintaining the validity of the generated counterfactuals.

Table 2 shows the counterfactuals generated for a loan applicant under 50 years old and at risk of financial distress. The results include CACTUS counterfactuals under both context-preserving and context-changing scenarios, as well as those generated by the baseline models. CACTUS generated more actionable counterfactuals compared to the ProtoCF models. ProtoCF-C suggested an income increase of 9,959 USD, which may be unrealistic for a young applicant. In contrast, CACTUS proposed a moderate adjustment in the context-preserving

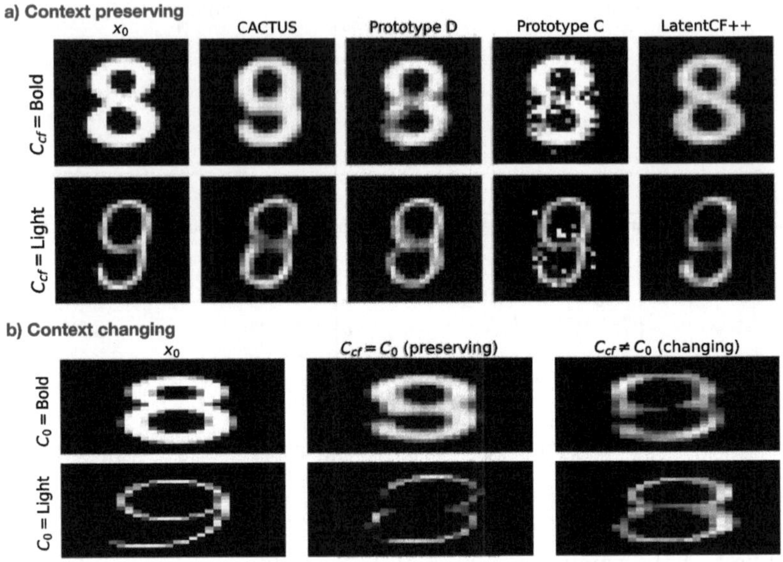

Fig. 4. CACTUS-generated counterfactuals on TMNIST. In (a), the first column shows original samples; the rest show context-preserving counterfactuals from CACTUS and the baseline. In (b), original samples appear in the first column, with context-preserving and context-changing counterfactuals in the second and third columns.

Table 2. Counterfactuals generated for the GiveMeCredit dataset under context $\mathbf{c_0}$ = Under 50, comparing two CACTUS-P and CACTUS-C scenarios (context-preserving and context-changing) with ProtoCF-D, ProtoCF-C, and LatentCF++. The red texts highlight the counterfactual changes, while the parentheses demonstrate the differences in these changes.

$\mathbf{c_0}$ = **Under 50**	x_0	**CACTUS-P**	**CACTUS-C**	**ProtoCF-D**	**ProtoCF-C**	**LatentCF++**
Revolving Utilization	0.0	0.0	0.0	0.0	0.0	0.0
# 30-59 Days Past Due	0.0	0.0	0.0	0.0	0.0	0.0
Debt Ratio	0.3	0.2 (-0.1)	0.2 (-0.1)	0.6 (+0.3)	0.0 (-0.3)	0.3
Monthly Income	4800	12070 (+7270)	13508 (+8708)	12203 (+7403)	14759 (+9959)	10009 (+5209)
# Credit Lines And Loans	8.0	7.0 (-1.0)	9.0 (+1.0)	13.0 (+5.0)	6.0 (-2.0)	9.0 (+1.0)
# 90 Days Late	0.0	0.0	0.0	0.0	0.0	0.0
# Real Estate Loans	1.0	1.0	1.0	1.0	3.0 (+2.0)	0.0 (-1.0)
# 60-89 Days Past Due	1.0	0.0 (-1.0)	0.0 (-1.0)	0.0 (-1.0)	1.0	0.0 (-1.0)
Predicted class	1.0	0.0	0.0	0.0	0.0	0.0

case (7, 270 USD) and a larger, yet plausible, increase under the context-changing scenario (8, 708 USD), consistent with the data distribution for older individuals and their associated socioeconomic patterns.

Moreover, the ProtoCF and LatentCF++ models tend to increase the number of credit lines – a feature more commonly associated with older individuals.

This causal relationship between the number of credit lines and age is precisely captured by CACTUS, which adds new credit lines only when the context is explicitly changed. This behavior is particularly useful for planning and supporting more informed decision-making processes.

6 Conclusions

We proposed CACTUS, a context-aware framework for counterfactual generation that can incorporate user-defined inputs for context-preserving and context-changing scenarios. Our experiments in binary classification with tabular and image data demonstrated that CACTUS outperformed state-of-the-art counterfactual methods in Context-LOF_{10} and in preserving initial context labels, while achieving competitive performance in validity, compactness, and proximity. An ablation study suggested a trade-off between validity and context loss, and the qualitative analysis supported our findings by showcasing the use of both context-preserving and context-changing scenarios in tabular and image prediction tasks. While this study focuses on a limited number of experimental domains, future work could involve broader settings and datasets, as well as an extended ablation study incorporating proximity coefficient γ. Additionally, our modeling approach can be extended to incorporate more objectives (e.g., class prototypes) that can improve the counterfactual learning outcomes. All our code to reproduce the experiments and the full results are publicly available at our supporting online repository: https://github.com/gsdpi/cactusCF.

Acknowledgments. This work was supported by Spanish National Research Agency (MCIN/AEI/10.13039/501100011033) under the Grant PID2020-115401GB-I00 and funded in part by the A3S Societal Innovation Project of the Digital Futures research center.

Disclosure of Interests. The authors have no competing interests to declare relevant to the content of this article.

References

1. Anwar, S., Griffiths, N., Bhalerao, A., Popham, T.J., Bell, M., Hellman, S.: CHILLI: a data context-aware perturbation method for XAI. In: ICML Workshop on Artificial Intelligence and Human-Computer Interaction, vol. 202. Honolulu, Hawaii, USA (2023)
2. Brughmans, D., Leyman, P., Martens, D.: Nice: an algorithm for nearest instance counterfactual explanations. Data Min. Knowl. Disc. **38**(5), 2665–2703 (2024). https://doi.org/10.1007/s10618-023-00930-y
3. Burgess, C.P., et al.: Understanding disentangling in β-vae. arXiv preprint arXiv:1804.03599 (2018)
4. Chai, J., Zeng, H., Li, A., Ngai, E.W.T.: Deep learning in computer vision: a critical review of emerging techniques and application scenarios. Mach. Learn. Appl. **6**, 100134 (2021). https://doi.org/10.1016/j.mlwa.2021.100134

5. Guidotti, R.: Counterfactual explanations and how to find them: literature review and benchmarking. Data Min. Knowl. Disc. (2022). https://doi.org/10.1007/s10618-022-00831-6
6. Hayashi, Y.: Emerging trends in deep learning for credit scoring: a review. Electronics **11**(19), 3181 (2022)
7. Ismail Fawaz, H., Forestier, G., Weber, J., Idoumghar, L., Muller, P.-A.: Deep learning for time series classification: a review. Data Min. Knowl. Disc. **33**(4), 917–963 (2019). https://doi.org/10.1007/s10618-019-00619-1
8. Jiang, N., Liu, X., Liu, H., Lim, E.T.K., Tan, C.W., Gu, J.: Beyond ai-powered context-aware services: the role of human-ai collaboration. Ind. Manag. Data Syst. **123**(11), 2771–2802 (2022)
9. Kanamori, K., Takagi, T., Kobayashi, K., Arimura, H.: DACE: distribution-aware counterfactual explanation by mixed-integer linear optimization. In: Proceedings of the Twenty-Ninth International Joint Conference on Artificial Intelligence, pp. 2855–2862. Yokohama, Japan (July 2020)
10. Karimi, A.H., Schölkopf, B., Valera, I.: Algorithmic recourse: from counterfactual explanations to interventions. In: Proceedings of the 2021 ACM Conference on Fairness, Accountability, and Transparency, pp. 353–362. ACM (2021). https://doi.org/10.1145/3442188.3445899
11. Kenny, E.M., Keane, M.T.: On generating plausible counterfactual and semi-factual explanations for deep learning. In: Proceedings of the AAAI Conference on Artificial Intelligence, vol. 35, no. 13, pp. 11575–11585 (2021)
12. Laugel, T., Lesot, M.J., Marsala, C., Detyniecki, M.: Issues with post-hoc counterfactual explanations: a discussion. In: ICML Workshop on Human in the Loop Learning (HILL 2019) (2019)
13. Lundberg, S.M., Lee, S.I.: A unified approach to interpreting model predictions. In: Advances in Neural Information Processing Systems, vol. 30. Curran Associates, Inc. (2017)
14. Magre, N., Brown, N.: Typography-MNIST (TMNIST): An MNIST-Style Image Dataset to Categorize Glyphs and Font-Styles (2022). http://arxiv.org/abs/2202.08112, arXiv:2202.08112 [cs]
15. Mahajan, D., Tan, C., Sharma, A.: Preserving causal constraints in counterfactual explanations for machine learning classifiers. CausalML Workshop, NeurIPS 2019 (2019)
16. Messmer, L.M., Reich, C., Abdeslam, D.O.: Context-aware machine learning: a survey. In: Arai, K. (ed.) Proceedings of the Future Technologies Conference (FTC) 2024, Volume 1. FTC 2024. LNNS, vol. 1154, pp. 252–272. Springer, Cham (2024). https://doi.org/10.1007/978-3-031-73110-5_17
17. Molnar, C.: Interpretable Machine Learning, 3 edn. (2025). https://christophm.github.io/interpretable-ml-book
18. Pawelczyk, M., Broelemann, K., Kasneci, G.: Learning model-agnostic counterfactual explanations for tabular data. In: Proceedings of the Web Conference 2020, pp. 3126–3132. New York, NY, USA (April 2020)
19. Looveren, A., Klaise, J.: Interpretable counterfactual explanations guided by prototypes. In: Oliver, N., Pérez-Cruz, F., Kramer, S., Read, J., Lozano, J.A. (eds.) ECML PKDD 2021. LNCS (LNAI), vol. 12976, pp. 650–665. Springer, Cham (2021). https://doi.org/10.1007/978-3-030-86520-7_40
20. Vellaichamy, S., Sharma, S., Albini, E., Brugere, I., Onigbanjo, A.: Context-aware SHAP for responsible feature selection. In: ACM ICAIF Workshop on Explainable AI in Finance (XAIFIN) (2024)

21. Verma, S., Boonsanong, V., Hoang, M., Hines, K., Dickerson, J., Shah, C.: Counterfactual explanations and algorithmic recourses for machine learning: a review. ACM Comput. Surv. **56**(12), 312:1–312:42 (2024)
22. Wachter, S., Mittelstadt, B., Russell, C.: Counterfactual Explanations Without Opening the Black Box: Automated Decisions and the GDPR. Technical report, Social Science Research Network (2017)
23. Wang, J., Bilgic, M.: Context-aware feature selection and classification. In: Proceedings of the Thirty-Second International Joint Conference on Artificial Intelligence (IJCAI-23), vol. 4, pp. 4317–4325 (August 2023). https://doi.org/10.24963/ijcai.2023/480, iSSN: 1045-0823
24. Wang, Z., Samsten, I., Mochaourab, R., Papapetrou, P.: Learning time series counterfactuals via latent space representations. In: Discovery Science, pp. 369–384 (2021). https://doi.org/10.1007/978-3-030-88942-5_29

Hubris Benchmarking with AmbiGANs: Assessing Model Overconfidence with Synthetic Ambiguous Data

Cátia Teixeira[1(✉)], Inês Gomes[1,3], Carlos Soares[1,2], and Jan N. van Rijn[3]

[1] Artificial Intelligence and Computer Science Laboratory, Faculty of Engineering, University of Porto, Porto, Portugal
up200808037@up.pt, catia.rds.teixeira@gmail.com, {ines.gomes,csoares}@fe.up.pt

[2] Fraunhofer Portugal AICOS, Porto, Portugal

[3] Leiden Institute of Advanced Computer Science, Leiden University, Leiden, The Netherlands
j.n.van.rijn@liacs.leidenuniv.nl

Abstract. The growing deployment of artificial intelligence in critical domains exposes a pressing challenge: how reliably models make predictions for ambiguous data without exhibiting overconfidence. We introduce hubris benchmarking, a methodology to evaluate overconfidence in machine learning models. The benchmark is based on a novel architecture, ambiguous generative adversarial networks (AmbiGANs), which are trained to synthesize realistic yet ambiguous datasets. We also propose the hubris metric to quantitatively measure the extent of model overconfidence when faced with these ambiguous images. We illustrate the usage of the methodology by estimating the hubris of state-of-the-art pre-trained models (ConvNext and ViT) on binarized versions of public datasets, including MNIST, Fashion-MNIST, and Pneumonia Chest X-ray. We found that, while ConvNext is on average 3% more accurate than ViT, it often makes excessively confident predictions, on average by 10% points higher than ViT. These results illustrate the usefulness of hubris benchmarking in high-stakes decision processes.

Keywords: Synthetic Data Generation · Overconfidence · Generative Adversarial Networks · Responsible Artificial Intelligence · Computer Vision

1 Introduction

Recently, the European Union High-Level Expert Group on Artificial Intelligence (AI) stated that trustworthy AI systems must be lawful, ethical, and robust [12]. With the growing use of machine learning and AI, responsible AI practices are essential for transparency and accountability [5]. Model evaluation

S. Džeroski et al. (Eds.): DS 2025, LNAI 16090, pp. 476–491, 2025.
https://doi.org/10.1007/978-3-032-05461-6_31

typically focuses on predictive performance metrics such as accuracy, precision, or recall. While high confidence in predictions is generally desirable, it is important to recognize when models may be unreliable due to inherent data uncertainty. In these cases, overconfident models can mislead decision-makers, posing risks in high-stakes applications. This issue is critical in fields such as medicine, where AI supports the medical expert in distinguishing ambiguous cases, such as a skin lesion or lung scan shadow, that could indicate either benign or malignant outcomes. Typical binary classification models produce output scores that are converted into predictions. In a well-calibrated model, scores should be near 0 or 1 for clear cases, while ambiguous instances should have scores around 0.5, reflecting an equal likelihood of belonging to either class.

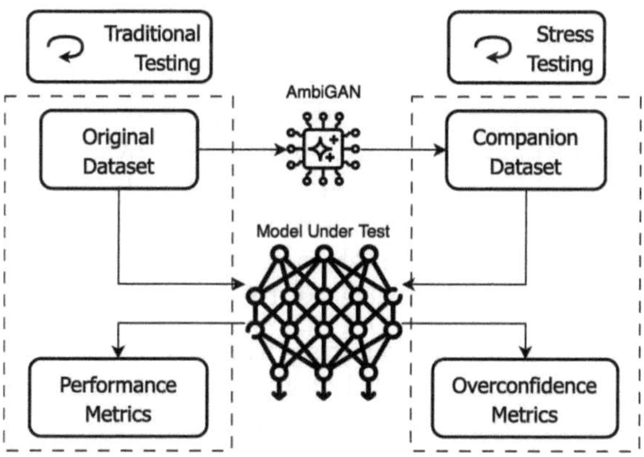

Fig. 1. Schematic view of how the hubris benchmarking methodology complements model testing: on the left side, we have the traditional testing with the original dataset and standard metrics; on the right, we have the stress testing with the companion dataset and overconfidence metrics.

This research proposes *hubris benchmarking*, a method to evaluate the overconfidence of a given classifier when facing ambiguous data. Given that ambiguous data may be scarce in the training set, the first step of our method is to generate a dataset of synthetic, ambiguous and realistic examples. Therefore, we propose a novel generative adversarial network (GAN) framework called ambiguous generative adversarial networks (*AmbiGAN*) to generate such images. To quantify the overconfidence, we propose a new metric called *hubris*, which measures the distance between the expected classifier output (ambiguity) and the true predictions. The metric proposed here is based on KullbackâĂŞLeibler (KL) divergence, but other similarity measures could be used as well. Hubris ranges from 0 to 1, where 0 indicates that the model perfectly distinguishes certain cases from uncertain ones. Therefore, the higher the hubris, the higher the model's overconfidence.

The proposed AmbiGAN is based on generative adversarial stress test networks (GASTeN) [9,27]—a GAN-based framework that generates synthetic realistic images close to the decision boundary of a given classifier. With AmbiGANs, we provide a generative framework that generates a universally ambiguous dataset, referred to as *companion dataset.* We consider a dataset universally ambiguous when different classifiers agree that a given data point has low confidence, as opposed to situations where we stress test models on samples generated specifically for that model [14]. This new companion dataset can test the hubris of machine learning models. We demonstrate the utility of our approach by creating companion datasets for subsets of three image classification benchmarks: MNIST [20], Fashion-MNIST [31], and Pneumonia Chest X-ray datasets [19]. We then evaluate two public, pre-trained models, ConvNext [22] and a Visual Transformer (ViT) [10]. Our results show that these models are overconfident when tested on our AmbiGAN-generated dataset, with hubris values close to 1.00. Most importantly, the results show how hubris benchmarking can provide an alternative perspective on evaluating models: while ConvNext is generally more accurate than ViT, it is also more overconfident, which could make it less interesting in some scenarios (e.g., medical diagnosis). These findings demonstrate how hubris benchmarking can be used to uncover model vulnerabilities by exposing its overconfidence in ambiguous scenarios. By providing a structured method to measure such vulnerabilities, this approach aids in building AI systems that are more transparent, reliable, and safer for deployment in high-stakes fields. The contributions in this article are the following:

1. the hubris benchmarking methodology, which can be applied to any binary image classification problem. It includes a novel hubris metric to evaluate the overconfidence of a model in ambiguous images;
2. the AmbiGAN architecture, which generates companion ambiguous datasets for binary classification purposes without the need for a human-in-the-loop;
3. an illustrative application of our methodology on three datasets, used to evaluate two pre-trained models: ConvNext [22] and ViT [10];
4. publicly available companion datasets for subsets of Pneumonia Chest X-ray[1], MNIST[2], and Fashion-MNIST[3];
5. AmbiGAN source code available for reproducibility and future usage[4].

2 Related Work

We review generative adversarial networks, focusing on those designed for stress testing, and discuss metrics for evaluating synthetic and ambiguous images.

2.1 Generative Adversarial Networks

Generative adversarial networks (GANs), introduced by Goodfellow *et al.* [15], model data distributions through a generator (G) and a discriminator (D) in

[1] https://huggingface.co/datasets/crdsteixeira/AmbiGAN-XRay.
[2] https://huggingface.co/datasets/crdsteixeira/AmbiGAN-MNIST.
[3] https://huggingface.co/datasets/crdsteixeira/AmbiGAN-Fashion.
[4] https://github.com/crdsteixeira/Hubris-AmbiGANs.

an adversarial setup. G produces synthetic samples to mimic real data, while D distinguishes real from generated samples [28]. Both are typically multilayer networks with convolutional and fully connected layers [8]. GANs aim for a Nash equilibrium where G generates data indistinguishable from the true distribution. This is formalized as a minimax optimization in Eq. 1, with D maximizing the value function $V(D, G)$ by classifying real samples $x \sim p_{\text{data}}$ and fake ones $G(z)$ with $z \sim p_z$, while G minimizes it by generating complex instances for D.

$$\begin{aligned}\mathscr{L}_{D,G}^{GAN} &= \min_G \max_D V(D, G) \\ &= \mathbb{E}_{x \sim p_{\text{data}}(x)}[\log D(x)] + \mathbb{E}_{z \sim p_z(z)}[\log(1 - D(G(z)))]\end{aligned} \tag{1}$$

Unlike models such as Boltzmann machines [1], GANs avoid Markov chains, improving efficiency. Still, issues remain with diversity, image quality, and training stability [29]. Variants such as WGAN [3], DCGAN [25], and InfoGAN [7] address these through improved architectures, adversarial [24], task-specific [11], and multi-objective loss functions [2].

2.2 Generative Adversarial Stress Test Networks

GASTeN [9] is a method that uses GANs to stress test a given model by generating ambiguous yet realistic inputs. It uses a DCGAN [25] to synthesize data near the classifier's decision boundary in binary classification, aiming to evaluate the performance and reliability of a model under ambiguous conditions. GASTeN follows a two-step training: first, pre-training a DCGAN; then, refining the generator using classifier (C) feedback through a modified loss, shown in Eq. 2. The equation includes a confusion distance term (*cd*), scaled by α, to measure the distance from the prediction to the decision threshold. Average confusion distance (ACD) quantifies the confusion distance (*cd*) of all the generated images.

$$\mathscr{L}_G^{GASTeN} = \mathscr{L}_G^{GAN} + \alpha \cdot cd(C(G(z))) \tag{2}$$

However, GASTeN often faced trade-offs between quality (measured by the Fréchet inception distance (FID)) and low confidence (measured by ACD) with instability and mode collapse during training. GASTeNv2 [27] addressed these by restoring the original DCGAN architecture, replacing earlier linear-layer modifications, and introducing a new loss. The updated loss combines GAN loss with a Gaussian negative log likelihood (GNLL) term, shown in Eq. 3. This term guides the generator to produce samples near the decision boundary (0.5) under the assumption of classifier calibration. The variance (σ^2) is treated as a hyperparameter, adjusting tolerance around the boundary. The updated loss improved both training stability and the ability to generate ambiguous examples.

$$\begin{aligned}\mathcal{L}_G^{GASTeNv2} &= \mathcal{L}_G^{GAN} + \alpha \cdot \mathcal{L}_{GNLL} \\ &= \mathcal{L}_G^{GAN} + \alpha \cdot \left(\frac{1}{2}\log(\sigma^2) + \frac{(C(G(z)) - 0.5)^2}{2\sigma^2}\right)\end{aligned} \tag{3}$$

2.3 Metrics for Synthetic Images

Synthetic image evaluation commonly uses FID [17] to measure image quality and realism. FID quantifies the similarity between real and generated data by comparing Gaussian feature statistics, capturing both realism and diversity [6,17]. An alternative approach is GIQA [16], which evaluates realism using two metrics: a quality score (QS), which measures perceptual fidelity, and a diversity score (DS), which measures sample variety. A higher quality score implies better visual quality; a higher diversity score indicates more diverse outputs. GIQA has been applied in generative frameworks for tasks such as image editing, segmentation, and synthesis [13,26,32], supporting more nuanced evaluation of image generation performance.

2.4 Ambiguous Images in Classification

Machine learning models rely on high-quality data, yet real-world scenarios often present ambiguous inputs, which are underrepresented in existing datasets [4,30]. Weiss *et al.* define true ambiguity as occurring when a single input yields nonzero probabilities for multiple classes: "x is truly ambiguous if and only if $P(c|x) > 0$ for more than one class c" [30]. Unlike model-specific ambiguity, true ambiguity is inherent to the data and independent of a classifier's decision boundary [21,30]. This aligns with aleatoric uncertainty, which arises from intrinsic data variability and cannot be reduced even with improved models [18,30]. In contrast, epistemic uncertainty stems from limited model knowledge and can be reduced through better training or data [18].

Various approaches explored the generation of ambiguous samples. GASTeN (Sect. 2.2) synthesizes ambiguity based on a model's decision boundary, linking it to model-specific uncertainty. In contrast, AMBIGUESS [30] constructs a model-agnostic dataset, ensuring that ambiguity arises from the input itself rather than a classifier. Similarly, AmbiguousMNIST [23] extends the standard MNIST dataset to include samples with multiple plausible labels, reflecting real-world ambiguity.

3 Hubris Benchmarking

Ensuring that AI systems meet the standards of transparency requires tackling a critical blind spot in traditional evaluation: how models handle ambiguity. Figure 2 shows a case of distinguishing between the handwritten digits "8" and "9". These images are inherently ambiguous, even for human observers, yet conventional empirical evaluation fails to assess how models respond to these uncertainties.

A typical binary classification model generates output scores that are subsequently converted into predictions. Ideally, these scores should be near the extremes (e.g., close to 0 or 1) when the label is clear, while for ambiguous cases, the scores should remain closer to 0.5, assuming proper model calibration.

Fig. 2. Ambiguous digits (with features from 9 and 8) generated with AmbiGAN.

At the same time, some models show high certainty for these ambiguous images. This is what is defined as model overconfidence. Without tools to measure and mitigate overconfidence in these scenarios, AI systems risk making unwarranted, high-certainty predictions.

3.1 Companion Datasets with Ambiguous Data

As illustrated in Fig. 1, model training and evaluation typically rely on standard datasets and metrics such as accuracy, precision, or recall. However, these metrics fail to capture model behavior in ambiguous scenarios. To address this, we generate synthetic, ambiguous images using AmbiGANs (see Sect. 4), forming a *companion dataset* that systematically challenges the model under test. This dataset enables the analysis of model predictions in ambiguous cases, revealing overconfidence not detected by traditional metrics. To quantify this, we introduce the *hubris* metric (Sect. 3.2), which measures the extent of overconfidence. These tools offer a complementary evaluation framework focused on model reliability in uncertain conditions.

3.2 Hubris Metric

We propose *hubris* as a metric to quantify model overconfidence in ambiguous samples. Hubris measures the distance between a model's prediction distribution, $\hat{y}$, and an ideal, unbiased reference prediction of 0.50. In binary classification, a balanced prediction of 0.50 ideally indicates no overconfidence since, in the case of an unbiased and calibrated model, it captures characteristics equally from both classes. In our approach, hubris leverages KL-Divergence to measure these distribution distances. *KL-Hubris* is defined in Eq. 4, with $\hat{y}$ being the prediction of the model under test. In KL-Hubris, *Ref* represents a Dirac distribution centered around 0.50, meaning any sampled value from *Ref* will always be exactly 0.50. U is a uniform distribution representing random predictions from 0 to 1. The result is scaled from 0.00 (no overconfidence) to 1.00 (total overconfidence), where lower KL-Hubris values indicate that the model treats ambiguous samples with predictions close to the ideal decision boundary.

$$\textit{KL-Hubris} = 1 - e^{-\left(\frac{KL(\hat{y}|Ref)}{KL(U|Ref)}\right)} \tag{4}$$

We define absolute hubris (H_A) as the measure of overconfidence relative to the ideal ambiguous output distribution centered at 0.50. Hubris can also measure overconfidence relative to any ambiguity estimation methodology, enabling comparisons of probability distributions across models. When used in this context, we refer to it as relative hubris (H_R), which will be used in Sect. 6.3.

4 Ambiguous Generative Adversarial Networks

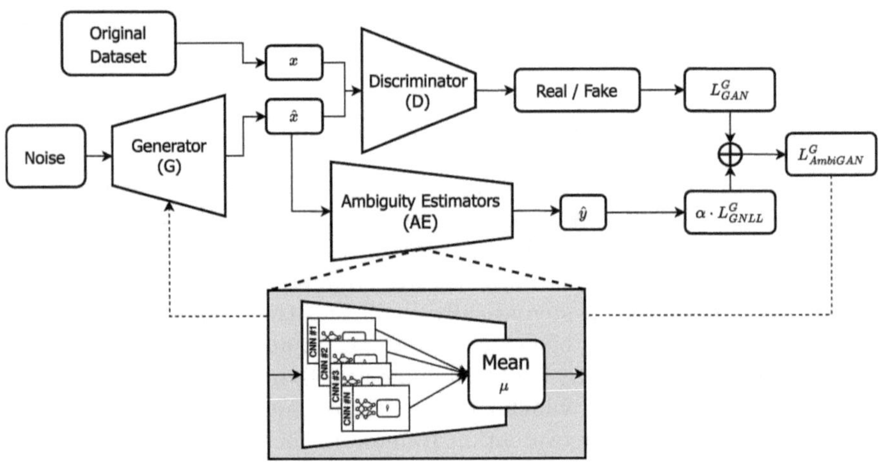

Fig. 3. AmbiGANs Architecture: a GAN-based architecture incorporating the ambiguity estimator (AE); AE averages the individual classifier's predictions; Generator loss combines both original GAN loss and Gaussian loss.

We propose hubris benchmarking, which assesses the overconfidence of models in datasets that contain only ambiguous data. However, these are limited in the standard datasets used to evaluate models in computer vision and other learning tasks. We address this by introducing ambiguous generative adversarial networks (*AmbiGANs*) to generate universally ambiguous synthetic samples.

We define *Universal Ambiguity* as equally challenging samples for machine learning models, i.e., where models predict equal scores for all labels (e.g., 0.5, assuming binary classification tasks). We approximate universal ambiguity by leveraging diversity among classifiers. We generate a set of diverse classifiers (by varying the architecture and training data). Samples that are ambiguous for all those classifiers are considered universally ambiguous. Hence, we implement this concept in AmbiGANs by integrating multiple diverse classifiers acting as the *ambiguity estimator* (AE), where the mean of the individual classifier's predictions is the final output prediction of the ambiguity estimator.

Figure 3 presents a summary of the proposed architecture. We maintain the approach of GASTeN (Sect. 2.2) regarding the multi-objective loss function, which minimizes both the standard GAN loss and Gaussian loss, to guide the generation to ambiguity. AmbiGANs balance realism and ambiguity, training the generator to produce realistic images that are universally ambiguous, according to the definition above. The ambiguity estimator output prediction is necessary to calculate the Gaussian loss of the generated images. The final loss propagated to the generator is the sum of the original GAN loss and Gaussian loss with a weighting factor (α), shown in Eq. 6.

$$\mathcal{L}_G^{AmbiGAN} = \mathcal{L}_G^{GAN} + \alpha \cdot \mathcal{L}_{GNLL} \tag{5}$$

$$= \mathcal{L}_G^{GAN} + \alpha \cdot \left(\frac{1}{2}\log(\sigma^2) + \frac{(AE(G(z)) - 0.5)^2}{2\sigma^2}\right) \tag{6}$$

5 Experimental Setup

To illustrate hubris benchmarking, we describe the AmbiGAN training process and companion dataset generation, followed by the evaluation of two pre-trained models. All experiments were conducted using a machine equipped with one Tesla T4 GPU.

5.1 AmbiGAN Training

Following the standard GASTeN training strategy (Sect. 2.2), AmbiGAN training is split into two stages. In the first stage, the generator and discriminator are pre-trained over 100 epochs for stability. In the second stage, over 50 epochs, the ambiguity estimator is introduced into the loss (Eq. 6, Sect. 4), influencing generator updates based on its predictions. The ambiguity estimator, frozen during the second stage, is built from 50 classifiers trained on the original dataset. To generalize ambiguity, classifiers vary in architecture (layer counts, feature sizes, and initializations) and are trained on non-overlapping data subsets. Final outputs use sigmoid activation to approximate universal ambiguity.

AmbiGAN architectures differ by dataset: for MNIST and Fashion-MNIST, the generator and discriminator both use 3 layers, with 512 features in their final and initial layers, respectively, and a latent space of 256. Due to larger images, a deeper 5-layer architecture is used for the Chest X-ray dataset, with 128 features and a latent space dimension of 512. Learning rates follow the ones proposed for DCGAN [25], set to 0.0002 for MNIST and Fashion-MNIST, and 0.001 for Chest X-ray. All models use Adam optimizer ($\beta_1 = 0.000$, $\beta_2 = 0.999$), and class balancing is enforced via random over-sampling. We tested α values of 0.5, 1.0, and 1.5 to control the trade-off between realism and ambiguity, and σ^2 values of 0.01 and 0.1 to set tolerance around the decision boundary. A lower variance leads to closer predictions of 0.5, while a higher one allows more variance. Three runs were conducted for each dataset, combining all α and σ^2 values, yielding six trained models per run. Hyperparameter selection relied on GASTeN metrics: classification accuracy, FID (image realism), and ACD (ambiguity). While lower values for both FID and ACD are ideal, FID is prioritized to ensure realism, provided ACD remains acceptable. Ambiguity estimator accuracy and ACD are calculated by averaging predictions across classifiers.

5.2 Companion Datasets

We generated companion datasets for MNIST [20], Fashion-MNIST [31], and Chest X-ray [19]. As in GASTeN, AmbiGANs are limited to binary classification [9], so we curated binary subsets: *9 vs. 8* in MNIST, and *Dress vs. T-Shirt* in Fashion-MNIST, selecting visually ambiguous class pairs for easier inspection. Each dataset subset includes grayscale 28×28 images, allowing for compact AmbiGAN architectures. The Chest X-ray dataset originally consisted of 640×640 RGB images labeled as Pneumonia or Normal. We down-scaled them to 128×128 pixels to match our model constraints. Unlike the other datasets, Chest X-ray is already binary but class-imbalanced.

Each companion dataset contains 5 000 generated samples, produced in three runs with different random seeds per dataset, totaling nine companion datasets. We made all companion datasets publicly available on HuggingFace.

To evaluate generation quality, we used FID, quality score, and diversity score, alongside visual inspection for realism, ambiguity, and GAN collapse. We compared each companion dataset against a baseline generated during the initial GAN training stage (before introducing the ambiguity estimator), which lacks deliberate ambiguity and serves as a reference. We also compared image quality with AmbiguousMNIST [23] (see Sect. 2.4) to benchmark against state-of-the-art ambiguity generation.

5.3 Evaluated Models

We used the generated companion datasets for hubris benchmarking of two pre-trained models, ConvNext [22] and ViT [10], originally designed for multi-class tasks. We fine-tuned the models for binary classification by modifying the final layer to output a single value, using a batch size of 64, a learning rate of 0.001, and the Adam optimizer ($\beta_1 = 0.9$, $\beta_2 = 0.999$). We first evaluated model accuracy on the original datasets. Then, using the hubris and ACD metrics, we assessed overconfidence on each companion dataset and on AmbiguousMNIST [23]. ACD was computed based on the model's output predictions.

We calculated absolute hubris (H_A), which compares model outputs to an ideal ambiguous distribution, and relative hubris (H_R), which compares outputs to an ambiguity estimator not used during AmbiGAN training. H_R was not computed for AmbiguousMNIST, as its ambiguity differs in nature. While our subsets focus on the ambiguity between specific pairs, AmbigousMNIST combines features from multiple classes across the dataset. For a fair comparison, we filtered AmbiguousMNIST to include only samples labeled with our subset.

6 Results and Discussion

Our experimental workflow consists of three phases: tuning AmbiGAN hyperparameters for each dataset, generating companion datasets, and evaluating pre-trained models.

6.1 AmbiGAN Training

In Table 1, we present the results with lower FID obtained during all runs for all datasets. The weight and variance for Gaussian loss used in the best run are also presented for each one.

Table 1. AmbiGAN results (for 2048 generated samples, averaged across 3 runs).

Dataset	Loss Params		ACD	FID
	Weight	Var		
MNIST (9 vs. 8)	0.50	0.10	0.08 ± 0.01	18.83 ± 4.23
Fashion-MNIST (Dress vs. T-Shirt)	1.50	0.10	0.06 ± 0.00	37.37 ± 0.77
Chest X-ray (Pneumonia vs. Normal)	1.50	0.10	0.08 ± 0.01	20.21 ± 2.41

The results highlight low ACD values ($\leq$0.08), indicating effective confusion generation, and FID scores ranging from 18.83 to 37.37, reflecting a balance between realism and ambiguity in the generated samples. These results demonstrate the adaptability of AmbiGAN architecture to diverse datasets while maintaining quality and ambiguity. ACD and FID had no significant deviation in all runs, and we saw no GAN collapse in any of the experiments performed.

6.2 Companion Datasets

After selecting the AmbiGAN variant with optimal hyperparameters for each dataset specified in Table 1, we generate the complete companion datasets for each of the three original datasets. Table 2 presents the image quality evaluation metrics for the companion datasets. The baseline metrics obtained during the AmbiGAN pre-training phase (before incorporating the ambiguity estimator) are also included for comparison. With this reference, we can analyze the compromise between realism and ambiguity. We aim to keep realism but also expect to improve ambiguity significantly.

While FID scores are higher in the companion datasets than the baseline, quality (QS) and diversity (DS) metrics remain consistent or improve slightly. This balance demonstrates AmbiGAN's ability to generate visually realistic and diverse samples while introducing ambiguity across datasets such as MNIST, Fashion-MNIST, and Chest X-ray. The increase in FID reflects the added variability introduced by the ambiguity estimator loss factor. This fact is caused

Table 2. Baseline and companion datasets metrics (for 5,000 generated samples) including FID, QS, and DS (Sect. 2.3), in comparison to the state-of-the-art dataset.

Companion Dataset	Baseline			Companion		
	FID	QS	DS	FID	QS	DS
Ambiguous-MNIST [23] (9 vs. 8)	–	–	–	74.40	0.83	0.80
AmbiGAN-MNIST (9 vs. 8)	3.19 ±0.06	0.82 ±0.03	0.78 ±0.03	17.67 ±0.18	0.83 ±0.01	0.79 ±0.00
AmbiGAN-Fashion (Dress vs. T-Shirt)	11.77 ±0.16	0.74 ±0.01	0.83 ±0.02	33.96 ±0.40	0.80 ±0.02	0.79 ±0.02
AmbiGAN-XRay (Pneumonia vs. Normal)	17.32 ±0.26	0.79 ±0.02	0.87 ±0.00	18.64 ±0.25	0.83 ±0.01	0.83 ±0.01

by the introduced variability in the new samples that reside within the decision boundary of the estimator. We also note that for AmbiGAN-Fashion and AmbiGAN-XRay, the DS decreases for the companion dataset when compared with the baseline. On the other hand, the QS increases for these companion datasets. This indicates that our companion ambiguous dataset has less diversity than the original dataset. For AmbiGAN-MNIST, both QS and DS increase, providing confidence that the generated samples are realistic, which we confirmed by manual inspection. However, in all cases, the differences are small, which indicates that the promotion of ambiguity does not reduce the quality of the generated data. Examples of generated images are presented in Fig. 4. Especially in those of AmbiGAN-MNIST and AmbiGAN-Fashion, which are images of common concepts, it is clear that they are realistic and ambiguous.

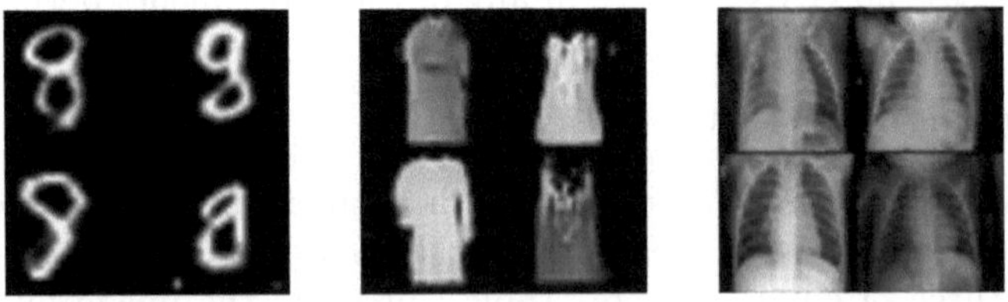

Fig. 4. Examples of generated companion datasets: in the left, AmbiGAN-MNIST (9 vs. 8); in the middle, AmbiGAN-Fashion (Dress vs. T-Shirt); in the right, AmbiGAN-XRay (Pneumonia vs. Normal).

When compared to AmbiguousMNIST [23] dataset, the AmbiGAN-MNIST companion dataset demonstrates significantly lower FID scores (17.67 vs. 74.40), indicating improved realism in the generated samples. Additionally, the quality (QS) and diversity (DS) are comparable (0.83 vs 0.83 and 0.79 vs. 0.80), suggesting that AmbiGAN maintains competitive levels of quality and diversity. A QS value around 0.8 indicates that the generated samples retain high visual realism, while a DS value near 0.8 reflects sufficient diversity to ensure a broad range of variability.

6.3 Hubris of ConvNext and ViT

We applied hubris benchmarking to pre-trained ConvNext and ViT models using the AmbiGAN-generated companion datasets and AmbiguousMNIST [23]. Results are summarized in Table 3.

Table 3. Pre-trained models evaluation with companion datasets. Accuracy of model with the original dataset is also presented.

Model	Companion Dataset	Accuracy	ACD	H_A^a	H_R^b
ConvNext [22]	Ambiguous-MNIST [23] (9 vs. 8)	99.95%	0.49 ±0.00	0.97 ±0.00	–
	AmbiGAN-MNIST (9 vs. 8)		0.48 ±0.00	0.97 ±0.00	0.91 ±0.00
	AmbiGAN-Fashion (Dress vs. T-Shirt)	98.30%	0.48 ±0.00	0.96 ±0.00	0.95 ±0.00
	AmbiGAN-XRay (Pneumonia vs. Normal)	95.38%	0.49 ±0.00	0.97 ±0.00	0.94 ±0.00
ViT [10]	Ambiguous-MNIST [23] (9 vs. 8)	99.07%	0.41 ±0.03	0.92 ±0.02	–
	AmbiGAN-MNIST (9 vs. 8)		0.39 ±0.02	0.89 ±0.03	0.79 ±0.03
	AmbiGAN-Fashion (Dress vs. T-Shirt)	93.97%	0.33 ±0.03	0.80 ±0.05	0.77 ±0.06
	AmbiGAN-XRay (Pneumonia vs. Normal)	90.63%	0.41 ±0.00	0.92 ±0.00	0.86 ±0.00

[a] Absolute hubris: KL-Hubris (ref = 0.50)
[b] Relative hubris: KL-Hubris (ref = Ambiguity estimator prediction)

Both models achieved high accuracy (>90%), with ConvNext outperforming ViT on MNIST (99.95%) and Chest X-ray (95.38%), confirming its superior predictive performance. However, hubris benchmarking reveals that accuracy alone can be misleading in ambiguous scenarios. ConvNext consistently showed high absolute hubris ($H_A \approx 1.00$) across all AmbiGAN datasets, indicating strong overconfidence. ViT had lower H_A on simpler datasets (0.89 on AmbiGAN-MNIST and 0.80 on AmbiGAN-Fashion) but approached ConvNext's hubris (0.92) on the more complex AmbiGAN-XRay. ACD values support this trend: ConvNext had values close to 0.50, while ViT achieved lower values (close to 0.36) on simpler datasets, reflecting less confident predictions. Relative hubris (H_R) further highlights overconfidence, particularly in ConvNext. Both models deviated significantly from the ambiguity estimator's more balanced outputs. This suggests that deeper architectures such as ConvNext may amplify confidence by reinforcing features across layers, leading to overconfident predictions in ambiguous cases. AmbiguousMNIST results mirror these findings. ConvNext reached $H_A = 0.97$, while ViT was slightly lower at 0.92, showing reduced overconfidence. These results highlight the limitations of relying solely on accuracy. While ConvNext appears superior by standard metrics, ViT's lower hubris scores suggest it may be more reliable in ambiguity-sensitive contexts. Companion datasets generated by AmbiGANs provide a valuable tool for revealing these differences.

6.4 Limitations

We identify some limitations that may hinder the conclusions drawn from this study. The image metrics used focus on assessing realism and quality but do not explicitly account for ambiguity, as they do not consider cases where a single image exhibits features from multiple realistic samples. In companion datasets, such as AmbiGAN-XRay, human labeling could further validate the realism and ambiguity of the generated images. Our approach to universal ambiguity assumes the ambiguity estimator represents all binary classifiers, which may be an oversimplification. Since models generalize differently, our estimator may not fully capture all their decision boundaries.

7 Conclusions

We introduce hubris benchmarking, a methodology to evaluate model overconfidence in machine learning models when faced with ambiguous data. The framework provides a new metric, hubris, to quantify how confidently models make predictions on ambiguous samples. This is relevant for applications such as medical diagnosis and autonomous driving, where overconfidence can be harmful.

Since ambiguous data is rarely available, we introduce AmbiGANs to generate realistic, ambiguous companion datasets. To enable systematic overconfidence evaluation, we generated companion datasets to MNIST, Fashion-MNIST, and Chest X-ray. Then, we benchmarked ConvNext and ViT models, finding that while ConvNext achieves 3% higher accuracy on average, it is also 10% points more overconfident than ViT.

Our results confirm that AmbiGANs effectively support hubris benchmarking. The provided codebase generalizes to other datasets for binary image classification, positioning hubris benchmarking as a practical standard for evaluating overconfidence in ambiguous settings.

Acknowledgments. This work was partially funded by projects AISym4Med and CRAI and LIACC. European Union under the Horizon Europe Framework Programme Grant Agreement No: 101095387; Agenda "Center for Responsible AI", nr. C645008882-00000055, investment project nr. 62, financed by the Recovery and Resilience Plan (PRR) and by European Union—NextGeneration EU. Funded by the European Union—NextGenerationEU; UID/00027 of the LIACC—Artificial Intelligence and Computer Science Laboratory—funded by Fundação para a Ciência e a Tecnologia, I.P./MCTES through the national funds. Views and opinions expressed are those of the author(s) only and do not necessarily reflect those of the European Union or the European Commission. Neither the European Union nor the European Commission can be held responsible for them. The computational resources of the Google Cloud Platform were provided by the project CPCA-IAC/AF/594904/2023.

Disclosure of Interests. The authors have no competing interests to declare that are relevant to the content of this article.

References

1. Ackley, D.H., Hinton, G.E., Sejnowski, T.J.: A learning algorithm for Boltzmann machines. Cogn. Sci. **9**(1), 147–169 (1985)
2. Albuquerque, I., Monteiro, J., Doan, T., Considine, B., Falk, T.H., Mitliagkas, I.: Multi-objective training of generative adversarial networks with multiple discriminators. In: Proceedings of the 36th International Conference on Machine Learning, pp. 202–211 (2019)
3. Arjovsky, M., Chintala, S., Bottou, L.: Wasserstein generative adversarial networks. In: Proceedings of the 34th International Conference on Machine Learning, pp. 214–223 (2017)
4. Aroyo, L., Paritosh, P.: Uncovering unknown unknowns in machine learning (2021). https://ai.googleblog.com/2021/02/uncovering-unknown-unknowns-in-machine.html, online; Google AI Blog
5. Arrieta, A.B., et al.: Explainable artificial intelligence (xai): concepts, taxonomies, opportunities and challenges toward responsible ai. Information Fusion **58**, 82–115 (2020)
6. Baraheem, S.S., Le, T., Nguyen, T.V.: Image synthesis: a review of methods, datasets, evaluation metrics, and future outlook. Artif. Intell. Rev. **56**, 10813–10865 (2023)
7. Chen, X., Duan, Y., Houthooft, R., Schulman, J., Sutskever, I., Abbeel, P.: Infogan: interpretable representation learning by information maximizing generative adversarial nets. Adv. Neural Inf. Process. Syst. **29**, 2172–2180 (2016)
8. Creswell, A., White, T., Dumoulin, V., Arulkumaran, K., Sengupta, B., Bharath, A.A.: Generative adversarial networks: an overview. IEEE Signal Process. Mag. **35**, 53–65 (2018)
9. Cunha, L., Soares, C., Restivo, A., Teixeira, L.F.: Gasten: generative adversarial stress test networks. In: Advances in Intelligent Data Analysis XXI — 21st International Symposium on Intelligent Data Analysis, pp. 91–102 (2023)
10. Dosovitskiy, A., et al.: An image is worth 16 × 16 words: transformers for image recognition at scale. CoRR (2021)
11. Dunn, I., Pouget, H., Melham, T.F., Kroening, D.: Adaptive generation of unrestricted adversarial inputs. CoRR (2019)

12. European Commission: Ethics guidelines for trustworthy AI (2019). https://digital-strategy.ec.europa.eu/en/library/ethics-guidelines-trustworthy-ai
13. Geng, Z., et al.: Instructdiffusion: a generalist modeling interface for vision tasks. In: Proceedings of the IEEE/CVF Conference on Computer Vision and Pattern Recognition, pp. 12709–12720 (2024)
14. Gomes, I., et al.: Finding patterns in ambiguity: interpretable stress testing in the decision boundary. In: IEEE/CVF Conference on Computer Vision and Pattern Recognition Workshops (CVPRW) (2024)
15. Goodfellow, I.J., et al.: Generative adversarial nets. In: Advances in Neural Information Processing Systems, pp. 2672–2680 (2014)
16. Gu, S., Bao, J., Chen, D., Wen, F.: GIQA: generated image quality assessment. In: Vedaldi, A., Bischof, H., Brox, T., Frahm, J.-M. (eds.) ECCV 2020. LNCS, vol. 12356, pp. 369–385. Springer, Cham (2020). https://doi.org/10.1007/978-3-030-58621-8_22
17. Heusel, M., Ramsauer, H., Unterthiner, T., Nessler, B., Hochreiter, S.: Gans trained by a two time-scale update rule converge to a local nash equilibrium. Adv. Neural Inf. Process. Syst. **30**, 6626–6637 (2017)
18. Hüllermeier, E., Waegeman, W.: Aleatoric and epistemic uncertainty in machine learning: an introduction to concepts and methods. Mach. Learn. **110**(3), 457–506 (2021)
19. Kermany, D.: Labeled optical coherence tomography (oct) and chest x-ray images for classification. Mendeley data (2018)
20. LeCun, Y., et al.: Handwritten digit recognition with a back-propagation network. Adv. Neural Inf. Process. Syst. **2**, 396–404 (1989)
21. Liu, Y., Feng, L., Wang, X., Zhang, S.: Deepboundary: a coverage testing method of deep learning software based on decision boundary representation. In: 22nd IEEE International Conference on Software Quality, Reliability, and Security, pp. 166–172 (2022)
22. Liu, Z., Mao, H., Wu, C.Y., Feichtenhofer, C., Darrell, T., Xie, S.: A convnet for the 2020s. In: Proceedings of the IEEE/CVF Conference on Computer Vision and Pattern Recognition, pp. 11976–11986 (2022)
23. Mukhoti, J., Kirsch, A., van Amersfoort, J., Torr;, P.H., Gal, Y.: Deterministic neural networks with inductive biases capture epistemic and aleatoric uncertainty. CoRR (2021)
24. Pan, Z., et al.: Loss functions of generative adversarial networks (gans): opportunities and challenges. IEEE Trans. Emerg. Top. Comput. Intell. **4**, 500–522 (2020)
25. Radford, A., Metz, L., Chintala, S.: Unsupervised representation learning with deep convolutional generative adversarial networks. In: 4th International Conference on Learning Representations (2016)
26. Tang, Z., Gu, S., Bao, J., Chen, D., Wen, F.: Improved vector quantized diffusion models. CoRR (2022)
27. Teixeira, C., Gomes, I., Soares, C., van Rijn, J.N.: GASTeNv2: generative adversarial stress testing networks with gaussian loss. In: Proceedings of the 23rd International Conference on Artificial Intelligence — EPIA 2024 (2024)
28. Wang, K., Gou, C., Duan, Y., Lin, Y., Zheng, X., Wang, F.: Generative adversarial networks: introduction and outlook. IEEE/CAA J. Autom. Sin. **4**, 588–598 (2017)
29. Wang, Z., She, Q.E., Ward, T.: Generative adversarial networks in computer vision: a survey and taxonomy. ACM Comput. Surv. **54**, 37:1–37:38 (2022)
30. Weiss, M., Gómez, A.G., Tonella, P.: Generating and detecting true ambiguity: a forgotten danger in dnn supervision testing. Empir. Softw. Eng. **28**, 146 (2023)

31. Xiao, H., Rasul, K., Vollgraf, R.: Fashion-mnist: a novel image dataset for benchmarking machine learning algorithms. CoRR (2017)
32. Yang, B., et al.: Paint by example: exemplar-based image editing with diffusion models. In: Proceedings of the IEEE/CVF Conference on Computer Vision and Pattern Recognition, pp. 18381–18391 (2023)

Reinforcement Learning, Bandit Optimization

Adversarial Bandit Optimization for Approximately Linear Functions

Zhuoyu Cheng[1(✉)], Kohei Hatano[1,2], and Eiji Takimoto[1]

[1] Joint Graduate School of Mathematics for Innovation, Kyushu University, Fukuoka, Japan
cheng.zhuoyu.874@s.kyushu-u.ac.jp, {hatano,eiji}@inf.kyushu-u.ac.jp
[2] RIKEN AIP, Tokyo, Japan

Abstract. We consider a bandit optimization problem for non-convex and non-smooth functions, where in each trial the loss function is the sum of a linear function and a small but arbitrary perturbation chosen after observing the player's choice. We give both expected and high probability regret bounds for the problem. Our result also implies an improved high-probability regret bound for the bandit linear optimization, a special case with no perturbation. We also give a lower bound on the expected regret.

Keywords: Bandit optimization · Normal barrier

1 Introduction

Bandit optimization is a sequential game between a player and an adversary. The game is played over T rounds, where T is a positive natural number called the horizon. The game is specified by a pair $(\mathcal{K}, \mathcal{F})$, where $\mathcal{K} \subseteq \mathbb{R}^d$ is a bounded closed convex set and $\mathcal{F} \subseteq \{f : \mathcal{K} \rightarrow \mathbb{R}\}$ is a function class. In each round $t \in [T]$, the player first chooses an action $x_t \in \mathcal{K}$ and the adversary chooses a loss function $f_t \in \mathcal{F}$, and then the player receives the value $f_t(x_t)$ as the loss. Note that f_t itself is unknown to the player. In this paper, we assume the adversary is oblivious, i.e., the loss functions are specified before starting the game[1]. The goal of the player is to minimize the regret

$$\sum_{t=1}^{T} f_t(x_t) - \min_{x \in \mathcal{K}} \sum_{t=1}^{T} f_t(x) \tag{1}$$

in expectation (expected regret) or with high probability (high-probability regret).

For convex loss functions, the bandit optimization has been extensively studied (see, e.g., [3,9,17]). $\mathcal{O}(d^{1/3}T^{3/4})$ regret bounds are shown by Flaxman et al. [10]. Lattimore [16] shows an information-theoretic regret bound $\widetilde{O}(d^{2.5}\sqrt{T})$

[1] We do not consider the case where the adversary is adaptive, i.e., it can choose the t-th loss function f_t depending on the previous actions $x_1, \ldots, x_{t-1}$.

S. Džeroski et al. (Eds.): DS 2025, LNAI 16090, pp. 495–509, 2025.
https://doi.org/10.1007/978-3-032-05461-6_32

for convex loss functions. For linear loss functions, Abernethy et al. [3] propose the SCRiBLe algorithm and give an expected regret bound $\mathcal{O}(d\sqrt{T}\ln T)$, achieving optimal dependence on T [8]. Lee et al. [17] propose SCRiBLe with lifting and increasing learning rates and show a high-probability regret bound $\widetilde{O}(d^2\sqrt{T})$.

Recently, non-convex functions are also getting popular in this literature. For example, Agarwal et al. [4] show a regret bound $\mathcal{O}(poly(d)T^{2/3})$ for smooth and bounded non-convex functions. Ghai et al. [12] propose algorithms with regret bounds $\mathcal{O}(poly(d)T^{2/3})$ under the assumption that non-convex functions are reparameterized as some convex functions.

In this paper, we investigate the bandit optimization problem for a class of non-convex and non-smooth loss functions. The function class consists of non-smooth and non-convex functions that are "close" to linear functions, in the sense that functions in the class can be viewed as linear functions with adversarial non-convex perturbations whose amount is up to ϵ. Bandit optimization for linear loss functions with stochastic noise (e.g., [1,5]) cannot be applied to our problem. Moreover, standard bandit linear optimization methods for gradient estimation, such as those based on self-concordant barrier regularizers [13], cannot be effectively applied to our problem. We propose a novel approach to analyzing high-probability regret, introducing a new method for decomposing regret.

1. When $\epsilon \neq 0$, we propose a modification of SCRiBLe with lifting and increasing learning rates [17] and leverage the properties of the ν-normal barrier [19] to prove its high probability regret bound $\widetilde{O}(d\sqrt{T} + \epsilon dT)$. We also derive its expected regret $\mathcal{O}(d\sqrt{T}\ln T + \epsilon dT)$.
2. When $\epsilon = 0$, this problem becomes bandit linear optimization, a special case with no perturbation. Compared to the result of [17], holding with probability $1-\gamma$, $\mathcal{O}(\ln^2(dT)d^2 \ln T\sqrt{T\ln\frac{\ln(dT)}{\gamma}})$, we use a different regret decomposition approach to achieve a better high-probability regret bound $\mathcal{O}(d\sqrt{T}\ln T + \ln T\sqrt{T\ln(\frac{\ln T}{\gamma})} + \ln(\frac{\ln T}{\gamma}))$.
3. We prove a lower bound $\Omega(\epsilon T)$, implying that our bounds are tight w.r.t. the parameter ϵ.

2 Related Work

Bandit linear optimization was first proposed by Awerbuch & Kleinberg [6], who achieved a regret bound of $\mathcal{O}(d^{3/5}T^{2/3})$ against oblivious adversaries. Later, McMahan & Blum [18] established a regret bound of $\mathcal{O}(dT^{3/4})$ when facing adaptive adversaries. A foundational approach in bandit optimization problems involves gradient-based smoothing techniques. Flaxman et al. [10] proposed a method for constructing an unbiased estimator of the loss function's gradient in the bandit setting. Abernethy et al. [3] introduced the SCRiBLe algorithm and achieved an expected regret bound of $\mathcal{O}(d\sqrt{T}\ln T)$ against oblivious adversaries. Bartlett et al. [7] proposed a high-probability regret bound of

$\mathcal{O}(d^{2/3}\sqrt{T \ln dT})$ under a special condition. Subsequently, Lee et al. [17] presented a high-probability regret bound $\widetilde{O}(d^2\sqrt{T})$ for both oblivious and adaptive adversaries. In recent work, Ito and Takemura [14,15] proposed a bandit linear algorithm that adapts to both stochastic and adversarial environments, achieving a regret bound of $\mathcal{O}(d\sqrt{T \ln T})$ in the adversarial setting. Rodemann et al. [21] established a connection between bandit problems and Bayesian blackbox optimization, offering theoretical foundations for regret bounds across both domains.

Unlike bandit convex optimization problems, which have been extensively explored and analyzed, bandit non-convex optimization problems introduce unique challenges due to the complexity of exploring and exploiting in a nonconvex area. Gao et al. [11] considered both non-convex losses and non-stationary data and established a regret bound of $O(\sqrt{T + poly(T)}$. Yang et al. [22] achieved a regret bound of $O(\sqrt{T \log T})$ for non-convex loss functions. However, they both require the loss functions to have smoothness properties, and our loss functions are neither convex nor smooth.

2.1 Comparison to Lee et al. [17]

Our approach builds upon Lee et al.'s work. Below, we highlight the key differences between our method and Lee et al.'s in the context of the oblivious bandit setting:

1. Simplified Regret Analysis: While Lee et al.'s regret analysis adds unnecessary complexity in the oblivious bandit setting, our approach offers a simplified analysis with clearer and more interpretable results.
2. Reduced Dependence on d: Lee et al.'s analysis results in a regret bound with greater dependence on d, whereas our method derives a bound with significantly reduced dependence on d (This distinction is demonstrated in the introduction and further illustrated in the case where $\epsilon = 0$).
3. Revised Generality of Problem Setting: Like the SCRiBLe algorithm [3], our approach is more general, treating bandit linear optimization as a special case within a broader problem framework.

3 Preliminaries

This section introduces some necessary notations and defines an ϵ-approximately linear function. Then we give our problem setting.

3.1 Notation

We abbreviate the 2-norm $\|\cdot\|_2$ as $\|\cdot\|$. For a twice differentiable convex function $\mathcal{R} : \mathbb{R}^d \to \mathbb{R}$ and any $x, h \in \mathbb{R}^d$, let $\|h\|_x = \|h\|_{\nabla^2\mathcal{R}(x)} = \sqrt{h^\top \nabla^2 \mathcal{R}(x) h}$, and $\|h\|_x^* = \|h\|_{(\nabla^2\mathcal{R}(x))^{-1}} = \sqrt{h^\top (\nabla^2\mathcal{R}(x))^{-1} h}$, respectively.

For any $v \in \mathbb{R}^d$, let $v^\perp$ be the space orthogonal to v. Let $\mathbb{S}_1^d = \{x \mid \|x\| = 1\}$. The vector $e_i \in \mathbb{R}^d$ is a standard basis vector with a value of 1 in the i-th position and 0 in all other positions. I is an identity matrix with dimensionality implied by context.

3.2 Problem Setting

Let $\mathcal{K} \subseteq \mathbb{R}^d$ be a bounded and closed convex set such that for any $x, y \in \mathcal{K}$, $\|x - y\| \leq D$. Furthermore, we assume that $\mathcal{K}$ contains the unit ball centered at the zero vector. Otherwise, we can apply an affine transformation to translate the center point of the convex set to the origin. Let $\mathcal{K}' = \{(x, 1) : x \in \mathcal{K}\}$. For any $\delta \in (0, 1)$, let $\mathcal{K}_\delta = \{x | \frac{1}{1-\delta} x \in \mathcal{K}\}$ and $\mathcal{K}'_\delta = \{(x, 1) : x \in \mathcal{K}_\delta\}$, respectively.

Definition 1. *A function $f : \mathcal{K} \to \mathbb{R}$ is ϵ-approximately linear if there exists $\theta_f \in \mathbb{R}^d$ such that $\forall x \in \mathcal{K}$, $|f(x) - \theta_f^\top x| \leq \epsilon$.*

For convenience, in the definition above, let $\sigma_f(x) = f(x) - \theta_f^\top x$, and we omit the subscript f of θ_f and σ_f if the context is clear. Note that $|\sigma(x)| \leq \epsilon$ for any $x \in \mathcal{K}$.

In this paper, we consider the bandit optimization $(\mathcal{K}, \mathcal{F})$, where $\mathcal{F}$ is the set of ϵ-approximately linear functions $f(x) = \theta^\top x + \sigma(x)$ with $\|\theta\| \leq G$. Bandit optimization for ϵ-approximately linear functions can be defined as the following statement. For every round $t = 1, .., T$, the player selects an action $x_t \in \mathcal{K}$, without knowing the loss function in advance. The environment, modeled as an oblivious adversary, chooses a sequence of linear loss vectors $\theta_1, \theta_2, ..., \theta_t$ before the interaction begins and independent of the player's actions. After selecting x_t, the adversary chooses a perturbation $\sigma_t(x_t)(|\sigma_t(x_t)| \leq \epsilon)$ and the player observes only the scalar loss $f_t(x_t)(= \theta_t^\top x_t + \sigma_t(x_t))$. The goal of the player is to minimize the regret $\sum_{t=1}^T f_t(x_t) - \min_{x \in \mathcal{K}} \sum_{t=1}^T f_t(x)$.

4 Main Results

In this section, we first introduce SCRiBLe with lifting, followed by presenting the main contributions of this paper with detailed explanations.

Algorithm 1 SCRiBLe with lifting

Input: T, parameters $\eta \in \mathbb{R}, \delta \in (0, 1)$, ν-normal barrier $\mathcal{R}$ on $con(\mathcal{K})$

1: Initialize: $x'_1 = \arg\min_{x' \in \mathcal{K}'} \mathcal{R}(x')$
2: **for** $t = 1, .., T$ **do**
3: let $\mathbf{A}_t = [\nabla^2 \mathcal{R}(x'_t)]^{-\frac{1}{2}}$
4: Draw μ_t from $\mathbb{S}_1^{d+1} \cap (\mathbf{A}_t e_{d+1})^\perp$ uniformly, set $y'_t = (y_t, 1) = x'_t + \mathbf{A}_t \mu_t$.
5: Play y_t, observe and incur loss $f_t(y_t)$. Let $g_t = d f_t(y_t) \mathbf{A}_t^{-1} \mu_t$.
6: Update $x'_{t+1} = \arg\min_{x' \in \mathcal{K}'} \eta \sum_{\tau=1}^t g_\tau^\top x' + \mathcal{R}(x')$
7: **end for**

4.1 SCRiBLe with Lifting

For the decision set $\mathcal{K}$ with a ν-normal barrier on $con(\mathcal{K})$, where $con(\mathcal{K}) = \{\mathbf{0}\} \cup \{(x, b) : \frac{x}{b} \in \mathcal{K}, x \in \mathbb{R}^d, b > 0\}$, we apply Algorithm 1 to ϵ-approximately linear functions. Recall $\mathcal{K}' = \{(x, 1) : x \in \mathcal{K}\}$.

We simplify SCRiBLe with lifting and increasing learning rates [17]. The algorithm performs original SCRiBLe [3] over a lifted decision set, using a ν-normal barrier $\mathcal{R}$ defined over the $con(\mathcal{K})$ (which always exists) as the regularizer to generate the sequence $x'_1, ..., x'_t$. It set $y'_t = x'_t + \mathbf{A}_t\mu_t$, where $\mathbf{A}_t = [\nabla^2\mathcal{R}(x'_t)]^{-\frac{1}{2}}$ and μ_t is uniformly sampled at random from $\mathbb{S}_1^{d+1} \cap (\mathbf{A}_t e_{d+1})^{\perp}$. Since μ_t is orthogonal to $\mathbf{A}_t e_{d+1}$, the last coordinate of $\mathbf{A}_t\mu_t$ is zero, ensuring that $y'_t = (y_t, 1)$ remains within $\mathcal{K}'$. The actual point played is still y_t. After playing y_t and observing $f_t(y_t)(= \theta_t^\top y_t + \sigma_t(y_t))$, it constructs the loss estimator the same way as SCRiBLe algorithm [3]: $g_t = df_t(y_t)\mathbf{A}_t^{-1}\mu_t$. Furthermore, it adopts the same update method as FTRL algorithm [13]: $x'_{t+1} = \arg\min_{x' \in \mathcal{K}'} \eta \sum_{\tau=1}^{t} g_\tau^\top x' + \mathcal{R}(x')$.

Compared with SCRiBLe with lifting and increasing learning rates [17], which focuses on bandit linear optimization, we do not use the increasing learning rates part, but retain the lifting. This preserves its advantages; for instance, the ν-normal barrier $\mathcal{R}$ always exists on $con(\mathcal{K})$ and we can leverage the properties of the ν-normal barrier. When the loss takes the form $f_t(y_t) = \theta_t^\top y_t$ (i.e., when $\epsilon = 0$), the analysis by Lee et al. [17] shows that the first d coordinates of g_t are indeed an unbiased estimator of θ_t. However, when $\epsilon \neq 0$ since g_t will always be influenced by $\sigma_t(y_t)$, g_t is no longer an unbiased estimator of θ_t.

Assumption 1. *Let* $x^* = \arg\min_{x \in \mathcal{K}} \sum_{t=1}^{T} f_t(x)$, $x_\delta^* = \Pi_{\mathcal{K}_\delta}(x^*)$, *and* $x_\delta^{*'} = (x_\delta^*, 1)$. *Assume that for a ν-normal barrier $\mathcal{R}$ on $con(\mathcal{K})$, the following inequality holds:* $\nabla\mathcal{R}(x'_t)^\top(x_\delta^{*'} - x'_t) \geq -\nu$.

Assumption 1 is introduced to enable the use of Lemma 4 in proving Lemma 8. Under this assumption, the theoretical upper bound can be established. However, when $\nabla\mathcal{R}(x'_t)^\top(x_\delta^{*'} - x'_t) < -\nu$, our method no longer admits the theoretical upper bound.

We present our main results: expected and high-probability regret bounds for the problem.

Theorem 1. *Assume Assumptions 1. The algorithm with parameters* $\eta = \frac{\sqrt{2\nu \log T}}{2d\sqrt{T}}$, $\delta = \frac{1}{T^2}$ *guarantees the following expected regret bound*

$$\mathbb{E}[\sum_{t=1}^{T} f_t(y_t) - \min_{x \in \mathcal{K}} \sum_{t=1}^{T} f_t(x)] \leq 4d\sqrt{2\nu T \log T} + \frac{GD}{T} + 2T\epsilon + dT\epsilon(2\nu + \sqrt{\nu}). \quad (2)$$

Theorem 2. *Assume Assumptions 1. The algorithm with parameters* $\eta = \frac{\sqrt{2\nu \ln T}}{2d\sqrt{T}}$, $\delta = \frac{1}{T^2}$ *ensures that with probability at least* $1 - \gamma$

$$\sum_{t=1}^{T} f_t(y_t) - \min_{x \in \mathcal{K}} \sum_{t=1}^{T} f_t(x) \leq 4d\sqrt{2\nu T \ln T} + \frac{GD}{T} + 2GD \ln \frac{C}{\gamma} + 2T\epsilon$$
$$+ C(1 + \epsilon)\sqrt{8T \ln \frac{C}{\gamma}} + Td\epsilon(2\nu + \sqrt{\nu})$$

where $C = \lceil \ln GD \rceil \lceil \ln((GD)^2 T) \rceil$.

4.2 Analysis

We primarily divide the regret into parts:

$$Regret = \underbrace{\sum_{t=1}^{T}(x'_t - h)^\top g_t}_{\text{REG-TERM}} + \underbrace{\sum_{t=1}^{T}(y'_t - x'_t)^\top \theta'_t}_{\text{DEVIATION-TERM}} + \underbrace{\sum_{t=1}^{T}(d\sigma_t(y_t)\mathbf{A}_t^{-1}\mu_t)^\top (x'_t - h)}_{\text{ERROR-TERM}}, \tag{3}$$

where $h \in \mathcal{K}'$, $\theta' = (\theta, 0)$. This differs from the decomposition proposed by Lee et al. [17], which divides the regret as follows:

$$Regret = \underbrace{\sum_{t=1}^{T}(x'_t - h)^\top g_t}_{\text{REG-TERM}} + \underbrace{\sum_{t=1}^{T}[{y'_t}^\top \theta'_t - {x'_t}^\top g_t + h^\top (g_t - \theta'_t)]}_{\text{DEVIATION-TERM}}, \tag{4}$$

where $h \in \mathcal{K}', \theta' = (\theta, 0)$. This means that, for an oblivious adversary, calculating the regret does not require considering the variance of the estimator g_t and θ_t, but only the variance between y_t and x_t. This difference is a key factor that enables us to achieve a better high-probability regret bound when $\epsilon = 0$. In addition, we need to account for an extra ERROR-TERM when $\epsilon \neq 0$.

The bound on the REG-TERM is identical to that in previous work [17], so we omit its explanation and first focus on the ERROR-TERM. We make the entire analysis hold in $\mathbb{R}^{d+1}$. The Cauchy-Schwarz inequality helps in deriving the bounds ERROR-TERM$\leq \|d\sigma_t(y_t)\mathbf{A}_t^{-1}\mu_t\|^*_{\nabla^2\mathcal{R}(x'_t)}\|x'_t - h\|_{\nabla^2\mathcal{R}(x'_t)}$ and $\|d\sigma_t(y_t)\mathbf{A}_t^{-1}\mu_t\|^*_{\nabla^2\mathcal{R}(x'_t)} \leq d\epsilon$. However, since the largest eigenvalue of $\nabla^2\mathcal{R}(x'_t)$ [19] could potentially approach infinity, the main challenge in bounding the ERROR-TERM lies in the difficulty of bounding $\|x'_t - h\|_{\nabla^2\mathcal{R}(x'_t)}$, has not been addressed in previous work. Lee et al. [17] derive a related inequality: $\|h\|_{\nabla^2\mathcal{R}(x'_t)} \leq -\|h\|_{\nabla^2\mathcal{R}(x'_{t+1})} + \nu\ln(\nu T + 1)$, but it is still too large and does not help in bounding $\|x'_t - h\|_{\nabla^2\mathcal{R}(x'_t)}$. We present a straightforward yet necessary Lemma 4, which helps to bound $\|h\|_{\nabla^2\mathcal{R}(x'_t)} \leq 2\nu$. It is clear that this bound is tighter than that of Lee et al. in terms of $\|h\|_{\nabla^2\mathcal{R}(x'_t)}$, which is one of the key reasons we are able to obtain a better high-probability regret bound when $\epsilon = 0$ and the adversary is oblivious. In addition, the properties of the ν-normal barrier tell us $\|x'_t\|_{\nabla^2\mathcal{R}(x'_t)} = \sqrt{\nu}$. With these two conditions, we can immediately deduce the bound $\|x_t - h\|_{\nabla^2\mathcal{R}(x'_t)}$ as $2\nu + \sqrt{\nu}$. This also implies that increasing learning rates are not required, as they are solely aimed at bounding $\|h\|_{\nabla^2\mathcal{R}(x'_t)}$ in Lee et al.'s paper [17].

Secondly, by obtaining the expected bound and high-probability bound for DEVIATION-TERM, we can derive the expected regret bound and high-probability regret bound, respectively. For the high-probability bound of DEVIATION-TERM, unlike SCRiBLe with lifting and increasing learning rates [17], which constrains the decision set from $\mathcal{K}'$ to $\mathcal{K}'_\delta$ to ensure that x'_t is never too close to the boundary (thus ensuring that the eigenvalues of $\mathbf{A}_t$ are bounded, especially for bounding $\|h\|_{\nabla^2\mathcal{R}(x'_t)}$). Our approach does not require x'_t to stay away from the boundary.

Furthermore, we do not need to bound the eigenvalues of $\mathbf{A}_t$, which gives us greater flexibility in choosing the value of δ (such as $\frac{1}{T^2}$), leading to a better upper bound for the regret.

Finally, we prove the lower bound of regret in Sect. 6.

5 Proof

This section introduces preliminaries on the ν-normal barrier, presents several essential lemmas and provides proofs of the main theorems. Omitted proofs are shown in the arxiv version[2].

5.1 ν-Normal Barrier

We introduce the ν-normal barrier, providing its definitions and highlighting several key properties that will be frequently used in the subsequent analysis.

Definition 2. *Let $\Psi \subseteq \mathbb{R}^d$ be a closed and proper convex cone and let $\nu \geq 1$. A function $\mathcal{R} : int(\Psi) \to \mathbb{R}$: is called a ν-logarithmically homogeneous self-concordant barrier (or simply ν-normal barrier) on Ψ if*

1. *$\mathcal{R}$ is three times continuously differentiable and convex and approaches infinity along any sequence of points approaching the boundary of Ψ.*
2. *For every $h \in \mathbb{R}^d$ and $x \in int(\Psi)$ the following holds:*

$$\sum_{i=1}^{d}\sum_{j=1}^{d}\sum_{k=1}^{d} \frac{\partial^3 \mathcal{R}(x)}{\partial x_i \partial x_j \partial x_k} h_i h_j h_k \leq 2\|h\|_x^3, \tag{5}$$

$$\|\nabla \mathcal{R}(x)^\top h\| \leq \sqrt{\nu}\|h\|_x, \tag{6}$$

$$\mathcal{R}(tx) = \mathcal{R}(x) - \nu \ln t, \forall x \in int(\Psi), t > 0. \tag{7}$$

Lemma 1 ([19,20]). *If $\mathcal{R}$ is a ν-normal barrier on Ψ, Then for any $x \in int(\Psi)$ and any $h \in \Psi$, we have*

$$\|x\|_x^2 = \nu, \tag{8}$$

$$\nabla^2 \mathcal{R}(x) x = -\nabla \mathcal{R}(x), \tag{9}$$

$$\|h\|_x \leq -\nabla \mathcal{R}(x)^\top h, \tag{10}$$

$$\nabla \mathcal{R}(x)^\top (h - x) \leq \nu. \tag{11}$$

Lemma 2 ([19]). *If $\mathcal{R}$ is a ν-normal barrier on Ψ, then the Dikin ellipsoid centered at $x \in int(\Psi)$, defined as $\{y : \|y - x\|_x \leq 1\}$, is always within Ψ. Moreover,*

$$\|h\|_y \geq \|h\|_x (1 - \|y - x\|_x) \tag{12}$$

holds for any $h \in \mathbb{R}^d$ and any y with $\|y - x\|_x \leq 1$.

[2] https://arxiv.org/abs/2505.20734.

Lemma 3 **([13]).** *Let* $\mathcal{R}$ *is a* ν*-normal barrier over* Ψ*, then for all* $x, z \in int(\Psi) : \mathcal{R}(z) - \mathcal{R}(x) \leq \nu \log \frac{1}{1-\pi_x(z)}$*, where* $\pi_x(z) = \inf\{t \geq 0 : x + t^{-1}(z - x) \in \Psi\}$.

ν-normal barrier plays a crucial role in addressing one of the key challenges in this problem: bounding $\|x'_t - h\|_{x'_t}$, where $h \in \mathcal{K}'$. Equation 8 give $\|x'_t\|_{x'_t} = \sqrt{\nu}$. Building upon Lemma 1, we introduce an effective Lemma 4 that aids in bounding $\|h\|_{x'_t}$.

Lemma 4. *If* $\mathcal{R}$ *be a* ν*-normal barrier for* $\Psi \subseteq R^d$*, and* $x \in int(\Psi)$*,* $h \in \Psi$ *satisfy* $\nabla\mathcal{R}(x)^\top (h - x) \geq -\nu$*, then*

$$\|h\|_x \leq 2\nu. \tag{13}$$

With the help of Eq. 8 and Lemma 4, it is easy to apply the triangle inequality to derive $\|x'_t - h\|_{x'_t} \leq \|x'_t\|_{x'_t} + \|h\|_{x'_t} \leq \sqrt{\nu} + 2\nu$, where $h \in \mathcal{K}'$.

5.2 Useful Lemmas

In addition to the properties of the normal barrier and its related lemmas, we also need to introduce some additional necessary lemmas.

Like Lemma 6 in the SCRiBLe algorithm [2], the next minimizer x'_{t+1} is "close" to x'_t and Lemma 5 implies $\|x'_{t+1} - x'_t\|_{x'_t} \leq 4d\eta$. This result will help us bound $g_t^\top (x'_t - h)$, where $h \in \mathcal{K}'$(see Lemma 7).

Lemma 5. $x'_{t+1} \in W_{4d\eta}(x'_t)$*, where* $W_r(x') = \{y \in \mathcal{K}' : \|y - x'\|_{x'} < r\}$.

This next lemma is based on Lemma B.9. [17], but due to the differences in the loss functions, $g_{t,i}$ is no longer an unbiased estimate of $\theta_{t,i}$, for $i \in [d]$. Lee et al. state that $\mathbb{E}_t[l_{t,i}] = \theta_{t,i}$, for $i \in [d]$, where $l_t = d(\theta_t, 0)(x'_t + \mathbf{A}_t\mu_t)\mathbf{A}_t^{-1}\mu_t$ [17]. We directly apply it to our analysis.

Lemma 6. *Let* $l_t = d(\theta_t, 0)(x'_t + \mathbf{A}_t\mu_t)\mathbf{A}_t^{-1}\mu_t$*. For Algorithm 1, we have* $\mathbb{E}_t[l_{t,i}] = \theta_{t,i}$*, for* $i \in [d]$.

The regret bound of FTRL algorithm [13] states that for every $u \in \mathcal{K}$, $\sum_{t=1}^T \nabla_t^\top x_t - \sum_{t=1}^T \nabla_t^\top u \leq \sum_{t=1}^T [\nabla_t^\top x_t - \nabla_t^\top x_{t+1}] + \frac{1}{\eta}[\mathcal{R}(u) - \mathcal{R}(x_1)]$, where ∇_t represents the gradient of the loss function f_t. In our adaptation, we replaced ∇_t with g_t and $\mathcal{K}$ with $\mathcal{K}'$. This modification does not fundamentally alter the original result. Since the update way $x'_{t+1} = \arg\min_{x' \in \mathcal{K}'} \eta \sum_{\tau=1}^t g_\tau^\top x' + \mathcal{R}(x')$ satisfied the condition of FTRL algorithm [13], we can apply Lemma 5.3. in [13] to Algorithm 1 as follow.

Lemma 7. *For Algorithm 1 and for every* $h \in \mathcal{K}'$*,* $\sum_{t=1}^T g_t^\top x'_t - \sum_{t=1}^T g_t^\top h \leq \sum_{t=1}^T [g_t^\top x'_t - g_t^\top x'_{t+1}] + \frac{1}{\eta}[\mathcal{R}(h) - \mathcal{R}(x'_1)]$.

The lemmas presented above are all intended to lead to the derivation of Lemma 8, which serves as a key result of this paper. Specifically, it provides a bound for $\sum_{t=1}^{T}\theta_t^\top x_t - \sum_{t=1}^{T}\theta_t^\top x^*$, where $x^* = \arg\min_{x\in\mathcal{K}}\sum_{t=1}^{T} f_t(x)$. Due to Lemma 8, we only need to consider $\sum_{t=1}^{T}\theta_t^\top y_t - \sum_{t=1}^{T}\theta_t^\top x_t$ when calculating the regret bound. This result plays a crucial role in deriving both expected and high-probability regret bounds.

Lemma 8. *For Algorithm 1, let* $f_t(x_t) = \theta_t^\top x_t + \sigma_t(x_t)$ *and* $x^* = \arg\min_{x\in\mathcal{K}}\sum_{t=1}^{T} f_t(x)$ *and we have*

$$\sum_{t=1}^{T}\theta_t^\top x_t - \sum_{t=1}^{T}\theta_t^\top x^* \leq 2\eta d^2 T + \frac{\nu\log(\frac{1}{\delta})}{\eta} + Td\epsilon(2\nu+\sqrt{\nu}) + \delta DGT. \tag{14}$$

With the help of Lemma 8, we are ready to present the proof of Theorem 1.

5.3 Proof of Theorem 1

Proof. Recall an ϵ-approximately linear function can be written as: $f(x) = \theta^\top x + \sigma(x)$. Thus, the regret of SCRiBLe with lifting algorithm

$$\begin{aligned}
\mathbb{E}[\sum_{t=1}^{T} f_t(y_t) - \sum_{t=1}^{T} f_t(x^*)] &= \mathbb{E}[\sum_{t=1}^{T}[\theta_t^\top y_t + \sigma_t(y_t)] - \sum_{t=1}^{T}[\theta_t^\top x^* + \sigma_t(x^*)]] \\
&= \mathbb{E}[\sum_{t=1}^{T}\theta_t^\top y_t - \sum_{t=1}^{T}\theta_t^\top x^*] + \mathbb{E}[\sum_{t=1}^{T}\sigma_t(y_t) - \sum_{t=1}^{T}\sigma_t(x^*)].
\end{aligned}$$

Firstly, we bound the front of the above equation,

$$\mathbb{E}[\sum_{t=1}^{T}\theta_t^\top y_t - \sum_{t=1}^{T}\theta_t^\top x^*] = \sum_{t=1}^{T}\mathbb{E}[\theta_t^\top y_t] - \sum_{t=1}^{T}\mathbb{E}[\theta_t^\top x_t] + \sum_{t=1}^{T}\mathbb{E}[\theta_t^\top x_t] - \sum_{t=1}^{T}\mathbb{E}[\theta_t^\top x^*].$$

From the Law of total expectation, we know

$$\begin{aligned}
\sum_{t=1}^{T}\mathbb{E}[\theta_t^\top y_t] - \sum_{t=1}^{T}\mathbb{E}[\theta_t^\top x_t] &= \sum_{t=1}^{T}\mathbb{E}[\theta_t^\top (y_t - x_t)] \\
&= \sum_{t=1}^{T}\mathbb{E}[\mathbb{E}_t[\theta_t^\top (y_t - x_t)]] \\
&= \sum_{t=1}^{T}\mathbb{E}[\theta_t^\top \mathbb{E}_t[(\mathbf{A}_t\mu_t)]] \\
&= \sum_{t=1}^{T}\mathbb{E}[\theta_t^\top \mathbf{0}] \\
&= \mathbf{0}.
\end{aligned}$$

Thus,

$$\mathbb{E}[\sum_{t=1}^{T}\theta_t^\top y_t - \sum_{t=1}^{T}\theta_t^\top x^*] = \mathbb{E}[\sum_{t=1}^{T}\theta_t^\top x_t - \sum_{t=1}^{T}\theta_t^\top x^*]. \tag{15}$$

From Lemma 8, we have

$$\begin{aligned}\mathbb{E}[\sum_{t=1}^{T}\theta_t^\top x_t - \sum_{t=1}^{T}\theta_t^\top x^*] &\leq \mathbb{E}[4\eta d^2 T + \frac{\nu\log(\frac{1}{\delta})}{\eta} + Td\epsilon(2\nu + \sqrt{\nu}) + \delta DGT] \\ &\leq 4\eta d^2 T + \frac{\nu\log(\frac{1}{\delta})}{\eta} + Td\epsilon(2\nu + \sqrt{\nu}) + \delta DGT.\end{aligned}$$

Since σ_t is chosen after knowing the player's action, it can cause as large a perturbation as possible. We using $\|\sigma_t(x)\| \leq \epsilon$ to bound $\sum_{t=1}^{T}\mathbb{E}[\sigma_t(y_t) - \sum_{t=1}^{T}\sigma_t(x^*)] \leq 2T\epsilon$ and combination of Lemma 8, we get

$$\begin{aligned}Regret &= \mathbb{E}[\sum_{t=1}^{T} f_t(y_t) - \sum_{t=1}^{T} f_t(x^*)] \\ &\leq 4d\sqrt{2\nu T\log T} + \frac{GD}{T} + Td\epsilon(2\nu + \sqrt{\nu}) + 2T\epsilon,\end{aligned}$$

where $\eta = \frac{\sqrt{2\nu\log T}}{2d\sqrt{T}}, \delta = \frac{1}{T^2}$.

5.4 Proof of Theorem 2

To establish the high-probability regret bound, we first introduce the necessary Lemma 9.

Lemma 9 (Theorem 2.2. in[17]). *Let $X_1, ..., X_T$ be a martingale difference sequence with respect to a filtration $F_1 \subseteq ... \subseteq F_T$ such that $\mathbb{E}[X_t \mid F_t] = 0$. Suppose $B_t \in [1, b]$ for a fixed constant b is F_t-measurable and such that $X_t \leq B_t$ holds almost surely. Then with probability at least $1 - \gamma$ we have*

$$\sum_{t=1}^{T} X_t \leq C(\sqrt{8Vln(C/\gamma)} + 2B^* ln(C/\gamma)), \tag{16}$$

where $V = \max\{1; \sum_{t=1}^{T}\mathbb{E}[X_t^2 \mid F_t]\}, B^* = \max_{t\in[T]} B_t$, *and* $C = \lceil logb \rceil \lceil log(b^2 T) \rceil$.

The analysis in [17] employs Lemma 9 to derive a high-probability bound for $\sum_{t=1}^{T}[y_t^\top\theta_t - x_t^\top g_t + u^\top(g_t - \theta_t)]$. In contrast, our approach defines $X_t = \theta_t^\top y_t - \theta_t^\top x_t$ and derives the high-probability bound for $\sum_{t=1}^{T}(\theta_t^\top y_t - \theta_t^\top x_t)$. This distinction in the application of Lemma 9 enables us to derive a tighter high-probability upper bound for bandit linear optimization.

With the support of Lemma 8 and Lemma 9, we are ready to prove Theorem 2.

Proof. Let $X_t = \theta_t^\top y_t - \theta_t^\top x_t$, then $\mathbb{E}_t[X_t] = \mathbb{E}_t[\theta_t^\top y_t - \theta_t^\top x_t] = 0$, $X_t = \theta_t^\top y_t - \theta_t^\top x_t \le \|\theta_t\| \|y_t - x_t\| \le GD$ and

$$\begin{aligned}
\mathbb{E}_t[X_t^2] &= \mathbb{E}_t[(\theta_t^\top y_t - \theta_t^\top x_t)^2] \\
&= \mathbb{E}_t[(\theta_t^\top y_t)^2 + (\theta_t^\top x_t)^2 - 2\theta_t^\top y_t \theta_t^\top x_t] \\
&= \mathbb{E}_t[(\theta_t^\top y_t)^2] + \mathbb{E}_t[(\theta_t^\top x_t)^2] - \mathbb{E}_t[2\theta_t^\top y_t \theta_t^\top x_t] \\
&= \mathbb{E}_t[(\theta_t^\top y_t)^2] - \theta_t^\top x_t \theta_t^\top x_t \\
&\le (1+\epsilon)^2.
\end{aligned}$$

Then,

$$\begin{aligned}
\sum_{t=1}^T f_t(y_t) - \sum_{t=1}^T f_t(x^*) &= \sum_{t=1}^T [\theta_t^\top y_t + \sigma_t(y_t)] - \sum_{t=1}^T [\theta_t^\top x^* + \sigma_t(x^*)] \\
&\le \sum_{t=1}^T \theta_t^\top y_t - \sum_{t=1}^T \theta_t^\top x^* + \sum_{t=1}^T \sigma_t(y_t) - \sum_{t=1}^T \sigma_t(x^*) \\
&\le \sum_{t=1}^T \theta_t^\top y_t - \sum_{t=1}^T \theta_t^\top x^* + 2T\epsilon \\
&= \sum_{t=1}^T \theta_t^\top y_t - \sum_{t=1}^T \theta_t^\top x_t + \sum_{t=1}^T \theta_t^\top x_t - \sum_{t=1}^T \theta_t^\top x^* + 2T\epsilon.
\end{aligned}$$

From Lemma 8, we know

$$\sum_{t=1}^T \theta_t^\top x_t - \sum_{t=1}^T \theta_t^\top x^* \le 4d\sqrt{2\nu T \ln T} + \frac{GD}{T} + Td\epsilon(2\nu + \sqrt{\nu}), \tag{17}$$

where $\eta = \frac{\sqrt{2\nu \ln T}}{2d\sqrt{T}}, \delta = \frac{1}{T^2}$. Then by Lemma 9,

$$\mathbb{P}(\sum_{t=1}^T (\theta_t^\top y_t - \theta_t^\top x_t) \le C(\sqrt{8V \ln(C/\gamma)} + 2B^* \ln(C/\gamma))) \ge 1 - \gamma, \tag{18}$$

where $V = (1+\epsilon)^2 T, B^* = b = GD$, and $C = \lceil \ln GD \rceil \lceil \ln((GD)^2 T) \rceil$. Combine everything to conclude the proof.

5.5 Application to Black-Box Optimization

From online to offline transformation, the result of this paper can also apply to black-box optimization for ϵ-approximately linear functions. This problem is important in that previous theoretical analyses for black-box optimization can only deal with linear/convex/smooth objectives in the adversarial environments (via bandit convex optimization). So, it is quite meaningful to clarify the possibility of the black-box optimization problems without such restrictions. In fact, our objective is not linear, nor smooth, even with a simple assumption.

Let $\hat{x}$ be the output of Algorithm 1. Then, it follows from Theorem 1 that $f(\hat{x}) - \min_{x\in\mathcal{K}} f(x) \leq \frac{4d\sqrt{2\nu \ln T}}{\sqrt{T}} + \frac{GD}{T^2} + d\epsilon(2\nu + \sqrt{\nu}) + 2\epsilon$. Additionally, we provide a lower bound 2ϵ for this problem (see Lemma 10). We can see that the difference between the lower bound and the upper bound is only $d\epsilon(2\nu + \sqrt{\nu})$ as T approaches infinity. This suggests the potential existence of "easier" settings between the adversarial environment and the standard stochastic environment, where better algorithms might be found. It also motivates us to explore these settings further.

6 Lower Bound

In this section, we show a lower bound of the regret. To do so, we consider a black-box optimization problem for the set $\mathcal{F}$ of ϵ-approximately linear functions $f : \mathcal{K} \to \mathbb{R}$. In the problem, we are given access to the oracle O_f for some $f \in \mathcal{F}$, which returns the value $f(x)$ given an input $x \in \mathcal{K}$. The goal is to find a point $\hat{x} \in \mathcal{K}$ such that $f(\hat{x}) - \min_{x\in\mathcal{K}} f(x)$ is small enough. Then, the following statement holds.

Lemma 10. *For any algorithm $\mathcal{A}$ for the black-box optimization problem for $\mathcal{F}$, there exists an ϵ-approximately linear function $f \in \mathcal{F}$ such that the output $\hat{x}$ of $\mathcal{A}$ satisfies*

$$f(\hat{x}) - \min_{x\in\mathcal{K}} f(x) \geq 2\epsilon. \tag{19}$$

Proof. Firstly, suppose that the algorithm $\mathcal{A}$ is deterministic. At iteration $t = 1, ..., T$, for any feedback $y_1, ..., y_{t-1} \in \mathbb{R}$, $\mathcal{A}$ should choose the next query point x_t based on the data observed so far. That is,

$$x_t = \mathcal{A}((x_1, y_1), ..., (x_{t-1}, y_{t-1})). \tag{20}$$

Assume that the final output $\hat{x}$ is returned after T queries to the oracle O_f. In particular, we fix the T feedbacks $y_1 = y_2 = \cdots = y_T = \epsilon$. Let $z \in \mathcal{K}$ be such that $z \notin \{x_1, ..., x_T, \hat{x}\}$. Then we define a function $f : \mathcal{K} \to \mathbb{R}$ is as

$$f(x) = \begin{cases} \epsilon, & x \neq z, \\ -\epsilon, & x = z. \end{cases} \tag{21}$$

The function f is indeed an ϵ-approximately linear function, as $f(x) = 0^\top x + \sigma(x)$, where $\sigma(x) = \epsilon$ for $x \neq z$ and $\sigma(x) = -\epsilon$ for $x = z$. Further, we have

$$f(\hat{x}) - \min_{x\in\mathcal{K}} f(x) \geq 2\epsilon. \tag{22}$$

Secondly, if algorithm $\mathcal{A}$ is randomized. It means each x_t is chosen randomly. We assume the same feedbacks $y_1 = y_2 = \cdots = y_T = \epsilon$. Let $X = \{x_1, ..., x_T, \hat{x}\}$. Then, there exists a point $z \in \mathcal{K}$ such that $P_X(z \in X) = 0$, since $\mathbb{E}_{z'}[P_X(z' \in X|z')] = P_{z',X}(z' \in X) = \mathbb{E}_X[P_{z'}(z' \in X|X)] = 0$, where the expectation on z' is defined w.r.t. the uniform distribution over $\mathcal{K}$. For the objective function f defined in (21), we have $f(\hat{x}) - \min_{x\in\mathcal{K}} f(x) \geq 2\epsilon$ while f is ϵ-approximately linear.

Theorem 3. *For any horizon $T \geq 1$ and any player, there exists an adversary such that the regret is at least $2\epsilon T$.*

Proof. We prove the statement by contradiction. Suppose that there exists a player whose regret is less than $2\epsilon T$. Then we can construct an algorithm for the black-box optimization problem from it by feeding the online algorithm with T feedbacks of the black-box optimization problem and by setting $\hat{x} = \min_{t \in [T]} f(x_t)$. Then,

$$f(\hat{x}) - \min_{x \in \mathcal{K}} f(x) \leq \frac{\sum_{t=1}^{T} f(x_t) - \sum_{t=1}^{T} \min_{x \in \mathcal{K}} f(x)}{T} < 2\epsilon,$$

which contradicts Lemma 10.

This lower bound indicates that $\Omega(\epsilon T)$ regret is inevitable for the bandit optimization problem for ϵ-approximately linear functions. We conjecture that the lower bound can be tightened to $\Omega(d\epsilon T)$, but we leave it as an open problem.

7 Experiments

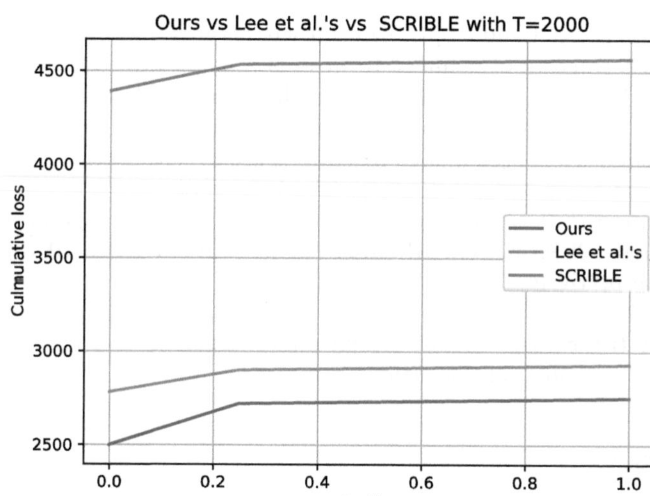

Fig. 1. Average cumulative loss of algorithms for artificial data sets. The blue line corresponds to the results of our algorithm, the yellow line represents the algorithm in [17], and the green line corresponds to the SCRiBLe algorithm [2] (Color figure online).

We conduct comparative experiments across a range of ϵ values: 0, 0.25, 0.5, 0.75, and 1, evaluating Algorithm 1, SCRiBLe [2], and SCRiBLe with lifting and increasing learning rates [17]. We generate the artificial data as follows. The input set is constructed such that each element $x \in \mathbb{R}^d$ satisfies $\|x\|_2 < D$. The loss vectors $\theta_1, ..., \theta_T$ are d-dimensional vectors randomly generated before the experiment, each with a Euclidean norm no greater than G. The ν-normal barrier is defined as $\mathcal{R}(x, b) = 400 \cdot \left(-\log \left(1 - \frac{\|x\|^2}{D^2} \right) - 2\nu \log b \right)$ and η is defined as $\frac{20\sqrt{2\log T}}{4d\sqrt{T}}$. In our experiments, we set $D = 3$, $G = 1$, $d = 5$, respectively. The perturbation is defined as $\sigma(y_t) = \epsilon \sin ((y_t^\top l)\pi)$, where l is a random vector in $\mathbb{R}^d$ (Note that this is not the worst-case perturbation, which may explain why the results do not increase significantly as ϵ grows). Each algorithm is run with a fixed time horizon of $T = 2000$. For each value of ϵ, we repeat the experiment 10 times for each algorithm and compute the average cumulative loss. The x-axis represents the value of ϵ, and the y-axis shows the average cumulative loss. As shown in the figure above, our method generally outperforms SCRiBLe. Compared to Lee et al.'s method, our approach achieves slightly better performance. This indicates that, when facing oblivious adversaries, not using increasing learning rates (i.e., dynamically adjusting the value of η) has little to no impact on the results (see Fig. 1).

8 Conclusion

In this work, we study bandit optimization with non-convex, non-smooth losses, where each loss consists of a linear term plus a small adversarial perturbation revealed after the action. We propose a new high-probability regret analysis and establish matching upper and lower bounds. Future work includes extending our results to broader loss classes, such as convex losses with perturbations.

Acknowledgments. This work was supported by WISE program (MEXT) at Kyushu University, JSPS KEKENHI Grant Numbers JP23K24905, JP23K28038, and JP24H00685, respectively.

Disclosure of Interests. The authors have no competing interests to declare that are relevant to the content of this article.

References

1. Abbasi-Yadkori, Y., Pál, D., Szepesvári, C.: Improved algorithms for linear stochastic bandits. Adv. Neural Inf. Process. Syst. **24** (2011)
2. Abernethy, J.D., Hazan, E., Rakhlin, A.: Competing in the dark: an efficient algorithm for bandit linear optimization. In: COLT, pp. 263–274 (2008)
3. Abernethy, J.D., Hazan, E., Rakhlin, A.: Interior-point methods for full-information and bandit online learning. IEEE Trans. Inf. Theory **58**(7), 4164–4175 (2012)
4. Agarwal, N., Gonen, A., Hazan, E.: Learning in non-convex games with an optimization oracle. In: Conference on Learning Theory, pp. 18–29. PMLR (2019)

5. Amani, S., Alizadeh, M., Thrampoulidis, C.: Linear stochastic bandits under safety constraints. Adv. Neural Inf. Process. Syst. **32** (2019)
6. Awerbuch, B., Kleinberg, R.D.: Adaptive routing with end-to-end feedback: distributed learning and geometric approaches. In: Proceedings of the Thirty-sixth Annual ACM Symposium on Theory of Computing, pp. 45–53 (2004)
7. Bartlett, P., Dani, V., Hayes, T., Kakade, S., Rakhlin, A., Tewari, A.: High-probability regret bounds for bandit online linear optimization. In: Proceedings of the 21st Annual Conference on Learning Theory-COLT 2008, pp. 335–342. Omnipress (2008)
8. Bubeck, S., Cesa-Bianchi, N., Kakade, S.M.: Towards minimax policies for online linear optimization with bandit feedback. In: Conference on Learning Theory, pp. 41–1. JMLR Workshop and Conference Proceedings (2012)
9. Dani, V., Kakade, S.M., Hayes, T.: The price of bandit information for online optimization. Adv. Neural Inf. Process. Syst. **20** (2007)
10. Flaxman, A.D., Kalai, A.T., Mcmahan, H.B.: Online convex optimization in the bandit setting: gradient descent without a gradient. In: Proceedings of the Sixteenth Annual ACM-SIAM Symposium on Discrete Algorithms, pp. 385 – 394 (2005). https://doi.org/10.5555/1070432
11. Gao, X., Li, X., Zhang, S.: Online learning with non-convex losses and non-stationary regret. In: International Conference on Artificial Intelligence and Statistics, pp. 235–243. PMLR (2018)
12. Ghai, U., Lu, Z., Hazan, E.: Non-convex online learning via algorithmic equivalence. Adv. Neural Inf. Process. Syst. **35**, 22161–22172 (2022)
13. Hazan, E., et al.: Introduction to online convex optimization. Found. Trends® Optim. **2**(3-4), 157–325 (2016)
14. Ito, S., Takemura, K.: Best-of-three-worlds linear bandit algorithm with variance-adaptive regret bounds. In: The Thirty Sixth Annual Conference on Learning Theory, pp. 2653–2677. PMLR (2023)
15. Ito, S., Takemura, K.: An exploration-by-optimization approach to best of both worlds in linear bandits. Adv. Neural Inf. Process. Syst. **36**, 71582–71602 (2023)
16. Lattimore, T.: Improved regret for zeroth-order adversarial bandit convex optimisation. Math. Stat. Learn. **2**, 311–334 (2020). https://doi.org/10.4171/msl/17
17. Lee, C.W., Luo, H., Wei, C.Y., Zhang, M.: Bias no more: high-probability data-dependent regret bounds for adversarial bandits and mdps. Adv. Neural Inf. Process. Syst. **33**, 15522–15533 (2020)
18. McMahan, H.B., Blum, A.: Online geometric optimization in the bandit setting against an adaptive adversary. In: Shawe-Taylor, J., Singer, Y. (eds.) COLT 2004. LNCS (LNAI), vol. 3120, pp. 109–123. Springer, Heidelberg (2004). https://doi.org/10.1007/978-3-540-27819-1_8
19. Nemirovski, A.: Interior point polynomial time methods in convex programming. Lecture Notes **42**(16), 3215–3224 (2004)
20. Nesterov, Y., Nemirovskii, A.: Interior-point polynomial algorithms in convex programming. SIAM (1994)
21. Rodemann, J., Jansen, C., Schollmeyer, G.: Reciprocal learning. Adv. Neural Inf. Process. Syst. **37**, 1686–1724 (2024)
22. Yang, L., Deng, L., Hajiesmaili, M.H., Tan, C., Wong, W.S.: An optimal algorithm for online non-convex learning. Proc. ACM Meas. Anal. Comput. Syst. **2**(2), 1–25 (2018)

A Contextual Bandit Algorithm for Recommending Item Sets with High Sum Diversity

Masaki Hashimoto and Atsuyoshi Nakamura(✉)

Graduate School of Information Science and Technology, Hokkaido University, Sapporo, Japan
{masakihashimoto,atsu}@ist.hokudai.ac.jp

Abstract. We study online recommendation of item sets that are high in both rating and diversity. We formalize our problem as a contextual bandit problem in which the reward of a recommended set is the rating sum of the member items plus the diversity of the set. In our problem setting, each item is represented as a feature vector (context) in a multi-dimensional metric space. Its rating is assumed to be the noisy value of some linear function of its feature vector, and the diversity of an item set is measured using the Sum diversity for the feature metric space. We develop an for this contextual bandit problem that guarantees a $O(\sqrt{n}\ln n) = \tilde{O}(\sqrt{n})$ upper bound on (1/2)-regret. According to our experimental results on a benchmark dataset, the proposed algorithm outperforms existing algorithms in terms of not only our combined reward but also the reward of rating sum only.

Keywords: diversity · recommender systems · contextual bandits

1 Introduction

In recent years, the use of internet-based services such as social networking platforms, e-commerce platforms, and video streaming platforms has been steadily increasing. Many of these services incorporate content recommendation features, which aim to enhance user satisfaction by leveraging user behavior data to personalize recommendations. Maintaining high user satisfaction is crucial, as it encourages continued use of the service. Therefore, providing appropriate recommendations is of significant importance.

A "good recommendation" is typically regarded as one that accurately reflects the user's preferences (i.e., items the user is likely to appreciate). However, several studies on recommendation diversity argue that focusing only on recommendation accuracy is insufficient, and some of these studies further suggest that introducing an appropriate level of diversity into the recommended items can enhance user satisfaction. As a result, diversity remains one of the key research themes in the field of recommender systems [4].

S. Džeroski et al. (Eds.): DS 2025, LNAI 16090, pp. 510–521, 2025.
https://doi.org/10.1007/978-3-032-05461-6_33

In this study, we address the problem of recommending a set of diverse items to users while considering their preferences by applying algorithms for the contextual bandit problem. Specifically, we assume that the user's preference for an item can be represented by a linear function of the item's feature vector $\boldsymbol{x}$, modeled as $\theta^T\boldsymbol{x}$. Based on this assumption, we develop a recommendation method of the C^{2}UCB algorithm framework [7]—which aims to maximize cumulative reward under a semi-bandit setting—to simultaneously achieve high accuracy and high diversity. To optimize the recommended item set, our algorithm uses the Sum diversity calculated by some distance measure over the feature vector space as the diversity measure.

In our bandit setting, the reward function $R_t(S)$ of item set S at time step t is defined as $\sum_{i\in S} r(i) + \lambda \sum_{\{i,j\}\subset S} l(\boldsymbol{x}(i), \boldsymbol{x}(j))$, where $r(i)$ and $\boldsymbol{x}(i)$ are the score and feature vector, respectively, of item i at time step t, and l is a distance over the feature vector space. The C^{2}UCB algorithm framework requires the set S that maximizes $R_t(S)$ among the available sets ($|S| = k$ in our setting for fixed positive integer k) for score upper confidence bounds $\{\hat{r}(i)\}_{i\in S}$ instead of true scores $\{r(i)\}_{i\in S}$. Unfortunately, this problem for our reward function $R_t(S)$ is known to be NP-hard [1], but a set S with $R_t(S) \geq (1/2)\max_S R_t(S)$ is known to be able to be found by a greedy algorithm. By using the greedy algorithm, our algorithm is guaranteed to achieve $O(\sqrt{n}\ln n) = \tilde{O}(\sqrt{n})$ regret in n-round setting with respect to 1/2-regret, which is defined as the difference of the expected cumulative reward from 1/2 times the maximum expected cumulative reward among the sequence of available sets.

According to our experimental results using the MovieLens 100K dataset, the proposed method outperforms C^{2}UCB algorithms using other reward functions not only in terms of our combined reward, which reflects both accuracy and diversity, but also in terms of the rating sum, which reflects accuracy alone. Notably, our algorithm performs better than C^{2}UCB algorithms using entropy regularizer as diversity measure [7], even in terms of the reward defined using entropy regularizer. This fact indicates the effectiveness of the item sets selected by our algorithm to learn the preference (rating) function.

Related Work

In the field of recommendation diversification, Ziegler et al. proposed a collaborative filtering approach that incorporates diversity, and conducted both online and offline studies on the impact of diversity in recommendation results [12]. For recommendation based on contextual bandit problem, Li et al. proposed the LinUCB algorithm for linear reward settings [5]. Furthermore, Qin et al. formulated a contextual combinatorial bandit problem and developed a new algorithm for recommending item sets, where the reward function includes a term designed to promote diversity based on their own metric [7].

2 Problem Setting

We address the contextual combinatorial bandit problem [7], in which each arm is represented by a feature vector, and a player repeatedly selects a set of arms based on their feature vectors and receives a corresponding reward. This problem is formulated as follows.

Let n be the number of time steps, and let m be the number of arms. We define $[i]$ as $[i] = \{1, 2, \ldots, i\}$ for any positive integer i. Let $\mathcal{S}_t \subseteq 2^{[m]}$ be the set of all available k-arm subsets at time step $t \in [n]$. We call each available k-arm subset a super arm. At each time step t, the player receives a set of m feature vectors $\{\boldsymbol{x}_t(1), \ldots, \boldsymbol{x}_t(m)\} \subseteq \mathbb{R}^d$ corresponding to the m arms, and selects a super arm $S_t \in \mathcal{S}_t$. The player then receives the scores $\{r_t(i)\}_{i \in S_t}$ associated with each arm contained in the super arm S_t. The score $r_t(i)$ of each arm $i \in [m]$ is assumed to be $r_t(i) = \boldsymbol{\theta}_*^\top \boldsymbol{x}_t(i) + \epsilon_t(i)$, where $\boldsymbol{\theta}_*$ is an unknown parameter vector and $\epsilon_t(i)$ is a zero-mean random variable.

The reward $R_t(S_t)$ obtained by the player by selecting a super arm S_t is defined as the sum of the scores $r_t(i)$ of arms i in S_t and a diversity $D(S_t, \boldsymbol{X}_t)$ of S_t for $\boldsymbol{X}_t = [\boldsymbol{x}_t(1), \ldots, \boldsymbol{x}_t(m)]$:

$$R_t(S_t) = \sum_{i \in S_t} r_t(i) + \lambda D(S_t, \boldsymbol{X}_t), \tag{1}$$

where λ is a positive parameter scaling the diversity term. In this study, we adopt Sum diversity SD_l based on some distance function l as the diversity measure D, which is defined as

$$\mathrm{SD}_l(S_t, \boldsymbol{X}_t) = \sum_{\{i,j\} \subseteq S_t} l(\boldsymbol{x}_t(i), \boldsymbol{x}_t(j)).$$

The player's objective is the maximization of the expected cumulative reward $\mathbb{E}[\sum_{t \in [n]} R_t(S_t)]$.

Define $f_{\boldsymbol{r},\boldsymbol{X}}(S)$ as

$$f_{\boldsymbol{r},\boldsymbol{X}}(S) = \sum_{i \in S} r(i) + \lambda D(S, \boldsymbol{X})$$

for $S \subseteq [m]$, $\boldsymbol{r} = \{r(i)\}_{i \in [m]}$ and $\boldsymbol{X} = [\boldsymbol{x}(1), \boldsymbol{x}(2), \ldots, \boldsymbol{x}(m)] \in \mathbb{R}^d$. Note that $\mathbb{E}[R_t(S_t)] = f_{\boldsymbol{r}_t^*, \boldsymbol{X}_t}(S_t)$, where $\boldsymbol{r}_t^* = \{\boldsymbol{\theta}_*^\top \boldsymbol{x}_t(i)\}_{i \in [m]}$. We assume that function $f_{\boldsymbol{r},\boldsymbol{X}}(\cdot)$ satisfies the following three properties:

A1: Monotonically non-decreasing with respect to $\boldsymbol{r}$

$$r(i) \le r'(i) \text{ for all } i \in [m] \Rightarrow f_{\boldsymbol{r},\boldsymbol{X}}(S) \le f_{\boldsymbol{r}',\boldsymbol{X}}(S).$$

A2: Lipschitz Continuous with respect to $\{r(i)\}_{i \in S}$

There exists $C > 0$ such that

$$|f_{\boldsymbol{r},\boldsymbol{X}}(S) - f_{\boldsymbol{r}',\boldsymbol{X}}(S)| \le C \sqrt{\sum_{i \in S} |r(i) - r'(i)|}.$$

A3: Existence of α-approximation oracle

There exists an oracle $\mathcal{O}_{\mathcal{S}}(\boldsymbol{r}, \boldsymbol{X})$ such that, for some $\alpha \leq 1$, given $\boldsymbol{r}$ and $\boldsymbol{X}$, $\mathcal{O}_{\mathcal{S}}(\boldsymbol{r}, \boldsymbol{X})$ returns $S \in \mathcal{S}$ that satisfies

$$f_{\boldsymbol{r},\boldsymbol{X}}(S) \geq \alpha f_{\boldsymbol{r},\boldsymbol{X}}(S_*),$$

where $S_* = \arg\max_{S \in \mathcal{S}} f_{\boldsymbol{r},\boldsymbol{X}}(S)$.

Using $f_{\boldsymbol{r},\boldsymbol{X}}(\cdot)$ that has the above three properties, we define α-regret α-Reg(n) of an algorithm that selects a super arm S_t at time step t as

$$\alpha\text{-Reg}(n) = \sum_{t \in [n]} \left(\alpha \max_{S \in \mathcal{S}_t} f_{\boldsymbol{r}_t^*, \boldsymbol{X}_t}(S) - f_{\boldsymbol{r}_t^*, \boldsymbol{X}_t}(S_t) \right),$$

where $\boldsymbol{r}_t^* = \{\boldsymbol{\theta}_*^\top \boldsymbol{x}_t(i)\}_{i \in [m]}$.

3 Algorithm

In this section, we explain the C²UCB algorithm proposed by Qin et al. [7] for solving the contextual combinatorial bandit problem and our version of the algorithm that uses Sum diversity as a diversity measure.

3.1 C²UCB Algorithm

C²UCB algorithm proposed by Qin et al. [7] is an extension of the LinUCB algorithm [5] to the set selection setting. The pseudocode of the C²UCB algorithm is shown in Algorithm 1. This algorithm repeatedly observes feature vectors, selects a set of arms, and receives a corresponding reward, aiming to maximize the expected cumulative reward $\mathbb{E}\left[\sum_{t \in [n]} R_t(S_t)\right]$ that can be represented by $\sum_{t \in [n]} f_{\boldsymbol{r}_t^*, \boldsymbol{X}_t}(S_t)$ using a function $f_{\boldsymbol{r},\boldsymbol{X}}(\cdot)$ satisfying Assumptions A1-3.

The following α-regret upper bound is guaranteed for the C²UCB algorithm.

Theorem 1 (Theorem 4.1 in [7]). *Assume that $||\theta_*|| \leq M$, $\|\boldsymbol{x}_t(i)\|_2 \leq 1$ and $r_t(i) \in [0, 1]$ for all $t \in [n]$ and $i \in [m]$. Given $0 < \delta < 1$, let $\alpha_t = \sqrt{d \log\left(\frac{1+tm/\lambda}{\delta}\right)} + \sqrt{\lambda} M$. Then, with probability at least $1 - \delta$, C²UCB satisfies*

$$\alpha\text{-Reg}(n) \leq C\sqrt{64nd\log(1 + nm/d\lambda)} \cdot \left(\sqrt{d\log((1 + nm/\lambda)/\delta)} + \sqrt{\lambda} M\right).$$

3.2 Monotone Submodular Function and Previous C²UCB Instance

First, we show the definition of a monotone submodular function because monotone-submodularity is an important property for a set function to have an efficiently-realizable α-approximation oracle for some small α.

Algorithm 1. C^2UCB algorithm [7]

Input: $\lambda_0, \alpha_1, \ldots, \alpha_n$

(Initialization) $\boldsymbol{V}_0 \leftarrow \lambda_0 \boldsymbol{I}_{d\times d}, \boldsymbol{b}_0 \leftarrow \boldsymbol{0}_d$

for $t = 1, \ldots, n$ **do**

  Receive $\boldsymbol{x}_t(1), \boldsymbol{x}_t(2), \ldots, \boldsymbol{x}_t(m)$ and $\mathcal{S}_t$.

  $\hat{\theta}_t \leftarrow \boldsymbol{V}_{t-1}^{-1} \boldsymbol{b}_{t-1}$

  for $i \in 1, \ldots m$ **do**

    $\bar{r}_t(i) \leftarrow \hat{\theta}_t^T \boldsymbol{x}_t(i)$

    $\hat{r}_t(i) \leftarrow \bar{r}_t(i) + \alpha_t \sqrt{\boldsymbol{x}(i)^T \boldsymbol{V}_t^{-1} \boldsymbol{x}(i)}$

  end for

  $S_t \leftarrow \mathcal{O}_{\mathcal{S}_t}(\hat{\boldsymbol{r}}_t, \boldsymbol{X}_t)$

  Play the super arm S_t and observe the scores $\{r_r(i)\}_{i \in S_t}$.

  $\boldsymbol{V}_t \leftarrow \boldsymbol{V}_{t-1} + \sum_{i \in S_t} \boldsymbol{x}_t(i) \boldsymbol{x}_t(i)^T$

  $\boldsymbol{b}_t \leftarrow \boldsymbol{b}_{t-1} + \sum_{i \in S_t} r_t(i) \boldsymbol{x}_t(i)$

end for

Definition 1. *Let U be a ground set. If a set function $f : 2^U \to \mathbb{R}$ satisfies*

$$f(S) + f(T) \geq f(S \cap T) + f(S \cup T)$$

for any subset $S, T \subseteq U$, then f is said to be submodular. Furthermore, if a submodular function $f : 2^U \to \mathbb{R}$ satisfies

$$f(S) \leq f(T)$$

for any subset $S \subseteq T \subseteq U$, then f is called a monotone submodular function.

For a monotone submodular function f, $(1 - 1/e)$-approximation solution $\hat{S}$ of S_* with $f(S_*) = \max_{S \in \mathcal{S}} f(S)$ is guaranteed by the output $\hat{S}$ of a greedy algorithm in the case that $\mathcal{S} = \{S \in 2^{[m]} \mid |S| \leq k\}$ [6].

C^2UCB algorithm, which is introduced in the previous subsection, can deal with various problems by changing the reward function $R_t(S_t)$ and the corresponding α-approximation oracle $\mathcal{O}_{\mathcal{S}_t}(\hat{\boldsymbol{r}}_t, \boldsymbol{X}_t)$. Qin et al. [7] employed the reward function $R_t(S_t)$ defined in Eq. (1) to evaluate a super arm S_t using an entropy regularizer h [8] as the diversity measure D. The entropy regularizer h is defined as

$$h(S_t, \boldsymbol{X}_t) = \frac{1}{2}|S_t| \log(2\pi e) + \frac{1}{2} \log \det(\boldsymbol{X}_t(S_t)^\top \boldsymbol{X}_t(S_t) + \sigma^2 \boldsymbol{I}),$$

where $\boldsymbol{X}_t(S_t)$ is the matrix whose columns are the feature vectors corresponding to the arms in the super arm S_t, and σ^2 is a scalar parameter. We let function $R_t^h(\cdot)$ denote the set function $R_t(\cdot)$ defined by Eq. (1) using the entropy regularizer h as D.

Since the set function $f_{r,\boldsymbol{X}}(S)$ defined by $f_{r,\boldsymbol{X}}(S) = \sum_{i \in S} r(i) + \lambda h(S, \boldsymbol{X})$ using the entropy regularizer h is monotone submodular when $\sigma^2 \geq 0.0586$, the greedy algorithm realizes a $(1 - \frac{1}{e})$-approximation oracle. Thus, in the setting with reward functions $\{R_t^h\}_{t \in [n]}$ and the oracle $\mathcal{O}_{\mathcal{S}_t}$ realized by the greedy algorithm for the monotone submodular function, the $(1 - \frac{1}{e})$-regret for C^2UCB is guaranteed to be $\tilde{O}(\sqrt{n})$ by Theorem 1.

3.3 Max-Sum k Diversification Problem

Our proposed algorithm is also an instance of the C^{2}UCB algorithm that uses reward function $R_t(\cdot)$ of Eq. (1) using Sum diversity SD_l for some distance function l as diversity measure D. Unfortunately, $R_t(\cdot)$ using SD_l does not satisfy submodularity due to the second term, thus the greedy algorithm is not guaranteed to output a $(1 - 1/e)$-approximation solution.

To address this issue, we consider an approximation algorithm for the following Max-Sum k diversification problem [1], MSkD problem for short.

Problem 1 (MSkD problem [1]*).* Let U be a metric space with a distance function $l(\cdot, \cdot)$, and let $f(\cdot)$ be a monotone submodular function defined on 2^U. Define $\phi(S)$ as

$$\phi(S) = f(S) + \lambda \sum_{\{u,v\}:u,v \in S} l(u, v)$$

for a constant $\lambda > 0$ and any $S \subseteq U$. Find a solution $S_* \subseteq U$ of the following problem:

$$\begin{aligned} &\text{maximize} && \phi(S) \\ &\text{subject to} && |S| = k. \end{aligned}$$

Algorithm 2 is the pseudocode of the greedy algorithm proposed by Borodin et al. [1] for the MSkD problem. This algorithm iteratively constructs a solution set S, starting from the empty set, by repeatedly adding the element that yields the largest marginal gain with respect to the function

$$\phi'(S) = \frac{1}{2} f(S) + \lambda \sum_{\{u,v\}:u,v \in S} l(u, v)$$

until the size of S reaches k. Note that the coefficient of $f(S)$ in $\phi'(\cdot)$ is not 1 but 1/2, thus $\phi'(\cdot)$ is different from the objective function $\phi(\cdot)$ of MSkD problem.

Algorithm 2. Greedy algorithm for MSkD problem [1]

$S \leftarrow \emptyset$
for $j = 1, \ldots, k$ **do**
  find $u \in U \setminus S$ maximizing $\phi'(S \cup \{u\}) - \phi'(S)$
  $S \leftarrow S \cup \{u\}$
end for
return S

As an approximation guarantee, the following theorem is known to hold for Algorithm 2.

Theorem 2 (Theorem 1 in [1]**).** *For the set $\hat{S} \subseteq U$ that is outputted by the greedy algorithm (Algorithm 2),*

$$\phi(\hat{S}) \geq \frac{1}{2} \max_{S \subseteq U : |S| = k} \phi(S)$$

holds.

3.4 Proposed Algorithm

Our proposed algorithm is the C^2UCB algorithm using SD_l for some distance function l in $\mathbb{R}^d$ as diversity measure D and using the greedy algorithm for MSkD problem as 1/2-approximation oracle.

In the case with $D = \mathrm{SD}_l$, the problem of maximizing $f_{\boldsymbol{r},\boldsymbol{X}}(S)$ among $S \in \mathcal{S} = \{S \in 2^{[m]} \mid |S| = k\}$ can be seen as an instance of Problem 1 for $U = \{\boldsymbol{x}(1), \ldots, \boldsymbol{x}(m)\}$ in $\mathbb{R}^d$ with distance function l, monotone submodular function $f(S) = \sum_{i:\boldsymbol{x}(i)\in S} r(i)$ with $r(i) \geq 0$, and $\phi(S) = f_{\boldsymbol{r},\boldsymbol{X}}(\{i \mid \boldsymbol{x}(i) \in S\}) = \sum_{i:\boldsymbol{x}(i)\in S} r(i) + \lambda \sum_{\{i,j\}:\boldsymbol{x}(i),\boldsymbol{x}(j)\in S} l(\boldsymbol{x}(i), \boldsymbol{x}(j))$. By Theorem 2, the greedy algorithm for MSkD problem can be used as 1/2-approximation oracle for $f_{\boldsymbol{r},\boldsymbol{X}}(S)$ with $D = \mathrm{SD}_l$. The pseudocode for the corresponding greedy algorithm is shown in Algorithm 3.

Algorithm 3. Greedy algorithm for MSkD problem with $U = \{\boldsymbol{x}(1, \ldots, \boldsymbol{x}(m))\} \subset \mathbb{R}^d$ and $f(S) = \sum_{i:\boldsymbol{x}(i)\in S} r(i)$.

Input: $\boldsymbol{r} \in \mathbb{R}^m, \boldsymbol{X} = [\boldsymbol{x}(1), \ldots, \boldsymbol{x}(m)] \in \mathbb{R}^{d\times m}$

$S \leftarrow \emptyset$
for $j = 1 \ldots k$ **do**
  for $i \in [m] \setminus S$ **do**
    $\delta_i^{(R)} \leftarrow \frac{1}{2} r(i)$
    $\delta_i^{(d)} \leftarrow \sum_{\ell\in S} l(\boldsymbol{x}(i), \boldsymbol{x}(\ell))$
  end for
  $i^* \leftarrow \arg\max_{i\in[m]\setminus S} \delta_i^{(R)} + \lambda \delta_i^{(d)}$
  $S \leftarrow S \cup \{i^*\}$
end for
return S

Note that Theorem 1 holds for our instance of C^2UCB with $\alpha = 1/2$. Thus, $1/2\text{-Reg}(n) = \tilde{O}(\sqrt{n})$ is guaranteed for the proposed algorithm.

4 Experiments

In this section, we demonstrate the effectiveness of the proposed method by comparing its performance with that of baseline approaches through movie recommendation simulations using the MovieLens 100K dataset.

4.1 Dataset

The MovieLens 100K dataset [2] used in our experiments is a benchmark dataset of movie ratings provided by GroupLens. It consists of 100,000 ratings from 943 users on 1,682 movies. Each data point is represented in the form that user u_i evaluates movie v_j with score $r_{i,j}$ ranging from 1 (lowest) to 5 (highest). Here, i and j denote the user ID and movie ID, respectively.

We use only the rating data of users who have rated more than 100 movies. The MovieLens 100K dataset contains 364 such users out of 943 total users, and we conduct movie recommendation experiments using bandit algorithms for these 364 users.

In our experiments, each movie is treated as an arm in the bandit algorithm, and its genre information is used as the feature vector of the arm. The MovieLens dataset includes 19 genres, and each movie belongs to at least one of them. For each movie i, we construct a 19-dimensional vector $\boldsymbol{x}'_i$ whose components correspond to genres:

$$x'_{i,j} = \begin{cases} 1 & \text{if movie i belongs to genre j, and} \\ 0 & \text{otherwise.} \end{cases}$$

The normalized vector $\boldsymbol{x}_i = \boldsymbol{x}'_i / \|\boldsymbol{x}'_i\|_2$ is used as the feature of arm i.

4.2 Comparison Algorithms

We compare the performance of our method with that of the following variations of the C^2UCB algorithm.

$\mathbf{C^2}$UCB[ER] Algorithm [7]**:** C^2UCB algorithm that uses the reward function $R_t^h(\cdot)$ with entropy regularizer diversity measure h, and uses the greedy algorithm for maximizing the corresponding monotone submodular function $f_{r,\mathbf{X}}(S)$ as a $(1-1/e)$-approximation oracle.

$\mathbf{C^2}$UCB[ND] Algorithm: C^2UCB algorithm that uses a reward function $R_t(\cdot)$ with no diversity term, that is, $R_t(S) = \sum_{i \in S} \hat{r}_t(i)$. An oracle for maximizing $f_{r,\mathbf{X}}(S) = \sum_{i \in S} r(i)$ among $S \in \mathcal{S}$ returns the set S_* that is composed of the arms i whose score $r(i)$ is within the top k, which can be calculated in $O(n \log k)$ time using the heap structure.

$\mathbf{C^2}$UCB[NDϵG] Algorithm: The ϵ-Greedy [9]-like C^2UCB[ND] algorithm that selects the set of k arms S by starting from $S = \emptyset$, repeatedly adding one arm to S from the set of the remaining arms $[m] \setminus S$. The added one arm is a randomly-selected arm with probability ϵ and the arm i with the highest estimated score $\hat{r}_t(i)$ with probability $1 - \epsilon$.

4.3 Evaluation Metrics

In addition to instantaneous and cumulative reward, we use precision and intra-list distance (ILD) [11,12] as evaluation metrics for the recommended movie set at each time step of the algorithms.

Precision: For each user u in the test data, we define the preferred movie set L_u as the set of movies that the user rated 4 or 5. Let $S_{t,u}$ be the set of items recommended to user u at time step t. Then, the precision $p_{t,u}$ of the recommendation for user u at time step t is defined as

$$p_{t,u} = \frac{|S_{t,u} \cap L_u|}{|S_{t,u}|}.$$

Intra-List Distance (ILD): Let $S_{t,u}$ be the set of items recommended to user u at time step t. The intra-list distance (ILD) of $S_{t,u}$ is defined using Jaccard distance between two sets $\{k \in [d] \mid x_{i,k} > 0\}$ and $\{k \in [d] \mid x_{j,k} > 0\}$ as

$$\frac{1}{|S_{t,u}|(|S_{t,u}| - 1)} \sum_{i \in S_{t,u}} \sum_{j \in S_{t,u}, i \neq j} \left(1 - \frac{|\{k \in [d] \mid x_{i,k}x_{j,k} > 0\}|}{|\{k \in [d] \mid x_{i,k} + x_{j,k} > 0\}|}\right).$$

Note that $x_{i,k} > 0 \Leftrightarrow$ (movieibelongs to genrek).

4.4 Settings

We conduct movie recommendation experiments for each of the 364 users who rated more than 100 movies, considering only the movies rated by the target user as candidate items. Let I_u denote the set of movies rated by user u. The experiments are carried out under two different scenarios:

Experiment 1: The algorithms recommend a set of 10 movies ($k = 10$) at each time step $t = 1, \ldots, 20$, under the condition that the same movie can be recommended multiple times. Thus, $\mathcal{S}_t = \{S \subset I_u \mid |S| = 10\}$ for user u and $t = 1, \ldots, 20$.

Experiment 2: The algorithms recommend a set of 5 movies ($k = 5$) at each time step $t = 1, \ldots, 20$, under the condition that each movie can be recommended at most once. Thus, $\mathcal{S}_t = \{S \subset I_u \mid |S| = 5\} \setminus \bigcup_{s<t} S_s$ for user u and $t = 1, \ldots, 20$.

The matrix initialization parameter λ_0 of C^{2}UCB algorithm (Algorithm 1) is set to $\lambda_0 = 10$, and the exploration parameter α_t at time step t is set to $\alpha_t = 3.0$ for all the algorithms. These parameters were selected by grid search among $(\lambda_0, \alpha_t) \in \{10, 20, 30, 40, 50\} \times \{1.0, 2.0, 3.0, 4.0, 5.0\}$ in terms of the precision of C^{2}UCB[ND] algorithm at $t = 20$ in the scenario of Experiment 1.

The diversity-scaling parameter is set to $\lambda = 0.01$ for the proposed algorithm, and $\lambda = 0.001$ for C^{2}UCB[ER] algorithm. These λ were selected by grid search among $\lambda \in \{0.001, 0.005, 0.01, 0.05, 0.1, 0.15, 0.20\}$ in terms of the average precision of the algorithms over 20 rounds in the scenario of Experiment 1.

The exploration-scaling parameter of the ϵ-greedy algorithm is set to $\epsilon = 0.1$.

The true score $r(i)$ of an arm i used as feedback is set to 1 if the corresponding movie is rated 4 or 5 by the target user, and 0 otherwise.

We evaluate the recommendation algorithms based on their performance over all time steps using precision and intra-list distance (ILD), in addition to the reward defined by Eq. (1). In the experiment, we use L_2 distance as distance l for Sum diversity in the proposed algorithm.

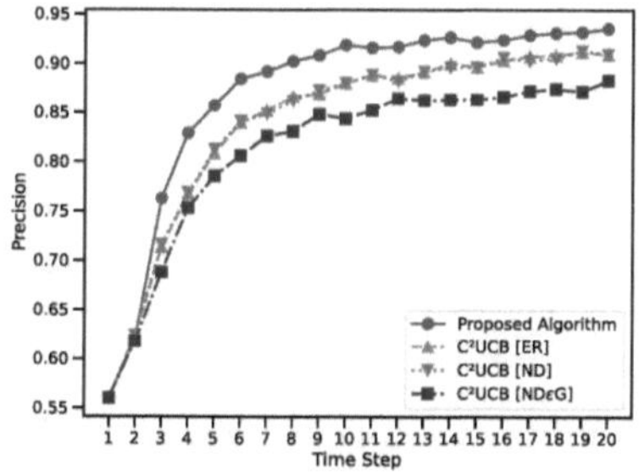

(a) Precision over time

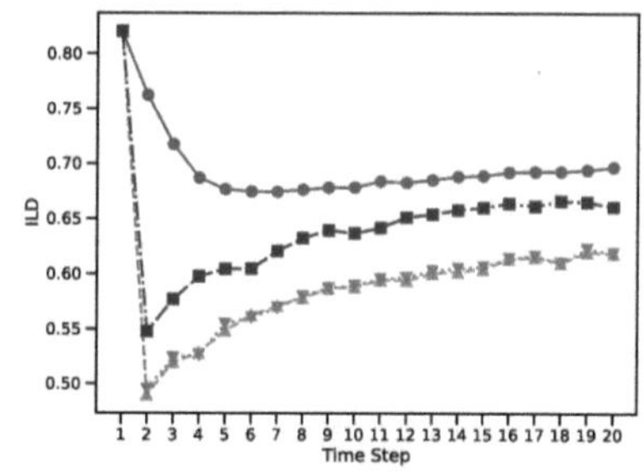

(b) Intra-list distance over time

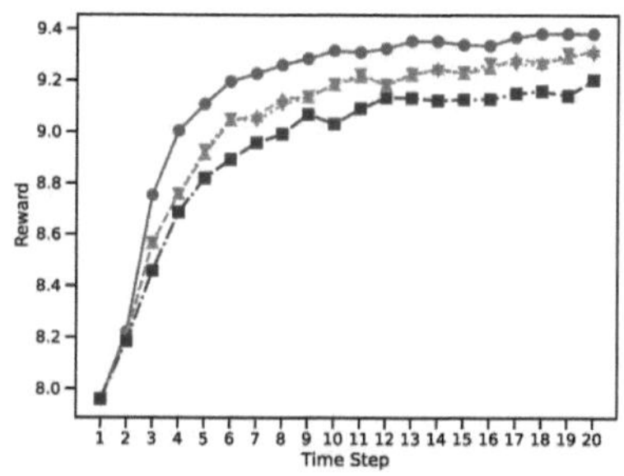

(c) Reward with entropy-regularizer

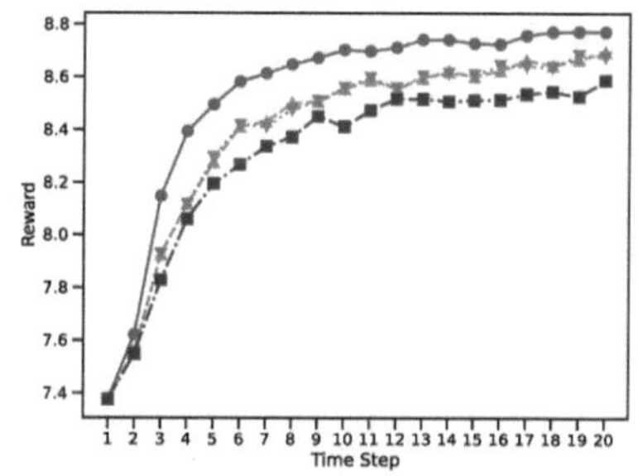

(d) Reward with Sum diversity

Fig. 1. Results of Experiment 1 for the 364 selected users. Each plot shows the average value over users at each time step.

4.5 Results

The results of Experiments 1 and 2 are shown in Figs. 1 and 2, respectively.

In terms of instantaneous and cumulative rewards, which are combined measures of precision and diversity, Figs. 1(c), 1(d), 2(c), and 2(d) demonstrate that the proposed algorithm achieves higher average reward than the other C^2UCB algorithms, including the C^2UCB[ER] algorithm. Note that C^2UCB[ER] selects a set of arms to maximize the reward with an entropy regularizer, but our algorithm performs better even in this evaluation measure. This fact indicates that arm sets selected using Sum diversity are better than those selected using the entropy regularizer to learn the parameter θ of the score function in this dataset.

Regarding precision, we can see that the proposed algorithm achieves the highest precision among all the algorithms at all the time steps through Fig. 1(a) and 2(a). In particular, it outperforms the C^2UCB[ND] algorithm, which does not consider the diversity of the selected arm set.

For the diversity measure, as shown in Fig. 1(b) and 2(b), our algorithm maintains higher intra-list distance (ILD) than the other algorithms. The arms in the set selected by our algorithm are more diverse than those in the set selected by C^2UCB[NDϵG] algorithm, which introduces randomness into the arm selection process.

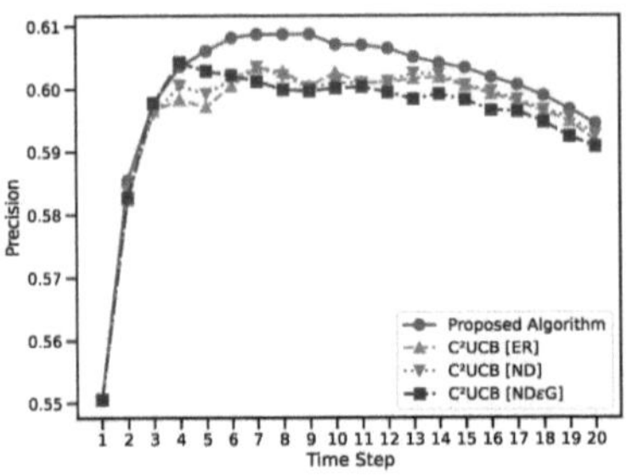

(a) Cumulative Average precision

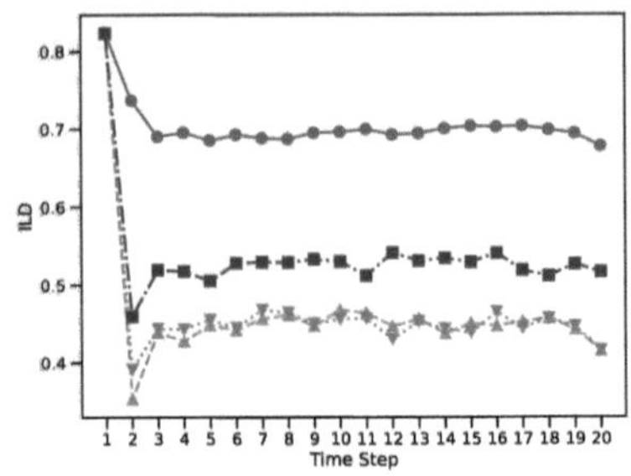

(b) Average Intra-list distance

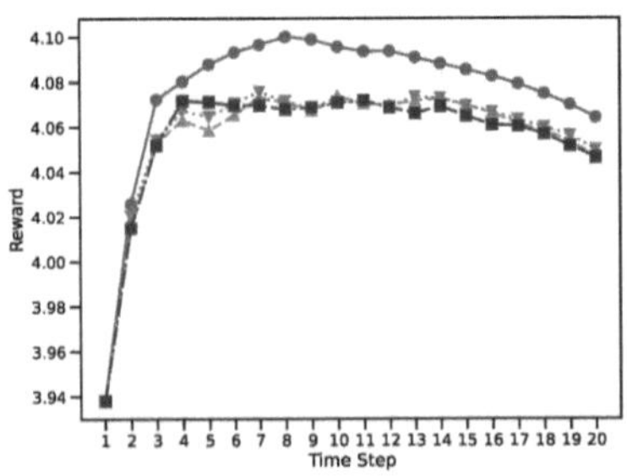

(c) Cumulative average reward with entropy-regularizer

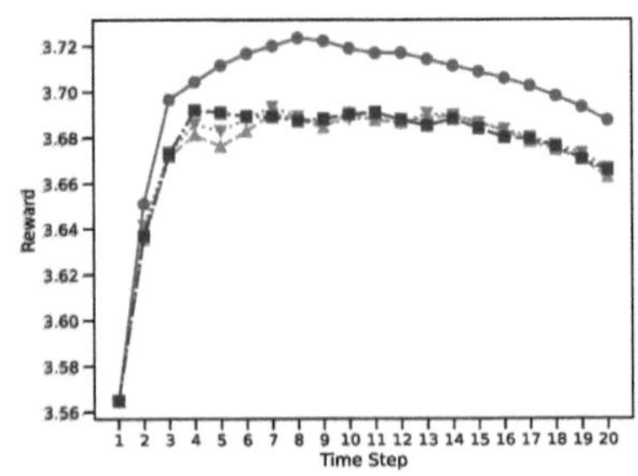

(d) Cumulative average reward with Sum diversity

Fig. 2. Results of Experiment 2 for the 364 selected users. Plots (a), (c), and (d) show the cumulative average values over time steps.

5 Conclusion and Future Work

In this paper, we proposed an online algorithm to recommend a set of items with high rating and diversity using a contextual bandit approach. Experiments on a benchmark dataset demonstrated that our method outperforms existing algorithms in terms of precision for high-rated items and diversity, in addition to the reward that are optimized in our algorithm.

As a rating prediction model, we adopt a linear function, which may limit the range of its application. One solution of this issue is to extend our algorithm to a kernelised version incorporating the techneque of KernelUCB [10] or GP-UCB [3] into our method.

Our experiment using a dense part of MovieLens 100K and genre information as context may not be enough to demonstrate the effectiveness of our algorithm. We are planning to conduct experiments on larger datasets with richer features.

Acknowledgments. This work was supported by JSPS KAKENHI Grant Number JP24H00685.

Disclosure of Interests. The authors have no competing interests to declare that are relevant to the content of this article.

References

1. Borodin, A., Lee, H.C., Ye, Y.: Max-sum diversification, monotone submodular functions and dynamic updates. In: Proceedings of the 31st ACM SIGMOD-SIGACT-SIGAI Symposium on Principles of Database Systems (PODS), pp. 155–166 (2012). https://doi.org/10.1145/2213556.2213580
2. Harper, F.M., Konstan, J.A.: The movielens datasets: history and context. ACM Trans. Interact. Intell. Syst. **5**(4) (2015). https://doi.org/10.1145/2827872
3. Krause, A., Ong, C.S.: Contextual gaussian process bandit optimization. In: Proceedings of the 25th International Conference on Neural Information Processing Systems, pp. 2447–2455. NIPS'11, Curran Associates Inc., Red Hook, NY, USA (2011)
4. Kunaver, M., Požrl, T.: Diversity in recommender systems – a survey. Knowl.-Based Syst. (KBS) **123**, 154–162 (2017). https://doi.org/10.1016/j.knosys.2017.02.009
5. Li, L., Chu, W., Langford, J., Schapire, R.E.: A contextual-bandit approach to personalized news article recommendation. In: Proceedings of the 19th International Conference on World Wide Web (WWW), pp. 661–670 (2010). https://doi.org/10.1145/1772690.1772758
6. Nemhauser, G., Wolsey, L., Fisher, M.: An analysis of approximations for maximizing submodular set functions-i. Math. Program. Ser. A and B **14**(1), 265–294 (1978). https://doi.org/10.1007/BF01588971
7. Qin, L., Chen, S., Zhu, X.: Contextual combinatorial bandit and its application on diversified online recommendation. In: Proceedings of the 2014 SIAM International Conference on Data Mining (SDM), pp. 461–469 (2014). https://doi.org/10.1137/1.9781611973440.53
8. Qin, L., Zhu, X.: Promoting diversity in recommendation by entropy regularizer. In: Proceedings of the Twenty-Third International Joint Conference on Artificial Intelligence (IJCAI), pp. 2698–2704 (2013)
9. Sutton, R.S., Barto, A.G.: Reinforcement Learning: An Introduction. A Bradford Book, Cambridge, MA, USA (2018)
10. Valko, M., Korda, N., Munos, R., Flaounas, I., Cristianini, N.: Finite-time analysis of kernelised contextual bandits. In: Proceedings of the Twenty-Ninth Conference on Uncertainty in Artificial Intelligence, pp. 654–663. UAI'13, AUAI Press, Arlington, Virginia, USA (2013)
11. Zhang, M., Hurley, N.: Avoiding monotony: improving the diversity of recommendation lists. In: Proceedings of the 2008 ACM Conference on Recommender Systems (RecSys), pp. 123–130 (2008). https://doi.org/10.1145/1454008.1454030
12. Ziegler, C.N., McNee, S.M., Konstan, J.A., Lausen, G.: Improving recommendation lists through topic diversification. In: Proceedings of the 14th international conference on World Wide Web (WWW), pp. 22–32 (2005). https://doi.org/10.1145/1060745.1060754

Explaining Anomalous Data with Reinforcement Learning

Simone Amirato(✉), Fabrizio Angiulli, and Fabio Fassetti

DIMES Department, University of Calabria, Rende, Italy
simone.amirato@dimes.unical.it,
{fabrizio.angiulli,fabio.fassetti}@unical.it

Abstract. Identifying the most informative features that discriminate between normal and anomalous data is a formidable challenge due to the enormous search space represented by the power set of the entire feature set. These features encode the explanation for the deviating behavior of selected observations and represent hidden knowledge of interest to analysts in many applications, such as cybersecurity, bioinformatics, and green AI, to name a few.

In this work, we propose a reinforcement learning-based algorithm to detect explanations from high-dimensional datasets. The algorithm takes as input an unlabeled collection of normal instances, a small set of labeled inlier examples, and a small set of anomalies to be explained. Separation between normal samples and anomalies is measured using a distance-based outlier score. By ranking features, our learning strategy isolates those that maximize separation accuracy.

Experimental results demonstrate that the size of the returned explanation is a small fraction of the full feature set, while the associated accuracy and subspace margin are of notably high quality.

Keywords: Explainable AI · Anomaly Explanation · Reinforcement Learning

1 Introduction

In this work, we address the *Anomaly Explanation problem*: given a population of normal data and a set of anomalous examples, the goal is to produce an interpretable explanation that justifies why the anomalies deviate from normal behavior.

We encode *explanations* as minimal feature subspaces where anomalies exhibit the most significant deviation. These subspaces highlight the most discriminative features, providing analysts with actionable insights in domains such as Cybersecurity, Bioinformatics, and Green AI.

However, identifying these features is challenging due to the combinatorial explosion of possible subsets in high-dimensional spaces. To tackle this, we propose the RLAX algorithm, for *Reinforcement Learning-based Anomaly eXplanation*, which formalizes the problem as a Markov Decision Process (MDP).

S. Džeroski et al. (Eds.): DS 2025, LNAI 16090, pp. 522–536, 2025.
https://doi.org/10.1007/978-3-032-05461-6_34

Specifically, given a measure of separation between normal and abnormal data, we define an *Explanation Gain* measure, which represents a trade-off between the separation accuracy achieved in a certain subspace and the subspace relative size. Our goal is thus that of detecting a feature set exhibiting the optimal *Gain*. Moreover, we define the *Subspace margin* as a measure of the separation between the densities associated with inliers and outliers.

RLAX explores the search space represented by all the subset of data features by taking as action the addition of a novel feature and by receiving as reward the corresponding variation of subspace margin. It follows the Proximal Policy Optimization (PPO) strategy to learn the optimal policy, thus taking advantage of its ability to work with large action spaces, a typical requirement for the setting considered here in which features to be selected represent actions to be taken. The outcome of the learning phase is eventually exploited to build a ranking among features, putting informative features first, from which the output anomaly explanation is obtained. As our main contribution, to the best of our knowledge, this is the first work to apply RL for anomaly explanation via feature selection. Experiments show that RLAX explanations are both concise (a tiny fraction of all features) and highly discriminative and that the associated accuracy and subspace margin are of remarkable quality, also in comparison with existing subspace selection methods for anomaly detection.

The rest of the work is organized as follows. Section 2 describes related work, Sect. 3 formalizes the anomaly explanation problem and describes the RLAX algorithm, Sect. 4 illustrates experimental results, and Sect. 5 draws conclusions and ends the work.

2 Related Works

Anomaly analysis involves two fundamental and complementary tasks: *anomaly detection* [5] and *anomaly explanation* [14,16,18], where the latter aims to determine the causes that make an instance anomalous. If an anomaly is detected by an algorithm, Explainable Anomaly Detection techniques aim at explaining the anomaly by making detector interpretable. Conversely, if an anomaly is identified by an expert, Explainable Anomaly Detection techniques aim at explaining why the given data instance is anomalous by extracting hidden knowledge from data. Our work deals with this second family and, then, in the following of the section we discuss on methods related to this task.

In particular, we focus on feature-based methods that provide explanations based on subspaces. We note that some methods for feature selection using reinforcement learning have been proposed in the context of classification [10, 11,20], however they are not designed for the anomaly explanation task.

The reader is referred to [14] for a survey on explainable anomaly detection. Among these methods, some do not rely on selection of interpretable or informative feature subsets, but instead project the data onto low-dimensional spaces. One representative example of this family of methods is SUOD [25], which employs random subspace projection techniques to generate diverse representations of the input space. These projections are used to build outlier ensembles. Similarly, [21] does not seek to explicitly select meaningful features, but performs outlier detection by combining random projection and sketching techniques. It estimates local density in randomly selected subspaces and aggregates this information to produce outlier scores. This family of techniques maps data in a feature-space different from the input one and, then, it cannot be used as explanation.

As for methods searching for subspace of the input feature-space and, thus, providing a set of "relevant" features interpretable as explanation, the seminal work [1] detects anomalies searching for subspaces in which the data density is exceptionally lower than the mean density of the whole data. Promising subspaces are detected by employing a technique based on genetic algorithms. Analogously, [22] presents the Z-score and the Isolation Path score (iPath), two dimensional-unbiased measures to avoid suffering high-dimensional distances. Although these methods work on the subspaces, they do not contemplate the presence of examples.

A related approach is proposed in [9], which formulates the outlier detection task as a subspace selection problem guided by both outlier and inlier examples. This method aims to identify the subspace where the separation between outliers and inliers is maximized. To efficiently explore the space of candidate subspaces a genetic algorithm is employed.

The work [19] introduces a method that integrates linear programming and heuristic search to improve detection in high-dimensional data. For each point, a local subspace that maximizes the sparsity of the current point with respect to its nearest neighbors is identified. Once a subset of feature is selected for each point, outlier score is computed using the Local Outlier Factor within the tailored subspace.

In [7] an ensemble-based outlier detection framework is proposed. Each ensemble identifies a subset of features by minimizing a ranking loss designed to separate outliers from inliers. A self-paced learning mechanism iteratively assigns weights to both features and samples, allowing the model to focus on easy instances and relevant attributes. The final outlier score is computed by aggregating the scores of all components of the ensemble through a weighted average with respect to the training loss.

HiCS [12] is a subspace-based outlier detection method that aims to identify high contrast subspaces, i.e., feature subsets where the local data distribution deviates significantly from the global distribution. The method uses statistical tests, particularly the Kolmogorov-Smirnov test, to assess the degree of contrast in candidate subspaces, and selects those that exhibit significant statistical dependence. Once such subspaces are identified, density-based outlier scoring method are applied to rank the anomalies.

The work [8] provides a mathematical formulation of the Multiple Views effect, casting subspace selection as a stochastic optimization problem. Based on this formulation, the authors introduce V-GAN, a generative model designed to solve the introduced optimization task. This approach allows the method to avoid an exhaustive search over the entire space of subsets of features.

In [13] the method SOD (Subspace Outlier Detection) is introduced, a subspace-based approach that detects outliers in varying subspaces of a high dimensional space. Instead of evaluating all features equally, SOD adaptively selects a relevant subspace for each data point based on its local neighborhood.

Some methods are focused on networks [24], some on sequence data [6,23] and some other methods are specifically designed for categorical data. One of the first methods has been proposed in [4], which addresses the task of identifying the outlying properties of given data instance. Although it does not explicitly perform feature selection technique per se, it focuses on extracting a subset of attributes that best account for the observed anomaly. CUFS [17] is a method, that deals with categorical data, performing unsupervised feature selection designed for outlier detection. It captures intra- and inter-feature dependencies through hierarchical value-feature coupling and maps them into a graph-based formulation. While it effectively filters irrelevant features in categorical settings, its applicability is restricted to non-numerical data.

M^2OE [3] is a method that assesses why a point is abnormal or not, by using a different kind of explanation. Specifically, it introduces a notion of transformation, where the explanation consists in identifying the minimal changes needed to turn an outlier instance into a normal one. Such method, like others [15], also work to contemporaneously explaining groups of outliers.

3 Problem Statement and Method

In this section we introduce RLAX algorithm, for Reinforcement Learning-based Anomaly eXplanation. The rest of the section is organized as follows: Sect. 3.1 provides the definition of Anomaly Explanation, Sect. 3.2 introduces the notion of Subspace Margin, Sect. 3.3 details the learning procedure, and Sect. 3.4 describes how the solutions are generated based on the output of the previous phase.

3.1 Anomaly Explanation

A *feature* f is an identifier with an associated domain $\mathbb{D}(f)$. The overall set of features is denoted by $F = \{f_1, \ldots, f_d\}$ and the associated *domain* $\mathbb{D}(F)$ is the cartesian production $\mathbb{D}(F) = \mathbb{D}(f_1) \times \ldots \times \mathbb{D}(f_d)$. A *subspace* S is a subset of F.

A *dataset* X on the set of features F is a set of elements x coming from $\mathbb{D}(F)$. Given a subspace S of F and an instance x of X, by x^S we denote the projection of x onto the subspace S, while by X^S we denote the set $\{x^S \mid x \in X\}$.

Assume a set E_{pos} of *outlier examples* (or *positive examples*) is given. We are interested in finding an *explanation* for the anomalous behavior of the instances in E_{pos}.

In this work, we encode explanations under the form of those minimal subspaces where the outlier examples deviate mostly from the dataset population.

Indeed, such an explanation makes intelligible the features characterizing the abnormality at hand.

Let *sep* be a measure of the separation quality between outlier and normal examples, that can be the subspace margin *margin* defined in the sequel of this section or any other different measure for evaluating classification accuracy, like the Area Under the ROC Curve (AUC) which is suitable for the unbalanced classes scenario typical of the anomaly detection task, to be employed as a measure of success in place of the subspace margin for a fair comparison with competitors aiming at optimizing other measures.

An *Anomaly Explanation* $\mathcal{E}$ of E_{pos} with respect to X is the subspace maximizing the value of *sep* over the dataset, that is to say the one exhibiting the better separation between normal and abnormal samples. More in the general, we consider the *Explanation Gain* measure, also called simply *Gain*, defined as

$$Gain_{\alpha}(S) = sep(S) - \alpha\frac{|S|}{|F|}$$

representing a trade-off between the separation accuracy guaranteed by the feature set S and its relative size $|S|/|F|$. Intuitively, putting one additional feature into the solution set should be compensated by a sufficiently large increase in the separation accuracy, increase which is precisely expressed by the quantity $\alpha/|F|$. Ideally $\alpha \leq 1$ and, if not otherwise stated, in the following we use the value $\alpha = 0.5$. We note that for $\alpha = 0$ the above measure is based only on the *sep* value. Thus, in the following we say that the best feature set is those maximizing the *Gain*, and refer to the *sep* measure also as $Gain_0$.

3.2 Subspace Margin

Given an instance x of X, the *anomaly score* $as(x)$ of x is defined as [2]:

$$as(x) = \frac{1}{K}\sum_{i=1}^{K} \text{dist}(x, nn_i(x)),$$

where $nn_i(x)$ denotes the i-th nearest neighbor of x in X according to some pre-defined distance measure. Thus, the anomaly score is given by the sum of the distances between x and its K nearest neighbors in the dataset. Its value provides an estimate of the dataset density in the neighborhood of the instance x. Intuitively, the higher the anomaly score associated with an instance, the more likely it is that the instance was generated by a distribution different from that of X. Such instances are called *outliers* or *anomalies.*

In the following, we assume a second set of examples is given, that is the set E_{neg} consisting of *inlier examples* (or *negative examples*), possibly singled out from the dataset X.

In order to measure the relevance of the subspace S as an explanation, next the concept of subspace margin is introduced. The *subspace margin* of the space

S with respect to the set of positive examples E_{pos} and the set of negative examples E_{neg} is

$$margin(S) = \frac{as(E_{pos}^S) - as(E_{neg}^S)}{\sqrt{|S|}},$$

where, with a little abuse of notation, the anomaly score $as(E)$ of a set of examples E is defined as the mean of the outlier scores associated with the elements of E. The subspace margin represents the difference between the anomaly scores of the positive and negative examples. Intuitively, the larger the subspace margin, the larger the separation between the densities associated with the two above mentioned set of examples. The denominator acts as a normalizing factor needed to compare distances coming from subspaces having different sizes.

Thus negative examples have a twofold usefulness: first of all they help to reduce the computational burden associated with estimating separation between normal and abnormal data densities and, moreover, they can be optionally pointed out by a domain expert in order to guide the search to focus on some characteristics of normal data of interest for the analysts. Finally, we note that if a set of negative examples is not available, a random selected subsample of the normal data X can always be reliably employed.

3.3 Learning Procedure

The RLAX algorithm, for Reinforcement Learning-based Anomaly eXplanation, formalizes the problem of detecting the anomaly explanation for a given set E_{pos} of negative examples as a reinforcement learning task.

In our formalization the notion of *state* S coincides with that of subspace. Specifically, we map the above problem into a Markovian Decision Process $\langle \mathcal{S}, A, T, R, \gamma \rangle$ where $\mathcal{S}$ is the set of states, that corresponds to the power set of the feature set F, and A is the set of actions, where an action a corresponds to adding a single feature to the current state.

We denote each action also as the feature that must be added, thus doing the action a in the state s produces the next state $s' = s \cup \{a\}$. Hence, the environment is deterministic and the probability $T(s, a, s')$ of transition from s to s' after doing a is always 1 for $s' = s \cup \{a\}$ and 0 otherwise. Clearly, in each state s, only the subset $A \setminus s$ of actions of A is available.

As for the *reward* R, we use the variation of the subspace margin between the increased subspace and the original subspace, that is

$$R(s, a, s \cup \{a\}) = margin(s \cup \{a\}) - margin(s).$$

We note that using the subspace margin during learning has two advantages. First of all, from the computational point of view it *can be more efficiently computed* with respect to measures, like the AUC, that require knowing the ranking of the anomaly scores associated with all the data, thus to compute the anomaly score of all examples and all dataset instances. Indeed, since the above scores have to be recomputed for each explored subspace, in the presence of large datasets the required temporal cost may become prohibitive. The margin instead can be more conveniently estimated by means of sampling procedures, like the one employed here exploiting positive and negative examples as representative samples of the abnormal and normal data populations respectively. Second, since the margin is a density estimation procedure, making decisions based on it enables the algorithm to be *agnostic to different quality measures* that the user could adopt to evaluate the returned explanation or that can be used to compare the quality of our explanations with that of alternative methods.

Each episode starts with the state $s_0 = \emptyset$, associated with the empty set of features, and proceeds by adding a selected feature a until a stopping criterion is met. The criterion for stopping current episode is the following: either (1) a maximum number t_{max} of steps is reached or (2) the reward remains negative for at least t_{neg} consecutive steps.

In order to learn the optimal policy we exploit the Proximal Policy Optimization (PPO) algorithm, belonging to the family of reinforcement learning algorithms that use policy gradient methods. These methods optimize the expected cumulative reward by finding a good policy under the form of a neural network parameterized policy, thanks to variants of stochastic gradient ascent with respect to the policy parameters. A peculiarity of these approaches is their ability to work with large and continuous action spaces. We prefer policy gradient methods over value based methods, like Deep Q-Networks (DQN), since in our setting the number of actions corresponds to the overall number of features, a number which is typically large in the application domain of interest for our technique and the latter methods are not well-suited to deal with large action spaces.

The learning phase stops after a total number of T steps is performed. Once the policy has been learned, the optimal subspace has to be obtained from it. To this aim, a criterion to select informative subspaces is needed. The idea is to exploit the outcome of the learning phase in order to build a *ranking* among features, putting informative features first, from which a set of candidate subspaces can be obtained.

3.4 Subspace Rankings and Solution

The first ranking we consider is the *MDR Ranking*, for empirical Mean of Discounted Rewards, defined as:

$$\mathrm{MDR}(a) = \frac{1}{c(a)} \sum_{k,t \;:\; a_t^{(k)}=a} \gamma^t r_t^{(k)},$$

where $a_t^{(k)}$ ($r_t^{(k)}$, resp.) denotes the action performed (reward obtained, resp.) at the t-th time step of the k-th episode, and $c(a)$ is the action count of a, that is the number of times a is performed during learning. According to the *MDR Ranking*, the first feature f_1 is that with the highest MDR value, the second feature f_2 is that with the second highest MDR value, and so on.

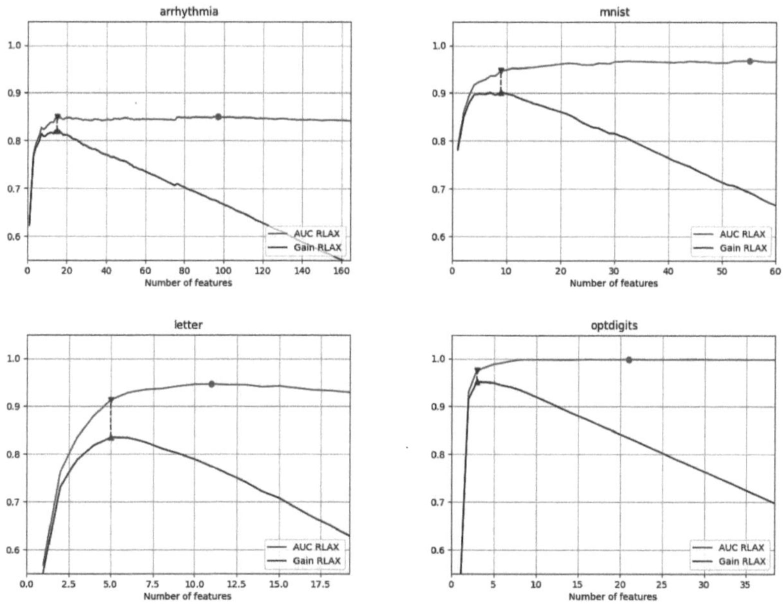

Fig. 1. Comparison between AUC and *Gain* with respect to the subspace dimension.(Color figure online)

The second ranking we consider, that we call *Policy Ranking*, can be derived from the learned policy π^*. According to this ranking, the i-th feature f_i ($1 \leq i \leq |F|$) is

$$f_i = \underset{f \in F \setminus \{f_1,\dots,f_{i-1}\}}{\arg\max} \pi^*(f \mid \{f_1, \dots, f_{i-1}\})$$

and, thus, the first feature f_1 is that with the highest probability $\pi^*(f \mid \emptyset)$.

Once a feature ranking $o = \langle f_1, \dots, f_d \rangle$ is fixed, a set $\mathcal{S}^o = \{S_1^o, \dots, S_d^o\}$ of d *candidate subspaces* is obtained, where $S_i^o = \{f_1, f_2, \dots, f_i\}$ ($1 \leq i \leq d$). Thus, according to each ranking o, we obtain the best explanation $\mathcal{E}^o$ as the subset of features S_i^o scoring the maximum gain

$$\mathcal{E}^o = \arg\max_{S_i^o} Gain(S_i^o)$$

The returned explanation $\mathcal{E}^*$ is the explanation scoring the maximum *Gain* among the considered rankings.

4 Experimental Results

In this section we present experiments conducted to assess the performances of RLAX[1]. In particular, experiments are organized as follows: Sect. 4.1 presents

[1] Code is available at https://tinyurl.com/FSWRLAX.

Table 1. Subspace dimensions where the best AUC and the best Gain is achieved and the AUCs associated with those subspaces

Dataset	Measure	Subspace Dim	Associated AUC
arrhythmia	Best AUC	97 of 274 (35.40%)	0.85044
	Best Gain	15 of 274 (5.47%)	0.84880
mnist	Best AUC	55 of 100 (55.00%)	0.96890
	Best Gain	9 of 100 (9.00%)	0.94705
letter	Best AUC	11 of 32 (34.38%)	0.94662
	Best Gain	5 of 32 (15.62%)	0.91333
optdigits	Best AUC	21 of 64 (32.81%)	0.99852
	Best Gain	3 of 64 (4.69%)	0.97483

the setting used for the experiments, Sect. 4.2 illustrates the behavior of the *Gain* and of the MDR and their benefits, Sect. 4.3 analyses the significance of the results of our method and Sect. 4.4 compares RLAX with the competitors.

4.1 Experimental Settings

In the experiments, we considered a number of datasets from the ODDS Repository[2], namely *arrhythmia* (274 features and 406 instances), *cardio* (21 features and 1675 instances), *ionosphere* (33 features and 245 instances), *letter* (32 features and 1520 instances), *mnist* (100 features and 6923 instances), *optdigits* (64 features and 5086 instances), *pendigits* (16 features and 6734 instances), *satellite* (36 features and 4419 instances), and *satimage-2* (36 features and 5752 instances).

For each of the above dataset, we obtained n_v variants by taking as E_{pos} and E_{neg} two samples consisting of n_e randomly picked outlier and inlier instances, and as X all the remaining inlier instances. Moreover, we performed n_r distinct runs of the same method on each of these variants. To summarize results on the various runs we considered the best explanations obtained, while to summarize experiments on the variants of the same dataset we averaged measures associated with best results. We used $n_e = 20$, $n_v = 10$, and $n_r = 10$ in the experiments reported in the following.

As for the hyper-parameters of our method, t_{max} has been set to the maximum between $\sqrt{d}$, with $d = |F|$, and 30, while t_{neg} is an hyper-parameter that can be conveniently set to a small integer, we left it to 3. The discounting rate γ has been set to 0.95, while the total number of steps T to 50,000.

Throughout all the experiments reported in the following we use as distance measure the Euclidean distance, and as measure *sep* of separation to evaluate and compare results the AUC (Area Under the ROC Curve).

[2] https://paperswithcode.com/dataset/odds.

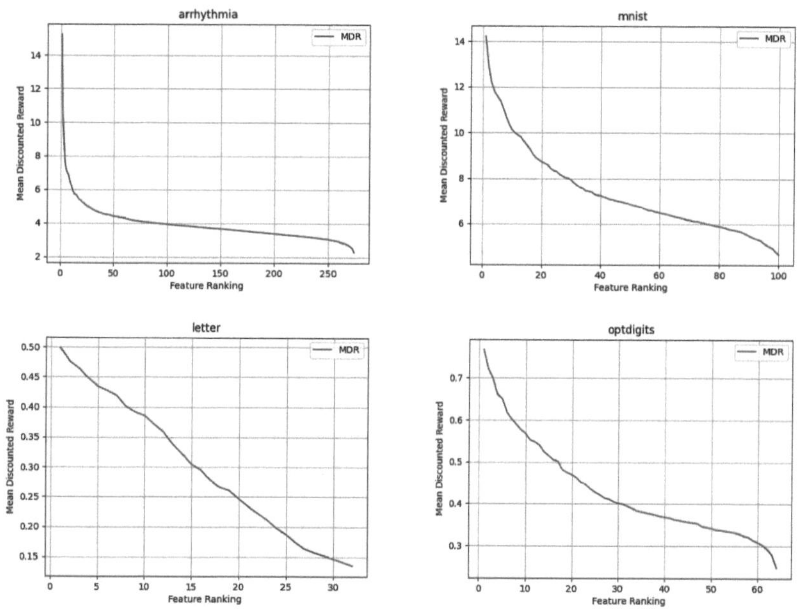

Fig. 2. Mean Discounted Reward trend as the number of considered features taken according to the sorting returned by RLAX increases.

4.2 Behavior of the Method

In this section we illustrate the peculiarities of the *Gain* measure providing experimental evidence of its benefits and discuss the behavior of the mean discounted reward and its advantages.

Figure 1 shows the curves of the AUC (red curve) and the *Gain* (blue curve) associated with the ranking of the features induced by RLAX on the datasets *arrhythmia*, *mnist*, *letter*, and *optdigits*. The curves of the other dataset are not reported since they show a similar trend. The best AUC is highlighted by a red circle, while the best gain by a blue triangle. Moreover, the AUC associated with the best gain is also highlighted by a second blue triangle. Specifically, values are reported in Table 1.

Clearly, the AUC associated with the best *Gain* cannot be higher than the optimal one. However, it can be noted that, despite being associated with a sensibly smaller number of features, the former value is very close to the latter one. Indeed, while the best AUC can be scored in a subspace consisting of more that the 30% of the whole feature set, the best *Gain* is often associated with a number of features representing only a few percent of the total.

Figure 2 reports the empirical Mean of Discounted Rewards (MDR) for the datasets *arrhythmia*, *mnist*, *letter*, and *optdigits*. Also in this case curves of the other dataset show a similar trend. The curves point out that, in most cases, the method learns a MDR distribution skewed to the right with respect to the number of features. This means the method has given importance to a small

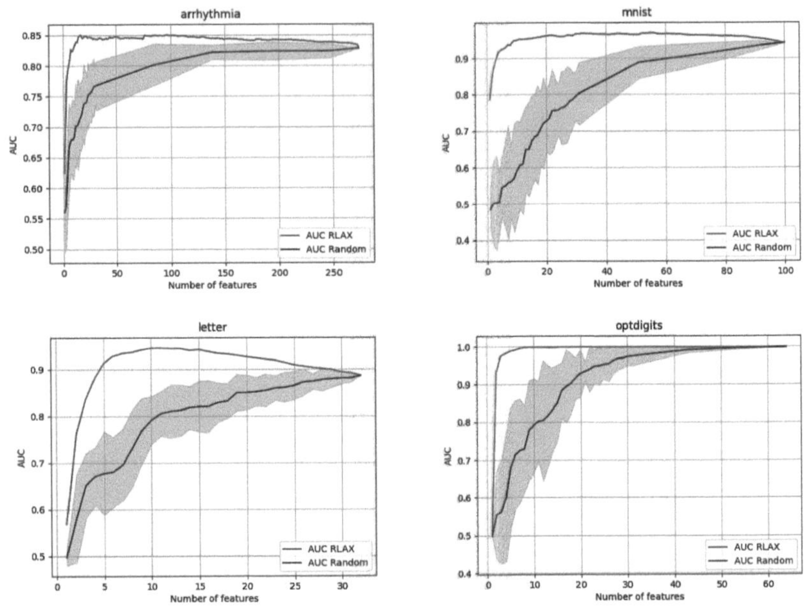

Fig. 3. Comparison between RLAX and random selection to assess the statistical significance of RLAX results, the blue area is associated with the standard deviation. (Color figure online)

selected subset of features, that are prioritized during candidate subspace generation. Since AUC curves (see Fig. 1) rise sharply for the first features, this behavior witnesses for the goodness of the selection strategy. We note that the MDR distribution for the *letter* dataset is less unbalanced, suggesting that the optimal AUC will be reached later on, as confirmed by the plot of Fig. 3.

4.3 Significance of the Results

In order to assess the significance of the results obtained by RLAX, we compared our method with the *Random* approach, which consists of randomly selecting a subset of features of a given size, assuming that each feature has an equal probability of being selected. Results of the random approach are averaged over all the runs in order to measure the expected quality associated with a random selection. Figure 3 compares the AUC of RLAX (red solid curve) with that of *random* (blue solid curve) on the datasets *arrhythmia*, *mnist*, *letter*, and *optdigits*. Curves associated with RLAX shown here are obtained by averaging, across the different variants of the same dataset, the best result achieved for each dimension, in order to provide a picture of its performances as a function of the subspace size.

It can be seen that the AUC of RLAX significantly deviates from the mean AUC associated with the random selection. To better understand the gap

Table 2. Comparison between RLAX and competitor on the subset of features scoring the best Gain, for each dataset the best *Gain* is highlighted in bold and the runner up is reported in italic

Method	arrhythmia			cardio			ionosphere		
	AUC	Gain	Dim	AUC	Gain	Dim	AUC	Gain	Dim
RLAX	0.876	**0.858**	11.0	*0.983*	**0.935**	*3.0*	0.926	**0.908**	*2.2*
Random	0.786	0.754	16.6	0.846	0.703	5.0	0.889	0.820	3.6
Genetic	*0.884*	0.653	127.0	0.970	0.670	12.6	*0.949*	0.606	22.6
HiCS	0.740	0.723	*9.8*	0.855	0.718	5.8	0.861	0.794	4.4
SOD	0.743	0.360	210.2	0.817	0.460	15.0	0.939	0.667	18.0
V-GAN	0.708	0.478	126.5	0.385	0.128	10.8	0.882	0.628	16.8

Method	letter			mnist			optdigits		
	AUC	Gain	Dim	AUC	Gain	Dim	AUC	Gain	Dim
RLAX	0.933	**0.867**	5.2	*0.979*	**0.958**	*5.2*	0.998	**0.984**	*2.8*
Random	0.812	0.665	8.4	0.820	0.714	20.2	0.955	0.819	16.4
Genetic	*0.948*	0.745	13.0	0.968	0.753	43.0	*1.000*	0.772	29.2
HiCS	0.704	0.634	*4.5*	0.638	0.596	8.4	0.789	0.741	6.1
SOD	0.916	0.663	16.2	0.864	0.474	78.0	0.749	0.449	38.4
V-GAN	0.485	0.230	16.3	0.569	0.322	49.4	0.495	0.241	32.6

Method	pendigits			satellite			satimage-2		
	AUC	Gain	Dim	AUC	Gain	Dim	AUC	Gain	Dim
RLAX	0.995	**0.939**	*2.8*	*0.956*	**0.931**	2.8	0.999	**0.985**	2.0
random	0.942	0.804	3.4	0.864	0.822	*2.0*	0.982	0.943	*1.8*
genetic	*0.999*	0.593	13.0	0.909	0.576	24.0	*1.000*	0.677	23.2
HiCS	0.932	0.778	4.9	0.865	0.800	4.7	0.978	0.912	4.7
SOD	0.903	0.653	8.0	0.844	0.633	15.2	0.813	0.604	15.0
V-GAN	0.691	0.435	8.2	0.822	0.564	18.6	0.955	0.697	18.6

between our results and those of the random approach, we have also included the standard deviation of the random results (corresponding to the shadowed area). There is no overlap also with the RLAX curve and these areas either, thus supporting the statistical significance of the selection performed by our algorithm.

4.4 Comparison with Competitors

We compared RLAX with V-GAN [8], SOD [13], HiCS [12], Genetic [9], and Random. As for the methods that output more than one subspace, we take the mean of the measure of interest over the output subspaces as the outcome of each single run.

Table 2 reports the AUC, the *Gain* and the subspace dimension associated with the best *Gain* scored in the various runs. We report in bold the best *Gain* achieved for a given dataset and in italic the best AUC and the best subspace dimension. We can note that for all datasets, our method outperforms all the competitors in terms of *Gain* thus witnessing that RLAX is able to suitably trading off AUC and subspace dimension. If we consider the column associated with AUC, we can note that in several cases RLAX is the running up method since for some datasets the Genetic method outputs a subspace scoring a slightly

Table 3. Comparison between RLAX and competitor on the subset of features scoring the best ROC-AUC, for each dataset the best ROC-AUC is highlighted in bold and the runner up is reported in italic

Method	arrhythmia			cardio			ionosphere		
	AUC	Gain	Dim	AUC	Gain	Dim	AUC	Gain	Dim
RLAX	**0.907**	*0.735*	95.0	**0.990**	*0.871*	*6.0*	0.944	0.553	26.8
Random	0.828	0.475	192.0	0.936	0.522	16.4	0.927	0.618	19.4
Genetic	0.886	0.652	128.2	0.974	0.651	13.6	0.955	0.586	24.4
HiCS	0.740	0.723	*9.8*	0.857	0.714	*6.0*	0.861	*0.794*	*4.4*
SOD	0.780	0.394	211.0	0.854	0.497	15.0	**0.975**	0.702	18.0
V-GAN	0.708	0.487	121.0	0.389	0.175	9.0	0.882	0.685	13.0

Method	letter			mnist			optdigits		
	AUC	Gain	Dim	AUC	Gain	Dim	AUC	Gain	Dim
RLAX	**0.969**	*0.816*	10.8	**0.986**	*0.763*	45.6	**1.000**	*0.916*	11.8
Random	0.887	0.462	26.2	0.934	0.479	90.0	0.996	0.566	54.0
Genetic	0.962	0.718	15.6	0.983	0.734	49.8	**1.000**	0.766	30.0
HiCS	0.705	0.634	*4.6*	0.638	0.596	*8.4*	0.789	0.741	*6.1*
SOD	0.960	0.710	16.0	0.913	0.523	78.0	0.806	0.509	38.0
V-GAN	0.487	0.284	13.0	0.568	0.343	45.0	0.495	0.277	28.0

Method	pendigits			satellite			satimage-2		
	AUC	Gain	Dim	AUC	Gain	Dim	AUC	Gain	Dim
RLAX	**1.000**	0.625	13.0	**0.958**	*0.930*	*3.0*	**1.000**	0.842	12.4
Random	0.999	0.499	15.0	0.887	0.670	14.6	**1.000**	0.633	25.4
Genetic	**1.000**	0.568	13.8	0.919	0.563	25.6	**1.000**	0.653	25.0
HiCS	0.933	*0.775*	*5.1*	0.865	0.800	4.7	0.978	*0.912*	*4.7*
SOD	0.948	0.698	8.0	0.871	0.662	15.0	0.968	0.760	15.0
V-GAN	0.692	0.473	7.0	0.816	0.621	14.0	0.954	0.760	14.0

higher AUC. However, it must be noticed that this is achieved at the expense of the subspace dimension which indeed results to be much larger. A similar behavior can be observed when analyzing the subspace dimension where, for some datasets, the HiCS algorithm (and in one case the Random algorithm) produces a slightly smaller subspace but at the cost of a significantly smaller AUC. Table 3 reports the AUC, the *Gain* and the subspace dimension associated with the best AUC associated to the various runs. We report in bold the best AUC achieved for a given dataset and in italic the best *Gain* and the best subspace dimension. We can note that in almost all cases, RLAX is able to detect a subspace scoring the highest AUC with respect to competitors. These AUCs are, nevertheless, higher than those associated with the best gain. Indeed, the subspaces scoring top AUCs are significantly large and this is suitably accounted by the gain which is sensitively smaller. Further, the table shows that among the methods achieving the highest AUC values, RLAX attains this performance while selecting the smallest subspace dimension.

5 Conclusions

In this work, we formalized the Anomaly Explanation problem as a feature selection task, aiming to identify minimal, interpretable subspaces that maximally discriminate anomalies from normal data. We proposed RLAX, a novel Reinforcement Learning (RL)-based framework that models the search for optimal explanations as a Markov Decision Process (MDP), leveraging Explanation Gain and Subspace Margin to quantify feature relevance. By adopting Proximal Policy Optimization (PPO), our method efficiently navigates the combinatorial feature space, addressing scalability challenges in high-dimensional datasets. Our key contributions include (*i*) *the first RL-based solution* for Anomaly Explanation—to our knowledge, RLAX is the first method to employ RL for this task; (*ii*) *feature selection as policy learning*, framing the problem as an MDP to optimize feature subset discovery; (*iii*) *interpretability and precision*, as RLAX generates concise explanations (small feature subsets) while achieving superior separation accuracy compared to existing subspace selection methods, as demonstrated experimentally. Future work includes extending the experimental evaluation to more challenging scenarios (e.g., dynamic or adversarial settings) and enhancing RLAX to incorporate domain-specific constraints (e.g., feature costs or dependencies) into the reward function.

Acknowledgments. We acknowledge the support of the PNRR project FAIR - Future AI Research (PE00000013), Spoke 9 - Green-aware AI, under the NRRP MUR program funded by the NextGenerationEU.

Disclosure of Interests. The authors have no competing interests to declare that are relevant to the content of this article.

References

1. Aggarwal, C.C., Yu, P.: Outlier detection for high dimensional data. In: Proceedings of the 2001 ACM SIGMOD International Conference on Management of data, Santa Barbara, CA, USA, 21–24 May 2001, pp. 37–46. ACM (2001)
2. Angiulli, F., Pizzuti, C.: Fast outlier detection in large high-dimensional data sets. In: Proceedings of the International Conference on Principles of Data Mining and Knowledge Discovery, pp. 15–26 (2002)
3. Angiulli, F., Fassetti, F., Nisticò, S., Palopoli, L.: Explaining outliers and anomalous groups via subspace density contrastive loss. Mach. Learn. **113**(10), 7565–7589 (2024)
4. Angiulli, F., Fassetti, F., Palopoli, L.: Detecting outlying properties of exceptional objects. ACM Trans. Database Syst. (tods) **34**(1), 1–62 (2009)
5. Chandola, V., Banerjee, A., Kumar, V.: Anomaly detection: a survey. ACM Comput. Surv. **41**(3) (2009)
6. Chandola, V., Banerjee, A., Kumar, V.: Anomaly detection for discrete sequences: a survey. IEEE Trans. Knowl. Data Eng. **24**(5), 823–839 (2012). https://doi.org/10.1109/TKDE.2010.235
7. Cheng, L., Wang, Y., Liu, X., Li, B.: Outlier detection ensemble with embedded feature selection. In: Proceedings of the AAAI Conference on Artificial Intelligence, vol. 34, pp. 3503–3512 (2020)
8. Cribeiro-Ramallo, J., et al.: Adversarial subspace generation for outlier detection in high-dimensional data. arXiv preprint arXiv:2504.07522 (2025)

9. Fassetti, F., Angiulli, F.: Finding distance-based outliers in subspaces through both positive and negative examples. In: ICAART (1), pp. 5–10 (2010)
10. Gaudel, R., Sebag, M.: Feature selection as a one-player game. In: Proceedings of the 27th International Conference on Machine Learning (ICML), 21–24 June Haifa, Israel, pp. 359–366 (2010)
11. Hazrati Fard, S.M., Hamzeh, A., Hashemi, S.: Using reinforcement learning to find an optimal set of features. Comput. Math. Appl. **66**(10), 1892–1904 (2013)
12. Keller, F., Muller, E., Bohm, K.: Hics: high contrast subspaces for density-based outlier ranking. In: 2012 IEEE 28th International Conference on Data Engineering, pp. 1037–1048. IEEE (2012)
13. Kriegel, H.-P., Kröger, P., Schubert, E., Zimek, A.: Outlier detection in axis-parallel subspaces of high dimensional data. In: Theeramunkong, T., Kijsirikul, B., Cercone, N., Ho, T.-B. (eds.) PAKDD 2009. LNCS (LNAI), vol. 5476, pp. 831–838. Springer, Heidelberg (2009). https://doi.org/10.1007/978-3-642-01307-2_86
14. Li, Z., Zhu, Y., Van Leeuwen, M.: A survey on explainable anomaly detection. ACM Trans. Knowl. Discov. Data **18**(1) (2023)
15. Macha, M., Akoglu, L.: Explaining anomalies in groups with characterizing subspace rules. Data Min. Knowl. Disc. **32**(5), 1444–1480 (2018). https://doi.org/10.1007/s10618-018-0585-7
16. Mokoena, T., Celik, T., Marivate, V.: Why is this an anomaly? Explaining anomalies using sequential explanations. Pattern Recogn. **121**, 108227 (2022)
17. Pang, G., Cao, L., Chen, L., Liu, H.: Unsupervised feature selection for outlier detection by modelling hierarchical value-feature couplings. In: 2016 IEEE 16th International Conference on Data Mining (ICDM), pp. 410–419. IEEE (2016)
18. Panjei, E., Gruenwald, L., Leal, E., Nguyen, C., Silvia, S.: A survey on outlier explanations. VLDB J. **31**(5), 977–1008 (2022)
19. Riahi-Madvar, M., Azirani, A.A., Nasersharif, B., Raahemi, B.: Subspace-based outlier detection using linear programming and heuristic techniques. Expert Syst. Appl. **207**, 117955 (2022)
20. Sali, R., Adewole, S., Akakpo, A.: Feature selection using reinforcement learning. CoRR (2021). https://arxiv.org/abs/2101.09460
21. Sathe, S., Aggarwal, C.C.: Subspace outlier detection in linear time with randomized hashing. In: 2016 IEEE 16th International Conference on Data Mining (ICDM), pp. 459–468. IEEE (2016)
22. Vinh, N.X., et al.: Discovering outlying aspects in large datasets. Data Min. Knowl. Disc. **30**(6), 1520–1555 (2016). https://doi.org/10.1007/s10618-016-0453-2
23. Wang, T., Duan, L., Dong, G., Bao, Z.: Efficient mining of outlying sequence patterns for analyzing outlierness of sequence data. ACM Trans. Knowl. Discov. Data (TKDD) **14**(5), 1–26 (2020)
24. Xu, H., et al.: Beyond outlier detection: interpreting outliers by attention-guided triplet deviation network. In: Proceedings of The Web Conference 2021, pp. 1328–1339 (2021)
25. Zhao, Y., Ding, X., Yang, J., Bai, H.: Suod: toward scalable unsupervised outlier detection. arxiv preprint arXiv:2002.03222 (2020)

Attackers Strike Back? Not Anymore—An Ensemble of RL Defenders Awakens for APT Detection

Sidahmed Benabderrahmane(✉) and Talal Rawhan

Division of Science, New York University NYUAD, Abu Dhabi, UAE
sidahmed.benabderrahmane@nyu.edu

Abstract. Advanced Persistent Threats (APTs) represent a growing menace to modern digital infrastructure. Unlike traditional cyberattacks, APTs are stealthy, adaptive, and long-lasting, often bypassing signature-based detection systems. This paper introduces a novel framework for APT detection that unites deep learning, reinforcement learning (RL), and active learning into a cohesive, adaptive defense system. Our system combines autoencoders for latent behavioral encoding with a multi-agent ensemble of RL-based defenders, each trained to distinguish between benign and malicious process behaviors.

We identify a critical challenge in existing detection systems: their static nature and inability to adapt to evolving attack strategies. To this end, our architecture includes multiple RL agents (Q-Learning, PPO, DQN, adversarial defenders), each analyzing latent vectors generated by an autoencoder. When any agent is uncertain about its decision, the system triggers an active learning loop to simulate expert feedback, thus refining decision boundaries. An ensemble voting mechanism, weighted by each agent's performance, ensures robust final predictions.

1 Introduction

1.1 Overview

Cybersecurity is undergoing a fundamental transformation as Advanced Persistent Threats (APTs) emerge as some of the most insidious forms of cyber attacks [1–3]. Unlike conventional threats, APTs are characterized by their stealth, persistence, and long-term objectives. They typically unfold over multiple phases—initial reconnaissance, exploitation, lateral movement, data exfiltration, and persistence—making them difficult to detect using traditional security systems [4,5].

The increased sophistication of APTs presents a profound challenge to defenders [6,7]. Signature-based Intrusion Detection Systems (IDS) fail to detect novel attack vectors, while classical anomaly detection systems struggle with high false positive rates and poor adaptability [8,9]. Moreover, APTs often mimic legitimate user behavior, leveraging living-off-the-land binaries (LOLBins) or hijacking system processes, further complicating detection.

S. Džeroski et al. (Eds.): DS 2025, LNAI 16090, pp. 537–552, 2025.
https://doi.org/10.1007/978-3-032-05461-6_35

To address these issues, there is growing interest in intelligent, learning-based defense mechanisms. Recent progress in machine learning has shown the potential of deep learning and reinforcement learning (RL) to model complex system behavior and adapt dynamically to evolving threats [10]. In particular, deep autoencoders have proven effective for reducing high-dimensional event data into compact latent representations that preserve essential patterns for anomaly detection [11].

Reinforcement learning introduces an agent-based paradigm wherein models learn optimal behaviors through interaction with an environment [12]. In the context of APT detection, RL agents can learn to classify process traces as benign or malicious by receiving feedback in the form of rewards. This approach is inherently adaptive, enabling defenders to adjust strategies based on new observations.

However, using a single RL agent is often insufficient. APTs exhibit diverse behaviors that challenge any single learning paradigm. Therefore, we propose an ensemble of heterogeneous RL agents, including Q-learning, PPO, DQN, and adversarially trained models, each contributing its strengths to the detection process. This ensemble not only increases robustness but also enables better generalization across varying attack scenarios.

To further refine the decision-making process, we integrate an active learning loop. When an RL agent encounters a sample about which it is uncertain—typically when outputs are near 0.5—it triggers a simulated query for expert labeling. This simulates real-world scenarios where human analysts are involved in labeling uncertain samples. By incorporating this feedback, agents are retrained iteratively to improve accuracy and reduce uncertainty.

1.2 Problem Statement and Objective

Advanced Persistent Threat (APT) detection remains a critical and unresolved challenge in cybersecurity. This difficulty stems from several factors: the rarity and subtlety of APT events in real-world datasets, the evolving strategies of attackers that induce concept drift over time, and the significant cost associated with false positives in sensitive operational environments.

To address these challenges, this paper introduces a multi-agent ensemble framework that integrates deep autoencoding, reinforcement learning, and active feedback loops. The framework is designed to extract latent patterns of system behavior through the use of autoencoders, while reinforcement learning agents are trained to identify APT activities using reward-driven learning paradigms. Additionally, an active learning mechanism is employed to concentrate model adaptation on uncertain and high-risk samples, effectively simulating expert feedback and accelerating learning.

The final decision process aggregates the outputs of multiple agents through a weighted ensemble strategy, ensuring robustness and adaptability. By unifying unsupervised feature extraction, multi-agent reinforcement learning, and uncertainty-aware querying, this work presents a resilient and continuously evolving defense system capable of countering sophisticated APTs.

2 Related Work

Reinforcement Learning (RL) has emerged as a powerful paradigm in machine learning, characterized by an agent that learns to take actions in an environment to maximize a cumulative reward [13,14]. The RL framework is formally defined by a Markov Decision Process (MDP), represented as a tuple $(\mathcal{S}, \mathcal{A}, P, R, \gamma)$, where: $\mathcal{S}$ is the set of states, $\mathcal{A}$ is the set of actions, $P(s'|s,a)$ defines the transition probabilities, $R(s,a)$ is the reward function, $\gamma \in [0,1]$ is the discount factor.

The goal is to learn a policy $\pi(a|s)$ that maximizes the expected return:

$$J(\pi) = \mathbb{E}_\pi\left[\sum_{t=0}^{\infty} \gamma^t R(s_t, a_t)\right]$$

2.1 Value-Based Methods: Q-Learning and DQN

Q-Learning is a model-free algorithm that learns the optimal action-value function $Q^*(s,a)$ using the Bellman equation:

$$Q(s_t, a_t) \leftarrow Q(s_t, a_t) + \alpha\left[r_t + \gamma \max_{a'} Q(s_{t+1}, a') - Q(s_t, a_t)\right]$$

Deep Q-Networks (DQN) extend Q-learning by using a neural network to approximate $Q(s,a;\theta)$, where θ are learnable parameters. The loss function is: $\mathcal{L}(\theta) = \left(r_t + \gamma \max_{a'} Q(s_{t+1}, a'; \theta^-) - Q(s_t, a_t; \theta)\right)^2$, where θ^- denotes a target network.

2.2 Policy-Based Methods: REINFORCE and PPO

In policy-based methods, we directly learn a parameterized policy $\pi_\theta(a|s)$. The REINFORCE algorithm updates the policy via gradient ascent on the expected return: $\nabla_\theta J(\theta) = \mathbb{E}_{\pi_\theta}\left[\nabla_\theta \log \pi_\theta(a|s) \cdot R_t\right]$

Proximal Policy Optimization (PPO) improves on REINFORCE by using a clipped surrogate objective: $L(\theta) = \mathbb{E}\left[\min\left(r(\theta)\hat{A}, \text{clip}(r(\theta), 1-\epsilon, 1+\epsilon)\hat{A}\right)\right]$ where $r(\theta) = \frac{\pi_\theta(a|s)}{\pi_{\theta_{old}}(a|s)}$ is the probability ratio and $\hat{A}$ is the advantage estimate.

2.3 Adversarial RL

Adversarial reinforcement learning introduces a perturbation-generating agent that seeks to fool the defender agent. This setting is often modeled as a minimax game: $\min_{\pi_D} \max_{\pi_A} \mathbb{E}[R(s, a_D, a_A)]$, where π_D is the defender's policy and π_A is the attacker's.

2.4 Multi-agent and Ensemble RL

Multi-agent reinforcement learning (MARL) extends single-agent RL to environments with multiple agents, which may be cooperative, competitive, or mixed. The joint policy $\pi = (\pi_1, \ldots, \pi_n)$ is learned to optimize a shared or individual reward signal.

Ensemble reinforcement learning combines predictions from multiple agents. In our work, we use weighted voting based on AUC performance:

$$P_{\text{ensemble}}(APT) = \sum_{i=1}^{N} w_i P_i(\text{APT}) \quad \text{with} \quad w_i \propto \text{AUC}_i$$

Few studies in cybersecurity combine these elements. Our contribution lies in synthesizing RL fundamentals with active learning feedback and ensemble predictions for robust, adaptive APT detection.

3 Proposed Method

Our proposed method introduces a multi-agent reinforcement learning (RL) framework combined with active learning and autoencoder-based latent encoding for robust and adaptive APT detection. The primary components and contributions of our architecture are as follows:

- We use a deep autoencoder to learn compact, latent representations of system behavior traces, allowing us to isolate meaningful behavioral anomalies.
- We train a set of heterogeneous RL agents (Q-Learning, PPO, DQN, Adversarial Agents), each making decisions independently based on latent features and AE reconstruction error.
- We implement an active learning loop that identifies uncertain predictions using softmax-based confidence margins, queries the Oracle (e.g. ground truth), and updates the RL agents accordingly.
- We introduce a weighted ensemble strategy that aggregates the decisions of all agents using their AUC scores as confidence weights.
- We propose a feedback refinement loop for the autoencoder by incorporating false-positive benign feedback, simulating analyst verification.

3.1 Autoencoder-Based Latent Representation

To efficiently detect Advanced Persistent Threats (APTs), which are often buried within high-dimensional system activity logs, we utilize an attention-based autoencoder AAE to derive compact and informative latent representations of process behavior. Let each input sample be a binary vector $x \in \{0, 1\}^d$, representing the presence or absence of specific system-level events or features. The autoencoder is a neural network composed of two parts: an encoder f_{enc} and a decoder f_{dec}.

$$z = f_{enc}(x; \theta_{enc}) \in \mathbb{R}^k, \quad \text{with } k \ll d \tag{1}$$

$$\hat{x} = f_{dec}(z; \theta_{dec}) \in \mathbb{R}^d \tag{2}$$

Here, z denotes the latent vector capturing compressed semantic information from x, and $\hat{x}$ is the reconstruction.

The reconstruction error, defined as the mean squared error (MSE) between the input and its reconstruction, is given by: $\mathcal{L}_{AE}(x, \hat{x}) = \frac{1}{d}\sum_{i=1}^{d}(x_i - \hat{x}_i)^2$. The autoencoder is trained to minimize this loss: $\min_{\theta_{enc},\theta_{dec}} \mathbb{E}_{x\sim\mathcal{D}}[\mathcal{L}_{AE}(x, f_{dec}(f_{enc}(x)))]$. The reconstruction error $\mathcal{L}_{AE}$ serves two purposes: (i) As an anomaly score: samples with high reconstruction error are likely anomalous. (ii) As a reward signal: to guide reinforcement learning agents (discussed in later sections). We denote the latent representation for the entire dataset as: $Z = \{z_i = f_{enc}(x_i)\}_{i=1}^{N}$, for N samples. These latent embeddings are used as the input state representations for the reinforcement learning agents, enabling them to operate in a reduced-dimensional, semantically meaningful space.

3.2 Multi-agent RL Architecture

To capture the sequential and interactive aspects of APT detection, we employ multiple reinforcement learning (RL) agents that act as autonomous defenders. Each agent receives a state derived from the AAE autoencoder latent space and outputs a prediction indicating whether the behavior is benign or malicious. We use multiple types of RL agents: Q-Learning, DQN, PPO, Adversarial Agent, Multiple Agents, and Active Adversarial Agents.

Let the environment be defined as a tuple $(\mathcal{S}, \mathcal{A}, R, \mathcal{T}, \gamma)$, where: $\mathcal{S}$: Set of states (AAE autoencoder latent vectors z). $\mathcal{A} = \{0, 1\}$: Actions ($0 =$ benign, $1 =$ APT). $R(s, a)$: Reward function based on prediction correctness and AAE error. $\mathcal{T}(s, a, s')$: State transition dynamics (stateless in this context). $\gamma \in [0, 1)$: Discount factor.

Each agent is trained on the latent space output of the autoencoder. The agents maintain or approximate a Q-function $Q(s, a)$ to guide decision-making:

$$Q^*(s, a) = \mathbb{E}\left[\sum_{t=0}^{\infty} \gamma^t R(s_t, a_t) \mid s_0 = s, a_0 = a\right] \tag{3}$$

Q-Learning Agent: Uses a tabular or neural approximation of $Q(s, a)$ updated via:

$$Q(s_t, a_t) \leftarrow Q(s_t, a_t) + \alpha\left[r_t + \gamma \max_a Q(s_{t+1}, a) - Q(s_t, a_t)\right] \tag{4}$$

DQN Agent: Approximates the Q-function with a deep network:

$$\mathcal{L}_{\text{DQN}} = \mathbb{E}_{(s,a,r,s')}\left[(r + \gamma \max_{a'} Q(s', a'; \theta^-) - Q(s, a; \theta))^2\right] \tag{5}$$

where θ and θ^- are the online and target network parameters.

PPO Agent: Optimizes a stochastic policy $\pi_\theta(a|s)$ using a clipped surrogate objective:

$$L^{\text{CLIP}}(\theta) = \mathbb{E}_t \left[\min\left(r_t(\theta)\hat{A}_t, \text{clip}(r_t(\theta), 1-\epsilon, 1+\epsilon)\hat{A}_t\right)\right] \tag{6}$$

where $r_t(\theta) = \frac{\pi_\theta(a_t|s_t)}{\pi_{\theta_{\text{old}}}(a_t|s_t)}$ and $\hat{A}_t$ is the estimated advantage.

Adversarial Agent: Trained under input perturbations δ to ensure robustness:

$$\max_{\delta \in \Delta} \mathcal{L}(Q(s+\delta, a), y), \quad \text{subject to } \|\delta\| \leq \epsilon \tag{7}$$

Multi-agent RL Architecture: To enhance detection capability and resilience, we adopt a multi-agent reinforcement learning (MARL) architecture wherein multiple heterogeneous RL agents operate in parallel, share training signals, and collaboratively improve their anomaly detection performance.

Let $\mathcal{A}_{RL} = \{A_1, A_2, \ldots, A_n\}$ denote the set of RL agents. Each agent A_i receives a state $s_i = [z_i, \epsilon_i]$, where $z_i \in \mathbb{R}^k$ is the latent vector from the autoencoder, and $\epsilon_i \in \mathbb{R}$ is the corresponding reconstruction error. Each agent aims to learn a policy $\pi_i(a|s)$ or Q-function $Q_i(s, a)$ to maximize its expected reward:

$$\mathbb{E}_{s \sim \mathcal{S}, a \sim \pi_i} \left[\sum_{t=0}^{\infty} \gamma^t R(s_t, a_t)\right] \tag{8}$$

We define a unified reward function across agents:

$$R(s, a) = \begin{cases} +1 + \beta \cdot \epsilon & \text{if correct prediction} \\ -1 + \beta \cdot \epsilon & \text{if benign misclassified} \\ -2 + \beta \cdot \epsilon & \text{if APT missed} \end{cases} \tag{9}$$

Here, $\beta \in \mathbb{R}^+$ is a tunable coefficient that modulates the influence of the reconstruction error ϵ on the reward signal. This parameter plays a crucial role in shaping the behavior of the RL agents. It allows the reward to reflect not only the correctness of the prediction but also the degree of anomaly suggested by the autoencoder. A higher β increases the agent's sensitivity to uncertain or anomalous states, thereby directing more learning focus toward high-risk areas. This mechanism encourages agents to assign greater importance to difficult cases that exhibit unusual behavior, effectively balancing classification performance with anomaly intensity.

This architecture provides redundancy, robustness, and the ability to learn complementary patterns, ultimately enhancing the detection of rare and stealthy APT behaviors.

3.3 Active Learning with Reward-Shaped Retraining

To improve sample efficiency, focus learning on ambiguous cases, and address the scarcity of labeled anomalous samples, we incorporate a targeted active learning loop into the MARL framework. At each inference step, an agent assesses its confidence level on the current input. If the output is deemed uncertain (e.g., softmax near 0.5 or Q-values close), the corresponding state is flagged and sent for simulated oracle labeling.

Uncertainty is computed using the margin criterion:

$$\text{uncertainty}(s) = |\pi(a = 1\ s) - 0.5| < \delta \quad \text{or} \quad |Q(s, 1) - Q(s, 0)| < \delta \tag{10}$$

Queried samples (s, y) are stored in an active replay buffer and used to fine-tune the corresponding agent. The training loss incorporates a reward modification based on the AE error: $R'(s, a) = R(s, a) + \lambda \cdot \mathcal{L}_{AE}(x, \hat{x})$. This encourages agents to prioritize high-error regions, effectively guiding learning toward suspicious behavior. The overall active learning routine is summarized in Algorithm 1.

Algorithm 1. Active Learning with Reward-Shaped Retraining

for each agent A_i **do**
  for each test sample s **do**
    Predict $Q(s, a)$ or $\pi(a|s)$
    if uncertain(s) **then**
      Simulate oracle to obtain y
      Add (s, y) to active buffer
    end if
  end for
  Retrain A_i using buffer with reward $R'(s, a)$
end for

This loop is executed at each feedback iteration. The impact is evaluated by computing model AUC before and after updates. This feedback-driven process helps refine decision boundaries in low-confidence areas, increasing true positive rates while reducing labeling cost.

3.4 Ensemble Decision Module

To consolidate the diverse predictions of the multi-agent RL system, we employ an ensemble strategy that aggregates the outputs of individual agents into a single final decision. This enhances robustness and allows leveraging the strengths of different models.

Let $\{A_1, A_2, ..., A_n\}$ be the set of agents, each producing a score or class prediction $y_i \in \{0, 1\}$. We explore following ensemble strategy:

1. Majority Voting:

$$y_{final} = \arg \max_{y \in \{0,1\}} \sum_{i=1}^{n} \mathbb{1}(y_i = y) \tag{11}$$

This decision fusion step capitalizes on diversity among agents and has been shown empirically to improve detection of stealthy APTs across datasets.

4 Results

4.1 Dataset Description

We use real-world APT datasets with binary process behavior vectors. Each vector represents the presence/absence of key system actions. APTs are rare and labeled via known attack stages (recon, exploit, persist, etc.). These cyber security data sources used in this paper come from the Defense Advanced Research Projects Agency (DARPA)'s `Transparent Computing TC`[1] program [15–20]. The aim of this program is to provide transparent provenance data of system activities and component interactions across different operating systems (OS) and spanning all layers of software abstractions. Specifically, the datasets include system-level data, background activities, and system operations recorded while APT-style attacks are being carried out on the underlying systems. Preserving the provenance of all system elements allows for tracking the interactions and dependencies among components. Such an interdependent view of system operations is helpful for detecting activities that are individually legitimate or benign but collectively might indicate abnormal behavior or malicious intent.

Here we specifically employ DARPA's data that has undergone processing conducted by the `ADAPT` (Automatic Detection of Advanced Persistent Threats) project's ingester [16–20]. The records come from four different source OS, namely Android (called in the TC program Clearscope), Linux (called Trace), BSD (called Cadets), and Windows (called Fivedirections or 5dir). For each system, the data comes from two separate attack scenarios: scenario 1 (also called Pandex) and scenario 2 (called Bovia), respectively. The processing includes ingesting provenance graph data into a graph database as well as additional data integration and deduplication steps. The final data includes a number of Boolean-valued datasets (data aspects), with each representing an aspect of the behavior of system processes as illustrated in Table 1. Each row in such a data aspect is a data point representing a single process run on the respective OS. It is expressed as a Boolean vector whereby a value of 1 in a vector cell indicates the corresponding attribute applies to that process.

For instance, in Table 1, the process with id `ee27fff2-a0fd-1f516db3d35f` has the following sequence of events:

`$<$/usr/sbin/avahi-autoipd, 216.73.87.152, EVENT_OPEN, EVENT_CONNECT, ...$>$.`

Specifically, the relevant datasets are interpreted as follows:

[1] https://gitlab.com/adaptdata

Table 1. Example of a boolean-valued dataset (data aspect) from the DARPA TC program. In each row, the boolean vector represents the list of features of the corresponding process.

	/usr/sbin/avahi-autoipd	216.73.87.152	EVENT_OPEN	EVENT_EXECUTE	EVENT_CONNECT	EVENT_SENDMSG	...
ee27fff2-a0fd-1f516db3d35f	1	1	1	0	1	0	...
b2e7e930-8f25-4242a52c5d72	0	1	0	1	1	1	...
07141a2a-832e-8a71ca767319	0	0	1	1	1	1	...
b4be70a9-98ac-81b0042dbecb	1	0	1	1	0	0	...
2bc3b5c6-9110-076710a13038	0	0	0	0	0	1	...
ad7716e0-8d59-5d45d1742211	1	1	0	1	0	1	...
...	...	...	...	...	...	...	...

- `ProcessEvent` (PE): Its attributes are event types performed by the processes. A value of 1 in `process[i]` means the process has performed at least one event of type i.
- `ProcessExec` (PX): The attributes are executable names that are used to start the processes.
- `ProcessParent` (PP): Its attributes are executable names that are used to start the parents of the processes.
- `ProcessNetflow` (PN): The attributes here represent IP addresses and port names that have been accessed by the processes.
- `ProcessAll` (PA): This dataset is described by the disjoint union of all attribute sets from the previous datasets.

Overall, with two attack scenarios, four OS (BSD, Windows, Linux, Android) and five aspects (PE, PX, PP, PN, PA), a total of forty individual datasets are composed, as illustrated in Fig. 1. They are described in Table 2 whereby the last column provides the number of attacks in each dataset. The substantially imbalanced nature of the datasets is clearly seen here. For each forensic configuration (OS×attack scenario×data aspect) we have the number of processes (instances) and the corresponding events (features). For instance, $Windows\_E1\_PE$ is the dataset represented by PE aspect belonging to Windows OS, produced during

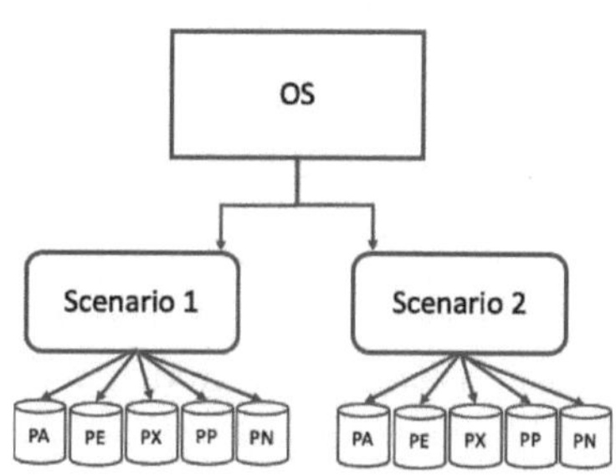

Fig. 1. Organization of the DARPA's TC datasets. Each OS undergoes two attack scenarios, each of which contains five data aspects ets. With four OS (BSD, Windows, Linux, Android), two attack scenarios, and five aspects (PE, PX, PP, PN, PA), a total of forty individual datasets are composed. Each forensic configuration (OS×attack scenario×data aspect) represents a single dataset [20].

the first attack scenario E1 (Pandex). It contains 17569 instances and 22 features with a total of 8 APTs anomalies (0.04%).

Table 2. Summary of the first source of 40 benchmark datasets belonging to DARPA's TC program for APT detection. A dataset entry (columns 4 to 8) is described by a number of rows (processes)/number of columns (attributes). For instance, with ProcessAll (PA) obtained from the second scenario using Linux, the dataset has 282104 rows and 6435 attributes with 46 APT attacks (0.01%).

	Scenario	Size	PE	PX	PP	PN	PA	$nb\_attacks$	$\%\frac{nb\_attacks}{nb\_processes}$
BSD	1	288 MB	76903/29	76698/107	76455/24	31/136	76903/296	13	0.02
	2	1.27 GB	224624/31	224246/135	223780/37	42888/62	224624/265	11	0.004
Windows	1	743 MB	17569/22	17552/215	14007/77	92/13963	17569/14431	8	0.04
	2	9.53 GB	11151/30	11077/388	10922/84	329/125	11151/606	8	0.07
Linux	1	2858 MB	247160/24	186726/154	173211/40	3125/81	247160/299	25	0.01
	2	25.9 GB	282087/25	271088/140	263730/45	6589/6225	282104/6435	46	0.01
Android	1	2688 MB	102/21	102/42	0/0	8/17	102/80	9	8.8
	2	10.9 GB	12106/27	12106/44	0/0	4550/213	12106/295	13	0.10

4.2 Results and Discussion

In this section, we present and analyze the performance of our proposed active adversarial multi-agent reinforcement learning (RL) framework for Advanced Persistent Threat (APT) detection. We structure the discussion around the following key aspects:

- *Evaluation Setup:* We evaluate our framework using real-world datasets containing labeled traces of benign and APT activities. The models are tested on a separate validation set after each active learning iteration. Evaluation metrics include Area Under the Curve (AUC), Precision, Recall and F1-score to assess classification performance and ranking quality.
- *Performance of Individual RL Agents:* We compare the effectiveness of Q-Learning, DQN, PPO, and an Adversarial Agent in detecting APT behavior. Each model is trained on latent representations derived from the autoencoder and retrained using active feedback. Their performance is tabulated over 50 iterations of active learning.
- *Active Learning Impact:* We analyze the influence of the active learning loop by tracking the AUC scores across iterations. Results show a consistent improvement in true positive rates and a reduction in false positives. Queried samples effectively refine the decision boundaries of the agents.
- *Ensemble Detection Strategy:* We evaluate the ensemble strategy with a majority voting and aggregation.

Evaluated Models. In our study, we evaluate the effectiveness of eight core anomaly detection models—seven based on reinforcement learning (RL) and one based on unsupervised reconstruction. These models are integrated into a broader multi-agent and ensemble-based architecture aimed at detecting Advanced Persistent Threats (APTs). Each model processes system behavior in the latent space produced by the AAE autoencoder and returns predictions indicating whether the behavior is benign or malicious.

- *Q-Learning Agent (Q-RL):* A classical value-based agent that updates its Q-values through tabular or function approximation using the Bellman equation. It serves as a baseline reinforcement learner with discrete state-action feedback.
- *Deep Q-Network (DQN):* An extension of Q-Learning that employs a deep neural network to approximate the Q-function. It uses experience replay and target networks to stabilize training in high-dimensional spaces.
- *Proximal Policy Optimization (PPO):* A policy-gradient agent that optimizes a clipped surrogate objective to ensure stable updates. PPO is known for balancing exploration and exploitation efficiently.
- *Rule-Based Autoencoder (AE-RL):* A baseline model that flags a process as malicious if its reconstruction error exceeds a predefined threshold. It serves as an unsupervised anomaly detector leveraging learned representations.
- *Adversarial Agent (AMARL):* A Q-learning-based model trained on both clean and perturbed inputs to ensure robustness against evasion attacks. The adversarial training process encourages the agent to remain stable under adversarial manipulations.
- *Active Learning Adversarial Agent (AAMARL):* To assess the impact of uncertainty-driven active learning, we simulate an iterative feedback loop over 50 rounds. During each iteration, the model identifies uncertain samples based on the softmax margin of its predictions. These samples are considered candidates for querying an oracle, which returns the ground truth labels. The newly labeled instances are then incorporated into the training set, enabling the model to refine its decision boundary incrementally.

  After each round of feedback and retraining, we evaluate the model on the full test set, computing the AUC and F1-score. This process yields a trajectory of performance metrics over the active learning timeline. The final reported AUC and F1 for the "Active Multi-Agent" configuration in Table 3 represent the average performance across all 50 iterations, thereby reflecting the cumulative benefit of integrating human feedback into the learning pipeline.
- *Meta Active Adversarial Multi-Agent Evaluation (EAAMARL):* In addition to individual agent retraining, we deploy active learning in the context of our proposed ensemble model, which combines predictions from Q-Learning, PPO, DQN, and adversarially trained agents. Over 50 active learning iterations, each agent selectively queries uncertain samples based on its softmax output margin. Ground truth labels for these samples are simulated as oracle feedback and integrated into each agent's experience replay buffer. As agents improve individually through this iterative process, their collective

predictions—aggregated using AUC-weighted voting—are updated accordingly. After each feedback round, the ensemble's decision is evaluated on the full test set using AUC and F1-score. The average values over the 50 iterations are reported as the final performance for the "EAAMARL" row in Table 3, highlighting the gains achieved by coordinated uncertainty-aware learning across multiple adversarially hardened agents.

Each of these models is integrated into an ensemble for improved robustness, and further enhanced using an active learning loop to address uncertainty in prediction and guide retraining dynamically.

Each model processes a latent vector obtained from the autoencoder and aims to classify whether the input behavior corresponds to a benign process or an APT. The agents were evaluated across 40 different datasets constructed by combining the following:

- *Operating Systems (4):* BSD, Windows, Linux, Android
- *Attack Scenarios (2):* Scenario 1, Scenario 2
- *Datasets (5):* PA, PE, PX, PN, PP

This results in a total of *40 datasets* allowing a comprehensive assessment of model performance across heterogeneous environments.

4.3 Discussion on Performance Accross Models

The results presented in Table 3 offer valuable insights into the comparative performance of reinforcement learning-based methods for Advanced Persistent Threat (APT) detection.

BSD: In both E1 and E2 scenarios, performance gradually improves from simple models like Q-RL and PPO toward the more advanced ensemble-based systems. Notably, in `BSD_E1_PA`, the EAAMARL model achieves the best AUC/F1 score of *0.91/0.50*, clearly outperforming standalone agents. Across all BSD datasets, EAAMARL consistently outperforms other models with average AUCs exceeding 0.85 and F1-scores above 0.35, showing its robustness to sparse signals typical of BSD traces. AAMARL also performs competitively, especially in PA dataset, indicating the value of active feedback loops.

Windows: Detection on Windows datasets exhibits moderate-to-high variance across models. For E1 datasets, `Windows_E1_PA` and `Windows_E1_PX` show top scores by EAAMARL at *0.91/0.41* and *0.89/0.39* respectively. Interestingly, AE-RL and MARL models are often surpassed by AMARL and AAMARL, emphasizing the benefits of adversarial robustness and active query refinement. In E2 scenarios, performance gains from MARL to EAAMARL are even more pronounced, suggesting stronger generalization under this setup.

Table 3. AUC and F1 Score Comparison Across Reinforcement Learning-Based APT Detection Models. Each entry reports the AUC and F1 score (*AUC/F1*) achieved on the corresponding dataset. The evaluated models include: Q-Learning (Q-RL), Proximal Policy Optimization (PPO), Deep Q-Network (DQN), Autoencoder-guided Reinforcement Learning (AE-RL), Multi-Agent Reinforcement Learning (MARL), Adversarial Multi-Agent Reinforcement Learning (AMARL), Active Adversarial Multi-Agent Reinforcement Learning (AAMARL), and Ensemble Active Adversarial Multi-Agent Reinforcement Learning (EAAMARL). The best-performing models are highlighted by consistently higher AUC and F1 scores across diverse datasets and attack scenarios.

Database (OS × Scenario × Dataset)			Q-RL	PPO	DQN	AE-RL	MARL	AMARL	AAMARL	EAAMARL
BSD	E1	PA	0.57/0.17	0.68/0.21	0.65/0.12	0.70/0.19	0.74/0.26	0.78/0.33	**0.85/0.42**	**0.91/0.50**
		PE	0.56/0.15	0.63/0.19	0.63/0.12	0.69/0.15	0.71/0.20	0.74/0.30	0.83/0.39	**0.92/0.50**
		PX	0.51/0.01	0.58/0.20	0.58/0.10	0.65/0.14	0.69/0.14	0.73/0.28	0.82/0.38	0.85/0.43
		PN	0.51/0.10	0.56/0.16	0.55/0.11	0.66/0.10	0.71/0.16	0.68/0.22	0.84/0.33	0.82/0.38
		PP	0.55/0.13	0.55/0.16	0.51/0.10	0.62/0.07	0.67/0.15	0.63/0.21	0.81/0.26	0.82/0.36
	E2	PA	0.56/0.13	0.63/0.19	0.64/0.12	0.69/0.17	0.69/0.16	0.76/0.28	0.83/0.33	0.88/0.45
		PE	0.55/0.08	0.60/0.19	0.60/0.11	0.65/0.09	0.72/0.15	0.74/0.22	**0.84/0.31**	**0.89/0.46**
		PX	0.51/0.01	0.57/0.15	0.55/0.08	0.66/0.10	0.73/0.16	0.68/0.19	0.82/0.32	0.86/0.37
		PN	0.50/0.02	0.53/0.10	0.54/0.07	0.63/0.08	0.68/0.13	0.68/0.21	0.82/0.28	0.83/0.40
		PP	0.51/0.08	0.51/0.11	0.52/0.08	0.62/0.07	0.67/0.12	0.68/0.20	0.81/0.27	0.82/0.30
Windows	E1	PA	0.54/0.19	0.55/0.12	0.58/0.12	0.67/0.11	0.72/0.17	0.69/0.29	**0.87/0.35**	**0.91/0.41**
		PE	0.56/0.02	0.55/0.12	0.56/0.12	0.64/0.08	0.69/0.14	0.69/0.29	0.83/0.30	0.88/0.36
		PX	0.53/0.08	0.52/0.16	0.54/0.10	0.65/0.09	0.70/0.15	0.70/0.23	0.85/0.34	0.89/0.39
		PN	0.57/0.19	0.56/0.15	0.52/0.09	0.63/0.08	0.68/0.13	0.69/0.19	0.82/0.30	0.85/0.35
		PP	0.53/0.08	0.53/0.11	0.50/0.06	0.61/0.07	0.67/0.12	0.68/0.18	0.80/0.25	0.81/0.32
	E2	PA	0.60/0.23	0.57/0.17	0.56/0.12	0.65/0.11	0.70/0.15	0.69/0.22	**0.84/0.37**	**0.88/0.37**
		PE	0.57/0.12	0.51/0.12	0.53/0.10	0.64/0.08	0.69/0.14	0.69/0.19	0.83/0.33	**0.89/0.39**
		PX	0.51/0.07	0.52/0.09	0.52/0.08	0.66/0.10	0.71/0.16	0.68/0.16	0.83/0.32	0.89/0.37
		PN	0.52/0.08	0.52/0.09	0.52/0.08	0.63/0.08	0.68/0.13	0.63/0.15	0.82/0.27	0.84/0.33
		PP	0.50/0.09	0.53/0.11	0.54/0.10	0.62/0.07	0.67/0.13	0.53/0.12	0.81/0.27	0.84/0.31
Linux	E1	PA	0.55/0.16	0.56/0.15	0.69/0.13	0.66/0.19	0.70/0.15	0.66/0.22	**0.85/0.39**	**0.89/0.36**
		PE	0.52/0.09	0.54/0.10	0.59/0.10	0.63/0.12	0.68/0.13	0.69/0.26	0.83/0.38	0.87/0.33
		PX	0.53/0.03	0.53/0.06	0.55/0.10	0.65/0.09	0.70/0.15	0.70/0.22	0.84/0.31	0.88/0.35
		PN	0.52/0.08	0.53/0.08	0.53/0.09	0.63/0.08	0.69/0.13	0.68/0.21	0.81/0.30	0.86/0.32
		PP	0.50/0.06	0.51/0.05	0.51/0.06	0.62/0.07	0.67/0.16	0.63/0.21	0.80/0.29	0.84/0.34
	E2	PA	0.56/0.10	0.57/0.08	0.58/0.09	0.66/0.13	0.70/0.10	0.72/0.21	**0.83/0.30**	**0.87/0.33**
		PE	0.55/0.14	0.54/0.15	0.55/0.11	0.63/0.08	0.73/0.16	0.70/0.23	0.82/0.38	0.86/0.31
		PX	0.52/0.08	0.52/0.08	0.52/0.12	0.65/0.09	0.69/0.15	0.69/0.22	0.80/0.30	0.81/0.35
		PN	0.53/0.12	0.52/0.13	0.52/0.08	0.63/0.08	0.68/0.16	0.62/0.21	0.81/0.27	0.82/0.32
		PP	0.51/0.08	0.50/0.09	0.51/0.09	0.62/0.07	0.67/0.12	0.60/0.19	0.80/0.25	0.81/0.30

Linux: Linux datasets reveal that simpler models (Q-RL, PPO) struggle to achieve both reasonable AUC and F1, with most scores hovering around 0.50–0.55. In contrast, MARL and AMARL agents gain momentum, especially in `Linux_E1_PA` and `Linux_E2_PE`, where AAMARL and EAAMARL score up to *0.89/0.36*. The autoencoder-based latent representation appears especially valuable in this OS, improving representation learning for RL agents. Overall, EAAMARL demonstrates the most consistent superiority in Linux experiments.

Android: Despite Android's more volatile behavioral traces, ensemble-based agents again lead the pack. `Android_E1_PA` and `Android_E2_PE` highlight this trend, with EAAMARL reaching *0.89/0.36* and *0.88/0.41*. Q-RL and PPO underperform significantly, with F1 scores often below 0.10, pointing to their poor recall under Android's dynamic process behaviors. AAMARL and EAAMARL show how adversarial training and active learning can collectively overcome these challenges.

Overall Trends. The general progression in performance across all OSes and scenarios is clear:

- **QRL, PPO, DQN**: Serve as lower baselines with poor sensitivity and low F1.
- **AE-RL and MARL**: Benefit from better latent representations and multi-agent learning.
- **AMARL and AAMARL**: Introduce robustness and adaptability, especially effective in dynamic or noisy environments.
- **EAAMARL**: Shows state-of-the-art performance in nearly every dataset, combining active learning, adversarial defense, and ensemble synergy.

These results support our hypothesis that combining diverse agents with active feedback and latent space modeling significantly enhances APT detection capability across heterogeneous environments.

5 Conclusion

In this work, we presented a novel architecture for Advanced Persistent Threat (APT) detection that integrates deep autoencoding, multi-agent reinforcement learning (RL), active learning, and ensemble decision-making. We addressed key challenges in the cybersecurity domain—such as the rarity of APT events, concept drift in attacker behavior, and high false-positive costs—by leveraging adaptive and interactive learning agents that evolve through uncertainty-driven feedback. Our proposed framework builds on multiple reinforcement learning paradigms, including Q-Learning, PPO, DQN, adversarial training, and multi-agent collaboration. These agents operate on latent representations extracted from autoencoders, allowing for compact, high-level system behavior abstraction. To further enhance detection sensitivity, we introduced an active learning loop that queries an oracle (simulated from ground truth) whenever uncertainty is detected. This mechanism enables dynamic labeling of ambiguous traces and accelerates agent retraining on the most informative samples. Through extensive evaluation across 40 datasets spanning four operating systems and multiple attack scenarios, we demonstrated that traditional single-agent RL methods perform poorly in isolation. However, performance significantly improves when agents collaborate (Multi-Agent RL), become adversarially trained (AMARL), and incorporate active learning (AAMARL). Our Ensemble Active Adversarial Multi-Agent RL (EAAMARL) framework achieved the best results, consistently outperforming all baselines in terms of AUC and F1 score. The promising performance of EAAMARL confirms the benefit of integrating multiple RL strategies with feedback-aware learning in cybersecurity. As future work, we aim to explore real-time deployment, transfer learning across domains, and the integration of symbolic reasoning or LLMs into the ensemble.

Disclosure of Interests. The authors declare that they have no conflict of interest.

References

1. Rebane, J.C.: The Stuxnet Computer Worm and Industrial Control System Security. Nova Science Publishers Inc., Hauppauge (2011)
2. Marczak, B., et al.: Hide and seek: tracking nso group's pegasus spyware to operations in 45 countries. Technical Report (2018)
3. Saad, G., et al.: Attribution is in the object: Using rtf object dimensions to track apt phishing weaponizers. Virus Bull. **12**, 1–2 (2020)
4. Ghafir, I., Prenosil, V.: Advanced persistent threat attack detection: an overview. Int. J. Adv. Comput. Netw. Secur. **4**(4), 5054 (2014)
5. Shackelford, S.J.: Protecting intellectual property and privacy in the digital age. Chapman Law Rev. **19**, 445 (2016)
6. Han, X., et al.: Provenance-based intrusion detection: opportunities and challenges. In: 10th USENIX TaPP Workshop (2018)
7. Jenkinson, G., et al.: Applying provenance in APT monitoring and analysis: practical challenges for scalable provenance. In: TaPP, pp. 16–21 (2017)
8. Salim, D.T., et al.: A systematic review for apt detection and effective cyber situational awareness conceptual model. Heliyon **9**(7), 17156 (2023)
9. Tankard, C.: Advanced persistent threats and how to monitor and deter them. Netw. Secur. **2011**(8), 16–19 (2011)
10. Nguyen, T.T., Reddi, V.J.: Deep reinforcement learning for cyber security. IEEE Trans. Neural Netw. Learn. Syst. **34**(8), 3779–3795 (2023)
11. Cengiz, E., Gök, M.: Reinforcement learning applications in cyber security: a review. Sakarya University J. Sci. **27**(2), 481–503 (2023)
12. Adawadkar, A.M.K., Kulkarni, N.: Cyber-security and reinforcement learning—a brief survey. Eng. Appl. Artif. Intell. **114**, 105116 (2022). https://www.sciencedirect.com/science/article/pii/S0952197622002512
13. Milani, S., Topin, N., Veloso, M., Fang, F.: Explainable reinforcement learning: a survey and comparative review. ACM Comput. Surv. **56**(7), 1–36 (2024)
14. Tang, C., Abbatematteo, B., Hu, J., Chandra, R., Martín-Martín, R., Stone, P.: Deep reinforcement learning for robotics: a survey of real-world successes. In: Proceedings of the AAAI Conference on Artificial Intelligence, vol. 39, no. 27, pp. 28 694–28 698 (2025)
15. Tran. computing. https://www.darpa.mil/program/transparent-computing
16. Berrada, G., et al.: Aggregating unsupervised provenance anomaly detectors. In: 11th USENIX TaPP Workshop. USENIX Association (2019)
17. Beerrada, G., et al.: A baseline for unsupervised advanced persistent threat detection in system-level provenance. FCGS **108**, 401–413 (2020)
18. Benabderrahmane, S., et al.: A rule mining-based advanced persistent threats detection system. In: Proceedings of IJCAI21, Canada, Montreal, pp. 3589–3596 (2021)

19. Cheney, J., Gombau, X., Berrada, G., Benabderrahmane, S.: Categorical anomaly detection in heterogeneous data using minimum description length clustering. CoRR arxiv:2006.07916 (2020)
20. Benabderrahmane, S., Hoang, N., Valtchev, P., Cheney, J., Rahwan, T.: Hack me if you can: aggregating autoencoders for countering persistent access threats within highly imbalanced data. Future Gener. Comput. Syst. **160**, 926–941 (2024). https://doi.org/10.1016/j.future.2024.06.050

Machine Learning Applications

A Novel AI Approach for the Diagnosis of Alzheimer's Disease from Multi-modal Incomplete Data

Veronica Buttaro[1,2(✉)], Giuseppe Lamanna[1], Donato Massaro[1,2,3], Claudio B. Caporusso[1], Gianvito Pio[2], Michelangelo Ceci[2], and Alzheimer's Disease Neuroimaging Initiative

[1] CoE R& D, Lutech SpA, Bari, Italy
veronica.buttaro@uniba.it
[2] Dept. of Computer Science, University of Bari Aldo Moro, Bari, Italy
[3] Department of Electrical and Information Engineering, Politecnico di Bari, Bari, Italy

Abstract. Alzheimer's Disease (AD) globally represents one of the most prevalent and devastating neurodegenerative disorders. Recent advances in biomarker identification, neuroimaging, and diagnostic technologies offer promising tools to detect AD before significant symptoms appear. Artificial intelligence (AI) systems represent one of the most powerful tools for this purpose, since they are revolutionizing AD detection and management by analyzing complex patterns across multi-modal data. In this context, this paper introduces VERONECA (VErsatile and RObust approach for NEuroimaging and Clinical data-based Alzheimer's diagnosis), a novel AI method based on an advanced multi-modal boosting approach which promotes diversity among modalities through multi-armed bandits. The method offers: *i)* state-of-the-art performance in AD diagnosis, through effective integration of heterogeneous data; *ii)* specialized learners tailored to work with specific peculiarities of each modality; and *iii)* a robust strategy for handling missing modalities. Our experiments show that VERONECA effectively leverages complex multi-modal data to diagnose AD. Moreover, its intrinsically modular design facilitates the seamless integration of emerging modalities as well as the adaptation to evolving clinical standards.

The code of the proposed method VERONECA is publicly available at https://github.com/code-paper-veroneca/veroneca.git

Keywords: Machine learning · Multi-modal learning · Boosting · Alzheimer's Disease diagnosis

V. Buttaro and G. Lamanna—Equal contributors.
Data used in preparation of this article were obtained from the Alzheimer's Disease Neuroimaging Initiative (ADNI) database (adni.loni.usc.edu). As such, the investigators within the ADNI contributed to the design and implementation of ADNI and/or provided data but did not participate in the analysis or writing of this paper. A complete listing of ADNI investigators can be found at: http://adni.loni.usc.edu/wp-content/uploads/how_to_apply/ADNI_Acknowledgement_List.pdf.

S. Džeroski et al. (Eds.): DS 2025, LNAI 16090, pp. 555–570, 2025.
https://doi.org/10.1007/978-3-032-05461-6_36

1 Introduction

Alzheimer's Disease (AD) is characterized by progressive neurodegenerative processes that significantly impact patients' quality of life and impose substantial economic burdens on healthcare systems [31]. Therefore, early and accurate diagnosis has emerged as an essential objective. However, the diagnosis of AD is particularly difficult, since early symptoms, such as mild memory loss or disorientation, are frequently misattributed to normal aging. Moreover, a single definitive diagnostic test is currently unavailable, and the current diagnostic protocols typically involve a complex combination of clinical evaluations, cognitive assessments, and neuroimaging techniques, each affected by its own limitations that can potentially introduce additional uncertainty [10].

These inherent limitations of traditional diagnostic approaches, coupled with the subjective nature of clinical interpretation, have created an urgent demand for more objective, standardized, and accessible diagnostic tools. In this respect, Artificial Intelligence (AI) tools can provide a relevant contribution to alleviate many of the limitations exhibited by conventional diagnostic methods. Early AI approaches for the diagnosis of AD relied on the analysis of a single data source (view), such as the results of neuropsychological assessments, or images collected through imaging techniques, such as Magnetic Resonance Imaging (MRI) [28]. However, recognizing that AD's complex pathophysiology manifests itself across multiple biological and cognitive perspectives, more recent works shifted toward integrating diverse data sources through multi-view and/or multi-modal learning frameworks [3]. This methodological evolution has been further accelerated by advances in deep learning, where sophisticated neural architectures have demonstrated remarkable performance in analyzing complex medical images and extracting subtle patterns from MRI data for the classification of AD stage [1].

This evolution represents a significant advancement in AD research, enabling the integration of heterogeneous data sources including structural and functional neuroimaging (MRI, fMRI, PET), electronic health records, genomic data, and neuropsychological assessments to develop more comprehensive diagnostic models: by leveraging multiple sources, possible represented through different modalities, it is possible to discover complex relationships that may otherwise be overlooked. However, the integration of heterogeneous data introduces several challenges: managing different noise levels and scaling properties, handling incomplete or partially missing data, maintaining computational efficiency while preserving diagnostic accuracy, and extracting meaningful cross-view feature representations that capture underlying disease characteristics.

Current data fusion strategies can generally be categorized as *early fusion* (direct combination of the input data), *intermediate fusion* (based on view/modality-specific feature extraction), and *late fusion* (based on the independent view processing, combining the outputs) [21], each with different solutions to representational challenges. While these approaches provided significant advancements in this field, a comprehensive solution capable of seamlessly handling heterogeneous, potentially incomplete, medical data while maintaining robust diagnostic performance, is still missing.

To address these limitations, we propose VERONECA (VErsatile and RObust approach for NEuroimaging and Clinical data-based Alzheimer's diagnosis), a novel multi-view and multi-modal method that leverages an advanced boosting mechanism based on adversarial multi-armed bandits, to dynamically and iteratively identify the optimal perspectives for AD diagnostic modeling in a *late fusion* fashion. VERONECA is also able to handle the possible incompleteness of the views, which is typical in medical data, and in providing a flexible architecture capable of integrating additional data views with different modalities, potentially achieving enhanced predictive performance.

The remainder of this paper is organized as follows: Sect. 2 discusses works related to the present paper; Sect. 3 describes the proposed method VERONECA; Sect. 4 describes our experimental setting and discusses the obtained results; finally, Sect. 5 concludes the paper and outlines future research directions.

2 Related Work

Traditional diagnosis of AD has evolved from relying solely on clinical symptoms and neuropsychological tests to incorporating sophisticated medical imaging, genetic and proteomic biomarkers. This advancement in medical research has also included AI methods, whose early approaches focused on a single view, limiting the full understanding of the disease [22].

The most common traditional machine learning (ML) techniques for the prediction of AD are Support Vector Machines (SVM), k-nearest neighbor (KNN), Decision Trees (DT), and Random Forests (RF) [12]. Besides their direct adoption on single-view (and, thus, uni-modal) data, feature extraction techniques have been integrated into more comprehensive frameworks. For example, in [13], the authors leverage dimensionality reduction techniques, like PCA and UMAP, to identify relevant features from magnetic resonance and positron emission tomography (PET) images of the OASIS dataset [24], before training ML models.

More recently, Deep Learning (DL) approaches have become widely adopted in the context of AD diagnosis, especially for analyzing medical images, biological networks or, more in general, unstructured data [20]. For example, in [1], the authors used three deep pre-trained neural networks (VGG19, Inception v3, and ResNet50) with transfer learning, with minimal training time from ImageNet, to classify AD stages using ADNI [23] fMRI data.

In another approach [29], the authors proposed an end-to-end AD diagnostic framework that combines a 2D Convolutional Autoencoder (2D-CAE) with a stacked Convolutional Neural Network (CNN). This two-components approach first uses the 2D-CAE to extract meaningful latent representations from brain MRI images for each disease class, then employs the stacked CNN to perform the actual classification based on these latent representations.

Moving toward multi-view approaches, existing methods can be mainly categorized as *early fusion*, *intermediate fusion*, and *late fusion* approaches, each offering distinct advantages and limitations [21].

In **early fusion** approaches, data from all perspectives are combined at the beginning, typically through concatenation or through a preliminary transformation, before learning a machine learning model. In this simple approach, machine learning methods cannot distinguish between features from different views. This technique is also not applicable in contexts where views are heterogeneous, i.e., have different modalities (e.g., images and tabular data). Another limitation is that, in order to adopt recent deep learning approaches, the possible view incompleteness must be handled through imputation methods to fill missing values. An example falling into this category is the work [18], where the authors propose a multiview learning approach that integrates miRNA expression profiles with electronic health records (EHR) to identify patients with various cancer-related conditions using, in advance, an imputation method to handle incomplete views.

Intermediate fusion approaches are based on a pre-processing phase applied to each view, that can also be specific for each modality, aiming to homogenize and possibly integrate data into a shared, intermediate, representation. For this purpose, autoencoder architectures are widely used to extract low-dimensional embeddings that capture essential features across multiple views, while reducing noise and redundancy. These compressed representations are considered robust for learning a predictive model. In [27], authors use stacked denoising auto-encoders to extract features from clinical and genetic data, and 3D-Convolutional neural networks for extracting embeddings from imaging data, using the concatenation of such extracted features as input to an ML method.

In **late fusion** methods, a model is trained from each view independently, which may also be based on different modalities. The output obtained through each view is then combined at the final stage of the pipeline to produce a final prediction. In this scenario, the emphasis shifts to the aggregation method, that may range from simple averaging or voting mechanisms to more complex techniques like meta-classifiers that integrate predictions from the different views. A relevant example is the work in [9], that represents a significant extension of the classical AdaBoost algorithm [30] to make it able to operate effectively in a multi-view learning setting. One of the most significant limitations of these approaches is that they can only work when samples are fully observed from all the considered views. A significant effort for the management of incomplete views was made in the method irBoost.SH [19], that was applied by the authors in the context of microbiome data analysis. However, it cannot handle heterogeneous data source, with different modalities (such as images and tabular data), since the hypothesis is that every view is analyzed through the same base learner.

The limitations exhibited by existing approaches highlight the need for a more versatile approach that can adapt to the practical challenges raised by the AD diagnosis: *i)* the possibility to analyze multiple data views; *ii)* the ability to handle the possible heterogeneity of the data (multiple modalities); *iii)* a proper approach to handle the incompleteness of the views, namely, instances (patients, in this domain) that are not necessarily represented according to all the available views. The approach VERONECA proposed in this paper exhibits all such characteristics.

3 The Proposed Method VERONECA

The approach we propose in this paper, called VERONECA, is based on the irBoost.SH method [19]. Specifically, we enable it to work with heterogeneous and multi-modal data that would possibly benefit from modality-optimized learners.

To clarify the terminology used throughout this work, we specify the distinction between ***views*** and ***modalities*** in multi-modal machine learning contexts. Modalities [16] refer to fundamentally different types of data sources or sensory channels (e.g., MRI imaging, genomic data, neuropsychological assessments), while views represent different representations or perspectives of the same instances. Multiple views can be derived from the same underlying data modality, possibly obtained using different preprocessing techniques applied to the same data sources (e.g., MRI scans with various feature extraction methods).

irBoost.SH is a multi-view boosting algorithm, based on multi-armed bandits, able to solve multi-class classification tasks considering possibly incomplete views. At each iteration, it identifies one winning view using adversarial multi-armed bandits and uses its predictions to update a shared instance weight distribution in a learning process based on boosting.

The main limitation of irBoost.SH is that every view is represented according to the same modality. Coherently, for each view, the approach based on adversarial multi-armed bandits learns a view-specific model by running the same machine learning method (i.e., Random Forest). This limitation turns out to be too restrictive in the analysis of multi-modal data since each modality possibly requires the adoption of learning algorithms with different peculiarities to take advantage of the data specificity. For instance, in the case of structured clinical data, more traditional classifiers such as RF or SVM can be appropriate, while unstructured data types require specialized architectures: medical images require convolutional neural networks (CNN), textual clinical notes require transformer-based language models, time-series physiological signals require recurrent neural networks, etc. VERONECA overcomes the limitations of irBoost.SH by handling any data modality that can be processed by appropriate machine learning models and, more generally, appropriate pipelines, transitioning from a purely multiview approach to a comprehensive multi-modal framework. Moreover, the integrated multi-modal approach is able to guarantee the interpretability advantages of the boosting framework, allowing the user to understand which data modalities contribute most significantly to the prediction (i.e., diagnostic decision).

VERONECA takes as input training data in the form $S = \{(x_i, y_i)\}_{i=1}^{n}$, where n is the number of instances, $y_i \in Y \subset \mathbb{N}^+$ is the label of instance x_i and each instance $x_i = x_i^{(1)}, x_i^{(2)}, ..., x_i^{(K)}$ is represented across K different views. For each view $j \in 1, 2, \ldots, K$, $x_i^{(j)} \in (\mathbb{R}^{q_j} \cup \{null\})$ represents the instance in a q_j-dimensional feature space.

Each instance x_i is also associated with its instance weight w_i^t, which is iteratively updated at iteration t. This weight is *shared* across multiple views in order to transfer knowledge across the views [7,19]. Initially, shared instance weights are uniformly distributed. For each iteration, the algorithm first trains weak classifiers for each view estimating their respective predictive performance (*edges*), then it probabilistically selects a winning view (rather than deterministically choosing the view with the highest edge). This probabilistic view selection incorporates principles from Adversarial Multi-armed Bandits (AMB), treating views as actions and using the Exponentially Weighted Average (EWA) forecaster to update view probabilities p_j^t (probability that the j-th view is selected at the iteration t). This approach encourages the selection of views with lower probabilities, thereby promoting diversity in the ensemble. After completing t iterations of this selection and weight update process, it constructs the final combined classifier by computing a weighted sum of all winning classifiers accumulated across the complete set of T iterations.

Formally, for each view $j \in 1, 2, \ldots, K$, we define N_j as the set of instances present in view j, while $N = \bigcup_{j=1}^{K} N_j$ represents the set of all instances. The algorithmic flow begins with the initialization of weights $w_i^1 = \frac{1}{n}$ for each instance $i \in N$ and probabilities $p_j^1 = e^{-\sigma\sqrt{K}}$ for each view $j \in 1, 2, \ldots, K$, where $\sigma > 0$ is a parameter that controls the initial exploration-exploitation balance. At each iteration t, the winning view, denoted as j^*, is selected according to the current probability distribution $q_j^t = (1-\gamma)\frac{p_j^t}{\sum_{k=1}^{K} p_k^t} + \frac{\gamma}{K}$, where $\gamma \in (0, 1]$ represents a mixing parameter that ensures minimum exploration of all views.

On the selected view, a weak classifier $h_t^{j^*}$ is trained using the current instance weights w_i^t. The classifier produces, for instances present in that view, both predicted labels $h_t^{j^*}(x_i^{j^*})$ and, for each possible label l, a class probability $v_{t,l}^{j^*}(x_i^{j^*})$. If one instance is not represented according to the view j^*, we assume that $h_t^{j^*}$ $(x_i^{j^*}) = 0$ and $v_{t,l}^{j^*}(x_i^{j^*}) = 0$, that is, the classifier predicts the value 0[1] with probability 0.

The computation of the edge is formally defined as:

$$\theta_t^{j^*} = 2 \cdot \sum_{i \in N} w_i \cdot (I[y_i = h_t^{j^*}(x_i^{j^*})] - 0.5) \tag{1}$$

where the agreement function $I[a = b] = 1$ if $a = b$, 0 otherwise.

This formulation assigns +1 to correct predictions and −1 to incorrect ones, scaled by instance weights. Importantly, instances missing from a particular view ($x_i \notin N_j$) receive a "null" prediction, which will never match their true label, thus not contributing positively to the edge calculation.

Based on this edge, both view probabilities and instance weights are updated following the exponentially weighted average forecaster principles from adversarial multi-armed bandits [2], after computing the so-called *reward.* The reward for the selected view j^* at iteration t is computed as $r_t^{j^*} = 1 - \sqrt{1 - (\theta_t^{j^*})^2}$, and then normalized by q_t^k, which encourages the views with low probability to be selected, promoting diversity. In general, the normalized reward for the view k $\hat{r}_t^k$ is defined as: $\hat{r}_t^k = r_t^k / q_t^k$ if $k = j^*$, 0 otherwise.

[1] Note that $Y \subset \mathbb{N}^+$, meaning that the class 0 does not correspond to any actual class.

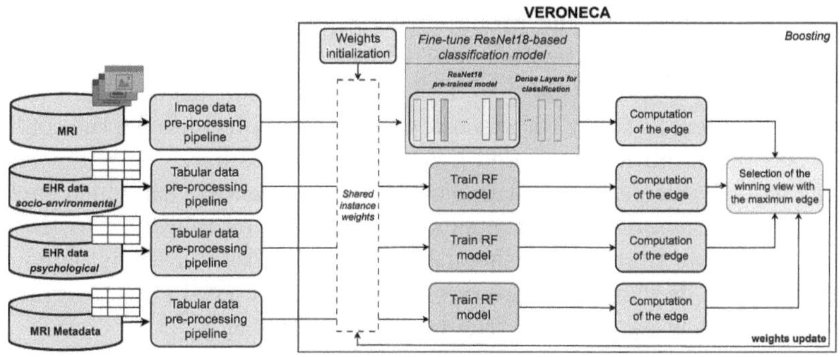

Fig. 1. Architecture of the variant VERONECA-Direct of the proposed method. The figure showcases an example with AD data across 4 views.

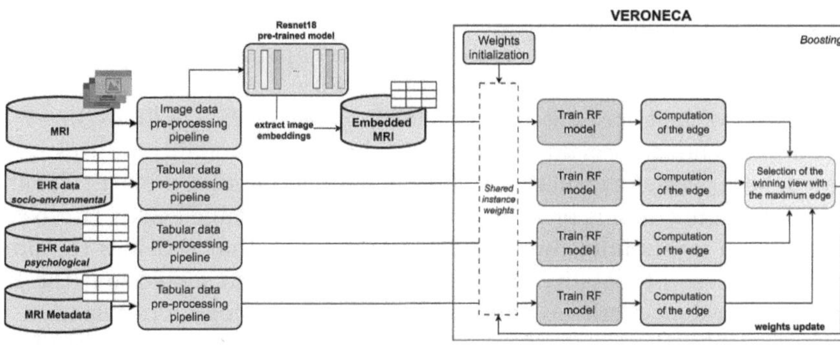

Fig. 2. Architecture of the variant VERONECA-Embedding of the proposed method. The figure showcases an example with AD data across 4 views.

The view probabilities are then updated as follows: $p^k_{t+1} = p^k_t \cdot e^{\sigma(\frac{1}{K}(\hat{r}^k_t + \frac{1}{\sqrt{KT}}))}$.

For the instance weight update, the algorithm computes a coefficient $\alpha_t = \frac{1}{2}\ln\left(\frac{1+\theta_t^{j^*}}{1-\theta_t^{j^*}}\right)$ and updates the weights as:

$$w_i^{(t+1)} = \frac{w_i^t}{Z_t^j} \cdot e^{-2\alpha_t(I[y_i = h_t^{j^*}(x_i^{j^*})]-0.5)} \tag{2}$$

where Z_t^j is a normalization factor such that $\sum_{i\in N} w_i^{(t+1)} = 1$. This update increases weights for misclassified examples while decreasing them for correctly classified ones, following the boosting paradigm for the multi-class setting.

The final prediction is produced as a weighted combination of all classifiers trained across the T iterations. For a test instance x, the algorithm computes:

$$H(x) = \arg\max_{l\in Y} \frac{\sum_{t=1}^{T} \alpha_t v_{t,l}^{j^*}(x^{j^*})}{C} \tag{3}$$

where $v_{t,l}^{j^*}(x^{j^*})$ represents the probability assigned to class l by the classifier trained at iteration t on view j^*, and C is a normalization constant chosen such that $\sum_{l \in Y} \sum_{t=1}^{T} \alpha_t v_{t,l}^{j^*}(x^{j^*}) = 1$. This formulation ensures that the final prediction considers not only the discrete class predictions but also the confidence scores (class probabilities) provided by each weak classifier.

A particularly interesting feature is that the frequency with which each view is selected as the *winner* can serve as an indicator of its relevance to the classification task: views that are consistently selected across iterations demonstrate greater utility, while views that are rarely chosen may contain less discriminative information. This is particularly true for small values of γ.

Figures 1 and 2 illustrate two possible variants of the VERONECA architecture. As previously discussed, each view can correspond to a distinct modality, preprocessing technique, and suitable representation (embedding). Additionally, each modality can be processed using a specific base machine learning algorithm, which is integrated into the boosting framework. We remind that in principle, we can plug any algorithm, provided that it is capable of returning both the predicted class and the associated class probability.

The first variant presented, henceforth called *VERONECA-Direct* (see Fig. 1), implements a specialized approach where each modality utilizes its own dedicated base learner: structured clinical data is handled by RF classifiers, while neuroimaging data is processed directly by a pre-trained CNN (ResNet18), followed by a 2-layers fully-connected neural network that provides the prediction and the class probability. The ResNet18 pretrained model and the fully-connected neural network are fine-tuned through a single optimization process. The second variant, henceforth called *VERONECA-Embedding* (see Fig. 2), employs a feature extraction approach: neuroimaging data are first embedded into a feature vector representation through the ResNet18 pre-trained CNN. All views are then processed using the same base learner in the boosting process. These two variants demonstrate that VERONECA offers a high degree of flexibility, enabling either the enforcement of homogeneous representations across modalities or the adoption of cross-modality end-to-end learning, in which the optimal representation is learned as an integral part of the boosting process.

Note that in this section we do not provide details on the preprocessing pipelines, which are instead specified in Sect. 4, after we describe the analyzed.

4 Experiments

In this section, we describe the experimental evaluation we performed to assess the effectiveness of the proposed method VERONECA for Alzheimer's disease diagnosis from multi-modal data.

The task we consider is a multi-class classification task where individuals can be classified according to the following classes: affected by Alzheimer's Disease (AD), affected by mild cognitive impairment (MCI), and control (CN).

In the following, we first describe the datasets, then outline the preprocessing pipelines, experimental setup, and finally present and discuss the results.

4.1 The Adopted Dataset

Data used in the preparation of this article were obtained from the Alzheimer's Disease Neuroimaging Initiative (ADNI) database. The current goals of ADNI include validating biomarkers for clinical trials, improving the generalizability of data by increasing diversity in the participant cohort, and to provide data concerning the diagnosis and the progression of Alzheimer's disease to the scientific community. For up-to-date information, see https://adni.loni.usc.edu/.

In our study, we consider the following views, based on different modalities:

- **MRI - Neuroimaging Data** (images): a dedicated image collection constructed comprising T1-weighted 3D brain MRI scans acquired with MPRAGE sequence [5] for high-resolution anatomical detail. The dataset includes DICOM images from participants aged 55–96 years across all ADNI study phases (ADNI-1, ADNI-GO, ADNI-2, ADNI-3).
- **EHR - Socio-enviromental** (tabular): includes information related to lifestyle, education level, social engagement, and environmental factors.
- **EHR - Psychological** (tabular): includes psychological tests administered to participants, such as the MMSE, clock drawing test [8], and other tools used to assess cognitive function and mental state.
- **MRI Metadata** (tabular): includes technical data and parameters from MRI scans (such as scanner type, resolution, and magnetic field strength), useful for analyzing brain imaging.

The dataset comprises 11,458 instances, categorized into three classes: Alzheimer's Disease (AD) with 2,449 instances, Mild Cognitive Impairment (MCI) with 4,989 instances, and Cognitively Normal (CN) with 4,020 instances. These instances correspond to a total of 2,431 unique individuals.

4.2 Data Preprocessing Pipelines

Both T1-weighted MRI images and the EHR data underwent rigorous preprocessing prior to analysis, as detailed below.

The T1-weighted (T1w) 3D brain MRI images from ADNI originate from various MRI scanners and undergo ADNI's standardized preprocessing pipeline, which includes corrections for gradient warping, B1 non-uniformity, and N3 intensity normalization. To prepare these images for machine learning analysis, we implemented an additional preprocessing workflow using the ANTsPy library [26] to ensure data consistency and reduce dimensionality. Despite ADNI's initial N3 correction, we applied N4ITK [25] bias field correction to further refine image homogeneity and ensure voxel intensities accurately reflect tissue properties rather than acquisition artifacts. Adaptive non-local means denoising [15] was then applied to reduce random noise while preserving critical edge information and structural details by averaging voxel intensities based on surrounding patch similarity. For standardization, all T1w images were registered to the MNI ICBM 152 T1-weighted template through affine registration to facilitate inter-subject comparisons. Finally, brain extraction was performed using a pre-trained

U-Net convolutional neural network [17] from the AntsPyNet package to isolate brain parenchyma from non-brain tissues including skull, scalp, and meninges.

For EHR data (available in the ADNIMERGE dataset) we implemented a preprocessing pipeline that involved systematic filtering steps: variables associated with dates, indexes, SW/HW versions were removed, along with those pointing to the diagnosis and any variables with more than 90% missing values.

After the pre-processing pipeline, while all instances are represented according to the *EHR-Socio-enviromentals* view, for 6 instances the *EHR-Psycological* view is missing, for 2731 instances the *MRI Metadata* view is missing, and for 10518 instances the *MRI - 3D images* is missing.

4.3 Experimental Setting

We evaluated model performance through a stratified 5-fold cross-validation, collecting precision, recall, and F1-score measures for each class. These measures were then averaged across the folds and the classes. In order to give proper importance to all the classes, the results are macro-averaged over the classes.

For tabular data, we adopted a Random Forest classifier as base learner, since widely adopted in biological applications [14], configured with 10 trees and a minimum leaf size of 60 instances. The number of trees and the minimum leaf size have been chosen to properly balance bias-variance in boosting-based ensemble learning, that is, to enforce the concept of boosting over weak learners.

For the neuroimaging modality, we adopt a CNN architecture based on ResNet18 [11], specifically employing the pre-trained implementation available in the MONAI package [6]. The ResNet18 network serves as a feature extractor to obtain dense embedding representations from the 3D MRI volumes for the *VERONECA-Embedding* variant (Fig. 2). Instead, for the *VERONECA-Direct* variant (Fig. 1), these embeddings are subsequently processed through fine-tuning of a neural network composed of two fully connected layers to produce the final predictions. This fine-tuning strategy is conducted using the Adam optimizer with a learning rate of 10^{-4}, a mini-batch size of 20, and 50 epochs.

We set the parameters $\sigma = 0.15$ and $\gamma = 0.3$ as suggested in [19], and performed $T = 25$ iterations of the boosting process.

To evaluate the effectiveness of VERONECA in capturing multi-modal information, we conducted comparative experiments against several uni-modal approaches, which play the role of baselines in our evaluation. In particular, for the baselines, we train a CNN-based classifier with the same architecture and training setup as the one implemented in *VERONECA-Direct* for the 3D MRI images. For all the tabular views, we train a RF classifier, also with the same architecture and training setup as the one adopted in VERONECA.

Since our framework falls under multi-modal classification methods, we also compared it against an intermediate fusion approach available in the literature. Specifically, we adapted the method in [27]. This method employs stacked autoencoders for EHR data and CNNs for MRI images, then concatenates the bottleneck embeddings from both modalities for classification using Random Forests. In our implementation, we used a stacked autoencoder for EHR data and the same pre-trained ResNet18 as in VERONECA for MRI images, concatenating the embeddings before applying an RF classifier for prediction. To ensure a fair comparison, we followed the original preprocessing steps, including replacing missing values with zeros to handle incomplete data. The purpose of this comparison is to show the possible advantage introduced by the multi-modal boosting approach in VERONECA with respect to another learning framework by employing the same base algorithms.

We also conducted experiments considering all tabular views (EHR - Socio-enviromental, EHR - Psycological and MRI Metadata) as a single view, obtained through concatenation. This setting enables the base learners to capture possible dependencies within the same modality, rather than leaving this task to the boosting algorithm. On the other hand, this choice can make the identification of the most relevant views less effective, limiting interpretability.

4.4 Results and Discussion

In Tables 1 and 2, we show the results obtained when considering all 4 views separately, and with all tabular views aggregated into a single view, respectively.

In the case of 4 views (see Table 1), we can see that VERONECA achieves F1 scores of 0.809 for the VERONECA-Direct variant and 0.806 for the VERONECA-Embedding variant, with substantial improvements over all the other approaches.

The results obtained by the uni-modal baselines highlight the importance of our multi-modal approach. In particular, the ResNet18 architecture processing 3D MRI images achieves moderate performance with an F1 score of 0.599, demonstrating reasonable but limited capability when relying solely on imaging data. In contrast, Random Forest applied to socio-environmental data or MRI metadata shows poor performance with an F1 score of 0.489 and 0.487, respectively. This indicates that these two views alone provide insufficient discriminative power for the classification task. This poor performance is particularly notable if we consider that the Random Forest implementation we used can naturally handle missing values. Indeed, the results suggest that if the proportion of missing values becomes excessive, as in the MRI Metadata view, even robust algorithms like Random Forest suffer significant performance degradation if not supported by complementary views. Instead, the single view related to psychological factors processed through Random Forest achieves significantly better performance with an F1 score of 0.796. This superior performance likely reflects both the inherent predictive value of psychological assessments and potentially more complete data collection in this domain.

The Intermediate Fusion method [27] performs poorly with an F1 score of 0.475, significantly behind VERONECA. This result highlights a critical limitation of traditional concatenation-based fusion methods when dealing with heterogeneous data modalities and missing values.

The analysis in Fig. 3 on the winning views across boosting iterations represents a way to better understand and interpret the results, emphasizing which view is mostly contributing to the predictive performance. In particular, Fig. 3

(left) reveals that, across 25 boosting iterations, the algorithm adaptively selects the winning views based on their contribution. The selection pattern highlights the dynamic contribution of each view, along the different folds. For example, in Folds 1 and 4, the view *EHR - psychological* appears frequently as the winner, indicating its strong relevance in that subset of instances, whereas in Fold 2, *MRI metadata* is chosen most frequently, suggesting a shift in view importance based on the underlying data distribution. This difference suggests that, for each fold, the boosting algorithm selects the most promising views to optimize the predictive performance. Nevertheless, the selection mechanism prevents the algorithm from being dominated by any single view, promoting diversity without sacrificing accuracy. This can be theoretically justified by the principles of Bayesian learning theory [4] and, more broadly, ensemble learning, both promoting diversity among multiple weak learners to enhance overall predictive accuracy.

Focusing on the results obtained with 2 views (see Table 2), we can observe the highest F1 score values, that reach 0.827 with VERONECA-Direct and 0.819 with VERONECA-Embedding. The uni-modal method *RF (EHR data + MRI metadata)* achieves a reasonably high F1 score of 0.800, while *Intermediate Fusion* reaches a F1 score of 0.781. The latter appears significantly higher than that obtained with 4 separate views (0.475), suggesting that aggregating homogeneous views allows the encoding part of these methods to leverage possible dependencies among features of originally different views. Similarly, VERONECA obtains better results with 2 views, with respect to those obtained with 4 views, possibly because the base learner adopted for *merged* tabular data (RF) could leverage possible dependencies among features of different views. However, this *artificial merge operation* is applicable only among views of the same modality.

In Fig. 3 (right), we show the frequency of selection of each view. In this case, the contribution of the views appear more balanced across the folds, except for Fold 2, where tabular data were selected for 20 out of 25 iterations. This confirms that, although VERONECA promotes diversity, when a view prominently performs better than the others for a given subset of instances, it is still properly exploited within the selection process.

Finally, when comparing the two variants of VERONECA, we observe that VERONECA-Direct constantly outperforms VERONECA-Embedding. This result is expected because in VERONECA-Direct the embedding of images is continuously optimized during the boosting process. In contrast, in VERONECA-Embedding the image embedding is fixed at the beginning and is not oriented to obtain optimal accuracy (the embedding phase is fully unsupervised).

Table 1. Results obtained with 4, where tabular data are aggregated into one view. The first 4 lines represent baselines, while the 3 three lines represent multi-modal approaches, including VERONECA. The best results are shown in bold.

	Method	Accuracy	Prec.	Recall	F1
Uni-modal	ResNet18 (MRI - 3D images)	0.608	0.677	0.616	0.599
	RF (EHR - Socio-enviromental)	0.506	0.533	0.477	0.489
	RF (EHR - Psychological)	0.784	0.802	0.791	0.796
	RF (MRI - metadata)	0.557	0.553	0.494	0.487
Multi-modal (4 views)	Intermediate Fusion [27]	0.488	0.485	0.472	0.475
	VERONECA-Direct	**0.802**	**0.818**	**0.803**	**0.809**
	VERONECA-Embedding	0.801	0.814	0.802	0.806

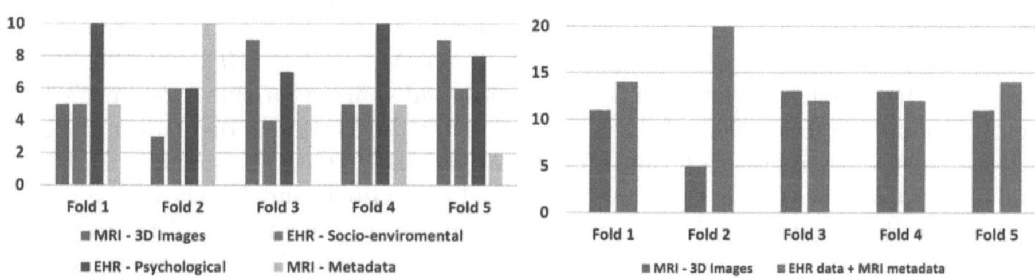

Fig. 3. Number of iterations of the boosting process of VERONECA with 4 views (on the left) and with 2 views (on the right), in which each view has been selected for each fold of the 5-fold cross validation.

Table 2. Results obtained with 2 views, where tabular data are aggregated into one view. The first 2 lines represent baselines, while the 3 three lines represent multi-modal approaches, including VERONECA. The best results are shown in bold.

	Method	Accuracy	Prec.	Recall	F1
Uni-modal	ResNet18 (MRI - 3D images)	0.608	0.677	0.616	0.599
	RF (EHR data + MRI metadata)	0.790	0.800	0.800	0.800
Multi-modal (2 views)	Intermediate Fusion [27]	0.772	0.785	0.778	0.781
	VERONECA-Direct	**0.822**	**0.832**	**0.823**	**0.827**
	VERONECA-Embedding	0.816	0.821	0.824	0.819

5 Conclusions

In this paper, we presented VERONECA, a novel multi-modal method for Alzheimer's Disease diagnosis. Key innovations of VERONECA include its ability to work on multi-view data, when represented according to different modalities, and its ability to work with incomplete views. Moreover, it can leverage optimized preprocessing and learning for specific modalities, integrating their output through a boosting-based multi-armed bandits approach.

Our experiments, performed on a real dataset provided by ADNI, shows the effectiveness of VERONECA in handling multi-view multi-modal data, in comparison with uni-modal methods and with existing multi-modal approaches.

For future work, we plan to extend our experiments to *i)* incorporate other modalities, such as genomic and transcriptomic data; *ii)* include additional datasets, also to enable the possibility to perform statistical tests; *iii)* consider alternative pre-trained models for each modality. Finally, we will work on the design of multi-modal explainability pipelines.

Acknowledgments. This work was partially supported by the European Union - NextGenerationEU through the Italian Ministry of University and Research, projects: CN3 RNA - "National Center for Gene Therapy and Drugs based on RNA Technology" (CUP: H93C22000430007), FAIR - "Future AI Research" (PE00000013, CUP: H97G22000210007 and CUP: B43D22000910004), and PRIN 2022 "BA-PHERD: Big Data Analytics Pipeline for the Identification of Heterogeneous Extracellular non-coding RNAs as Disease Biomarkers" (2022XABBMA, CUP: H53D23003690006), under the NRRP MUR Program. Data collection and sharing for the Alzheimer's Disease Neuroimaging Initiative (ADNI) is funded by the National Institute on Aging (National Institutes of Health Grant U19AG024904). B.V., C.C., L.G. and M.D. thank Anni Domenica Arias, Ivan Palmisano, Leonardo Di Gioia, and Francesca Laera for helpful scientific discussions. Moreover, they acknowledge the support of Barbara Alicino and Giuseppe Ieva for enabling this research within Lutech CoE R&D in Bari.

References

1. Abed, M.T., Fatema, U., Nabil, S.A., Alam, M.A., Reza, M.T.: Alzheimer's disease prediction using convolutional neural network models leveraging pre-existing architecture and transfer learning. In: ICIEV & icIVPR, pp. 1–6 (2020)
2. Auer, P., Cesa-Bianchi, N., Freund, Y., Schapire, R.E.: The nonstochastic multiarmed bandit problem. SIAM J. Comput. **32**(1), 48–77 (2002)
3. Baltrušaitis, T., Ahuja, C., Morency, L.P.: Multimodal machine learning: a survey and taxonomy. IEEE Trans. Pattern Anal. Mach. Intell. **41**(2), 423–443 (2018)
4. Bishop, C.M.: Pattern Recognition and Machine Learning (Information Science and Statistics), 1 edn. Springer, Heidelberg (2007)
5. Brant-Zawadzki, M., Gillan, G.D., Nitz, W.R.: Mp rage: a three-dimensional, t1-weighted, gradient-echo sequence-initial experience in the brain. Radiology **182**(3), 769–775 (1992)
6. Cardoso, M.J., Li, W., Brown, R., et al.: Monai: an open-source framework for deep learning in healthcare (2022). https://arxiv.org/abs/2211.02701
7. Ceci, M., Pio, G., Kuzmanovski, V., Džeroski, S.: Semi-supervised multi-view learning for gene network reconstruction. PLOS ONE **10**(12), 1–27 (2015)
8. Davis, D.H., Creavin, S.T., Noel-Storr, A., et al.: Neuropsychological tests for the diagnosis of alzheimer's disease dementia and other dementias: a generic protocol for cross-sectional and delayed-verification studies. Cochrane Datab. Syst. Rev. (3), CD010460 (2013)
9. Elfwing, S.: Neural information processing. Theory and algorithms, p. 215 (2010)

10. Grimmer, T., et al.: Visual versus fully automated analyses of 18f-fdg and amyloid pet for prediction of dementia due to alzheimer disease in mild cognitive impairment. J. Nucl. Med. **57**(2), 204–207 (2016)
11. Hara, K., Kataoka, H., Satoh, Y.: Can spatiotemporal 3d cnns retrace the history of 2d cnns and imagenet? (2018). https://arxiv.org/abs/1711.09577
12. Kavitha, C., Mani, V., Srividhya, S., Khalaf, O.I., Tavera Romero, C.A.: Early-stage alzheimer's disease prediction using machine learning models. Front. Public Health **10**, 853294 (2022)
13. Lazli, L.: Machine learning classifiers based on dimensionality reduction techniques for the early diagnosis of alzheimer's disease using magnetic resonance imaging and positron emission tomography brain data. In: International Meeting on Computational Intelligence Methods for Bioinformatics and Biostatistics, pp. 117–131 (2021)
14. Li, Y., Wu, F.X., Ngom, A.: A review on machine learning principles for multi-view biological data integration. Brief. Bioinform. **19**(2), 325–340 (2018)
15. Manjón, J.V., Coupé, P., Martí-Bonmatí, L., Collins, D.L., Robles, M.: Adaptive non-local means denoising of mr images with spatially varying noise levels. J. Magn. Reson. Imaging **31**(1), 192–203 (2010)
16. Parcalabescu, L., Trost, N., Frank, A.: What is multimodality? arXiv preprint arXiv:2103.06304 (2021)
17. Ronneberger, O., Fischer, P., Brox, T.: U-net: convolutional networks for biomedical image segmentation. In: Navab, N., Hornegger, J., Wells, W.M., Frangi, A.F. (eds.) MICCAI 2015. LNCS, vol. 9351, pp. 234–241. Springer, Cham (2015). https://doi.org/10.1007/978-3-319-24574-4_28
18. Rosa, D., Pellicani, A., Pio, G., D'Elia, D., Ceci, M.: Exploiting microrna expression data for the diagnosis of disease conditions and the discovery of novel biomarkers. In: ISMIS 2024, pp. 77–86. Springer, Heidelberg (2024). https://doi.org/10.1007/978-3-031-62700-2_8
19. Simeon, A., Radovanović, M., Lončar-Turukalo, T., Ceci, M., Brdar, S., Pio, G.: Multi-class boosting for the analysis of multiple incomplete views on microbiome data. BMC Bioinf. **25**(1), 188 (2024)
20. Song, J., Zheng, J., Li, P., Lu, X., Zhu, G., Shen, P.: An effective multimodal image fusion method using mri and pet for alzheimer's disease diagnosis. Front. Digital Health **3**, 637386 (2021)
21. Stahlschmidt, S.R., Ulfenborg, B., Synnergren, J.: Multimodal deep learning for biomedical data fusion: a review. Brief. Bioinf. **23**(2), bbab569 (2022)
22. Tang, X., Guo, Z., Chen, G., et al.: A multimodal meta-analytical evidence of functional and structural brain abnormalities across alzheimer's disease spectrum. Ageing Res. Rev. 102240 (2024)
23. Toga, A.W., Crawford, K.L.: The alzheimer's disease neuroimaging initiative informatics core: a decade in review. Alzheimer's Dementia **11**(7), 832–839 (2015)
24. Tuan, P.M., Phan, T.L., Adel, M., Guedj, E., Trung, N.L.: Autoencoder-based feature ranking for alzheimer disease classification using pet image. Mach. Learn. Appl. **6**, 100184 (2021)
25. Tustison, N.J., et al.: N4itk: improved n3 bias correction. IEEE Trans. Med. Imaging **29**(6), 1310–1320 (2010)
26. Tustison, N.J., Cook, P.A., Holbrook, A.J., et al.: The antsx ecosystem for quantitative biological and medical imaging. Sci. Rep. **11**(1), 9068 (2021)
27. Venugopalan, J., Tong, L., Hassanzadeh, H.R., Wang, M.D.: Multimodal deep learning models for early detection of alzheimer's disease stage. Sci. Rep. **11**(1), 3254 (2021)

28. Williams, J.A., Weakley, A., Cook, D.J., Schmitter-Edgecombe, M.: Machine learning techniques for diagnostic differentiation of mild cognitive impairment and dementia. In: Workshops at the 27th AAAI Conference on AI, vol. 21 (2013)
29. Yagis, E., De Herrera, A.G.S., Citi, L.: Convolutional autoencoder based deep learning approach for alzheimer's disease diagnosis using brain mri. In: 2021 IEEE CBMS, pp. 486–491. IEEE (2021)
30. Ying, C., Qi-Guang, M., Jia-Chen, L., Lin, G.: Advance and prospects of adaboost algorithm. Acta Automatica Sinica **39**(6), 745–758 (2013)
31. Zhang, H., Zhou, W., Zhang, D.: Direct medical costs of parkinson's disease in southern china: a cross-sectional study based on health insurance claims data in guangzhou city. Int. J. Environ. Res. Public Health **19**(6), 3238 (2022)

Exploring Curriculum Learning for Languages: Lessons from Regular Language Tasks

Vanessa Toborek[1,3](✉), Florian Seiffarth[1,3], Sebastian Müller[1,3], Tamás Horváth[1,2,3], and Christian Bauckhage[1,2,3]

[1] University of Bonn, Bonn, Germany
toborek@cs.uni-bonn.de
[2] Fraunhofer IAIS, Sankt Augustin, Germany
[3] Lamarr Institute for Machine Learning and Artificial Intelligence, Bonn, Germany

Abstract. Despite its intuitive appeal, the effectiveness of data-level curriculum learning (CL) remains debated, mainly due to the absence of unambiguous notions of sample difficulty in real-world tasks. As a step towards a better understanding of the effective use of different curriculum strategies in natural language learning, we study CL in the context of regular languages, where both ground truth and sample difficulty can be precisely defined using deterministic finite automata. We consider two natural measures of difficulty: a data-driven metric based on input length and a task-specific metric derived from the automaton's structure. Training RNNs and LSTMs across ten regular language classification tasks, we find that CL is not just beneficial but, in some cases, essential for generalisation. Surprisingly, straightforward data-driven curricula outperform more complex task-specific strategies, with the most successful approaches oversampling the shorter lengths early in training.

Keywords: curriculum learning · data difficulty · regular languages

1 Introduction

Curriculum learning (CL), inspired by observations of how children acquire knowledge [7], organises the training of a model by ordering the data points according to some notion of difficulty. The aim is to improve a model's generalisation or accelerate convergence during training [2,17,19]. While intuitively appealing and empirically successful in certain areas of natural language processing (NLP) [1,11,13,15], its overall effectiveness remains a matter of debate [18]. Moreover, even across the settings where CL proves beneficial, there is little consensus on what makes a training sample "easy" or "hard" – largely because the structure of the target language is typically only partially observable. Consequently, curriculum strategies rely on heuristics or proxy measures.

Motivated by these challenges, we investigate two key questions: (1) Can curriculum learning effectively improve performance on language learning tasks?

S. Džeroski et al. (Eds.): DS 2025, LNAI 16090, pp. 571–586, 2025.
https://doi.org/10.1007/978-3-032-05461-6_37

If so, (2) how do *task-specific* and *data-driven* notions of difficulty compare in their ability to guide learning? By task-specific information, we mean knowledge derived from the underlying data-generating process (e.g. the grammar defining the target language), while data-driven information refers to features that are directly observable from the training data (e.g. sentence length).

To answer these questions, we focus on *regular languages*, the simplest class in the Chomsky hierarchy, which are succinctly represented by deterministic finite automata (DFAs). Formal languages – particularly regular languages – offer several advantages to study CL before extending it to the more complex domain of NLP. First, regular languages are theoretically well-defined and unambiguous – unlike natural language data, which often suffers from ambiguity and label noise. These properties allow us to isolate and analyse the specific impact of CL. Second, DFAs provide full (task-specific) control over the target language and its difficulty levels, enabling a systematic generation of training instances with precisely defined difficulty. The availability of ground-truth structure also allows for rigorous evaluation of CL's effectiveness. Third, regular languages support controlled examination whether models genuinely generalise beyond the training set – one of the core goals of CL. Finally, this idealised setting serves as a clean "proof of concept", yielding valuable insights that can guide the design of effective curricula in more complex NLP tasks.

Our work operates within the widely adopted yet often loosely defined paradigm of global, data-level CL, where the order of training examples shapes the training process [17,19]. We formalise and systematically evaluate this setup in a highly controlled environment to assess how different notions of difficulty and scheduling strategies affect learning outcomes. Notably, this necessitated an extensive hyperparameter search, as we found model performance to be surprisingly brittle – even when training on short sequences. Our *lessons learned* can be summarised as follows:

(i) For certain regular languages, CL is not just beneficial but *essential* for successful generalisation.
(ii) Surprisingly, input length – a simple, data-driven difficulty measure – outperforms more sophisticated task-specific curricula.
(iii) Scheduling strategies that oversample shorter input sequences early in training achieve the best results. We further validate this lesson by showing that restricting training to very short sequences can improve no-curriculum performance.
(iv) In contrast, anti-curricula (i.e. training in reverse order) often underperform even the no-curriculum baselines.

Our results, though based on a formal setting, have broader implications for NLP. They suggest that data-driven difficulty measures, such as input length, can be more effective than task-specific ones and that sampling strategies should account for specific properties of the input distribution – such as the long tail of shorter input sequences observed in our work.

The remainder of the paper is structured as follows: Sect. 2 offers a concise review of related work. Section 3 formalises our approach, while Sect. 4 details our experimental setup. In Sect. 5 we present and discuss our results. Finally, Sect. 6 concludes the paper.

2 Related Work

The Chomsky hierarchy [4] classifies formal languages by their generative complexity, ranging from regular languages at the lowest level to recursively enumerable languages at the highest. Regular languages are generated by deterministic finite-state automata and are characterised by their structural simplicity. Still, they are not trivial for neural networks to learn and are used to describe the bounds of a model's reasoning capabilities [3,5].

Curriculum learning originates from observations in cognitive science: children tend to learn in a structured manner, beginning with simpler concepts before progressing to more complex ones [7]. This phenomenon gained popularity in machine learning when it was demonstrated that ordering training samples by some notion of difficulty can improve both convergence speed and generalisation in neural networks, independent of the input domain [2]. While theoretical guarantees for the effectiveness of CL primarily exist in the context of convex optimisation [20], there remains a gap in the understanding of how curricula affect training in more complex, non-convex learning settings like neural networks.

In NLP, data-level CL has been applied to a wide range of tasks, including language modelling [11,16], neural machine translation [15], and question answering [12]. However, in global data-level curriculum learning, the choice of difficulty metric is often task-specific or guided by intuition. For example, curriculum strategies have used cross-domain similarity in machine translation [14], or Levenshtein distance in text simplification [1]. Further, input length is a widely used, though somewhat naive, measure, employed in tasks such as machine translation [15], language modelling [11,16], and code execution [21].

Overall, it remains unclear which difficulty metrics are most effective across different tasks and model settings. Our work contributes to this effort by using regular languages as a controlled test bed to study the interaction between difficulty measures, sampling strategies, and learning outcomes. The regular language setting allows us to investigate CL with access to ground-truth task structure.

3 Regular Languages and Curriculum Learning

This section presents our benchmark tasks and describes our CL framework. We begin by formalising the regular language tasks used for training and evaluation. We then detail our curriculum design approach, comprising difficulty measures and pacing functions, which enables systematic variation of training regimes across experiments.

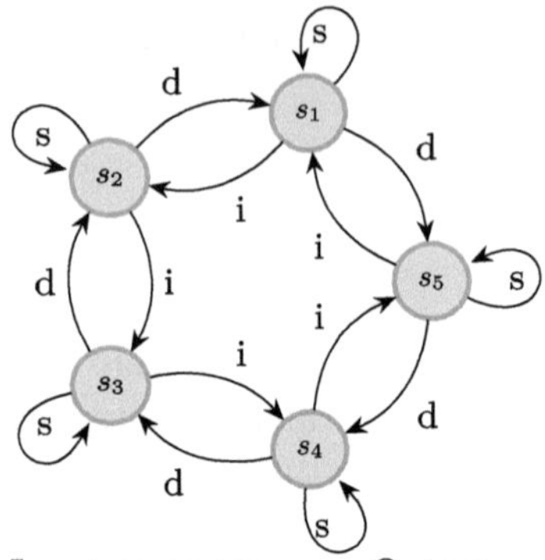

Input sequence	Output
isdd	s_5
iiss	s_3

Fig. 1. Cycle Navigation: Automaton and example sequences using symbols stay (s), increase (i), and decrease (d) actions.

Table 1. Overview of all ten tasks and the description of their underlying automaton.

Name	$\lvert\Sigma\rvert$	$\lvert Q\rvert$	$\lvert C\rvert$
Modular Arithmetic	15	5	5
Cycle Navigation	3	5	5
355 Random	3	5	5
Even Pairs	2	5	2
Parity Check	2	2	2
Cycle Navigation S	3	3	3
333 Random	3	3	3
252 Random	2	5	2
FirstSymbol	2	5	2
LastSymbol	2	2	2

3.1 Regular Languages

Regular languages, the simplest class in the Chomsky hierarchy, can be compactly represented by deterministic finite-state automata (DFAs), making them well-suited for the development of task-specific curriculum metrics. Traditionally, DFAs divide the set of words over the input alphabet into two blocks: those accepted by the automaton and those rejected. In our definition, we generalise this by allowing partitions into an arbitrary finite number of classes: A DFA is a tuple $A = (Q, \Sigma, \delta, q_0, C, L)$, where Q is a finite set of states, Σ is the alphabet, specifying the finite set of possible input symbols, $\delta : Q \times \Sigma \rightarrow Q$ is the transition function, $q_0 \in Q$ is the initial state, C is a finite set of output labels, and $L : Q \rightarrow C$ is a labelling function. For recognition tasks, $C = \{0, 1\}$, while for classification tasks, C can contain more than two classes. The set of finite strings over Σ is denoted by Σ^*. For any word $\mathbf{x} \in \Sigma^*$, the class of $\mathbf{x}$ is determined solely by the state reached from the initial state after processing every symbol of $\mathbf{x}$.

Using this formalism, we specify each task in our benchmark with a compact, automaton-based representation. Table 1 lists ten regular language problems that vary in structural characteristics such as the size of the input alphabet ($|\Sigma|$), the number of states ($|Q|$), and the number of output labels ($|C|$). The problems Modular Arithmetic, Cycle Navigation, Even Pairs, and Parity Check are standard benchmarks [3,5] (see Fig. 1 for the Cycle Navigation automaton and example sequences). To extend our analysis, we create simpler variants of them by either reducing the number of states (Cycle Navigation S) or by randomly assigning edge values, resulting in random DFAs with the same formal specifications in terms of the number of symbols, states, and classes (252, 355 and 333 Random). The task FirstSymbol (resp. LastSymbol) involves a simpler regular language, where the goal is to learn a language consisting of all strings that start (resp. end) with one particular symbol of Σ, say a. Together with 252 Random, Even Pairs, and Parity Check they represent recognition problems.

3.2 Curriculum Design

As outlined in Sect. 1, all curriculum strategies in this work adhere to the global data-level CL paradigm that influences the training process by the order in which samples are presented. To formalise such a strategy, we now introduce the necessary notions and notation. For any $n \in \mathbb{N}$, $[n]$ denotes the set $\{1, 2, \ldots, n\}$. Let the *training dataset* be $\mathcal{D} = \{X_i\}_{i=1}^N$ with $X_i = (\mathbf{x}_i, y_i)$, where $\mathbf{x}_i$ is a string over Σ (i.e. $\mathbf{x}_i \in \Sigma^*$) and $y_i \in C$ denotes its class label. Given $\mathcal{D}$, the goal is to learn a model $f_\theta : \Sigma^* \to C$, parametrised by θ, that predicts the true class label for every word in Σ^*. Training is conducted over M steps, using mini-batches $\mathbb{B}_1, \ldots, \mathbb{B}_M$, with $\mathbb{B}_t \subset \mathcal{D}$ for all $t \in [M]$. The goal of a curriculum strategy is to specify how the mini-batches are constructed from $\mathcal{D}$.

To define the curriculum strategies evaluated in this work, we adopt the convention of decomposing curriculum design into two core components: (i) partitioning the training data into an ordered sequence of so-called "difficulty buckets" and (ii) specifying the mini-batch scheduling. This separation allows for greater flexibility in curriculum design. We now describe (i) and (ii) in detail.

(i) Partitioning $\mathcal{D}$ into Difficulty Buckets. This component is designed to partition $\mathcal{D}$ into a sequence of meaningful buckets, ordered according to a predefined difficulty measure. While there are many ways to define difficulty-based partitions in general, our setting naturally lends itself to two distinct approaches. The first is data-driven and based on input length, the other is task-specific and based on the set of automaton states used to generate each example. These partitions differ in both structure and granularity, offering complementary perspectives on the notion of difficulty.

Input Length. This ordered partition relies on the difficulty measure defined by the *length* of the examples. It naturally induces an ordered partition $\langle S_1, \ldots, S_K \rangle$ of $\mathcal{D}$, where, for each $k \in [K]$, S_k contains all training examples of length l_k, with $l_k < l_{k+1}$ and l_1 being the smallest length present in $\mathcal{D}$.

DFA State Sets. This ordered partition is based on the set of states of the DFA that are visited during the processing of a training example. More precisely, for a DFA $A = (Q, \Sigma, \delta, q_0, C, L)$ and $(\mathbf{x}, y) \in \mathcal{D}$, let

$$m_{\text{state}}(\mathbf{x}) = \{q \in Q : q = \hat{\delta}(q_0, \mathbf{x}'),\ \mathbf{x}' \text{ is a prefix of } \mathbf{x}\} \ ,$$

where $\hat{\delta} : \Sigma^* \times Q \to Q$ is the extended transition function of A. An ordered partition $\langle S_1, \ldots, S_K \rangle$ of $\mathcal{D}$ is defined as follows: for all $(\mathbf{x}, y), (\mathbf{x}', y') \in \mathcal{D}$, $(\mathbf{x}, y), (\mathbf{x}', y') \in S_k$ for some k if and only if $m_{\text{state}}(\mathbf{x}) = m_{\text{state}}(\mathbf{x}')$, and for all $1 \le i < j \le K$, $|m_{\text{state}}(\mathbf{x}_u)| \le |m_{\text{state}}(\mathbf{x}_v)|$ for all $(\mathbf{x}_u, y_u) \in S_i, (\mathbf{x}_v, y_v) \in S_j$.

The above definitions partition the training data into disjoint buckets based on the difficulty function m. In the finest-grained case, each unique value of m defines a separate bucket. However, coarser groupings are also allowed, where instances in the same bucket may differ in $m(\mathbf{x})$. This flexibility supports a trade-off between curriculum granularity and scheduling control.

(ii) Mini-Batch Scheduling. The second component, defining a particular curriculum strategy, involves two key elements for each mini-batch: specifying the set of difficulty buckets from which the mini-batch elements are sampled, and defining the strategy for random sampling from the selected buckets. Given an ordered partition $\mathcal{S} = \langle S_1, \ldots, S_K \rangle$ of $\mathcal{D}$, we consider the following scheduling strategies for constructing the mini-batches $\mathbb{B}_1, \ldots, \mathbb{B}_M$:

Baseline. As a baseline, we consider i.i.d. uniform sampling from the full dataset $\mathcal{D}$, completely disregarding the buckets in $\mathcal{S}$.

Curriculum. This is a monotonic scheduling strategy that progresses through the buckets in their order in $\mathcal{S}$. More precisely, the bucket from which a mini-batch is sampled is determined by a *pacing function* $g : [M] \to [K]$ that is surjective onto $[K]$ (i.e. for all $k \in [K]$, there exists $t \in [M]$ such that $g(t) = k$) and monotone non-decreasing (i.e. $g(t) \leq g(t')$ for all $t < t'$). This definition of g yields K consecutive training phases. In each phase, all mini-batches in that phase are sampled exclusively and uniformly at random from a single bucket of $\mathcal{S}$, beginning with the "easiest" one (i.e. S_1).

Anti-Curriculum. This is the reverse of the curriculum strategy defined above. More precisely, given a pacing function $g_c : [M] \to [K]$ for a curriculum strategy, the pacing function g of the corresponding anti-curriculum strategy is defined by $g(t) = g_c(M - t + 1)$ for all $t \in [M]$. Thus, training begins with the most "difficult" examples selected from S_K and gradually moves towards the easiest ones. Like the curriculum strategy, this is a monotonic scheduling strategy, but in this case, the pacing function g is monotone decreasing.

Single Bucket. For each $t \in [M]$, we select an index k_t from $[K]$ independently and uniformly at random and construct $\mathbb{B}_t$ by uniform sampling from S_{k_t}. This strategy corresponds to the training in [5].

Uniform over Buckets. For all $t \in [M]$, approximately the same number of examples is sampled from each bucket of $\mathcal{S}$ to construct a balanced mini-batch $\mathbb{B}_t$ (i.e. $\frac{|\mathbb{B}_t \cap S_k|}{|\mathbb{B}_t|} \approx \frac{1}{K}$ for all $k \in [K]$). This ensures that each bucket in $\mathcal{S}$ is equally represented in each mini-batch. Examples from each of the K buckets are selected uniformly at random. [3] achieve a similar data distribution due to their data generation process.

4 Experimental Setup

Having defined the formal structure of regular languages and introduced the details of our curriculum designs, we now describe the full experimental setup. This includes the generation of training and test data, the choice of model architectures, and the results of our extensive hyperparameter search. We begin by detailing how input sequences are sampled and labelled using the DFA representations defined in the previous section.

4.1 Data Generation

Following [3,5], we generate 100 million training samples per task, approximately uniformly distributed across input sequences of length $[1, 40]$. Since the number of possible sequences decreases drastically for shorter lengths, we adjust sampling with respect to the input length. In a first step, we sample exhaustively without repetition for lengths ≤ 18 and randomly generate 500 000 samples otherwise. Next, we label the samples using the finite state automaton. For validation, we sample 1 000 sequences uniformly at random from the training data. The test data consists of sequences of lengths in the range $[41, 500]$. In particular, for each length we generate 10 samples, resulting in 4 600 test samples for each problem. This setup allows a clear test of generalisation: a model has learned the problem's underlying structure if it can generalise well beyond the input lengths encountered during training.

Number of Difficulty Buckets. As described in Sect. 3.2, the number of buckets can be chosen to be smaller than the dataset's maximum. We performed preliminary experiments on the Cycle Navigation task with $[2, 4, 5, 10, 20, 40]$ buckets. Two buckets consistently failed to support learning. While LSTM performance remained stable across higher bucket counts, the RNN showed erratic drops in the final stages of training with intermediate bucket counts. This motivated our choice of 20 buckets as a stable configuration for the data-driven difficulty measure. For the task-specific difficulty measure, we use the maximum number of buckets per task.

4.2 Models and Hyperparameter Sensitivity

We conduct our experiments using two recurrent neural network architectures: vanilla RNNs [6] and LSTMs [10], both of which have been shown to be capable of learning regular language tasks [3,5].[1] An important and somewhat surprising observation in our study is the extreme sensitivity of both architectures to the hyperparameter configurations – an issue largely overlooked in prior work. Despite the apparent simplicity of the tasks, with training sequences as short as

[1] Transformer models were also evaluated during preliminary experimentation; however, they were excluded from the final analysis due to persistently unstable performance across a wide range of hyperparameter configurations.

two symbols, the vast majority of configurations *fail* to achieve perfect training accuracy. This revealed an unexpectedly rugged optimisation landscape, where effective hyperparameter combinations were rare and difficult to find. To gain a deeper understanding of this behaviour, we conducted a hyperparameter search across a wide range of configurations.

4.3 Hyperparameter Search

We perform a grid search focused on three different tasks: Even Pairs, Parity Check, and Cycle Navigation. To keep the hyperparameter search computationally feasible, we strictly limit the maximum sequence length for each task to 5 and evaluate whether a model configuration is able to learn the limited number of sequences by heart (i.e. perfect training accuracy of 1.0).

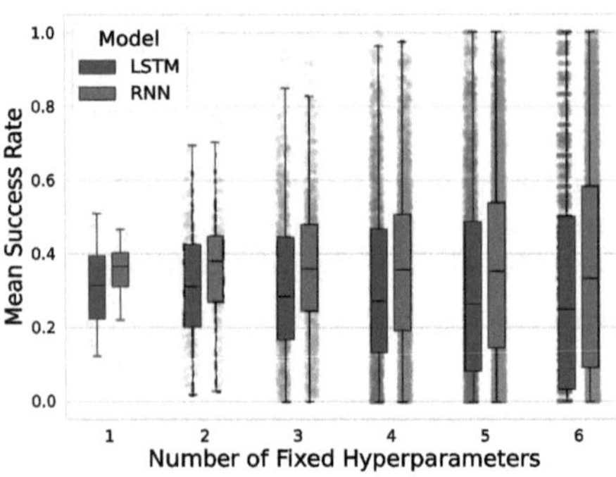

Fig. 2. Mean success rate of the hyperparameter configurations as a function of the number of fixed hyperparameters. At least five must be fixed for perfect accuracy across all settings.

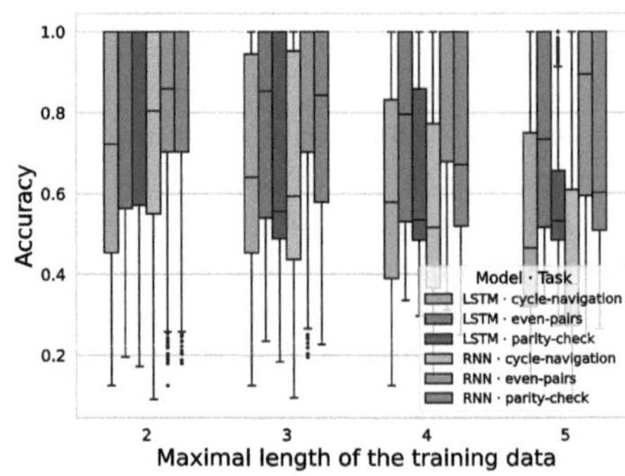

Fig. 3. Training accuracy of hyperparameter configurations by task and maximum training sequence length.

We vary batch size {64, 128, 256}, learning rate $\{10^{-4}, 10^{-3}\}$, loss function {Cross-Entropy, MSE}, hidden size {64, 128, 256}, and output activation {None, tanh, ReLU}, and consider five weight initialization schemes: PyTorch default, Glorot uniform/normal [8], and He uniform/normal [9]. For the RNNs we also consider two nonlinearities {ReLU, tanh}. Each configuration is run with three different random seeds. Initial experiments showed that models struggled with sequences of length ≤ 5, prompting us to test limits at $\{2, 3, 4, 5\}$. For each of the three different tasks we thus train 12 960 RNN and 6 480 LSTM models. To adjust for the varying batch sizes, each model has a budget of 2^{15} samples. We train the models without a curriculum using the Adam optimizer.

As a first step, we investigate how many hyperparameters must be fixed to consistently achieve a perfect success rate of 1.0 within the allocated data budget (see Fig. 2). We find that both architectures require fixing at least five of the above hyperparameters to reliably achieve perfect training performance across sequence lengths and tasks. Next, we examine how the results vary for different training sequence lengths. As shown in Fig. 3, performance on the Cycle Navigation task degrades rapidly with increasing input length – starting from length three, both RNN and LSTM architectures exhibit configurations that fail to fit the task. Results for Even Pairs appear fairly robust: for each sequence length, at least 25% of configurations reach perfect accuracy. Regarding Parity Check, RNNs handle the task generally more consistently compared to LSTMs.

4.4 Final Setup

The above grid search shows that the *optimal* hyperparameters are quite stable across architectures. Thus, we use the same hyperparameter configuration for all architectures in our experiments: a learning rate of 10^{-4}, the MSE loss function, the tanh activation function, no output activation, the He uniform initialisation [9], a hidden dimension of 256, and a batch size of 128. Notably, MSE loss consistently outperforms cross-entropy loss in all tested scenarios.

We train all tasks for 10 000 updates (i.e. mini-batches), except for Modular Arithmetic, Cycle Navigation, and Cycle Navigation S, which require 100 000 updates. To account for variability, each experiment is repeated with five different random initialisations. For each training run, we measure three aspects: (1) final generalisation accuracy, evaluated on the test set described in Sect. 4.1, (2) convergence speed, quantified by the average number of training updates needed to reach perfect accuracy on the test dataset, and (3) training stability, assessed by the variance across runs and sudden performance drops during training. This allows us to compare not only final accuracy but also how reliably and efficiently CL strategies enable learning across different tasks.

5 Results and Discussion

Drawing on experiments conducted across ten regular language tasks and two neural architectures, we address the following research questions in this section:

1. Does curriculum learning improve generalisation in regular language tasks?
2. How do DFA state sets (task-specific) compare to input length (data-driven) as a difficulty measure for guiding curriculum learning?
3. What role do the scheduling strategies play in shaping learning dynamics?
4. How does anti-curriculum compare to curriculum and baseline training?

5.1 Effectiveness of Curriculum Learning

Table 2 summarises test accuracies for all models and CL strategies, highlighting how CL affects generalisation across the ten tasks. Overall, CL improves test accuracy across many tasks, especially when using the data-driven difficulty measure input length. Length-based strategies consistently outperform those using

Table 2. Mean test accuracy for all models and strategies. We encode the standard deviation across random seeds: ☼ for $= 0.00$, ☽ for < 0.10, and ϟ for ≥ 0.10.

			Input Length				DFA State Sets			
	Problems	no-curr	curr	single	uniform	anti	curr	single	uniform	anti
RNN	Modular Arithmetic	1.00☼	1.00☼	1.00☼	1.00☼	0.64ϟ	0.60☽	0.97☽	0.96☽	0.27☽
	Cycle Navigation	0.19☼	0.51ϟ	0.73ϟ	0.74ϟ	0.20☽	0.20☽	0.21☽	0.29☽	0.20☼
	355 Random	1.00☼	1.00☼	1.00☼	1.00☼	0.98☽	1.00☼	1.00☼	1.00☼	0.39ϟ
	Even Pairs	0.51☼	1.00☼	1.00☼	1.00☼	0.60ϟ	0.50☼	0.49☼	0.49☼	0.51☼
	Parity Check	0.50☼	1.00☼	1.00☼	1.00☼	0.50☽	0.50☼	0.50☼	0.50☼	0.50☼
	Cycle Navigation S	0.33☼	1.00☼	1.00☼	1.00☼	0.33☼	0.33☼	0.33☼	0.34☽	0.33☼
	333 Random	1.00☼	1.00☼	1.00☼	1.00☼	1.00☼	1.00☼	1.00☼	1.00☼	0.85ϟ
	252 Random	1.00☼	1.00☼	1.00☼	1.00☼	0.93☽	1.00☼	1.00☼	1.00☼	0.62☽
	FirstSymbol	0.50☼	1.00☼	1.00☼	1.00☼	0.50☼	0.50☼	0.50☼	0.61ϟ	0.49☼
	LastSymbol	1.00☼	1.00☼	1.00☼	1.00☼	1.00☼	1.00☼	1.00☼	1.00☼	1.00☼
LSTM	Modular Arithmetic	1.00☼	1.00☼	1.00☼	1.00☼	1.00☼	1.00☼	1.00☼	1.00☼	0.34☽
	Cycle Navigation	0.19☼	0.65☽	0.57☽	0.64☽	0.20☽	0.46☽	0.37☽	0.46☽	0.20☼
	355 Random	1.00☼	1.00☼	1.00☼	1.00☼	0.88☽	1.00☼	1.00☼	1.00☼	0.79☽
	Even Pairs	1.00☼	1.00☼	1.00☼	1.00☼	1.00☼	0.50☼	1.00☼	1.00☼	0.51☼
	Parity Check	0.50☼	1.00☼	1.00☼	1.00☼	0.50☼	0.51☼	0.51☼	0.50☼	0.50☼
	Cycle Navigation S	0.34☼	0.74ϟ	0.63☽	0.64☽	0.33☽	0.34☼	0.45☽	0.51☽	0.34☼
	333 Random	1.00☼	1.00☼	1.00☼	1.00☼	1.00☽	1.00☼	1.00☼	1.00☼	0.92☽
	252 Random	1.00☼	1.00☼	1.00☼	1.00☼	0.90☽	0.99☽	0.84☼	1.00☼	0.63☼
	FirstSymbol	1.00☼	1.00☼	1.00☼	1.00☼	0.79ϟ	0.50☼	1.00☼	1.00☼	0.50☼
	LastSymbol	1.00☼	1.00☼	1.00☼	1.00☼	1.00☼	1.00☼	1.00☼	1.00☼	1.00☼

the task-specific DFA state sets, and – importantly – never degrade performance (except in the case of anti-curriculum). Compared to the stable performance of the no-curriculum baseline (which is stable regardless of the final training performance), CL strategies can introduce more variability. This variability is more pronounced with the task-specific difficulty measure and on tasks like Cycle Navigation, where instability occurs regardless of the model or strategy used.

Looking at model-specific patterns, we find that RNNs are able to learn several tasks without any curriculum support, including all three random automata, the trivial LastSymbol, and, perhaps surprisingly, Modular Arithmetic. LSTMs extend the list to further include Even Pairs and FirstSymbol, suggesting they are more capable of learning under unstructured training. However, both models consistently struggle with Cycle Navigation, which remains the only task for which no CL strategy or architecture achieves perfect accuracy. Interestingly, the LSTM architecture is particularly sensitive to Cycle Navigation and its smaller variant, often failing to converge even when trained with CL strategy.

These findings highlight that task difficulty is not directly correlated with formal complexity. For instance, tasks 355 and 333 Random (random automata with the same formal complexity as Cycle Navigation and Cycle Navigation S, respectively) are far easier to learn. The difference lies in structural properties like edge distribution and transition patterns. This becomes especially clear when comparing Modular Arithmetic and Cycle Navigation: although the latter can be framed as a simplified version of the former (using only additions and subtractions of 1 s and 0 s), Cycle Navigation proves significantly harder for neural networks to learn. To explore this further, we evaluated a larger version of Cycle Navigation with ten states and output classes. Despite its higher formal complexity, this variant achieved much better test accuracy (over 60% even without a curriculum), implying that learnability does not depend on task compactness.

In summary, CL does help: across several tasks, successful generalisation is achieved only when curriculum strategies are applied. A prerequisite for success is the right choice of both the difficulty measure and the mini-batch scheduling strategy.

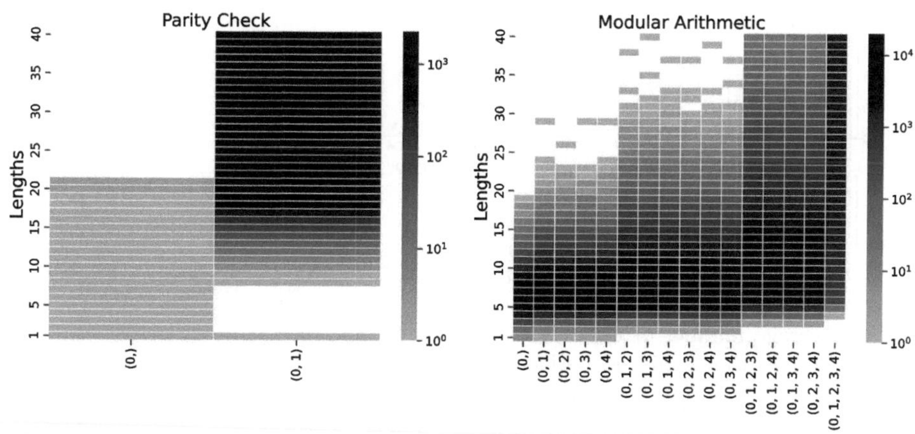

Fig. 4. Distribution of sequence lengths within buckets defined by the task-specific difficulty measure. The x-axis indicates the DFA state sets of samples in each bucket. Here, no bucket contains only short sequences, preventing their oversampling.

5.2 Data-Driven vs. Task-Specific Measures

A key aspect of our study is the comparison of difficulty measures: one grounded in the underlying task structure – specifically, DFA state sets – the other a simpler, data-driven alternative such as input length. Regular languages offer an ideal setting in which we can access a compact, ground-truth representation of the task in form of a DFA. However, our results show that this structure-aware notion of difficulty does not translate into better performance. In fact, it often underperforms compared to curricula based on input length.

Notably, CL strategies using task-specific difficulty measures not only fail to improve generalisation but can actively *hinder* learning. This effect is most evident in tasks already learnable without a curriculum, such as Modular Arithmetic with RNNs and Even Pairs or FirstSymbol with LSTMs. This is further supported by a broader comparison across tasks: for RNNs, task-specific performs worse than data-driven in six out of ten tasks; for LSTMs, in five.

One likely explanation lies in the distribution of input lengths across the task-specific buckets. As shown in Fig. 4, these buckets replicate the overall distribution of the dataset rather than isolating short sequences. This limits the opportunity to oversample shorter, simpler examples. Additionally, the task-specific difficulty measure can be too coarse: for example, in the Parity Check task only two DFA states yield just two buckets. This lack of structure leads to performance that is indistinguishable from the no-curriculum baseline. Lastly, LSTMs appear more sensitive to scheduling choices. In four tasks (Cycle Navigation, Even Pairs, Parity Check, and FirstSymbol), curriculum scheduling underperforms compared to single and uniform sampling.

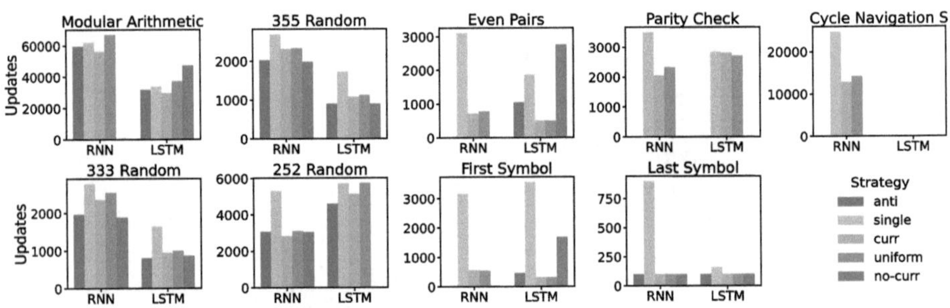

Fig. 5. Average number of updates to convergence for each strategy across all tasks. Only strategies that achieved perfect test accuracy are shown (note: no strategy for the LSTM on Cycle Navigation S converged perfectly).

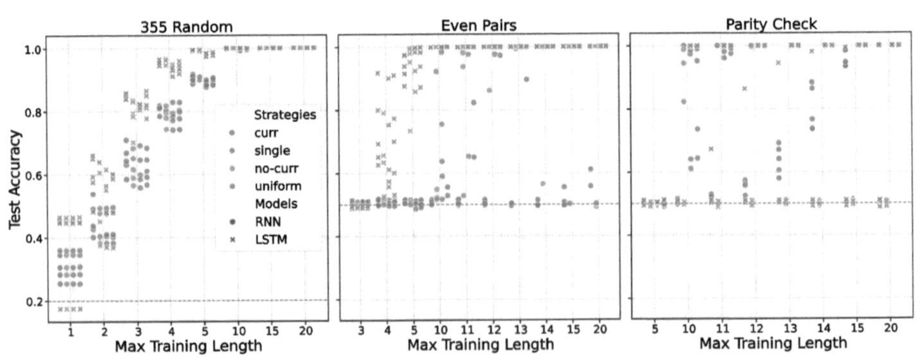

Fig. 6. Effect of limiting training sequence length on generalisation with the data-driven difficulty measure. The dashed line marks the random baseline. For many tasks, high test accuracy is achievable with training lengths well below the maximum.

Together, these findings highlight a key insight: access to the task's underlying structure does not guarantee a beneficial curriculum. In this respect, input length proves to be not only simpler but also more reliable.

5.3 Mini-batch Scheduling Dynamics

Curriculum learning is often guided by the intuition that a monotonic progression in difficulty supports effective training. Yet, our results suggest a different dynamic is at play. As shown in Table 2, curriculum, single, and uniform scheduling yield comparable test performance across tasks, despite their different approaches. This challenges the assumed importance of difficulty progression as such. Instead, we hypothesise that the benefit of CL in our setting stems from the *oversampling* of short, simple sequences early in training. These scheduling strategies provide similar advantages because they emphasise short sequences either by design or through repeated early exposure. As shown in Fig. 5, curriculum schedules typically lead to slower convergence compared to single or uniform. This is expected, as early training is limited to short inputs.

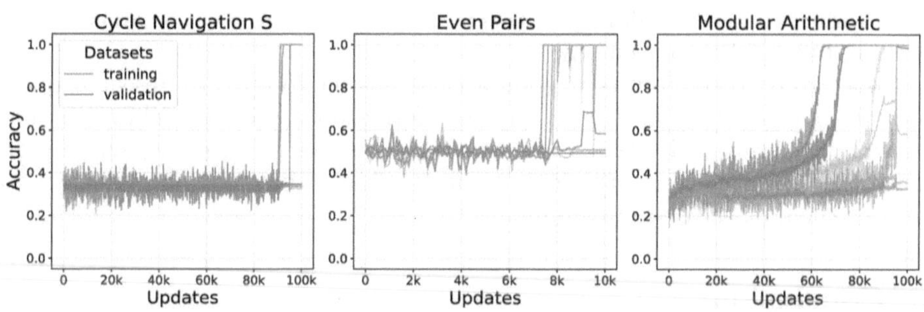

Fig. 7. Training and validation accuracy for the data-driven anti-curriculum across all seeds. Models fail to learn until the short input sequences are introduced. Modular Arithmetic shows notable sensitivity to random initialisation.

Generalisation from Short Sequences. To validate the hypothesis on the importance of short input sequences, we trained models on progressively increasing subsets of the training data with the maximum length varying from 1 up to 20, while evaluating performance on unseen test sequences as before. Figure 6 reveals that many tasks can be learned to near-perfect accuracy using only very short sequences. Interestingly, under these constraints, the no-curriculum baseline achieves better performance with the RNN. For example, for Even Pairs (lengths 11, 12) and Parity Check (lengths 1012), limiting the training length enables generalisation that does not emerge when training across the full length range. As the optimal sequence lengths for generalisation in no-curriculum setups are unknown in advance, this emphasises the benefit of CL strategies that, by design, prioritise early exposure to shorter sequences.

Taken together, our findings suggest that the scheduling strategy shapes learning dynamics not by enforcing a monotonic difficulty order, but by controlling early exposure to certain types of inputs.

5.4 Anti-curriculum and Failure Cases

As a sanity check, we included a deliberately adverse condition: the anti-curriculum strategy (see Sect. 3.2) is expected to impede learning if CL is beneficial.

Indeed, anti-curriculum generally performs poorly across tasks and models. As shown in Table 2, in several cases it even underperforms the no-curriculum baseline. For RNNs, this includes Modular Arithmetic, 355 Random, and 252 Random. For LSTMs, the same pattern appears in 355 Random, 252 Random, and also FirstSymbol. Furthermore, where curriculum helps, anti-curriculum often harms: Cycle Navigation, Even Pairs, Parity Check, Cycle Navigation S, and FirstSymbol (RNN), and Cycle Navigation, Parity Check, and Cycle Navigation S (LSTM). The training curves in Fig. 7 further illustrate this effect. Models often remain near random performance until shorter sequences are introduced later in training. At this point they begin to recover, albeit rarely reaching full validation accuracy.

Collectively, these results provide a clear answer to our last research question: anti-curriculum consistently underperforms both the curriculum and no-curriculum baselines. It emphasises that the presence of short sequences is in some cases essential for learning to occur at all and that it is not sufficient to only see them at the end of training.

6 Conclusion

We studied curriculum learning in the controlled setting of regular language tasks, focusing on how different difficulty measures and scheduling strategies affect learning. Using ten diverse tasks and two widely used recurrent architectures, we found that CL can substantially improve generalisation for some tasks – but only when designed carefully. Irrespective of model architecture, constructing the curriculum using input length, a straightforward data-driven measure, consistently outperformed the task-specific alternatives. Scheduling strategies that oversample short sequences early in training proved most effective, while anti-curricula often degraded performance. We further substantiate the importance of shorter lengths by showing that the baseline training also benefits from restricted input sequence length.

Overall, these results open the door for exploring curriculum learning in more complex settings, such as NLP. A promising step towards this direction is to identify which observable distributional features in natural language could serve as effective difficulty measures – especially since our findings suggest that good curricula do not require access to task-specific, unobservable information.

Acknowledgments. This publication was supported by the Open Access Publication Fund of the University of Bonn.

References

1. Agrawal, S., Carpuat, M.: An imitation learning curriculum for text editing with non-autoregressive models. In: ACL (2022). https://aclanthology.org/2022.acl-long.520/
2. Bengio, Y., Louradour, J., Collobert, R., Weston, J.: Curriculum learning. In: ICML (2009). https://dl.acm.org/doi/10.1145/1553374.1553380
3. Butoi, A., Khalighinejad, G., Svete, A., Valvoda, J., Cotterell, R., DuSell, B.: Training neural networks as recognizers of formal languages. In: ICLR (2025). https://openreview.net/forum?id=aWLQTbfFgV
4. Chomsky, N.: On certain formal properties of grammars. Inf. Control **2**(2) (1959). https://doi.org/10.1016/S0019-9958(59)90362-6
5. Delétang, G., et al.: Neural networks and the chomsky hierarchy. In: ICLR (2023). https://openreview.net/forum?id=WbxHAzkeQcn
6. Elman, J.L.: Finding structure in time. Cogn. Sci. **14**(2) (1990). https://doi.org/10.1016/0364-0213(90)90002-E
7. Elman, J.L.: Learning and development in neural networks: the importance of starting small. Cognition **48**(1) (1993). https://doi.org/10.1016/0010-0277(93)90058-4
8. Glorot, X., Bengio, Y.: Understanding the difficulty of training deep feedforward neural networks. In: AISTATS (2010). http://proceedings.mlr.press/v9/glorot10a.html
9. He, K., Zhang, X., Ren, S., Sun, J.: Delving deep into rectifiers: surpassing human-level performance on imagenet classification. In: ICCV (2015). https://doi.org/10.1109/ICCV.2015.123
10. Hochreiter, S., Schmidhuber, J.: Long short-term memory. Neural Comput. **9**(8) (1997). https://doi.org/10.1162/neco.1997.9.8.1735
11. Nagatsuka, K., Broni-Bediako, C., Atsumi, M.: Pre-training a BERT with curriculum learning by increasing block-size of input text. In: RANLP (2021). https://aclanthology.org/2021.ranlp-1.112/
12. Liu, C., He, S., Liu, K., Zhao, J.: Curriculum learning for natural answer generation. In: IJCAI (2018). https://doi.org/10.24963/ijcai.2018/587
13. Maharana, A., Bansal, M.: On curriculum learning for commonsense reasoning. In: NAACL (2022). https://aclanthology.org/2022.naacl-main.72/
14. Mohiuddin, T., Koehn, P., Chaudhary, V., Cross, J., Bhosale, S., Joty, S.: Data selection curriculum for neural machine translation. In: Findings EMNLP (2022). https://aclanthology.org/2022.findings-emnlp.113/
15. Platanios, E.A., Stretcu, O., Neubig, G., Poczos, B., Mitchell, T.M.: Competence-based curriculum learning for neural machine translation. In: NAACL (2019). https://aclanthology.org/N19-1119/
16. Ranaldi, L., Pucci, G., Massimo Zanzotto, F.: Modeling easiness for training transformers with curriculum learning. In: RANLP (2023). https://aclanthology.org/2023.ranlp-1.101/
17. Soviany, P., Ionescu, R.T., Rota, P., Sebe, N.: Curriculum learning: a survey. Int. J. Comput. Vision **130**(6) (2022). https://doi.org/10.1007/s11263-022-01611-x

18. Surkov, M., Mosin, V., Yamshchikov, I.: Do data-based curricula work? In: Proceedings of the Third Workshop on Insights from Negative Results in NLP (2022). https://aclanthology.org/2022.insights-1.16/
19. Wang, X., Chen, Y., Zhu, W.: A survey on curriculum learning. IEEE Trans. Pattern Anal. Mach. Intell. **44**(9) (2021). https://ieeexplore.ieee.org/document/9392296
20. Weinshall, D., Amir, D.: Theory of curriculum learning, with convex loss functions. J. Mach. Learn. Res. **21**(222) (2020). http://jmlr.org/papers/v21/18-751.html
21. Zaremba, W., Sutskever, I.: Learning to execute. arXiv preprint (2015). http://arxiv.org/abs/1410.4615

Exploring Recommendations Attacks Through Blind Optimization

Julio Corona[1(✉)], Rafael Teixeira[1,2], Mário Antunes[1,2], and Rui L. Aguiar[1,2]

[1] Instituto de Telecomunicações, Universidade de Aveiro, Aveiro, Portugal
{jcamejo,rafaelgteixeira,mario.antunes,ruilaa}@av.it.pt
[2] DETI, Universidade de Aveiro, Aveiro, Portugal

Abstract. Numerous studies have demonstrated various techniques for creating adversarial samples to exploit inherent or flawed behaviors in intelligent algorithms. However, many of these methods are intricate and may necessitate model- or data-specific information. This work introduces a novel approach to crafting adversarial samples, utilizing blind optimization algorithms combined with efficient population initialization methods. We apply this approach to evaluate the resilience of well-known recommendation algorithms in adversarial scenarios. Specifically, we conducted experiments using two recommendation systems featured in the Netflix Prize competition. Our primary objective was to inflate the ratings of a target movie, increasing the likelihood of it being recommended, by creating adversarial interactions (adversarial samples) within the network. Our evaluation results reveal that while the tested recommendation algorithms exhibit resilience to blind adversarial attacks, optimization-based methods can induce moderate shifts in predicted ratings, exposing subtle but exploitable vulnerabilities.

Keywords: Adversarial Machine Learning · Recommender Systems · Differential Evolution

1 Introduction

Nowadays, the use of (AI)-powered systems has proliferated across various systems and applications, enabling humans and systems to perform actions that were once considered impossible due to their complexity.

All computer systems, including intelligent systems [9], have their flaws and vulnerabilities that require adjustments to ensure system correctness and integrity. Developers typically introduce these flaws during the development phase. However, the inherent behavior of specific algorithms can be exploited, posing a threat to the security environment of these systems [17].

The online market for entertainment through audio-visual content, such as movies, series, and documentaries, is highly attractive and involves numerous entities and resources. Due to its growth, leading platforms like Netflix, HBO,

S. Džeroski et al. (Eds.): DS 2025, LNAI 16090, pp. 587–601, 2025.
https://doi.org/10.1007/978-3-032-05461-6_38

and Amazon Prime have implemented various algorithms to enhance the user experience, including recommendation systems that suggest movies to users based on their ratings of other movies [21]. For movie publishers, the likelihood of recommendation is crucial, as it determines the success or failure of a movie in terms of viewership. Therefore, any manipulation of the recommendation system can lead to platform instability due to manipulated recommendations, where the recommended movies do not accurately reflect user preferences [10]. The primary aim of this work is to assess the feasibility of a specific entity investing in this type of attack to gain an unfair advantage as a movie owner by maximizing the likelihood of a particular movie being recommended.

Our experiments are based on models and data from the Netflix Prize challenge, as this setup provides a controlled and reproducible environment for evaluating adversarial strategies. The attack targets two well-known recommendation algorithms used in the challenge, namely **BellKor** and **Funk-SVD**. These algorithms assign scores of 1 to 5 for individual movies. The goal is to increase the average movie rating (across all users) generated by the recommendation algorithm. The optimal solution should result in the highest possible rating for every user on the platform. Detailed information about the attack corpus is provided in Sect. 3. In practical terms, the idea is to construct adversarial interactions using Adversarial Machine Learning (AML) to maximize the overall movie ratings for all users by exploiting inherent properties of the recommendation algorithm. However, instead of using the primary algorithms to craft adversarial examples, as discussed in Sect. 2, we employ black-box optimization algorithms, specifically DE and Grey Wolf Optimization (GWO), as described in Sect. 4.

The main contributions of this work are:

- Introducing a novel attack strategy targeting two recommendation systems involved in the Netflix Prize contest [1] - **BellKor** & **Funk-SVD**.
- Conducting a comprehensive experimental study that assesses the effectiveness and limitations of both **Influential Users** and **Bot Network** attacks, under multiple optimizers and dataset configurations, revealing the inherent robustness of the tested recommendation algorithms to blind adversarial manipulation.

The remainder of this document is organized as follows. Section 2 reviews related work on attacks against Machine Learning (ML) systems and discusses algorithmic approaches for generating adversarial samples. Section 3 details the proposed attack methodology targeting two recommendation systems, where adversarial examples are crafted using metaheuristic blind optimization algorithms. The evaluation of the proposed attacks is presented in Sect. 5, followed by the conclusions in Sect. 6.

2 Background

This section presents the state of the art in attacks against ML systems, focusing on their strategies and real-world implications. Given the widespread adoption of ML, ensuring the security of these systems is critical, mainly due to their black-box nature and susceptibility to manipulation.

The main types of attacks include:

- **Adversarial Machine Learning (AML)**: Slight input perturbations lead to misclassification [3].
- **Data Poisoning (DP)**: Injecting malicious data into training sets [7].
- **Online Adversarial (OA)**: Feeding false data in online learning setups [16].
- **Distributed Denial of Service (DDoS)**: Overloading the model with high-complexity inputs [15].

This work concentrates on the AML approach due to its relevance to our scenario. Other methods, although effective, assume access levels beyond our scope.

AML resides at the intersection of ML and cybersecurity, where adversaries craft **adversarial examples**—input perturbations that retain original semantics but mislead the model:

$$F(x') = T \tag{1}$$

Such vulnerabilities often stem from flawed design assumptions, particularly the reliance on fixed data distributions [6]. Attackers exploit distributional shifts or direct access to training data to compromise models.

Various algorithms exist for generating adversarial samples, including:

1. Gradient Ascent [22]
2. Fast Gradient Sign Method (FGSM) [8]
3. Jacobian based saliency map attack (JSMA) [19]
4. Carlini-Wagner [4]

Common AML strategies include **evasion attacks** at test time and **poisoning attacks** during training [2,11,14,18]. These typically require knowledge about the model (white/grey box attacks).

In contrast, this work explores a black-box approach using blind optimization techniques such as DE and GWO to generate adversarial examples with minimal system assumptions.

3 Attack Definition

This section outlines the proposed attack and its objectives. The goal is to manipulate a recommendation system to increase the likelihood of recommending specific audiovisual content (for simplicity, referred to as a movie hereafter). Most recommendation algorithms, including the two used in this work, are vulnerable to data poisoning through the insertion of malicious samples, which can result in unintended recommendations. The attack aims to generate such samples to boost the recommendation of a target movie. The decision to inject adversarial samples directly into the training dataset is grounded in the principles of data poisoning attacks. This approach is particularly relevant for recommendation systems that rely heavily on historical user-item interactions, as even a small set of manipulated entries can significantly bias the learned representations of users and items. By evaluating the impact of such poisoned data, we aim to expose potential vulnerabilities in the model's learning process and demonstrate how attackers might exploit these flaws to influence recommendations without access to model internals.

We begin by presenting the dataset used as the foundation for this work in Subsect. 3.1, followed by a description of the recommendation algorithms employed—*BellKor's Pragmatic Chaos* and *Funk-SVD*. Subsection 3.3 details the two attack strategies explored, while Sect. 4 explains the attack's internal mechanics and its limitations.

3.1 Dataset

To train and evaluate the recommendation algorithms, we needed a dataset with the required information, so we used a sample from the MovieLens 25M Dataset[1]. It is a well-known dataset that has been exhaustively used in recommendation systems.

This dataset describes 5-star rating activity from MovieLens, a movie recommendation service. It comprises 25,000,095 ratings from 162,541 users to 62,423 different movies. The ratings were gathered between January 09, 1995, and November 21, 2019. The ratings are made on a 5-star scale, with half-star increments (0.5 stars - 5.0 stars). Timestamps represent seconds since midnight Coordinated Universal Time (UTC) of January 1, 1970. The data provided in this dataset has been normalized to meet the demands of each algorithm as part of the pre-processing.

3.2 BellKor's Pragmatic Chaos and Funk-SVD

Understanding how these algorithms compute recommendation scores is crucial for contextualizing the attack models.

Both algorithms rely on user interaction data structured as follows:

- **Movie ID**: Integer value that identifies a specific movie
- **User ID**: Integer value that identifies a specific user
- **Rating**: Float value representing the rating score of a movie by a user. (0.5 scale values)
- **Timestamp**: Integer value that represents the timestamp of a user's evaluation

During the training process, both of these algorithms accept as input the movie ID, user ID, and rating (the timestamp value is only used in the BellKor algorithm), allowing batch processing to accelerate model training.

After the training process (model creation), the algorithm is ready to infer a user's rating of a new movie. In this phase, the algorithm predicts the estimated

[1] https://grouplens.org/datasets/movielens/.

rating a user would assign to an unseen movie, based on their previous rating records. The overall architecture of the recommendation system, including both training and inference stages, is illustrated in Fig. 1.

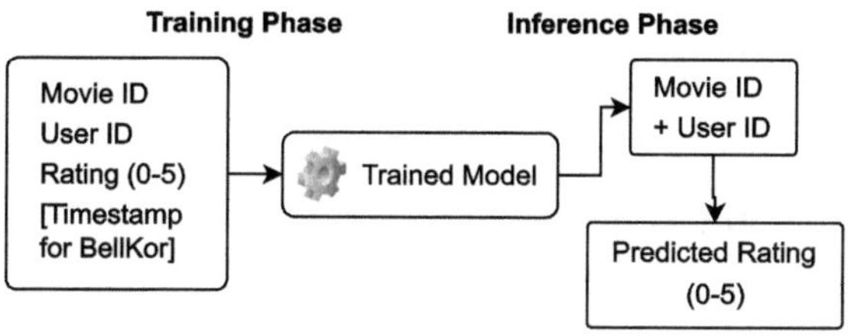

Fig. 1. Recommendation system architecture.

The rating values are displayed as a numeric value between 0 and 5 (with a 0.5 value increment), where the higher the value, the greater the likelihood of a recommendation, and vice versa. The 5-star-rated movies are the ones that users like the most.

As introduced earlier, both of these recommendation mechanisms utilize the user's historical ratings to make an informed guess about a new movie and therefore recommend films that best suit the user's preferences. The difference between them relies on the use of an extra parameter (in the BellKor algorithm) - the **timestamp** parameter, which is used to address the main temporal effects, including the fact that the popularity of a movie may change over time and the possibility that a user will change his or her rating over time.

For further details about these recommendation algorithms, we recommend checking the following literature: BellKor [12] & Funk-SVD [13].

3.3 Attack Strategies

Two attack strategies are considered in this work: **Influential Users** and **Bot Network**. In both cases, blind optimization algorithms are used to identify the most effective manipulation points for increasing the likelihood of recommending the target movie.

Influential Users. This strategy aims to identify the n most influential users, referring to those whose ratings have the most significant impact on the recommendation system's predictions. The attack increases the overall average predicted rating of the target movie by persuading or simulating these users to assign maximum ratings (five stars). The optimal outcome is to identify the smallest possible n capable of significantly elevating the target movie's predicted ratings (Fig. 2).

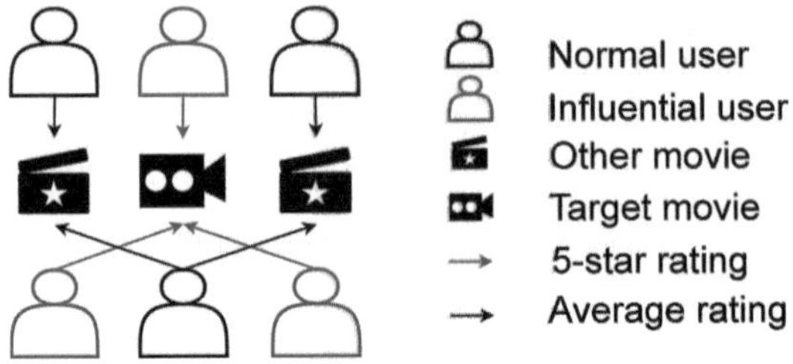

Fig. 2. Influential Users attack scheme.

Bot Network. In this approach, a network of n bots is created. Each bot first rates a set of highly popular movies (with high average ratings and many user interactions), and subsequently rates the target movie with five stars. By associating the target movie with popular items, the bots indirectly increase its likelihood of recommendation (Fig. 3). Similar to the previous strategy, the objective is to determine the minimal effective n.

4 Optimization Algorithm

The attack formulation described in this work requires identifying an optimal set of n objects (users or movies) to maximize the manipulation's effectiveness. This process involves exploring a combinatorial search space where each candidate solution represents a possible subset of users or movies to be manipulated. To address this optimization problem, we adopt blind (black-box) metaheuristic algorithms that rely exclusively on the evaluation of a fitness function. This fitness function measures the attack's effectiveness by computing the target movie's average predicted rating after adversarial samples have been injected.

In this work, two different optimization algorithms are evaluated: DE and GWO. While these algorithms are known to require significant computational resources and careful hyperparameter tuning, they remain among the most flexible and effective tools for optimization tasks where the objective function is non-differentiable or expensive to evaluate, as in our case, where each iteration involves retraining the recommendation model. Additionally, both methods operate under black-box assumptions, without requiring internal access to the recommendation models.

4.1 Differential Evolution

DE [5] is a population-based evolutionary algorithm designed for continuous optimization problems. It iteratively refines a population of candidate solutions by applying mutation, crossover, and selection operators.

For each generation, DE perturbs candidate solutions using scaled differences between randomly selected individuals within the population, followed by recombination and evaluation based on the fitness function. This process allows DE to balance exploration and exploitation efficiently across the search space.

In the context of this attack, DE explores which users (for the Influential Users attack) or which movies (for the Bot Network attack) should be selected to maximize the target movie's predicted rating. Its black-box nature allows DE to operate without requiring any knowledge of the recommendation model's internal structure, gradients, or training procedures.

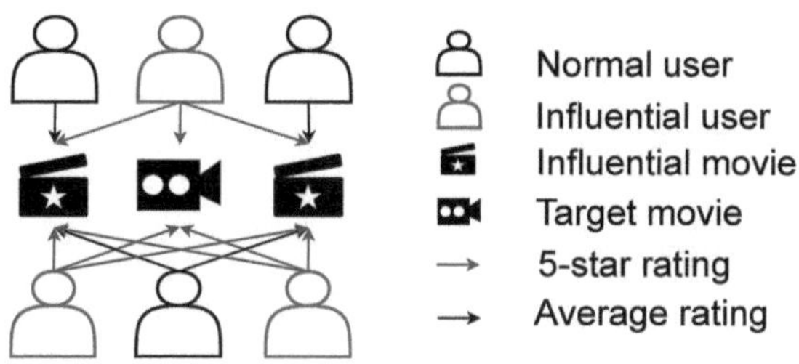

Fig. 3. Bot Network attack scheme.

4.2 Grey Wolf Optimization

GWO [20] is a nature-inspired metaheuristic that simulates the leadership hierarchy and hunting strategy of grey wolves in nature. In GWO, candidate solutions are ranked into four levels: alpha, beta, delta, and omega, which guide the search process through leadership-based position updates.

During each iteration, wolves update their positions by encircling and following the best-known solutions, with exploration and exploitation phases controlled by a dynamic weighting mechanism. This hierarchy-driven search behavior allows GWO to effectively balance diversification and intensification when navigating complex search spaces.

Similar to DE, GWO is applied here to identify the most effective set of users or movies to target, relying only on the evaluation of the fitness function and not requiring any access to the underlying recommendation model.

4.3 Population Initialization

The performance of metaheuristic algorithms strongly depends on the quality and diversity of the initial population, particularly in complex and high-dimensional search spaces. Poor initialization may lead to premature convergence or inefficient exploration. To address this challenge, we adopt the OBLESA initialization strategy [23], which combines multiple mechanisms to improve population diversity and accelerate convergence.

OBLESA operates in two stages. First, a random initial population is generated, and opposition-based learning is applied to produce complementary solutions, effectively expanding the initial coverage of the search space. Second, the

Empty-Space Search Algorithm is employed to guide additional candidate solutions into under-explored regions, improving diversity by explicitly targeting sparse areas within the search space. This phase uses a Lennard-Jones potential function to iteratively reposition artificial agents toward regions of lower sample density.

After both stages, the augmented candidate pool is pruned by selecting the best-performing individuals, based on fitness evaluations, to compose the final initial population. This enriched population is then used to initialize both DE and GWO algorithms, ensuring a more comprehensive exploration from the start of the optimization process.

4.4 Attack Workflow

This section outlines the algorithmic procedures implemented for both attack strategies. At its core, each attack's workflow defines its objective function.

Both attacks follow a similar logic: the adversarial samples are injected into the training dataset by appending customized entries to the original data. The average predicted score for the target movie across all users is calculated to assess the effectiveness of these new entries.

The objective function for the **Influential Users** approach is detailed in Fig. 1. In this case, the additional samples are generated by pairing the normalized User IDs produced by the optimization algorithm with the target Movie ID (lines 3âĂŞ4). The current timestamp is also included in the BellKor algorithm.

Algorithm 1: Objective function in Influential Users

Input: de_params, dataset, all_users, target_movie_id
Output: objective_value

1 recommender ← BellKor() or FunkSVD();
2 attack_entries ← [ ];
3 **for** *user_id* **in** *de_params* **do**
4   attack_entries.append(user_id, target_movie_id, 5, now_timestamp);
5 modified_dataset ← dataset + attack_entries;
6 recommender.train(modified_dataset);
7 ratings ← [ ];
8 **for** *user_id* **in** *all_users* **do**
9   predicted_rating ← recommender.predict(user_id, target_movie_id, now_timestamp);
10   ratings.append(predicted_rating);
11 mean_rating ← mean(ratings);
12 objective_value ← 5 - mean_rating;
13 **return** *objective_value*;

In case of the Bot Network attack, as it's exposed in algorithm 2, the sample creation is implemented in two main interaction points:

1. The first one is responsible for creating an interaction between the bot and the target movie by combining the bot's ID and the target movie's ID in the same object (lines 3–4).
2. The second one creates an interaction between the bot and the "influential movies", by taking the normalized parameters from the optimization algorithm (Movie IDs) and combining them with the bot IDs (lines 5–6).

Once again, if the BellKor algorithm is used as the recommendation system, the *current_timestamp* parameter is also utilized.

Algorithm 2: Objective function in Bot Network

Input: de_params, dataset, all_users, bots_id
Output: objective_value

```
1  recommender ← BellKor() or FunkSVD();
2  attack_entries ← [ ];
3  for bot_id in bots_id do
4      attack_entries.append(bot_id, target_movie_id, 5, now_timestamp);
5      for movie_id in de_params do
6          attack_entries.append(bot_id, movie_id, 5, now_timestamp);

7  modified_dataset ← dataset + attack_entries;
8  recommender.train(modified_dataset);

9  ratings ← [ ];
10 for user_id in all_users do
11     predicted_rating ← recommender.predict(user_id, target_movie_id,
        now_timestamp);
12     ratings.append(predicted_rating);

13 mean_rating ← mean(ratings);
14 objective_value ← 5 - mean_rating;
15 return objective_value;
```

Regarding restrictions, the main limitation is the high computational cost associated with the training phase of recommendation algorithms. These systems are not typically designed to be retrained from scratch frequently, which is required in every iteration of the optimization algorithm. Another constraint relates to the practical feasibility of executing this attack in a real-world application:

Influential Users Attack. After identifying the top n influential users, it is necessary to persuade them to assign a 5-star rating, and potentially to watch the target movie, since some platforms may invalidate ratings for unwatched content.

Bot Network. For this attack type, the main restriction involves movie duration. Each bot must interact with n movies from the network, and the rating timestamps should align with the expected viewing time. Bots should not rate all movies at once, as watching multiple movies in a short period is unrealistic. Bots must mimic human behavior.

5 Attack Evaluation

After describing the attack implementation, this section presents the evaluation results obtained from multiple executions. The experiments were conducted using two distinct adversarial attack strategies, **Influential Users** and **Bot Network**, each assessed under three optimization approaches: Differential Evolution (DE), Grey Wolf Optimization (GWO), and a random (baseline) search. To ensure a fair and robust comparison, the random strategy was executed using five different random seeds: 1, 3, 5, 7, and 11. These seeds were selected to provide a diverse representation of the search space while ensuring reproducibility. As a comprehensive baseline, we report the best, worst, and average outcomes of the random search, capturing its performance variability and offering a reference point for comparing the optimization-based approaches. The results enable the evaluation of both the effectiveness of the attacks and the vulnerability of the targeted recommendation systems.

The experiments were conducted using different configurations of input parameters:

- **Dataset dimension**: Number of rows selected from the MovieLens 25M Dataset[2], ranging from 10,000 to 40,000 in increments of 10,000 ([10,000, 20,000, ..., 40,000]).
- **n_ratio**: Percentage of selected objects (users or movies). For the **Influential Users** attack, this corresponds to a proportion of users; for the **Bot Network** attack, it corresponds to a proportion of movies. The absolute number n of objects is calculated as $n = n\_ratio \times \#objects\_in\_dataset$, with values ranging from 0.1 to 0.3 in increments of 0.05 ([0.1, 0.15, 0.2, 0.25, 0.3]).

In all figures, the x-axis represents the dataset size, and the y-axis reflects the average predicted rating of the target movie across all users. Each reported value corresponds to the highest mean rating obtained among the tested n values for each configuration.

5.1 Influential Users Attack: Optimization Vs Random

The results for the **Influential Users** attack are summarized in Figs. 4 and 5. Across all dataset dimensions, both recommendation algorithms show limited sensitivity to the applied attacks.

[2] https://grouplens.org/datasets/movielens/.

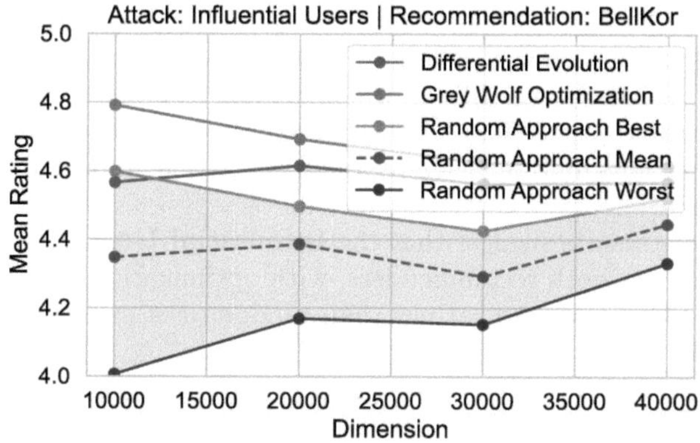

Fig. 4. Influential Users attack: BellKor—Differential Evolution vs Grey Wolf Optimization vs Random.

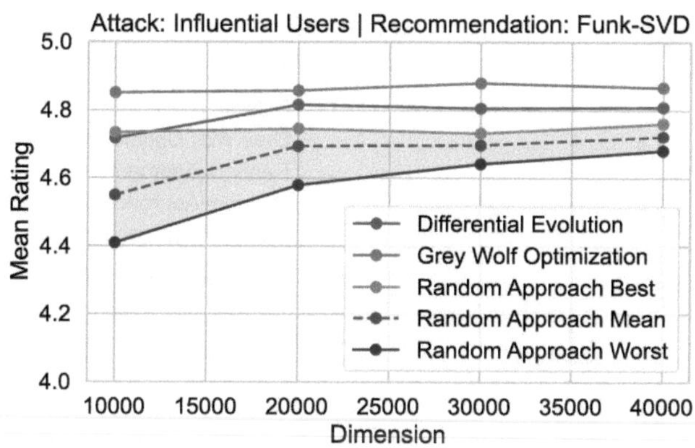

Fig. 5. Influential Users attack: Funk-SVD—Differential Evolution vs Grey Wolf Optimization vs Random.

For BellKor, the average target movie rating under DE optimization reached a mean of 4.58, while GWO achieved a slightly higher mean of 4.68. The Random approach yielded an average of the mean results across seeds of 4.36, with the highest observed mean reaching 4.51 and the lowest dropping to 4.17. The differences across optimization methods indicate that optimization-based attacks consistently yield slightly higher average predicted ratings for the target movie compared to random approaches. However, the overall effect remains moderate across all dataset sizes.

For Funk-SVD, the observed behavior is similar. DE yielded a mean rating of 4.79, while GWO performed slightly better with a mean of 4.86. The Random approach produced an average of the mean results across seeds of 4.67, with individual runs yielding mean values ranging from 4.58 to 4.74. Once again, optimization showed a slight advantage over the random baseline.

Interestingly, Funk-SVD consistently produced slightly higher target movie ratings than BellKor across all configurations. This is likely explained by Funk-SVD's lack of temporal weighting, which makes it more susceptible to new, highly rated samples, regardless of their recency. In contrast, BellKor incorporates timestamp information, thereby dampening the impact of injected user ratings over time.

Overall, these results suggest that the **Influential Users** attack has limited effectiveness against both recommenders, with optimization algorithms yielding only slightly higher predicted ratings than random user selection.

5.2 Bot Network Attack: Optimization Vs Random

The results for the **Bot Network** attack are presented in Figs. 6 and 7. In this scenario, both recommendation algorithms demonstrate moderate vulnerability; however, the advantage of optimization over random search remains limited.

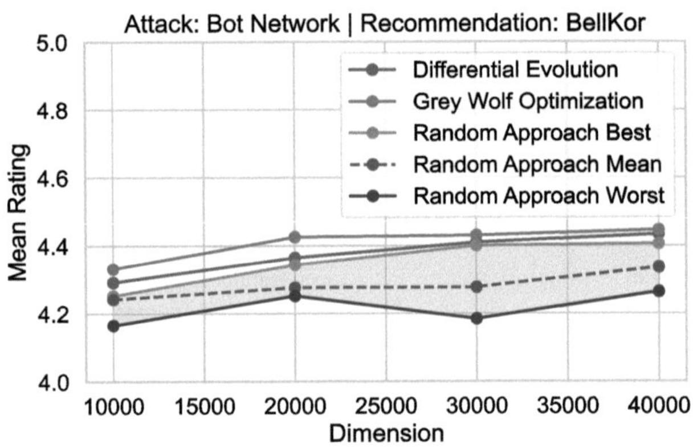

Fig. 6. Bot Network attack: BellKor—Differential Evolution vs Grey Wolf Optimization vs Random.

For BellKor, the DE optimizer achieved a mean target rating of 4.38, while GWO performed slightly better with a mean of 4.41. The Random approach, averaged across multiple seeds, yielded a mean target rating of 4.28, with individual runs producing values ranging from 4.22 (worst) to 4.35 (best). The differences among the three approaches are small and mostly fall within the standard deviation range, indicating that optimization techniques do not provide substantial improvements over random selection.

For Funk-SVD, a similar pattern is observed. The DE optimizer obtained a mean rating of 4.25, GWO achieved 4.26, and the Random approach yielded a mean of 4.22 when averaged across seeds, with values ranging from 4.20 to 4.24. While BellKor appears marginally more susceptible than Funk-SVD under this attack, the performance gap between optimization-based and random strategies remains very small.

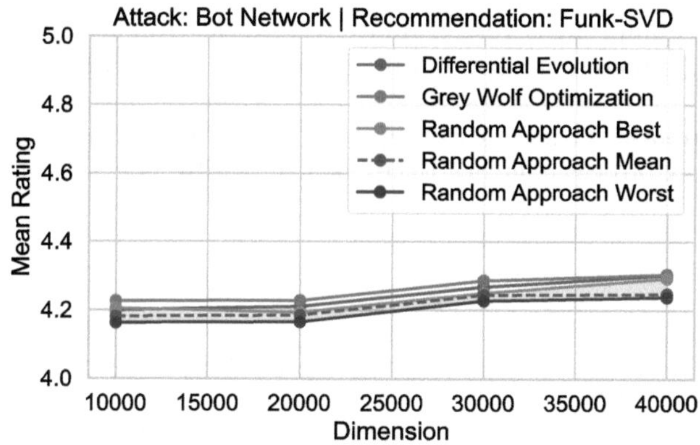

Fig. 7. Bot Network attack: Funk-SVD—Differential Evolution vs Grey Wolf Optimization vs Random.

Overall, these results confirm that the **Influential Users** attack achieves slightly higher rating increases than the **Bot Network** strategy, particularly for Funk-SVD. However, across both recommendation algorithms, blind optimization techniques offer little advantage over random sampling, further reinforcing the robustness of these systems to black-box adversarial attacks.

6 Conclusions

This study explored the feasibility of manipulating recommendation systems using adversarial samples generated through blind optimization strategies under two attack models: **Influential Users** and **Bot Network**. The objective was to increase the predicted rating of a target movie without requiring internal access to the recommendation models.

The empirical results demonstrate that both BellKor and Funk-SVD exhibit substantial resilience against adversarial sample injection. Although the **Influential Users** attack achieved slightly higher manipulation levels than the **Bot Network** attack, particularly for Funk-SVD, the overall effectiveness of both attacks remained limited, and optimization techniques consistently offered only marginal benefits over random sampling.

While Funk-SVD showed a somewhat higher susceptibility to manipulation—likely due to its absence of temporal modeling—BellKor's integration of timestamp information contributed to its additional robustness. These results reinforce the value of incorporating temporal or contextual features to enhance resistance against certain classes of adversarial attacks.

Although the tested recommendation algorithms demonstrated robustness under the evaluated blind black-box conditions, these findings do not fully generalize to real-world system security. Future work should investigate alternative attack strategies, broader experimental configurations, and combined hybrid attacks to more comprehensively evaluate real-world vulnerabilities, particularly in domains where recommendation bias could yield financial or political gains.

Acknowledgements. This work was supported by FCT - Fundação para a Ciência e Tecnologia, I.P. by project reference UIDB/50008: Instituto de Telecomunicações.

References

1. Bennett, J., Lanning, S., et al.: The netflix prize. In: Proceedings of KDD Cup and Workshop, vol. 2007, p. 35 (2007)
2. Biggio, B., et al.: Evasion attacks against machine learning at test time. In: Blockeel, H., Kersting, K., Nijssen, S., Železný, F. (eds.) ECML PKDD 2013. LNCS (LNAI), vol. 8190, pp. 387–402. Springer, Heidelberg (2013). https://doi.org/10.1007/978-3-642-40994-3_25
3. Biggio, B., Roli, F.: Wild patterns: ten years after the rise of adversarial machine learning. Pattern Recogn. **84**, 317–331 (2018). https://doi.org/10.1016/j.patcog.2018.07.023
4. Carlini, N., Wagner, D.: Towards evaluating the robustness of neural networks. In: 2017 IEEE Symposium on Security and Privacy (SP), pp. 39–57 (2017).https://doi.org/10.1109/SP.2017.49
5. Das, S., Suganthan, P.N.: Differential evolution: a survey of the state-of-the-art. IEEE Trans. Evol. Comput. **15**(1), 4–31 (2011). https://doi.org/10.1109/TEVC.2010.2059031
6. Duddu, V.: A survey of adversarial machine learning in cyber warfare. Def. Sci. J. **68**(4), 356 (2018). https://doi.org/10.14429/dsj.68.12371
7. Goldblum, M., et al.: Dataset security for machine learning: data poisoning, backdoor attacks, and defenses. IEEE Trans. Pattern Anal. Mach. Intell. (2022).https://doi.org/10.1109/tpami.2022.3162397
8. Goodfellow, I.J., Shlens, J., Szegedy, C.: Explaining and harnessing adversarial examples. arXiv preprint arXiv:1412.6572 (2014)
9. Himeur, Y., Sohail, S.S., Bensaali, F., Amira, A., Alazab, M.: Latest trends of security and privacy in recommender systems: a comprehensive review and future perspectives. Comput. Secur. **118**, 102746 (2022). https://doi.org/10.1016/j.cose.2022.102746
10. Himeur, Y., Sohail, S.S., Bensaali, F., Amira, A., Alazab, M.: Latest trends of security and privacy in recommender systems: a comprehensive review and future perspectives. Comput. Secur. **118**, 102746 (2022)
11. Kloft, M., Laskov, P.: Online anomaly detection under adversarial impact. In: Teh, Y.W., Titterington, M. (eds.) Proceedings of the Thirteenth International Conference on Artificial Intelligence and Statistics. Proceedings of Machine Learning Research, vol. 9, pp. 405–412. PMLR, Chia Laguna Resort, Sardinia (2010). https://proceedings.mlr.press/v9/kloft10a.html
12. Koren, Y.: The bellkor solution to the netflix grand prize. Netflix Prize Document. **81**(2009), 1–10 (2009)
13. Koren, Y., Bell, R., Volinsky, C.: Matrix factorization techniques for recommender systems. Computer **42**(8), 30–37 (2009)
14. Liang, B., Li, H., Su, M., Bian, P., Li, X., Shi, W.: Deep text classification can be fooled. CoRR arxiv:1704.08006 (2017)

15. Mallikarjunan, K.N., Muthupriya, K., Shalinie, S.M.: A survey of distributed denial of service attack. In: 2016 10th International Conference on Intelligent Systems and Control (ISCO). IEEE (2016).https://doi.org/10.1109/isco.2016.7727096
16. Mladenovic, A., et al.: Online adversarial attacks. In: International Conference on Learning Representations (2022)
17. Mobasher, B., Burke, R., Bhaumik, R., Williams, C.: Toward trustworthy recommender systems. ACM Trans. Internet Technol. **7**(4), 23 (2007). https://doi.org/10.1145/1278366.1278372
18. Mozaffari-Kermani, M., Sur-Kolay, S., Raghunathan, A., Jha, N.K.: Systematic poisoning attacks on and defenses for machine learning in healthcare. IEEE J. Biomed. Health Inform. **19**(6), 1893–1905 (2015). https://doi.org/10.1109/JBHI.2014.2344095
19. Papernot, N., McDaniel, P., Jha, S., Fredrikson, M., Celik, Z.B., Swami, A.: The limitations of deep learning in adversarial settings. In: 2016 IEEE European Symposium on Security and Privacy (EuroS P), pp. 372–387 (2016). https://doi.org/10.1109/EuroSP.2016.36
20. Sharma, I., Kumar, V., Sharma, S.: A comprehensive survey on grey wolf optimization. Rec. Adv. Comput. Sci. Commun. (Formerly: Rec. Patents Comput. Sci.) **15**(3), 323–333 (2022)
21. Sharma, N., Dutta, M.: Movie recommendation systems: a brief overview. In: Proceedings of the 8th International Conference on Computer and Communications Management, pp. 59–62 (2020)
22. Shi, Y., Wang, S., Han, Y.: Curls & whey: boosting black-box adversarial attacks. In: Proceedings of the IEEE/CVF Conference on Computer Vision and Pattern Recognition (CVPR) (2019)
23. Zhang, X., Antunes, M., Estro, T., Zadok, E., Mueller, K.: Smart starts: accelerating convergence through uncommon region exploration. arXiv e-prints pp. arXiv–2505 (2025)

Author Index

S. Džeroski et al. (Eds.): DS 2025, LNAI 16090, pp. 603–604, 2025.
https://doi.org/10.1007/978-3-032-05461-6

MIX
Papier aus verantwortungsvollen Quellen
Paper from responsible sources
FSC® C105338

If you have any concerns about our products,
you can contact us on
ProductSafety@springernature.com

In case Publisher is established outside the EU,
the EU authorized representative is:
Springer Nature Customer Service Center GmbH
Europaplatz 3, 69115 Heidelberg, Germany

Printed by Libri Plureos GmbH
in Hamburg, Germany